D1091546

STP 1020

Fracture Mechanics: Perspectives and Directions (Twentieth Symposium)

Robert P. Wei and Richard P. Gangloff, editors

ASTM
1916 Race Street
Philadelphia, PA 19103

ASTM Publication Code Number (PCN: 04-010200-30)
ISBN: 0-8031-1250-5
ISN: 1040-3094

NOTE

The Society is not responsible, as a body,
for the statements and opinions
advanced in this publication.

Peer Review Policy

Each paper published in this volume was evaluated by three peer reviewers. The authors addressed all of the reviewers' comments to the satisfaction of both the technical editor(s) and the ASTM Committee on Publications.

The quality of the papers in this publication reflects not only the obvious efforts of the authors and the technical editor(s), but also the work of these peer reviewers. The ASTM Committee on Publications acknowledges with appreciation their dedication and contribution of time and effort on behalf of ASTM.

Printed in Ann Arbor, MI
November 1989

Foreword

The Twentieth National Symposium on Fracture Mechanics was held on 23–25 June 1987 at Lehigh University, Bethlehem, Pennsylvania. ASTM Committee E-24 on Fracture Testing was the sponsor of this symposium. Robert P. Wei, Lehigh University, and Richard P. Gangloff, University of Virginia, served as coeditors of this publication. Robert P. Wei also served as chairman of the symposium.

Contents

Overview

Fracture mechanics forms the basis of a maturing technology and is used in quantifying and predicting the strength, durability, and reliability of structural components that contain cracks or crack-like defects. First utilized in the late 1940s to analyze catastrophic fractures in ships, the fracture mechanics approach found applications and increased acceptance in the aerospace industries through the late 1950s and early 1960s. Much of the early work was spearheaded by Dr. George R. Irwin and his co-workers at the U.S. Naval Research Laboratory and was nurtured through a special technical committee of ASTM, chaired by Dr. John R. Low. Over the past 20 years, fracture mechanics has undergone major development and has become an important subdiscipline in solid mechanics and an enabling technology for materials development, component and system design, safety and life assessments, and scientific inquiries. The contributions are now utilized in the design and analysis of chemical and petrochemical equipment, fossil and nuclear power generation systems, marine structures, bridges and transportation systems, and aerospace vehicles. The fracture mechanics approach is being used to address all of the major mechanisms of material failure; namely, ductile and cleavage fracture, stress corrosion cracking, fatigue and corrosion fatigue, and creep cracking. From its origin in glass and high strength metallic materials, the approach is currently applied to most classes of materials; including metallic materials, ceramics, polymers, composites, soils, and rocks.

The first National Symposium on Fracture Mechanics was organized by Professor Paul C. Paris, and was held on the campus of Lehigh University in June 1967. The National Symposium has gained prominence and international recognition and serves as an important international forum for fracture mechanics research and applications under the sponsorship of ASTM Committee E-24 on Fracture Testing. It has been held annually since 1967, with the exception of 1977. The growth of the National Symposium has paralleled the development and utilization of fracture mechanics. Landmark papers and Special Technical Publications have resulted from this Symposium series. It is appropriate that this, the 20th anniversary meeting of the National Symposium, be held again at Lehigh University and that the proceedings be archived in an ASTM book.

At this anniversary, following from two decades of intense and successful developments, it is appropriate and timely to conduct an introspective examination of the field of fracture mechanics and to define directions for future work. The Organizing Committee, therefore, set the following goals for the 20th National Symposium on Fracture Mechanics, *Fracture Mechanics: Perspectives and Directions*:

1. To provide perspective overviews of major developments in important areas of fracture mechanics and of associated applications over the past two decades.

2. To highlight directions for future developments and applications of fracture mechanics, particularly those needed to encompass the nontraditional areas.

To achieve the stated goals, the technical program was organized into the following six sessions:

(a) Analytical Fracture Mechanics
(b) Nonlinear and Time Dependent Fracture Mechanics
(c) Microstructure and Micromechanical Modeling
(d) Fatigue Crack Propagation
(e) Environmentally Assisted Cracking
(f) Fracture Mechanics of Nonmetals and New Frontiers

This Special Technical Publication accurately adheres to the objectives and approach of the Symposium. The twelve invited review papers, organized topically in the order of their presentation in one section, provide authoritative and comprehensive descriptions of the state of the art and important challenges in each of the six topical areas. The worker new to the field will be able to survey current understanding through the use of these seminal contributions. The thirty-one contributed papers, organized topically in a separate section, provide reports of current research. These papers are of particular importance to fracture mechanics researchers.

Although each manuscript was subjected to rigorous peer reviews in accordance with ASTM procedures, the authors of invited review papers were encouraged to respond thoughtfully to the reviewers comments and suggestions, but were granted considerable latitude to exercise their judgment on the final manuscript. This action was taken by the Editors to preserve the personal (*vis-à-vis*, a consensus) perspective of the individual experts, and to accurately reflect agreements and differences in opinion on the direction of future research. The invited papers, therefore, need to be read in this context. The opinions expressed and positions taken by the individual authors are not necessarily endorsed by the author's peers, the Editors, or the ASTM.

The review papers document the significant progress achieved over the past two decades of active research in fracture mechanics. Collectively, the authors provide compelling arguments for the need of continued development and exploitation of this technology, and insights on the challenges that must be faced. Some of the specific challenges are as follows:

1. On the analytical front, we must expand upon the effort to integrate continuum fracture mechanics analyses with the microscopic processes which govern local fracture at the crack tip.
2. In the area of advanced heterogeneous materials, fracture mechanics methods must be further developed and applied to describe novel failure modes. Claims of high performance for these materials must be supported by quantitative and scalable characterizations of fracture resistance that is relevant to specific applications.
3. In the area of subcritical crack growth (for both fatigue and sustained-load crack growth in deleterious environments and at elevated temperatures), the gains in understanding from multidisciplinary (mechancis, chemistry, and materials science) research must be reduced to practical life prediction methodologies. The critical issues of formulating mechanistically based procedures that enable the extrapolation of short-term laboratory data in predicting long-term service performance (that is, from weeks to decades) must be addressed.

4. In the area of education, we must better inform engineering students and practitioners on the interdisciplinary nature and intricacies of the material failure problem, whether by subcritical crack growth or by catastrophic defect-nucleated fracture. We must also continue to develop and to communicate governing ASTM standards to the engineering community.

This volume demonstrates that the existing fracture mechanics foundation is well positioned to meet these challenges over the next decade.

Professors Paul C. Paris and George R. Irwin provided important insights during the closing of the symposium and at the Conference Banquet. The banquet provided an opportunity for the awarding of the first ASTM E-24 Fracture Mechanics Medals to Professors Irwin and Paris.

We gratefully acknowledge the contributions of the Symposium Organizing Committee: R. Badaliance (NRL), T. W. Crooker (NASA Headquarters), F. Erdogan (Lehigh University) and R. H. Van Stone (GE-Evandale), and of the Session Chairmen: R. Badaliance, R. J. Bucci (ALCOA), S. C. Chou (AROD), F. Erdogan, J. Gilman (EPRI), R. J. Gottschall (DOE/BES), D. G. Harlow (Lehigh), C. Hartley (NSF), R. Jones (EPRI), R. C. Pohanka (ONR), A. H. Rosenstein (AFOSR), A. J. Sedriks (ONR), D. P. Wilhelm (Northrop); the assistance of the Local Committee: Terry Delph, Gary Harlow, Ron Hartranft, and Gary Miller; the hospitality of Lehigh University; and especially the skill and devotion of the Symposium Secretary, Mrs. Shirley Simmons.

We particularly acknowledge the work of our many colleagues who participated as authors, as speakers, and in the technical review process; the support of the ASTM staff; and the able editorial assistance provided by Helen Hoersch and her colleagues.

Financial support by the Office of Naval Research is gratefully acknowledged. All of the funds were used to provide matching support to graduate students across the United States so that they can participate in this introspective review of fracture mechanics. Nearly 30 students participated, and all of them expressed their appreciation for the opportunity to attend.

Robert P. Wei
Lehigh University, Bethlehem, PA.

Richard P. Gangloff
University of Virginia, Charlottesville, VA.

PART I

Invited Papers

Analytical Fracture Mechanics

George C. Sih[1]

Fracture Mechanics in Two Decades

REFERENCE: Sih, G. C., "**Fracture Mechanics in Two Decades,**" *Fracture Mechanics: Perspectives and Directions (Twentieth Symposium), ASTM STP 1020*, R. P. Wei and R. P. Gangloff, Eds., American Society for Testing and Materials, Philadelphia, 1989, pp. 9–28.

ABSTRACT: A brief historical and technical perspective precedes emphasizing the need to understand some of the fundamental characteristics of fracture. What the state of the art was two decades ago is no longer adequate in the era of modern technology. Observed mechanisms of failure at the atomic, microscopic, and macroscopic scale will continue to be elusive if the combined interaction of space/time/temperature interaction is not considered. The resolution of analysis, whether analytical or experimental or both, needs to be clearly identified with reference to local and global failure. Microdamage versus macrofracture is discussed in connection with the exchange of surface and volume energy, which is inherent in the material damage process. This gives rise to dilatation/distortion associated with cooling/heating at the prospective sites of failure initiation. Analytical predictions together with experimental results are presented for the compact tension and central crack specimens.

KEY WORDS: surface and volume energy, change of volume with surface, dilatation and distortion, cooling and heating, energy dissipation, material damage, space/time/temperature interaction, thermal/mechanical effects, crack initiation and growth

The rapid advance of technology in the past two decades has substantially altered the performance limits and reliability objectives dealing with the application of advanced materials. More and more of the conventional metals, whose mechanical and failure behavior are characterized by macroparameters in an homogeneous fashion, are being replaced by multi-phase materials such as composites that reflect a complex dependence on their constituents or microstructure. Past methodologies [*1–3*] which relied on a single-parameter characterization are no longer adequate as the new materials become more application-specific. New concepts are needed to replace the old ones, making this communication on fracture mechanics quite timely.

Fracture mechanics became a recognized discipline after World War II because of the inability of continuum-mechanics theories to address failure by unexpected fracture, a situation that occurs less frequently as the trade-off between strength and fracture toughness is now better understood. Research activities have fallen into two categories: *material science* and *continuum mechanics*. The former seeks to look at damage from a microscopic or atomistic viewpoint or both, determining what happens to the atoms and grains of a solid, which is beyond the scope of this discussion. The latter attempts to formalize the results of macroscopic experiments without probing very deeply into the origin and physics of how failure initiates. It would be desirable to have a *unique approach* such that the hierarchy of the physical damage mechanisms, each dominant over a certain range of load-time history

[1] Professor of mechanics and director of the Institute of Fracture and Solid Mechanics, Packard Laboratory No. 19, Lehigh University, Bethlehem, PA 18015.

can be assessed quantitatively when material or geometry or both are changed. This goal will be emphasized.

The translation of data collected from specimens with or without a crack to the design of larger structural components has been problematic. It is the common practice to employ both uniaxial and fracture data for predicting structural behavior. This is an oversupply[2] of input data and introduces arbitrariness and inconsistency into the analysis. Linear elastic fracture mechanics (LEFM) based on the toughness parameter K_{Ic} or G_{Ic} merely addresses a go and no-go situation. It implies a unique amount of energy release for a small crack extension that triggers rapid fracture. Such a concept obviously has no room in situations where a crack grows slowly at first and then rapidly. The irresistible urge to characterize ductile fracture for the development of small specimen tests without an understanding of the underlying physics and principles has hindered progress. Among the two leading candidates were the crack opening displacement (COD) [*6*] and J-integral approach [*7*]. The COD measurements, being sensitive to changes in strain rate, triaxiality of stress, specimen size and geometry, etc., served little or no useful purpose in design. Application of the J-integral caused a great deal of confusion because the idea applies only to elastic deformation and the same symbol J has been used [*7,8*] to represent the variations of areas under the load-extension versus crack length curve that include the effect of permanent deformation. What should be remembered is that the formalism of J precludes the distinction[3] of energy used in permanent deformation and crack extension that are interwoven in the experimental data. It would, therefore, be totally misleading to interpret J as the crack driving force except in the elastic case for then it is identically equal to G. The experimental data [*8*] did not yield a linear relation between J and crack growth. Alternate forms of dJ/da were suggested [*9*] and resulted in nonlinear crack growth resistance curves.[4]

The Charpy V-Notch (CVN) impact test is another empirical method for collecting data on dynamic fracture with no consideration given to rate effect.[5] It has been used for establishing the nil ductility temperature (NDT). The rate at which energy is actually used to create dynamic fracture must be isolated from other forms of energy dissipation such as plastic deformation, acoustic emission, etc., in order to obtain a reliable assessment of the time-dependent failure process. Current research [*13*] has not yet recognized that dynamic fracture is inherently load dependent and cannot be characterized by a single parameter such as K_{ID}. One of the major shortcomings of the conventional approaches is that they failed to separate the fracture energy from other forms of energy dissipation. This is why COD, J, CVN, K_{ID}, etc., are all sensitive to change in loading and specimen geometry and size. In this regard, they can hardly be claimed as fracture toughness parameters; much less, as material constants. A detailed discussion of their limitations can be found in Refs *14* and *15*.

The empirical "4th power law" [*16*] on relating the crack growth rate da/dN and change

[2] There are no difficulties to predict the fracture behavior of cracked specimens by using uniaxial data only [*4,5*].

[3] The separation of energy dissipated by permanent deformation and crack extension is not additive. This was not recognized by those who attempted to formally include the so-called plastic deformation term in the J-integral approach.

[4] Data collected on precracked polycarbonate specimens [*10*] showed that the dJ/da = const. condition was not satisfied. This was pointed out and discussed in Ref *11*. The strain energy density factor S, when plotted against crack growth, did yield a linear relationship such that specimen size and loading rate effects can then be easily resolved by interpolation.

[5] The Charpy impact energy normally specified in foot-pounds is not sufficient to describe dynamic fracture. The *rate* of energy used to initiate dynamic fracture as distinguished from that dissipated in plastic deformation and other forms is the relevant quantity [*12*].

in Mode I stress-intensity factor ΔK continues to dominate the literature on fatigue. Such a correlation was found to be invalid when moisture effects were accounted for [*17*]. A multitude of the so-called fatigue crack propagation laws have been proposed to correct separately for the mean stress, specimen thickness, temperature, crack size, crack opening, or closure effects. None of them had any theoretical basis. The majority applied elasticity to a process that is inherently dissipative. Thresholds in ΔK were found to disappear when the same crack growth data were replotted against change in energy release rate, ΔG. The influence of mean stress on da/dN reversed in trend for some metals. Special treatments were suggested for "small or short cracks" as data showed considerable scatter on the da/dN versus ΔK plot. It was only discovered later that ΔK was simply the wrong parameter to use [*18*]. The phenomenon of crack retardation due to occasional overload in fatigue has not been explained adequately.

Contrived explanation and data gathering in the absence of theoretical support are destined to fall by the wayside. Unless the basic fundamentals of energy dissipation associated with material damage are understood and quantified, fracture mechanics will not withstand the rigors of progress. Verbal or elegant, or both, mathematical descriptions of already known and observed events cannot be considered original research for they merely serve as a means of bookkeeping. *Predictive capability* is needed such that simple test data can be used to forecast the behavior of more complex systems regardless of whether they are loaded monotonically, repeatedly, or dynamically. A unified approach for addressing material damage at the atomic, microscopic, and macroscopic scale levels is long overdue. With this objective in mind, a few selected topics are chosen to illustrate some of the fundamental aspects of the fracture process that are not commonly known.

Micro- and Macro-Mechanics of Fracture

Fracture[6] is a process that involves the creation of free surface at the microscopic and macroscopic scale level. It entails a hierarchy of failure modes, each associated with a certain range of stress, strain rate, temperature, and material type. Examinations are frequently traced to the ways with which damages are influenced by the microstructure. The empirical results have been couched in terms of void growth, cleavage failure, brittle fracture, ductile fracture, etc. Although observations can be made on failure mechanisms, it has been very difficult to construct a quantitative theory of failure or damage that can relate the microscopic entities to the useful macroscopic variables. These difficulties can, in retrospect, be identified with the inability of theories such as elasticity, plasticity, etc., to account for the *irreversible* nature of thermal/mechanical interactions that are inherent in the fracture process. These effects are assumed to be mutually exclusive in classical mechanics that invoke the independence of surface and volume energy or the decoupling of mechanical deformation and thermal fluctuation.[7] Addition of the different energy forms leaves out coupling effects that are, in themselves, problematic. Quantitative assessment of the fracture process cannot be achieved without an understanding of the underlying physics.

Surface Energy Density

Prior to modeling of the fracture process, it would be instructive to specify the resolution of the analysis with reference to defect size and microstructure detail. Figures 1*a* to 1*c*

[6] In this communication, *fracture* pertains to material discontinuities at the microscopic and macroscopic level. Imperfections at the atomic level should be referred to as vacancies, dislocations, etc.

[7] Isothermal condition can only be realized conceptually in the limit as disturbances or changes become infinitesimally small.

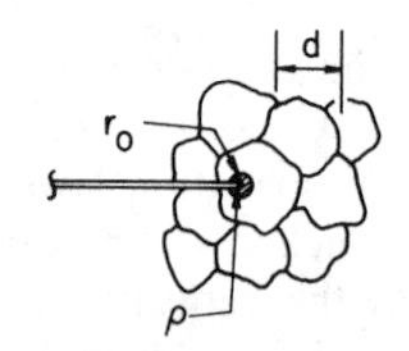

(a) Crack Tip In A Grain: $\rho << d$

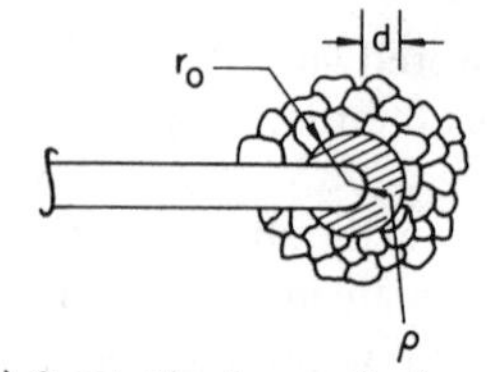

(b) Crack Tip Among Grains: $\rho \simeq d$

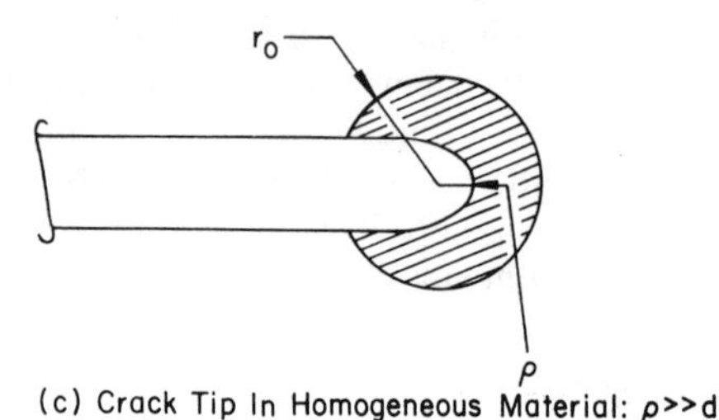

(c) Crack Tip In Homogeneous Material: $\rho >> d$

FIG. 1—*Size of crack tip compared with material microstructure.*

compare the mean linear dimension d of the grains in a polycrystal to the mean radius of curvature of a crack given by ρ. In the first case given by Fig. 1*a*, the crack tip is located in a single grain and the basic element would be submicroscopic in size, requiring information on the properties within a single grain because $\rho \ll d$. When $\rho \simeq d$, as indicated in Fig. 1*b*, the core region[8] radius r_0 would have to be increased accordingly and a microelement consisting of the average properties of the grains should suffice. The assumption of material homogeneity can be justified as ρ becomes many times greater than d. Fig. 1*c*. The dimension r_0 is then comparable with the macroelement size. It is the scale level at which damage is being analyzed that determines the size of r_0.

Past discussions on crack growth initiation [*21–23*] have focused attention on the energy required to extend a unit area of crack extension which is generally designated by the symbol γ and known as the specific surface energy. This quantity is, in fact, the surface energy density dW/dA. The global instability of a precracked solid shown in Fig. 2 can be predicted by considering the failure of a microelement or macroelement ahead of the crack. The distinction between the creation of a unit macrocrack and microcrack surface is not only necessary, but must be assessed accordingly when computing the specific surface energy

[8] On physical grounds, it is necessary to consider a ligament distance or core region of radius r_0 around the crack front which served as a measure of the resolution of analysis within which failure will not be considered [*19*]. Such an assumption can be justified even at atomic level where a dislocation free zone [*20*] prevails directly ahead of the crack tip, with r_0 being the order of an angstrom.

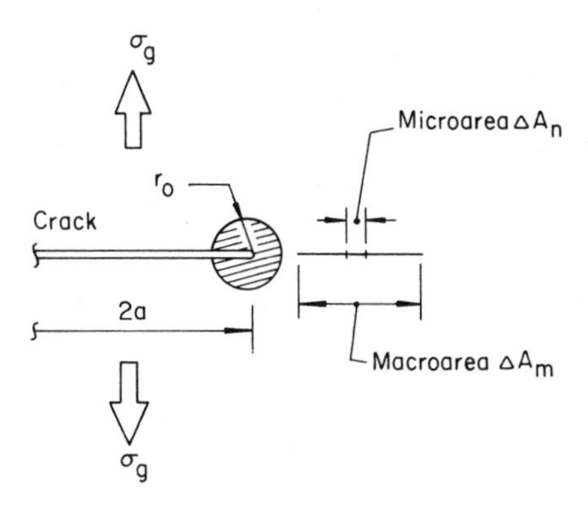

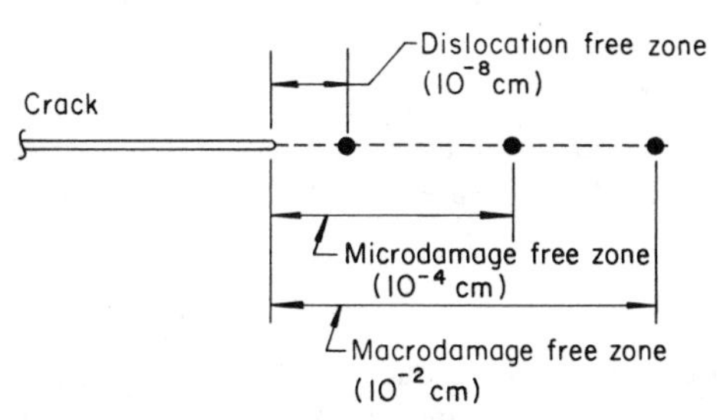

FIG. 2—*Macro- and micro-element ahead of crack.*

dW/dA. In what follows, $(dW/dA)_m$ or $(dW/dA)_n$ will be used to denote energy needed to create a unit surface area at the macroscopic or microscopic scale. The ratio of $(dW/dA)_m$ to $(dW/dA)_n$ can be two to three orders in magnitude depending on ΔA_m and ΔA_n. If the rate of energy dissipation associated with the formation of a unit free surface is assumed to occur so quickly that the failure of a local microelement or macroelement coincides with the breaking of the specimen, then the critical uniform stress σ_g applied over a panel with a central crack of length $2a$ is given by[9]

$$\sigma_g = \sqrt{\frac{E}{\pi a}} \sqrt{(dW/dA)_m} \tag{1}$$

where E is the Young's modulus. It was shown in Refs *24* and *25* that Eq 1 can also be expressed in terms of $(dW/dA)_n$, the energy required to create a unit-free micro surface, that is,

$$\sigma_g = \sqrt{\frac{E}{\pi a}} \sqrt{\left(\frac{E}{e}\right)\left(\frac{d}{\rho}\right)} \sqrt{(dW/dA)_n} \tag{2}$$

[9] For simplicity, let it be understood that the Young's modulus E in Eq 1 stands for $E/(1 - \nu^2)$ with ν being the Poisson's ratio such that the stress state ahead of the crack is plane strain.

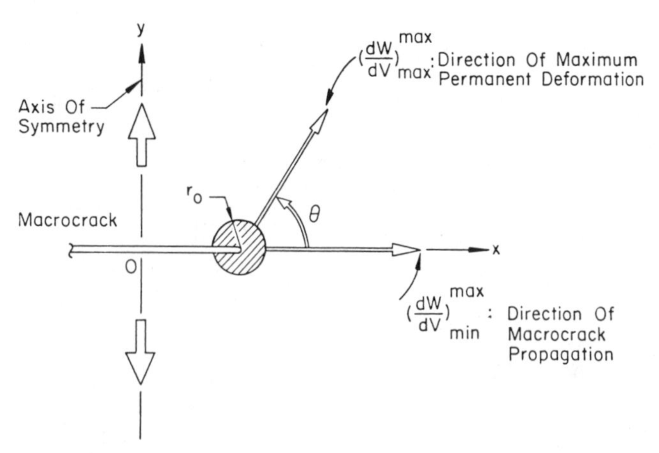

FIG. 3—*Stationary values of volume energy density function associated with damage by deformation or fracture or both.*

in which e is the microscopic modulus and d a microscopic characteristic length. The factor[10] $\sqrt{Ed/e\rho}$ provides the scale shift that relates the macro- and microquantities.

Volume Energy Density

Sudden fracture is a rare event. More frequently, failure initiates locally and there prevails a period of slow crack growth as the material undergoes shape change or permanent deformation. Such a behavior is commonly known as ductile fracture. It is a strain-rate-sensitive process that depends inherently on the load time history. The consideration of specific surface energy or dW/dA alone can no longer adequately describe the rate dependent crack growth process because energy per unit volume dW/dV also changes nonuniformly with time. The quantity dW/dV will, in general, vary from one location in the solid to another. The peaks and valleys or the maxima and minima of dW/dV are intimately associated with the local and global instability of the system. Initial imperfections, say microcracks or microvoids in a solid, tend to enhance the onset of instability. A subcritical stage of local disturbance, however small, must necessarily prevail prior to reaching the critical condition at large.

Excessive distortion and dilatation are two of the most common modes of failure. They can be associated with the maximum and minimum values of dW/dV. Illustrated schematically in Fig. 3 is a two-dimensional macrocrack subjected to tension in the y-direction. Among many of the minima of dW/dV, the maximum of $(dW/dV)_{min}$ is assumed to coincide

[10] This factor was taken as unity in Ref *23*. It is not justified to arbitrarily add γ_p onto γ_g and then drop γ_g on the basis that $\gamma_p >> \gamma_g$, that is,

$$\sigma_g = \sqrt{\frac{2E}{\pi a}}\sqrt{\gamma_g + \gamma_p} \simeq \sqrt{\frac{2E}{\pi a}}\sqrt{\gamma_p}$$

The so-referred-to γ_p is along the prospective path of macrocrack growth and should be distinguished from the macroyielding off to the side of the crack, in which case the macrocrack may grow very slowly at first prior to the onset of rapid fracture. This removes the ambiguity raised in Ref *23* for associating the brittle fracture of low carbon steel with the large amount of $(dW/dA)_n$ or γ_p that is many times larger than the elastic work $(dW/dA)_m$ or γ_g. The energy per unit of macrosurface area should be clearly distinguished from that per unit of microsurface area. See Refs *24* and *25* for more details.

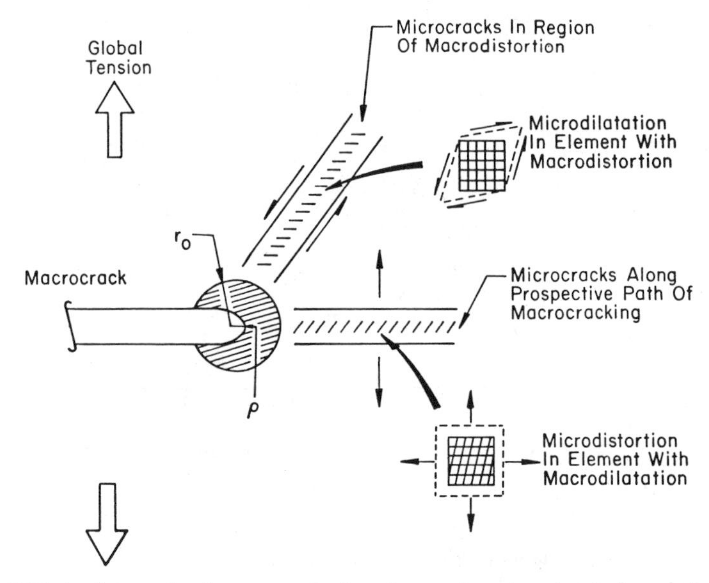

FIG. 4—*Schematic of macro- and micro-damage in region ahead of crack tip.*

with macrocrack growth and the largest of $(dW/dV)_{max}$ with the direction of maximum permanent deformation. The former can be associated with excessive dilatation while the latter with excessive distortion. This is consistent with the intuition that large volume change leads to fracture and shape change to permanent deformation.[11]

The physical interpretation of the stationary values of the volume energy density function can be best illustrated in Fig. 4. According to the criterion, the macroelements along the path of prospective macrocracking experience macrodilatation and microdistortion. This accounts for the creation of slanted microcracks prior to macrocracking along planes normal to the applied tension. In the same vein, the macroelements off to the sides are subjected to macrodistortion and microdilatation. They are responsible for permanent deformation and creation of microcracks. The interplay between distortion and dilatation is an inherent part of material behavior. They vary in proportion depending on the load history and thermal fluctuation. The dilatational effect tends to dominate along the path of macrocracking while the distortional effect governs permanent macrodeformation. Moreover, macrocracking is initiated by microdistortion and permanent macrodeformation by microdilatation. These damages were observed in Ref *23*.

Energy Dissipation and Irreversibility

Permanent deformation prevails when a solid is stretched or compressed beyond the linear range, as illustrated by the uniaxial stress and strain curve in Fig. 5. For a multiaxial stress state, the equivalent uniaxial stress σ_ξ and strain ϵ_ξ on the plane of homogeneity can be used instead of those in Fig. 5 without loss of generality. The definition of this plane will be given subsequently. The rate of change of area under the σ versus ϵ curve gives the change of dW/dV as a function of time. The shaded area *oypq* denoted by $(dW/dV)_p$ represents the

[11] The separation of dW/dV into its dilatational and distortional components, of course, can be carried out only by assuming a linear relationship between stress and strain. No such separation can be made as *a priori* when the response becomes nonlinear.

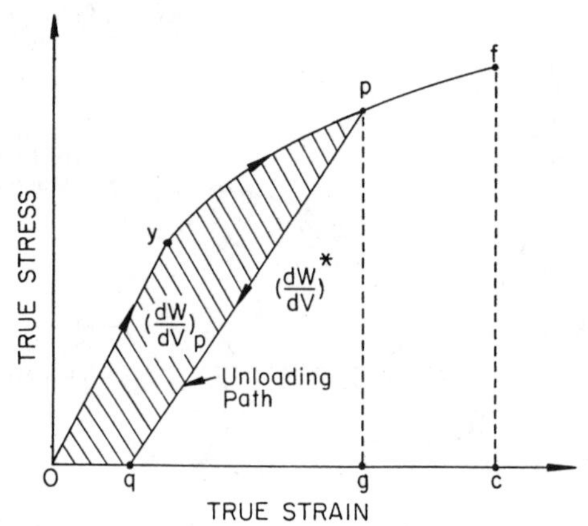

FIG. 5—*Schematic of true stress and true strain.*

energy dissipated by permanent deformation, while the area *qpg* given by

$$\left(\frac{dW}{dV}\right)^* = \left(\frac{dW}{dV}\right) - \left(\frac{dW}{dV}\right)_p \tag{3}$$

is the available energy density that is released when a unit of macrovolume fails as $(dW/dV)^*$ becomes critical. The unloading path may or may not be parallel with *oy*, the initial path of loading. Its slope depends on the loading rate. Final failure corresponds to *p* reaching the point *f* with the strain at *c*.

The dissipated energy density function $(dW/dV)_p$ in Eq 3 must be positive definite and is a manifestation of the irreversible exchange of surface and volume energy density [*26*]

$$\left(\frac{dW}{dA}\right)_i = \left(\frac{dV}{dA}\right)_i \frac{dW}{dV}, \qquad i = \xi,\eta,\zeta \tag{4}$$

where $i = \xi,\eta,\zeta$ denote a system of orthogonal coordinates. Both $(dW/dA)_i$ and $(dV/dA)_i$ are components of vector quantities which can change with the orientation of the plane on which they are defined. Without loss in generality, Eq 4 can be related to the uniaxial data, while dW/dV can be computed only from a knowledge of the strain ϵ_ξ on the *plane of homogeneity*[12]

$$\frac{dW}{dV} = \iint \lambda\left(\frac{dV}{dA}\right)_\xi d\epsilon_\xi d\epsilon_\xi \tag{5}$$

with λ being a proportionality parameter. It is obtained by letting dV/dA to be proportional to the slope of the stress and strain curve $d\sigma/d\epsilon$. The expended portion of the energy in

[12] This is the plane [*26*] on which the surface energy density $(dW/dA)_i$ in the three orthogonal directions $i = \xi,\eta,\zeta$ are equal such that $(dW/dA)_\xi = (dW/dA)_\eta = (dW/dA)_\zeta$ and $(dV/dA)_\xi = (dV/dA)_\eta = (dV/dA)_\zeta$ can be equated to dW/dA and dV/dA for the uniaxial case. This provides a one-to-one correspondence of energy states of elements in a three-dimensional system to those under uniaxial stress states without invoking any simplifying assumptions.

dW/dV or $(dW/dV)_p$ is associated with thermal agitation and their relation is given by [*15*]

$$\frac{\Delta\theta}{\theta} = -\frac{\Delta\sigma\Delta\epsilon}{\Delta(dW/dV)_p} \tag{6}$$

The negative sign stands for work done on the system, while $\Delta\sigma$ and $\Delta\epsilon$ represent the increment of stress and strain, respectively. In Eq 6, θ is given in deg K and equals the classical temperature T only when the dissipation energy $(dW/dV)_p$ takes place all in the form of heat. This, of course, need not be the case in general. Note that because $\Delta\sigma/\Delta\epsilon = \lambda(\Delta V/\Delta A)$, the temperature change $\Delta\theta$ can be directly related to the incremental change of volume with surface $\Delta V/\Delta A$, a quantity assumed to vanish in classical mechanics which is responsible for the decoupling of mechanical deformation and thermal fluctuation.

It should be noted that the invocation of separate fracture criteria with continuum-mechanics theories, such as elasticity and plasticity, have been known to introduce inconsistencies, arbitrariness, and unrealistic conditions.[13] A case in point is the assumption in plasticity that the uniaxial data coincide with the effective stress and effective strain, leaving out the effect of dilatation. This cannot be justified because a crack has been known to extend along the path where dilatation dominates (Fig. 4). Moreover, the local strain rates and strain rate history change from element to element in the region ahead of the crack, whereas plasticity assumes the same stress and strain response everywhere. Even more serious is the neglect of temperature change during crack initiation and growth, which can seriously affect the resulting stress and strain field.

Crack Initiation and Growth: Thermal/Mechanical Interaction

Alteration of temperature affects the thermal and mechanical properties of solids. Because energy dissipates in an irreversible manner, dilatational and distortional effects are not additive and they inherently control the heat transfer process. That is, dilatation enhances cooling while distortion leads to heating. This interaction cannot be ignored when discussing crack initiation and growth.

Cooling/Heating

As the region ahead of the crack is highly dilated, cooling is expected to occur even if the load increases monotonically. The period between cooling and heating can be significant. This phenomenon[14] is related to the rate change of volume with surface, dV/dA. It has been predicted theoretically in a compact tension specimen made of 1020 steel loaded at a displacement rate of $\dot{u} = 0.051$ cm/min. Temperature measurements were also made and the results matched with those obtained from the theory [*29*].

[13] Contrary to experimental observation, the analysis in Ref *27* assumes that local yielding occurs uniformly in a circular region around the crack tip. Regardless of the scale level, the local crack tip region contains two distinct planes, one pertaining to maximum distortion and the other to maximum dilatation. This gives rise to a nonhomogeneous stress field and precludes the representation of the local stress field amplitude by a single parameter such as the plastic intensity factor proposed in Ref *28*.

[14] Intuitions developed from theories that assume $dV/dA = 0$ are obviously not valid. Moreover, any attempts made to concoct the cooling/heating behavior in the classical continuum mechanics theories would be inconsistent with the assumption just mentioned. The thermoviscoplasticity theory proposed in Ref *30* for determining the cooling and heating of uniaxially stretched specimens is suspect.

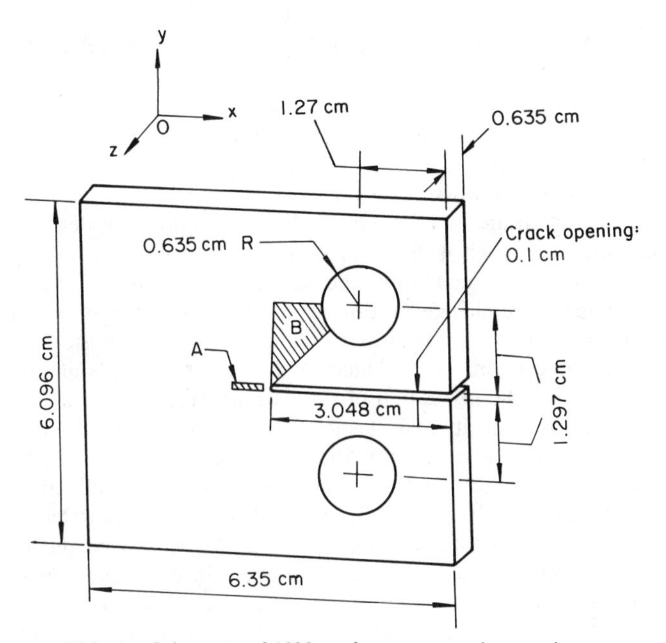

FIG. 6—*Schematic of 1020 steel compact tension specimen.*

More specifically, refer to the specimen configuration in Fig. 6. Since there is no need for an *a priori* knowledge of the stress and strain response, the energy density theory [*26*] requires only a knowledge of the initial slope of the stress and strain curve $(d\sigma/d\epsilon)_0$ = 198.71×10^3 MPa for the 1020 steel. The magnitude of the first-time step is taken to be sufficiently small so that all elements can be assumed to have the same stress and strain. After that, all stress and strain states will be derived individually in steps of $\Delta t = 24$ s. In two dimensions, damage in the thickness direction is assumed to be independent of the variables in the plane of the specimen. For the specimen thickness in Fig. 6 and past experience, the appropriate value of $(dV/dA)_z$ is 3.44. Large dV/dA corresponds to thin plates, while an infinitely thick plate is approached in the limit as $dV/dA \rightarrow 1$. From a knowledge of the equivalent uniaxial strain ϵ_ξ on the plane of homogeneity, the equivalent uniaxial stress σ_ξ can be found from

$$\sigma_\xi = \iint \lambda \left(\frac{dV}{dA}\right)_\xi d\epsilon_\xi \tag{7}$$

in which $\lambda = 22.742$ MPa/cm is obtained from the initial slope of the 1020 steel stress and strain curve. Figure 7 displays the stress and strain response of Elements A and B that are located directly ahead and off to the side of the crack, respectively. The locations of these elements are shown in Fig. 6. In comparison with the base material which would have been used in the theory of plasticity for every element regardless of the load-time history, Element A experiences a higher strain rate, while Element B experiences a lower strain rate. Exhibited in Fig. 7 is the time-history of the energy density dissipation function $(dW/dV)_p$ for a local spot 1 mm in diameter in the immediate vicinity of the crack tip. The curve rises very slowly for small time, which is indicative of the gradual occurrence of irreversibility. An order of magnitude jump in $(dW/dV)_p$ occurs between $t = 144$ and 192 s. This dramatic change may

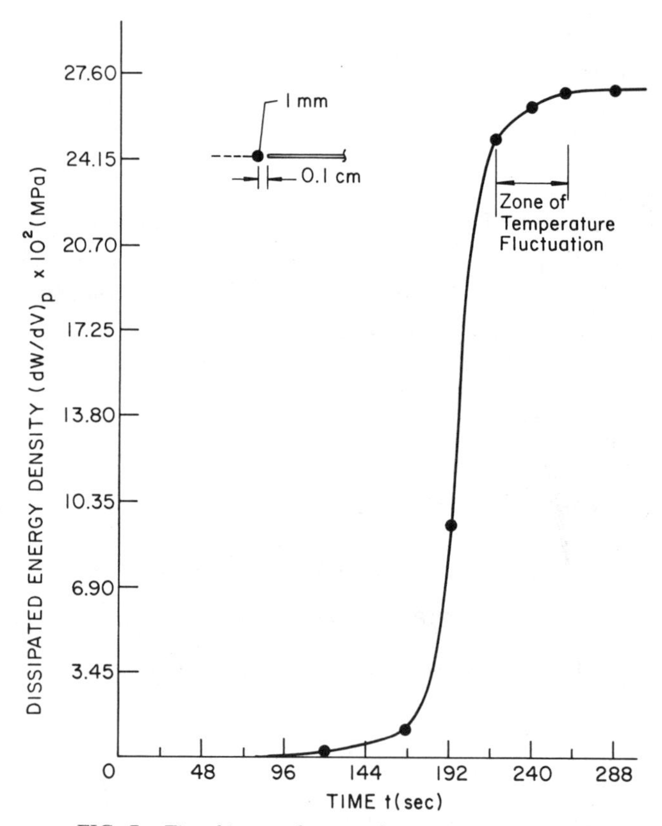

FIG. 7—*Time history of energy dissipation ahead of crack.*

be regarded as an engineering approximation of the onset of *irreversibility*. This is not the commonly referred to yield point[15] because during this time interval, the stress and strain already respond nonlinearly, as shown by curve labelled Element A in Fig. 8. Note that $(dW/dV)_p$ indeed increases monotonically with time.

Once $(dW/dV)_p$ is known, the temperature θ can be found from Eq 6, with $\Delta\sigma$ and $\Delta\epsilon$ corresponding to those on the plane of homogeneity. For the same location ahead of the crack tip, Fig. 9 gives a plot of θ-θ_0 against time, with θ_0 being the ambient temperature. The solid and open circles refer, respectively, to the theoretical and experimental results, which agreed extremely well for the 1020 steel, particularly in the cooling range that lasted for more than 3 min before heating starts. This delay cannot be ignored or neglected. Even more important is the second fluctuation in the local temperature predicted by the energy density theory and detected by experimental measurement. This is clearly indicated in Fig. 9 and occurs between t = 216 and 264 s and implies a possible change in the material microstructure near the crack tip region. It corresponds to a change in curvature of the *H*-function[16] versus time plot, a quantity that plays the role of the classical entropy function

[15] The yield point is a notion established empirically with no theoretical support and the 0.2% offset procedure is equally unsatisfying.

[16] The *H*-function in the energy density theory is given by the relation, $\Delta H = -\Delta(dW/dV)_p/\theta$. It reduces to the classical Second Law of Thermodynamics when $\Delta(dW/dV)_p$ equals to $\Delta\theta$, that is, heat exchange and $\theta = T$.

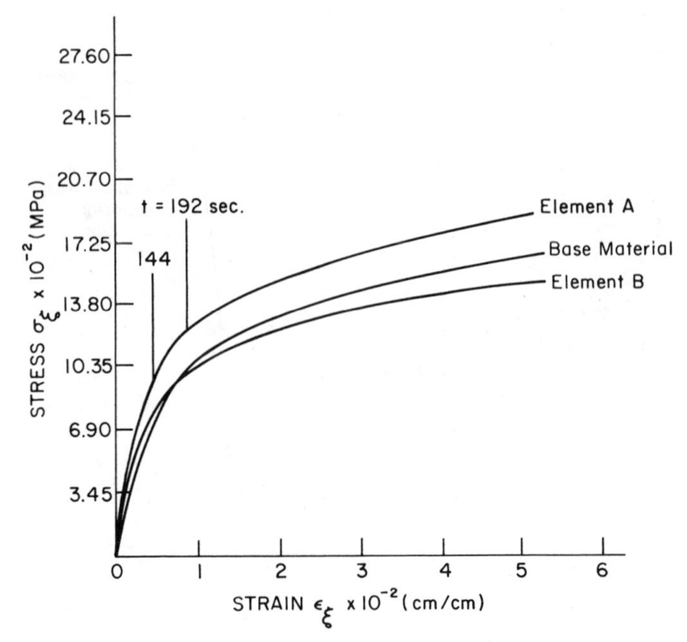

FIG. 8—*Stress and strain response on plane of homogeneity.*

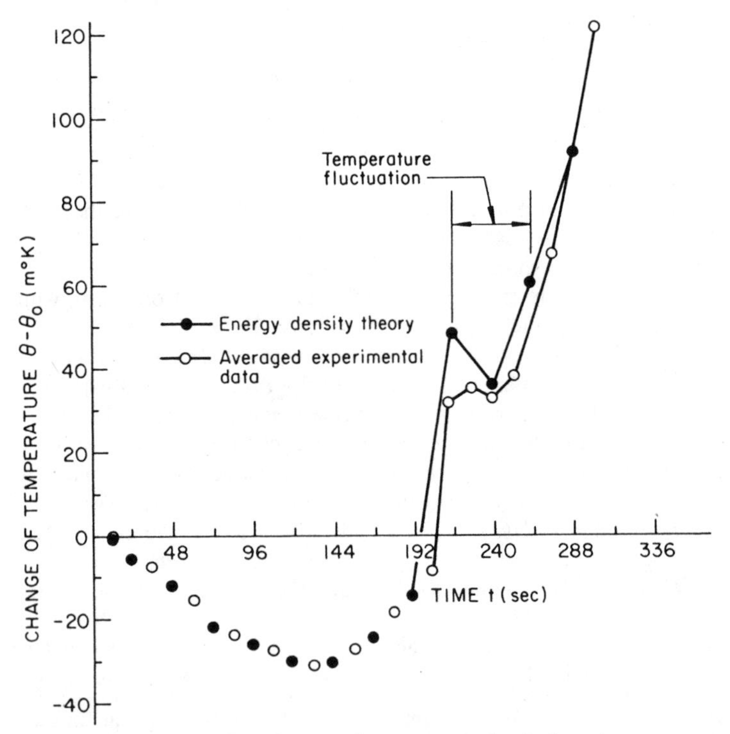

FIG. 9—*Time history of temperature ahead of crack.*

but has a much more general interpretation. Such a feature is reminiscent of the behavior of the entropy function associated with phase transformation in classical thermodynamics.

Damage-Free Zone

It is now common knowledge that failure always initiates at a finite distance ahead of the crack. There prevails a crack-tip ligament that would appear to be undamaged only because theory and experiment are necessarily limited in resolution. Even at the atomic scale level, a dislocation-free zone ahead of the crack has been detected [*20*] in single crystals of stainless steel, copper, and aluminum subjected to tension and cyclic loadings. What it means is that electron micrograph does not have the resolution for detecting sub-atomic disturbance or damage. Dislocations should not be regarded as the basic mechanism of material. There are obviously infinite numbers of even smaller scale level of imperfections and disorders that have yet to be discovered and quantified in terms of the material damage process. A ligament free of damage can thus be defined at each scale level. This is illustrated in Fig. 10. A microdamage-free zone ahead of the crack simply defines the resolution of analyses and experiments carried out at the macroscopic scale level.

The size of the macrodamage free zone has been estimated in Ref *19* for a compact tension specimen 5 by 6 m with a crack 1 m in length. It is loaded with a displacement rate of $\dot{u}$ = 0.02 cm/min and is made of structural steel with an initial modulus equal to that of the specimen in Fig. 6. A damage-free zone can be identified from a plot of the dissipation energy density $(dW/dV)_p$ against the distance r ahead of the crack for different time as shown in Fig. 11. The dissipated energy is negligibly small for the first 60 s of loading. Significant increase in $(dW/dV)_p$ is detected for t near or greater than 80 s, as indicated in Fig. 11. There prevails a distinct zone with lineal dimension of approximately 4.75×10^{-3} cm within which the dissipation energy $(dW/dV)_p$ is undetectable from the continuum mechanics analysis. This is referred to as the *macrodamage-free zone*. Obviously, this does not imply that there is no damage at the microscopic or atomic level or both. In fact, such a zone prevails at all scales as illustrated in Fig. 10, for it represents nothing more than the resolution of observation.

Crack Growth Characteristics

The interaction of surface tractions $\overset{n}{\underset{\sim}{T}}$ with body force or local inertia $\rho\ddot{\underset{\sim}{u}}$ with ρ being the density and dot representing differentiation with time is not negligible in regions where the local strain rates are high. This is precisely the situation ahead of the crack. If $\underset{\sim}{n}$ denotes the unit normal vector on an element of surface where $\overset{n}{\underset{\sim}{T}}$ is applied, then, according to the

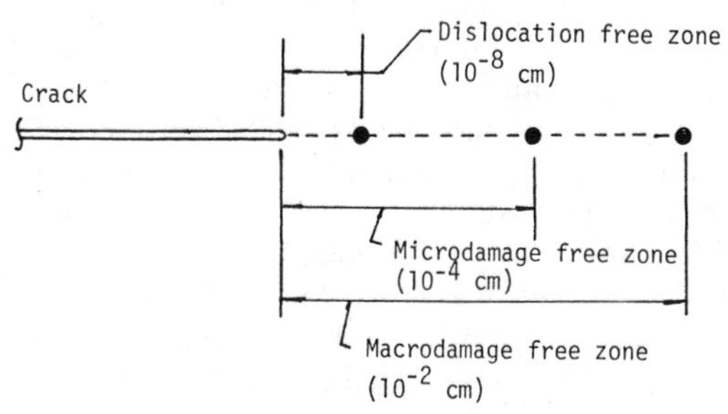

FIG. 10—*Scaling of damage-free zone.*

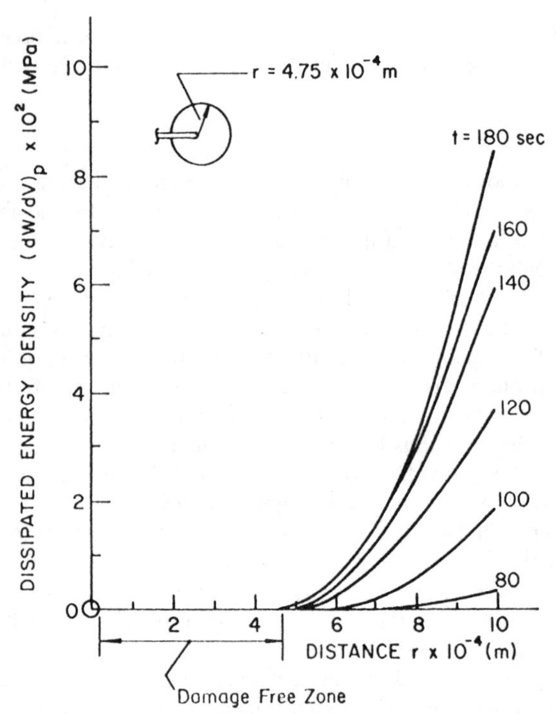

FIG. 11—*Variations of dissipated energy density versus distance from the crack tip.*

energy density theory, the following relation must be satisfied

$$\overset{n}{T}_i = \sigma_{ji}n_j + \rho\ddot{u}_i\left(\frac{dV}{dA}\right)_n \tag{8}$$

where σ_{ij} are the stress components. Even when the globally applied load is static, the product of the local inertia $\rho\ddot{\underset{\sim}{u}}$ and $(dV/dA)_n$ is of the same order as $\overset{n}{\underset{\sim}{T}}$ and $\underset{\sim}{\sigma}$ because $(dV/dA)_n$ attains high values near the crack. This effect is completely ignored in all classical continuum-mechanics theories and becomes more important as the loading rate or crack growth rate or both are increased.

The crack growth characteristics obtained from the energy density theory deviate significantly from those predicted from the incremental theory of plasticity. Large and finite deformation are considered in both analyses and have been made [*31*] for the problem of a central crack of length $2a$ in a rectangular panel 25.4 cm wide and 50.8 cm high. Equal and opposite uniform stress σ is applied incrementally normal to the crack at a rate of $\dot{\sigma}$ = 4.14 $\times$ 10^3 MPa/s. For a common structural steel, the initial slope of the stress and strain curve can be taken as $(d\sigma/d\epsilon)_0$ = 207 $\times$ 10^3 MPa. Using an asterisk to denote the available portion of the volume energy density as defined in Eq 3, the decay of $(dW/dV)^*$ with the distance r along the prospective path of crack growth for the first time increment t = 6.67 $\times$ 10^{-2} s with σ = 276 MPa is displayed in Fig. 12. From a knowledge of the stress

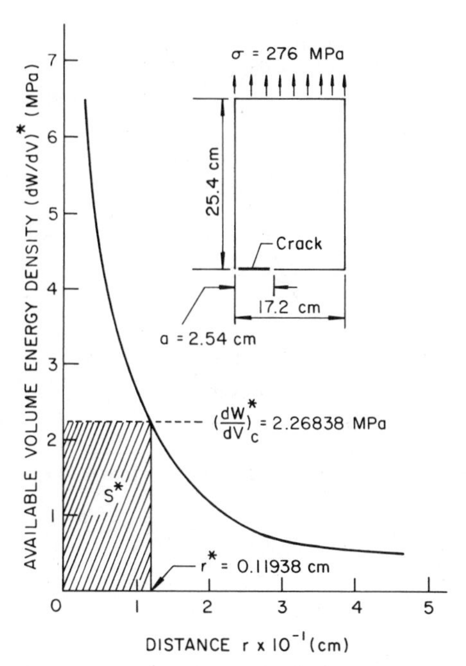

FIG. 12—*Decay of available volume energy density for the first increment of crack growth at* t = *6.67* $\times$ *10*$^{-2}$ *s.*

and strain time history, the critical value of $(dW/dV)_c^* = 2.2684$ MPa is found.[17] This yields a $r^* = 0.1194$ cm for the first increment of crack growth. An energy density factor S^* which represents the cross-hatched area in Fig. 12 can be obtained[18]

$$S^* = r^* \left(\frac{dW}{dV}\right)_c^* = 27.080 \times 10^2 \text{ N/m} \tag{9}$$

It has the same units as the energy release rate quantity in the conventional theory of LEFM, but the physical and mathematical implications are entirely different. Because the stress and strain state ahead of the crack changes for each load or time increment, all the quantities in Eq 9 are functions of time, that is

$$\left(\frac{dW}{dV}\right)_c^* = \frac{S^*(t)}{r^*(t)} \tag{10}$$

In other words, the crack driving force S^* in the energy density theory is loading rate dependent. The variations of S^* with crack growth for eight time steps are plotted in Fig.

[17] The available surface energy density $(dW/dA)^*$, in fact, also reaches its critical value of $(dW/dA)_c^* = 0.7533$ MPa $\sqrt{\text{m}}$. This, however, yields a microdamage characteristic length of $r^* = 0.30 \times 10^{-2}$ cm which is negligible in comparison with the macrocrack growth increment of 0.1194 cm as predicted from $(dW/dV)_c^*$. Therefore, it is not necessary to restructure the material elements.

[18] The average value for seven crack growth steps is 27.160×10^2 N/m. Refer to the results in Fig. 13.

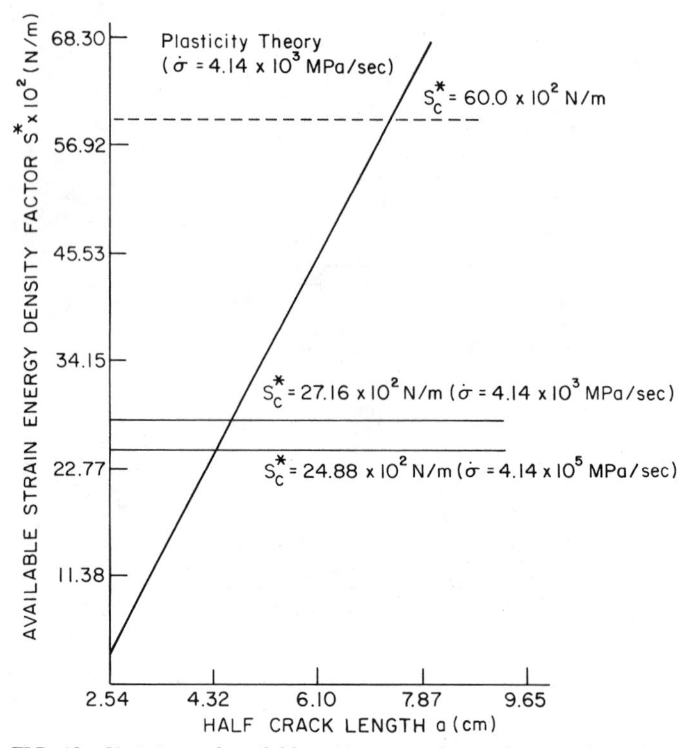

FIG. 13—*Variations of available strain energy density factor with crack growth.*

13. The available strain energy density remained constant during crack growth. A lower value of $S^* = 24.880 \times 10^2$ N/m is obtained when the loading rate is increased two orders of magnitude to $\dot{\sigma} = 4.14 \times 10^5$ MPa/s. This is, in essence, the crack growth resistance curve. The slanted line represents data obtained from the incremental theory of plasticity for $\dot{\sigma} = 4.14 \times 10^3$ MPa/s, which staisfies the condition dS^*/da = const. with a critical value of $S_c^* = 60.0 \times 10^2$ N/m. This differed substantially from that of the energy density theory. Large variances are also reflected in the local stresses. Illustrated in Fig. 14 are plots of the stress component σ_{yy} on an element next to the crack tip for the energy density and plasticity theory at $t = 6.67 \times 10^{-2}$ s where crack growth has not occurred. For distances r less than the half crack length $a = 2.54$ cm, the difference is significant.

Conclusion

Technological advancement cannot rely on general consensus and standardization, else it will be short-lived. Fracture mechanics is no exception and has remained stagnant for many years. Old ideas need to be modified and replaced by new ones with emphases placed on understanding the fundamental entity of the fracture process. Piecemeal attempts are application-specific and lack predictive capability. There is no reason why failure by monotonic and fatigue loading cannot be predicted directly from uniaxial data. Empirical approaches and unsupported tests are costly and uninformative. They can no longer be justified in modern times, when high-resolution computers can be readily used to make predictions.

Numerical analyses that incorporate the classical continuum-mechanics theories will not

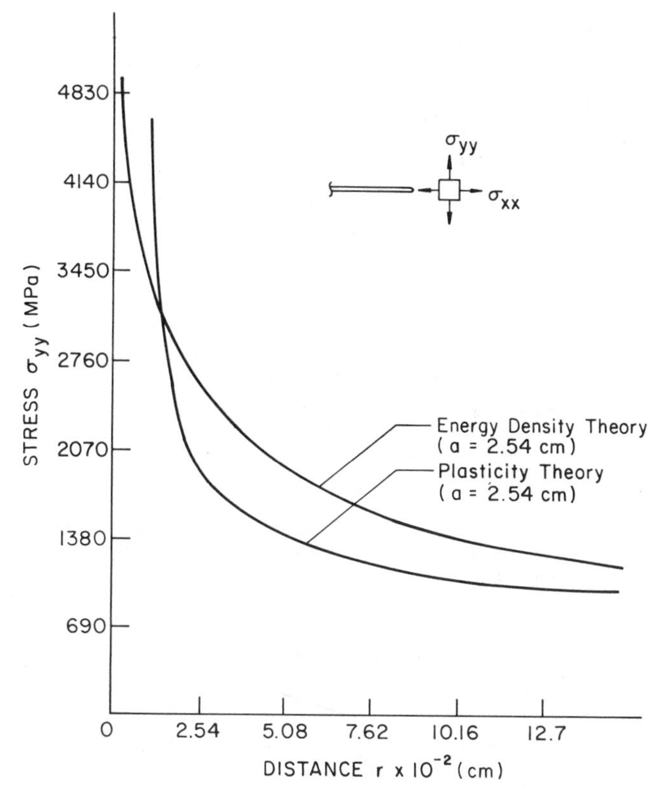

FIG. 14—*Variations of stress component* σ_{yy} *with distance prior to crack growth at* t = *6.67* × *10*$^{-2}$ *s.*

succeed in developing a generalized model of material damage. Finite-element computations, when employed to solve elasticity problems, for instance, will always yield finite values of dV/dA, which invalidates the original assumption that $dV/dA \rightarrow 0$ in the derivation of elasticity equations. This inconsistency introduces large errors in regions near the crack tip where dV/dA undergoes high oscillation. This, in retrospect, explains why the classical approaches cannot consistently assess material damage at the atomic, microscopic, and macroscopic level. The transient character of size/time/temperature interaction has eluded those working in fracture mechanics up to this date. Failure modes observed at one scale level may differ from that seen at another level, and they are further complicated by changes in loading rates. It is anticipated that dilatation/distortion and cooling/heating tend to flip-flop as the scale level of observation is altered; depending, of course, also on the time response. The fundamentals of this alternating mechanism are discussed in Ref *32* and will only be mentioned briefly in relation to the state of affairs near the macrocrack tip illustrated in Figs. 15*a* and 15*b*. Partial agreements with experiments have been obtained for the uniaxial specimens.[19] At an element along the path of possible crack growth at $\theta = 0$ deg in Fig.

[19] The combination of size/time/temperature data are selected arbitrarily for this discussion. Actual values for uniaxial tensile and compressive specimens have been obtained and can be found in Ref *33*.

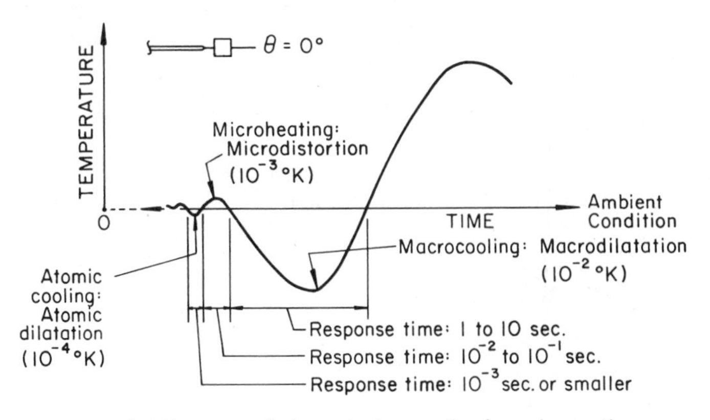

(a) Response of element along path of crack growth

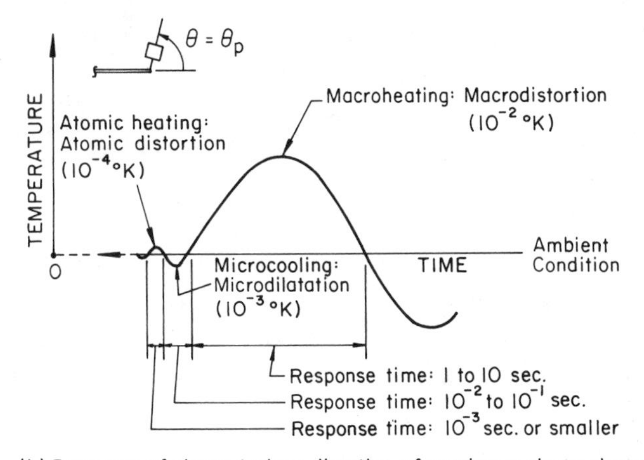

(b) Response of element along direction of maximum shape change

FIG. 15—*Size/time/temperature response of element ahead of crack at different scale level.*

15*a*, what appears to be macrocooling[20] and macrodilatation for response time of 1 to 10 s corresponds to microheating and microdistortion when the same event is viewed within the time interval of 10^{-2} to 10^{-1} s. Disturbances at the atomic scale refer to even smaller response time and temperature fluctuation. The situation for an element in the direction of excessive distortion at $\theta = \theta_p$ in Fig. 15*b* behaves opposite to that in Fig. 15*a*. What was macrocooling now becomes macroheating. The same applies to the micro- and atomic-scale together with the corresponding time response and temperature. Scaling of size/time/temperature is being assessed quantitatively such that seemingly different behavior of the same fracture process when viewed at the atomic, microscopic, and macroscopic level can be related uniquely. This interaction will become better understood as more actual examples and results are made available.

[20] Macro refers to dimensions of 10^{-3} to 10^{-2} cm; micro to 10^{-4} to 10^{-5} cm; and atomic to 10^{-6} to 10^{-8} cm. The size/time/temperature scale can shift depending on the loading or local strain rate.

References

[1] *Linear Fracture Mechanics*, G. C. Sih, R. P. Wei and F. Erdogan, Eds., Envo Publishing Co., Bethlehem, PA, 1974.
[2] Pellini, W. S., "Principles of Structural Integrity Technology," Office of Naval Research, Arlington, VA, 1976.
[3] *Plane-Strain Crack Toughness Testing of High-Strength Metallic Materials, ASTM STP 410*, W. F. Brown, Jr. and J. E. Srawley, Eds., American Society for Testing and Materials, Philadelphia, 1966.
[4] Sih, G. C. and Tzou, D. Y., "Mechanics of Nonlinear Crack Growth: Effects of Specimen Size and Loading Step," *Proceedings on Modelling Problems in Crack Tip Mechanics*, J. T. Pindera, Ed., Martinus Nijhoff Publishers, Amsterdam, the Netherlands, 1984, pp. 155–169.
[5] Sih, G. C. and Chen, C., "Non-Self-Similar Crack Growth in an Elastic-Plastic Finite Thickness Plate," *Journal of Theoretical and Applied Fracture Mechanics*, Vol. 3, No. 2, 1985, pp. 125–139.
[6] Wells, A. A., "Application of Fracture Mechanics at and Beyond General Yielding," *British Welding Journal*, Vol. 10, 1963, pp. 563–569.
[7] Begley, J. A. and Landes, J. E., "The J Integral as a Fracture Criterion," *Fracture Toughness, ASTM STP 514*, American Society for Testing and Materials, Philadelphia, 1972, pp. 1–20.
[8] Bucci, R. J., Paris, P. C., Landes, J. E., and Rice, J. R., "J Integral Estimation Procedure," *Fracture Toughness, ASTM STP 514*, American Society for Testing and Materials, Philadelphia, 1972, pp. 40–69.
[9] Shih, C. F., DeLorenzi, H. G., and Andrews, W. R., "Studies on Crack Initiation and Stable Crack Growth," *Elastic-Plastic Fracture Mechanics, ASTM STP 668*, American Society for Testing and Materials, Philadelphia, 1979, pp. 65–120.
[10] Bernstein, H. L., "A Study of J-Integral Method using Polycarbonate," AFWAL-TR-82-4080, Air Force Wright Aeronautical Laboratories, Wright-Patterson Air Force Base, OH, Aug. 1982.
[11] Sih, G. C. and Tzou, D. Y., "Crack Extension Resistance of Polycarbonate Material," *Journal of Theoretical and Applied Fracture Mechanics*, Vol. 2, No. 3, 1984, pp. 220–234.
[12] Sih, G. C., and Tzou, D. Y., "Dynamic Fracture Rate of Charpy V-Notch Specimen," *Journal of Theoretical and Applied Fracture Mechanics*, Vol. 5, No. 3, 1986, pp. 189–203.
[13] *Proceedings of Workshop on Dynamic Fracture*, California Institute of Technology, Pasadena, CA, Feb. 1983.
[14] Sih, G. C., "Fracture Mechanics of Engineering Structural Components," *Fracture Mechanics Methodology*, G. C. Sih and L. Faria, Eds., Martinus Nijhoff Publishers, Amsterdam, the Netherlands, 1984, pp. 35–101.
[15] Sih, G. C., "Outlook on Fracture Mechanics," *The Mechanism of Fracture*, V. S. Goel, Ed., *Proceedings of the Annual American Society of Metal Conference*, Salt Lake City, UT, 2–6 Dec. 1985, pp. 1–16.
[16] Paris, P. C., "The Growth of Cracks Due to Variations in Load," Ph.D. dissertation, Department of Mechanics, Lehigh University, Bethlehem, PA, 1962.
[17] Wei, R. P., "Contribution of Fracture Mechanics to Subcritical Crack Grwoth Studies," *Linear Fra ture Mechanics*, G. C. Sih, R. P. Wei, and F. Erdogan, Eds., Envo Publishing Co., Bethlehem, PA, 1974, pp. 287–302.
[18] Vecchio, R. S., and Hertzberg, R. W., "A Rationale for the Apparent Anomalous Growth Behavior of Short Fatigue Cracks," *Journal of Engineering Fracture Mechanics*, Vol. 22, No. 6, 1985, pp. 1049–1060.
[19] Sih, G. C. and Tzou, D. Y., "Heating Preceded by Cooling Ahead of Crack: Macrodamage Free Zone," *Journal of Theoretical and Applied Fracture Mechanics*, Vol. 6, No. 2, 1986, pp. 103–111.
[20] Ohr, S. M., Horton, J. A., and Chung, S. J., "Direct Observations of Crack Tip Dislocation Behavior During Tensile and Cyclic Deformation," *Defects, Fracture and Fatigue*, G. C. Sih and J. W. Provan, Eds., Martinus Nijhoff Publishers, Amsterdam, the Netherlands, 1982, pp. 3–15.
[21] Griffith, A. A., "The Theory of Rupture," *Proceedings*, 1st International Congress for Applied Mechanics Delft, the Netherlands, 1924, pp. 55–63.
[22] Irwin, G. R., "Analysis of Stresses and Strains Near the End of a Crack Traversing a Plate," *Journal of Applied Mechanics*, Vol. 24, 1957, pp. 361–364.
[23] Orowan, E., "Energy Criteria of Fracture," *Welding Research Supplement*, Vol. 34, 1955, pp. 157s–160s.
[24] Sih, G. C., "Some Basic Problems in Fracture Mechanics and New Concepts," *International Journal of Engineering Fracture Mechanics*, Vol. 5, No. 2, 1973, pp. 365–377.
[25] Sih, G. C., "Introductory Chapters of Vol. I to Vol. VII," *Mechanics of Fracture*, G. C. Sih, Ed., Martinus Nijhoff Publishers, Amsterdam, the Netherlands, 1972–1982.

[*26*] Sih, G. C., "Mechanics and Physics of Energy Density and Rate of Change of Volume with Surface," *Journal of Theoretical and Applied Fracture Mechanics*, Vol. 4, No. 3, 1985, pp. 157–173.
[*27*] Hutchinson, J. W., "Plastic Stress and Strain Fields at a Crack Tip," *Journal of the Mechanics and Physics of Solids*, Vol. 16, 1968, pp. 337–342.
[*28*] Hilton, P. D. and Hutchinson, J. W., "Plastic Intensity Factors for Cracked Plates," *Journal of Engineering Fracture Mechanics*, Vol. 3, 1971, pp. 435–451.
[*29*] Sih, G. C., Tzou, D. Y., and Michopoulos, J. G., "Secondary Temperature Fluctuation in Cracked 1020 Steel Specimen Loaded Monotonically," *Journal of Theoretical and Applied Fracture Mechanics*, Vol. 7, No. 2, 1987, pp. 79–87.
[*30*] Cernocky, E. P. and Krempl, E., "A Theory of Thermoviscoplasticity for Uniaxial Mechanical and Thermal Loading," *Journal de Mechanique Appliquee*, Vol. 5, No. 3, 1981, pp. 293–321.
[*31*] Tzou, D. Y. and Sih, G. C., "Thermal/Mechanical Interaction of Subcritical Crack Growth in Tensile Specimens," *Journal of Theoretical and Applied Fracture Mechanisms*, Vol. 10, No. 1, pp. 59–72.
[*32*] Sih, G. C., "Thermal/Mechanical Interaction Associated with the Micromechanisms of Material Behavior," *Institute of Fracture and Solid Mechanics Technical Report*, Lehigh University, Bethlehem, PA, Feb. 1987.
[*33*] Sih, G. C. and Chao, C. K., "Scaling of Size/Time/Temperature Associated with Damage of Uniaxial Tensile and Compressive Specimens," *Journal of Theoretical and Applied Mechanisms*, forthcoming.

James R. Rice[1]

Weight Function Theory for Three-Dimensional Elastic Crack Analysis

REFERENCE: Rice, J. R., **"Weight Function Theory for Three-Dimensional Elastic Crack Analysis,"** *Fracture Mechanics: Perspectives and Directions* (*Twentieth Symposium*), *ASTM STP 1020,* R. P. Wei and R. P. Gangloff, Eds., American Society for Testing and Materials, Philadelphia, 1989, pp. 29–57.

ABSTRACT: Recent developments in elastic crack analysis are discussed based on extensions and applications of weight function theory in the three-dimensional regime. It is shown that the weight function, which gives the stress intensity factor distribution along the crack front for arbitrary distributions of applied force, has a complementary interpretation: It characterizes the variation in displacement field throughout the body associated, to first order, with a variation in crack-front position. These properties, together with the fact that weight functions have now been determined for certain three-dimensional crack geometries, have allowed some new types of investigation. They include study of the three-dimensional elastic interactions between cracks and nearby or emergent dislocation loops, as are important in some approaches to understanding brittle versus ductile response of crystals, and also the interactions between cracks and inclusions which are of interest for transformation toughening. The new developments further allow determination of stress-intensity factors and crack-face displacements for cracks whose fronts are slightly perturbed from some reference geometry (for example, from a straight or circular shape), and those solutions allow study of crack trapping in growth through a medium of locally nonuniform fracture toughness. Finally, the configurational stability of cracking processes can be addressed: For example, when will an initially circular crack, under axisymmetric loading, remain circular during growth?

KEY WORDS: fracture mechanics, elasticity theory, weight functions, stress intensity factors, dislocation emission, crack-defect interactions, configurational stability, crack trapping

Bueckner introduced the concept of "weight functions" for two-dimensional elastic crack analysis in 1970 [*1*]. His weight functions satisfy the equations of linear elastic displacement fields, but they equilibrate zero body and surface forces and have a stronger singularity at the crack tip than would be admissible for an actual displacement field. The worklike product of an arbitrary set of applied forces with the weight function gives the crack-tip stress intensity factor induced by those forces. Bueckner's contribution led to what is now a vast literature on two-dimensional elastic crack analysis. One of the earliest works of that literature was a 1972 paper by the writer [*2*] which showed that weight functions could be determined by differentiating known elastic displacement field solutions with respect to crack length. It was also shown [*2*] that knowledge of a two-dimensional elastic crack solution, as a function of crack length, for any one loading enables one to determine directly the effect of the crack on the elastic solution for the same body under any other loading system.

The subject here is three-dimensional weight-function theory. Foundations of the three-dimensional theory were given independently by the writer, in the Appendix of Ref *2*, based

[1] Professor, Division of Applied Sciences and Department of Earth and Planetary Sciences, Harvard University, Cambridge, MA 02138.

on displacement field variations associated to first order with an arbitrary variation in position of the crack front, and in a review by Bueckner [*3*], based on three-dimensional solutions of the elastic displacement field equations that equilibrate null forces and that have arbitrary distributions of strength of a normally inadmissible singularity along the crack front. Bueckner refers to such fields as "fundamental fields."

Since 1985, there has been a surge of interest in the three-dimensional theory. That recent work, to be discussed, has allowed new types of three-dimensional crack investigations, including crack tip interactions with dislocations and other defects, stress analysis for perturbed crack shapes, crack-front trapping in growth through heterogeneous solids, and the configurational stability of crack shape during growth. However, three-dimensional weight function theory had a rather quiet first 13 years or so. Notable developments in that period include Besuner's [*4*] 1974 observation that the formulation based on crack-front variation [*2*] could be applied to determine certain weighted averages of K_I (tensile mode stress intensity factor) along the front of an arbitrarily loaded elliptical crack, by differentiating a known solution with respect to parameters describing the ellipse (Bueckner [*3*] had earlier used the same approach to construct some examples of his fundamental singular fields). Also, Parks and Kamenetzky [*5*] outlined a three-dimensional finite-element procedure for calculating numerically the variation of elastic displacement fields with crack-front position that are necessary to determine the three-dimensional weight function by the procedure of Ref *2*. In a 1977 paper [*6*], Bueckner determined fundamental fields for tensile half-plane and circular cracks for distributions of singularity strength that vary trigometrically with distance along the crack front. He also used that approach to rederive known results for the stress intensity factor distribution induced by a pair of wedge-opening point forces acting on the crack surfaces.

In 1985, the writer [*7*] pointed out the relation between three-dimensional weight function concepts and the determination of tensile-mode stress intensity factors along crack fronts whose locations are perturbed slightly from some simple reference geometry, and used such results to address the configurational stability of crack front shape during quasi-static crack growth. He also solved directly for the Mode 1 weight function for a half-plane crack in a full space, by determining the three-dimensional elastic field variations to first order for an arbitrary variation of crack front location, and generalized the three-dimensional theory to arbitrary mixed-mode conditions in the manner briefly reviewed in the next section. A related paper [*8*] pointed out how to use weight function concepts to describe the three-dimensional elastic interaction between crack tips and dislocation loops or zones of shape transformation, and Gao and Rice [*9*] developed the perturbation approach of Ref *7* to determine also the shear-mode stress intensity factors along the front of a generally loaded half-plane crack when that front is slightly perturbed from a straight line.

In a significant recent paper, Bueckner [*10*] completed the determination of weight functions for all three modes for the half-plane crack and further determined them for a "penny-shaped" circular crack. Also, Gao and Rice [*11,12*] applied the crack shape perturbation method of Ref *7* to determine tensile-mode stress intensity factors along crack fronts whose shapes are moderately perturbed from circles, dealing with the respective cases of near-circular cracks in full spaces and near-circular connections (that is, external cracks) bonding elastic half-spaces. They also note [*11*], and compare their methods to, a much earlier but apparently little known paper by Panasyuk [*13*], which directly derived a first-order perturbation solution for a near-circular crack (see also the 1981 review by Panasyuk et al. [*14*]). Gao and Rice [*11,12*] used their results to determine conditions for configurational stability in the growth of cracks with initially circular fronts under axisymmetric loading, as will be discussed subsequently. By using shear-mode results of Bueckner [*10*], Gao [*15*] solved for shear-mode intensity factors along a slightly noncircular shear crack and used the results to determine, to first-order accuracy, the shape of a shear loaded crack having constant energy release rate along its front. Rice [*16*] applied the crack front perturbation analysis to address some elementary problems in crack front trapping by tough obstacles in growth through heterogeneous microstructures. Also, Anderson and Rice [*17*] applied the methods of Ref *8* to evaluate the three-dimensional stress field and energy of a prismatic dislocation loop emerging from a half-plane crack tip, and studies of this type were recently extended by Gao [*18*] and Gao and Rice [*19*] to general shear dislocation loops. Sham [*20*] recently gave a new finite-element procedure for three-dimensional weight-function determination in bounded solids, as an alternative to the virtual crack extension method of Ref *5*.

Since weight functions are interpretable as intensity factors induced by arbitrarily located point forces, they can sometimes also be extracted from the existing literature on three-dimensional elastic crack analysis. That is too extensive to summarize here, but the reader is referred to the review by Panasyuk et al. [*14*] and also to the recent work of Fabrikant [*21*,*22*], which gives general solutions for arbitrarily loaded circular cracks.

Theory of Three-Dimensional Weight Functions

Background and Notation

For background, Fig. 1*a* shows a local coordinate system along a three-dimensional elastic crack front. Axes of the local system are labelled to agree with mode number designations for stress intensity factors $K_\alpha(\alpha = 1,2,3)$. Thus, at small distance ρ' ahead of the tip, on the prolongation of the crack plane, the stress components σ_{11}, σ_{12}, σ_{13} have the asymptotic

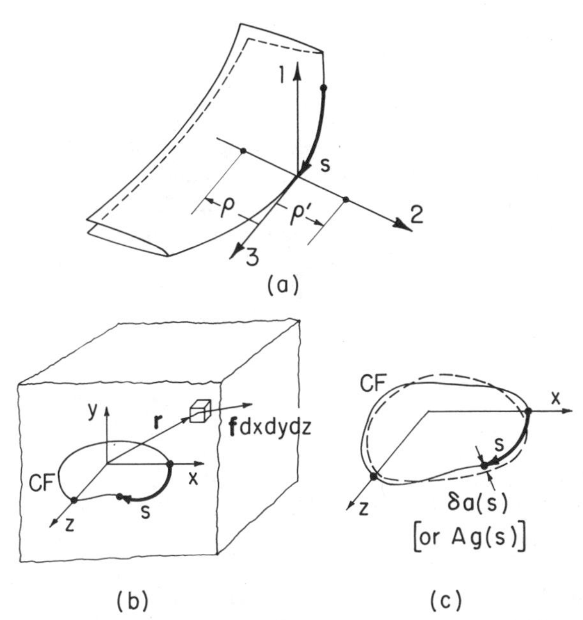

FIG. 1—(a) *Local coordinates along front of three-dimensional crack; numbering of axes corresponds to stress intensity modes.* (b) *Loaded solid with planar crack on* y = *0; CF denotes crack front, arc length* s *parameterizes locations along CF, vector* **r** *denotes position in body.* (c) *Advance of crack front normal to itself by* δa(s); *advance sometimes labelled* Ag(s) *where* A *is amplitude and* g(s) *a fixed function.*

form

$$\sigma_{1\alpha} \sim K_\alpha/\sqrt{2\pi\rho'} \tag{1}$$

Similarly, at small distance ρ behind the tip (that is, along the -2 axis) one has the asymptotic form of displacement discontinuities Δu_1, Δu_2, Δu_3 between the upper and lower crack surfaces of

$$\Delta u_\alpha \sim 8\Lambda_{\alpha\beta}K_\beta\sqrt{\rho/2\pi} \tag{2}$$

Here there is summation on repeated Greek indices over 1, 2, 3, associated with the local coordinate system at the point of interest along the crack front. (Repeated Latin indices, as appear later, are to be summed over directions x, y, z of a fixed-coordinate system as in Fig. 1*b*.) The matrix $\Lambda_{\alpha\beta}$ is given by

$$[\Lambda_{\alpha\beta}] = \frac{1}{2\mu}\begin{bmatrix} 1-\nu & 0 & 0 \\ 0 & 1-\nu & 0 \\ 0 & 0 & 1 \end{bmatrix} \tag{3}$$

for an isotropic material (μ = shear modulus, ν = Poisson ratio); for the general anisotropic material, $\Lambda_{\alpha\beta}$ remains symmetric but not necessarily diagonal, and is proportional to the inverse of a prelogarithmic energy factor matrix arising in the expression for self energy of a straight dislocation line with same direction as the local crack-front tangent [*23,24*]. Also, the energy G released per unit area of crack advance at the crack-front location considered is

$$G = \Lambda_{\alpha\beta}K_\alpha K_\beta \tag{4}$$

Figure 1*b* shows an elastic solid with a planar crack on $y = 0$. The crack front is denoted as CF, and arc length s parameterizes position along CF. The cracked body is loaded by some distribution of force vector $\mathbf{f} = \mathbf{f}(\mathbf{r})$ per unit volume. Here $\mathbf{r}$ is the position vector relative to the fixed x, y, z system, that is, $\mathbf{r} = (x,y,z)$, and $\mathbf{f}$ has cartesian components denoted by f_j where $j = x, y, z$. Also, in cases involving loading by a distribution of imposed stresses, or tractions, on the surface of the body, it will be convenient to regard those surface tractions as a singular layer of body force. Thus, the work-like product of the entire set of applied loadings with any vector field $\mathbf{u} = \mathbf{u}(\mathbf{r})$ will generally be written as

$$\int_{\text{Body}} \mathbf{f}(\mathbf{r}) \cdot \mathbf{u}(\mathbf{r})\, dxdydz \text{ or } \int_{\text{Body}} f_j(\mathbf{r})u_j(\mathbf{r})\, dxdydz$$

where the integral of $\mathbf{f} \cdot \mathbf{u}$ over the "Body" is to be interpreted as an integral of $\mathbf{f} \cdot \mathbf{u}$ over the interior of the body (with surface layer excluded) plus an integral of $\mathbf{T} \cdot \mathbf{u}$ over the surface of the body, including crack surfaces, where $\mathbf{T}$ is the vector of imposed surface tractions.

In addition, it will be assumed in general that the body considered is restrained against displacement over some part of its surface so that it can sustain arbitrary force distributions. This requirement can be disregarded if, in the intended applications, the actual imposed loadings are self-equilibrating.

Weight Functions and Their Properties

The weight functions $\mathbf{h}_1$, $\mathbf{h}_2$, $\mathbf{h}_3$ are three vector functions of position $\mathbf{r}$ in the body and locations s along CF: $\mathbf{h}_\alpha = \mathbf{h}_\alpha(\mathbf{r};s)$. One such vector function is associated with each crack tip stressing mode at location s, the mode being indicated by the value of subscript α on $\mathbf{h}_\alpha$. The vector functions $\mathbf{h}_\alpha$ have cartesian components $h_{\alpha j}(\alpha = 1,2,3;\ j = x,y,z)$, so that altogether there are nine scalar functions involved.

The weight functions have two properties, now outlined, as introduced in the developments via Refs *2* and *7*. Either of the first or the second property may be taken to define the weight function, and then the other property may be derived from that one by basic elasticity and fracture mechanics principles.

The first property is that the stress intensity factors induced at location s along the crack front, by arbitrary loading of the body (Fig. 1*b*) are given by

$$K_\alpha(s) = \int_{\text{Body}} \mathbf{h}_\alpha(\mathbf{r};s) \cdot \mathbf{f}(\mathbf{r})\, dxdydz \tag{5}$$

Thus $h_{\alpha j}(\mathbf{r};s)$ gives the mode α intensity factor induced at location s along CF by a unit point force in the j direction at $\mathbf{r}$.

The second property is that if, under fixed applied loadings, the crack front is advanced normal to itself, in the plane $y = 0$, by an amount $\delta a = \delta a(s)$, variable along CF as in Fig. 1*c*, the associated change in the displacement field $\mathbf{u}(\mathbf{r})$ is

$$\delta\mathbf{u}(\mathbf{r}) = 2\int_{\text{CF}} \Lambda_{\alpha\beta}(s)\mathbf{h}_\alpha(\mathbf{r};s)K_\beta(s)\delta a(s)\, ds \tag{6}$$

to first order in $\delta a(s)$. Thus $h_{\alpha j}(\mathbf{r},s)$, when weighted with $2\Lambda_{\alpha\beta}K_\beta$, gives the increase of displacement component u_j at $\mathbf{r}$ per unit enlargement of crack area near s [note that $\delta a(s)$ ds is an element of area].

To state the second property, Eq 6, more precisely, as well as to aid certain derivations, the following alternative is useful: Let $g(s)$ be an arbitrary but, once chosen, fixed dimensionless function of position along CF. Then a family of crack-front locations, with parameter A, may be defined by advancing the original crack front, CF, normal to itself by amount $A\, g(s)$. That is, the increment labelled $\delta a(s)$ in Fig. 1*c* is now understood as $A\, g(s)$. The loading is regarded as fixed so that the displacement field associated with this family of crack-front locations may be written as $\mathbf{u} = \mathbf{u}(\mathbf{r},A)$. Then the statement that Eq 6 holds to first order in $\delta a(s)$ is equivalent to

$$\left.\frac{\partial\mathbf{u}(\mathbf{r},A)}{\partial A}\right|_{A=0} = 2\int_{\text{CF}} \Lambda_{\alpha\beta}(s)\mathbf{h}_\alpha(\mathbf{r};s)K_\beta(s)g(s)\, ds \tag{7}$$

Since the growth increment is written as $A\, g(s)$, this equation corresponds, of course, to writing $\delta a(s) = \delta A\, g(s)$ in Eq 6, which is then required to hold to first order in δA.

Example: Mode 1 Weight Function for the Half-Plane Crack

The writer solved [7] for the Mode 1 weight function for a half-plane crack, denoted here as $\mathbf{h}_1(\mathbf{r};s)$. As shown in Fig. 2, s now denotes the z-coordinate of the location of interest

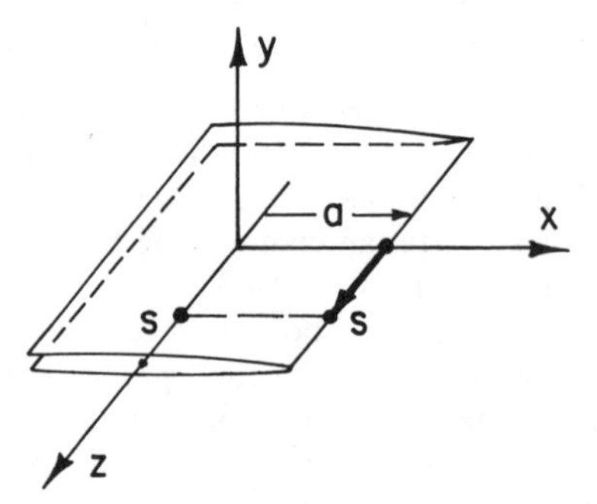

FIG. 2—*Half-plane crack with straight front in infinite solid.*

along the crack front, and the front is at $x = a$ on $y = 0$. The derivation was accomplished using the second property, Eqs 6 or 7, to define $\mathbf{h}_1$. The relevant elasticity equations were solved directly for $\partial \mathbf{u}(\mathbf{r}, A)/\partial A$, at $A = 0$, for arbitrary growth functions $g(s)$ and arbitrary Mode 1 loadings [hence arbitrary $K_1(s)$], and the solution was put in the form of Eq 7 to identify $\mathbf{h}_1$. The results are

$$h_{1y} = H - [1/2(1 - \nu)]y\partial H/\partial y$$

$$\{h_{1x}, h_{1z}\} = -[1/2(1 - \nu)]\{\partial/\partial x, \partial/\partial z\}\left[yH - (1 - 2\nu)\int_y^\infty H\,dy\right] \quad (8)$$

where

$$H = \frac{\text{Im}[(x - a + iy)^{1/2}]}{\pi\sqrt{2\pi}[(x - a)^2 + y^2 + (z - s)^2]} \quad (9)$$

and $i = \sqrt{-1}$, Im means "imaginary part of," and the branch cut for the ½ power term is along the crack.

The solution for $\mathbf{h}_1$ could also have been developed by using Fourier analysis together with some fundamental fields given by Bueckner [6], having a cos ωz variation along the crack front, to construct what he has called a fundamental field with a point of concentration. Essentially, his fundamental fields would have to be superposed over all ω by weighting each with the Fourier transform at frequency ω of a Dirac function, centered at $z = s$, and integrating over all ω to obtain $\mathbf{h}_1(\mathbf{r};s)$.

As remarked, Bueckner later derived [10] the full set of three weight functions for the half-plane crack and for the circular crack. From his work, the function defined by the integral in Eq 8, with $a = s = 0$, is

$$\frac{1}{\pi\sqrt{2\pi}}\int_y^\infty \frac{\text{Im}(x + iy)^{1/2}\,dy}{x^2 + y^2 + z^2} = \text{Re}\left\{\frac{1}{2\pi\sqrt{\pi}\zeta}\ln\left[\frac{q + \zeta}{q - \zeta}\right]\right\} \quad (10)$$

where $\zeta = (x + iz)^{1/2}$ and $q = \text{Re}[\sqrt{2}(x + iy)^{1/2}]$, and all the weight functions for the half-plane crack may be expressed in terms of linear differential operation on the complex function whose real part appears on the right in Eq 10.

Derivation of Second Property from First

Assume that the $\mathbf{h}_\alpha(\mathbf{r};s)$ are defined primitively by their first property, Eq 5, that is, $h_{\alpha j}(\mathbf{r};s)$ is defined as the mode α intensity factor at location s on CF due to a unit point force in the j direction at position $\mathbf{r}$. The writer's derivation [2,7] of the second property is outlined, and modestly recast, here. Let U be the strain energy of the cracked solid in Fig. 1*b* and let

$$V = -\int_{\text{Body}} \mathbf{f}(\mathbf{r}) \cdot \mathbf{u}(\mathbf{r})\, dxdydz \tag{11}$$

be the potential energy of the applied forces (regarded as fixed), which induce intensity factors $K_\alpha(s)$ along the crack front.

Consider a family of crack shapes defined by advancing CF normal to itself by $Ag(s)$, where, again, $g(s)$ is arbitrary but fixed once chosen, and the advance process corresponds to Fig. 1*c* with the label $\delta a(s)$ replaced by $A\, g(s)$. Observe that the area elements swept out in incremental change δA in A can be written as

$$\delta(\text{area}) = p(A,s)\, \delta A\, g(s)\, ds \tag{12}$$

where $p(A,s)$ is a function dependent on the curvature of CF at s but need not be written out here since we will, in the end, only need its value at $A = 0$, at which $p(0,s) = 1$.

Also, suppose that in addition to the given load system, an arbitrary point force $\mathbf{F}$ is applied to the body at $\mathbf{r}$, where the displacement is $\mathbf{u}(\mathbf{r})$ or, more fully, $\mathbf{u}(\mathbf{r}, A,\mathbf{F})$. (Formally, $\mathbf{u}$ is unbounded at $\mathbf{r}$ when $\mathbf{F}$ differs from zero, but we shall shortly be setting $\mathbf{F} = 0$. To keep things finite, we may distribute $\mathbf{F}$ uniformly over a small sphere [7] of radius ϵ about $\mathbf{r}$, interpret $\mathbf{u}(\mathbf{r})$ as the average over that same sphere, and later let $\epsilon \rightarrow 0$ after setting $\mathbf{F} = 0$.) Thus the total stress intensity factors $\hat{K}_\alpha = \hat{K}_\alpha(s, A,\mathbf{F})$ have the form, when $A = 0$,

$$\hat{K}_\alpha(s,0,\mathbf{F}) = K_\alpha(s) + \mathbf{h}_\alpha(\mathbf{r},s) \cdot \mathbf{F} \tag{13}$$

Now, by the definition of the elastic energy release rate G, and the relation between increments of work and energy, one must have

$$\delta U(A,\mathbf{F}) = -\delta V(A,\mathbf{F}) + \mathbf{F} \cdot \delta\mathbf{u}(\mathbf{r}, A,\mathbf{F}) - \int_{\text{CF}} G(s, A,\mathbf{F})[p(A,s)\, \delta A\, g(s)\, ds] \tag{14}$$

for arbitrary variations of $\mathbf{F}$ and A, where U is the strain energy of the cracked body. Thus

$$\delta[\mathbf{F} \cdot \mathbf{u} - V - U] = \mathbf{u}(\mathbf{r}, A,\mathbf{F}) \cdot \delta\mathbf{F} + \left\{\int_{\text{CF}} G(s, A,\mathbf{F}) p(A,s) g(s)\, ds\right\} \delta A \tag{15}$$

and, evidently, the right side must be a perfect differential in $\delta\mathbf{F}$ and δA. Thus their coefficients must satisfy the Maxwell-like reciprocal relations

$$\frac{\partial \mathbf{u}(\mathbf{r},A,\mathbf{F})}{\partial A} = \frac{\partial}{\partial \mathbf{F}}\left\{\int_{CF} G(s,A,\mathbf{F})p(A,s)g(s)\,ds\right\} \tag{16}$$

Since, by Eq 4, $G = \Lambda_{\alpha\beta}\hat{K}_\alpha\hat{K}_\beta$, this means that

$$\frac{\partial \mathbf{u}(\mathbf{r},A,\mathbf{F})}{\partial A} = 2\int_{CF} \Lambda_{\alpha\beta}(s)\,\frac{\partial \hat{K}_\alpha(s,A,\mathbf{F})}{\partial \mathbf{F}}\,\hat{K}_\beta(s,A,\mathbf{F})p(A,s)g(s)\,ds \tag{17}$$

One now sets $A = 0$, recalling that then $p(0,s) = 1$ and, from Eq 13

$$\partial \hat{K}_\alpha(s,0,\mathbf{F})/\partial\mathbf{F} = \mathbf{h}_\alpha(\mathbf{r};s) \tag{18}$$

Setting $\mathbf{F} = 0$ also, in which case $\hat{K}_\beta(s,0,0) = K_\beta(s)$, the left and right sides of Eq 17 coincide with those of Eq 7, thus providing the desired proof of the second property enunciated for the weight functions.

A New Derivation of the Second Property

Consider an arbitrary location s along CF. Relative to the local coordinate system there, Fig. 1*a*, let us move along the negative 2 axis (that is, perpendicular to CF, into the crack zone) a small distance ρ and, at that site apply a force pair $\mathbf{Q}$ to the upper crack surface and $-\mathbf{Q}$ to the lower. Let us now apply the elastic reciprocal theorem to the load system just described and to another system consisting of a point force $\mathbf{F}$ at $\mathbf{r}$. The latter causes intensity factors $\mathbf{h}_\alpha(\mathbf{r};s)\cdot\mathbf{F}$ at location s and hence, by Eq 2, the relative crack surface displacements $\Delta\mathbf{u}^F$ induced by the force $\mathbf{F}$ at distance ρ, very near to location s along CF, is

$$\Delta u_\alpha^F = 8\sqrt{\rho/2\pi}\Lambda_{\alpha\beta}(s)h_{\beta j}(\mathbf{r},s)F_j \tag{19}$$

By the elastic reciprocal theorem, the work product $Q_\alpha\Delta u_\alpha^F$ equals the product $F_j u_j^Q(\mathbf{r})$, where $u_j^Q(\mathbf{r})$ denotes the j direction displacement induced at $\mathbf{r}$ by the pair of point forces $\mathbf{Q}$ and $-\mathbf{Q}$ at (small) distance ρ from location s along CF.

Thus another characterization of the weight functions which emerges is that

$$u_j^Q(\mathbf{r}) = 8\sqrt{\rho/2\pi}Q_\alpha\Lambda_{\alpha\beta}(s)h_{\beta j}(\mathbf{r},s) \tag{20}$$

is the j direction displacement induced at $\mathbf{r}$ by the force pair near CF. The only sense in which ρ is assumed small in this derivation is that terms of order higher than $\sqrt{\rho}$, in the expression for $\Delta\mathbf{u}^F$ induced at distance ρ from CF by force $\mathbf{F}$ at $\mathbf{r}$, must be negligible by comparison to $\sqrt{\rho}$. Note that the components Q_α in Eq 20 are referred to the local 1, 2, 3 coordinate system at s, just as are those of Δu_α^F in Eq 19.

Now consider the process of crack advance by $\delta a(s)$, as in Fig. 1*c*. The variation $\delta\mathbf{u}(\mathbf{r})$ in displacement at $\mathbf{r}$ can be calculated as the effect of removing the stresses of type

$$\sigma_{1\alpha} = K_\alpha(s)/\sqrt{2\pi\rho'} \tag{21}$$

which acted before enlargment. This manner of addressing the effects of crack enlargement is similar to Panasyuk's [13] approach to the perturbed circular crack. Stress removal is equivalent to placing pairs of infinitesimal forces

$$Q_\alpha = [K_\alpha(s)/\sqrt{2\pi\rho'}]\, d\rho' ds \tag{22}$$

on the crack faces at distance $\rho = \delta a(s) - \rho'$ from the new crack tip. Each such force causes the displacement u_j at $\mathbf{r}$ identified above as $u_j^\rho(\mathbf{r})$, and thus the net displacement variation $\delta u_j(\mathbf{r})$ due to the considered crack advance is

$$\delta u_j(\mathbf{r}) = \int_{CF}\int_0^{\delta a(s)} \left[8\sqrt{\frac{\delta a(s) - \rho'}{2\pi}}\, \frac{K_\alpha(s)}{\sqrt{2\pi\rho'}}\, \Lambda_{\alpha\beta}(s) h_{\beta j}(\mathbf{r};s) \right] d\rho' ds \tag{23}$$

The integral on ρ' is elementary, and one readily confirms that this equation agrees with Eq 6, providing the alternate derivation. As stated, this derivation assumes that $\delta a(s)$ is everywhere positive. It is not hard to modify it when $\delta a(s)$ is negative (in those zones one applies infinitesimal forces Q_α to create, rather than remove, the appropriate near-tip stresses $\sigma_{1\alpha}$), and thus to make the derivation fully general.

The reciprocal interpretation of the three-dimensional weight functions in Eq 20 has also been noticed by Bueckner (private communication). It generalizes an interpretation given by Paris et al. [25] in the two-dimensional case.

Variation of Green's Function with Change of Crack Front Position

The Green's function $G_{jk}(\mathbf{r},\mathbf{r}')$ for an elastic body is defined by the property

$$u_j(\mathbf{r}) = \int_{Body} G_{jk}(\mathbf{r},\mathbf{r}') f_k(\mathbf{r}')\, dx'dy'dz' \tag{24}$$

and, naturally, the Green's function depends on the position of the crack and varies with change of that position. Letting $\delta G_{jk}(\mathbf{r},\mathbf{r}')$ be that variation, it is seen by the second property of the weight functions, when the $K_\beta(s)$ in Eq 6 is expressed by use of the first property, Eq 5, that

$$\delta G_{jk}(\mathbf{r},\mathbf{r}') = 2\int_{CF} \Lambda_{\alpha\beta}(s) h_{\alpha j}(\mathbf{r},s) h_{\beta k}(\mathbf{r}',s)\delta a(s)\, ds \tag{25}$$

to first order in $\delta a(s)$ when the crack front is advanced, as in Fig. 1*c*.

This emphasizes the remarkable information content of the weight functions. While primitively they have the relatively humble role of describing only the distribution of stress intensity factors induced along the crack front by arbitrary point forces, they turn out to relate to the Green's function and thus to the entire displacement field induced throughout the body by such point forces. In fact, if the weight functions are known for a sequence of crack-front positions, corresponding to introduction of the crack and enlargement to its present size, then $G_{jk}(\mathbf{r},\mathbf{r}')$ can be calculated directly from the weight functions, by integrating $\delta G_{jk}(\mathbf{r},\mathbf{r}')$ of Eq 25, provided that the initial $G_{jk}(\mathbf{r},\mathbf{r}')$ is known for the uncracked solid.

For example, consider a half-plane crack with tip position at a, as in Fig. 2, or a circular crack of radius a (Fig. 3*b*). Then $G_{jk} = G_{jk}(\mathbf{r},\mathbf{r}';a)$ and, by letting $\delta a(s)$ be uniform in s,

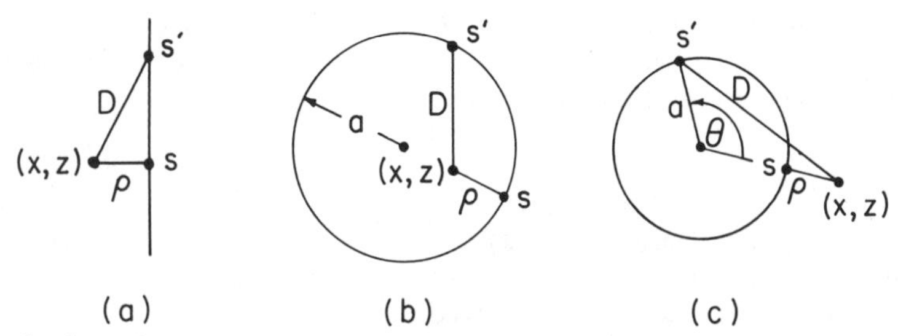

FIG. 3—*Cracks in the plane* y = 0 *and notation to describe crack face weight functions for* (a) *half-plane crack,* (b) *circular crack, and* (c) *circular connection (externally cracked).*

and dividing both sides of Eq 25 by that uniform value, the left side becomes $\partial G_{jk}(\mathbf{r},\mathbf{r}';a)/\partial a$. Thus, by knowing the classical Kelvin G_{jk} for an uncracked full-space and adding to it the integral of $\partial G_{jk}/\partial a$, from $-\infty$ to a for the half-plane crack or from 0 to a for the circular crack, one obtains G_{jk} for the cracked solid.

Sometimes it is not necessary to know all three weight functions to calculate what will serve as a Green's function for the class of loadings actually experienced by a cracked solid. For example, suppose our concern is exclusively with loading systems that produce pure Mode 1 along the crack front. In that case, when $\delta G_{jk}(\mathbf{r},\mathbf{r}')$, or its integral over some crack introduction sequence, is actually multiplied by $f_k(\mathbf{r}')$ and integrated over all volume elements $dx'dy'dz'$ of the body, the product $h_{\beta k}f_k$ integrates to zero, by Eq 5, for all load systems of the class considered when $\beta = 2$ or 3. It then suffices, for the pure Mode 1 load systems which are considered, to write (for example, for an isotropic solid)

$$\delta G_{jk}(\mathbf{r},\mathbf{r}') = \frac{1-\nu}{\mu}\int_{CF} h_{1j}(\mathbf{r};s)h_{1k}(\mathbf{r}';s)\delta a(s)\,ds \tag{26}$$

The concepts outlined here have been used to derive the Green's function or, related to it, the expression for relative crack surface displacement $\Delta\mathbf{u}$ under general loadings, for half-plane [*7*] and circular [*11,15*] cracks and circular connections [*12*].

Relation to Bueckner's Concept of Fundamental Fields

Bueckner's approach to the three-dimensional theory may be summarized as follows: Let $\mathbf{v}(\mathbf{r})$ be a fundamental field, that is, a solution to the Navier displacement equations of three-dimensional elasticity, equilibrating null applied loading. In general, no such field other than $\mathbf{v} = 0$ (or $\mathbf{v}$ = rigid motion) would exist, for such is clearly a solution of the elasticity equations for null loading and, by the uniqueness theorem, no other type of solution could exist. However, the fundamental fields lie outside the scope of fields covered by the uniqueness theorem, since the fundamental fields, to be useful, must have unbounded strain energy. In fact, such fields have displacements which become infinite as $1/\sqrt{\rho}$ near the crack front, and hence stresses and strains which become infinite as $1/\rho\sqrt{\rho}$. Their strength at location s along the crack front is characterized in terms of the discontinuity $\Delta\mathbf{v}$ between upper and lower crack surfaces by the Bueckner strength function

$$B_\alpha(s) = \lim_{\rho\to o}\left[\sqrt{\pi\rho/2}\,\Delta v_\alpha(s,\rho)\right] \tag{27}$$

where now the $\Delta\mathbf{v}$ are referred to the local coordinates (Fig. 1*a*) at s and $\Delta v_\alpha(s,\rho)$ means Δv_α at distance ρ along the -2 axis through the location along CF at arc length s.

In terms of these fundamental fields and their strength distributions around CF, Bueckner's basic result is that

$$\int_{CF} B_\alpha(s)K_\alpha(s)\,ds = \int_{Body} \mathbf{v}(\mathbf{r}) \cdot \mathbf{f}(\mathbf{r})\,dxdydz \tag{28}$$

The proof is as follows: $\mathbf{v}(\mathbf{r})$ can be regarded as an unobjectional elastic displacement field for a cracked solid from which we exclude a small cylindrical tube, say, of radius ρ, along CF. The stresses associated with $\mathbf{v}$ equilibrate zero body and surface force everywhere except along the tube surface, where tractions $\mathbf{T}$ of order $1/\rho\sqrt{\rho}$ must be applied to maintain displacements $\mathbf{v}$.

We now apply the elastic reciprocal theorem to the pair of fields consisting of the fundamental field $\mathbf{v}(\mathbf{r})$ just described and the actual displacement field $\mathbf{u}(\mathbf{r})$ induced by the applied forces $\mathbf{f}(\mathbf{r})$. Thus, the work of the forces of the $\mathbf{v}$ field (that is, of the tractions $\mathbf{T}$, of order $1/\rho\sqrt{\rho}$ along the tube) on the $\mathbf{u}$ displacements equals the work of the forces $\mathbf{f}$, and of tractions on the tube resulting from the $\mathbf{u}$ field, on the $\mathbf{v}$ displacements. We can let $\rho \to 0$ in the two work expressions. The work of $\mathbf{f}$ is plainly given by the right side of Eq 28, and it should appear plausible that the limit of the works of the tube tractions is given by the left side, since the $\mathbf{u}$ field near CF is proportional to the K's times $\sqrt{\rho}$. Thus $\mathbf{T} \cdot \mathbf{u}$ is of order $1/\rho$ along the tube, as is the work on $\mathbf{v}$, and the $1/\rho$ gets cancelled out when we integrate over the surface of the tube, so there is a well-defined limit as $\rho \to 0$. Of course, the strengths $B_\alpha(s)$ have been so defined in Eq 27 that the tube-surface work terms combine to what is written on the left of Eq 28.

Consider a limiting fundamental field which may be said to have a point of concentration at location s', that is, for which the strength distribution is

$$B_\alpha(s) = \delta^K_{\alpha\beta}\delta^D(s - s') \tag{29}$$

where $\delta^K_{\alpha\beta}$ is the Kronecker-δ and $\delta^D(\ldots)$ the Dirac-δ. Then by comparing the result of Eqs 28 to 5, giving the first property of the weight functions, it is evident that the limiting fundamental field described is just

$$\mathbf{v}(\mathbf{r}) = \mathbf{h}_\beta(\mathbf{r};s') \tag{30}$$

Since the field $\mathbf{u}(\mathbf{r})$ created by general applied loadings satisfies the Navier displacement equations of elasticity, so also must $\delta\mathbf{u}(\mathbf{r})$ of Eq 6 [and $\partial\mathbf{u}(\mathbf{r},A)/\partial A$ of Eq 7]. Further, since both $\mathbf{u}(\mathbf{r})$ and $\mathbf{u}(\mathbf{r}) + \delta\mathbf{u}(\mathbf{r})$ equilibrate the same system of loadings, $\delta\mathbf{u}(\mathbf{r})$ and $\partial\mathbf{u}(\mathbf{r},A)/\partial A$ satisfy the displacement equations of elasticity corresponding to null loading. One therefore suspects that not only $\mathbf{h}_\alpha(\mathbf{r};s)$, but also every field of type $\partial\mathbf{u}(\mathbf{r},A)/\partial A$ meets the requirements to be a Bueckner fundamental $\mathbf{v}(\mathbf{r})$ field. It is easy to confirm that $\partial\mathbf{u}(\mathbf{r},A)/\partial A$ has a singularity of the appropriate order, $1/\sqrt{\rho}$, near the crack front so that it is indeed a candidate $\mathbf{v}(\mathbf{r})$. The Bueckner strength distribution $B_\alpha(s)$ associated with $\partial\mathbf{u}(\mathbf{r},A)/\partial A$ is readily determined, either by examining the field near $\rho = 0$ or by substituting $\partial\mathbf{u}(\mathbf{r},A)/\partial A$ as expressed by Eq 7 into Eq 28 for $\mathbf{v}(\mathbf{r})$ and then using Eq 5. Either way, one finds that

$$B_\alpha(s) = 2\Lambda_{\alpha\beta}(s)K_\beta(s)g(s) \tag{31}$$

We have just seen that every field $\partial \mathbf{u}(\mathbf{r}, A)/\partial A$, corresponding to variation of crack front location under fixed load, provides a Bueckner fundamental field $\mathbf{v}(\mathbf{r})$. The converse applies too: Every Bueckner fundamental field $\mathbf{v}(\mathbf{r})$ can be identified with a field $\partial \mathbf{u}(\mathbf{r}, A)/\partial A$ for a suitably loaded cracked solid with suitable choice of growth function $g(s)$. This is true because $\Lambda_{\alpha\beta}$ is invertible, and hence every distribution of $B_\alpha(s)$ corresponds, by Eq 31, to an equivalent distribution of products $K_\beta(s)g(s)$ in the description of crack growth under fixed load. Since by suitable choice of body-force distribution it is possible to make the $K_\beta(s)$ vary in any desired manner around the crack front, this confirms that every $\mathbf{v}(\mathbf{r})$ has a $\partial \mathbf{u}(\mathbf{r}, A)/\partial A$ representation. Thus Bueckner's class of fundamental fields and the writer's class of fields generated by incremental crack growth are identical.

Crack-Face Weight Functions, Some Examples, and Symmetry Properties

For applications to perturbations of crack shape, and for some other purposes, there is no need to know the full-field weight functions $\mathbf{h}_\alpha(\mathbf{r};s)$ for positions $\mathbf{r}$ throughout the entire body. Rather, it suffices to know their jumps across the crack plane, that is, to know the crack-face weight functions defined by

$$\mathbf{k}_\alpha(x,z;s') = \mathbf{h}_\alpha(x,0^+,z;s') - \mathbf{h}_\alpha(x,0^-,z;s') \tag{32}$$

for all positions (x,z) within the crack zone and locations s' along CF. The properties of the $\mathbf{k}_\alpha$ are analogous to those enunciated for the $\mathbf{h}_\alpha$ earlier. First, if the loading consists of tractions $\mathbf{T} = \mathbf{T}(x,z)$ per unit area on the upper crack face, and $-\mathbf{T}$ on the lower, then the intensity factors at location s' along the crack front are

$$K_\alpha(s') = \int_{\text{crack}} \mathbf{k}_\alpha(x,z;s') \cdot \mathbf{T}(x,z)\, dxdz \tag{33}$$

Of course, general loadings can always be reduced to this case by the well-known superposition procedure. Second, if the crack front location is altered by $\delta a(s)$, as in Fig. 1*c*, under fixed loading conditions then the variation of the crack-surface displacement discontinuity is

$$\delta[\Delta\mathbf{u}(x,z)] = 2\int_{\text{CF}} \Lambda_{\alpha\beta}(s')\mathbf{k}_\alpha(x,z;s')K_\beta(s')\delta a(s')\, ds' \tag{34}$$

to first order in $\delta a(s)$. These are the specific forms of Eqs 5 and 6 for crack-face loading and crack-face displacement discontinuities.

Some examples of crack-face weight functions are cited now for cracks in unbounded isotropic solids. First one may observe that symmetry requires

$$k_{1x} = k_{1z} = k_{2y} = k_{3y} = 0 \tag{35}$$

in all such cases. Thus, for tensile (Mode 1) loadings, one needs only $k_{1y}(x,z;s')$. The half-plane crack, circular crack of radius a, and circular connection of radius a (external annular crack of infinite outer radius) are shown in Figs. 3*a*, *b*, and *c*, respectively. Choose a point (x,z) within the crack space, let $\rho = \rho(x,z)$ be the shortest distance to the crack front, and $D = D(x,z;s')$ be the distance from (x,z) to location s' along the crack front. Then for the half-plane crack (for examples, see Refs *6* and *7*)

$$k_{1y}(x,z;s') = \sqrt{2\rho}/\pi\sqrt{\pi}\, D^2 \tag{36}$$

and for the circular crack [*6,11,13*]

$$k_{1y}(x,z;s') = \sqrt{\rho(2\alpha - \rho)}/\pi\sqrt{\pi a}\, D^2 \tag{37}$$

When the two half-spaces joined by the circular connection are restrained against displacement at infinity [*12*]

$$k_{1y}(x,z;s') = \sqrt{\rho(2a + \rho)}/\pi\sqrt{\pi a}\, D^2 \tag{38}$$

When the restraints at infinity are removed so as to allow free translation in the y direction, free rotation, or both (that is, completely unrestrained), various terms must be added to the expression for k_{1y} just given. For example, in the completely unrestrained case, the terms [*12*]

$$(\pi a)^{-3/2}\{\cos^{-1}[1/(1 + \rho/a)][1 + 3(1 + \rho/a)\cos\theta] + [\sqrt{\rho(2a + \rho)}/(a + \rho)]\cos\theta\}$$

must be added, where the angle $\theta = \theta(x,z;s')$ is identified in Fig. 3*c*.

Shear mode crack face weight functions, k_{2x}, k_{2z}, k_{3x}, and k_{3z} are given in Refs *7* and *9* for the half-plane crack and in Ref *15* for the circular crack.

Let us limit attention to pure Mode 1 conditions in homogeneous isotropic solids for the rest of this section. Consider a general crack shape as in Fig. 4, choose two locations s and s' along the crack front, locate a point (x,z) by moving into the crack zone a small perpendicular distance ρ from s, and a point (x',z') by moving a small distance ρ' from s'.

Guided by the above examples, we observe that $k_{1y}(x,z;s')/\sqrt{\rho(x,z)}$ has a well-defined limit as $\rho \rightarrow 0$, that is, as (x,z) approaches the location s along the crack front. Let us therefore introduce the general representation

$$k_{1y}(x,z;s') = \frac{\sqrt{2\rho(x,z)}W(x,z;s')}{\pi\sqrt{\pi}\, D^2(x,z;s')} \tag{40}$$

where $W(x,z;s')$ has a well-defined limit, denoted by

$$W(s,s') = \lim_{\rho(x,z)\rightarrow 0} [W(x,z;s')] \tag{41}$$

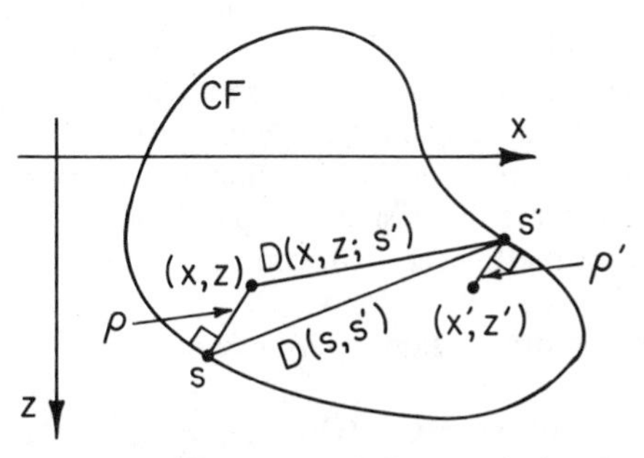

FIG. 4—*General crack shape, for discussion of crack face weight function and symmetry properties.*

as (x,z) approaches location s. We use $D(s,s')$ to denote the corresponding limit of $D(x,z;s')$; it is the distance from location s to location s' (Fig. 4). For example

$$W(s,s') = 1 \tag{42}$$

for the cases of the half-plane crack and the circular crack, and also for the circular connection when points at infinity are restrained against displacement, whereas

$$W(s,s') = 1 + D^2(s,s')[1 + 6 \cos \theta(s,s')]/a^2 \tag{43}$$

for the circular connection with unrestrained displacements at infinity.

Consider now the opening displacement $\Delta u_y^Q(x',z')$ induced by force Q_y in the y (or 1) direction on the upper crack face, and $-Q_y$ on the lower crack face, at (x,z). By Eqs 20 and 32

$$\Delta u_y^Q(x',z') = 8\sqrt{\rho(x,z)/2\pi}[(1 - \nu)/2\mu]k_{1y}(x',z';s)Q_y \tag{44}$$

and by the reciprocal theorem this must be the same as the opening displacement induced at (x,z) when the same force pair is applied at (x',z'). Thus

$$\sqrt{\rho(x,z)}\, k_{1y}(x',z';s) = \sqrt{\rho'(x',z')}\, k_{1y}(x,z;s') \tag{45}$$

and dividing this equation by $\sqrt{\rho\rho'}$ and then letting $\rho \rightarrow 0$ and $\rho' \rightarrow 0$ (recall that Eq 20 is exact in this limit), we obtain the symmetry property

$$W(s',s) = W(s,s') \tag{46}$$

Some Applications of Three-Dimensional Weight-Function Theory

Guided by the two-dimensional theory, it might be thought that the main application of three-dimensional weight-function theory is to determine stress intensity factors for a given crack geometry under a variety of different loading conditions. There is evidently much to be done on such applications, but here the focus will be on some new types of investigation allowed by the theory.

Crack Tip Interaction with Transformation Strains and Dislocations

This class of applications is of interest for the micromechanics of fracture. The three-dimensional weight-function concepts enable one to develop useful relations for study of the effects of particles of different elastic moduli, or of particles that undergo a stress-induced shape transformation (as in transformation toughening), or of smaller-scale micro-cracks, on the stress intensity factors and stress fields near a crack tip. Similarly, they allow one to address various three-dimensional dislocation interactions with crack tips, including development of models for three-dimensional loop emission from stressed crack tips.

The starting point is with the Eshelby [*26*] concept of a distribution of stress-free transformation strain $\epsilon_{pq}^T(\mathbf{r})$. If we consider a solid of elastic moduli $C_{jkpq}(\mathbf{r})$, then the stress field in the presence of this strain distribution is assumed to be given by

$$\sigma_{jk}(\mathbf{r}) = C_{jkpq}(\mathbf{r})[u_{p,q}(\mathbf{r}) - \epsilon_{pq}^T(\mathbf{r})] \tag{47}$$

Here $C_{jkpq} = C_{pqjk} = C_{kjpq} = C_{jkqp}$, and if $L = L(\mathbf{r})$, then L,j for $j = (x,y,z)$ denotes $\partial L/\partial(x,y,z)$. Equilibrium in the absence of any actual body force requires that $\sigma_{jk,j} = 0$ and this equation, using Eq 42, is the same as the equilibrium equation in terms of $\mathbf{u}(\mathbf{r})$ for a solid having zero transformation strain but being subject to a distribution of "effective" body force

$$f_k^{\text{eff}}(\mathbf{r}) = -[C_{jkpq}(\mathbf{r})\epsilon_{pq}^T(\mathbf{r})],j \tag{48}$$

(and also effective surface tractions when ϵ_{pq}^T is nonzero along the crack faces or external surfaces).

Thus the writer [*8*] showed that the intensity factors induced at location s along CF by a distribution of $\epsilon_{pq}^T(\mathbf{r})$ in some region V are

$$K_\alpha(s) = \int_V h_{\alpha k,j}(\mathbf{r};s)C_{jkpq}(\mathbf{r})\epsilon_{pq}^T(\mathbf{r})\; dxdydz \tag{49}$$

General Somigliana dislocations, that is, surfaces S on which a displacement jump $\Delta\mathbf{u}(\mathbf{r})$ is prescribed for $\mathbf{r}$ on S, can also be considered as a limiting case. Thus [*8*]

$$K_\alpha(s) = \int_S h_{\alpha k,j}(\mathbf{r};s)C_{jkpq}(\mathbf{r})N_p(\mathbf{r})\Delta u_q(\mathbf{r})\; dS(\mathbf{r}) \tag{50}$$

where $\mathbf{N}$ is the local normal to S, chosen so that $\Delta\mathbf{u}$ is the difference between $\mathbf{u}$ on the side of S towards which $\mathbf{N}$ points and $\mathbf{u}$ on the other side.

This formula requires a modification, outlined in Refs *17* to *19* for the case of the half-plane crack, to by-pass formally divergent integrals for locations s along any segment of CF which happens to lie in the dislocated surface S or along a part of its border. The latter is particularly an issue for a microcrack [representable as some distribution, albeit unknown, of $\Delta\mathbf{u}(\mathbf{r})$] or dislocation loop emanating from the crack tip. In the simplest case, when the crack can be regarded as a half-plane crack with straight front, as in Fig. 2, and when the surface S is a planar zone emanating from the crack tip, the fix-up is simple. We replace $\Delta\mathbf{u}(x,y,z)$ in Eq 50 by $\Delta\mathbf{u}(x,y,z) - \Delta\mathbf{u}(0,0,s)$, where the location of interest along the crack front is at $z = s$ and the crack tip is assumed to be at $a = 0$ (Fig. 2), and we extend the integral over the entire half-plane emanating from the tip that contains S [*17*]. This is equivalent to subtracting the effect of a constant displacement, $\Delta\mathbf{u}(0,0,s)$, over that entire half-plane, which has no effect on the stress field.

Stress Field—In the types of micromechanical applications mentioned, one generally wants to know the stress field induced in V by the transformation strain, or on S by the dislocation. For example, if we use $\epsilon_{pq}^T(\mathbf{r})$ as an artifact, to represent the effect of an inclusion of different moduli than its surroundings (the Eshelby [*26*] procedure), then the ϵ_{pq}^T distribution over the inclusion region V should, in principle, be chosen so that the local values of σ_{jk}, due to ϵ_{pq}^T and to the external loading, relate to the local values of $u_{j,k}$ within V in a manner compatible with the actual constitutive relation (not necessarily linear) for the inclusion material. This clearly poses a formidable problem, but one which would be tractible in a more approximate form, for example, if ϵ_{pq}^T were taken as locally uniform in V, or in some subregions of V, and the constitutive relation were required to be satisfied only at the centroid of V, or at centroids of its subregions. Similarly, when S represents a microcrack one would, in principle, choose $\Delta\mathbf{u}(\mathbf{r})$ on S so that the traction stresses $N_j\sigma_{jk}$, due to that $\Delta\mathbf{u}$ distribution and to the external loading, vanish on S. Equally, for calculating the overall energy and alteration of Peach-Koehler (configurational) forces along a near-tip (Volerra, constant $\Delta\mathbf{u}$) dislocation loop, the stresses are required.

The stress field due to transformation strain or dislocation can be represented as follows for the half-plane crack with tip along the z axis (that is, with $a = 0$ in Fig. 2). Let $K_\alpha(s,a)$ denote the intensity factor distribution calculated from Ref *49* or *50* when the tip is at $x = a$. Also, let $\mathbf{h}_\alpha(x - a,y,z - s)$ denote the weight functions $\mathbf{h}_\alpha(\mathbf{r};s)$; this is the form they take for the half-plane crack in a homogeneous material (for example, see Eqs 8 and 9). Then [*8*]

$$\sigma_{pq}(\mathbf{r}) = \sigma^0_{pq}(\mathbf{r}) + 2\Lambda_{\alpha\beta}C_{jkpq}\int_{-\infty}^{0}\int_{-\infty}^{+\infty} h_{\alpha k,j}(x - a,y,z - s)K_\beta(s,a)\, ds da \tag{51}$$

Here $\sigma^0_{pq}(\mathbf{r})$ is the stress field which that same distribution of transformation strain or dislocation would induce in an infinite, uncracked solid. When the crack tip contains segments which lie in or on the border of a dislocated surface S, certain subtraction procedures, analogous to subtracting $\Delta\mathbf{u}(0,0,s)$ as explained earlier, have to be used to avoid formally divergent terms. For example, as discussed in Refs *17* to *19*, when S is a dislocated surface emanating from the crack tip, both σ^0_{pq} and the integral in Eq 51 contribute singular stress terms, of order $1/\rho'$ near the tip (Fig. 1*a*), which cancel one another to leave the proper $1/\sqrt{\rho'}$ singularity.

Although the methods outlined in this section have great potential for the three-dimensional micromechanics of phenomena at crack tips, it should be cautioned that the integrations involved are formidable. Equation 51 for a dislocation loop effectively involves integration over four variables, two to get the K_α from Eq 50 and two more displayed explicitly in Eq 51. Also, the subject is young, the Mode 1 weight function being published only in 1985 [*7*] and the Mode 2 and 3 functions only in 1987 [*10*], and only rather simple applications have been made thus far. Some of these are now summarized, all involving isotropic solids.

Dilatant Shape Transformation—Suppose that a transformation strain distribution corresponding to pure dilatation, $\epsilon^T_{pq}(\mathbf{r}) = \delta_{pq}\theta(\mathbf{r})/3$, occurs in V. Then from Eq 49 and Eqs 8 and 9, K_1 along the crack front is given by [*8*]

$$K_1(s) = \frac{2\mu(1 + \nu)}{3(2\pi)^{3/2}(1 - \nu)}\int_V \frac{\cos(\phi/2)}{\rho^{1/2}D^2}\left[1 - \frac{8\rho^2}{D^2}\sin^2\left(\frac{\phi}{2}\right)\right]\theta(\mathbf{r})\, dxdydz \tag{52}$$

where (notation of Fig. 2) $\tan\phi = y/(x - a)$, $\rho^2 = (x - a)^2 + y^2$, and $D^2 = \rho^2 + (z - s)^2$. This reproduces known results from two-dimensional models [*27,28*] developed in the literature on transformation toughening of ceramics, when $\theta(x,y,z)$ reduces to $\theta(x,y)$.

Microcrack Ahead of Half-Plane Crack—Let the microcrack surface S lie on $y = 0$ and assume that S lies entirely in the region $x > 0$. Here the (main) crack tip is at $x = 0$ (that is, $a = 0$ in Fig. 2) so the microcrack is separated from the main crack. The problem is assumed to have Mode 1 symmetry so that the y (and only) component of displacement discontinuity on S is $\Delta u_y(x,z)$.

Then the intensity $K_1(s)$ induced along the half-plane crack tip by the opening distribution can be evaluated from Eq 50 and has the representation [8]

$$K_1(s) = \frac{\mu}{(2\pi)^{3/2}(1-\nu)} \int_S \frac{\Delta u_y(x,z)\,dxdz}{\sqrt{x}[x^2 + (z-s)^2]} \tag{53}$$

When $\Delta u_y \geq 0$, $K_1(s)$ is everywhere positive, so microcrack opening directly ahead of the main crack increases K_1 everywhere along the crack front. Of course, to this K_1 must be added the K_1 due to the externally applied loading (in the absence of the microcrack), and the actual presently unknown Δu_y distribution on S will be directly proportional to the intensity of that loading, so that Eq 53, like the equation for σ_{yy} below, is only a result that is useful as part of the process of constructing a fuller solution.

The stress integral of Eq 51 gives in this case [8], for $x > 0$

$$\sigma_{yy}(x,0,z) = \sigma^o_{yy}(x,0,z) + \frac{\mu}{4\pi^2(1-\nu)\sqrt{x}} \int_S \frac{Q(x,x',z-z')\Delta u_y(x',z')\,dx'dz'}{\sqrt{x'}[(x-x')^2 + (z-z')^2]} \tag{54}$$

where

$$Q = 1 - P\tan^{-1}(1/P) \quad \text{and} \quad P = 2\sqrt{xx'}/\sqrt{(x-x')^2 + (z-z')^2} \tag{55}$$

[so that the integrand is nonsingular when (x',z') coincides with (x,z)] and [29]

$$\sigma^o_{yy}(x,0,z) = \frac{\mu}{4\pi(1-\nu)} \int_S \frac{(x'-x)\Delta u_{y,x}(x',z') + (z'-z)\Delta u_{y,z}(x',z')}{[(x-x')^2 + (z-z')^2]^{3/2}}\,dx'dz' \tag{56}$$

gives the stress induced by the same (unknown) opening on S in an uncracked full space. Anderson and Rice [17] show how to modify these expressions when a border of S coincides with a segment of the crack tip. Again, the stresses induced by applied loadings, in the absence of the microcrack, must be added to those given here.

The present formulation leads to a singular integral equation for $\Delta u_y(x,z)$ on S, analogous to those developed as a starting point in numerical treatments of three-dimensional crack problems in simpler full or half-plane geometries [29–31] and similar methods of discretization can be employed for numerical solution.

Dislocation Loop Emanating from Tip—When $\Delta u_y(x,z) = b$ (a constant; Volterra dislocation) over a region S (on $y = 0$) with border along the crack tip, we have the problem of a prismatic dislocation loop emanating from the crack tip. The two-dimensional elastic interaction between a crack tip and a parallel dislocation line has been worked out in great generality [32,33]. Such interactions are of interest in studies of intrinsic cleavability of solids, where an attempt is made to estimate the critical combination of intensity factors at which a dislocation nucleates from the crack tip [32,34]. Thus far, the three-dimensional aspects of such dislocation emission have been treated only approximately by assuming that the elastic self-energy of an emergent loop is just half of that for the corresponding "full loop" (emergent loop plus its mirror image relative to the crack tip). The problem is of interest for shear dislocations [18,19] emerging from the tip on slip planes that are generally inclined to the crack plane; the prismatic version [17], emerging in that plane, provides a more readily addressed first case. We review here only the half-circular emergent loop, over $x^2 + z^2 \leq R^2$, $x \geq 0$.

The self-energy of a full circular loop (in an uncracked body) is [35]

$$U^{\text{full loop}} = 2\pi R A_0 \ln(8R/e^2 R_0) \tag{57}$$

where

$A_0 = \mu b^2/4\pi(1 - \nu)$ for a prismatic dislocation,
R_0 = core cut-off, and
e = natural logarithm base.

The self-energy U of the emergent loop therefore may be calculated as

$$U - \frac{1}{2} U^{\text{full loop}} = -(b/2) \int_S [\sigma_{yy}(x,0,z) - \sigma_{yy}^{\text{full loop}}(x,0,z)]\, dxdz \tag{58}$$

which Anderson and Rice [17] show to have the form $\pi R A_0$ times a constant, which they calculate by using the previous formulae for σ_{yy}, and write as $\ln(m)$. Thus

$$U = \pi R A_0 \ln(8mR/e^2 R_0) \tag{59}$$

for the emergent loop. It is found [17] that $m = 2.21$ for the half-circular loop. Calculations for a rectangular emergent loop show that the analogously defined (with πR replaced by loop outer perimeter) value of m is 1.92 when the loop and its image form a square full loop, that m increases towards a maximum of 2.27 when the rectangle is elongated from that shape by a factor of about 4 in the direction parallel to the crack tip, and that with greater elongation m diminishes towards its limit $m = 2$ for the infinitely elongated rectangular loop (that is, for the two-dimensional, or line, dislocation [17]).

More recent work [18,19] has used Bueckner's [10] shear-mode weight functions to evaluate an analogously defined m, based on $A_0 = (2 - \nu)\mu b^2/8\pi(1 - \nu)$, for a half-circular shear dislocation loop emerging from the tip. When it is on the same plane as the crack, this gives $m = 2.35$ when $\mathbf{\Delta u}$ is in the x direction, and $m = 1.82$ when $\mathbf{\Delta u}$ is in the z direction. For the highly elongated loop (two-dimensional limit), $m = 2$ independently of the direction of $\mathbf{\Delta u}$ when the slip plane is the crack plane. However, m decreases, in different ways for different directions of $\mathbf{\Delta u}$ in the slip plane, when the slip plane is rotated (about the crack tip) relative to the crack plane [17,19].

The energy of an emergent loop is an important quantity in the theory of dislocation emission and its competition with cleavage decohesion at a crack tip [32,34]. The fact that $m > 1$ for typical cases means that the approximation of estimating self-energy as half that for a full loop has tended to underestimate (by $1/\sqrt{m}$) the K needed for dislocation nucleation.

Variation of Crack Shape

A general crack-front shape is shown in Fig. 5, and advance normal to itself by a distribution $\delta a(s)$ is indicated, as well as the distance function $D(s_1,s)$. The focus here is on setting up the formalism for calculating variations in the K_α along a crack front, both to first-order accuracy in the function $\delta a(s)$ and, in principle, exactly by integrating results for a sequence of infinitesimal advances $\delta a(s)$. For simplicity, attention is limited to Mode 1 conditions in isotropic solids.

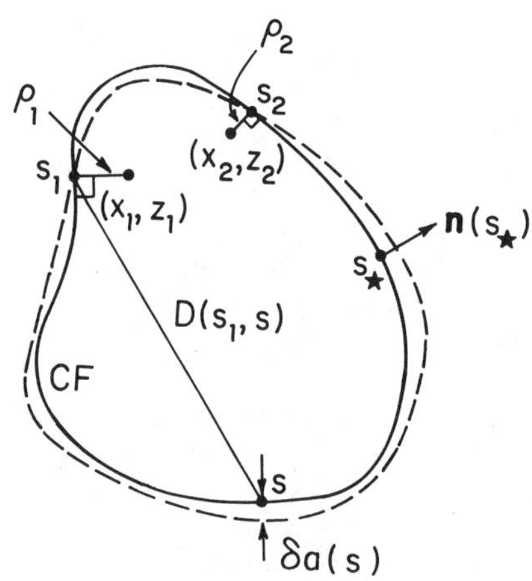

FIG. 5—*Variation of crack front location, for discussion of corresponding variations in* $K_I(s_I)$ *and* $W(s_I,s_2)$.

Note that $\delta a(s) = 0$ when $s = s_1$ or s_2 in Fig. 5. These are important locations because it is possible to work out the first-order variations $\delta K_1(s_1)$, $\delta K_1(s_2)$ and $\delta W(s_1,s_2)$, where W is the function of Eqs 40 to 43 and Eq 46. At least this is so when $d[\delta a(s)]/ds$ exists at s_1 and s_2. Naturally, one is also interested in the δK_1 and δW associated with locations where $\delta a(s) \neq 0$, and a procedure will be explained shortly for calculating those.

Consider a point (x_1, z_1) at small distance ρ_1 from CF along its perpendicular at s_1. By Eqs 2 and 3

$$\Delta u_y(x_1, z_1) \sim [4(1 - \nu)/\mu] K_1(s_1)\sqrt{\rho_1/2\pi} \tag{60}$$

Hence the variation in Δu_y is

$$\delta[\Delta u_y(x_1, z_1)] \sim [4(1 - \nu)/\mu] \delta K_1(s_1)\sqrt{\rho_1/2\pi} \tag{61}$$

to first order in $\delta a(s)$, where $\delta K_1(s_1)$ is the corresponding first-order variation at location s_1 along CF. The perpendicular to the new crack front (dashed line, Fig. 5) at location s_1 no longer passes through (x_1, z_1), but misses it by a distance $\rho_1 \delta\theta_1$ measured parallel to CF, where $\delta\theta_1 = d[\delta a(s)]/ds$ at $s = s_1$. This effect may be included in the analysis, recognizing that $\delta[\Delta u_y]$ in Eq 61 should, strictly, be replaced by its value at (x_1, z_1) plus $\rho_1 \delta\theta_1$ times the gradient of Δu_y in the direction parallel to CF. However, that modification gives a term of order $\rho_1\sqrt{\rho_1}\,\delta\theta_1$, the same order in ρ_1 as those already deleted on the right in Eqs 60 and 61, and all these will disappear shortly when we divide Eq 61 by $\sqrt{\rho_1}$ and let $\rho_1 \to 0$.

From Eqs 3, 34, and 40, one now observes by the second property of the weight functions that

$$\delta[\Delta u_y(x_1, z_1)] = \frac{1 - \nu}{\mu} \frac{\sqrt{2\rho_1}}{\pi\sqrt{\pi}} \int_{CF} \frac{W(x_1, z_1; s)}{D^2(x_1, z_1; s)} K_1(s) \delta a(s)\, ds \tag{62}$$

Now Eq 61 is used and both sides of Eq 62 are divided by $\sqrt{\rho_1}$, after which one lets $\rho_1 \to 0$ [that is, $(x_1 z_1)$ approaches location s_1 along CF]. Since $W(x_1, z_1; s)$ smoothly approaches

$W(s_1,s)$, Eq 41, in this limit, and since $\delta a(s)$ vanishes at s_1, the limit of the integral exists as a principal value (PV) integral. Thus

$$\delta K_1(s_1) = \frac{1}{2\pi} PV \int_{CF} \frac{W(s_1,s)}{D^2(s_1,s)} K_1(s)\delta a(s)\, ds \tag{63}$$

to first order in $\delta a(s)$ when $\delta a(s_1) = 0$. This generalizes to arbitrary crack front shapes results given in Refs *7*, *11*, and *12* for specific shapes.

Note that $W(s_1,s)$ is known for simple crack shapes, Fig. 3 and Eqs 42 and 43. Thus, Eq 63 is ready for use, provided that $\delta a(s_1) = 0$.

How does one calculate $\delta K_1(s_*)$ at a location like s_* in Fig. 5, where $\delta a(s_*) \neq 0$? In general, the following procedure solves the problem: We represent the given $\delta a(s)$ as the sum of two functions

$$\delta a(s) = \delta_* a(s) + [\delta a(s) - \delta_* a(s)] \tag{64}$$

where $\delta_* a(s) = \delta a(s)$ at $s = s_*$ so that the second function, in brackets, vanishes as required for validity of Eq 63, and where $\delta_* a(s)$ is a simple motion of CF for which it is straightforward to calculate independently $\delta_* K_1(s_*)$, the variation of K_1 corresponding to $\delta_* a(s)$. Thus

$$\delta K_1(s_*) = \delta_* K_1(s_*) + \frac{1}{2\pi} PV \int_{CF} \frac{W(s_*,s)}{D^2(s_*,s)} K_1(s)[\delta a(s) - \delta_* a(s)]\, ds \tag{65}$$

As examples, we may take for $\delta_* a(s)$ a rigid translation of CF of amount $\mathbf{n}(s_*)\delta a(s_*)$, so that

$$\delta_* a(s) = \mathbf{n}(s) \cdot \mathbf{n}(s_*)\delta a(s_*) \tag{66}$$

where $\mathbf{n}(s)$ is the unit outer normal (in the x,z plane) to CF at location s. For a finite-size crack in an unbounded solid under remotely uniform tension, such translation gives $\delta_* K_1(s_*) = 0$. Alternately, as in Refs *11* and *12* for circular CF, one may sometimes take a self-similar scaling of the crack shape by expansion relative to any convenient location of the x,z cordinate origin, which does not lie along the tangent line to CF at s_*. In that case, $\delta_* a(s)$ has the form $\mathbf{r}(s) \cdot \mathbf{n}(s)\delta\lambda$, where here $\mathbf{r}$ is the position vector along CF, and we choose $\delta\lambda$ to make $\delta_* a(s_*) = \delta a(s_*)$. Thus

$$\delta_* a(s) = \mathbf{r}(s) \cdot \mathbf{n}(s)\delta a(s_*)/\mathbf{r}(s_*) \cdot \mathbf{n}(s_*) \tag{67}$$

Yet a third alternative for choosing $\delta_* a(s)$ is provided by a rigid rotation of CF about any point which does not lie on the perpendicular to CF at s_*.

This discussion has focused on growth from the solid line CF in Fig. 5 to the dashed line. However, often the given reality in a problem is the actual (dashed line) crack front. Then the solid line CF is just an artifact that we introduce and are free to locate arbitrarily, subject to the restrictions that we must know how to determine K_1 along CF and that we want it to be close enough to the actual shape to justify use of the first order perturbation formulae.

Thus, if we want to know K_1 at the location marked s_1 along the dashed crack front in Fig. 5, we simply position the (arbitrarily chosen) solid-line CF to pass through that location, as has been done for the figure. Examples follow shortly.

Variation of W—Equation 63, or its relative, Eq 65, provides a solution for $\delta K_1(s_*)$ provided that one knows $W(s_*,s)$. W is known for simple crack shapes, Eqs 42 and 43. However, one cannot consider yet Eq 65 as providing an infinitesimal δK_1 distribution, associated with infinitesimal $\delta a(s)$, which can be integrated over a sequence of successive crack shapes, starting at a simple shape, to give K_1 for a general crack shape, because a procedure for calculating the evolution of $W(s_*,s)$ has not been given.

Such a procedure can be obtained by applying Eq 63 to loading by a pair of unit y-direction point forces on the crack faces at (x,z). Thus the $K_1(s)$ appearing in Eq 63 becomes $k_{1y}(x,z;s)$ of Eq 32, given for simple crack shapes by Eqs 36 to 38 and, in general form, by Eq 40. We choose a $\delta a(s)$ which, as in Fig. 5, vanishes at two points s_1 and s_2, for which we wish to know $W(s_1,s_2)$. [If the given $\delta a(s)$ does not already so vanish, it can be modified to do so by a procedure like the one in Eq 64, combining any appropriate pair of translations, rotation, and scaling.] We choose (x,z) as the point (x_2,z_2) at distance ρ_2 along the perpendicular to CF at location s_2 in Fig. 5. Then

$$\delta k_{1y}(x_2,z_2;s_1) = \frac{1}{2\pi} PV \int_{CF} \frac{W(s_1,s)}{D^2(s_1,s)} k_{1y}(x_2,z_2;s)\delta a(s)\, ds \tag{68}$$

to first order in $\delta a(s)$ where, from Eq 40,

$$k_{1y}(x_2,z_2;s) = \frac{\sqrt{2\rho_2}\,W(x_2,z_2;s)}{D^2(x_2,z_2;s)} \tag{69}$$

Thus by dividing by $\sqrt{\rho_2}$ and then letting $\rho_2 \to 0$, we obtain the new result

$$\delta W(s_2,s_1) = \frac{D^2(s_2,s_1)}{2\pi} PV \int_{CF} \frac{W(s_1,s)W(s_2,s)}{D^2(s_1,s)D^2(s_2,s)} \delta a(s)\, ds \tag{70}$$

to first order in $\delta a(s)$, when $\delta a(s_1) = \delta a(s_2) = 0$.

In principle, this equation lets us begin with any simple crack shape for which W is known (for example, as $W = 1$ along a circular crack) and to sum by integration a sequence of δW associated with infinitesimal $\delta a(s)$ to calculate W for an arbitrary crack shape. To make the procedure work, one must find simple functions of type $\delta_* a(s)$ in Eq 64 above, for which the associated $\delta_* W$ is readily computed, and which have two disposable degrees of freedom to make $\delta a(s) = \delta_* a(s)$ at locations corresponding to all possible pairs of arguments of $W(s_*,s_*')$. One then rewrites Eq 70 analogously to how Eq 63 is rewritten as Eq 65. Fortunately, the simple $\delta_* a(s)$ functions can be provided for cracks in unbounded solids by combining translations, rotation, or scaling, all of which then cause zero change $\delta_* W$ in the function W associated with a fixed pair of phase features along CF.

While the idea is simple, the algebra necessary to write out fully the $\delta_* a(s)$ is complex and is not pursued here, except in the simple variants of Eqs 72 and 76 to follow. The important point is that $W(s,s')$ can be determined, in principle, for complex crack shapes by integration, and once $W(s,s')$ is known through a sequence of crack shapes, $K_1(s)$ due to arbitrary loadings is likewise determined, in principle, by integration. That is, Eqs 63 or 65 and 70 can be regarded as a pair of equations which allow us to solve for K along a complex crack geometry by integrating over a sequence of crack shapes, beginning with a simple one. While that is of great theoretical interest, it remains to be seen if the formulation outlined, when implemented numerically, offers any advantage over more customary methods of three-dimensional crack analysis.

First-Order Expressions for K_I *Along Perturbed Crack Fronts*

Half-Plane Crack—Consider a half-plane crack with front lying along the curve $x = b(z)$, Fig. 6*a*, where $b(z)$ differs slightly from constancy. To calculate $K_1(z_*)$, that is, the K_1 at the point along the crack front whose coordinates (x,z) are $[b(z_*), z_*]$, we choose for CF the straight crack front, as in Fig. 2, with a identified as $b(z_*)$. Let $K_1^o[z;a]$ denote the K_1 distribution that the applied loads would induce along a straight crack with tip at $x = a$, as in Fig. 2. We regard $K_1^o[z,a]$ as a known function of its two arguments. Thus, with the choice of CF made above, we identify $\delta K(z)$ as $K_1(z) - K_1^o[z;b(z_*)]$, and thus have [7]

$$K_1(z_*) = K_1^o[z_*;b(z_*)] + \frac{1}{2\pi} \text{PV} \int_{-\infty}^{+\infty} \frac{K_1^o[z;b(z_*)]}{(z - z_*)^2} [b(z) - b(z_*)] \, dz \qquad (71)$$

to first order in $b(z) - b(z_*)$. An analogous expression, accurate to first order, based on Eq 34 was also given [7] for the crack-face opening $\Delta u_y(x,z)$.

The shear modes are addressed in a similar manner, but involve somewhat more com-

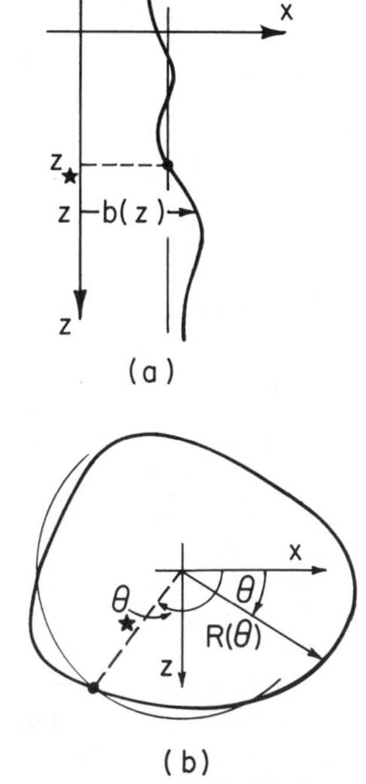

FIG. 6—(a) *Half-plane crack with perturbed front.* (b) *Perturbed circular crack or circular connection.*

plicated expressions because they couple together. The expressions analogous to Eq 71 for K_2 and K_3 are given in Ref *9*.

We may also record a first-order expression, derived from Eq 70 with $W = 1$, for the half-plane crack with tip at $x = b(z)$. Translation and rotation have been used to make vanish the terms corresponding to $\delta a(z)$ at z_1 and z_2. Thus

$$W(z_1,z_2) = 1 + \frac{(z_1 - z_2)^2}{2\pi} \text{PV} \int_{-\infty}^{+\infty} \frac{1}{(z - z_1)^2(z - z_2)^2} \times \left[b(z) - \frac{b(z_2) - b(z_1)}{2(z_2 - z_1)} (2z - z_1 - z_2) - \frac{b(z_2) + b(z_1)}{2} \right] dz \quad (72)$$

to first order in the deviation of $b(z)$ from constancy. A little analysis shows that knowing W to first order enables one to determine $K(z)$ to second order, although this is not pursued further here.

As an example, suppose that

$$b(z) = a_0 + B\cos(2\pi z/\lambda) \quad (73)$$

where $\lambda > 0$ and, for validity of first-order expressions, we assume $2\pi B/\lambda \ll 1$. Then, when the external loadings cause $K_1^o[z;a]$ to be independent of z, and then denoted by $K_1^o[a]$, we obtain

$$K_1(z) = K_1^o[a_0 + B\cos(2\pi z/\lambda)][1 - (\pi B/\lambda)\cos(2\pi z/\lambda)] \quad (74)$$

Expanded consistently to first order in B, this is

$$K_1(z) = K_1^o[a_0] + \left\{ \frac{dK_1^o[a_0]}{da_0} - \frac{\pi}{\lambda} K_1^o[a_0] \right\} B\cos(2\pi z/\lambda) \quad (75)$$

The function $W(z_1,z_2)$ associated to first order with this choice of $b(z)$ is

$$W(z_1,z_2) = 1 + \frac{2\pi B}{\lambda} \left[\frac{\sin\eta_2 - \sin\eta_1}{\eta_2 - \eta_1} - \frac{\cos\eta_2 + \cos\eta_1}{2} \right] \quad (76)$$

where $\eta = 2\pi z/\lambda$ and, as remarked, this would actually allow expansion of $K_1(z)$ to order B^2.

Slightly Noncircular Crack—Figure 6*b* shows a slightly noncircular crack, corresponding to radius $R = R(\theta)$ as measured from the adopted coordinate system. Thus, let $K_1^o[\theta;a]$ be the intensity factor which the given applied loadings would induce around a circular crack of radius, a (as in Fig. 3*b*). To calculate $K_1(\theta_*)$, we take CF to be the circle of radius $a = R(\theta_*)$. Thus, Eq 63 with $\delta K_1(\theta_*)$ identified as $K_1(\theta_*) - K_1^o[\theta_*, R(\theta_*)]$ leads to

$$K_1(\theta_*) = K_1^o[\theta_*;R(\theta_*)] + \frac{1}{8\pi} \text{PV} \int_0^{2\pi} \frac{K_1^o[\theta;R(\theta_*)][R(\theta)/R(\theta_*) - 1]}{\sin^2[(\theta - \theta_*)/2]} d\theta \quad (77)$$

to first order in $R(\theta) - R(\theta_*)$, as in Ref *11*.

If we consider axisymmetric loading so that $K_1{}^o[\theta;a]$ is independent of θ, hence denoted $K_1{}^o[a]$, and consider the perturbation

$$R(\theta) = a_0 + B \cos n\theta \tag{78}$$

(where n is a positive integer and $nB/a_0 \ll 1$), then, when consistently expanded to first order in B, Eq 77 gives a result identical to Eq 75 when we replace z with $a_0\theta$ and identify λ as $2\pi a_0/n$.

Gao and Rice [*11*] tested the range of validity of Eq 77 by using it to calculate K_1 along an elliptical crack and found good agreement with the exact solution up to 2:1 aspect ratios. Comparable accuracy was found for displacements Δu_y estimated by the first order formula based on Eq 34. The estimate of K_1 was found to remain good up to aspect ratios approaching 5:1 when the reference circular CF was given a radius equal to that of the minor semi-axis of the ellipse (maximum inscribed circle), with the center of the reference circle then being shifted, in general, from that of the ellipse to make $\delta a = 0$ at a location of interest.

Corresponding results for shear loading of cracks perturbed from a circle are given by Gao [*15*]. An alternate, approximate approach to planar tensile cracks of arbitrary noncircular and nonelliptical shape has been given by Fabrikant [*36*].

Slightly Noncircular Connection—Now let a connection, joining two half-spaces, occupy the region shown in Fig. 6*b*, with boundary $R = R(\theta)$. Let $K_1{}^o[\theta;a]$ be the intensity factor which the given loadings induce when the connection is circular, of radius a, as in Fig. 3*c*. As already suggested by Eqs 42 and 43, in this case the results depend on the conditions of restraint (or its lack) that one assumes for the remote displacements in going from the circular reference shape to the actual shape. Thus [*12*]

$$K_1(\theta_*) = K_1{}^o[\theta_*;R(\theta_*)] + \frac{1}{8\pi}\,\mathrm{PV}\int_0^{2\pi} \frac{W(\theta_*,\theta)K_1{}^o[\theta;R(\theta_*)][1 - R(\theta)/R(\theta_*)]}{\sin^2[(\theta - \theta_*)/2]}\,d\theta \tag{79}$$

to first order in $R(\theta) - R(\theta_*)$, where

$$W(\theta_*,\theta) = 1 + 4\sin^2[(\theta - \theta_*)/2][1 + 6\cos(\theta - \theta_*)] \tag{80}$$

from Eq 43 when remote points are unrestrained, where the $6\cos(\theta - \theta_*)$ in the final bracket is dropped when remote points cannot rotate but can move in the y direction, where the 1 in the final bracket is dropped when remote points on a y axis through the center of the reference circle cannot displace in the y direction but can rotate, and from Eq 42 the entire final bracket is dropped when remote points are constrained against any motion.

For axisymmetric loading, in the sense that $K_1{}^o[\theta;a]$ is independent of θ, and denoted $K_1{}^o[a]$, the perturbation

$$R(\theta) = a_0 - B \cos n\theta \tag{81}$$

(minus sign, compared to Eq 78, so that positive B corresponds to crack growth) leads to

$$K_1(\theta) = K_1{}^o[a_0] - \left\{\frac{dK_1{}^o[a_0]}{da_0} + \frac{n_1}{2a_0}K_1{}^o[a_0]\right\} B \cos n\theta \tag{82}$$

Here, in dealing with remote displacement restraint cases in the order given after Eq 80, one has the following: unrestrained, $n_1 = -3$ when $n = 1$ and $n_1 = n + 2$ otherwise; restrained against rotation only, $n_1 = n + 2$; restrained against y direction displacement along axis only, $n_1 = -5$ when $n = 1$ and $n_1 = n$ otherwise; totally restrained, $n_1 = n$.

Finite-Element Analog—Following deKoning and Lof [37], in three-dimensional finite-element studies of cracked solids, the crack-front position is specified by corner-node positions for the string of elements along CF, and a stress intensity factor $K_1^{(1)}$, $K_1^{(2)}$, . . . , $K_1^{(m)}$ may be associated with each such node, the latter regarded as being defined [37] as in Eq 60 from the calculated near-tip opening, Δu_y, along a perpendicular to CF at the associated corner node. Parameters, a_1, a_2, . . . , a_m characterize crack front position; they vanish along CF and correspond to outward shifts of the corner-node positions along perpendiculars to CF. The linearized form

$$K_1^{(i)} = [K_1^{(i)}]^{(0)} + [\partial K_1^{(i)}/\partial a_j]^{(0)}a_j \tag{83}$$

is now considered, where the superscript (0) means that the quantity is evaluated with all $a_i = 0$, that is, for the crack front along CF.

The $\partial K_1^{(i)}/\partial a_j$ are calculated from $\partial(\Delta u_y)/\partial a_j$ at fixed distances from the moving crack tip nodes, and these displacement derivatives at $a_i = 0$ are calculated directly from the inverse of the stiffness matrix for the unperturbed crack geometry and from certain quantities that are calculated in the stiffness derivative procedure [37]. In fact, displacement derivatives calculated in this way form the discretized weight functions in the Parks and Kamenetzky [5] finite element formulation, so that the deKoning and Lof procedure seems to be a discretized version of the steps presented here for calculating K_1 to first order along perturbed crack fronts.

Application to Crack Trapping—Crack trapping arises in brittle crack advance through solids of locally heterogeneous fracture toughness. The front advances nonuniformly and has segments which are trapped, at least temporarily, by contact with tough obstacles whose K_{1c} exceeds the local K_1.

In an idealized model of the process the crack may be considered planar, as here, but with a wavy front as in Figs. 6*a* and 6*b*. Important problems are to predict, in a statistical sense, the configuration of the crack front and the far-field K_1 or load level which just enables growth through the heterogeneous fracture resistance. An elementary approach [16] can be based on the first order perturbation results of Eqs 71, 77, 79, and 83. This will not suffice for all problems. For example, with sufficiently tough local obstacles, crack-front segments will tend to bow out between obstacles and ultimately join with neighboring bows, to advance forward as a single crack whose surfaces remain bridged by uncracked obstacles left behind. Description of such processes lie outside the range of the first-order perturbation and will require the integration of perturbation effects along a sequence of crack shapes as discussed following Eq 70.

For cases with sufficiently tame crack fronts, amenable to the first-order perturbation approach, Eqs 71, 77, and 79 become singular integral equations for crack-front locations $b(z)$ or $R(\theta)$. Each point along the crack front is either active (K_1 = local K_{1c}) or trapped ($K_1 <$ local K_{1c}). In general, if we start with an initial half-plane crack front and increase the load, $K_1(z)$ is known as a function of $b(z)$ (on which the local K_{1c} depends) in active segments which are taking part in growth; $b(z)$ is unknown there. Also, $b(z)$ is known where the crack front remains trapped, while $K_1(z)$ is unknown there. Thus, Eq 71 can be regarded as an integral equation whose solution gives crack-front shape $b(z)$. Complications are that the active and trapped zone locations are not known *a priori,* that a previously active zone may later become trapped, and that the problem will not always have a solution (for example, when conditions are met for part of the crack front to jump forward dynamically, either to a new array of trapping obstacles or as the start of final fracture).

The simplest case is for the half-plane crack with $K_1^o[z,a]$ independent of z and, for the size of excursion $b(z)$ considered, of a, too. Hence it is written simply as K_1^o, effectively dependent on applied load only, and then Eq 71 becomes, after integration by parts [9]

$$K_1(z_*)/K_1^o = 1 + \frac{1}{2\pi}\,\mathrm{PV}\int_{-\infty}^{+\infty} \frac{db(z)/dz}{z - z_*}\,dz \tag{84}$$

This is of the same form as for plane stress of a thin elastic sheet occupying the x,z plane, where loading of the sheet is by remotely uniform stress σ_{xx}^o and its entire z axis can open up by an x-direction displacement gap of distribution $\Delta(z)$, hence creating nonuniform stress $\sigma_{xx}(z)$ at points along the z axis. In that case, $\sigma_{xx}(z)/\sigma_{xx}^o$ corresponds to $K_1(z)/K_1^o$ in Eq 84, and $(1 + \nu)\mu\Delta(z)/\sigma_{xx}^o$ to $b(z)$. Thus, prescribing $K_1(z)$ along active portions of the crack front is equivalent to prescribing $\sigma_{xx}(z)$ in the plane stress problem over the same segment of z axis, which is then regarded as a cracked segment. Prescribing $b(z)$ along the trapped portion of crack front corresponds to prescribing the opening gap $\Delta(z)$ over that segment of the z axis.

As a simple example [16], suppose the segment $-H < z < +H$ is active, with $K_1 = K_{1c}$, but that segment $H < z < 2L - H$ is completely trapped with $b(z) = 0$. Let these active and trapped zones alternate periodically, with period $2L$, and symmetry relative to $z = 0$ (middle of an active zone) and $z = L$ (middle of a trap). The problem is analogous to that of a periodic infinite array of collinear cracks in plane stress [38]. Thus, the solution, for example, for the average b_{pen} of crack penetration $b(z)$ over an active zone like $-H < z < +H$, is

$$b_{pen} = (4L^2/\pi H)(1 - K_{1c}/K_1)\ell n[1/\cos(\pi H/2L)] \tag{85}$$

When $K_{1c} = 1.5K_{1c}$, this equation predicts $b_{pen}/2H = 0.49$ when $(L - H)/L$, the line fraction of traps, is 0.1 and $b_{pen}/2H = 0.29$ when the line fraction is 0.5. The maximum penetration $b(0)$ for these two cases is only about 25% above b_{pen}.

Configurational Stability in Quasi-Static Crack Growth

Suppose that a tensile crack grows quasistatistically (say, by $R = 0$ fatigue or sustained load stress corrosion) under elastic fracture-mechanics conditions, such that the rate of crack growth is an increasing function of K_1. Further, suppose that the nature of the loading is such that K_1 is uniform along the crack front initially, and that a solution exists to the combined equations of elasticity and quasistatic crack growth such that crack can grow while maintaining a uniform K_1 along its front.

Such would be the case for homogeneous materials containing half-plane cracks with initially straight fronts, loaded under plane-strain conditions (K_1 independent of z), and also for homogeneous materials with circular cracks or connections under axisymmetric loading (K_1 independent of θ). In these respective cases, a solution exists such that the crack continues to grow with a straight or circular front.

Here we review conditions for such crack fronts to be configurationally stable. That is, if the crack front starts off slightly deviated from that ideal shape, will those deviations grow or diminish as the crack enlarges? Considering trigometric deviations as in Eqs 73, 78, and 81, stability requires that K_1 be smallest where the crack has advanced most and largest where the crack has advanced least, for then the corresponding variations in the growth rate will tend to diminish the amplitude of the crack front fluctuations. Conversely, if K_1 is greatest where the crack has advanced most, and smallest where least, then the variation in growth rates causes the fluctuations in crack front position to grow in amplitude.

Thus, since B times the cosine term measures advance of the crack front in Eqs 73, 78, and 81, for configurational stability we require that the coefficient of B times the cosine term in the resulting expressions for K_1 be negative. Thus, assuming $K_1{}^o[a_0] > 0$ and letting

$$Q(a_0) = \{dK_1{}^o[a_0]/da_0\}/K_1{}^o[a_0] \tag{86}$$

one has from Eqs 75 and 82 that for configurational stability

$$Q(a_0) < \pi/\lambda \tag{87}$$

for a half-plane crack subject to perturbations of wavelength $\lambda(>0)$ [7],

$$2a_0Q(a_0) < n \tag{88}$$

for a circular crack with perturbation having $n(>0)$ wavelengths on its circumference ($\lambda = 2\pi a_0/n$) [*11*], and

$$-2a_0Q(a_0) < n_1 \tag{89}$$

for a circular connection with perturbations having n wavelengths on the circumference [*12*], where n_1 is the function of n, dependent on conditions of remote displacement restraint, given after Eq 82. When the inequalities are reversed, Eqs 87, 88, and 89 become conditions for configurational instability.

These expressions show that growth with negative Q for the half-plane and circular crack is always configurationally stable. Also, there is inevitably stability against perturbations of short wavelength λ or high n. Thus, if there is to be configurational instability, it will tend to develop at long wavelengths or low n.

For an edge crack of depth a_0 in a remotely stressed half space, $Q = 1/2a_0$, and thus Eq 87 predicts configurational stability unless λ is so great as to exceed $2\pi a_0$. This could be important for initially short cracks, but the value of λ is so great compared to crack length that no reliance can then be put on predictions based on the half-plane model.

For circular cracks under remote tension, $2a_0Q = 1$ so that by Eq 88 there is neutral stability when $n = 1$ (corresponding to translational shift of the crack front) and stability for all higher n. For a crack loaded by pressure [*11*], $p = p(R) > 0$ dependent on distance R from the center, there is stability to all n if $dp/dR < 0$. If $dp/dR > 0$, the translational shift mode ($n = 1$) is unstable, and if p increases rapidly enough with R (as happens for p = constant $\times$ R), higher n become unstable too, such as $n = 2$ corresponding to an elliptical-like growth mode Note that these unstable growth modes represent effects, presumed to be small at least initially, superposed on the basic axisymmetric growth mode.

Similarly, for the circular connection under remote axially symmetric tensile loading [*12*], there is stability when $n > 1$. However, the $n = 1$ translational shift mode is neutrally stable [*12*] when the remote points are constrained against rotation (this means that the crack grows as a circle but does not necessarily remain of fixed center). For loading only by a remotely imposed force, centered on the circular connection, with no restraint against remote rotation, the translational shift mode is unstable. The crack initially tends to grow in a circle but of shifting center, and this induces an unfavorable moment relative to the shifting center which aggravates the effect and ultimately leads to strongly noncircular growth [*12*].

Acknowledgments

This study was supported by the Office of Naval Research under contract N00014-85-K-0405 with Harvard University and contract N00014-86-K-0753 through sub-agreement VB 38639-0 with Harvard from the University of California. I am grateful to Drs. H. F. Bueckner, V. I. Fabrikant, and H. Gao for comments on the original manuscript.

References

[*1*] Bueckner, H. F. *Zeitschrift für Angewandte Mathematik und Mechanik,* Vol. 50, 1970, pp. 529–545.
[*2*] Rice, J. R., *International Journal of Solids and Structures,* Vol. 8, 1972, pp. 751–758.
[*3*] Bueckner, H. F. in *Mechanics of Fracture I: Methods of Analysis and Solution of Crack Problems,* G. C. Sih, Ed., Noordhoff, Leyden, the Netherlands, 1973, pp. 239–314.
[*4*] Besuner, P. M. in *Mechanics of Crack Growth, ASTM STP 590,* American Society for Testing and Materials, Philadelphia, 1974, pp. 403–419.
[*5*] Parks, D. M. and Kamenetzky, E. A., *International Journal of Numerical Methods in Engineering,* Vol. 14, 1979, pp. 1693–1706.
[*6*] Bueckner, H. F. in *Fracture Mechanics and Technology,* Vol. 2, G. C. Sih and C. L. Chow, Eds., Sijthoff and Noordhoff, Amsterdam, the Netherlands, 1977, pp. 1069–1107.
[*7*] Rice, J. R., *Journal of Applied Mechanics,* Vol. 52, 1985, pp. 571–579.
[*8*] Rice, J. R., *International Journal of Solids and Structures,* Vol. 21, 1985, pp. 781–791.
[*9*] Gao, H. and Rice, J. R., *Journal of Applied Mechanics,* Vol. 53, 1986, pp. 774–778.
[*10*] Bueckner, H. F., *International Journal of Solids and Structures,* Vol. 23, 1987, pp. 57–93.
[*11*] Gao, H. and Rice, J. R., *International Journal of Fracture,* Vol. 33, 1987, pp. 155–174.
[*12*] Gao, H. and Rice, J. R., *Journal of Applied Mechanics,* Vol. 54, 1987, pp. 627–643.
[*13*] Panasyuk, V. V., *Dopovidi Academii Nauk Ukrainskoi RSR* (in Ukranian), No. 2, 1962, pp. 891–895.
[*14*] Panasyuk, V. V., Andrejkiv, A. E., and Stadnik, M. M., *Engineering Fracture Mechanics*, Vol. 14, 1981, pp. 245–260.
[*15*] Gao, H., *International Journal of Solids and Structures,* Vol. 24, 1988, pp. 177–193.
[*16*] Rice, J. R., in *Analytical, Numerical and Experimental Aspects of Three-Dimensional Fracture Processes,* A. Rosakis et al., Eds., American Society of Mechanical Engineers AMD-Vol. 91, 1988, pp. 175–184.
[*17*] Anderson, P. M. and Rice, J. R., *Journal of the Mechanics and Physics of Solids,* Vol. 35, 1987, pp. 743–769.
[*18*] Gao, H., *Journal of the Mechanics and Physics of Solids,* Vol. 37, 1989, pp. 133–154.
[*19*] Gao, H. and Rice, J. R., *Journal of the Mechanics and Physics of Solids,* Vol. 37, 1989, pp. 155–174.
[*20*] Sham, T.-L., *International Journal of Solids and Structures,* Vol. 23, 1987, pp. 1357–1372.
[*21*] Fabrikant, V. I., *Philosophical Magazine A,* Vol. 56, 1987, pp. 191–207.
[*22*] Fabrikant, V. I., *Advances in Applied Mechanics,* 1989, in press.
[*23*] Stroh, A. N., *Philosophical Magazine,* Vol. 3, 1958, pp. 625–646.
[*24*] Barnett, D. M. and Asaro, R. J., *Journal of the Mechanics and Physics of Solids,* Vol. 20, 1972, pp. 353–366.
[*25*] Paris, P. C., McMeeking, R. M., and Tada, H. in *Cracks and Fracture, ASTM STP 601,* American Society for Testing and Materials, Philadelphia, 1976, pp. 471–489.
[*26*] Eshelby, J. D., *Proceedings of the Royal Society* (*London*), Vol. A241, 1957, pp. 376–396.
[*27*] McMeeking, R. M. and Evans, A. G., *Journal of the American Ceramics Society,* Vol. 65, 1982, pp. 242–246.
[*28*] Budiansky, B., Hutchinson, J. W., and Lambropoulos, J. C., *International Journal of Solids and Structures,* Vol. 19, 1983, pp. 337–355.
[*29*] Weaver, J., *International Journal of Solids and Structures,* Vol. 13, 1977, pp. 321–330.
[*30*] Bui, H. D., *Journal of the Mechanics and Physics of Solids,* Vol. 25, 1977, pp. 29–39.

[31] Murakami, Y. and Nemat-Nasser, S., *Engineering Fracture Mechanics,* Vol. 17, 1983, pp. 193–210.
[32] Rice, J. R. and Thomson, R., *Philosophical Magazine,* Vol. 29, 1974, pp. 73–97.
[33] Rice, J. R. in *Fundamentals of Deformation and Fracture,* B. A. Bilby, K. J. Miller, and J. R. Willis, Eds., Cambridge University Press, Cambridge, U.K., 1985, pp. 33–56.
[34] Anderson, P. M. and Rice, J. R., *Scripta Metallurgica,* Vol. 20, 1986, pp. 1467–1472.
[35] Hirth, J. P. and Lothe, J., *Theory of Dislocations,* McGraw-Hill, New York, 1968.
[36] Fabrikant, V. I., *Philosophical Magazine A,* Vol. 56, 1987, pp. 175–189.
[37] DeKoning, A. U. and Lof, C. J. in *Proceedings of the Third International Conference on Numerical Methods in Fracture Mechanics,* A. R. Luxmoore and D. R. J. Owen, Eds., Pineridge Press, Swansea, U.K., 1984, pp. 195–203.
[38] Koiter, W. T., *Ingenieur Archiv,* Vol. 28, 1959, pp. 168–172.

Nonlinear and Time-Dependent Fracture Mechanics

John W. Hutchinson[1] *and Viggo Tvergaard*[2]

Softening Due to Void Nucleation in Metals

REFERENCE: Hutchinson, J. W. and Tvergaard, V., **"Softening Due to Void Nucleation in Metals,"** *Fracture Mechanics: Perspectives and Directions (Twentieth Symposium), ASTM STP 1020,* R. P. Wei and R. P. Gangloff, Eds., American Society for Testing and Materials, Philadelphia, 1989, pp. 61–83.

ABSTRACT: The mechanics of void nucleation in an elastic-plastic solid stressed into the plastic range is studied with emphasis on the contribution of nucleation to softening of overall stress-strain behavior. Results for nucleation of an isolated spherical void in an infinite matrix under triaxial remote stressing are used to predict overall stress-strain behavior when interaction between voids is negligible. Calculations for nucleation of a void at a rigid spherical particle in a cylindrical cell model the simultaneous nucleation of a uniform distribution of voids. A strong dependence of the nucleation process on the matrix material specification is found when results based on isotropic hardening are contrasted with those based on kinematic hardening. At issue is the magnitude of the softening contribution due to void nucleation. This issue and the role of nucleation in promoting flow localization are discussed.

KEY WORDS: cell model, dilatation, isolated, isotropic, J_2-flow theory, kinematic hardening, macroscopic, matrix strain hardening, nucleation, plasticity, softening, strain, stress, triaxiality, void

Nucleation of voids in a plastically deforming metal has two consequences. Most obvious is the generation of damage, which ultimately will lead to failure of a ductile metal after further straining. A less obvious but equally important consequence is the reduction in macroscopic strain-hardening capacity due to the nucleation process itself, independent of the subsequent growth of the void. The shedding of the load carried by a particle when a void nucleates by interface debonding between the particle and the matrix or by particle cracking causes a redistribution of stress and strain in the matrix, which alters the overall stress-strain behavior of the material. The consequences of nucleation in offsetting matrix strain hardening can be dramatic. There is evidence that some high-strength steels undergo shear localization simultaneously, or almost simultaneously, with the onset of void nucleation at second-phase particles.

This paper focuses on the effect of void nucleation on macroscopic stress-strain behavior. We introduce the subject by displaying one set of results from a numerical calculation carried out in the section on Cell Model Calculations. An axisymmetric cell model is used to model the simultaneous nucleation of voids at a uniform distribution of rigid spherical particles. The cell, which is shown in Fig. 1, is constrained such that the ends remain planar and the lateral surface remains cylindrical. The macroscopic true stresses S and T are obtained as averages of the local normal stresses over the respective surfaces. Prior to nucleation, S and T are increased monotonically with a fixed ratio, T/S, and the distribution of stress and

[1] Professor, Division of Applied Sciences, Harvard University, Cambridge, MA 02138.
[2] Professor, Department of Solid Mechanics, The Technical University of Denmark, Lyngby, Denmark.

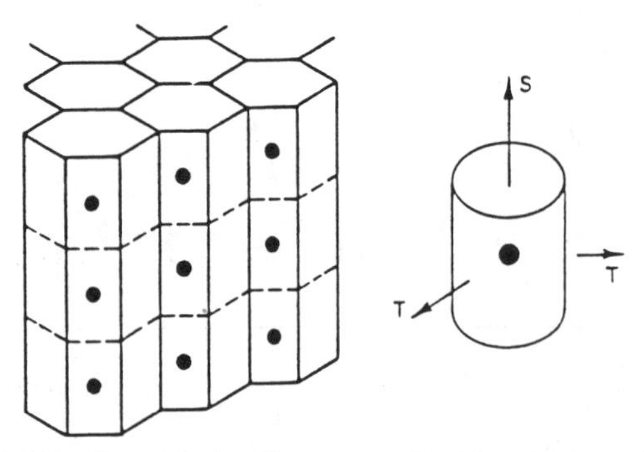

FIG. 1—*Cylindrical cell model of uniform array of void-nucleating spherical particles.*

strain in the cell is calculated. Then at a given macroscopic true stress state, the void in the cell is "nucleated" by incrementally reducing the traction across the particle/matrix interface to zero.

The overall stress-strain behavior shown in Fig. 2 was calculated for nucleation under constant S and T at two stress levels (other possibilities are analyzed in the section noted above). The horizontal segments display the amount of overall strain which occurs during nucleation. In this example, the volume fraction of the spherical particle is 1%, the stress triaxiality is $T/S = 0.5$, and the multiaxial stress dependence of the matrix material is described by J_2-flow theory. Included in Fig. 2 for reference are the stress-strain behavior in the absence of nucleation and the stress-strain behavior of the cell containing a void which at the start of straining has a 1% volume fraction. The strain following nucleation at finite stress is only slightly less than corresponding strain when the void is present from the start, and following nucleation the two curves are similarly close. By contrast, the growth in volume of the void during nucleation, ΔV, has a very strong history dependence. This can be seen in Fig. 3, where the volume growth, normalized by the particle volume V_0 and the

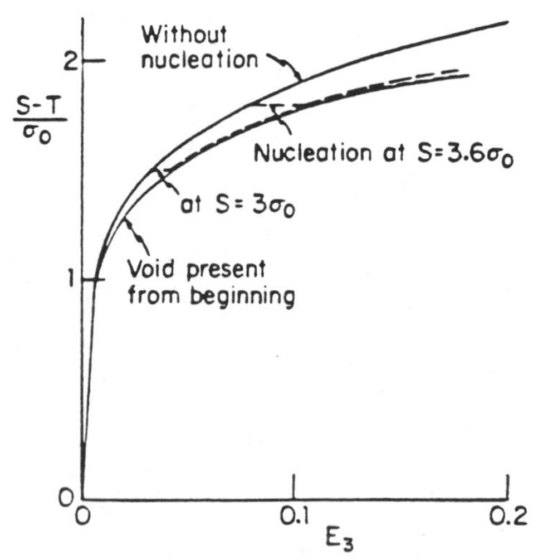

FIG. 2—*Overall stress-strain behavior for nucleation at constant stress for rigid spherical particles in a matrix material governed by* J_2*-flow theory* ($\rho = 0.01$, $\epsilon_0 = 0.004$, $n = 5$, $\nu = 0.3$).

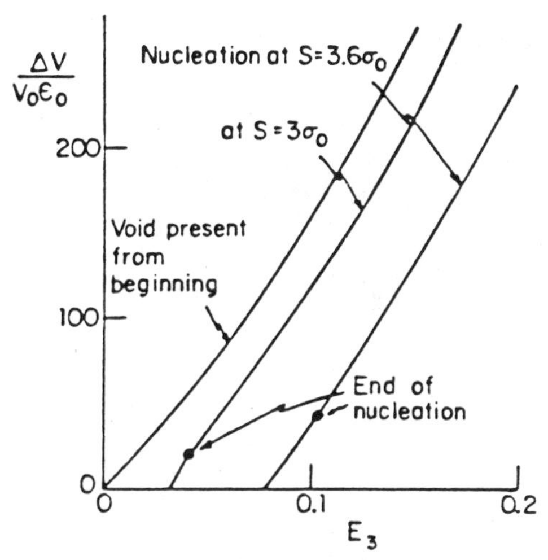

FIG. 3—*Normalized volume increase as a function of overall strain for nucleation at constant stress in a* J_2*-flow theory matrix.*

initial yield strain ϵ_0, is plotted as a function of the macroscopic logarithmic strain. The segment of the curve ending in the solid dot corresponds to the void growth during nucleation, which is far less than that experienced by a void present from the start.

The first part of the paper deals with the basic problem of nucleation of a spherical void in an infinite matrix under remote triaxial stressing. Application of the solution to this problem to predict the effect of nucleation on macroscopic stress-strain behavior is discussed. The second part of the paper presents an analysis of the cylindrical cell model described above. In each part, the role of the multiaxial stress characterization of the matrix material is explored by using both J_2-flow theory (isotropic hardening) and kinematic hardening. Nucleation as predicted by the Gurson model [*1*] is related to the results of the present paper in the section on Effect of Nucleation.

Nucleation of an Isolated Void and Its Effect on Macroscopic Stress-Strain Behavior

Before analyzing in detail the mechanics of void nucleation, we first outline the nature of the results expected and show how these can be used to predict the effect of void nucleation on stress-strain behavior.

Consider nucleation of a single void at a spherical particle of unit volume in an infinite matrix stressed into the plastic range by proportional application of a remote stress $\mathbf{\Sigma}$. Imagine that the interface between the particle and the matrix debonds, either partially or completely, when the remote stress reaches $\mathbf{\Sigma}$. This nucleation event causes a redistribution of stress, some growth of the void and additional straining in the matrix. Denote the average extra strain that occurs as a result of nucleation of the void, over and above what occurs in the absence of nucleation, by $\Delta\mathbf{E}$, such that $\Sigma_{ij}\Delta E_{ij}$ is the extra work done by the remote stress due to nucleation. For proportional stressing to $\mathbf{\Sigma}$, $\Delta\mathbf{E}$ is an isotropic function of $\mathbf{\Sigma}$, assuming the unstressed matrix is isotropic. Approximate results for $\Delta\mathbf{E}$ are given in the next section.

Let $\mathbf{M}$ denote the current overall (diagonally symmetric) incremental compliances of a macroscopic element of material subject to macroscopic stress $\mathbf{\Sigma}$ when no particles nucleate voids. Thus the macroscopic strain increment is

$$\dot{E}_{ij} = M_{ijkl}\dot{\Sigma}_{kl} \tag{1}$$

in the absence of nucleation. Now suppose that during the stress increment $\dot{\mathbf{\Sigma}}$ voids nucleate in the material element corresponding to a volume fraction increment $\dot{\rho}$. If the nucleated voids are sufficiently widely spaced so that their interaction can be ignored (for example, $\dot{\rho}$ sufficiently small), then in the presence of nucleation

$$\dot{E}_{ij} = M_{ijkl}\dot{\Sigma}_{kl} + \dot{\rho}\Delta E_{ij} \tag{2}$$

With $\mathbf{L} = \mathbf{M}^{-1}$ as the incremental moduli of the material element in the absence of nucleation, it follows from Eq 2 that in the presence of nucleation

$$\dot{\Sigma}_{ij} = L_{ijkl}\dot{E}_{kl} - \dot{\rho}\Delta\Sigma_{ij} \tag{3}$$

where

$$\Delta\Sigma_{ij} = L_{ijkl}\Delta E_{kl} \tag{4}$$

The quantity $\dot{\rho}\Delta\mathbf{\Sigma}$ can be interpreted as the average stress drop due to nucleation of an increment of void volume fraction $\dot{\rho}$ relative to the stress in the absence of nucleation at the same macroscopic strain. Figure 4 displays the schematic interpretation of $\dot{\rho}\Delta\mathbf{E}$ and $\dot{\rho}\Delta\mathbf{E}$ relative to the overall stress-strain curves with and without nucleation.

The focus in this paper is on the first nucleation of voids in a void-free material, but the above discussion also applies, at least approximately, to subsequent nucleation adding to voids nucleated earlier in the stress history. Then, $\mathbf{L}$ and $\mathbf{M}$ correspond to incremental moduli and compliances in the presence of a volume fraction ρ of voids but without nucleation of additional voids $\dot{\rho}$. The extra average strain ΔE due to nucleation of the isolated void should in general account for interaction with preexisting voids.

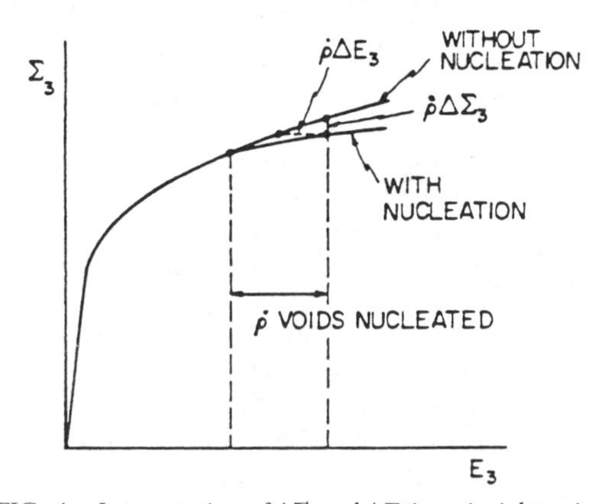

FIG. 4—*Interpretation of* $\Delta\mathbf{\Sigma}$ *and* $\Delta\mathbf{E}$ *in uniaxial tension.*

Extra Strain, ΔE, and Stress Drop, ΔΣ, Due to Nucleation of an Isolated Spherical Void

In this section the nucleation of an isolated spherical void in an infinite block of material is modeled as a mechanics problem. The solution procedure is given in the Appendix, and approximate recipes for **ΔE** and **ΔΣ** are presented in this section.

An infinite block of material is stressed into the plastic range by proportional application of remote stress **Σ**. The void is nucleated from a fictitious "particle" which deforms uniformly with the matrix prior to nucleation. Thus at the onset of nucleation the block of material is in a uniform state of stress **Σ** and the "particle" is taken to be spherical with unit volume. The nucleation process is modeled as a plasticity problem in which tractions across the particle/matrix interface at the onset, $T_i^0 = \Sigma_{ij} n_j$, are incrementally reduced to zero. Specifically, traction rates, $-\dot{\lambda}_0 T_i^0$, are applied to the interface, with $\lambda_0 = 0$ coinciding with the onset of nucleation and $\lambda_0 = 1$ with completion. The tractions on the interface are reduced to zero quasi-statically and uniformly. In many instances, the actual debonding process is likely to involve a dynamic interfacial separation by progressive cracking, which is not modeled here. A more detailed treatment of the debonding process is given by Needleman [2]. Contact between the "particle" and the nucleating void is ignored, but does not occur in any case in which the remote stress has modest triaxiality. We consider also the possibility that the nucleation process occurs under proportionally increasing remote stress $\dot{\lambda}_\infty \mathbf{\Sigma}$ simultaneously with traction-rates $-\dot{\lambda}_0 T_i^0$ on the interface.

The void nucleation problem just described is a small strain plasticity problem. The sequence of incremental problems in the void nucleation process does not lead to either large geometry changes of the void or large strain changes anywhere in the material surrounding the void. For our purposes here we take the hardening level in the matrix, as measured by the tangent modulus E_t of the effective stress-strain curve, to be constant during nucleation. In doing so, it is imagined that the strain changes during nucleation are small compared to the strain at the onset of nucleation such that only very small changes in E_t would occur during nucleation, and these are neglected. Results will be presented for both J_2-flow theory (isotropic hardening) and kinematic hardening to give some indication of how strongly the predictions are influenced by the matrix material description. Some influence is certainly expected since the stress changes in the vicinity of the void which occur during nucleation are distinctly nonproportional, and thus the material characterized by isotropic hardening should offer more resistance to plastic straining than the kinematic hardening material.

Let σ_0 be the initial tensile yield stress of the material, E its Young's modulus, and $\epsilon_0 = \sigma_0/E$ the tensile strain at initial yield. For computational convenience, we have taken the material to be incompressible. For J_2-flow theory the increment in the stress deviator, s, is

$$\dot{s}_{ij} = \frac{2}{3} E\dot{\epsilon}_{ij} - (E - E_t) s_{ij} s_{kl} \dot{\epsilon}_{kl} / \sigma_e^2 \tag{5}$$

for loading ($\sigma_e = (\sigma_e)_{\max}$ and $s_{kl}\dot{\epsilon}_{kl} \geq 0$) and

$$\dot{s}_{ij} = \frac{2}{3} E\dot{\epsilon}_{ij} \tag{6}$$

for elastic unloading. Here $\sigma_e = (3 s_{ij} s_{ij}/2)^{1/2}$ is the effective stress and, as already mentioned, E_t is the tangent modulus associated with the remote stress state **Σ**. For kinematic hardening theory based on the shifted J_2-invariant, yield is specified by $(3\hat{s}_{ij}\hat{s}_{ij}/2)^{1/2} = \sigma_0$ where $\hat{s} =$

$s - \boldsymbol{\alpha}$ and $\boldsymbol{\alpha}$ is the deviator specifying the center of the yield surface. For plastic loading ($\hat{s}_{ij}\dot{\epsilon}_{ij} > 0$),

$$\dot{s}_{ij} = \frac{2}{3} E\dot{\epsilon}_{ij} - (E - E_t)\hat{s}_{ij}\hat{s}_{kl}\dot{\epsilon}_{kl}/\sigma_0^2 \tag{7}$$

and

$$\dot{\alpha}_{ij} = E_t\hat{s}_{ij}\hat{s}_{kl}\dot{\epsilon}_{kl}/\sigma_0^2 \tag{8}$$

while Eq 6 holds for elastic increments with $\dot{\boldsymbol{\alpha}} = 0$.

The remote stressing is taken to be axisymmetric with respect to the 3-axis such that the nonzero components of $\boldsymbol{\Sigma}$ are

$$\Sigma_{33} = S, \quad \Sigma_{22} = \Sigma_{11} = T \tag{9}$$

Denote the remote mean and effective stresses by Σ_m and Σ_e so that

$$\Sigma_m = \frac{1}{3}(S + 2T) \text{ and } \Sigma_e = |S - T| \tag{10}$$

Let δE_N denote the increase in the remote axial strain component E_{33} which is prescribed to occur during the nucleation process (that is, the strain change associated with $\dot{\lambda}_\infty\boldsymbol{\Sigma}$). The solution procedure is given in the Appendix. The results of the calculations are now reported.

J_2-*Flow Theory Results*

We begin by an example in Fig. 5 based on J_2-flow theory which shows the evolution during nucleation of both the total dilatation of the void, ΔE_{kk}^T, and the dilatation just due to nucleation, ΔE_{kk}, for three different choices of $\delta E_N/\epsilon_0$ where $\epsilon_0 = \sigma_0/E$ is the elastic strain at yield. The dilatation due to nucleation, ΔE_{kk}, is the total dilatation with the contribution due to δE_N (that is, due to $\dot{\lambda}_\infty\boldsymbol{\Sigma}$) subtracted off. Note that ΔE_{kk} is essentially independent of δE_N and thus is, indeed, meaningfully identified as the contribution due to nucleation. The normalized quantities $\Delta E_{kk}^T/(\Sigma_m/E)$ and $\Delta E_{kk}/(\Sigma_m/E)$ are also found to be essentially independent of Σ_m/E and of E_t/E when it is small (that is, $E_t/E < 0.1$); they do, however, depend on triaxiality, $X \equiv \Sigma_m/\Sigma_e$, as shown below.

The significance of considering different values of δE_N is that, in an actual nucleation process, this value is determined by the mechanism of debonding together with the way in which the external loading is applied. Since no particular debonding mechanism is studied here, the relevant value of δE_N cannot be determined. Therefore it is of interest to extract the part of the macroscopic behavior that is essentially independent of δE_N.

An important feature of the process is the relatively small dilatation due to nucleation. Were the process a linearly elastic one (in an incompressible elastic matrix), then

$$\Delta E_{kk} = (9/4)\Sigma_m/E \tag{11}$$

The dilatation contribution ΔE_{kk} at the end of nucleation in Fig. 5 is only about twice this elastic value. This feature stems from the distinctly nonproportional stressing in the vicinity of the void during nucleation and the resistance of the material to plastic deformation needed to enlarge the void. By contrast, if the void were nucleated this way in a nonlinear elastic material (for example, a J_2-deformation theory material), its enlargement would be inde-

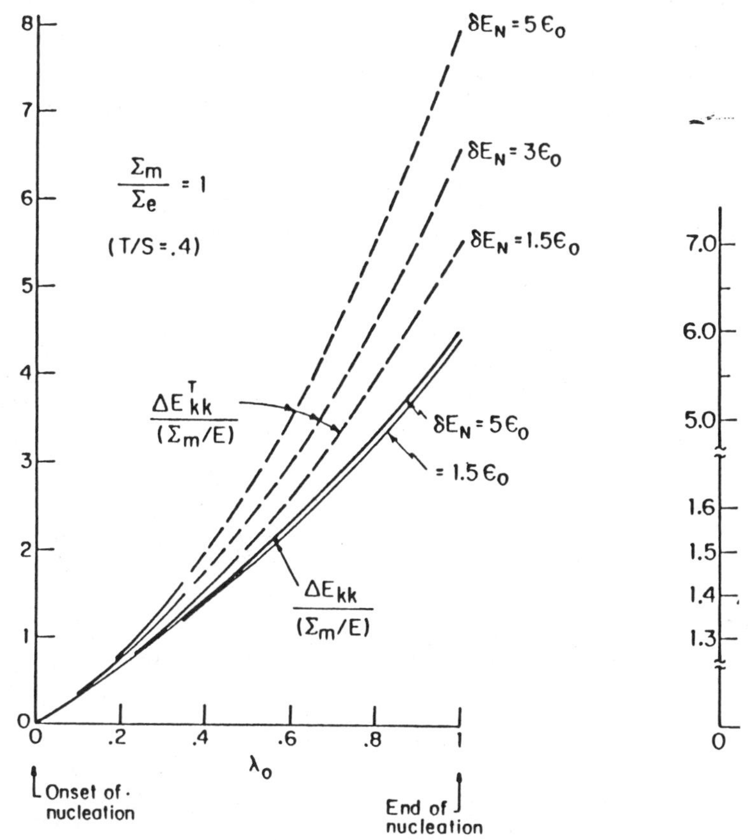

FIG. 5—*Normalized dilatation contribution for nucleation of a spherical void in an infinite matrix governed by* J_2*-flow theory* ($E_t/E = 0.02$, $S/\sigma_0 = 5$).

pendent of the history and would depend strongly on the strain at nucleation. The present results for the dilatation are essentially independent of prior plastic strain.

The deviatoric part of ΔE is much larger than the dilatation resulting from a redistribution of stress and strain throughout the matrix. It is inversely proportional to E_t rather than E. Given that ΔE must have an isotropic dependence of $\mathbf{\Sigma}$, we write for the contribution due just to nucleation

$$\Delta E_{ij} = F(X)\Sigma'_{ij}/E_t + G(X)\Sigma_m\delta_{ij}/E \tag{12}$$

where $\mathbf{\Sigma}'$ is the deviator of $\mathbf{\Sigma}$, $\Sigma_m = \frac{1}{3}\Sigma_{kk}$, $\Sigma_e = (3\Sigma'_{ij}\Sigma'_{ij}/2)^{1/2}$ and $X = \Sigma_m/\Sigma_e$. For axisymmetric stressing with $S \geq T$, Eq 12 is a general representation of the stress dependence, but under general remote stressing Eq 12 will only be valid if one can neglect dependence on the third invariant of $\mathbf{\Sigma}$ and on $\Sigma'_{ip}\Sigma'_{pj}$. The functions F and G also depend implicitly on E_t/E and δE_N, but our numerical calculations indicate that they vary by less than 2% for E_t/E in the range from 0.01 to 0.1. Figure 6 displays the dependence of F and G on δE_N for the case $X = 1$ ($T = 0.4S$). The results for the triaxiality dependence of F and G shown in Fig. 7 are computed with $\delta E_N/\epsilon_0 = 10$, but as seen in Fig. 6, the dependence on δE_N is very weak. Included in Fig. 7 are predictions for F which derive from the Gurson [*1*] model, which will be discussed in the section on Effect of Nucleation.

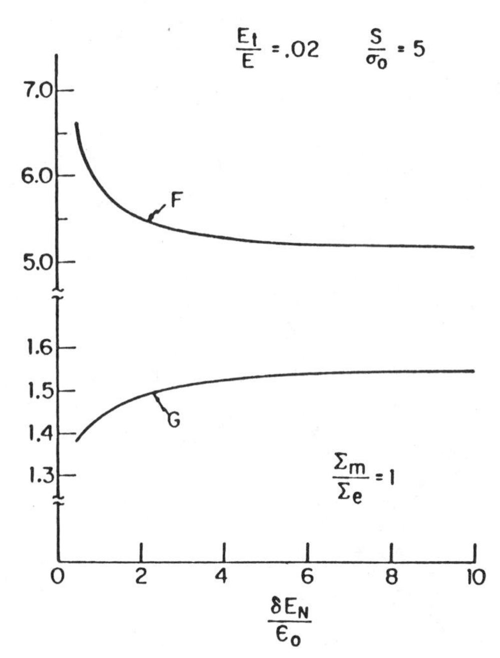

FIG. 6—*Dependence of* F *and* G *on* $\delta E_N/\epsilon_0$ *for nucleation in a matrix governed by* J_2*-flow theory.*

As mentioned earlier, interaction between the "particle" and the void surface is ignored in the calculations once nucleation starts. Inspection of the numerical solution indicates that the void surface pulls away from the particle at every point if $T/S > 0.1$ and δE_N is not large. Thus, only the values of F and G at very low triaxialities ($X < 0.44$) would be changed by a calculation which accounted for constraint of the particle on the void deformation.

Kinematic Hardening Results

For proportional stressing histories, the two plasticity theories (Eqs 5 and 7) coincide, but the isotropic hardening material offers more resistance to plastic flow under nonproportional histories than does its kinematic counterpart. As already mentioned, the nucleation process involves distinctly nonproportional stressing near the void, and thus it is expected that the curvature of the yield surface will affect ΔE.

The calculations of F and G were repeated using the kinematic hardening description of the matrix material, Eqs 6 to 8. In this case, there is a strong dependence on $Y \equiv \Sigma_e/\sigma_0$ as well as on the triaxiality measure $X = \Sigma_m/\Sigma_e$. For kinematic hardening Eq 12 is rewritten as

$$\Delta E_{ij} = F(X,Y)\Sigma'_{ij}/E_t + G(X,Y)\Sigma_m\delta_{ij}/E \tag{13}$$

Plots of F and G as functions of X are shown in Fig. 8 for three levels of $Y = \Sigma_e/\sigma_0$. For Y just above unity, corresponding to nucleation before the material hardens appreciably, the results for kinematic hardening are only slightly larger than those for isotropic hardening, as would be expected. For nucleation at larger values of Y, the strain contribution ΔE predicted by the kinematic theory is significantly larger than the isotropic hardening result, by factors as much as two or three.

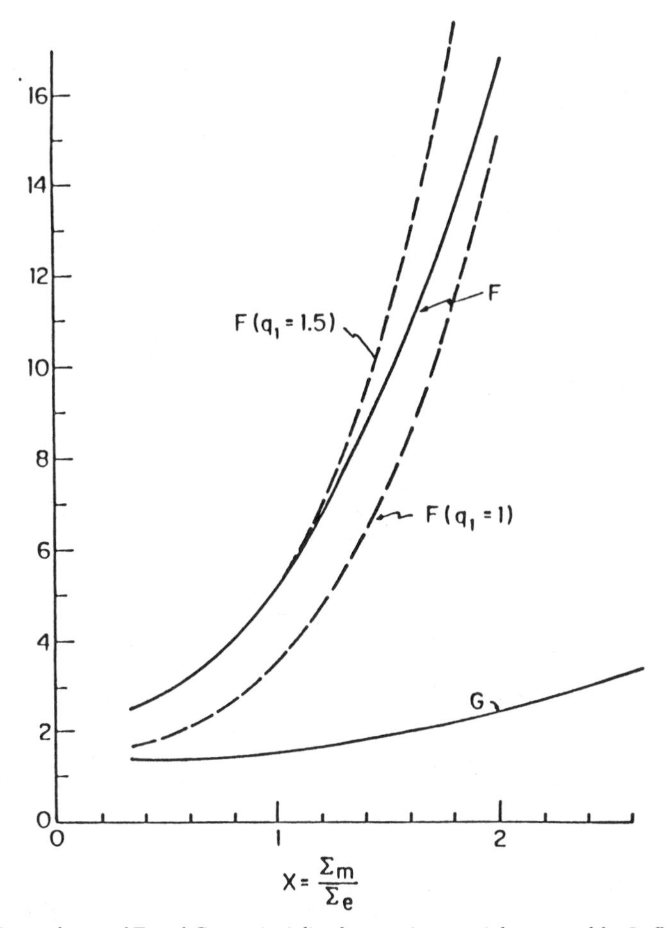

FIG. 7—*Dependence of* F *and* G *on triaxiality for matrix material governed by* J_2*-flow theory.*

As will be even clearer from the cell model results in the next section, the enhancement of the strain produced by nucleation in a kinematic hardening material over that in an isotropic hardening material is a major effect. In other problem areas, such as plastic instability phenomena, where there are significant differences between the predictions based on these two material models, the isotropic hardening model invariably tends to be overly stiff compared with experimental observations. The issue here is not fully reversed loading and Bauschinger effects; rather, it is continued loading under nonproportional stress histories. The kinematic theory reflects, albeit crudely, the high curvature or possibly even a corner, which develops at the loading point of the subsequent yield surface.

To calculate $\mathbf{\Delta\Sigma}$ defined by Eq 4, assume that $\mathbf{\Delta E}$ is given approximately by Eq 12 or 13 even when Poisson's ratio ν is not ½. Then by Eqs 4 and 12, one obtains

$$\Delta\Sigma_{ij} = \frac{2}{3} F\Sigma'_{ij} + \frac{1}{1 - 2\nu} G\Sigma_m\delta_{ij} \tag{14}$$

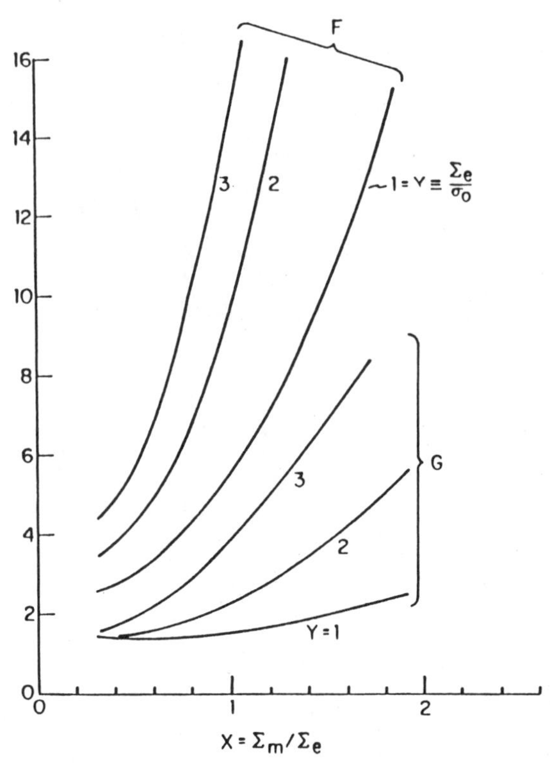

FIG. 8—*Dependence of* F *and* G *on triaxiality and* Σ_e/σ_0 *for matrix material governed by kinematic hardening theory.*

when $E_t \ll E$. This result also holds for kinematic hardening with the respective F and G values. The dilatational part of the stress drop due to nucleation at a given macroscopic strain necessarily becomes ill-defined for an incompressible matrix material.

Cell Model Calculations of Overall Stress-Strain Behavior as Influenced by Void Nucleation

The effect of a uniform distribution of spherical particles that nucleate voids simultaneously is studied by numerical solution of the axisymmetric model problem illustrated in Fig. 1. Here, the particles are assumed to be rigid, which means that in contrast to the calculation in the previous section, the stress state in the matrix material is not uniform prior to nucleation. As in the previous calculations, nucleation is simply modeled by releasing the displacements of the matrix material on the particle-matrix interface and incrementally reducing the corresponding surface tractions to zero.

Finite strains are accounted for in these cell model calculations. A Lagrangian formulation of the field equations is used, with reference to a cylindrical coordinate system, in which x^1 is the radius, x^2 is the circumferential angle, and x^3 is the axial coordinate. The displacement components on the reference base vectors are denoted u^i, where $u^2 \equiv 0$ by the assumption of axisymmetry. The Lagrangian strains are given by

$$\eta_{ij} = \frac{1}{2}(u_{i,j} + u_{j,i} + u^k_{,i}u_{k,j}) \tag{15}$$

where $(\ )_{,i}$ denotes covariant differentiation in the reference configuration. The contravariant components τ^{ij} of the Kirchhoff stress tensor on the deformed base vectors are related to the Cauchy stress tensor σ^{ij} by

$$\tau^{ij} = \sqrt{G/g}\,\sigma^{ij} \tag{16}$$

where g and G are the determinants for the metric tensors g_{ij} and G_{ij} in the reference configuration and the current configuration, respectively.

The finite strain generalization of J_2-flow theory used here has been discussed in detail in Ref *3*. The incremental stress-strain relationship is of the form

$$\dot{\tau}^{ij} = L^{ijkl}\dot{\eta}_{kl} \tag{17}$$

with the tensor of instantaneous moduli given by

$$L^{ijkl} = \frac{E}{1+\nu}\left\{\frac{1}{2}(G^{ik}G^{jl} + G^{il}G^{jk}) + \frac{\nu}{1-2\nu}G^{ij}G^{kl} - \beta\frac{3}{2}\frac{E/E_t - 1}{E/E_t - (1-2\nu)/3}\frac{s^{ij}s^{kl}}{\sigma_e^2}\right\} - \frac{1}{2}\{G^{ik}\tau^{jl} + G^{jk}\tau^{il} + G^{il}\tau^{jk} + G^{jl}\tau^{ik}\} \tag{18}$$

Here, the value of β is 1 or 0 for plastic yielding or elastic unloading, respectively, and the tangent modulus E_t is the slope of the uniaxial true stress versus natural strain curve at the stress level σ_e. The uniaxial stress-strain behavior is represented by a piecewise power law

$$\epsilon = \begin{cases} \dfrac{\sigma}{E}, & \text{for } \sigma \leq \sigma_0 \\ \dfrac{\sigma_0}{E}\left(\dfrac{\sigma}{\sigma_0}\right)^n, & \text{for } \sigma > \sigma_0 \end{cases} \tag{19}$$

where σ_0 is the uniaxial yield stress and n is the strain-hardening exponent. The values of these parameters are taken to be $\sigma_0/E = 0.004$ and $n = 5$, and, furthermore, elastic compressibility is accounted for, taking Poisson's ratio $\nu = 0.3$.

The finite strain generalization of kinematic hardening theory is analogous to the above equations. The full formulation has been given in Ref *4* and will not be repeated here.

In the numerical solution, equilibrium is based on the incremental principle of virtual work, and the boundary conditions, specified in terms of the nominal surface tractions T^i, are

$$\dot{u}^3 = 0, \quad \dot{T}^1 = \dot{T}^2 = 0, \quad \text{at } x^3 = 0 \tag{20}$$

$$\dot{u}^3 = \dot{U}_{\text{III}}, \quad \dot{T}^1 = \dot{T}^2 = 0, \quad \text{at } x^3 = B_0 \tag{21}$$

$$\dot{u}^1 = \dot{U}_{\text{I}}, \quad \dot{T}^2 = \dot{T}^3 = 0, \quad \text{at } x^1 + A_0 \tag{22}$$

$$\left.\begin{aligned} \dot{u}^i &= 0 \text{ before nucleation} \\ \dot{T}^i &= 0 \text{ after nucleation} \end{aligned}\right\} \quad \text{at } (x^1)^2 + (x^3)^2 = R_0^2 \tag{23}$$

The two constants $\dot{U}_{\text{I}}$ and $\dot{U}_{\text{III}}$ are displacement increments, and the ratio $\dot{U}_{\text{I}}/\dot{U}_{\text{III}}$ is calculated in each increment such that there is a fixed prescribed ratio between the macroscopic true stresses T and S (see also Ref 5).

The initial geometry of the region analyzed is shown in Fig. 9, where R_0 is the inclusion radius, A_0 is the initial radius of the cylindrical body analyzed, and $2B_0$ is the inclusion spacing along the cylinder axis. In the cases analyzed, the initial geometry is specified by $B_0/A_0 = 1$ and $R_0/A_0 = 0.2466$, corresponding to a volume fraction 0.01 of particles. The mesh used for the finite element solutions is shown in the figure, where each quadrilateral consists of four triangular elements.

Figure 2 shows the stress $(S - T)/\sigma_0$ versus the average axial logarithmic strain E_3 for two cases, where nucleation takes place under constant macroscopic stress, at $S = 3.0\sigma_0$ and $S = 3.6\sigma_0$, respectively, while $T/S = 0.5$. The matrix material follows J_2-flow theory. For comparison, the behavior of matrix material with the bonded rigid particle without nucleation and the behavior when the void is present from the beginning are also shown in the figure. Prior to nucleation, the macroscopic stress-strain curve with a rigid particle differs from that of the matrix material by less than 1%. After nucleation the stress level for a given value of the strain E_3 is nearly reduced to the level found when the void is present from the beginning. Figure 3 shows the corresponding growth ΔV of the void after nucleation,

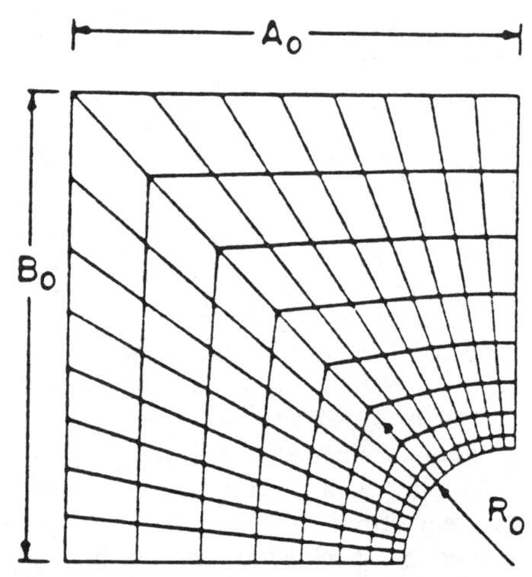

FIG. 9—*Initial geometry and finite-element grid for cylindrical cell model.*

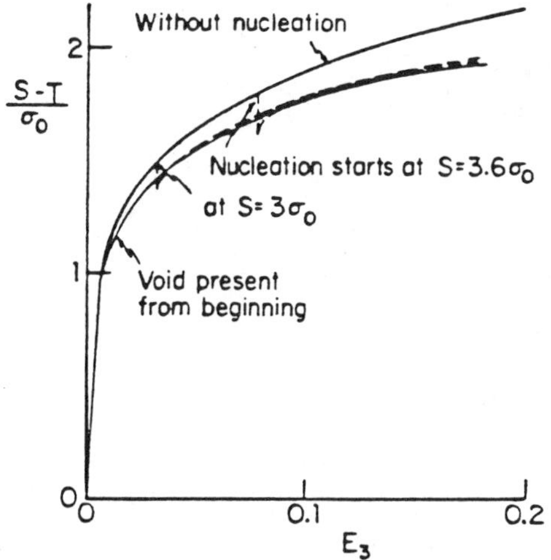

FIG. 10—*Overall stress-strain behavior for nucleation at constant strain from rigid spherical particles in a matrix material governed by* J_2*-flow theory* (ρ = *0.01,* ϵ_0 = *0.004,* n = *5,* ν = *0.3*).

normalized by the particle volume V_0 and the initial yield strain $\epsilon_0 = \sigma_0/E$. The solid dot on two of the curves indicates the point where nucleation is completed (that is, the tractions on the particle-matrix interface have been relaxed to zero). At these positions the growth is far less than that of a void present from the beginning.

Figures 10 and 11 show the same computations repeated with the only difference that here nucleation takes place under a constant macroscopic axial strain E_3, but still such that the stress ratio $T/S = 0.5$ remains constant. Here the macroscopic stresses decay during nucleation, to values somewhat below the curve for a void present from the beginning, and simultaneously a great deal of elastic unloading occurs in the matrix material near the voids. However, after some subsequent stretching these unloading regions disappear, and the

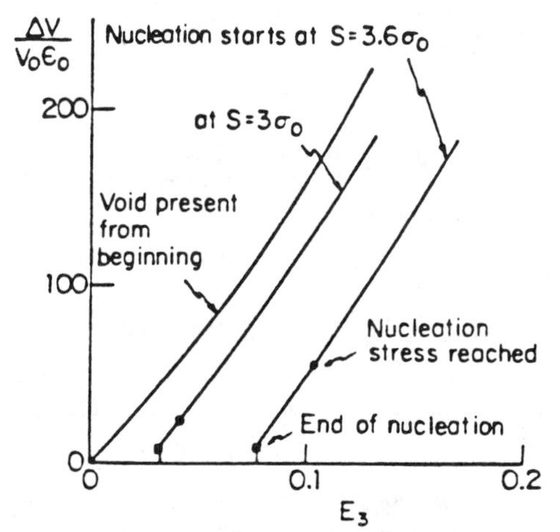

FIG. 11—*Normalized volume increase as a function of overall strain for nucleation at constant strain in a* J_2*-flow theory matrix.*

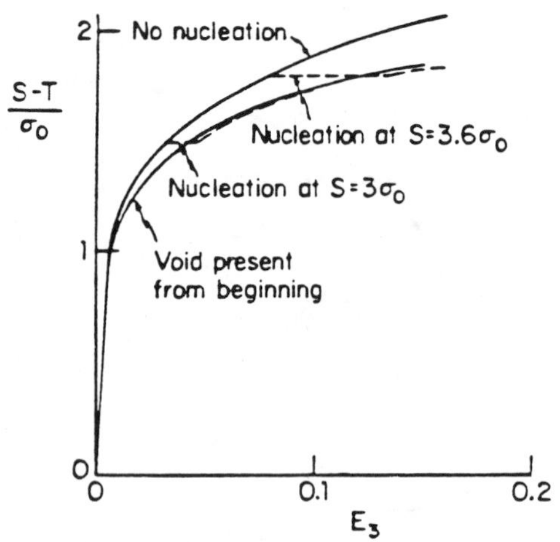

FIG. 12—*Overall stress-strain behavior for nucleation at constant stress from rigid spherical particles in a matrix material governed by kinematic hardening theory* (ρ = *0.01*, ϵ_0 = *0.004*, n = *5*, ν = *0.3*).

strains at which the macroscopic stress for the onset of nucleation is reached again are nearly identical to those at the end of nucleation in Figs. 2 and 3.

Kinematic hardening theory has been used to analyze the same cases, as illustrated in Figs. 12–15. Clearly, kinematic hardening has a strong influence on the predictions in the present cases, as would be expected based on Fig. 8, since the material has high hardening and the relevant values of Σ_e/σ_0 are 1.5 and 1.8, respectively. Figures 12 and 13 show that the stress-strain curves after nucleation remain below that corresponding to a void present from the beginning, in contrast to the behavior found for J_2-flow theory. Thus, the mac-

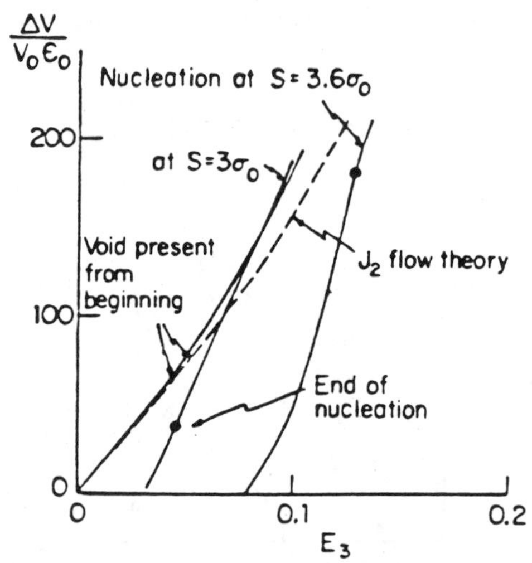

FIG. 13—*Normalized volume increase as a function of overall strain for nucleation at constant stress in a matrix material governed by kinematic hardening theory. For reference, the prediction for a matrix governed by* J_2*-flow theory is included as a dashed line curve for the case when the void is present from the start of straining.*

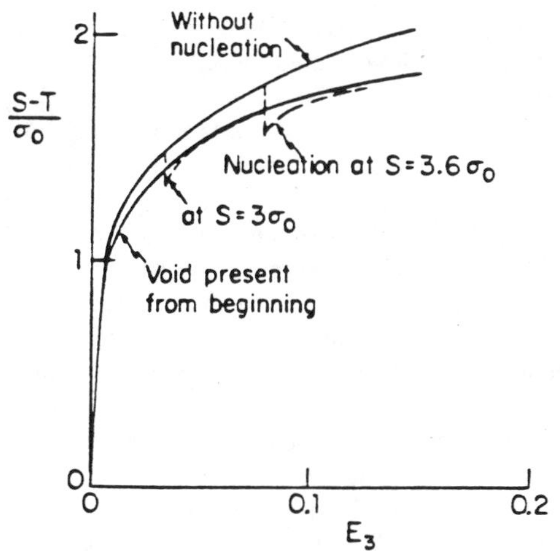

FIG. 14—*Overall stress-strain behavior for nucleation at constant strain from rigid spherical particles in a matrix material governed by kinematic hardening theory* (ρ = *0.01*, ϵ_0 = *0.004*, n = *5*, ν = *0.3*).

roscopic strain increment resulting from nucleation is significantly larger according to kinematic hardening theory. Figures 13 and 15 show that kinematic hardening also predicts a larger void growth during nucleation. For completeness, the comparison between the dilatation predicted by the two matrix hardening rules is included in Fig. 13 for the case in which the void is present from the start of straining. Initially, there is essentially no difference between the two predictions since the straining is everywhere nearly proportional; however, the two sets of predictions diverge as straining progresses due to nonproportionality associated with the finite expansion of the void.

The results for ΔE_{ij} obtained by these cell model calculations may be compared with Eq 2, using Eqs 12 and 13. Here, $\dot{E}_{ij}$ are interpreted as logarithmic strain increments, and $\dot{\Sigma}_{ij}$ are interpreted as true stress increments. The computations illustrated in Figs. 2 and 12 are

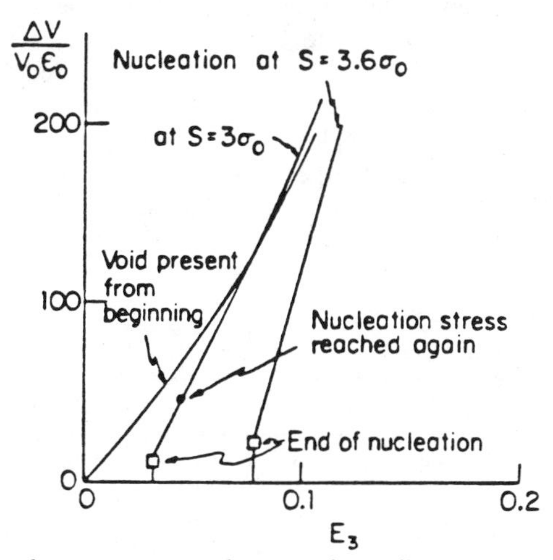

FIG. 15—*Normalized volume increase as a function of overall strain for nucleation at constant strain in a matrix material governed by kinematic hardening theory.*

most conveniently used for this comparison, since here $\dot{\Sigma}_{ij} = 0$ during nucleation. Then, with $T/S = 0.5$ prescribed in all the calculations, $\Sigma_m/\Sigma_e = 4/3$ is constant, and the values of Σ_e/σ_0 at nucleation are either 1.5 or 1.8 in the cases considered. Table 1 shows the values of ΔE_3 and ΔE_1 obtained by the cell model computations, according to Eq 2 with $\dot{\rho} = 0.01$, and the values of the functions F and G computed from Eqs 12 and 13.

The two J_2-flow theory calculations give values of F that are not too much higher than the value $F = 8.0$ given in Fig. 7, whereas the values found for G are significantly larger than the value 1.7. The results found for kinematic hardening agree qualitatively with Fig. 8, since both F and G are increased relative to the J_2-flow theory results. For the lower stress level, $\Sigma_e/\sigma_0 = 1.5$, the increases are of the same order of magnitude as those found in Fig. 8. For the larger stress level $\Sigma_e/\sigma_0 = 1.8$, the values of both F and G are much higher than those in Fig. 8.

For the values of F and G calculated by the cell model, it should be noted that the volume fraction increment $\dot{\rho} = 0.01$ is rather large. The value of the tangent modulus E_t in Eqs 12 and 13 is taken to be that of uniformly stressed matrix material at the stress level corresponding to the onset of nucleation; but due to the power-hardening relation (Eq 19) used here, this will only give a good approximation for small values of $\dot{\rho}$. The accuracy is least good for kinematic hardening at the higher stress level $\Sigma_e/\sigma_0 = 1.8$, since here the maximum stress-carrying capacity is nearly reached at the point where nucleation ends.

Effect of Nucleation as Predicted by the Gurson Isotropic Hardening Model

A prototype constitutive relation modeling void nucleation and growth has been proposed by Gurson [*1*], and this model is probably the most complete and widely used model of its type. Gurson's theory is endowed with a yield condition, a flow law, a measure of void volume fraction, a rule for nucleating voids, and a law for evolution of the void volume fraction. Its yield surface was derived from approximate solutions to a volume element of perfectly-plastic material containing a void, and it was extended to strain-hardening materials under the assumption of isotropic hardening. The stress-strain behavior of the void-free material is part of the specification of the model. With no voids present the model reduces to the classical isotropic hardening theory based on the von Mises invariant, J_2-flow theory.

Here the main equations governing the Gurson model will be briefly stated, and the quantity ΔE introduced in the section on Nucleation of an Isolated Void will be identified. We will also attempt to bring out the effect of nucleation on macroscopic strain behavior as predicted by the model. A more complete specification of the model can be found in the papers by Saje, Pan, and Needleman [*6*] and Needleman and Rice [*7*], who particularly emphasize the role of nucleation in offsetting strain hardening and in promoting flow localization.

TABLE 1—*Values of* F *and* G *computed using the cell model for* ΔE_3 *and* ΔE_1.

	ΔE_3	ΔE_1	X	Y	E_t/E	F	G
Fig. 2	0.973	−0.446	4/3	1.5	0.0395	9.3	3.4
Fig. 2	2.62	−1.22	4/3	1.8	0.0191	10.2	6.3
Fig. 12	1.391	−0.614	4/3	1.5	0.0395	13.2	6.8
Fig. 12	5.55	−2.37	4/3	1.8	0.0191	20.9	28.4

Gurson's yield function involves two material state parameters, the void volume fraction ρ and a measure of the current flow stress of the matrix material σ. With $\mathbf{\Sigma}$ as the macroscopic stress and with $\mathbf{\Sigma}'$, Σ_m and Σ_e defined as before, the yield condition is

$$\Phi(\mathbf{\Sigma}, \sigma, \rho) = \left(\frac{\Sigma_e}{\sigma}\right)^2 + 2q_1\rho \cosh\left(\frac{3q_2}{2}\frac{\Sigma_m}{\sigma}\right) - 1 - q_1^2\rho^2 = 0 \tag{24}$$

The factors q_1 and q_2 were introduced in Ref *8* to bring the yield function into better agreement with numerical results for periodic arrays of spherical voids. Gurson's original proposal employed $q_1 = q_2 = 1$, while the suggestion in Ref *8* was $q_1 = 3/2$ and $q_2 = 1$. See Ref *5* for a discussion of the current status of the yield function in comparison with experimental data and micro-mechanical calculations.

In addition to the yield condition, the following equations are postulated for plastic loading

$$\dot{E}_{ij}^p = \dot{\lambda}\partial\Phi/\partial\Sigma_{ij} \tag{25}$$

$$\Sigma_{ij}\dot{E}_{ij}^p = (1 - \rho)\sigma\dot{\sigma}[1/E_t(\sigma) - 1/E] \tag{26}$$

$$\dot{\rho} = \dot{\rho}_{\text{growth}} + \dot{\rho}_{\text{nucleation}} \tag{27}$$

$$\dot{\rho}_{\text{growth}} = (1 - \rho)\dot{E}_{kk}^p \tag{28}$$

$$\dot{\rho}_{\text{nucleation}} = A(\sigma, \Sigma_m)\dot{\sigma} + B(\sigma, \Sigma_m)\dot{\Sigma}_m \tag{29}$$

Normality is invoked in Eq 25; the condition for continued yielding, $\dot{\Phi} = 0$, allows one to determine $\dot{\lambda}$ as

$$\dot{\lambda} = \sigma\left(\frac{1}{E_t} - \frac{1}{E}\right)\left(\frac{\partial\Phi}{\partial\Sigma_{ij}}\dot{\Sigma}_{ij} + \dot{\rho}\frac{\partial\Phi}{\partial\rho}\right) \div \left(-\frac{\partial\Phi}{\partial\sigma}\frac{\partial\Phi}{\partial\Sigma_{mn}}\Sigma_{mn}\right) \tag{30}$$

Equation 26 equates the macroscopic plastic work rate to the plastic work rate in the matrix, where $E_t(\sigma)$ is the tangent modulus of the effective stress-strain curve of the matrix at σ. Equation 27 separates the increase in void volume fraction into a contribution due to growth of previously nucleated voids (Eq 28) and a contribution due to nucleation of new voids (Eq 29). Several nucleation rules of the form (Eq 29) have been proposed [*6,7*], but will not be detailed here.

The strain contribution due to nucleation, $\dot{\rho}\Delta E_{ij}$ in Eq 2, as predicted by the Gurson model is readily determined from the foregoing equations. For the first voids nucleated (when $\rho = 0$) at $\mathbf{\Sigma}$, the result for the void of unit volume is

$$\Delta E_{ij} = \frac{3}{2}q_1\frac{1}{E_t}\cosh\left(\frac{3q_2}{2}\frac{\Sigma_m}{\Sigma_e}\right)\Sigma_{ij}' \tag{31}$$

assuming $E_t \ll E$. Thus, when cast in the form of Eq 12, the Gurson model gives

$$F = \frac{3}{2}q_1\cosh(3q_2X/2) \text{ and } G = 0 \tag{32}$$

and this prediction is compared with the isotropic hardening results in Fig. 7 using both $q_1 = 3/2$ and $q_1 = 1$, in each case with $q_2 = 1$. The absence of any dilatational contribution ($G = 0$) at first nucleation is a consequence of the normality assumption (Eq 25) invoked for the model. While not strictly correct, the dilatational contribution derived in the section on Nucleation of an Isolated Void is generally much smaller than the deviatoric contribution.

Nucleation is included in the Gurson model in a highly coupled manner. The quantitative effects of nucleation on macroscopic behavior are not transparent in the model. The fact that the model is in good agreement with the micro-mechanical calculation of ΔE for isotropic hardening lends confidence to the model.

The effect of nucleation as specified by the Gurson model is transparent in the case of pure shear. Since $\Sigma_m = 0$ in pure shear, the change in ρ is due entirely to nucleation. With Σ_{12} as the macroscopic shear stress and with $\tau \equiv \sigma/\sqrt{3}$ as the equivalent shear stress in the matrix material, the yield condition (Eq 24) implies (with $q_1 = 1$)

$$\Sigma_{12} = (1 - \rho)\tau \tag{33}$$

Then, with

$$\dot{\gamma}^p \equiv \sqrt{3}\dot{\epsilon}^p = \sqrt{3}\dot{\sigma}(1/E_t - 1/E)$$

as the equivalent shear strain rate in the matrix, Eqs 26 and 33 give the macroscopic shear strain rate as simply

$$2\dot{E}^p_{12} = \dot{\gamma}^p$$

independent of ρ. Thus the relation between the macroscopic shear stress-strain curve and the corresponding matrix curve is exceptionally simple as sketched in Fig. 16. The relation between Σ_{12} and E^p_{12} depends only on the current value of ρ, independent of when the voids have been nucleated.

Under other stress histories, the post-nucleation state is not so simply related to the history where voids have been present from the start since void growth itself has a strong history dependence. Under proportional stressing the following statement quite closely reflects the Gurson model prediction. The deviatoric part of the strain following nucleation of ρ volume fraction of voids is almost the same as the deviatoric strain if the voids were present from the start and had grown to a current void volume fraction ρ. In other words, under proportional stressing, the deviatoric macroscopic strain at a given current void volume fraction ρ is essentially independent of whether the voids were present from the start or whether they were nucleated late in the history.

Conclusions

Relatively small amounts of void nucleation can significantly affect macroscopic hardening behavior. Moreover, the longer nucleation is delayed generally the larger will be its softening contribution. For example, in uniaxial tension from Eq 12 or 13,

$$\dot{\rho}\Delta E_3 = \frac{2}{3}\dot{\rho}F\Sigma_3/E_t \tag{34}$$

if the small dilatational component is neglected. Delaying nucleation increases Σ_3 and decreases E_t, thereby increasing the strain contribution due to nucleation. The tangent modulus of the macroscopic stress-strain curve in the presence of the nucleation, E_t^N, is related to the tangent modulus in the absence of nucleation, E_t, by

$$\frac{1}{E_t^N} = \frac{\dot{E}_3}{\dot{\Sigma}_3} = \frac{1}{E_t}\left[1 + \frac{\dot{\rho}}{\dot{\Sigma}_3}\frac{2}{3}F\Sigma_3\right] \tag{35}$$

Regarding ρ as a function of E_3 and replacing $\dot{\rho}/\dot{\Sigma}_3$ with $(d\rho/dE_3)/E_t^N$, one obtains

$$\frac{E_t^N}{E_t} = 1 - \frac{2}{3}F\frac{d\rho}{dE_3}\frac{\Sigma_3}{E_t} \tag{36}$$

This formula reveals that the macroscopic hardening rate as measured by E_t^N is not only diminished by delayed nucleation, as just discussed, but also by an increased rate of nucleation as measured by $d\rho/dE_3$ and by triaxial effects through F. As discussed by Needleman and Rice [7], the macroscopic hardening rate can become negative at rates of nucleation which are not excessively large. In uniaxial tension, $2F/3$ from the J_2-flow theory calculation is about 2. A typical value of Σ_3/E_t is about 1, in which case E_t^N will be negative if $d\rho/dE_3 \cong$ ½. In other words, a burst of nucleation giving a 1% volume fraction of voids over a 2% range of strain will produce a negative overall strain hardening rate over this range. Such bursts of nucleation are destabilizing, leading to flow localization on the macroscopic scale.

A separate issue which has surfaced in the present study is the unusually strong sensitivity of the predictions to the choice of multiaxial plasticity law for the matrix material, that is, to isotropic or kinematic hardening. The F-factor in Eqs 34 and 36, as computed assuming kinematic hardening, can be as much as two or three times the corresponding value computed assuming isotropic hardening. It is an open question at this point as to which plasticity law gives the more realistic representation of matrix behavior in this application. Based on experience with other applications involving nonproportional stressing where large differ-

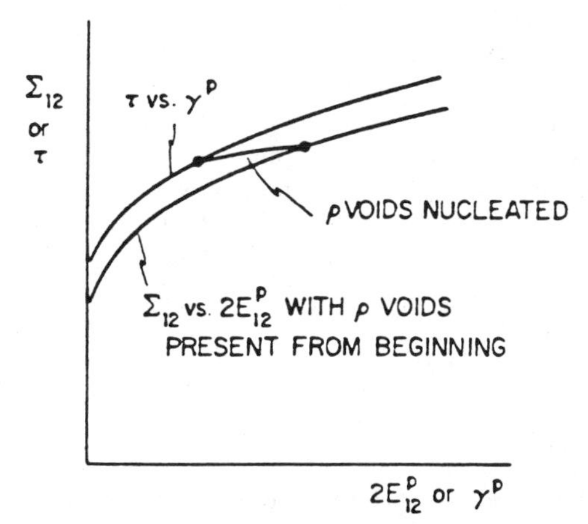

FIG. 16—*Effect of nucleation on shear stress-strain behavior according to Gurson theory.*

ences between predictions on the two plasticity laws are found, the predictions assuming isotropic hardening are likely to underestimate the strain contribution while the kinematic hardening predictions may be more realistic. If so, bursts of nucleation at finite strain are even more destabilizing than one would infer from the Gurson model for example.

The remarkable thing about the kinematic hardening results in Fig. 12 is that at a given stress, the strain subsequent to delayed nucleation is greater than when the void is present from the beginning. This effect seems counterintuitive since even a nonlinear elastic solid would only experience the same strain following delayed nucleation as when the void is present from the start. The effect can be understood only in terms of the nonproportional stressing in the vicinity of the void during nucleation and the reduced resistance to plastic flow associated with the high curvature of the kinematic hardening yield surface.

Acknowledgment

The work of JWH was supported in part by the Materials Research Laboratory under Grant NSF-DMR-83-16979, by the National Science Foundation under Grant NSF-MSM-84-16392, and by the Division of Applied Sciences, Harvard University. The work of VT was supported by the Danish Technical Research Council through Grant 16-4006.M.

APPENDIX

Method for Calculating ΔE

To formulate a minimum principle for the displacement rates, let

$$w(\mathbf{s}, \dot{\boldsymbol{\epsilon}}) = \frac{1}{2}\dot{\sigma}_{ij}\dot{\epsilon}_{ij} \tag{37}$$

be the stress dependent strain-rate potential of the matrix material evaluated using either Eqs 5 and 6 for J_2-flow theory or Eqs 6 and 7 for kinematic hardening theory. For the moment, suppose the void is nucleated in a spherical block of material of radius R (subsequently R will be allowed to become infinite), which is stressed into the plastic range by uniform tractions $\Sigma_{ij}n_j$ applied to its surface A_R, where n is the outward unit normal to A_R. As discussed in the section on Nucleation of an Isolated Void, during nucleation traction rates $-\dot{\lambda}_0 T_i^0$ are applied to the surface A_0 of the unit void being nucleated, simultaneously with traction rates $\dot{\lambda}_\infty \Sigma_{ij}n_j$ applied on A_R. For the finite region with outer radius R, the actual displacement rates minimize

$$W = \int_V w(\mathbf{s}, \dot{\boldsymbol{\epsilon}})\, dV + \dot{\lambda}_0 \int_{A_0} T_i^0 \dot{u}_i \, dA - \dot{\lambda}_\infty \int_{A_R} \Sigma_{ij} n_j \dot{u}_i \, dA \tag{38}$$

where V is the region exterior to A_0 and interior to A_R.

The principle must be modified such that the functional remains bounded when $R \rightarrow \infty$. To this end, let $\dot{\boldsymbol{\epsilon}}^\infty$ be the strain rate at infinity associated with $\dot{\lambda}_\infty \Sigma_{ij}$ and let $\dot{\mathbf{u}}^\infty$ be a displacement rate field such that

$$\dot{\epsilon}_{ij}^{\infty} = \frac{1}{2}(\dot{u}_{i,j}^{\infty} + u_{j,i}^{\infty})$$

Following procedures similar to those given for visco-plastic behavior in Ref 9, one can show that Eq 38 can be replaced by the modified functional of the additional displacement rates

$$W = \int_V [w(\mathbf{s}, \dot{\boldsymbol{\epsilon}}) - w(\mathbf{s}, \dot{\boldsymbol{\epsilon}}^{\infty}) - \dot{\Sigma}_{ij}\dot{\tilde{\epsilon}}_{ij}]\, dV + \int_{A_0} (\dot{\lambda}_{\infty}\Sigma_{ij}n_j + \dot{\lambda}_0 T_i^0)\dot{\tilde{u}}_i\, dA \tag{39}$$

where n points into the void on A_0 and the additional rate quantities are defined by

$$\dot{\tilde{\boldsymbol{\epsilon}}} = \dot{\boldsymbol{\epsilon}} - \dot{\boldsymbol{\epsilon}}^{\infty} \text{ and } \dot{\tilde{\mathbf{u}}} = \dot{\mathbf{u}} - \dot{\mathbf{u}}^{\infty} \tag{40}$$

The modified function is minimized by the actual additional displacement rates. Moreover, the modified functional remains well-conditioned as $R \rightarrow \infty$ for all fields for which $\dot{\tilde{\boldsymbol{\epsilon}}}$ decays faster than $r^{-3/2}$.

The desired extra strain contribution, ΔE, due to nucleation of the void is obtained for the *finite problem* by integrating over the nucleation history the incremental contribution

$$\Delta\dot{E}_{ij} = \frac{1}{2}\int_{A_R} (\dot{\tilde{u}}_i^0 n_j + \dot{\tilde{u}}_j^0 n_i)\, dA \tag{41}$$

where $\dot{\tilde{\mathbf{u}}}^0$ is that part of the additional field due to just the traction rate $-\dot{\lambda}_0 T_i^0$ on A_0. As discussed in the section on Nucleation of Isolated Voids, the contribution due to $\dot{\lambda}_{\infty}\Sigma_{ij}n_{ij}$ on A_R is not included. The two contributions to the additional field are easily separated as discussed below. Equation 41 applies for the finite region but cannot be used for the limit solution with $R = \infty$. An alternative means of calculating $\Delta\mathbf{E}$, which does apply in the limit, makes use of the reciprocal theorem and an auxiliary solution. For the auxiliary solution, let $\dot{\Sigma}_{ij}^A n_j$ be applied on A_R with zero traction rates on A_0 and let $\dot{u}_i^A$ be the associated displacement-rate field calculated using the same distribution of the incremental moduli as in the incremental nucleation problem itself. By reciprocity

$$\dot{\Sigma}_{ij}^A \Delta\dot{E}_{ij} = \int_{A_R} \dot{\Sigma}_{ij}^A \dot{\tilde{u}}_i^0 n_j\, dA = -\int_{A_0} \dot{\lambda}_0 T_i^0 \dot{u}_i^A\, dA \tag{42}$$

since $\dot{\tilde{u}}_i^0$ is that part of the solution associated with zero traction rates on A_R. The integral over A_0 in Eq 42 is readily calculated in the limit $R \rightarrow \infty$, and the individual components of $\Delta\dot{\mathbf{E}}$ can be computed by making several appropriate choices for $\dot{\mathbf{\Sigma}}^A$.

Since the solution has axial symmetry with respect to the x_3-axis, let r and θ be radial and azimuthal coordinates with θ measured from the x_3-axis. The additional velocity fields in the incompressible matrix are generated from a velocity potential according to

$$\dot{\tilde{u}}_r = -r^{-2}(\sin\theta)^{-1}(\Psi\sin\theta)_{,\theta}, \quad \dot{\tilde{u}}_\theta = r^{-1}\Psi_{,r} \tag{43}$$

The velocity potential used in the calculations was

$$\Psi = a_0 \cot \theta + \sum_{k=2,4,\ldots} \sum_{j=1,2,3,\ldots} a_{kj} r^{-j+1} [P_k(\cos \theta)]_{,\theta} \tag{44}$$

where the a's are amplitude factors which were chosen to minimize Eq 39 and $P_k(x)$ is the Legendre polynomial of degree k. The lead term, $a_0 \cot \theta$, is the spherically symmetric contribution.

Denote the set of free amplitude factors by A_i, $i = 1, N$ and introduce the notation

$$\dot{\mathbf{u}} = \sum_{i=1}^{N} \dot{A}_i \mathbf{u}^{(k)}, \quad \dot{\boldsymbol{\epsilon}} = \sum_{i=1}^{N} \dot{A}_i \boldsymbol{\epsilon}^{(k)} \tag{45}$$

The functional (Eq 39) becomes

$$W = \frac{1}{2} \sum_{i=1}^{N} \sum_{j=1}^{N} M_{ij} \dot{A}_i \dot{A}_j + \sum_{i=1}^{N} B_i \dot{A}_i \tag{46}$$

where

$$M_{pq} = \int_V L_{ijkl} \epsilon_{ij}^{(p)} \epsilon_{kl}^{(q)} \, dV \tag{47}$$

and

$$B_p = \int_{A_0} (\dot{\lambda}_\infty \Sigma_{ij} n_j + \dot{\lambda}_0 T_i^0) u_i^{(p)} \, dA + \int_V [L_{ijkl} - L_{ijkl}^\infty] \dot{\epsilon}_{ij}^\infty \epsilon_{kl}^{(p)} \, dV \tag{48}$$

Here $\mathbf{L}$ are the incremental moduli at a given point and $\mathbf{L}^\infty$ are the incremental moduli at infinity. The stress and the incremental moduli are updated at each incremental step of the solution allowing for the possibility of elastic unloading or plastic loading. The equations for the increments of the amplitude factors follow immediately from Eq 46 as

$$\sum_{j=1}^{N} M_{ij} \dot{A}_j = -B_i \quad i = 1, N \tag{49}$$

The volume integrals in Eqs 47 and 48 were evaluated numerically using 10×10 Gauss-type formulas over the domain of r and θ. The surface integrals in Eqs 48 and 42 were evaluated analytically. The auxiliary problem and the problem for $\dot{\mathbf{u}}^0$ are obtained from Eq 49 simply by changing the B-vector. For the auxiliary problem $\dot{\lambda}_\infty \boldsymbol{\Sigma}$ is replaced by $\dot{\boldsymbol{\Sigma}}^A$ and $\dot{\lambda}_0$ is set to zero; for the problem for $\dot{\mathbf{u}}^0$, $\dot{\lambda}_\infty$ (and $\dot{\boldsymbol{\epsilon}}^\infty$) is set to zero. The strain contribution due to nucleation of the void is readily calculated from Eq 42. The calculations for F and G reported in the body of the paper were carried out using the same seven free amplitudes as in Ref *9*, corresponding to a_0 and a_{kj} with $k = 2, 4$ and $j = 1, 2, 3$ in Eq 44.

References

[*1*] Gurson, A. L., *Journal of Engineering Materials and Technology,* Vol. 99, 1977, pp. 2–15.
[*2*] Needleman, A., "A Continuum Model for Void Nucleation by Inclusion Debonding," *Journal of Applied Mechanics,* Vol. 54, 1987, pp. 525–532.

[3] Hutchinson, J. W. in *Numerical Solution of Nonlinear Structural Problems,* AMD-Vol. 6, R. F. Hartung, Ed., American Society of Mechanical Engineering, 1973, pp. 17–30.
[4] Tvergaard, V., *International Journal of Mechanical Sciences,* Vol. 20, 1978, pp. 651–658.
[5] Tvergaard, V., *Journal of the Mechanics and Physics of Solids,* Vol. 35, 1987, pp. 43–60.
[6] Saje, M., Pan, J., and Needleman, A., *International Journal of Fracture,* Vol. 19, 1982, pp. 163–182.
[7] Needleman, A. and Rice, J. R. in *Mechanics of Sheet Metal Forming,* D. P. Koistinen and N.-M. Wang, Eds., Plenum Publishing Company, New York, 1978, pp. 237–265.
[8] Tvergaard, V., *International Journal of Fracture,* Vol. 18, 1982, pp. 237–252.
[9] Budiansky, B., Hutchinson, J. W., and Slutsky, S. in *Mechanics of Solids,* H. G. Hopkins and M. J. Sewell, Eds., Pergamon Press, New York, 1982, pp. 13–46.

L. B. Freund[1]

Results on the Influence of Crack-Tip Plasticity During Dynamic Crack Growth

REFERENCE: Freund, L. B., **"Results on the Influence of Crack-Tip Plasticity During Dynamic Crack Growth,"** *Fracture Mechanics: Perspectives and Directions (Twentieth Symposium), ASTM STP 1020,* R. P. Wei and R. P. Gangloff, Eds., American Society for Testing and Materials, Philadelphia, 1989, pp. 84–97.

ABSTRACT: Dynamic fracture processes in structural materials are often described in terms of the relationship between a measure of the crack driving force and the crack-tip speed. In this paper, ongoing research directed toward establishing a basis for such a relationship in terms of crack-tip plastic fields is described. In particular, the role of material inertia on a small scale as it influences the perceived fracture resistance of a rate-independent material is discussed. Also, the influence of material strain rate sensitivity on the development of crack-tip plastic deformations, and the implications for cleavage propagation and arrest in a material that can undergo a fracture mechanism transition is considered. The discussion is concluded with mention of a few outstanding problems in the study of dynamic fracture, including some recent experimental evidence that the traditional crack-tip characterization viewpoint of fracture mechanics may be inadequate under very high rate loading conditions.

KEY WORDS: dynamic fracture, dynamic crack propagation, elastic-plastic fracture, high strain rate fracture, crack arrest

Consider growth of a crack in an elastic-plastic material under conditions that are essentially two-dimensional. The process depends on the configuration of the body in which the crack grows and on the details of the applied loading, in general. However, if the region of active plastic flow is confined to the crack-tip region, and if the elastic fields surrounding the active plastic zone are adequately described in terms of an elastic stress-intensity factor, then it is commonly assumed that the prevailing stress-intensity factor controls the crack-tip inelastic process. With this point of view, the stress-intensity factor provides a one-parameter representation of the input into the crack-tip zone. The viewpoint mimics the small-scale-yielding hypothesis of elastic-plastic fracture mechanics [*1*], but the basis for it in the study of rapid crack growth is less well established.

Experimental data on rapid crack growth in elastic-plastic materials are commonly interpreted on the basis of an extension of the Irwin crack growth criterion in those cases in which a stress-intensity factor field exists. If the applied stress-intensity factor is K_a, then a common constitutive assumption is that there exists a material parameter or material function, say $K_m(v, T)$, depending on crack speed, and possibly on temperature, such that the crack grows with $K_a = K_m$. Indeed, in the jargon of fracture dynamics, such a condition provides an equation of motion for the position of the crack tip as a function of time.

Rapid crack growth in metals subjected to quasistatic loading or stress wave loading of modest intensity seems to follow this constitutive assumption to a sufficient degree so that

[1] Professor, Division of Engineering, Brown University, Providence, RI 02912.

a systematic study of some of its physical underpinnings is warranted. Thus, in recent years considerable effort has been devoted to developing models to explain the reasons why K_m depends on crack speed, and possibly on temperature, as it does for real materials. The approach is quite straightforward. It is assumed that a crack grows at some speed v in an elastic-plastic or elastic-viscoplastic material under the action of the input K_a applied remotely from the crack-tip region. The potentially large stresses near the edge of the crack are relieved through inelastic deformation in an active plastic zone, and a permanently deformed layer is left in the wake of the active plastic zone along each crack face as the crack advances through the material. A solution of this problem is then obtained for arbitrary K_a and v in the form of stress and deformation fields that satisfy the field equations in some sense. With this solution in hand, a crack growth criterion motivated by the physics of the process may be imposed on the solution to yield a relationship between K_a and v that must be satisfied for the crack to steadily advance. This relationship is, in fact, the material function $K_m(v, T)$ for the model problem.

The way in with K_m depends on v, T, and other system parameters depends critically on the details of the fracture separation process, that is, whether it is a void nucleation and ductile hole growth mechanism or a cleavage mechanism, whether there is a strain rate induced elevation of flow stress or not, whether the material strain hardens significantly or flows with little hardening, and so on. The processes that must be analyzed are inherently nonlinear, but a few general results have been obtained for dynamic crack growth in elastic-plastic materials. Some recent studies are described in the sections to follow.

Rate-Independent Material Response

Experimental data on the dependence of dynamic fracture toughness versus crack speed for American Iron and Steel Institute (AISI) 4340 steel and other materials that are commonly considered as elastic-plastic in their bulk mechanical response have some common features. Here, attention is limited to situations in which crack growth occurs by the single mechanism of void nucleation and ductile hole growth to coalescence, and in which the extent of plasticity is sufficiently limited to permit interpretation of the fields surrounding the crack tip region on the basis of a stress-intensity factor. The toughness is found to be relatively insensitive to variations in speed for very low speeds (less than 20% of the shear wave speed) but to increase dramatically with crack speed for greater speeds. The speed dependence of the surrounding elastic field is not nearly great enough to account for this dependence, so an explanation must be sought in the plastically deforming region itself. The most likely reasons for this toughness-speed behavior are material inertia and material rate sensitivity. While the strain rate in the crack tip region is necessarily very high, the same general behavior has been observed in materials that are relatively rate insensitive in their bulk response up to strain rates of 10^3 s^{-1}, so the role of material inertia has been examined separately from the role of rate effects in research on this matter. The general idea is to generate a theoretical fracture toughness versus crack speed relationship to determine the role of inertia on the scale of crack-tip plastic zone on the observed dynamic crack growth response.

A rough estimate of the conditions under which material inertia has a significant influence on the development of fields within the active plastic zone is obtained as follows. For steady quasistatic growth of a crack in the plane strain opening mode in an elastic-ideally plastic material, Rice, Drugan, and Sham [2] have constructed an asymptotic field consisting of a constant state region ahead of the crack tip, followed by a fan sector with singular plastic strain, an elastic unloading region, and finally, a small plastic reloading zone along the crack flanks. Within the region of singular plastic strain, the distribution of particle velocity and shear strain in crack tip polar coordinates is

$$\dot{u}_r, \dot{u}_\theta \sim v\epsilon_0 \ln\left(\frac{r_0}{r}\right), \epsilon^p_{r\theta} \sim \epsilon_0 \ln\left(\frac{r_0}{r}\right) \tag{1}$$

where

ϵ_0 = tensile yield strain,
r_0 = maximum extent of the plastic zone, and
r = the radial distance from the crack tip.

If an expression for the kinetic energy density and the stress work density are derived for this deformation field, then the ratio of the kinetic energy density to the stress work density as a function of r is

$$KE/SW \approx 10 \frac{v^2}{c^2} \ln\left(\frac{r_0}{r}\right) \tag{2}$$

where $c = \sqrt{E/\rho}$ is an elastic wave speed. Based on this simple estimate, it might be expected that inertial effects will be significant when the ratio in Eq 2 is greater than one-tenth. For example, if $v/c = 0.1$, then the ratio is greater than one-tenth if $r/r_0 \leq 0.3$.

In retrospect, this estimate has a feature that could have been interpreted as a warning that asymptotic analysis of this problem would have some subtle difficulties. The estimate suggests that for any nonzero v/c there is a range of r/r_0 for which material inertial effects are important, but that the size of that region diminishes very rapidly as v/c approaches zero. Indeed, the estimate suggests that inertial effects are important only over a region for which $(r/r_0)^{v/c} \ll 1$, which is similar in form to the restriction on the domain of validity of the dynamic asymptotic solution given below in Eq 4.

The steady-state growth of a crack at speed v in the antiplane shear mode, or Mode III in fracture mechanics terminology, under small-scale yielding conditions was analyzed by Freund and Douglas [*3*] and by Dunayevsky and Achenbach [*4*]. The field equations governing this process include the equation of momentum balance, the strain-displacement relations, and the condition that the stress distribution far from the crack edge must be the same as the near tip stress distribution in a corresponding elastic problem. For elastic-ideally plastic response of the material, the stress is assumed to lie on the Mises yield locus, a circle of radius τ_0 in the plane of rectangular stress components, and the stress and strain are related through the incremental Prandtl-Reuss flow rule. The material is linearly elastic with shear modulus μ in regions where the stress state does not satisfy the yield condition.

With a view toward deriving a theoretical relationship between the crack tip speed and the imposed stress intensity factor required to sustain this speed according to a critical plastic strain crack growth criterion, attention was focussed on the strain distribution on the crack line within the active plastic zone and on the influence of material inertia on this stress distribution. It was found that the distribution of shear strain on this line, say $\gamma_{yz}(x, 0)$ in crack tip rectangular x, y coordinates, could be determined *exactly* in terms of the plastic zone size r_0 in the parametric form

$$\gamma_{yz}(x, 0) = \frac{\mu}{\tau_0}\left\{1 - \left(\frac{1 - m^2}{2m^2}\right) \ln\left(\frac{1 - m^2h^2}{1 - m^2}\right)\right\}$$
$$x = r_0 \frac{I(-h)}{I(m)}, \; I(t) = \int_0^{(1-t)/(1+t)} \frac{s^{(1-m)/2m}}{(1 + s)} ds \tag{3}$$

where $m = v/c_s$ and c_s is the elastic shear wave speed. While the integral $I(t)$ has a representation in terms of elementary functions only for very special values of its argument, it is easily evaluated by numerical methods for any nonzero value of m.

The exact result (Eq 3) resolved a long standing paradox concerning mode III crack tip fields. Rice [1] showed that the near tip distribution of strain $\gamma_{yz}(x, 0)$ for steady growth of a crack under equilibrium conditions was singular as $\ln^2(x/r_0)$ as $x/r_0 \to 0$. On the other hand, Slepyan [5] showed that the asymptotic distribution for any $m > 0$ was of the form $(m^{-1} - 1)\ln(x/r_0)$ as $x/r_0 \to 0$. These two features could be verified by examining the behavior of the exact solution for dynamic growth Eq 3 under the condition $m \to 0$ for any nonzero value of x/r_0 and under the condition that $x/r_0 \to 0$ for any nonzero value of m, respectively. The resolution of the paradox was found, however, in the observation that Slepyan's asymptotic solution is valid only if

$$(x/r_0)^{2m/(1+m)} \ll 1 \tag{4}$$

Thus, the apparent inconsistency arises from the fact that the asymptotic result due to Slepyan is valid over a region that becomes vanishingly small as $m \to 0$.

Graphs of the plastic strain distribution on the crack line in the active plastic zone are shown in Fig. 1 for $m = 0, 0.3, 0.5$. The plastic strain is singular in each case, as has already been noted. The most significant observation concerns the influence of material inertia on the strain distribution. An increase in crack speed results in a substantial reduction of the level in plastic strain for a fixed fractional distance from the crack tip to the elastic-plastic boundary. Therefore, if a local ductile crack growth criterion is imposed, then it would appear that the fracture resistance or toughness would necessarily increase with increasing crack tip speed. To quantify this idea, the fracture criterion proposed by McClintock and

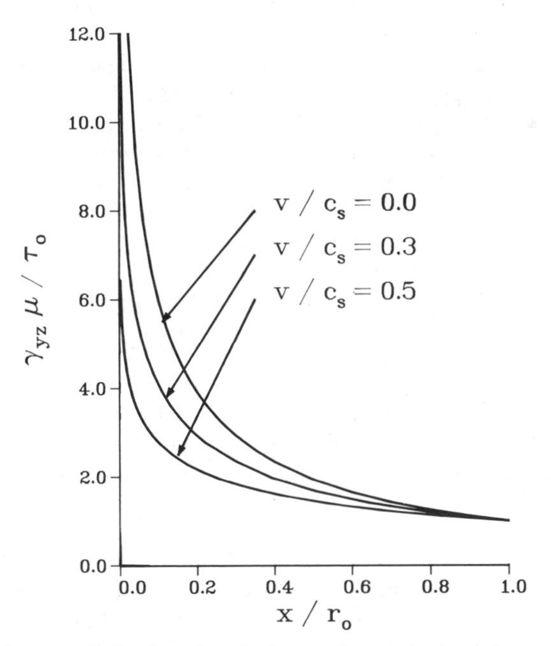

FIG. 1—*Total strain on crack line in active plastic zone for steady dynamic growth of Mode III crack in elastic-ideally plastic material from Eq 3.*

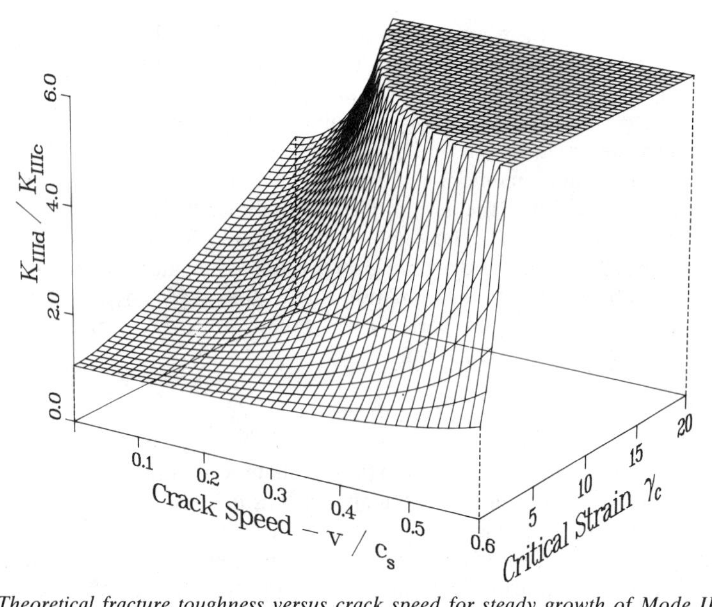

FIG. 2—*Theoretical fracture toughness versus crack speed for steady growth of Mode III crack according to critical plastic strain at characteristic distance criterion, for levels of critical plastic strain from* $\gamma_c = 0$ *to* $\gamma_c = 20\tau_0/\mu$. *The plateau is an artifice arising from truncation of the surface at a convenient level.*

Irwin [*6*] was adopted. According to this criterion, a crack will grow with a critical value of plastic strain at a point on the crack line at a characteristic distance ahead of the tip. The crack will not grow for levels of plastic strain at this point below the critical level, and levels of plastic strain greater than the critical level are inaccessible. To make a connection between the plastic strain in the active plastic zone and the remote loading, a relationship between the size of the plastic zone and the remote applied stress intensity factor is required. This can be provided only through a complete solution of the problem, which was obtained for the case of Mode III by Freund and Douglas on the basis of a full-field numerical solution of the governing equations. The resulting theoretical fracture toughness K_{IIId} versus crack speed is shown in Fig. 2 for continuous variation of the critical plastic strain from $\gamma_c = 0$ to $\gamma_c = 20\tau_0/\mu$. The critical distance has been eliminated in favor of K_{IIIc}, the level of applied stress intensity required to satisfy the same criterion for a stationary crack in the same material under equilibrium conditions. The variable intercept at $m = 0$ indicates an increasing amount of plasticity with increasing critical plastic strain, and the intercept values correspond to the so-called steady state toughness values of the theory of stable crack growth, that is, with the plateau level of the resistance curve.

The plot in Fig. 2 illustrates some typical features. The ratio of K_{IIId}/K_{IIIc} is a monotonically increasing function of crack speed m for fixed critical strain, and this function takes on large values for moderate values of m. Although there is no unambiguous way to associate a terminal velocity with these results, they suggest a maximum attainable velocity well below the elastic wave speed of the material. It is emphasized that the variation of toughness with crack speed in Fig. 2 is due to inertial effects alone. The material response is independent of rate of deformation, and the crack growth criterion that is enforced involves no characteristic time. If inertial effects were neglected, the calculated toughness would be *completely independent* of speed. The question of the influence of material rate sensitivity on this relationship is a separate issue.

The equivalent plane strain problem of dynamic crack growth in an elastic-ideally plastic material has not been so fully developed. However, a numerical calculation leading to a fracture toughness versus crack speed relationship, analogous to Fig. 2, has been described by Lam and Freund [7]. They adopted the critical crack tip opening angle growth criterion and derived results for Mode I on the basis of the Mises yield condition and J_2 flow theory of plasticity that are quite similar in general form to those shown for Mode III. The nature of the elastic-plastic fields deep within the active plastic zone were difficult to discern from the finite element results, and an analytical study of the asymptotic field was undertaken by Leighton, Champion, and Freund [8] in order to examine this feature. It was found that the plastic strain components had bounded limiting values at the crack tip for any nonzero crack speed $m = v/c_s$, but that these limits depended on crack speed as $1/m$. Both Slepyan [5] and Achenbach and Dunayevsky [9] reported studies of this problem in which they took elastic compressibility into account. They were able to extract solutions valid very close to the crack tip in the limit of vanishing crack speed. It should be noted that in both studies, the Tresca yield condition was used together with the Mises flow rule, so that the fields described are consistent with normality of the plastic strain rate to the yield surface only in the limit of incompressibility. In addition, both studies imposed restrictions on the out-of-plane deformation that arise from assuming normality of the plastic strain rate to the Tresca yield surface. In a study of the same problem in the incompressible limit, Gao and Nemat-Nasser [10] reported a solution with jumps in stress and particle velocity across radial lines emanating from the crack tip for all crack speeds between zero and the elastic shear wave speed, where the jump magnitudes were subject to the appropriate jump conditions. It is shown in Ref 8, however, that if the sequence of deformation states throughout the jump must be admissible plastic states, consistent with the theory of mechanical shocks, then discontinuities in the angular variation of stress and particle velocity components around the crack edge can be ruled out.

Some data on the dynamic fracture toughness of metals during crack growth are available. Rosakis, Duffy, and Freund [11] used the optical shadow spot method in reflection mode to infer the prevailing stress intensity factor during rapid crack growth in 4340 steel hardened to $R_C = 45$. This is a relatively strain-rate-insensitive material with very little strain hardening, so that the material may presumably be modeled as elastic-ideally plastic. The observed toughness varied little with crack speed for speeds up to about 600 to 700 m/s, and thereafter the toughness increased sharply with increasing crack tip speed. The general form of the toughness versus speed data was similar to the theoretical prediction based on the numerical simulation reported by Lam and Freund [7], lending support to the view that material inertia on the scale of the crack tip plastic zone has an important influence on the perceived dynamic fracture toughness. Similar data were reported by Kobayashi and Dally [12], who made photoelastic measurements of the crack tip stress field by means of a birefringent coating on the specimen. Data on crack propagation and arrest in steels were also reported by Dahlberg, Nilsson, and Brickstad [13].

Viscoplastic Material Response

An estimate of the plastic strain rate near the tip of an advancing crack may be obtained as follows. Suppose that the yield stress in shear is τ_y and that the elastic shear modulus is μ, so that the yield strain is τ_y/μ. As a rough estimate of the plastic strain rate, consider the yield strain divided by the time required for the crack tip to traverse a region that is the size of the active plastic zone at speed v. Following McClintock and Irwin [6], if the size of the plastically deforming region is interpreted as the largest extent in an elastic field of the locus of points on which the maximum shear stress is τ_y, then the estimate of strain rate is

$$(\dot{\gamma}^p)_{est} \sim 7v\tau_y^3/\mu^2 G \tag{5}$$

where energy release rate G is the characterizing parameter for the elastic field. Clearly, for rapid growth of a crack in a low toughness material, the strain rate estimate can be enormous, well in excess of 10^6 s^{-1}. Some results concerned with fairly large plastic strains (compared to elastic strains) under high rate conditions are discussed in this section, and the case of viscoplastic crack growth with small plastic strains is considered in the next section.

Consider steady crack growth in an elastic-plastic material for which the flow stress depends on the rate of deformation. The particular material model known as the over-stress power law model has been considered by Lo [*14*], Brickstad [*15*], and a number of other authors. According to this idealization, the plastic strain rate in simple shear $\dot{\gamma}^p$ depends on the corresponding shear stress τ through

$$\dot{\gamma}^p = \dot{\gamma}_t + \dot{\gamma}_0\{(\tau - \tau_t)/\mu\}^n \quad \text{for} \quad \tau \geq \tau_t \tag{6}$$

where $\dot{\gamma}_t$ is the threshold strain rate for this description, or the plastic strain rate when $\tau = \tau_t$. The description also includes the elastic shear modulus μ, the viscosity parameter $\dot{\gamma}_0$, and the exponent n. A common special case is based on the assumption that the slow loading response of the material is elastic-ideally plastic and that *all* inelastic strain is accumulated according to Eq 6. For this case, $\dot{\gamma}_t = 0$ and τ_t is the slow loading flow stress τ_0. For other purposes, it is assumed that Eq 6 provides a description of material response only for high plastic strain rates, in excess of the transition plastic strain rate $\dot{\gamma}_t$ and for stress in excess of the corresponding transition stress level τ_t. For low or moderate plastic strain rates, the variation of plastic strain rate with stress is weaker than in Eq 6, and a common form for the dependence is (see also Ref *16*)

$$\dot{\gamma}^p = g_1(\tau)\exp\{-g_2(\tau)\} \tag{7}$$

where g_1 and g_2 are algebraic functions. The marked difference between response at low or moderate plastic strain rates and at high strain rates may be due to a change in fundamental mechanism of plastic deformation with increasing rate, or it may be a structure induced transition. For present purposes, it is sufficient to regard the difference as an empirical observation. The two forms of constitutive laws in Eqs 6 and 7 can lead to quite different results in analysis of crack-tip fields and, indeed, the form Eq 6 leads to fundamentally different results for different values of the exponent n.

Lo [*14*] extended some earlier work on the asymptotic field for steady quasistatic crack growth in an elastic-viscoplastic material by Hui and Riedel [*17*] to include inertial effects. In both cases, the multiaxial version of Eq 6 with $\dot{\gamma}_t = 0$ and $\tau_t = \tau_0$ was adopted to describe inelastic response, with no special provision for unloading. They showed that for values of the exponent n less than 3, the asymptotic stress field is the elastic stress field. For values of n greater than 3, on the other hand, Lo constructed an asymptotic field including inertial effects having the same remarkable feature of complete autonomy found by Hui and Riedel, that is, it revealed no dependence on the level of remote loading. For steady antiplane shear Mode III crack growth, Lo found the radial dependence of the inelastic strain on the crack line ahead of the tip to be

$$\gamma_{yz}^{p}(x, 0) \approx (n - 1)(v/\dot{\gamma}_0 x)^{1/(n-1)} T_L(v/c_s) \tag{8}$$

where the dependence of the amplitude factor T_L on crack speed is given graphically by Lo, who also analyzed the corresponding plane strain problem. Note that as $n \rightarrow \infty$, the plastic strain singularity becomes logarithmic. The full-field solution for this problem under small scale yielding conditions was determined numerically by Freund and Douglas [18]. The numerical results showed a plastic strain singularity much stronger than for the rate independence case, and it appeared from the numerical results that the domain of dominance of the asymptotic field within the crack tip plastic zone expanded with increasing crack-tip speed. These observations are consistent with Eq 8.

High Strain Rate Crack Growth

A particularly interesting class of dynamic fracture problems are those concerned with crack growth in materials that may or may not experience rapid growth of a sharp cleavage crack, depending on the conditions of temperature, stress state and rate of loading. These materials may fracture by either a brittle or ductile mechanism on the microscale, and the focus of work in this area is on establishing conditions for one or the other mode to dominate. The phenomenon is most commonly observed in ferritic steels. Such materials show a dependence of flow stress on strain rate, and the strain rates experienced by a material particle in the path of an advancing crack are potentially enormous. Consequently, the mechanics of rapid growth of a sharp macroscopic crack in an elastic-viscoplastic material that exhibits a fairly strong variation of flow stress with strain rate has been of interest in recent years. The general features of the process as experienced by a material particle on or near the fracture path are straight forward. As the edge of a growing crack approaches, the stress magnitude tends to increase there due to the stress concentrating effect of the crack edge. The material responds by flowing at a rate related to the stress level in order to mitigate the influence of the crack edge. It appears that the essence of cleavage crack growth is the ability to elevate the stress to a critical level before plastic flow can accumulate to defeat the influence of the crack tip. In terms of the mechanical fields near the edge of an advancing crack, the rate of stress increase is determined by the elastic strain rate, while the rate of crack-tip blunting is determined by the plastic strain rate. Thus, an equivalent observation is that the elastic strain rate near the crack edge must dominate the plastic strain rate for sustained cleavage. It is implicit in this approach that the material is intrinsically cleavable, and the question investigated is concerned with the way in which work can be supplied to the crack-tip region.

The problem has been studied from this point of view by Freund and Hutchinson [19]. They adopted the constitutive description (Eqs 6 and 7) with $n = 1$. This is indeed a situation for which the near tip elastic strain rate dominates the plastic strain rate. Through an approximate analysis, conditions necessary for a crack to run at high velocity in terms of constitutive properties of the material, the rate of crack growth, and the overall crack driving force were extracted under small yielding conditions.

Consider the crack gliding along through the elastic-viscoplastic material under plane strain conditions. At points far from the crack edge, the material remains elastic and the stress distribution is given in terms of the applied stress-intensity factor K_I. Equivalently, the influence of the applied loading may be specified by the rate of mechanical energy flow into the crack-tip region from remote points G, and these two measures are related by means of

$$G = \frac{1 - \nu^2}{E} A(v) K_{\mathrm{I}}^2 \tag{9}$$

where ν and E are the elastic constants of an isotropic solid and A is a universal function of the instantaneous crack speed v. The function has the properties that $A(0) = 1$, $A'(0) = 0$ and $A(v) \rightarrow \infty$ as $v \rightarrow c_r$. For points near the crack edge, the potentially large stresses are relieved through plastic flow, and a permanently deformed but unloaded wake region is left behind the advancing plastic zone along the crack flanks. For material particles in the outer portion of the active plastic zone, the rate of plastic straining is expected to be in the low or moderate strain rate range, whereas for particles close to the crack edge, the response is modeled by the constitutive law (Eq 6) with $n = 1$. Because of elastic rate dominance, the stress distribution within this region has the same spatial dependence as the remote field but with a stress-intensity factor *different* from the remote-stress-intensity factor. The crack tip stress intensity factor, say K_{Itip}, is assumed to control the cleavage growth process. The influence of the remote loading is screened from the crack tip by the intervening plastic zone, and the main purpose of the analysis is to determine the relationship between the remote loading and the crack-tip field. For present purposes, it is assumed that the crack grows as a cleavage crack with a fixed level of local energy release rate, say G_{tip}^c. The question then concerns the conditions under which enough energy can be supplied remotely to sustain the level of energy release rate G_{tip}^c at the crack tip.

The matter of relating the applied G to G_{tip}^c was pursued by enforcing an overall energy rate balance. The balance may be cast into the form

$$G_{\mathrm{tip}}^c = G - \frac{1}{v} \int_A \sigma_{ij} \dot{\epsilon}_{ij} \, dA - \int_{-h}^{h} U_e^* \, dy \tag{10}$$

where

A = area of the active plastic zone in the plane of deformation,
h = thickness of the plastic wake far behind the crack tip, and
U_e^* = residual elastic strain energy density trapped in the remote wake.

This relation simply states that the energy being released from the body at the crack tip is the energy flowing into the crack tip region reduced by the energy dissipated through plastic flow in the plastic zone, and further reduced by the energy trapped in the wake due to incompatible plastic strains. The expression is exact.

Through several approximations, the complete energy balance (Eq 10) was reduced in Ref *19* to the simple form

$$G/G_{\mathrm{tip}}^c = 1 + D(m) P_c \tag{11}$$

where the dimensionless parameter P_c is

$$\dot{\gamma}_0 \sqrt{\mu\rho}\, G_{\mathrm{tip}}^c (1 + 2\dot{\gamma}_t \mu / \dot{\gamma}_0 \tau_t) / 3\tau_t^3$$

and $D(m)$ is a dimensionless function of crack-tip speed $m = v/c_r$ and ρ is the material mass density. P_c is a monotonically increasing function of temperature for steels with values in the range from about 0 to 10 as temperature varies from 0 K to about 400 K. The function $D(m)$ is asymptotically unbounded as $m \to 0$ and $m \to 1$, and it has a minimum at an intermediate crack tip speed. The applied crack tip driving force, say G_S, is related to the crack-tip energy release rate by $G_S = G/(1 - v/c_r)$ for a semi-infinite crack in an otherwise unbounded body, and this relationship is adopted here as an approximation. A graph of G_S/G^c_{tip} is shown in Fig. 3 in the form of a surface over the crack speed-temperature plane.

The graph in Fig. 3 gives the locus of combinations G_S, v, T for which steady-state propagation of a sharp crack can be sustained. The implication is that if a cleavage crack can be initiated for a combination G_S, v, T that is *above* the surface, then the crack will accelerate to a state on the stable branch of the surface (that is, the side with increasing G_S at fixed temperature). If the driving force diminishes as the crack advances, or if the local material temperature increases as the crack advances, then the state combination will move toward the minimum point on the surface at the local temperature. If the driving force is further decreased, or if the temperature is further increased, then growth of a sharp cleavage crack cannot be sustained according to the model. The implication is that the crack will arrest abruptly from a fairly high speed, and a plastic zone will then grow from the arrested crack.

Of special significance is the observation that, at any given temperature, the variation of required driving force with crack speed has an absolute minimum, say G^*_S/G^c_{tip}. This implies that, according to this model, it is impossible to sustain cleavage crack growth at that temperature with a driving force below this minimum. Thus, this minimum as a function of temperature may be interpreted as the variation of the so-called arrest toughness for the material with material temperature. This minimum is plotted against temperature for the case of mild steel in Fig. 4.

Further crack growth beyond the first arrest is possible if either a ductile growth criterion can be met or if cleavage can be reinitiated through strain hardening in the evolving plastic zone. The details of the model have been refined through full numerical solution of the problem [*20*], but the essential features have not changed with more precise analysis. A

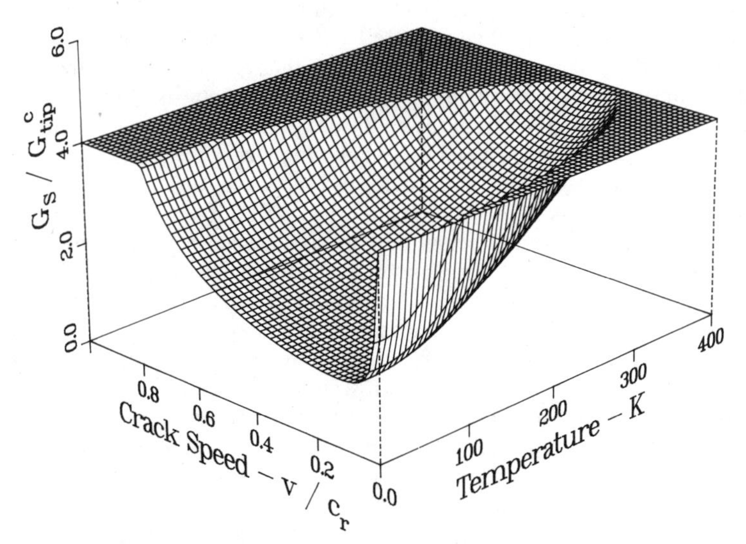

FIG. 3—*A surface representing conditions on applied crack-tip driving force* G_S*, crack tip speed* v *and temperature that correspond to steady crack propagation, as predicted by Eq. 11. The plateau is an artifice arising from truncation of the surface at a convenient level.*

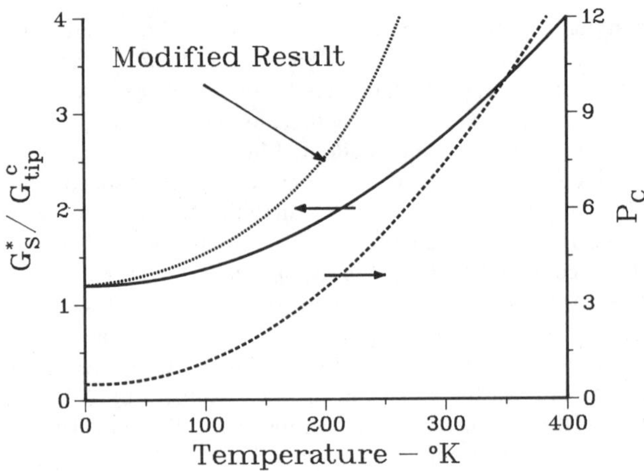

FIG. 4—*The minimum driving force* G_S^* *needed to drive the crack dynamically as a function of temperature. The solid line corresponds to Fig. 3, and the dashed line from a modified result in Ref* 21.

modification of the basic model was proposed by Mataga, Freund, and Hutchinson [*21*] which provides a description in better agreement with full-field numerical solutions than the model outlined above. In the original development [*20*], it was assumed that the plastic dissipation was completely controlled by the near tip stress-intensity factor field, say K_{tip}. However, this stress-intensity factor, which is asymptotically correct, must give way to the far-field stress stress-intensity factor K with increasing distance from the crack edge. It was observed in Ref *21* that the estimate of plastic dissipation was improved significantly, at least in comparison to finite element simulations, if the plastic dissipation was estimated on the basis of a "mean" stress-intensity factor $\sqrt{K_{tip}K}$. The graph of arrest toughness versus temperature for parameters corresponding to mild steel is also shown in Fig. 4.

Important experiments on crack propagation and arrest in steel specimens are currently being carried out by deWit and Fields [*22*]. Their specimens are enormous, single-edge notched plates loaded in tension. The growing crack thus experiences an increasing driving force as it advances through the plate. A temperature gradient is also established in the specimen so that the crack grows from the cold side of the specimen toward the warm side. Based on the presumption that the material becomes tougher as the temperature is increased, the crack also experiences increasing resistance as it advances through the plate. The specimen material is A533B pressure vessel steel, which is both very ductile and strain-rate sensitive. In the experiments, the fracture initiates as a cleavage fracture and propagates at high speed through the specimen into material of increasing toughness. The crack then arrests abruptly in material whose temperature is *above* the nil-ductility temperature for the material based on Charpy tests. A large plastic zone grows from the arrested crack edge, and cleavage crack growth is occasionally reinitiated. The essential features of the experiment appear to be consistent with the model of high strain rate crack growth outlined in Ref *19*, and this model appears to provide a conceptual framework for interpretation of the phenomenon. An analysis of rapid crack growth in a rate dependent plastic solid has also been carried out by Brickstad [*15*] in order to interpret some experiments on rapid crack growth in a high-strength steel. It is noted that the work of deWit and Fields [*22*] is part of the Nuclear Regulatory Commission's Heavy Section Steel Technology Program, which sustains an integrated dynamic fracture mechanics effort involving experiments, computation, and material characterization.

Concluding Remarks

The results described in the preceding sections reflect some progress toward discovery of the role played by crack-tip plastic fields in establishing conditions for rapid advance of a crack in an elastic-plastic material. Understanding of this issue is far from complete, and a few of the open questions that could be profitably pursued are identified in this concluding section. For example, much of the modeling that has resulted in a detailed description of crack-tip elastic-plastic fields has been based on the assumption that the fields are steady as seen by a crack-tip observer. This approach overlooks all transient aspects of the process. The picture of the way in which a crack-tip plastic zone develops in a cracked, rate-sensitive structural material under the action of stress wave loading is not clear, but the question is important in the sense that these fields determine whether or not the crack will advance. The same issue appears to be at the heart of the cleavage initiation process in steels, but on a microstructural scale. Here, the sudden cracking of carbides or other brittle phases due to incompatible plastic strains provides a nucleation mechanism, and the question is whether or not these dynamic microcracks penetrate into the adjacent ferrite as sharp cracks. The answer seems to hinge on the way in which plastic strains develop near the carbide-ferrite interface due to the appearance of the microcracks in the brittle phases.

The transients of the arrest of a cleavage crack in a structural material are also unclear at this point. A running crack appears to arrest because conditions for the continuous reinitiation of cleavage cannot be maintained [*23*]. In terms of the model discussed in the previous section, arrest occurs because conditions for elastic rate dominance of the local field cannot be maintained. However, the model does not provide information on the process thereafter. It appears from the experiments reported by deWit and Fields [*22*] that arrest is quite abrupt, that a large plastic zone grows from the crack edge following arrest of the cleavage crack, that the crack may grow subsequently in a ductile mode, and that cleavage may be reinitiated at a later stage. It is not clear if the cleavage reinitiation is due to a rate effect or to a combination of strain hardening and constraint in the interior portions of the specimen.

Modeling of plasticity effects in dynamic crack growth has been restricted to two-dimensional systems, for the most part. It is likely that a number of three-dimensional effects are of sufficient importance to warrant further investigation. For example, crack propagation studies are often carried out with plate specimens. For such specimens, the transition from plane-stress conditions in regions far from the crack tip compared with plate thickness to plane-strain or generalized plane-strain conditions near the crack edge is not clear. Yang and Freund [*24*] suggest that plane-stress conditions prevail only for points beyond about one-half the plate thickness from the crack edge for elastic deformations. Out-of-plane inertia is of potential importance in these three-dimensional fields, but this effect has not been investigated to date. Furthermore, the roles of ductile shear lips at the free surfaces or of ductile ligaments left behind a cleavage crack as it advances through a structural metal are not clear at this time. In a study of fracture initiation in dynamically loaded specimens of a ductile material by Nakamura, Shih, and Freund [*25*], it was shown that these three-dimensional effects are potentially very significant.

Virtually all of the foregoing discussion has focused on issues from the traditional fracture mechanics point of view, that is, the use of a single parameter to characterize the mechanical state of a dominant crack in a stressed body. This section is concluded, however, with mention of somewhat speculative interpretations of recent work that suggest a departure from the traditional fracture mechanics viewpoint. Some exciting new data on fracture initiation and crack growth in a 4340 steel in a very hard condition were recently reported by Ravichandran and Clifton [*26*]. Through a modification of the plate impact apparatus, they were able to examine fracture initiation and crack growth of a few millimetres for the plane-strain situation of a semi-infinite crack in an unbounded body subject to plane wave loading, at least for a microsecond or two. At the lowest testing temperature reported, the cracks grew as cleavage cracks. Based on optical measurements of the surface motion of the specimen and comparison with detailed elastic-viscoplastic calculations, it appeared that the cracks grew more nearly at constant velocity crack than with a fixed level of energy release rate or stress intensity factor. This observation is similar to that made by Ravi-Chandar and Knauss [27], who studied crack growth in the brittle polymer Homalite-100. In both cases, this observation was made in situations where the load was suddenly applied and the load level was very intense compared with the minimum load necessary to induce fracture in the same situation. The results suggest that the one-parameter characterization of the crack-tip conditions may not be adequate to describe fracture response under such conditions.

Acknowledgments

It is a pleasure to acknowledge ongoing collaboration on various aspects of plasticity effects in dynamic fracture with J. W. Hutchinson of Harvard University. Research on plasticity effects in dynamic fracture is supported at Brown University by the Office of Naval Research through contracts N00014-87-K-0481, the Army Research Office through contract DAAG29-85-K-0003, and the Brown University NSF Materials Research Group through grant DMR-8714665. This support is gratefully acknowledged.

References

[1] Rice, J. R., "Mathematical Analysis in the Mechanics of Fracture," *Fracture—Vol. 2,* H. Liebowitz, Ed., Academic Press, 1968, pp. 191–311.
[2] Rice, J. R., Drugan, W. J., and Sham, T. L., "Elastic-Plastic Analysis of Growing Cracks," *Fracture Mechanics, (Twelfth Conference), ASTM STP 700,* American Society for Testing and Materials, Philadelphia, 1980, pp. 189–219.
[3] Freund, L. B. and Douglas, A. S., "The Influence of Inertia on Elastic-Plastic Antiplane Shear Crack Growth," *Journal of the Mechanics and Physics of Solids,* Vol. 30, 1982, pp. 59–74.
[4] Dunayevsky, V. and Achenbach, J. D., "Boundary Layer Phenomenon in the Plastic Zone Near a Rapidly Propagating Crack Tip," *International Journal of Solids and Structures,* Vol. 18, 1982, pp. 1–12.
[5] Slepyan, L. I., "Crack Dynamics in an Elastic-Plastic Body," *Mechanics of Solids,* Vol. 11, (English translation of *Mechanika Tverdogo Tela,* 1976, pp. 126–134.
[6] McClintock, F. A. and Irwin, G. R., "Plasticity Aspects of Fracture Mechanics," *Fracture Toughness Testing and Its Applications, ASTM STP 381,* American Society for Testing and Materials, Philadelphia, 1965, pp. 84–113.
[7] Lam, P. S. and Freund, L. B., "Analysis of Dynamic Growth of a Tensile Crack in an Elastic-Plastic Material," *Journal of the Mechanics and Physics of Solids,* Vol. 33, 1985, pp. 153–167.
[8] Leighton, J. T., Champion, C. R., and Freund, L. B., "Asymptotic Analysis of Steady Dynamic Crack Growth in an Elastic-Plastic Material," *Journal of the Mechanics and Physics of Solids,* Vol. 35, 1987, pp. 541–563.
[9] Achenbach, J. D. and Dunayevsky, V., "Fields Near a Rapidly Propagating Crack-Tip in an Elastic-Plastic Material," *Journal of the Mechanics and Physics of Solids,* Vol. 29, 1981, pp. 283–303.
[10] Gao, Y. C. and Nemat-Nasser, S., "Dynamic Fields Near a Crack Tip in an Elastic Perfectly Plastic Solid," *Mechanics of Materials,* Vol. 2, 1983, pp. 47–60.
[11] Rosakis, A. J., Duffy, J., and Freund, L. B., "The Determination of Dynamic Fracture Toughness of AISI 4340 Steel by the Shadow Spot Method," *Journal of the Mechanics and Physics of Solids,* Vol. 32, 1984, pp. 443–460.
[12] Kobayashi, T. and Dally, J. W., "Dynamic Photoelastic Determination of the a-K Relation for

4340 Steel," *Crack Arrest Methodology and Applications, ASTM STP 711,* G. T. Hahn and M. F. Kanninen, Eds., American Society for Testing and Materials, Philadelphia, 1979, pp. 189–210.

[*13*] Dahlberg, L., Nilsson, F., and Brickstad, B., "Influence of Specimen Geometry on Crack Propagation and Arrest Toughness," *Crack Arrest Methodology and Applications, ASTM STP 711,* G. T. Hahn and M. F. Kanninen, Eds., American Society for Testing and Materials, Philadelphia, 1980, pp. 89–108.

[*14*] Lo, K. K., "Dynamic Crack-Tip Fields in Rate Sensitive Solids," *Journal of the Mechanics and Physics of Solids,* Vol. 31, 1983, pp. 287–305.

[*15*] Brickstad, B., "A Viscoplastic Analysis of Rapid Crack Propagation Experiments in Steel," *Journal of the Mechanics and Physics of Solids,* Vol. 31, 1983, pp. 307–327.

[*16*] Frost, H. J. and Ashby, M. F., *Deformation-Mechanism Maps,* Pergamon Press, Oxford, 1982.

[*17*] Hui, C. Y. and Riedel, H., "The Asymptotic Stress and Strain Field Near the Tip of a Growing Crack Under Creep Conditions," *International Journal of Fracture,* Vol. 17, 1981, pp. 409–425.

[*18*] Freund, L. B. and Douglas, A. S., "Dynamic Growth of an Antiplane Shear Crack in a Rate-Sensitive Elastic-Plastic Material," *Elastic-Plastic Fracture: Second Symposium. Volume 2, ASTM STP 803,* C. F. Shih and J. Gudas, Eds., American Society for Testing and Materials, Philadelphia, 1983, pp. 5–20.

[*19*] Freund, L. B. and Hutchinson, J. W., "High Strain-Rate Crack Growth in Rate-Dependent Plastic Solids," *Journal of the Mechanics and Physics of Solids,* Vol. 33, 1985, pp. 169–191.

[*20*] Freund, L. B., Hutchinson, J. W., and Lam, P. S., "Analysis of High Strain Rate Elastic-Plastic Crack Growth," *Engineering Fracture Mechanics,* Vol. 23, 1986, pp. 119–129.

[*21*] Mataga, P. A., Freund, L. B., and Hutchinson, J. W., "Crack Tip Plasticity in Dynamic Fracture," *Journal of the Physics and Chemistry of Solids,* Vol. 48, 1987, pp. 985–1005.

[*22*] deWit, R. and Fields, R., "Wide Plate Crack Arrest Testing," *Nuclear Engineering and Design,* Vol. 98, 1987, pp. 149–155.

[*23*] Irwin, G. R., "Brittle-Ductile Transition Behavior in Reactor Vessel Steels," *Proceedings,* WRSI Meeting, Gaithersburg, MD, Oct. 1986, NUREG/CP-0082, Vol. 2, U.S. Nuclear Regulatory Commission, Feb. 1987, pp. 251–272.

[*24*] Yang, W. and Freund, L. B., "Transverse Shear Effects for Through-Cracks in an Elastic Plate," *International Journal of Solids and Structures,* Vol. 21, 1985, pp. 977–994.

[*25*] Nakamura, T., Shih, C. F., and Freund, L. B., "Three Dimensional Transient Analysis of a Dynamically Loaded Three Point Bend Ductile Fracture Specimen," *Nonlinear Fracture Mechanics, ASTM STP 995,* American Society for Testing and Materials, Philadelphia, 1988, pp. 217–241.

[*26*] Ravichandran, G. and Clifton, R. J., "Dynamic Fracture Under Plane Wave Loading," Brown University Report, 1986.

[*27*] Ravi-Chandar, K. and Knauss, W. G., "An Experimental Investigation into Dynamic Fracture: III. On Steady State Crack Propagation and Crack Branching," *International Journal of Fracture,* Vol. 26, 1984, pp. 141–154.

Microstructure and Micromechanical Modeling

Hermann Riedel[1]

Creep Crack Growth

REFERENCE: Riedel, H., **"Creep Crack Growth,"** *Fracture Mechanics: Perspectives and Directions* (*Twentieth Symposium*), *ASTM STP 1020*, R. P. Wei and R. P. Gangloff, Eds., American Society for Testing and Materials, Philadelphia, 1989, pp. 101–126.

ABSTRACT: On the background of the historical evolution of the subject area, the current knowledge on creep crack growth is reviewed. The discussion is grouped around the C^* integral, which is the appropriate load parameter for describing creep crack growth in ductile materials. Crack growth by coalescence with grain boundary cavities is modeled and the lifetimes of cracked specimens are calculated by integrating the resulting laws for crack growth. Limitations to C^* arise from the initial elastic-plastic material response, from primary creep, from crack-tip blunting and from widespread cavitation damage. All these bounds are shown on a load parameter map. This is a diagram in the stress-time plane indicating the regimes in which different load parameters are applicable: K_I for predominantly elastic deformation; J if plasticity plays a role; C_h^* for primary creep; and C^* for steady-state creep. Tertiary creep is included in a damage-mechanics analysis of creep crack growth. The final section discusses a few three-dimensional aspects of fracture mechanics in general and of creep crack growth in particular.

KEY WORDS: creep crack growth, high temperatures, grain boundary cavities, damage mechanics, three-dimensional aspects

A component operating in the creep regime may fail by the slow extension of a macroscopic crack. Slow stable cracking at elevated temperatures is called "creep crack growth" if it occurs under more or less constant load, and "creep-fatigue crack growth" under cyclic loading conditions. Since corrosive effects often play a role, it may be justified to speak of high-temperature stress corrosion cracking in some cases. The present paper focuses the attention on the mechanics and the mechanisms of creep crack growth.

Usually, but not necessarily, creep crack growth occurs along grain boundaries by the formation and coalescence of grain boundary cavities. Cavitation may be confined to a small zone near the tip of the growing main crack ("small-scale damage"). Then, the lifetime will be determined by crack growth and a fracture-mechanics approach appears to be appropriate to deal with the problem. In other cases, the whole specimen may have developed cavities before the crack grows markedly. Then failure occurs by more or less homogeneous cavitation and the crack plays no particular role for the lifetime.

Because creep crack growth has been investigated for almost 20 years, only a fraction of the published literature can be referenced here. More complete reference lists can be found in previous reviews [*1–8*].

The Early Years of Creep Crack Growth Testing

The history of research on creep crack growth could have started a few years earlier than it actually did. In the 1960s, major advances had been made in the theory of elastic-plastic

[1] Head of the Department Applications of Materials at High Temperatures, Fraunhofer-Institut für Werkstoffmechanik, Wöhlerstraße 11, D-7800 Freiburg, Federal Republic of Germany.

fracture mechanics. In particular, the relevance of the J-integral for crack analysis had been recognized [9,10], the Hutchinson, Rice, and Rosengren (HRR) crack-tip fields had been derived [11,12], and the role of crack-tip blunting had been explored [13]. For theoreticians, who were aware of Hoff's analogy [14], it was obvious that these discoveries could be applied directly to viscously creeping bodies. Hoff's analogy states that a nonlinear elastic body obeying a material law $\epsilon = f(\sigma)$, and a nonlinear viscous body characterized by $\dot{\epsilon} = f(\sigma)$, develop the same stress field when subjected to the same loads. The elastic strain field corresponds to the viscous strain-rate field, and a viscous analogue to the J-integral exists, which has become known as C^* [15–17]. But in the 1960s, these ideas were not directly pursued, since creep crack growth had only started to be recognized as a problem of practical significance, and this knowledge had not yet spread within the scientific community.

In the early 1970s, the first papers appeared which reported slow stable cracking under elevated-temperature creep conditions [18–22]. To analyze the measured crack growth rates, some of the early workers tried to use the linear-elastic stress intensity factor, K_I. They were aware of the fact that K_I could only be the appropriate load parameter under predominantly elastic, or small-scale yielding, conditions. This is illustrated by Fig. 1, which shows two specimens after crack growth tests under sustained load. It cannot be expected that the linear-elastic stress intensity factor can be applied to the heavily deformed specimen of the chromium-molybdenum steel. But crack growth in the Nimonic alloy was accompanied by no detectable nonelastic deformation, and hence, K_I should be applicable. Similarly, Siverns and Price [22] obtained a good correlation between K_I and the crack growth rate in an embrittled 2¼ Cr-1Mo steel (heat treated to simulate the coarse-grained, heat-affected zone of a weld), which exhibited little nonelastic deformation during crack growth. In more ductile materials, crack growth is accompanied by extensive creep of the specimen. Before the C^* integral had become known among experimentalists, many workers tried to correlate crack growth rates in ductile materials with the net section stress, σ_{net}, or with a reference stress, σ_{ref}. Harrison and Sandor [19], for example, found a good correlation of crack growth rate with σ_{ref} for a small set of data on a turbine rotor steel.

However, it should be noted that there is no theory which suggests using σ_{net} or σ_{ref} as a parameter correlating crack growth rates in different specimens. It may be useful to compare rupture lifetimes (but not crack growth rates) of cracked, notched, and smooth specimens on the basis of the net section stress. In this connection, the terms *notch strengthening* and *notch weakening* have been introduced to indicate that a cracked or notched specimen

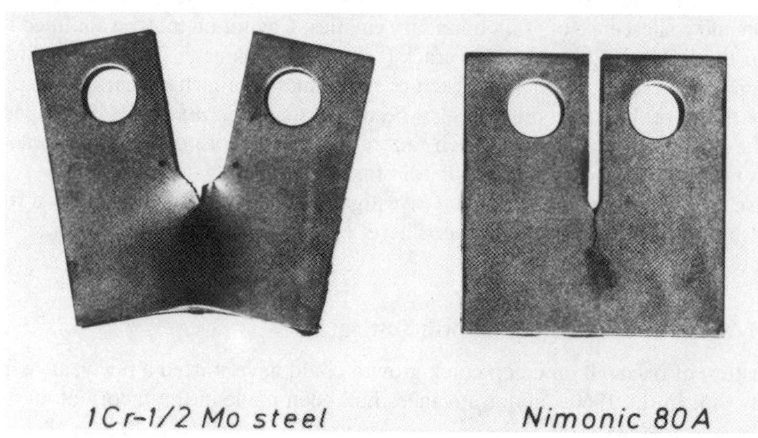

FIG. 1—*Two specimens after creep crack growth at 535°C (CrMo steel) and 650°C (Nimonic 80 A).*

sustains a given net section stress for a longer or shorter time, respectively, than a smooth specimen. (In the case of notch strengthening, the triaxial constraint, which retards creep flow, dominates, whereas notch weakening occurs if the stress concentration at the crack or notch causes premature crack growth and failure.) However, these terms are vague and specimen-geometry dependent, and should be used with care as an empirical guideline for characterizing the creep ductility or brittleness of materials.

Although σ_{net} and σ_{ref} have no theoretical basis in relation to creep crack growth rates, many data were evaluated using these parameters (for a summary, see Ref *23*). The question of which parameter should be used has important practical consequences, since, as Riedel and Rice [*24*] point out, the lifetime of a cracked component may be under- or overestimated by orders of magnitude if an inappropriate load parameter is employed to transfer crack growth rates from laboratory test specimens to the component. However, the theoretical basis for discussing that question was lacking in the early 1970s.

The C^* Integral

An important step towards a well-founded theory of creep crack growth was the introduction of C^* by Landes and Begley [*15*], by Ohji, Ogura, and Kubo [*16*], and by Nikbin, Webster, and Turner [*17*]. Other notations for C^* are $\dot{J}$, J^*, and J'.

The General Idea

The use of C^*, like that of other fracture mechanics parameters, relies on the following set of arguments:

1. In nonlinear viscous materials characterized by an arbitrary stress/strain-rate relation, $\dot{\epsilon} = f(\sigma)$, the contour integral C^* defined in Fig. 2 is independent of the choice of the path. The proof of the path independence of C^* is completely analogous to that of J [*9,10*].
2. The numerical value of C^* depends on the applied loading, the specimen geometry and the specific form of the law describing the material's nonlinear-viscous behavior. Practical guidelines on how to determine C^* from the second formula in Fig. 2 are given by Landes and Begley [*15*]. For the special case of power-law viscous materials, more convenient expressions for C^* will be presented in the next subsection.
3. The same C^* which is measured far away from the crack tip at the load line must characterize the deformation field near the crack tip, since C^* is path independent.
4. The asymptotic field near the crack tip has a unique form independent of the specimen geometry and is unequivocally determined once C^* is specified. A power-law viscous material is a convenient example to illustrate this (see the next subsection).

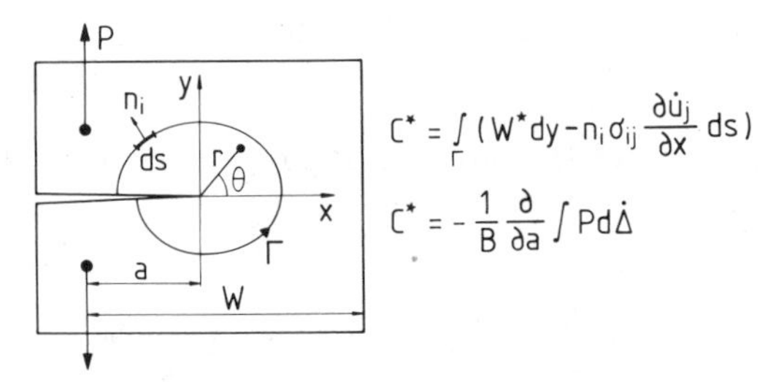

FIG. 2—*Definition of the contour integral* C^*.

5. Hence, creep crack growth rates in nonlinear viscous specimens of different sizes and shapes should be the same, if the applied loadings lead to the same C^*.
6. This remains true even if the material deviates from nonlinear viscous behavior, as long as these deviations are confined to a near-tip zone that is sufficiently small. Such a deviation may result, for example, from grain-boundary cavitation near the crack tip (see the section on damage mechanics).

Since a nonlinear viscous description represents a reasonable approximation to the behavior of metallic alloys under creep conditions, C^* has been applied successfully for describing creep crack growth in numerous materials (for example, [*15,25–31*]). However, since viscous behavior corresponds to steady-state creep, a description by C^* does not include primary and tertiary creep, nor the elastic-plastic material response. The limitations arising from these effects will be discussed later in this paper.

Specific Formulas for Power-Law Viscous Materials

In power-law materials, characterized in uniaxial tension by

$$\dot{\epsilon} = A\,\sigma^n \tag{1}$$

the stress field is proportional to the applied load, while the strain rate field scales with the nth power and C^* with the $(n + 1)$th power of the load [*32*]. Due to Hoff's analogy [*14*], the tables for J calculation given in the Electric Power Research Institute (EPRI) plastic fracture handbook [*33*] can be applied to calculate C^* as well. For compact specimens, for example, convenient forms adopted from Ref *33* are

$$C^* = A\,(W - a)\,\sigma_{\text{ref}}^{n+1}\,h_1\,(a/W,n) \tag{2a}$$

$$= \sigma_{\text{net}}\,\dot{\Delta}\,g_2\,(a/W,n) \tag{2b}$$

where

a = crack length
$W - a$ = ligament width, and
$\dot{\Delta}$ = load-line deflection rate.

The dimensionless function h_1 has been obtained from finite-element calculations and has been tabulated [*33*], and g_2 is related to h_1 by

$$g_2 = \frac{W - a}{a}\,\frac{h_1}{1.455\,\eta\,h_3} \tag{3}$$

where h_3 is also tabulated in Ref *33*. Polynomial approximations for h_1, h_3, and g_2 are given in Ref *8*. The reference stress is defined here as

$$\sigma_{\text{ref}} = \frac{P}{1.455\,\eta\,B(W - a)} \tag{4}$$

Here, P is load, B is specimen thickness, and

$$\eta = \left[\left(\frac{2a}{W-a}\right)^2 + \frac{4a}{W-a} + 2\right]^{1/2} - \frac{2a}{W-a} - 1 \tag{5}$$

In general, the reference stress is defined such that an ideally plastic body yields when σ_{ref} reaches the yield stress. The factor 1.455, which appears in Eqs 3 and 4, is for plane strain and is replaced by 1.071 for plane stress. For other specimen geometries, Eq 2 retains its general form, but the geometrical functions h_1 and g_2 assume different values.

The asymptotic stress field in power-law viscous material at small distances, r, from the crack tip can be taken from work on power-law elastic material using Hoff's analogy. This so-called HRR field [*11,12*] has the form

$$\sigma_{ij} = \left(\frac{C^*}{I_n\, A\, r}\right)^{1/(n+1)} \tilde{\sigma}_{ij}(\theta) \tag{6}$$

The dimensionless factor I_n and the angular function $\tilde{\sigma}_{ij}(\theta)$ are tabulated or given graphically in Refs *11*, *12*, and *34*. As Eq 6 shows, the asymptotic field is indeed independent of the specimen geometry and the loading, except through C^*.

Crack Growth Rate as a Function of C*

After C^* had been established theoretically and experimentally as a load parameter correlating crack growth rates in viscously creeping materials, the question arose as to how fast a crack grows in response to any given applied value of C^*. The answer to this question requires a knowledge not only of the macroscopic deformation fields but also of the micromechanism of failure directly at the crack tip. Starting from the observation that local failure often occurs by grain-boundary cavitation, several workers [*35–39*] modeled creep crack growth, as is illustrated in Fig. 3. Cavities ahead of the main crack grow either by the classical diffusive growth mechanism (Hull-Rimmer [*40*]) or by a strain-controlled mechanism, for example the constrained diffusive mechanism proposed by Dyson [*41,42*]. Further, it is assumed that the stress and strain-rate fields felt by the cavities are the HRR fields which move with the moving crack tip and are not disturbed by the presence of the cavities. The crack must grow at such a rate that at a microstructural distance, x_c, ahead of its tip, the cavities are just about to link up. This requirement leads to an integral equation for the crack growth rate, $\dot{a}$, as a function of the amount of crack growth, Δa, which has occurred since the beginning of the test. Here, solutions are reported for strain-controlled cavitation, since these solutions were found to agree with measured crack growth rates. In this case the crack is predicted to start growing after the incubation time

$$t_i = \left(\frac{I_n\, A x_c}{C^*}\right)^{n/(n+1)} \frac{\epsilon_f}{A\,[\tilde{\sigma}_e(0)]^n} \tag{7}$$

where ϵ_f is the critical strain for local failure at a distance x_c ahead of the crack tip, and $\tilde{\sigma}_e(0)$ is the value of the dimensionless angular function appearing in the HRR field for the equivalent stress directly ahead of the crack tip. The initial growth rate is found to be

$$\dot{a} = \frac{n+1}{n}\frac{x_c}{t_i} \tag{8}$$

For large growth increments, $\Delta a \gg x_c$, the solution approaches the limiting behavior [35,39]

$$\dot{a} = \frac{\pi[\tilde{\sigma}_e(0)]^n A^{1/(n+1)}}{\sin[\pi n/(n+1)]\,\epsilon_f}\left(\frac{C^*}{I_n}\right)^{n/(n+1)}[(\Delta a)^{1/(n+1)} - \beta x_c^{1/(n+1)}] \tag{9a}$$

$$= \frac{\pi\, x_c/t_i}{\sin[\pi n/(n+1)]}[(\Delta a/x_c)^{1/(n+1)} - \beta] \tag{9b}$$

where the numerical factor β has values of $\beta = 0.85$, 0.90, 0.95, and 1 for $n = 4$, 5, 7.5 and ∞, respectively.

Experimental results reported by Riedel and Wagner [31] on a 1Cr-1/2Mo steel are shown in Fig. 4. In a log-log plot, the $\dot{a}(C^*)$ relation exhibits a slope of $n/(n+1)$, as predicted by Eqs 8 and 9. Further, the data are only moderately temperature dependent, which agrees also with theory. Note that in Eqs 7 to 9, the strongly temperature-dependent factor A (from Norton's creep law) is raised to the small power $1/(n+1)$ (n is typically 4 or greater), so that a moderate temperature dependence results. Finally, it should be noted that Eq 9 predicts a dependence of $\dot{a}$ on Δa. This is so because each volume element of the material has experienced a different stressing history before the crack tip arrives. For large $\Delta a/x_c$, this dependence of $\dot{a}$ on Δa is weak, and the data in Fig. 4 are taken from the later stages of the experiments. In the early stages, however, the dependence on Δa leads to a steep increase of the crack growth rate, as shown in Fig. 5. This behavior has been observed frequently [25–30]. Figure 5, in particular, shows results of Detampel [43] on a 2¼Cr-1Mo steel tested at 540°C under constant load. The first 25% of the test is not shown, since results in this earlier portion are determined by transients, which will be discussed later. For between about 25 and 85% of the lifetime, the crack growth rate exhibits an increase, which is ascribed to the dependence on Δa, while C^* is still nearly constant. Later, C^* increases as a consequence of the crack extension. Only after about 85% of the lifetime, has the experimental curve merged into the line having a slope of $n/(n+1)$.

In addition, the tests of Riedel and Wagner [31] showed that the low-alloy ferritic steel exhibits the same crack growth rates in air and in a mixture of argon and hydrogen.

The Lifetime of a Cracked Component

In practice, lifetimes of cracked components can be calculated by integrating the crack growth laws over time, provided that the $\dot{a}(C^*)$ relation has been determined experimentally and that the C^* values at the crack in the structure have been calculated. The latter task sometimes requires finite-element calculations, but in other cases, solutions can be inferred from tabulated solutions for similar geometries.

If one wishes to neglect the dependence of $\dot{a}$ on Δa, the measured data can often be represented by $\dot{a} = \gamma\, C^{*q}$, where γ and q are adjustable parameters. Time integration gives the lifetime

$$t_f = \frac{1}{\gamma}\int_{a_i}^{a_f}\frac{da}{[C^*(a)]^q} \tag{10}$$

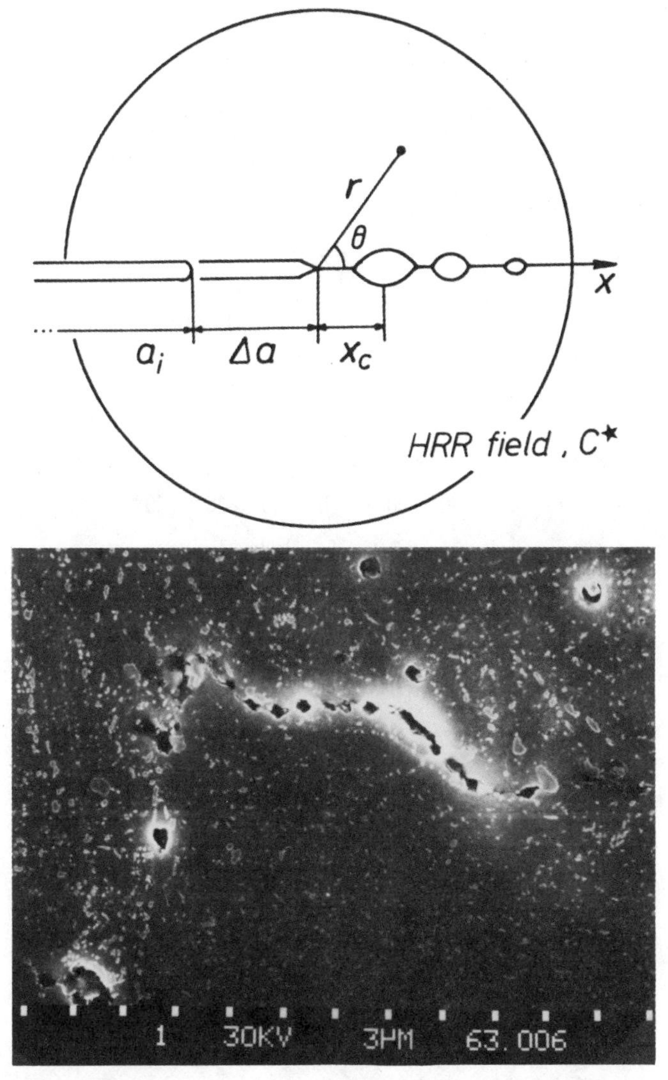

FIG. 3—*Creep crack growth by grain boundary cavitation:* (top) *schematic:* (bottom) *cavities ahead of a crack in a CrMo steel* [31].

where a_i is the initial crack length and a_f is the crack length at which failure occurs. Usually the precise value of a_f does not have much of an influence on the value of t_f. The integral in Eq 10 must be evaluated numerically in most cases.

Neglecting the dependence of $\dot{a}$ on Δa sometimes yields an overly conservative estimate of the lifetime. As a first approximation for including this dependence, it suffices to start from the approximation for large Δa, Eq 9, with $\beta = 0$. Then Eq 9 takes the form $\dot{a} = \gamma(A\Delta a)^{1/(n+1)}C^{*n/(n+1)}/\epsilon_f$, where γ can be considered an adjustable dimensionless parameter, or its theoretical value can be obtained by comparison with Eq 9*a*. Integration of this growth

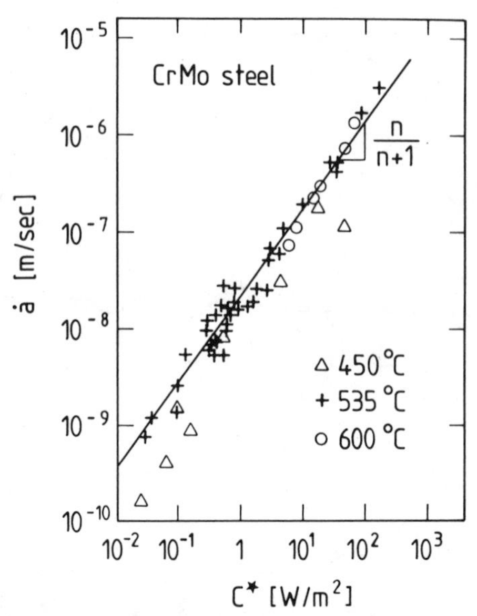

FIG. 4—*Crack growth rates in 1Cr-1/2Mo steel at various temperatures. Solid line: Eq 9. From Ref* 31.

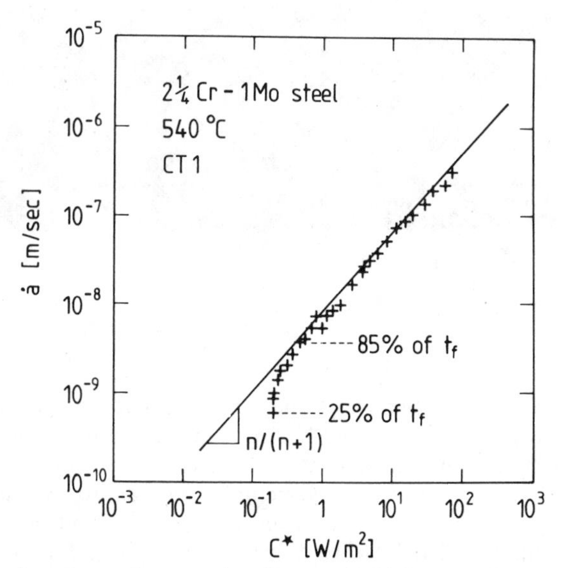

FIG. 5—*Increase of crack growth rate at virtually constant* C* *between 25% and 85% of the lifetime, supposedly caused by a dependence of* ȧ *on* Δa. *From Ref* 43.

law, together with Eq 2*a* for C^*, gives

$$t_f = \frac{F}{A(\sigma_{\mathrm{ref},i})^n} \tag{11}$$

where $\sigma_{\mathrm{ref},i}$ is the reference stress for the initial crack length and F is an abbreviation for

$$F = \frac{\epsilon_f}{\gamma}\int_0^{x_i}\left[\frac{x\,\eta(x)}{x_i\,\eta(x_i)}\right]^n \frac{dx}{(x_i - x)^{1/(n+1)}(x\,h_1)^{n/(n+1)}} \tag{12}$$

and $x = 1 - a/W$ with the initial value $x_i = 1 - a_i/W$. The integral must be evaluated numerically. Results will be shown later (Fig. 7), after transient effects have been taken into account.

Comparison with a Reference Stress Concept

In the formula for C^*, Eq 2, the reference stress was introduced merely as an auxiliary quantity. The load or the net section stress could have been used as well. But the reference stress has also been suggested to control the lifetime under conditions when damage spreads across the whole specimen and the ligament fails like a more or less homogeneously stressed specimen. For strain-controlled fracture, the reference-stress concept predicts

$$t_f = \frac{C_{MG}}{A(\sigma_{\mathrm{ref},i})^n} \tag{13}$$

where C_{MG} $(= \dot{\epsilon}_s t_f)$ is the Monkman-Grant product obtained in uniaxial tension.

Hence the same dependence of the lifetime on stress is predicted irrespective of whether one assumes that failure occurs by crack growth, Eq 11, or by homogeneously distributed cavitation, Eq 13. The two expressions differ only in the numerical factor, F versus C_{MG}. Webster, Smith, and Nikbin [*44*] suggest that among these two failure modes the one which gives the smaller lifetime predominates. Failure by crack growth tends to predominate for long cracks, whereas homogeneous cavitation is preferred, if the initial crack is small. This is a useful first approach to discuss the question, within what limits creep rupture of cracked specimens can be described by C^*, and when it is more appropriate to view the specimen as homogeneously damaged. This question will be taken up again in a later section on continuum damage mechanics. As Webster, Smith, and Nikbin [*44*] also note, an uncertainty regarding the appropriateness of plane strain or plane stress leads to a great uncertainty in the predicted lifetime. This problem will be discussed in a separate section on three-dimensional aspects.

Effects of Elasticity in Creep Crack Growth

Further progress in the understanding of creep crack growth has been made when it was recognized that the description by the stress intensity factor and that by the C^*-integral were limiting cases contained in an elastic/nonlinear viscous material law, and when the transition from the short-time elastic behavior to long-time creep was analyzed.

First attempts to describe the behavior of cracked bodies of elastic/nonlinear viscous materials employed Dugdale-type models [*45–47*]. Such models, however, do not represent the long-time, C^*-controlled limit correctly and are not described here. The next step was a Mode III analysis [*48*] in close analogy to the development of elastic-plastic fracture mechanics. In the sequel, a Mode I analysis will be described, which was found independently by Riedel and Rice [*24*] and by Ohji, Ogura, and Kubo [*49*].

Ductile Material: The Specimen Reaches Steady-State Creep After a Transient

Riedel and Rice [*24*] have analyzed the transient from the initial elastic stress field to steady-state creep in a body with a stationary crack. The material is characterized by $\dot{\epsilon} = \dot{\sigma}/E + A\sigma^n$ in uniaxial tension, where the first term describes elastic straining with E = Young's modulus. During the transient, a creep zone grows around the crack tip. For short times after load application, the creep zone size increases with time according to $r_{cr} \propto K_I^2(AEt)^{2/(n-1)}$, while the stress near the crack tip relaxes according to $\sigma \propto [K_I^2/(AEt)]^{1/(n+1)}$. After long times, the creep zone spreads out through the whole specimen, that is, the elastic/nonlinear viscous material approaches its purely nonlinear viscous limit, and the stress field becomes a steady-state field.

In the whole time domain, the stress field well inside the creep zone has an HRR-type singularity, $r^{-1/(n+1)}$, but C^* in Eq 6 is replaced by a time-dependent quantity $C(t)$ which approaches C^* for long times. A useful interpolation formula which reproduces the short- and the long-time limits, and which was checked against finite-element results [*50*] is

$$C(t) = (t_1/t + 1)\, C^* \tag{14}$$

Here, t_1 is the characteristic time of the transient

$$t_1 = \frac{J}{(n+1)\, C^*} \tag{15}$$

For elastic/nonlinear viscous materials, is $J = K_I^2(1 - \nu^2)/E$, but Eq 15 above retains its form if the initial material response is elastic-plastic or fully plastic [*8*].

Equation 14 does not only describe the transient behavior of the stress field but also of the crack growth rate [*8*], if in Eq 9 C^* is replaced by $C(t)$. Due to the decrease in the stress field, expressed by the t_1/t-term in Eq 14, the crack decelerates during the transient. Later, when C^* increases as a consequence of crack growth, the crack accelerates. This behavior will be shown later, after primary creep has been taken into account.

Saxena and Landes [*51*] and Saxena [*52*] (using a slightly different definition) have proposed a parameter called C_t, which is designed to approximate the crack-tip parameter $C(t)$. For short times, C_t does not exactly have the same behavior as $C(t)$ [*8*]. Its advantage is that it is easily measurable, whereas Eq 14 for $C(t)$ relies to a great extent on theory.

Brittle Material: Crack Growth Under Small-Scale Creep Conditions

If a cracked specimen fractures at times that are short compared to the characteristic time for stress redistribution, t_1 from Eq 15, the material is called brittle. In other words, the crack grows so fast that it overtakes the creep zone. At growing cracks, a new type of crack-tip field develops. The character of this field depends on the value of the stress exponent n [*53*]. For $n < 3$, elastic strain rates dominate near the crack tip and the stress and strain fields have a $1/\sqrt{r}$-singularity. This case was pursued by Hart [*54*]. For $n \geq 3$, the new type of singularity

$$\sigma_{ij} \propto \left(\frac{\dot{a}}{EAr}\right)^{1/(n-1)} \tag{16}$$

$$\epsilon_{ij} \propto \frac{1}{E}\left(\frac{\dot{a}}{EAr}\right)^{1/(n-1)} \tag{17}$$

was obtained by Hui and Riedel [*53*]. Hence, this field is sometimes called an HR field. It should be noted that the HR field is the asymptotic field at growing cracks in elastic/nonlinear viscous materials both in the K_I-controlled limit and the C^*-controlled limit. In the latter case, however, the validity of the HR field is usually confined to an extremely small region (often smaller than atomic dimensions), so that it can generally be neglected. But under small-scale creep conditions, the HR field plays a role.

The HR field has surprising properties. A few workers have investigated its interplay with other singular fields (the HRR field and the elastic singular field) [*53,55–59*]. The results are somewhat complicated and will not be reported here. In particular, if the stress analysis is combined with a critical-strain criterion (or any other related criterion) in order to obtain crack growth rates, a complicated and sometimes irregular growth behavior results [*55,59*]. The simplest case is that of large K_I (compared to $E\epsilon_f\sqrt{x_c}$). In this case, the range of validity of the HR field goes to zero and the steady-state crack growth rate becomes [*8,53*]

$$\dot{a} = \frac{2}{n-2}\frac{Ax_c}{\epsilon_f}\left[\frac{(1-2\nu)K_I}{\sqrt{2\pi x_c}}\right]^n \tag{18}$$

A power-law relation of the form $\dot{a} \propto K_I^n$ has in fact been reported frequently as an experimental result. Figure 6, for example, shows crack growth rates in the nickel-base alloy Nimonic 80A [*31*]. In these tests, the characteristic time t_1 was typically a few hundred years. Correspondingly, the specimens, one of which was shown in Fig. 1, exhibited no

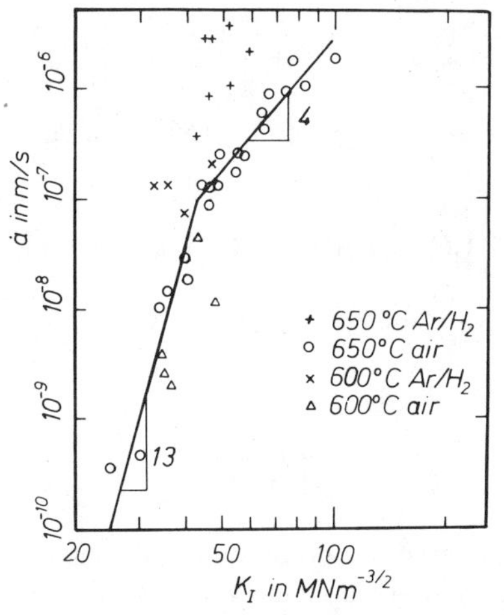

FIG. 6—*Crack growth rates in Nimonic 80A in air and in a Ar/H_2 mixture. The line with slope 13 represents Eq 18. From Ref* 31.

nonelastic deformation, and K_I was the appropriate load parameter. The stress exponent of the material at 650°C was found to be $n = 13$. At small growth rates, the $\dot{a}(K_I)$ relation shown in Fig. 6 in fact exhibits a slope of 13 in a log-log representation. At greater $\dot{a}$, the slope becomes smaller. The reasons for this have not been investigated. As an important qualitative result, Fig. 6 also shows an influence of the atmosphere on the crack growth rate. Although oxygen is usually considered to be an aggressive agent for nickel-base alloys, the crack growth rate in an argon-hydrogen mixture (which has a low oxygen partial pressure) was higher than in air.

Primary Creep

In connection with creep crack growth, primary creep has been modeled by either a time hardening law [*24*], which is the simplest and least realistic constitutive description, or by a strain-hardening law [*60*], or by a hardening/recovery model using an internal variable [*61*], which is the most realistic model. Here, it suffices to consider a strain-hardening law of the form

$$\dot{\epsilon} = A_1\, \sigma^{m(1+p)}\, \epsilon^{-p} \tag{19}$$

The hardening exponent is typically $p = 2$. Then Andrade's primary-creep law $\epsilon \propto t^{1/3}$ results upon time-integration of Eq 19 for constant stress. If both, primary and secondary creep are to be included, one adds the right-hand sides of Eqs 1 and 19.

If besides primary and secondary creep, elastic strains are also admitted, the deformation fields in a cracked body develop as follows [*60*]. At short times after load application, a primary-creep zone surrounded by the initial elastic stress field grows around the crack tip. Within the primary-creep zone, a secondary-creep zone develops. If the primary-creep stage of the material is pronounced, the primary-creep zone spreads through the whole specimen while the secondary-creep zone is still small. The characteristic time for the transition from the initial elastic-plastic state to extensive primary creep is given by

$$t_1 = \frac{1}{m+1}\left(\frac{J}{C_h^*}\right)^{1+p} \tag{20}$$

where C_h^* is the primary-creep analogue to C^* [*60*]. The equation corresponding to Eq 2 for C^* is

$$C_h^* = [A_1(1+p)]^{1/(1+p)}\, (W - a)\, \sigma_{\mathrm{ref}}^{m+1}\, h_1(a/W,m) \tag{21}$$

The transition from extensive primary to extensive secondary creep occurs with the characteristic time

$$t_2 = \left(\frac{C_h^*}{(1+p)C^*}\right)^{(p+1)/p} \tag{22}$$

If, on the other hand, the material has no pronounced primary creep stage, the secondary-creep zone overtakes the primary-creep zone while both are still small. Subsequently, the secondary-creep zone grows in an elastic field as described in the preceding section.

Well inside the secondary-creep zone, the crack-tip field at a stationary crack has the HRR-type form of Eq 6, but with a time-dependent amplitude $C(t)$ instead of the steady-state value C^*. As an interpolation for the stress field and for the crack growth rate, it has been suggested [8,62] to replace C^* in Eqs 6 and 9 by

$$C(t) = [t_1/t + (t_2/t)^{p/(1+p)} + 1]\, C^* \tag{23}$$

Here, t_2 is given by Eq 22 and t_1 by Eq 15. Depending on the relative magnitude of t_1, t_2, and the lifetime t_f, the transients by elastic straining and by primary creep may be more or less important.

Two experimental examples were analyzed in the literature [8,62], both of which showed that primary creep plays a role for creep crack growth in creep-resistant chromium-molybdenum-vanadium steels. Here, a test on a $2\frac{1}{4}$Cr-1Mo steel at 540°C is described [8]. A CT1-specimen with 20% side grooves and $a/W = 0.5$ is loaded to 20.12 kN. The material parameters are $A = 1.3 \times 10^{-24}$, $n = 7.3$, $A_1 = 7 \times 10^{-29}$, $m = n/3$, $p = 2$, $E = 150$ GPa, $\nu = 0.3$ with dimensions of MPa and s in A and A_1.

The load parameters were calculated from Eqs 2*a* and 21 to be $C^* = 7.2 \times 10^{-2}$ W/m^2 and $C_h^* = 965$ $J/(m^2s^{1/3})$. The J-integral was determined from the initial specimen response using the Merkle-Corten formula [63] to be $J = 18.7$ kJ/m². The linear-elastic value of J would have been 7 kJ/m², that is, instantaneous plasticity contributed substantially to J. From these values of the load parameters, the characteristic times were calculated as $t_1 = 8.7$ h from Eq 15, and $t_2 = 83$ h from Eq 22. The lifetime of the specimen was $t_f = 596$ h, so that in this example, the transients are expected to play a role during the first 20% of the test.

To illustrate the effect of the transients on the crack growth behavior, Eq 9 (with $x_c = 0$) for the crack growth rate was integrated as in Eq 12, but with C^* replaced by the approximation for $C(t)$ given in Eq 23. Figure 7 shows the result for the crack length as a function of time in comparison with measured data. The solid line was calculated using the values for t_1 and t_2 estimated above, while the dotted line was computed with $t_2 = 0$, that is, neglecting primary-creep effects. For the dashed line, $t_1 = t_2 = 0$ was assumed, that is,

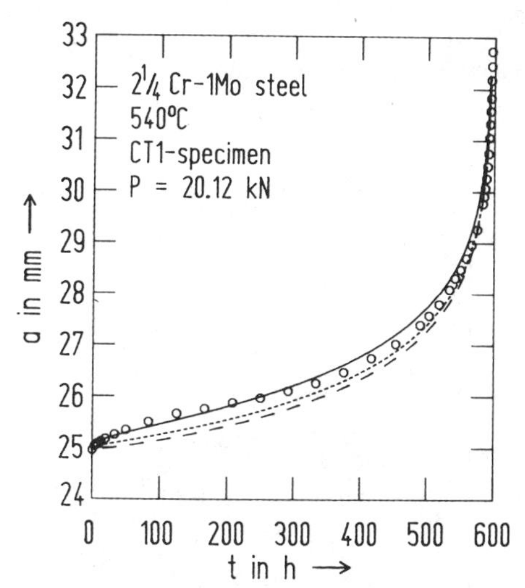

FIG. 7—*Evolution of crack length in a CT-specimen. Circles: Measured data. Solid line: primary creep and elastic-plastic strain included. Dotted line: primary creep neglected. Dashed line: only secondary creep considered* [8].

all transients were neglected. As the comparison with the data shows, the initial behavior can only be described if primary creep is taken into account. It should be noted that all theoretical lines have been normalized by adjusting the value of the critical strain, ϵ_f, in Eq 9 to give the experimental lifetime. The fit required $\epsilon_f = 1.82\%$, 2.05%, and 3.28% for the dashed, the dotted, and the solid line, respectively.

Load Parameter Maps

The regimes in the time-stress plane in which different load parameters are valid can be represented conveniently on a load parameter map. Figure 8 schematically shows an example. The abscissa is the stress which is applied to the cracked specimen. Preferably the reference stress is used, although any other measure of the applied stress could be used as well. When a specimen is loaded to a small value of σ_{ref}, the instantaneous response is elastic, while at high σ_{ref}, the specimen becomes fully plastic. Hence at low loads and short times, the stress-intensity factor, K_I, is the load parameter which characterizes the crack-tip fields, while at higher loads and short times, the J-integral dominates. (Of course, J comprises K_I.) If the load is maintained, a creep zone grows around the crack tip and finally leads to a transition to either primary or secondary creep of the whole specimen. The lines on the load parameter maps represent the characteristic times for the transitions. For example, the transition from elastic behavior to primary creep occurs at the time $t_1 \propto (J/C_h{}^*)^{1+p}$ given by Eq 20. In the small-scale yielding limit is $J \propto \sigma_{ref}{}^2$ and, from Eq 21, $C_h{}^* \propto \sigma_{ref}{}^{m+1}$. Hence this transition time is represented on a log-log plot by a line with slope $-(m - 1)(1 + p)$. The intercept of that line with the axis depends on the material parameters A_1 and E on the specimen geometry, but not on the absolute specimen size. Analogously, the other lines in Fig. 8 can be constructed.

Figure 8 also shows a regime at small stresses in which linearly viscous creep (or diffusion creep) predominates. In this range, K_I and C^* are equivalent and are related by $C^* = (3/4)\, A_d\, K_I{}^2$, where A_d is the coefficient of the law for linear-viscous creep, $\dot{\epsilon} = A_d\sigma$ [8].

Further, the range of validity of J and C^* is bounded on the right by a line, t_b, which indicates excessive crack-tip blunting. In the rate-independent regime, this line represents the relation $J/\sigma_y = a/25$, which is commonly used for compact-type (CT) specimens in elastic-plastic fracture mechanics. Blunting in the viscous regime will be considered in the following section.

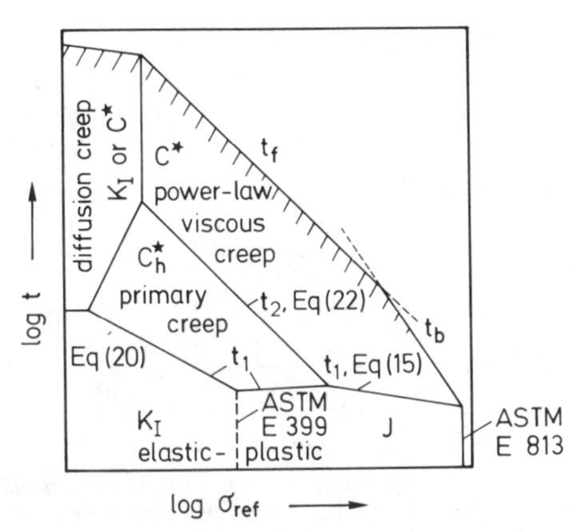

FIG. 8—*Load parameter map (schematic).*

Finally, Fig. 8 shows the time to fracture, t_f. The hatched band below t_f indicates that grain boundary damage has spread over the whole ligament of the specimen, thus invalidating a nonlinear viscous description and the use of C^*. The width of the hatched band will be investigated in a section on damage mechanics.

Limitations to C^* Due to Crack-Tip Blunting

If one uses C^* as a crack-growth-rate correlating parameter, one relies on the assumption that crack tips in all kinds of specimens are encompassed by the same type of singular stress field (for example, the HRR field in power-law viscous materials). In reality, however, the singular field is always perturbed near the crack tip, for example, by crack-tip blunting, by cavitation damage, or by any other deviation from the solid mechanics equations for non-linear viscous materials. Nevertheless, C^* remains a valid parameter, as long as the perturbed zone is small enough to be surrounded by a field which is still governed by the singular stress field to within a desired accuracy. Then, whatever happens in the perturbed zone is controlled by the singular field. This guarantees a unique cracking behavior in all specimen geometries.

The analysis of crack-tip blunting in elastic-plastic or fully plastic materials [*13,64–66*] has shown that the HRR field has a finite range of approximate validity between the disturbances by blunting on one side and by the outer specimen geometry on the other, if the crack-tip opening displacement, δ_t, is smaller than a certain fraction of the crack length or ligament width:

$$\delta_t \leq \frac{a}{2M} \text{ or } \frac{W - a}{2M} \tag{24}$$

The factor M depends on the hardening exponent, the specimen geometry, and the desired accuracy. For practical purposes, $M = 25$ is suggested for compact-type specimens, whereas center-cracked tension specimens require a much higher value, $M = 200$, due to the limited range of validity of the HRR field in this specimen geometry [*65*]. In rate-independent fracture mechanics, Eq 24 is often formulated in terms of J which is related to δ_t by $J = 2\sigma_y\delta_t$, where σ_y is yield stress.

The results obtained in rate-independent fracture mechanics can be directly applied here. Only the crack-tip opening displacement needs to be calculated for viscous materials. In a power-law viscous material, the crack-tip stress field was given in Eq 6, and the corresponding displacement rate field is

$$\dot{u}_j = (Ar)^{1/(n+1)} \left(\frac{C^*}{I_n}\right)^{n/(n+1)} \tilde{u}_j(\theta) \tag{25}$$

where the nondimensional angular function $\tilde{u}_j(\theta)$ was tabulated by Shih [*34*]. The profile of a stationary crack is obtained by setting $\theta = \pm\pi$ and taking $u_\theta = \dot{u}_\theta\, t$ (for constant C^*). Using the definition of δ_t with the ±30-deg lines, which is illustrated in Fig. 9, leads to

$$\delta_t = \left[\frac{2|\tilde{u}_\theta(\pi)|^{1+1/n}}{(\tan 30°)^{1/n}\, I_n}\right] A^{1/n}\, C^*\, t^{1+1/n} \tag{26}$$

The term in brackets is equal to 1.25 and 1.11 for $n = 5$ and 7, respectively. If the crack starts growing when a critical strain is reached, the crack growth initiation time is given by

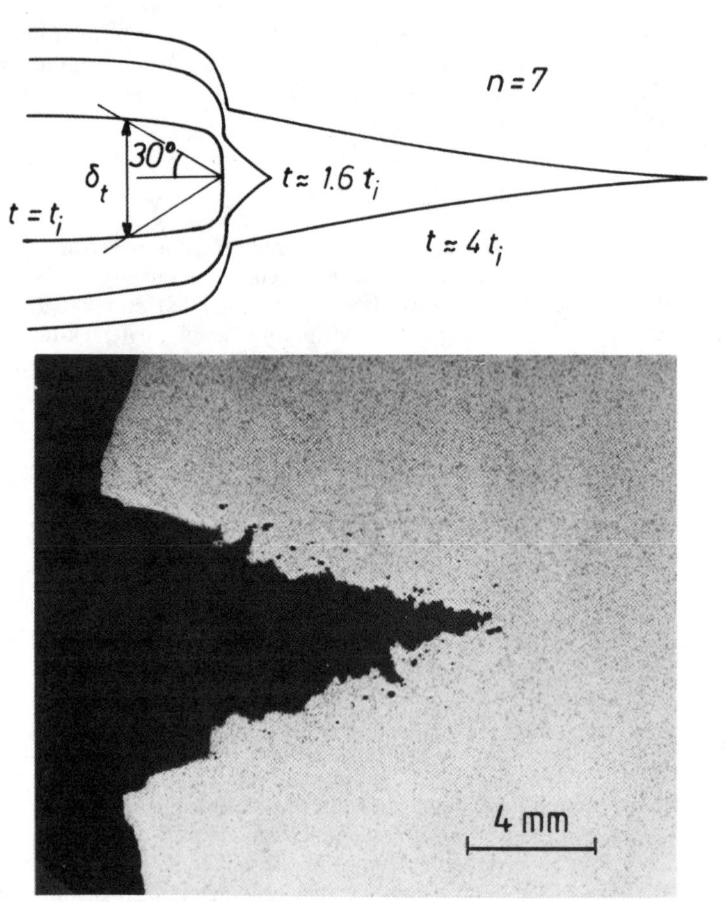

FIG. 9—*Evolution of crack-tip shape during creep crack growth:* (top) *calculated for* n = 7, ϵ_f = *1%:* (bottom) *experimental in 2¼Cr-1Mo steel tested at 540°C.*

Eq 7. The corresponding crack opening is obtained by substituting t_i for t in Eq 26,

$$\delta_t = 452\ \epsilon_f^{1+1/n}\ x_c \tag{27}$$

where the numerical factor is valid for $n = 7$. If $\epsilon_f = 1\%$ and $x_c = 10$ μm, Eq 27 gives $\delta_t = 23$ μm. This is sufficiently small compared to typical specimen dimensions to satisfy Eq 24.

After growth initiation, the crack shape continues to change in a manner shown in Fig. 9. These crack profiles were calculated by replacing r in Eq 25 by $\Delta a - x$, where Δa is the crack growth increment and x is the coordinate in the crack-growth direction with $x = 0$ at the initial crack tip position. Then Eq 25 is integrated over the time, the time differential is replaced by $da/\dot{a}$, and the integral extends from 0 (if $x < 0$) or from x (if $x > 0$) to the current Δa. Since the crack growth rate, $\dot{a}$, is not known in closed analytic form, the procedure is carried out using the limiting formulas, Eqs 8 and 9. For $x > 0$, there results for small and large Δa, respectively,

$$u_\theta = \frac{\tilde{u}_\theta \, \epsilon_f}{[\tilde{\sigma}_e(0)]^n} \frac{n}{n + 2} x_c \left(\frac{\Delta a - x}{x_c}\right)^{(n+2)/(n+1)} \tag{28}$$

$$u_\theta = \frac{\tilde{u}_\theta \, \epsilon_f}{[\tilde{\sigma}_e(0)]^n} \frac{1}{\pi} \sin \frac{\pi n}{n + 1} \int_x^{\Delta a} \left(\frac{a - x}{a}\right)^{1/(n+1)} da \tag{29}$$

Figure 9 was calculated for $n = 7$ and $\epsilon_f = 1\%$. Since the large-Δa limit starts to become accurate at rather large Δa, which cannot be shown in Fig. 9, the profile for $t = 4t_i$ represents an estimate based on interpolation. As Fig. 9 shows, the growing crack develops a relatively sharp tip characterized by $u_\theta \propto (\Delta a - x)^{(n+2)/(n+1)}$. The tangent of the half crack opening angle is given by $u_\theta(x = 0)/\Delta a$, which approaches the value $13\epsilon_f$ for $n = 7$ in the large Δa-limit.

Figure 9 also shows an experimental example for the profile of a growing crack. To fit the observed opening angle with the above formulas requires $\epsilon_f = 3\%$. This agrees with the value that had been obtained by adjusting the theoretical lifetime in Fig. 7. In uniaxial tension, the critical strain of 2¼Cr-Mo steel is greater than 3%, but in the triaxial HRR field, ϵ_f is expected to be lower than in uniaxial tension. In the experimental example in Fig. 9, the growing crack has developed a shape which differs significantly from an idealized mathematically sharp crack for which the HRR field has been derived. What this deviation from the idealized shape means for the range of approximate validity of the HRR field is not quantitatively clear. A very acute crack opening angle allows for valid C^* testing up to considerable amounts of crack growth, while for a less acute angle, the applicability of C^* will be restricted.

Damage Mechanics Limitations to C^* by the Growth of a Process Zone

The zone around the crack tip in which the nonlinear viscous fields are perturbed by cavitation damage, or by any other microscopic damage mechanism, is called the process zone. In plane-strain fracture at room temperature, the process zone is of the same order of magnitude as the crack-tip opening displacement [*13*]. In creep crack growth, this may be different.

Riedel [*67*] has modeled the evolution of the process zone by the damage mechanics equations of Kachanov and Rabotnov. Although for the crack analysis the three-dimensional form of the equations [*68*] is needed, it suffices here to consider the uniaxial form. An internal variable, called the damage parameter ω, is introduced, which varies from $\omega = 0$ for the virgin material to $\omega = 1$ at fracture. It is assumed to increase according to the kinetic law

$$\dot{\omega} = D \frac{\sigma^\chi}{(1 + \phi)(1 - \omega)^\phi} \tag{30}$$

(where D, χ, and ϕ are material parameters) and to enhance the strain rate as a function of stress as

$$\dot{\epsilon} = A \frac{\sigma^n}{(1 - \omega)^n} \tag{31}$$

In a cracked body, ω is, of course, a field which varies from point to point. Directly at the crack tip, the stress concentration leads to rapid growth of ω from 0 to 1, so that the crack starts growing. In other words, crack growth is a feature of the solutions of Eqs 30 and 31 with no additional local failure criterion being needed.

Mathematically, the evolution of the stress, strain, and damage fields can be described in the following way. Directly upon load application at time $t = 0$, the stress field of a nonlinear viscous material with $\omega = 0$ develops. The field at the crack tip, especially, is the HRR field. For small, but finite times, the HRR field is perturbed by damage accumulation near the crack tip, but as long as this disturbed process zone is small enough compared to the crack length, the HRR field retains its validity in an annular zone surrounding the growing process zone. Hence the HRR field can serve as the remote boundary condition for the evolution of the fields in the process zone, and the crack can be considered to be of an infinite length compared to the process-zone size. This is the well-known boundary layer approach which has been used to analyze the small-scale yielding limit in elastic-plastic fracture and which is applied here to the small-scale damage limit. With the HRR field being both the initial condition at $t = 0$ and the remote boundary condition at $r \rightarrow \infty$, Eqs 30 and 31, together with the equilibrium and compatibility conditions, have similarity solutions, as standard arguments of dimensional consistency show [*67*].

These similarity solutions directly give the crack growth rate. Apart from a numerical factor, the result is equivalent to Eq 9 with $x_c = 0$. (The damage mechanics equations contain no structural length.) Hence, no serious error has been introduced in the previous analysis by neglecting the effect of damage on the stress distribution.

In the damage mechanics analysis, stress redistribution by damage is taken into account and the stress in the process zone as obtained from a finite-element analysis [*67*] is shown in Fig. 10. The crack-tip position initially lay at $r = 0$ and the crack has grown by an amount Δa. Hence the stress component $\sigma_{\theta\theta}$, which is shown in Fig. 10 and which is normal to the crack, is zero between $r/\Delta a = 0$ and 1. Far away from the crack tip, the stress must approach

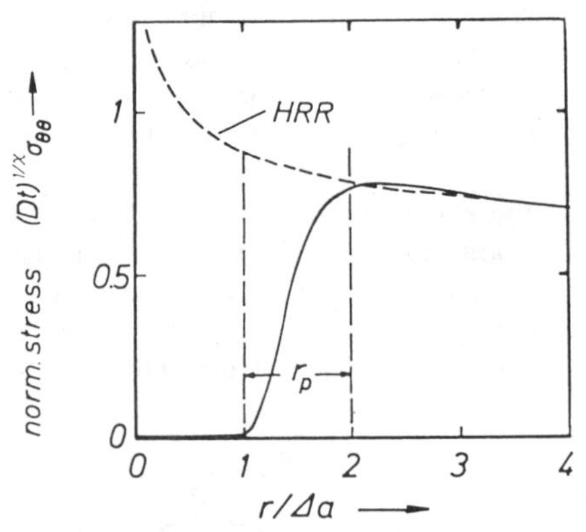

FIG. 10—*Stress distribution in the process zone ahead of the crack tip. Finite-element calculation from Ref* 67.

the HRR field in the small-scale damage limit. The process zone can be defined as the zone in which the stress deviates markedly from the HRR field. Due to the similitude of the small-scale damage solutions, Δa and the process zone size must grow in a fixed proportion. In the example shown in Fig. 10, the two are approximately equal.

The process zone plays the same role in bounding the range of validity of C^* as does the crack-opening displacement in relation to J or C^*. Only as long as the process zone is smaller than, say, $2.5\Delta a/M$ with M as in Eq 24, has the HRR field a range of sufficient accuracy and is C^* applicable. This condition for valid C^*-testing can be expressed as

$$\Delta a \leq (5 \text{ to } 15)\% \text{ of } a \text{ or } W - a \tag{32a}$$

For typical test specimen configurations this corresponds to

$$t \leq (60 \text{ to } 90)\% \text{ of } t_f \tag{32b}$$

Experiments [*43,62*] seem to indicate that the above conditions are unnecessarily restrictive. Experimentally crack growth rates in different specimens can be correlated by C^* up to greater crack growth increments and lifetime fractions than permitted by Eq 32.

If, on the other hand, the range of validity of the small-scale damage limit and hence of C^*, had been found to be restricted to very small values of $\Delta a/a$ and t/t_f, crack growth in a structure and its lifetime could not be predicted using C^*. In that case, a separate finite-element analysis of each cracked structure would be required based on the continuum damage equations [*69*]. This is not only difficult and expensive, but it can also lead to seriously nonconservative estimates of lifetime in cases like the following. If corrosive processes occur at the crack tip, the damage mechanics equations cannot be expected to capture the enhancement of the crack growth rate by corrosion. On the other hand, a creep crack growth test done in the same environment reflects the corrosive effects, and the prediction for the cracked structure will be correct provided that C^* has a sufficient range of validity.

The Third Dimension in Fracture Mechanics

It is not always sufficient to consider the limiting cases of plane strain and plane stress, which correspond to the idealized situations of infinitely thick and infinitely thin specimens. For example, if C^* is calculated from the applied load using the EPRI plastic fracture handbook [*33*], the results for plane strain and plane stress can differ by more than an order of magnitude. This would lead to a similar uncertainty in lifetime prediction, and this is often unacceptable.

Another problem is whether the crack tip region experiences plane-strain or a plane-stress conditions. Since some of the micromechanisms of crack extension sensitively depend on the state of stress, it is necessary to ensure that the situation in the test specimen and the cracked structure are comparable. These two questions, and a few others, will be discussed next.

The Load/Displacement Response and C*

It is plausible and it has been demonstrated frequently that the load/displacement response of the standard, parallel-sided specimens falls between plane strain and plane stress. In fracture mechanics tests, including creep crack growth tests, the use of side-grooved specimens has become common. Their load/displacement behavior has been studied using three-dimensional finite element calculations [*70,71*].

TABLE 1—*Normalized compliance,* $E\Delta B_{tot}/P$, *of compact specimens without or with 25% side grooves.*

a/w	Plane Stress	Plane Strain	3-D, No Side Grooves	3-D, 25% Side Grooves
0.5	37.0	33.7	36.3	38.2
0.6	63.4	57.7	60.9 [*70*]	63.6 [*70*]
			62.1 [*72*]	63.3 [*71*]

In the linear elastic case, the compliances calculated for plane stress and plane strain differ by a factor $(1 - \nu^2)$. This difference is often insignificant. In Table 1, the plane limits are compared with results of three-dimensional finite-element calculations [*70*] for parallel-sided and side-grooved CT-specimens.

In nonlinear elastic (or viscous) materials, the difference in load-line deflection (or deflection rate) between plane strain and plane stress is of the order 0.75^n. For large n, say for $n = 7$, this means that plane-strain displacements are an order of magnitude smaller than for plane stress. Similarly, the plane-stress and plane-strain formulas for C^*, Eq 2*a*, differ by a factor of the order 0.75^{n+1}. To reduce this uncertainty, three-dimensional finite element calculations have been performed [*71*]. One of the results—the load/displacement curve of a side-grooved CT-specimen—is shown in Fig. 11. An elastic-plastic material is considered here with a hardening law in the plastic range $\epsilon = \alpha\epsilon_0(\sigma/\sigma_0)^n$ with $\alpha = 0.5$ (erroneously given as $\alpha = 0.05$ in Ref *71*; private communication by C. F. Shih), $\epsilon_0 = \sigma_0/E$, $\sigma_0 = 414$ MPa, $E = 207\,000$ MPa, $n = 10$. In the fully plastic limit, this material is analogous to a power-law viscous material. For comparison, results for the fully plastic limit from the EPRI handbook [*33*] are also shown as the straight lines. All of them are for plane

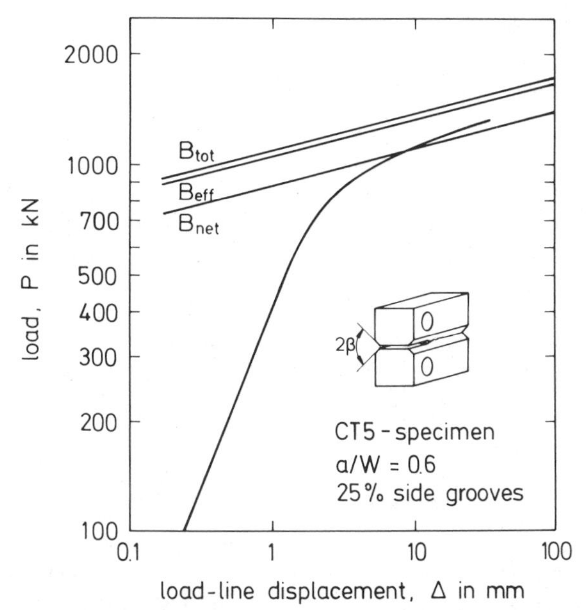

FIG. 11—*Load/displacement response of a side-grooved CT-specimen. Comparison of 3-D finite-element calculation* [71] *with fully plastic plane-strain results (straight lines, Ref* 33).

strain. They differ by the specimen thickness which the load is referred to, that is, either the gross (total) thickness, B_{tot}, or the net thickness between the side grooves, B_{net}, or the effective thickness defined as $B_{eff} = B_{tot} - (B_{tot} - B_{net})^2/B_{tot}$ [70]. As Fig. 11 shows, the use of the net thickness approximates the three-dimensional calculation best. The lines for plane stress lie below those for plane strain and are not shown here.

Hence, if C^* is to be determined for side-grooved specimens, it is recommended either to use Eq 2*a* with the stress calculated as load per net thickness or, if the deflection rate has been measured, to use Eq 2*b*, which is relatively insensitive to the choice of plane strain or plane stress.

Three-Dimensional Stress Distribution Ahead of a Crack

Figure 12 qualitatively shows the stress distribution on the ligament ahead of a straight crack front in a parallel-sided specimen. This picture has been proposed in Ref *8*. To be definite, the material is assumed to be nonlinear viscous and to obey a power law, $\dot{\epsilon} = A\sigma^n$, but the main conclusions of the following do not depend on that assumption.

First one notes that asymptotically near a mathematically sharp crack, the stress field must approach the plane-strain HRR field. This is so since the transverse strain component, ϵ_{33}, must not become singular in order to avoid an infinite displacement, $u_3 = \int \epsilon_{33}\, dz$. Since ϵ_{33} remains bounded, while the in-plane components of strain are singular as $\epsilon \propto r^{-n/(n+1)}$, ϵ_{33} is negligible, that is, plane-strain conditions are approached asymptotically.

At the vertex where the crack front intersects the specimen surface, plane-strain conditions cannot be maintained, and consequently the distance from the crack front over which the plane-strain field is valid shrinks to zero. At the specimen surface, the boundary condition $\sigma_{3i} = 0$ must be satisfied, but this does not mean, as is often believed, that the field is the plane-stress field. (What is usually called a plane-stress field, additionally requires that all derivatives $\partial/\partial z$ are zero.) Rather, the vertex field exhibits a new type of singularity which has not yet been investigated for power-law material. A vertex in linear elastic material, however, has been studied by Benthem [*72,73*]. His results are worth mentioning here since, due to the elastic-viscous analogy, they comprise the linear viscous case ($n = 1$). The results for arbitrary n are expected to exhibit the same qualitative features. Benthem shows that the stress singularity has the general form $\sigma \propto \rho^{-s}$ where ρ is the distance from the vertex with an exponent s that depends on Poisson's ratio ν. For incompressible material, which

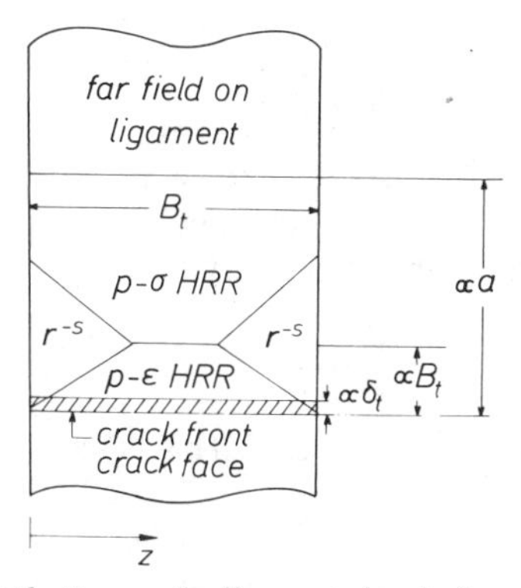

FIG. 12—*Schematic stress distribution on the ligament ahead of a crack in a thin specimen;* p − ε = *plane strain,* p − σ = *plane stress;* r^{-s} = *vertex field; the hatched zone is disturbed by blunting and by the process zone. From Ref* 8.

is of most interest here, since viscous creep is considered to be incompressible, he obtains $s = 0.332$. For $\nu = 0.3$, $s = 0.452$. Further, he shows that the vertex field approaches the plane-strain crack-tip field close to the crack front, as one would expect.

At larger distances from the crack front (where the specimen thickness B is the characteristic length to compare with), the deformation fields approach plane-stress conditions, if the in-plane dimensions of the specimen—crack length, a, and ligament width, W-a—are large enough compared to the thickness. This is so since the through-the-thickness stress component is $\sigma_{33} = 0$ at the surface and of the order $\sigma_{33} \approx B\, \partial\sigma/\partial r$ in the center of the specimen (since the equilibrium condition demands $\partial\sigma_{33}/\partial z \approx \partial\sigma/\partial r$), while the in-plane components, σ, are of the order $\sigma \propto r^{-1/(n+1)}$. This means that at greater distances from the crack tip $\sigma_{33}/\sigma \rightarrow 0$. For thin specimens, there is a region in which the plane-stress HRR field dominates, before at even larger distances (the crack length being the characteristic length to compare with), the far field is established, which depends on the specimen geometry.

Near the crack tip, a zone, which is represented by the hatched area in Fig. 12, must be exempt from the description in terms of singular fields. In this zone, the singular fields are perturbed by blunting or by damage formation as described in the previous section. If this process zone is smaller than the range of validity of the plane-strain HRR field, damage accumulation occurs under plane-strain conditions. To evaluate the condition for plane strain in the process zone, one notes that the range of the plane-strain field scales with specimen thickness. Based on an analysis by Yang and Freund [*74*], Riedel [*8*] estimates the range of the plane-strain field to be about 1/8 of the specimen thickness. The process-zone size in creep crack growth was found to be of the order of the crack growth increment, Δa. Hence, plane-strain conditions near the crack tip are maintained as long as

$$\Delta a \leq B/8 \tag{34}$$

Incidentally, the same argument when applied to rate-independent fracture mechanics, where the process zone size is about $5\delta_t$, leads to

$$B \geq 20\, J/\sigma_y \tag{35}$$

This rationalizes the ASTM Test Method for J_{Ic}, a Measure of Fracture Toughness (E 813-81) for plane-strain fracture testing, where a factor of 25 is suggested instead of 20.

The Distribution of K$_I$, J, *or* C* *Along the Crack Front*

In specimens of finite thickness, the stress varies as one moves along the crack front. The linear elastic case is considered first. As a consequence of the ρ^{-s}-stress singularity at the vertex, the coefficient of the $1/\sqrt{r}$-crack-tip singularity, which is called the local stress intensity factor, goes to zero at the specimen surface as

$$K_I(z) \propto z^{1/2-s} \tag{36}$$

In linear elasticity is $J = K_I^2(1 - \nu^2)/E \propto z^{1-2s}$. This behavior is shown by the dashed line in Fig. 13 for the case of $\nu = 0.3$ in which $s = 0.452$. For comparison, the solid line represents the result of a three-dimensional finite-element calculation [*71*], which apparently exhibits the correct behavior near the vertex. Further, from the same source, the distribution of J in elastic-plastic material is shown when the load, P, exceeds the plastic limit load, P_L, by 25%. As this example shows, the distribution of J is less homogeneous in the fully plastic

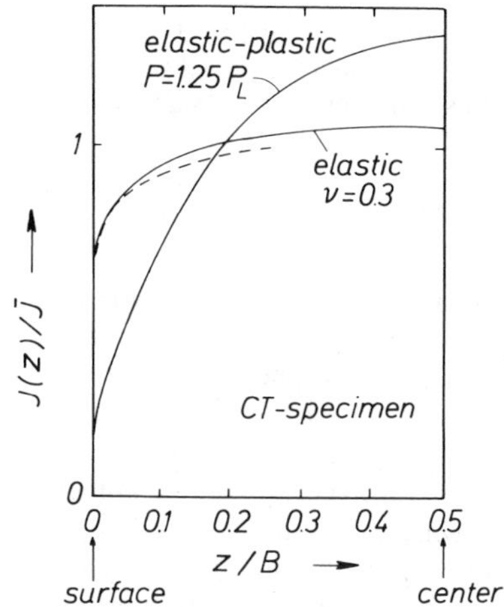

FIG. 13—*Local* J-*integral along crack front, normalized by the average value* $\bar{J}$. *Solid line: finite-element results* [71]; *dashed line: vertex solution for* $\nu = 0.3$.

case than in the elastic case. As a consequence of the viscous-elastic analogy, a similar picture could be drawn for the variation of C^* in nonlinear viscous materials.

In Fig. 13, the J-integral is normalized by its thickness average, $\bar{J}$. This average value is what is measured at the load points of a specimen [*8,75*].

Crack Tunneling and Its Suppression by Side Grooves

Since J (or C^*) is higher in the center of a specimen with a straight crack front than at the specimen surface, the crack will tend to develop a thumbnail-shaped front, which tunnels ahead in the center of the specimen.

For linear elastic material with $\nu = 0.3$, Bazant and Estenssoro [*76*] find that the vertex singularity depends on the angle under which the crack front hits the free surface. When the crack-front direction deviates by 11 deg from the normal on the surface, the vertex singularity becomes an inverse-square-root singularity. This is a prerequisite for a constant K_I-value along the crack front. In fact, three-dimensional finite-element calculations [*77*] show that K_I is constant if the crack tunnels such that an angle of 10 to 15 deg is established. Further, fatigue cracks growing under linear elastic conditions actually develop a surface angle in this range.

In nonlinear material (plastic or viscous), crack tunneling is much more pronounced than in linear elastic material, which is plausible from the behavior of J shown in Fig. 13. Correspondingly, three-dimensional finite-element calculations [*78*] show that only after a considerable amount of tunneling, a state of constant J (or C^*) along the crack front is reached in nonlinear material.

Since tunneling of crack fronts and the formation of shear lips, which often accompanies tunneling, renders crack growth tests difficult to evaluate, it has been suggested to use side-grooved specimens (see Fig. 11). It is obvious that side grooves will counteract the tendency to form shear lips. Further, at the vertex where the crack front intersects the (sharp) root of the side groove, a stress singularity develops, supposedly again of the general form $\sigma \propto \rho^{-s}$, with s depending on the included side-groove angle 2β, on the "tunneling angle" and on the stress exponent n. No solutions for s are known to the author, but finite-element calculations [*71,78,79*] and experimental experience suggest that an included side-groove angle of 2β = 45 deg is favorable for a uniform distribution of J (or C^*) along a straight crack front. The depth of the side grooves is usually chosen between 20% and 25% of the half specimen thickness. Then straight crack fronts are obtained both in pre-fatigue cracking and in creep crack growth.

Future Perspectives

Many aspects of creep crack growth are now well understood, but there are also unresolved problems. Corrosive effects, for example, are very important in superalloys, but a complete understanding is lacking of how the atmosphere interacts with the crack-tip region. Another problem is the large crack-opening angle which occurs in ductile materials at growing cracks. It is not yet clear to what extent a large crack-opening angle bounds the applicability of C^*. This problem has been partially resolved after the completion of the manuscript (see Ref *80*). Three-dimensional problems, although understood in principle, remain at least a computational problem. Creep crack growth under small-scale yielding conditions exhibits surprising features, which are a consequence of the properties of the HR field.

In relation to practical applications, several problems arise which need attention. Very often, the loading history contains loading cycles so that creep-fatigue crack growth must be considered. Another problem is that cracks are often formed preferentially in weldments or in the heat-affected zone. A fracture mechanics theory for a creeping inhomogeneous material, like a weld, has not been developed and little testing has been done. In highly creep-resistant materials, small cracks may gain an importance, similarly to the situation in fatigue failure.

Finally, it remains a task of practical significance to determine in a given situation if failure will occur by C^*-controlled crack growth or by homogeneous cavitation of the cross section.

References

[*1*] Van Leeuwen, H. P., *Engineering Fracture Mechanics*, Vol. 9, 1977, pp. 951–974.
[*2*] Ellison, E. G. and Harper, M. P., *Journal of Strain Analysis*, Vol. 13, 1978, pp. 35–51.
[*3*] Pilkington, R., *Metal Science*, Vol. 13, 1979, pp. 555–564.
[*4*] Speidel, M. O. in *Advances in Fracture Research, Proceedings*, 5th International Conference on Fracture, Vol. 6, D. Francois et al., Eds., Pergamon Press, Oxford, 1981, pp. 2685–2704.
[*5*] Sadananda, K. and Shahinian, P., *Engineering Fracture Mechanics*, Vol. 15, 1981, pp. 327–342.
[*6*] Riedel, H. in *Subcritical Crack Growth Due to Fatigue, Stress Corrosion and Creep*, L. H. Larsson, Ed., Elsevier Applied Science Publishers, Barking, U.K., 1984, pp. 449–467.
[*7*] Riedel, H. in *Flow and Fracture at Elevated Temperatures*, R. Raj, Ed., American Society for Metals, Metals Park, OH, 1985, pp. 149–177.
[*8*] Riedel, H., *Fracture at High Temperatures*, Springer-Verlag, Berlin, 1987.
[*9*] Rice, J. R., *Transactions*, American Society of Mechanical Engineers, *Journal of Applied Mechanics*, Vol. 35, 1968, pp. 379–386.
[*10*] Rice, J. R. in *Fracture: An Advanced Treatise*, H. Liebowitz, Ed., Vol. 2, Academic Press, New York, 1968, pp. 191–311.
[*11*] Hutchinson, J. W., *Journal of the Mechanics and Physics of Solids*, Vol. 16, No. 1, 1968, pp. 13–31.
[*12*] Rice, J. R. and Rosengren, G. F., *Journal of the Mechanics and Physics of Solids*, Vol. 16, No. 1, 1968, pp. 1–12.
[*13*] Rice, J. R. and Johnson, M. A. in *Inelastic Behavior of Solids*, M. F. Kanninen et al., Eds., McGraw-Hill, New York, 1970, pp. 641–671.
[*14*] Hoff, N. J. *Quarterly of Applied Mathematics*, Vol. 12, 1954, pp. 49–55.
[*15*] Landes, J. D. and Begley, J. A. in *Mechanics of Crack Growth, ASTM STP 590*, American Society for Testing and Materials, Philadelphia, 1976, pp. 128–148.
[*16*] Ohji, K., Ogura, K., and Kubo, S. *Transactions*, Japanese Society of Mechanical Engineers, Vol. 42, 1976, pp. 350–358.

[17] Nikbin, K. M., Webster, G. A., and Turner, C. E. in *Cracks and Fracture (9th Conference), ASTM STP 601*, American Society for Testing and Materials, Philadelphia, 1976, pp. 47–62.
[18] Siverns, M. J. and Price, A. T., *Nature*, Vol. 228, 1970, pp. 760–761.
[19] Harrison, C. B. and Sandor, G. N., *Engineering Fracture Mechanics*, Vol. 3, 1971, pp. 403–420.
[20] Robson, K., "Creep Crack Growth in Two Carbon Steels at 450°C," presented at the International Conference on Creep Resistance in Steels, Verein Deutscher Eisenhüttenleute, Düsseldorf, 1972.
[21] Thornton, D. V., "Creep Crack Growth Characteristics of Certain Ferritic Steels," presented at the International Conference on Creep Resistance in Steels, Verein Deutscher Eisenhüttenleute, Düsseldorf, 1972.
[22] Siverns, M. J. and Price, A. T., *International Journal of Fracture*, Vol. 9, 1973, pp. 199–207.
[23] van Leeuwen, H. P., "The Application of Fracture Mechanics to the Growth of Creep Cracks," Advisory Group for Aerospace Research and Development, Report No. 705, 1983.
[24] Riedel, H. and Rice, J. R. in *Fracture Mechanics: Twelfth Conference, ASTM STP 700*, P. C. Paris, Ed., American Society for Testing and Materials, Philadelphia, 1980, pp. 112–130.
[25] Koterazawa, R. and Mori, T., *Transactions*, American Society of Mechanical Engineers, *Journal of Engineering Materials and Technology*, Vol. 99, 1977, pp. 298–305.
[26] Taira, S., Ohtani, R., and Kitamura, T., *Transactions*, American Society of Mechanical Engineers, *Journal of Engineering Materials and Technology*, Vol. 101, 1979, pp. 154–161.
[27] Saxena, A. in *Fracture Mechanics: Twelfth Conference, ASTM STP 700*, P. C. Paris, Ed., American Society for Testing and Materials, Philadelphia, 1980, pp. 131–151.
[28] Ohji, K., Ogura, K., Kubo, S., and Katada, J. in *Proceedings of the International Conference on Engineering Aspects of Creep*, Vol. 2, The Institution of Mechanical Engineers, London, 1980, pp. 9–16.
[29] Smith, D. J. and Webster, G. A., *Journal of Strain Analysis*, Vol. 16, 1981, pp. 137–143.
[30] Maas, E. and Pineau, A., *Engineering Fracture Mechanics*, Vol. 22, No. 2, 1985, pp. 307–325.
[31] Riedel, H. and Wagner, W. in *Advances in Fracture Research, Proceedings*, 6th International Conference on Fracture, Vol. 3, S. R. Valluri et al., Eds., Pergamon Press, Oxford, 1985, pp. 2199–2206.
[32] Ilyushin, A. A., *Prikladnaia Matematika i Mekhanika*, Vol. 10, 1946, p. 347.
[33] Kumar, V., German, M. D., and Shih, C. F., "An Engineering Approach for Elastic-Plastic Fracture Analysis," Report NP-1931 on Project 1237-1 for Electric Power Research Institute, Palo Alto, CA, 1981.
[34] Shih, C. F., "Tables of the Hutchinson-Rice-Rosengren Singular Field Quantities," Brown University Report MRL E-147, Providence, RI, 1983.
[35] Riedel, H. in *Creep in Structures*, A. R. S. Ponter and D. R. Hayhurst, Eds., Springer-Verlag, Berlin, 1981, pp. 504–519.
[36] Bassani, J. L. in *Creep and Fracture of Engineering Materials and Structures*, B. Wilshire and D. R. J. Owen, Eds., Pineridge Press, Swansea, U.K., 1981, pp. 329–344.
[37] Bassani, J. L. and Vitek, V. in *Proceedings*, 9th U.S. National Congress on Theoretical and Applied Mechanics, L. B. Freund and C. F. Shih, Eds., 1982, pp. 127–133.
[38] Wilkinson, D. S. and Vitek, V., *Acta Metallurgica*, Vol. 30, 1982, pp. 1723–1732.
[39] Hui, C. Y. and Banthia, V., *International Journal of Fracture*, Vol. 25, 1984, pp. 53–67.
[40] Hull, D. and Rimmer, D. E., *Philosophical Magazine*, Vol. 4, 1959, pp. 673–687.
[41] Dyson, B. F., *Metal Science*, Vol. 10, 1976, pp. 349–353.
[42] Dyson, B. F., *Canadian Metallurgical Quarterly*, Vol. 18, 1979, pp. 31–38.
[43] Detampel, V., "An Investigation of Creep Crack Growth in Creep-Resistant Pipe Steels" (in German), Ph.D. thesis, RWTH Aachen, Germany, 1987.
[44] Webster, G. A., Smith, D. J., and Nikbin, K. M. in *Proceedings*, International Conference on Creep, Tokyo, Japanese Society of Mechanical Engineers, 1986, pp. 303–308.
[45] Vitek, V., *International Journal of Fracture*, Vol. 13, 1977, pp. 39–50.
[46] Riedel, H., *Materials Science and Engineering*, Vol. 30, 1977, pp. 187–196.
[47] Ewing, D. J. F., *International Journal of Fracture*, Vol. 14, 1978, pp. 101–112.
[48] Riedel, H., *Zeitschrift für Metallkunde*, Vol. 69, No. 12, 1978, pp. 755–760.
[49] Ohji, K., Ogura, K., and Kubo, S., *Journal of the Society of Materials Science*, Japan, Vol. 29, No. 320, 1980, pp. 465–471.
[50] Ehlers, R. and Riedel, H. in *Advances in Fracture Research, Proceedings*, 5th International Conference on Fracture, Vol. 2, D. Francois et al., Eds., Pergamon Press, Oxford, 1981, pp. 691–698.
[51] Saxena, A. and Landes, J. D. in *Advances in Fracture Research, Proceedings*, 6th International Congress on Fracture, Vol. 6, P. Rama Rao et al., Eds., Pergamon Press, Oxford, 1985, pp. 3977–3988.

[52] Saxena, A. in *Fracture Mechanics: Seventeenth Volume, ASTM STP 905*, J. H. Underwood et al., Eds., American Society for Testing and Materials, Philadelphia, 1986, pp. 185–201.
[53] Hui, C. Y. and Riedel, H., *International Journal of Fracture*, Vol. 17, No. 4, 1981, pp. 409–425.
[54] Hart, E. W., *International Journal of Solids and Structures*, Vol. 16, 1980, pp. 807–823.
[55] Riedel, H. and Wagner, W. in *Advances in Fracture Research, Proceedings*, 5th International Conference on Fractures, Vol. 2, D. Francois et al., Eds., Pergamon Press, Oxford, 1981, pp. 683–688.
[56] Hui, C. Y. in *Elastic-Plastic Fracture: Second Symposium, Volume 1—Inelastic Crack Analysis, ASTM STP 803*, C. F. Shih and J. P. Gudas, Eds., American Society for Testing and Materials, Philadelphia, 1983, pp. 573–593.
[57] Hawk, D. E. and Bassani, J. L., *Journal of the Mechanics and Physics of Solids*, Vol. 34, 1986, pp. 191–212.
[58] Hui, C. Y. and Wu, K.-C., *International Journal of Fracture*, Vol. 31, 1986, pp. 3–16.
[59] Wu, F.-H., Bassani, J. L., and Vitek, V., *Journal of the Mechanics and Physics of Solids*, Vol. 34, 1986, pp. 455–475.
[60] Riedel, H., *Journal of the Mechanics and Physics of Solids*, Vol. 29, 1981, pp. 35–49.
[61] Kubo, S. in *Elastic-Plastic Fracture: Second Symposium, Volume I—Inelastic Crack Analysis, ASTM STP 803*, C. F. Shih and J. P. Gudas, Eds., American Society for Testing and Materials, Philadelphia, 1983, pp. 594–614.
[62] Riedel, H. and Detampel, V., *International Journal of Fracture*, Vol. 33, 1987, pp. 239–262.
[63] Merkle, J. G. and Corten, H. T., *Transactions*, American Society of Mechanical Engineers, *Journal of Pressure Vessel Technology*, Series J, Vol. 98, No. 4, Nov. 1984, pp. 286–292.
[64] McMeeking, R. M., *Journal of the Mechanics and Physics of Solids*, Vol. 25, 1977, pp. 357–381.
[65] McMeeking, R. M. and Parks, D. in *Elastic-Plastic Fracture, ASTM STP 668*, J. D. Landes, J. A. Begley, and G. A. Clarke, Eds., American Society for Testing and Materials, Philadelphia, 1979, pp. 175–194.
[66] Hutchinson, J. W. and Paris, P. C. in *Elastic-Plastic Fracture, ASTM STP 668*, J. D. Landes, J. A. Begley, and G. A. Clarke, Eds., American Society for Testing and Materials, Philadelphia, 1979, pp. 37–64.
[67] Riedel, H. in *Fundamentals of Deformation and Fracture*, B. A. Bilby, K. J. Miller, and J. R. Willis, Eds., Cambridge University Press, Cambridge, 1985, pp. 293–309.
[68] Hayhurst, D. R. and Leckie, F. A. in *Mechanical Behaviour of Materials*, IV, Vol. 2, J. Carlsson and N. G. Ohlsson, Eds., Pergamon Press, Oxford, U.K., 1983, pp. 1195–1212.
[69] Hayhurst, D. R., Brown, P. R., and Morrison, C. J., *Philosophical Transactions of the Royal Society London A*, Vol. 311, 1984, pp. 131–168.
[70] Shih, C. F., deLorenzi, H. G., and Andrews, W. R., *International Journal of Fracture*, Vol. 13, 1977, pp. 544–548.
[71] deLorenzi, H. G. and Shih, C. F., *International Journal of Fracture*, Vol. 21, 1983, pp. 195–220.
[72] Benthem, J. P., *International Journal of Solids and Structures*, Vol. 13, 1977, pp. 479–492.
[73] Benthem, J. P., *International Journal of Solids and Structures*, Vol. 16, 1980, pp. 119–130.
[74] Yang, W. and Freund, L. B., *International Journal of Solids and Structures*, Vol. 21, 1985, pp. 977–994.
[75] Budiansky, B. and Rice, J. R., *Journal of Applied Mechanics*, Vol. 40, 1973, pp. 201–203.
[76] Bazant, Z. P. and Estenssoro, L. F., *International Journal of Solids and Structures*, Vol. 15, 1979, pp. 405–426.
[77] Smith, A. P., Towers, O. L., and Smith, I. J. in *Proceedings of the Third International Conference on Numerical Methods in Fracture Mechanics*, A. R. Luxmoore and D. R. J. Owen, Eds., Pineridge Press, Swansea, U.K., 1984, pp. 205–217.
[78] Kikuchi, M. and Miyamoto, H. in *Proceedings of the Third International Conference on Numerical Methods in Fracture Mechanics*, A. R. Luxmoore and D. R. J. Owen, Eds., Pineridge Press, Swansea, U.K., 1984, pp. 249–260.
[79] Kikuchi, M. and Miyamoto, H., *International Journal of Pressure Vessels and Piping*, Vol. 16, 1984, pp. 1–16.
[80] Riedel, H., *Proceedings*, 7th International Conference on Fracture, Vol. 2, K. Salama et al., Eds., Pergamon Press, Oxford, 1989, pp. 1495–1523.

Ali S. Argon[1]

The Role of Heterogeneities in Fracture

REFERENCE: Argon, A. S., **"The Role of Heterogeneities in Fracture,"** *Fracture Mechanics: Perspectives and Directions (Twentieth Symposium), ASTM STP 1020,* R. P. Wei and R. P. Gangloff, Eds., American Society for Testing and Materials, Philadelphia, 1989, pp. 127–148.

ABSTRACT: Heterogeneities play a dual role in the mechanical behavior of materials. They alter the stiffness and plastic resistance of a solid, but also provide interfaces across which stresses are concentrated due to the accumulating deformation-induced misfit that can hasten cavitation and lead to eventual fracture. This paper discusses conditions of cavitation around hard particles in low-temperature ductile fracture of metals and alloys and in intergranular fracture during creep. Also, analogous, but complementary, roles of compliant particles and heterogeneity phases in the craze plasticity of glassy polymers are also surveyed.

KEY WORDS: heterogeneities, hard particles, composite particles, ductile fracture, creep fracture, crazing, toughened polymers

Heterogeneities play an important dual role in the mechanical behavior of materials. In intrinsically ductile inorganic solids, on the one hand, they modulate the usual inelastic behavior, often in a profound manner, by enforcing local redundant deformations necessary to maintain compatibility between them and the matrix. On the other hand, they introduce interfaces into the solid across which deformation-induced stresses can be concentrated to result in cavitation, either there, or inside the heterogeneities themselves, by cracking them. In intrinsically brittle solids, a similar dual role is also observed, where the heterogeneities can either serve to initiate supercritical cracks that can propagate in the cleavage mode with very low energy consumption or, alternatively, reinforce or even toughen the solid by crack-tip shielding or diversion. In glassy amorphous polymers, where crazing can introduce an element of dilatational transformation and a potential for toughness, heterogeneities play equally important roles in either the initiation of the crazes, governing their growth, and eventually their fracture.

Here, we consider the heterogeneities from a micromechanical, continuum point of view, which usually is proper for purposes of understanding fracture. We point up, however, instances where this point of view becomes stretched, and where more mechanistic non-continuum views are also necessary.

Fracture in Homogeneous Ductile Solids

An initially crack-free homogeneous ductile solid fails in tension or extensional deformation only by first undergoing deformation localization, where no internal cavitation or cracking is involved, but where, by a continuous process of deformation, the part fails under decreasing load by rupturing at a pin-point or chisel edge. The mechanics of such macroscopic

[1] Quentin Berg Professor, Department of Mechanical Engineering, Massachusetts Institute of Technology, Cambridge, MA 02139.

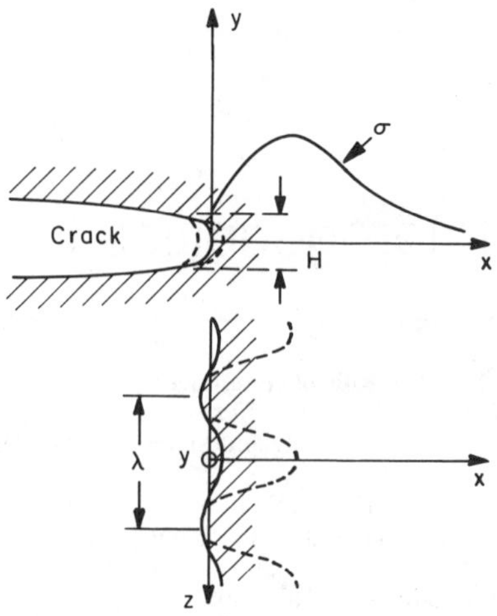

FIG. 1—*Convolution of a concave, blunted crack-tip interface in a ductile nonhardening, but strain rate sensitive solid, such as amorphous metal.*

deformation localization in bars and sheets has been developed in detail [*1*] and is not a subject of our discussion here. Conventional fracture with less than complete reduction of area is impossible in such materials if deformation resistance remains low or if large negative pressures cannot be built up by severe flow inhomogeneities. This is because the homogeneous cavitation strength of a pure solid without defects is equal to its ideal cohesive strength and cannot be reached under the above conditions.

A homogeneous nonhardening, but nonlinearly viscous ductile solid containing a sharp crack, however, can undergo a ductile fracture process with minimal apparent reduction of area through the development of a crack tip convolution process that is a variant of a ubiquitous meniscus instability in fluids studied first by Taylor [*2*] and later by many others [*3–5*].

As is well known, a sharp crack in a nonhardening ductile solid will blunt under Mode I deformation. This not only wipes out any initial elastic stress concentration, but through a logarithmic spiral plastic strain field produces a concentration of negative pressure in the interior, at a distance somewhat larger than the radius of curvature of the blunted crack [*6,7*]. The resulting positive gradient of negative pressure in the plastic flow field ahead of the blunted crack makes the concave interface of the crack unstable to lateral perturbations of all wave lengths, where the only stabilizing element is the surface tension. As shown in Fig. 1, considering the concave crack front deformation fields as a variant of a meniscus in a nonlinear fluid, advancing under a suction gradient in a flow channel of height H, a simple perturbation analysis gives the most rapidly growing principal lateral wave length λ as

$$\lambda = 2\sqrt{\sqrt{3}\,\frac{H\chi}{\eta_0}\left(\frac{2}{\dot{\epsilon}_e}\right)^n} \tag{1}$$

where χ is the surface tension and η_0, $\dot{\epsilon}_e$, and n are related to the equivalent tensile deformation resistance Y_e by the nonlinear viscosity law of

$$Y_e = \frac{3}{2}\,\eta_0\dot{\epsilon}_e^{\,n} \tag{2}$$

Considering further the perturbed fluid meniscus as built-into the logarithmic spiral strain field, with the dimension H as the critical crack opening displacement (CCOD), where the advancing parallel ridges of the perturbed field at the crack interface are just undergoing rupture, one obtains for the CCOD and the fracture toughness, K_{Ic}

$$\text{CCOD} = C(n)\frac{\chi}{Y_e} \therefore K_{Ic} = \sqrt{\frac{\text{CCOD}\ \pi Y_e E}{\alpha}} \tag{3a,b}$$

where

$C(n)$ = a numerical constant, a function only of the power exponent n,
Y_e = plastic resistance in tension,
E = Young's modulus, and
α = a numerical constant of about 3.

In a typical case of a rather nonlinear (plastic, nonhardening) solid, such as an amorphous metal alloy with $n = 0.1$, $E = 140$ GPa, $\chi = 2$ J/m^2, $Y_e = 2.4$ GPa, $C(n) = 2 \times 10^3$ [5], the CCOD, the principal ridge wave length λ, and the fracture toughness, K_{Ic}, are found to be

$$\text{CCOD} = 0.48\ \mu\text{m} \tag{4a}$$

$$\lambda = 0.2\ \mu\text{m},\ K_{Ic} = 10.56\ \text{MPA}\ \sqrt{\text{m}} \tag{4b}$$

These computed values agree very well with the measured K_{Ic} of 9.50 MPa $\sqrt{\text{m}}$ and principal wave length of ridge spacing λ of 0.36 μm on the fracture surface of this amorphous metal alloy. A typical fracture surface of this type in a strip of a palladium-silicon ($Pd_{80}Si_{20}$) amorphous alloy is shown in Fig. 2 [8].

Amorphous metallic alloys are particularly well suited for this type of fracture because they are quite homogeneous and exhibit no strain hardening, but exhibit regular nonlinear viscous behavior.

Ductile Fracture in Metals

That metals fracture prematurely by a process of internal necking on a very fine scale was postulated first by Joseph Henry of electromagnetism fame, in 1855 [9], who advised that wire drawing and rolling were preferable to stretching in tension. In more recent times, Tipper [10] identified the source of ductile fracture of plastically deformed metals to be second-phase particles. As we now know, these offer ready interfaces that can decohere or fracture internally under the deformation-induced internal stresses arising from the differences in deformation resistance between matrix and particle.

This observation has been of great technological importance because it has clarified the dual role of hard particles, first, as raisers of plastic resistance, but also as providers of sites for early cavitation that can hasten fracture. Through careful control of particle size, shape, and interface quality, alloys can be made both quite deformation resistant and tough. Here, we will concern ourselves only with the aspects of cavitation from particle interfaces, since the suppression of this has a profound effect on increasing overall toughness for a given volume fraction of particles. There have been other suggestions as alternatives to second-phase particles as sources of ductile cavitation in metals, such as vacancy coalescence or cavitation under high stress in heavily dislocated regions of metal [11]. Although such

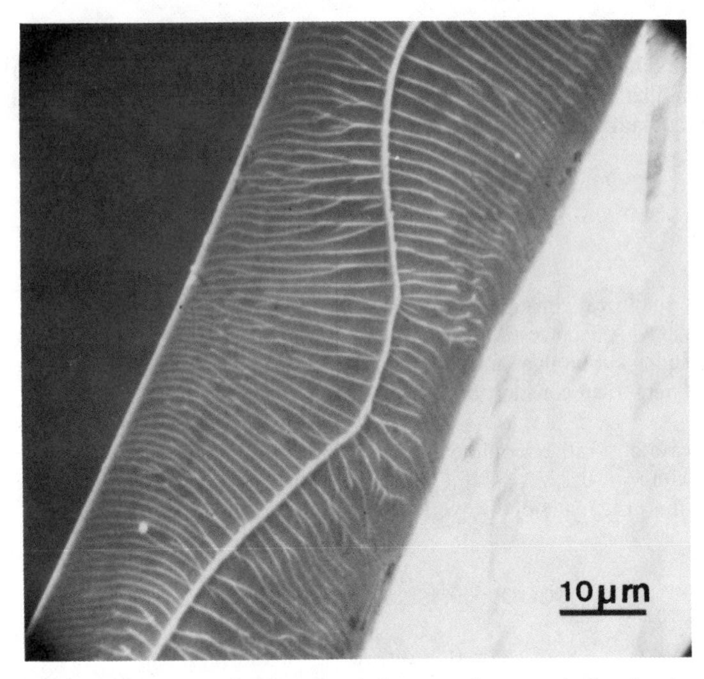

FIG. 2—*Fracture surface of a thin sample of a $Pd_{80}Si_{20}$ amorphous metal alloy showing parallel ridge form of ruptures, which is the signature of the interface convolution mechanism of separation* (*from Ref* 8, *courtesy of Elsevier*).

homogeneous cavitation cannot be entirely ruled out in the presence of very high triaxial stresses, the evidence for this has not been favorable, since most such reports have been on very thin films studied with transmission electron microscopy (TEM), where stress triaxiality must be quite small. On the other hand, careful experiments on the origins of ductile cavities in high-purity aluminum [*12*] and detailed examination of alternative claims [*13*] have established quite conclusively that second-phase particles are indeed the predominant origins of ductile fracture cavities. Figure 3 shows a typical example of cavitation from iron carbide (Fe_3C) particles in spheroidized medium carbon steel.

Since most ductile metals have yield stresses that are only a small fraction of the cohesive strength of particle interfaces, plastic deformation is necessary to concentrate stress on interfaces or inside particles through the buildup of deformation induced misfit between particles and matrix arising from their very different plastic resistances. This problem has been considered by a number of investigators [*14–20*]. Since cavity initiating particles are generally irregular in shape, an exact analysis of the problem of generating high stresses between a nondeformable particle and the deforming matrix is not possible.

By considering the particles as rigid and the surrounding ductile matrix as either a non-hardening or a linearly hardening continuum, Argon and coworkers [*16,17*] have obtained approximate solutions both for the maximum interface stress for equiaxed particles and for internal concentration of stress inside elongated particles in a pure shear field. As a first-order approximation, they considered any triaxial component of stress in the distant field to be directly additive to the stress produced by the shear field to result in the following

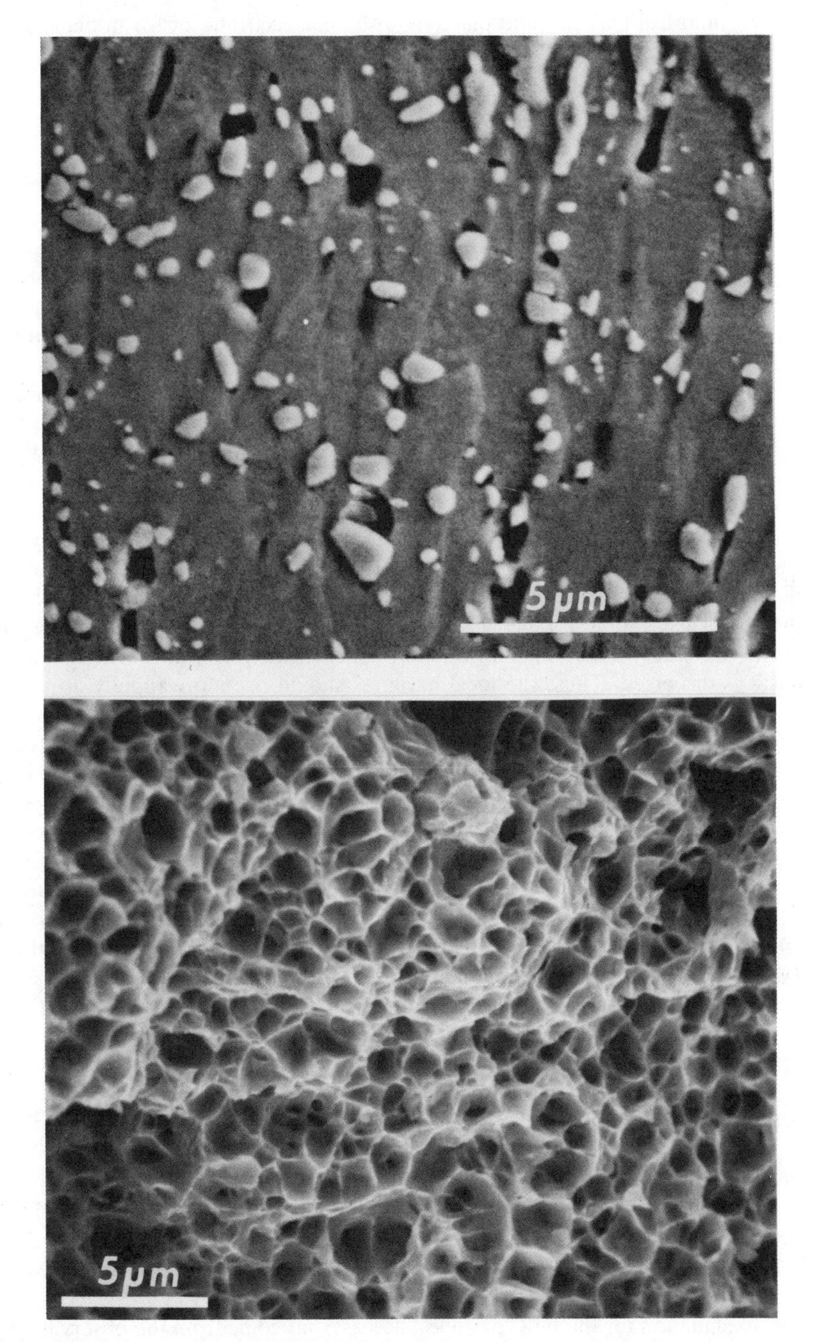

FIG. 3—*Ductile fracture in spheroidized 1045 Steel at room temperature:* (top) *cavitation around Fe_3C particles in central region of a necked bar, revealed on an etched longitudinal section;* (bottom) *resulting dimples on the fracture surface.*

relations for cylindrical particles and platelets with an aspect ratio of ℓ/h in plane strain:

$$\sigma_{rr} = \sigma + Y_e(\bar{\epsilon}^p) \qquad \text{(for cylinders)} \tag{5}$$

$$\sigma_{zz} = \sigma + \frac{\sqrt{3}}{2} Y_e(\bar{\epsilon}^p) \left[1 + \frac{2\pi(\ell/h)}{3[\ell n(\ell/h) - C]} \right] \qquad \text{(for platelets)} \tag{6}$$

In Eqs 5 and 6, σ_{rr} is the plastic misfit induced maximum radial stress acting across the interface of the cylindrical particle under the maximum stress at the center of the elongated particle considered aligned parallel to the maximum principal tension; $Y_e(\bar{\epsilon}^p)$ is the tensile deviatoric deformation resistance in the region of the particle, had the particle not been present; σ is the mean normal stress (negative pressure) in the field; and C is a numerical constant, which is typically 0.15. Both Eqs 5 and 6 are rather approximate and are based on the condition that although the plastic resistance of the matrix surrounding the particle will increase with strain hardening, an eventual steady state must exist, in which the local interface stress relates to the resistance of the surrounding field by a constant multiple leading to a constant stress concentration factor.

The analysis leading to Eqs 5 and 6 and its assumptions have been criticized by several investigators. Goods and Brown [*18*] have maintained that a continuum analysis does not apply for small particles, because experiments and simple kinematical analyses show that the rate of excess dislocation storage with local strain is inversely proportional to the particle radius in the secondary zones of plastic strain concentration around particles where the plastic misfit is dispersed. Thus, the local hardness should rise more rapidly around small particles, and these should come apart earlier—an expectation quite contrary to any experimental observations of cavitation, where particle size effects have been reported.

Thomson and Hancock [*19*], on the other hand, have concluded from their finite-element analyses of elastic and rigid particles in three dimensions that for even quite large plastic strains, a steady state appears not to become established, and that the stress concentration factor relating the distant plastic resistance to the peak interface traction steadily rises. This is not supported by a new numerical analysis of Wilner [*20*] using a nonlinear variational method, which demonstrates that a steady state with a well-defined stress concentration factor for the interface stress does indeed become established and depends only on the stress triaxiality and the exponent of the nonlinear power law. This analysis also shows, as might be suspected, that the negative pressure is not merely additive to the stresses of the shear field, but affects the local inhomogeneous flow field itself. Using Wilner's [*20*] analysis to reevaluate the results of the extensive cavity nucleation experiments of Argon and Im [*21*], we find that the interface strengths for equiaxed particles of: Fe_3C in spheroidized 1045 steel; copper-chromium particles in copper, and titanium carbide (TiC) particles in unaged (VM300) maraging steel are, respectively, 1.89 ± 0.19 GPa, 1.10 ± 0.11 GPa, and 1.89 ± 0.05 GPa. These interface strengths are 8.9×10^{-3} times the Young's modulus for the two cases of the steel matrix, and $8.3 \times 10^{-3}E$ for the copper matrix, respectively. They are also within a few percent of the earlier answers obtained by the approximate analysis based on Eq 5. These normalized strengths are somewhat lower than what might be expected from the strength of incoherent interfaces in pure crystalline matter.

A partial explanation for the Fe_3C particles, at least, has come from the observations that a large fraction of the decohered particles were on grain boundaries and that exposure to hydrogen enhanced the fraction that causes cavitation [*22*]. Thus, a certain reduction in the intrinsic interface strength due to segregation of embrittling impurities of sulfur, phosphorus, arsenic, and antimony, commonly found in all steels as trace impurities, can be suspected [*23*].

The criteria for cavity nucleation due to the buildup of deformation-induced interface stress are independent of the scale of the particles in both continuum analyses and analyses involving punched dislocations dispersing the plastic strain misfit over secondary plastic zones [*16*]. Two particle size effects, nevertheless, are relevant to the decohesion problem. First, very small particles cannot nucleate a cavity, because the elastic strain energy stored around them is insufficient for producing the surface energy of the cavity, or, alternatively and more simply stated, their elastic-strain-induced displacement misfit across the interface is less than the critical decohesion displacement of the interface. On the other side of the size scale, some reports [*24*] show that large particles are more prone to cavity nucleation. As stated already above, this has been explained [*16*] as a particle interaction effect, where in a randomly dispersed field of particles with broad size distribution, a significant probability exists for two larger-than-average particles to be neighbors at a spacing less than the average spacing, resulting in an elevation of interface stress due to interpenetration of secondary plastic zones around the particles. The apparent particle size effect in spheroidized medium carbon steel is of this type and should be recognized more as a local volume fraction variation effect [*16,21*].

Some authors [*14,18,25,26*] in both the mechanics and materials literature have advocated strain criteria for cavity nucleation either on fundamental grounds or for operational ease as damage criteria in deformation analyses. Clearly, when the local stress criterion for decohesion or particle cracking is known, it can be as readily recast into a critical local strain criterion. However, this would be of limited utility for problems with different initial conditions of plastic resistance and deformation paths differing from simple radial loading.

Once cavities have been nucleated, their growth to final fracture or to a point where rapid shear localization sets in has been analyzed by numerous investigators [*25,27–31*]. All of these analyses show that cavity growth is dramatically enhanced by the presence of a sustainable negative pressure component. They all indicate, however, that quasi-homogeneous growth of cavities seriously overpredicts the total strain to fracture and that actual ductile fracture, even in an initially crack free part, must involve early cavitational localization leading to the formation and propagation of a macro-crack. Although some severe localization of cavitation into narrow planar zones of so-called void sheets [*32,33*] have been widely observed, the conditions of their formation is less well understood. The contributing causes to such behavior must include deformation localization due to strain softening arising from microstructure degradation, local clustering of particles with weak interfaces due to earlier deformation processing, and finally, the destabilizing effect on the deformation by the cavity nucleation itself [*34*].

Intergranular Cavitation in Creeping Metals

At elevated temperatures where the plastic deformation in metals becomes much more temperature dependent because of the presence of recovery effects, the ductile cavitation process discussed above initiated from particles dispersed in the matrix is still present in relatively unaltered form when the strain rates are high enough, so that grain boundaries can transmit shear tractions fully [*21*]. For small strain rates, however, when grain boundary sliding displacements become comparable to overall grain elongations to produce significant relaxation of grain boundary shear tractions, the cavitation process becomes localized almost entirely to grain boundaries, giving rise to grain boundary fracture, as shown in Fig. 4.

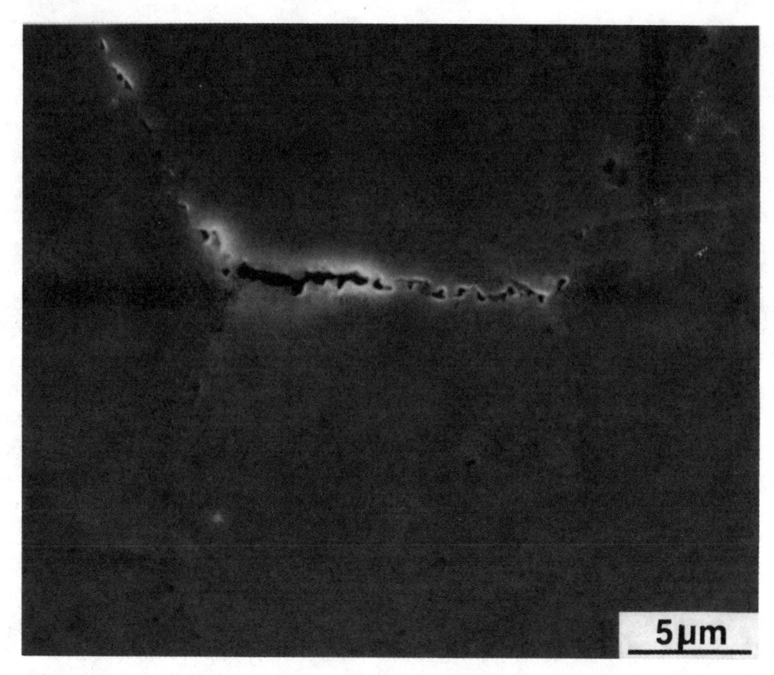

FIG. 4—*Development of an intergranular wedge crack from the accelerated growth and linkage of cavities near a triple junction of grains in a creeping 304 stainless steel (from Ref* 35, *courtesy of Pergamon Press).*

In intergranular cavitation, all aspects that were important in low-temperature ductile fracture are also present in modified form. Here too, the total cavitation process must be divided into two successive steps of nucleation and growth. The nucleation of intergranular cavities is no less difficult than nucleation of cavities from particle interfaces. It too requires important stress concentration, which in this case comes from grain boundary particles [*35*]. Detailed analysis shows that the rate of cavity nucleation $\dot{\rho}$ on stressed interfaces of grain boundary particles can be given as

$$\dot{\rho} = \dot{\rho}_0 \exp(-\Delta G^*/kT) \tag{7a}$$

$$\dot{\rho}_0 = (2\pi r^* D_b \delta/\Omega^{4/3}) \exp(\sigma_n \Omega/kT) \tag{7b}$$

$$\Delta G^* = \frac{4}{3}\frac{\psi^2 \chi_s^{\ 3}}{\sigma_n^{\ 2}} F_v(\psi) \tag{7c}$$

$$r^* = 2\psi^2 \chi_s/\sigma_n \tag{7d}$$

where r^* and ΔG^* are the radius and free energy of the critical size cavity at its saddle point configuration, nucleated under a local normal stress σ_n at a temperature T, and where $D_b\delta$ is the linear diffusive conductance of a grain boundary, χ_s the surface free energy, ψ the dihedral half-angle at the apex of the cavity, Ω the atomic volume, and $F_v(\psi)$ is a constant affected by the shape of the cavity governed by whether it is along a flat boundary, a curved boundary, or at a corner of grains, it is unity for a flat boundary, but can become as small as 0.5 for some reentrant grain corners.

The normal stress σ_n, that is needed to cavitate the solid along the interfaces of grain boundary particles, has been obtained by Argon and coworkers [36,37] to be

$$\frac{\sigma_n}{\tau_{gb}} = k = C(m)(\ell/p)\frac{(1-\lambda)}{(1-\sqrt{2}\Lambda/p)^{1-\lambda}}\left(\frac{p}{\sqrt{2}\Lambda}\right)^{\lambda} \quad (8)$$

where

τ_{gb} = average grain boundary shear traction, supported exclusively by particles of size p at an average spacing of ℓ,

λ = range exponent of the concentrated interface stresses at the apex of the particle [36], and

$C(m)$ = a numerical constant.

The values of λ and $C(m)$, which are dependent on the dihedral angle of the particle, and the creep exponent m, have been evaluated by Lau et al. [36] for many relevant geometries and power exponents. The length Λ is a critical diffusion length, over which diffusional smoothing effectively levels down stress singularities. It is given as

$$\Lambda = (\delta D_b \Omega \sigma_e / kT\dot{\epsilon}_e)^{1/3} \quad (9)$$

where σ_e and $\dot{\epsilon}_e$ are the equivalent stress and equivalent minimum creep rates, respectively. Evaluation of Eqs 8 and 9 for typical cases of creeping alloys with $m = 5$ have given stress concentration factors of 10 to 20 [37].

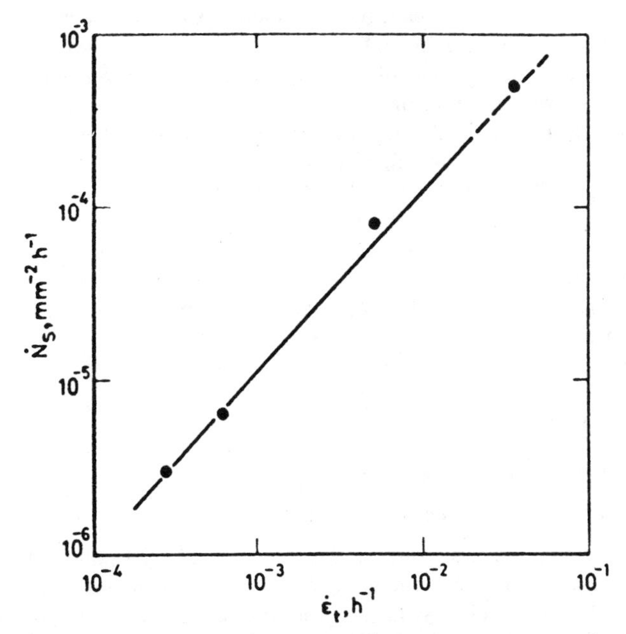

FIG. 5—*Linear dependence of intergranular cavity nucleation rate on creep strain rate in iron (from Ref 37, courtesy of Pineridge Press).*

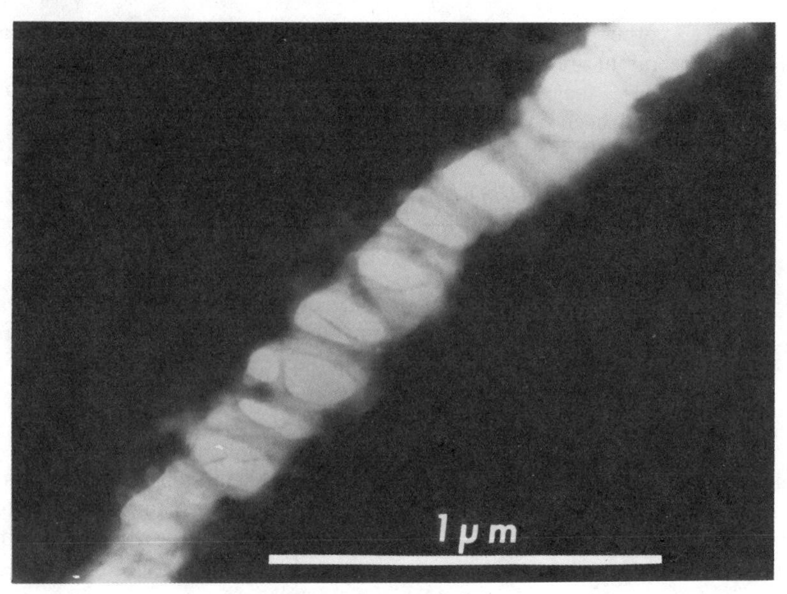

FIG. 6—*Microstructure of a craze in PS at room temperature. Note drawn polymer tufts bridging the craze flanks together (from Ref* 49, *courtesy of Taylor & Francis, Ltd.).*

While the details of the cavity nucleation indicated in Eqs 7 to 9 are quantitatively satisfactory, they are difficult to verify experimentally, since the critical cavity sizes r^* at nucleation are too small and quite outside the range of most routine methods of direct detection. When they can be detected by convenient techniques, such as scanning electron microscopy, their rate of appearance reflects more their rate of growth through a threshold of detectability, rather than a rate of nucleation. In fact, the rate of cavity appearance correlates quite well with the rate of background creep as shown in Fig. 5. This has been explained by the facts that nucleation occurs only during stochastic transients of grain boundary sliding and that the overall rate of appearance of such transients must be proportional by compatibility to the background creep rate [*37*].

Once nucleated, the rate of growth of cavities on stressed grain boundaries is governed by a combination of conditions of diffusional transport of point defects along grain boundaries, and power law creep. These growth mechanisms have been developed in detail by a number of investigators [*38–45*] and are much better understood than the nucleation process. They have been reviewed recently by Argon [*37*] and will not be discussed here further. The predictions of these mechanisms agree quite well with experimental observations [*44,45*].

Dilatational Plasticity in Polymers

Crazing and Fracture in Homogeneous Glassy Polymers

The ductile cavitation processes discussed above for crystalline solids have interesting and important counterparts in both homogeneous and heterogeneous polymers. Perhaps the most intriguing of these parallels is in the process of crazing exhibited by glassy polymers, which in itself plays a dual role as both the ingredient of toughness and the element of fracture. It has been known for a long time that glassy polymers, such as polymethyl methacrylate (PMMA) and polystyrene (PS), tend to be brittle at room temperature and that this brittleness has something to do with the phenomenon of crazing, which they exhibit. It had been observed in the course of studies of this effect that the application of a tensile stress well below general yield results in the formation of crazes on surfaces of stressed parts, which look like cracks but are actually load-bearing. Eventually one of these crazes initiates rapid and brittle fracture [*46*]. In a pioneering study of electron microscopy, Kambour and Holik [*47*] demonstrated that crazes indeed differed from cracks, in that tiny tufts of drawn polymer only a few tens of nanometers in diameter connect the two flanks of the craze and make it a traction transmitting planar discontinuity, behaving in many aspects as a lenticular dilatational transformation. Figure 6 shows an electron micrograph of the interior

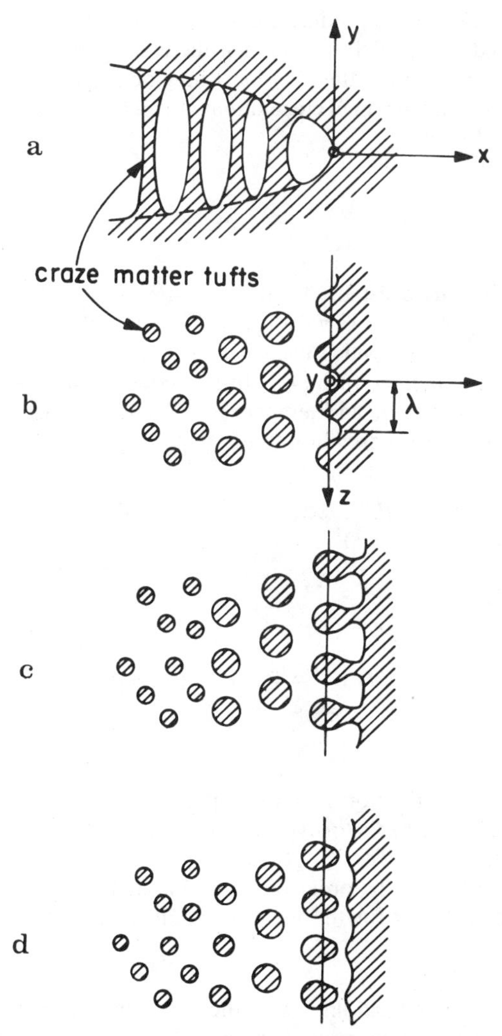

FIG. 7—*Sketch of a sequence in the advance of a craze front generating by the interface convolution process and repeated pinch-off of ridges the craze matter tufts shown in Fig. 6:* (a) *side view;* (b,c,d) *top view* (*from Ref* 55, *courtesy of Springer*).

structure of a typical craze in PS. The revelation of this fibrillar internal structure of a craze put in perspective the earlier observations by Bucknall and Smith [*48*] that the profuse formation of crazes from composite rubbery particles under stress, makes high-impact polystyrene (HIPS) tough.

While the mechanism of nucleation of crazes from molecular level heterogeneities remains obscure, the mechanism of growth of crazes under stress has been explained successfully as being a variant of the interface convolution process discussed above for the quasi-continuous planar ductile cavitation of amorphous metals [*49*]. As in the case of an amorphous metal, a sharp crack in a glassy polymer will blunt by local crack-tip plastic flow and produce the very same crack-tip plastic flow field with the same meniscus instability. Since glassy polymers, however, do not rupture upon large strain drawing, but undergo strong orientation hardening due to the presence of molecular entanglements, the ridges between the convolutions merely stretch out and produce the tufty craze matter by repeatedly pinching off the ridges, as sketched in Fig. 7. In this model, the craze growth rate is simply the rate at which material elements entering the very local flow field process volume at the craze tip stretch out to establish a fully-formed craze tuft at an opening displacement δ at the concave convoluted interface, that is

$$\frac{da}{dt} = \alpha\delta\dot{\epsilon} \tag{10}$$

where α is a proportionality constant, δ is the craze-tip opening displacement, scaling principally with the ratio χ/Y_e of the surface tension χ to the plastic resistance Y_e in tension, and $\dot{\epsilon}$ is the plastic strain rate, which itself is an exponential function of the stress acting across the craze and the temperature [*49*]. Introduction of the specific dependences of δ and $\dot{\epsilon}$ on distant tensile stress σ_∞ and temperature T, and considering the character of the fibrillar craze structure, gives a growth rate of

$$\frac{da}{dt} = \frac{D}{\lambda_n\sigma_\infty/\hat{Y}} \exp\left\{-\frac{B}{RT}\left[1 - \left(\frac{\lambda_n\sigma_\infty}{\hat{Y}}\right)^{5/6}\right]\right\} \tag{11}$$

with

$$D = \frac{\beta\chi\dot{\epsilon}_0}{\hat{Y}} \therefore \hat{Y} = 0.190\ \mu \tag{12a,b}$$

where

λ_n = effective extension ratio of the craze tufts,
B = activation energy scale factor typically of the order of 100 kJ/mole,
μ = shear modulus,
β = numerical constant of the order of 1.7×10^3, and
$\dot{\epsilon}_0$ = pre-exponential factor for the inelastic strain rate being typically of the order of $10^{13}s^{-1}$ [*49*].

Figure 8 compares the theoretical lines with the experimental data for the growth rate of crazes in PS and PMMA at either room temperature or −20°C (253 K). The temperature-dependent shifts are well within the ranges predicted by the theoretical model.

Clearly, the stable filling of a glassy polymer with crazes can mean a very handsome level

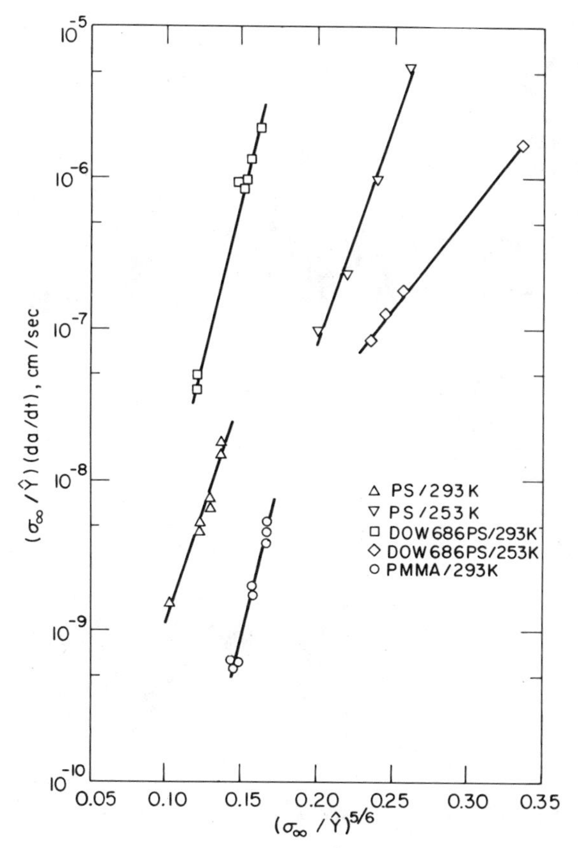

FIG. 8—*Stress dependence of craze growth rate in two PS samples at two different temperatures and PMMA at room temperature comparing experimental results with theory (from Ref* 49, *courtesy of Taylor & Francis, Ltd.).*

of dilatational plasticity—albeit at a relatively low stress level in comparison with what would be required to produce distortional plasticity. Unfortunately, crazes can turn into cracks when the delicate craze matter ruptures locally. The rupture of craze matter is nearly always initiated from inorganic dust particles entrapped in the polymer, when such particles are "acquired" by the growing craze in either its forward growth or in its widening toward both sides as it draws out additional solid polymer from the flanks. The discontinuity in the craze matter that is formed when the inorganic particle is torn loose from it can constitute a supercritical flaw in the craze. An event of this type, which is about to fracture a craze, is shown in Fig. 9. Normally, in unmodified single-phase glassy polymers, the stress that initiates and grows crazes is in excess of 30 MPa, for which many entrapped inorganic dust particles become supercritical flaws when they are incorporated into the craze. Several possibilities exist to counteract early fracture of craze matter from dust particles. Elimination of the offending particles by processing of polymers in super-clean environments is only marginally effective [*51*]. Much more effective ways include neutralization of the effect of the particles. One such way is by regularly modulating the polymer with very finely dispersed rubbery copolymer phases, which can be made to cavitate in a three-dimensional cellular manner and which effectively isolates the offending particles. Another and equally effective

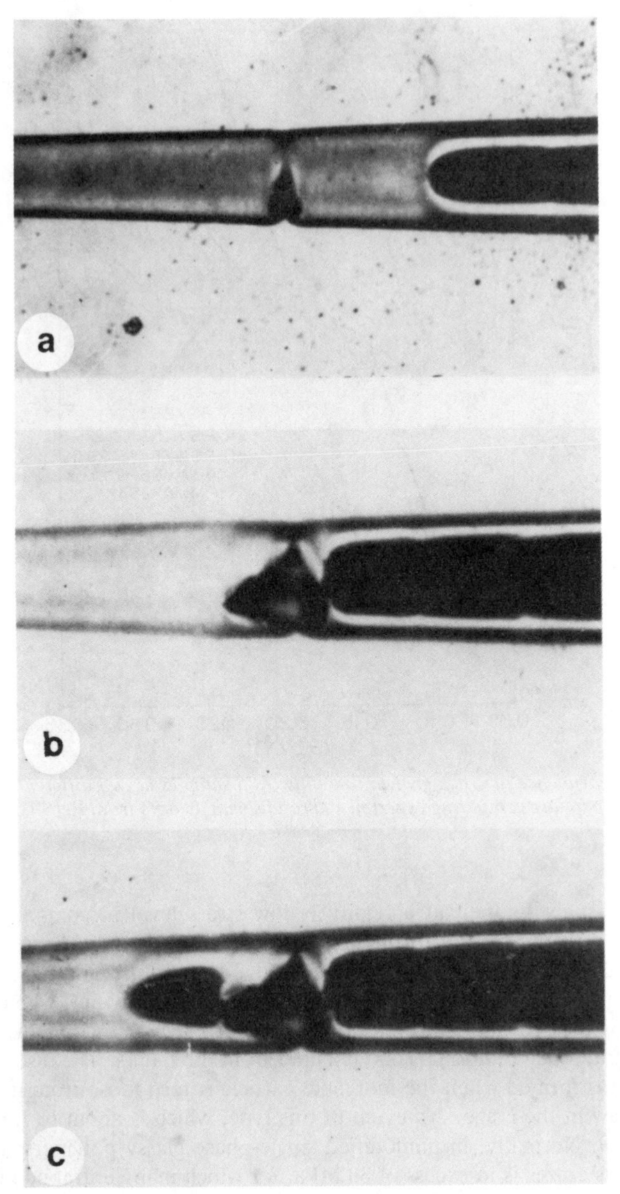

FIG. 9—*Sequence in the fracture of a craze:* (a) *an advance cavitation event from a dust particle ahead of the main crack in a craze;* (b,c) *fracture spreading from the advance cavitation site joins the main crack* (*from Ref* 50, *courtesy of the Royal Society of London*).

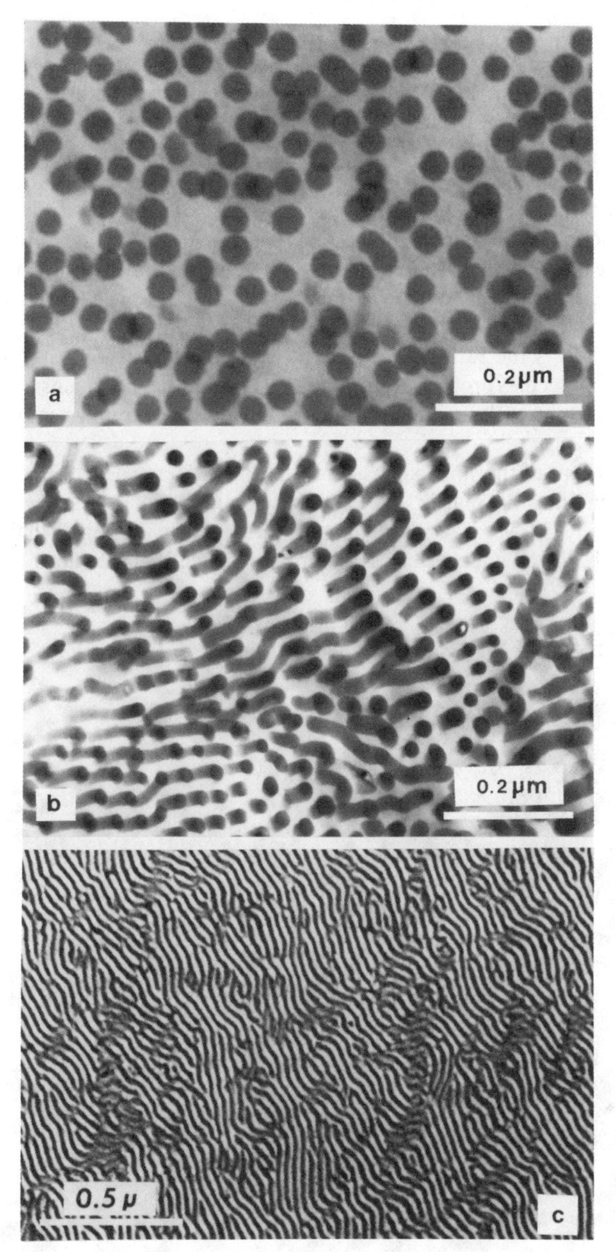

FIG. 10—*Di-block copolymers of PS and PB:* (a) *spheres of PB in PS;* (b) *rods of PB in PS;* (c) *parallel lamallae of PB and PS.*

way is to lower the stress for initiation and growth of crazes substantially, so that the usual set of entrapped particles become sub-critical flaws.

Crazing in Block Copolymers

Block copolymers are composed of chain molecules, with two or more separate but closely controlled lengths of linearly attached segments of different chemical nature, such as PS and polybutadiene (PB). When such block copolymers are solidified from the melt or are prepared from solution, they undergo a very regular type of phase separation, in which the entrapped minority phase is dispersed as spheres, rods, or platelets in the majority phase as the volume fraction of the minority phase increases. We will discuss here briefly only the unusual mechanical properties of diblock copolymers of PS and PB, in which the minority phase is PB rubber. Figure 10 shows examples of equilibrium phase forms of such a set of block copolymers, starting with rubbery domains in the form of spheres and progressing to rods and to platelets as the ratio of the length of the PB block segment to that of the PS block segment on molecules progressively increases. The rubbery domains in the micrographs appear black as a result of diffusing into them relatively electron-opaque osmium tetroxide molecules.

Since the coefficient of thermal expansion of the PB rubber is nearly 3.5 times larger than that of the surrounding PS matrix, the rubbery domains are in a state of high negative pressure of typically 30 MPa at room temperature, which is roughly half of their ideal cavitation strength. When a tensile stress is applied to such PS base block copolymers, they

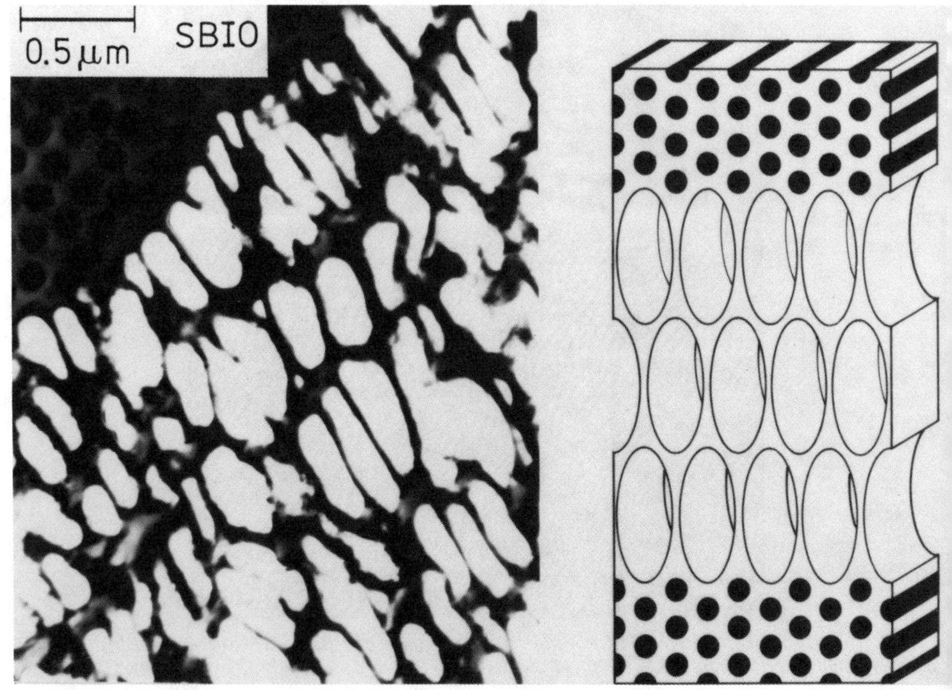

FIG. 11—*Craze in a PB/PS di-block copolymer with rod morphology showing drawn cell walls of PS after cavitation of the PB rods. The associated drawing illustrates the process of cavitation (from Ref 55, courtesy of Springer).*

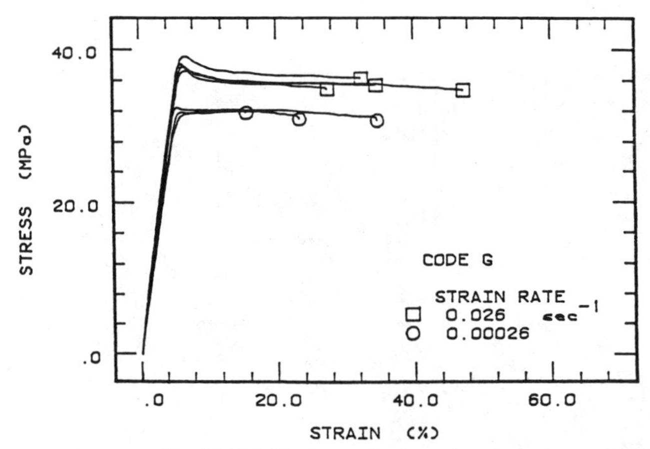

FIG. 12—*Stress-strain curve of a PB/PS block copolymer with spherical morphology PB domains, having a volume fraction of 0.11.*

craze quite readily by regularly cavitating the very small rubbery domains, which first turns the material into a cellular spongy solid, followed by the plastic drawing of the PS cell walls, as shown in the drawing and associated micrograph of a block copolymer of rod morphology (Fig. 11). The remarkably regular morphology of the cellular craze matter of only tens of nanometer dimensions and possibly the associated better wetting of dust particle interfaces by the ubiquitous PB rubber appear to render effectively the particles neutral at the same high craze flow stress levels of homogeneous PS, while providing a capacity for a large strain to fracture. The stress-strain curve in Fig. 12 shows an example of a block copolymer with spherical PB domains, while Fig. 13 shows a typical craze in such a polymer composed of a planar zone of cavitated spherical rubbery domains and plastically stretched cell walls of PS. The kinetics of growth of such crazes by particle cavitation in a plane have been discussed in detail elsewhere [*52,53*] and will not be repeated here.

Crazes Initiated by Compliant Particles

The second way of imparting toughness to glassy polymers is by introducing into them composite compliant particles by blending block copolymer phases into the glassy polymer matrix. Four prominent examples are shown in Fig. 14. Of these, the one appearing in Fig. 14*a* is the block copolymer with the designation of KRO-1, having a topologically continuous stiff PS phase; that in Fig. 14*b* is a conventional high-impact PS (HIPS) particle with a topologically continuous, but minority phase of PB rubber; that in Fig. 14*c* is a particle composed of a set of concentric spherical shells (CSS) of PS and PB, while those shown in Fig. 14*d* are small particles of homogeneous (PB) rubber. The effect on the stress-strain curve of the first three of these is shown in Fig. 15, in comparison with the rather brittle behavior of pure PS. The effect of the small rubber particles is rather similar to the effect of the HIPS particles, even though their volume fraction is only about 1% of the latter. Clearly, as the particles become increasingly more potent in initiating crazes and as the craze flow stress drops, the strain to fracture increases, as does the overall toughness. The principal cause for the improving behavior is the increasing compliance of the particles (or their lowered modulus). This trend in the average particle moduli is shown in Table 1, as calculated

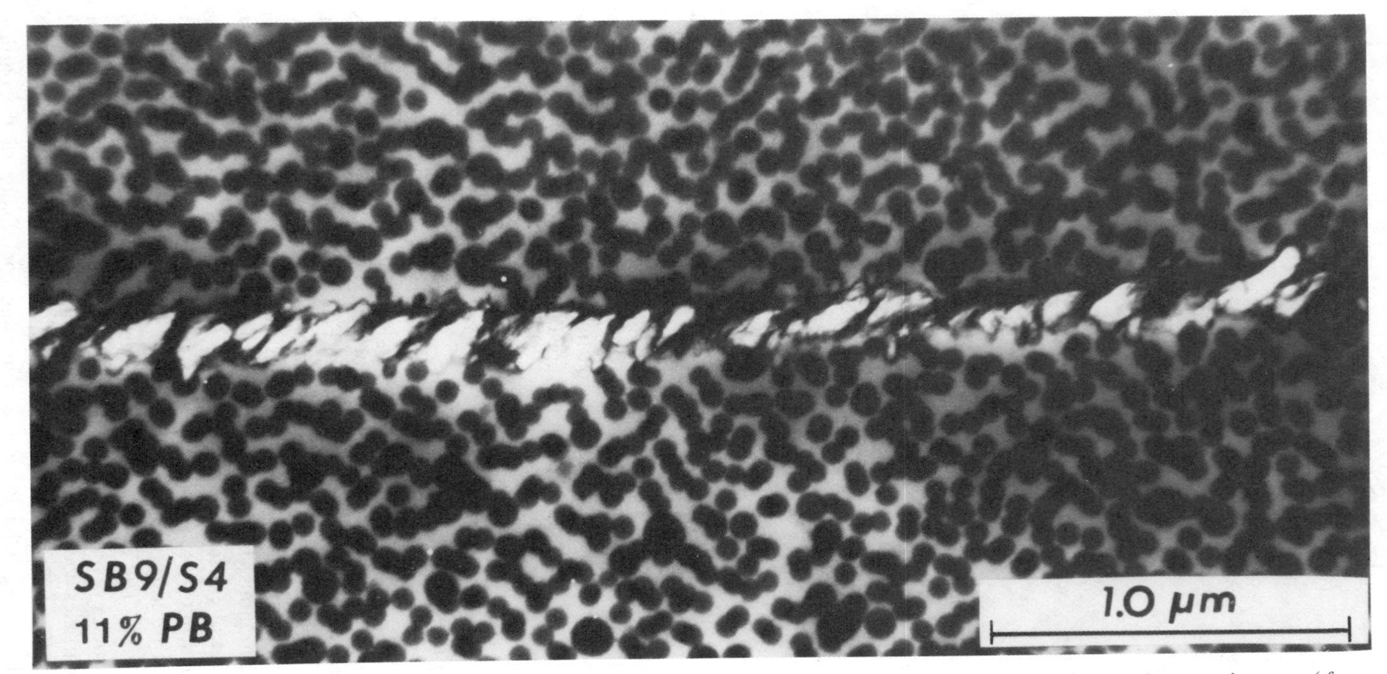

FIG. 13—*Craze in a PB/PS block copolymer with spherical morphology PB domains, cavitating in a planar zone making up the craze (from Ref* 53, *courtesy of Taylor & Francis, Ltd.).*

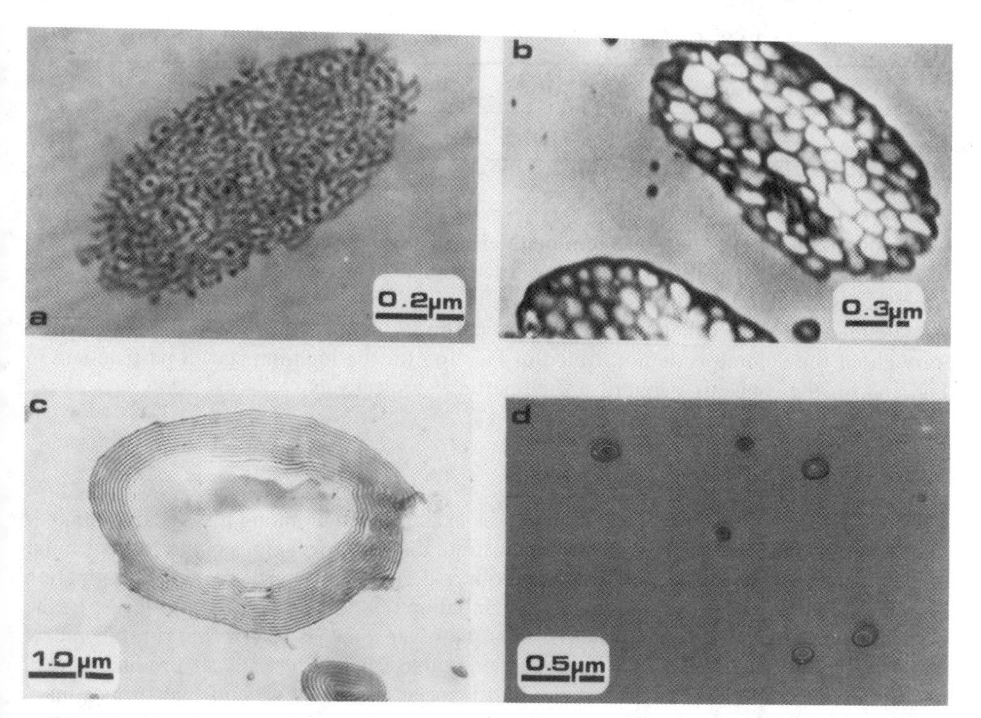

FIG. 14—*Morphologies of four different block copolymer particles blended into a PS matrix:* (a) *KRO-1 resin particle of tortous PB rods in PS:* (b) *high-impact PS particles;* (c) *particles of alternating concentric spherical shells of PB and PS;* (d) *pure PB particles* (*from Ref* 55, *courtesy of Pergamon Press*).

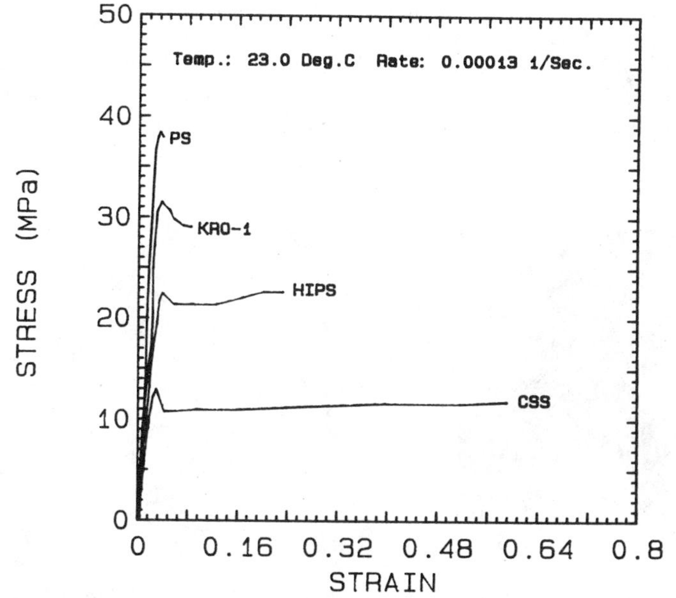

FIG. 15—*Stress-strain curves of PS containing three of the four particles* (a,b,c) *shown in Fig. 14* (*from Ref* 55, *courtesy of Pergamon Press*).

TABLE 1—*Effective Young's moduli of particles, GPa* [*55*].

KRO-1 Particle	HIPS	CSS	PB
2.400	0.333	0.138	0.047

by various analytical or finite-element method (FEM) models; it parallels the reduced flow stresses in Fig. 15. The elastic properties of such composite particles and their craze initiating effectiveness have been discussed in detail elsewhere [*54,55*] and are too involved to be repeated here. The fact that these compliant particles are very effective in initiating crazes throughout the volume is demonstrated in Fig. 16*a* for the high-impact PS particle and in Fig. 16*b* for the concentric spherical shell particle, respectively.

Discussion

The several topics that have been chosen as examples from among a very large number of other possible cases should serve to demonstrate that heterogeneities play a rather similar and fundamental dual role in governing both a desirable property of high deformation resistance and an undesirable side effect of initiating fracture from their interfaces. In the case of the polymers with composite particles, however, they also show that careful control of particle properties can result in controlled stable cavitation that can produce a very handsome level of dilatational plasticity and associated toughness, and that proper management of dilatation by cavitation can be quite advantageous.

Acknowledgment

The researches that led to the content of this communication have been supported by the Center for Materials Science and Engineering at M.I.T. throughout the past decade and a half, under several NSF/MRL grants, of which DMR 84-18718 has been the most recent. Additional support has also come from the Monsanto Chemical Company, the Mobil Chemical Company, and the Allied Corporation, in the form of doctoral and post-doctoral fellowships.

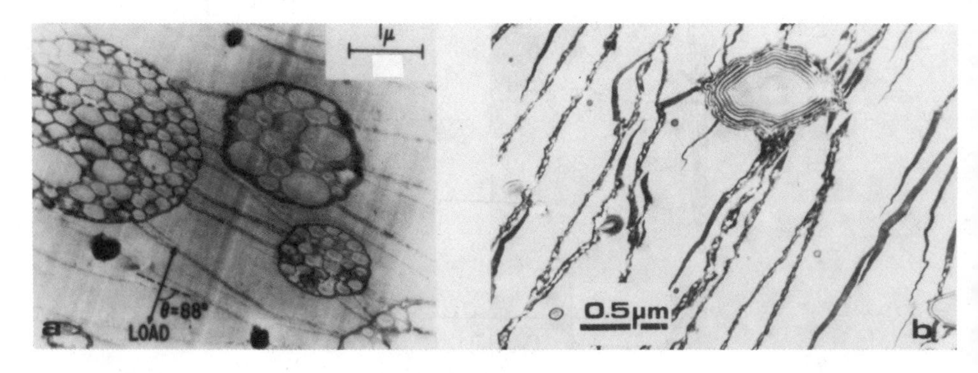

FIG. 16—*Crazes emanating from:* (a) *HIPS particles;* (b) *CSS particles* (*from Ref* 55, *courtesy of Pergamon Press*).

References

[1] *Mechanics of Sheet Metal Forming,* D. P. Koistinen and N. M. Wang, Eds., Plenum Press, New York, 1978.
[2] Taylor, G. I., *Proceedings,* Royal Society, London, Vol. A201, 1950, pp. 192–196.
[3] Pearson, J. R. A., *Journal of Fluid Mechanics,* Vol. 7, 1960, pp. 481–500.
[4] Pitts, E. and Greiller, J., *Journal of Fluid Mechanics,* Vol. 11, 1961, pp. 33–51.
[5] Argon, A. S. and Salama, M. M., *Materials Science and Engineering,* Vol. 23, 1976, pp. 219–230.
[6] McClintock, F. A. in *Physics of Strength and Plasticity,* A. S. Argon, Ed., Massachusetts Institute of Technology Press, Cambridge, MA, 1969, pp. 307–326.
[7] Rice, J. R. and Johnson, M. A. in *Inelastic Behavior of Solids,* M. F. Kanninen et al., Eds., McGraw-Hill, New York, 1970, pp. 641–670.
[8] Megusar, J., Argon, A. S., and Grant, N. J., *Materials Science and Engineering,* Vol. 38. 1979, pp. 63–72.
[9] Backofen, W. A., *Metallurgical Transactions,* Vol. 4, 1973, pp. 2679–2699.
[10] Tipper, C. F., *Metallurgia,* Vol. 39, 1948, pp. 133–137.
[11] Lyles, R. L. and Wilsdorf, H. G. F., *Acta Metallurgica,* Vol. 23, 1975, pp. 269–277.
[12] Miller, D. R. and Besag, F. M. C. in *Proceedings,* 1st International Conference on Fracture (Sendai), T. Yokobori et al., Eds., Japan Society for Strength and Fracture of Materials, Sendai, Vol. 2, 1966, pp. 711–722.
[13] Thomson, A. W. and Weihrauch, P. F., *Scripta Metallurgica,* Vol. 10, 1976, pp. 205–210.
[14] McClintock, F. A. in *Ductility,* American Society for Metals, Metals Park, OH, 1968, pp. 255–277.
[15] Ashby, M. F., *Philosophical Magazine,* Vol. 14, 1966, pp. 1157–1178.
[16] Argon, A. S., Im, J., and Safoglu, R., *Metallurgical Transactions,* Vol. 6A, 1975, pp. 825–837.
[17] Argon, A. S., *Journal of Engineering Materials and Technology,* Vol. 98, 1976, pp. 60–68.
[18] Goods, S. H. and Brown, L. M., *Acta Metallurgica,* Vol. 27, 1979, pp. 1–15.
[19] Thomson, R. D. and Hancock, J. W., *International Journal of Fracture,* Vol. 24, 1984, pp. 209–228.
[20] Wilner, B., *Journal of the Mechanics and Physics of Solids,* Vol. 36, 1988, pp. 141–165.
[21] Argon, A. S. and Im, J., *Metallurgical Transactions,* Vol. 6A, 1975, pp. 839–851.
[22] Cialone, H. and Asaro, R. J., *Metallurgical Transactions,* Vol. 10A, 1979, pp. 367–375.
[23] Fishmeister, H. F., Navarra, E., and Easterling, K. E., *Metal Science,* Vol. 6, 1972, pp. 211–215.
[24] Palmer, I. G. and Smith G. C. in *Oxide Dispersion Strengthening,* Gordon and Breach, New York, 1968, pp. 253–290.
[25] Needleman, A. and Rice, J. R. in *Mechanics of Sheet Metal Forming,* D. P. Koistinen and N. M. Wang, Eds., Plenum Press, New York, 1978, pp. 237–267.
[26] Fisher, J. R. and Gurland, J., *Metal Science,* Vol. 15, 1981, pp. 193–202.
[27] McClintock, F. A., *Journal of Applied Mechanics,* Vol. 35, 1968, pp. 363–372.
[28] Rice, J. R. and Tracy, D. M., *Journal of the Mechanics and Physics of Solids,* Vol. 17, 1969, pp. 201–217.
[29] Budiansky, B., Hutchinson, J. W., and Slutsky, S. in *Mechanics of Solids,* H. G. Hopkins and M. J. Sewell, Eds., Pergamon Press, Oxford, 1981, pp. 13–45.
[30] McMeeking, R. M. in *Chemistry and Physics of Fracture,* R. M. Latanision and R. H. Jones, Eds., Martinus Nijhoff, Boston, 1987, pp. 91–128.
[31] Tvergaard, V. and Needleman, A., *Acta Metallurgica,* Vol. 32, 1984, pp. 157–169.
[32] Rogers, H. C. in *Ductility,* American Society for Metals, Metals Park, OH, 1968, pp. 31–61.
[33] Cox, T. B. and Low, J. R., *Metallurgical Transactions,* Vol. 5A, 1974, pp. 1457–1470.
[34] Hutchinson, J. W. in *Constitutive Relations and their Physical Basis,* S. I. Anderson et al., Eds., Riso National Laboratory, Roskilde, Denmark, 1987, pp. 95–105.
[35] Chen, I. W. and Argon, A. S., *Acta Metallurgica,* Vol. 29, 1981, pp. 1321–1333.
[36] Lau, C. W., Argon, A. S., and McClintock, F. A. in *Elastic-Plastic Fracture: Second Symposium, Volume I—Inelastic Crack Analysis, ASTM STP 803,* C. F. Shih and J. P. Gudas, Eds., American Society for Testing and Materials, Philadelphia, 1983, pp. I-551–I-572.
[37] Argon, A. S. in *Recent Advances in Creep and Fracture of Engineering Materials and Structures,* B. Wilshire and R. J. Owen, Eds., Pineridge Press, Swansea, U.K., 1982, pp. 1–52.
[38] Chuang, T. J., Kagawa, K. I., Rice, J. R., and Sills, L. B., *Acta Metallurgica,* Vol. 27, 1979, pp. 265–284.
[39] Needleman, A. and Rice, J. R., *Acta Metallurgica,* Vol. 28, 1980, pp. 1315–1332.
[40] Rice, J. R., *Acta Metallurgica,* Vol. 29, 1981, pp. 675–681.
[41] Beere, W. and Speight, M. V., *Metal Science,* Vol. 12, 1978, pp. 172–176.

[*42*] Chen, I. W. and Argon, A. S., *Acta Metallurgica,* Vol. 29, 1981, pp. 1759–1768.
[*43*] Pharr, G. M. and Nix, W. D., *Acta Metallurgica,* Vol. 27, 1979, pp. 1615–1631.
[*44*] Goods, S. H. amd Nix, W. D., *Acta Metallurgica,* Vol. 26, 1978, pp. 739–752.
[*45*] Stanzl, S. E., Argon, A. S., and Tschegg, E. K., *Acta Metallurgica,* Vol. 31, 1983, pp. 833–843.
[*46*] Bartenev, G. M. and Zuyer, Y. S., *Strength and Failure of Visco-elastic Materials,* Pergamon Press, Oxford, 1968, pp. 96–104.
[*47*] Kambour, R. P. and Holik, A. S., *Journal of Polymer Science,* Part A-2, Vol. 7, 1969, pp. 1393–1403.
[*48*] Bucknall, C. B. and Smith, R. R., *Polymer,* Vol. 6, 1965, pp. 437–446.
[*49*] Argon, A. S. and Salama, M. M., *Philosophical Magazine,* Vol. 36, 1977, pp. 1217–1234.
[*50*] Doyle, M. J., Maranci, A., Orowan, E., and Stork, S. T., *Proceedings,* Royal Society (London), Vol. A329, 1972, pp. 137–151.
[*51*] Yang, A. C. M., Kramer, E. J., Kuo, C. C., and Phoenix, L., *Macromolecules,* Vol. 18, 1986, pp. 2010–2019.
[*52*] Argon, A. S., Cohen, R. E., Gebizlioglu, O. S., and Schwier, C. E. in *Advances in Polymer Science,* H. H. Kausch, Ed., Springer, Berlin, Vol. 52/53, 1983, pp. 276–334.
[*53*] Schwier, C. E., Argon, A. S., and Cohen, R. E., *Philosophical Magazine,* Vol. 52, 1985, pp. 581–603.
[*54*] Boyce, M. E., Argon, A. S., and Parks, D. M., *Polymer,* Vol. 28, 1987, pp. 1680–1694.
[*55*] Argon, A. S., Cohen, R. E., and Gebizlioglu, O. S. in *Mechanical Behavior of Materials—V,* M. G. Yau et al., Eds., Pergamon Press, Oxford, Vol. 1, 1987, pp. 3–14.

Fatigue Crack Propagation

Keisuke Tanaka[1]

Mechanics and Micromechanics of Fatigue Crack Propagation

REFERENCE: Tanaka, K., **"Mechanics and Micromechanics of Fatigue Crack Propagation,"** *Fracture Mechanics: Perspectives and Directions (Twentieth Symposium), ASTM STP 1020,* R. P. Wei and R. P. Gangloff, Eds., American Society for Testing and Materials, Philadelphia, 1989, pp. 151–183.

ABSTRACT: The present paper reviews major developments in mechanics and micromechanics of fatigue crack propagation at an ambient temperature over the past two decades, and suggests directions of future developments. Mechanics of plastic deformation and closure of fatigue cracks under cyclic loading will first be presented, then the propagation behavior of long and short cracks will be described. Emphasis will be placed on the experimental as well as the theoretical aspects of the mechanical treatment of fatigue crack propagation.

KEY WORDS: fatigue (materials), fracture mechanics, crack propagation, small cracks, crack closure, threshold conditions, micromechanics

The fatigue fracture of materials is caused by the nucleation and propagation of cracks. The understanding of the growth law of a fatigue crack is essential to predict and control the progress of the fatigue cracking process. The rate of fatigue crack propagation is governed by a loading parameter which is primarily a function of the applied stress and crack length.

$$da/dN = f(\sigma, a) \tag{1}$$

Once the growth law of a fatigue crack is determined, we can compute the life of engineering components by integration from the initial to final crack length

$$N_f = \int_{a_i}^{a_f} da/f(\sigma, a) \tag{2}$$

Figure 1 illustrates the crack size scale for initial and final cracks [*1*]. The initial crack length varies from 1 μm to about 1 mm, depending on defects existing in materials. The final crack length is above tens of millimeters. The distinction between long and short cracks is sometimes taken as about 1 mm in length [*2,3*].

The main subject of mechanics of fatigue crack propagation is to establish the equation of crack propagation as a function of a loading parameter under complex service conditions. The factors which influence the crack propagation rate are material microstructures, component geometry, loading conditions, environments and temperatures.

Since Paris [*4,5*] first applied the fracture mechanics to the problem of fatigue crack propagation, the stress-intensity factor (SIF) has been widely used as a loading parameter

[1] Associate professor, Department of Engineering Science, Kyoto University, Kyoto 606, Japan.

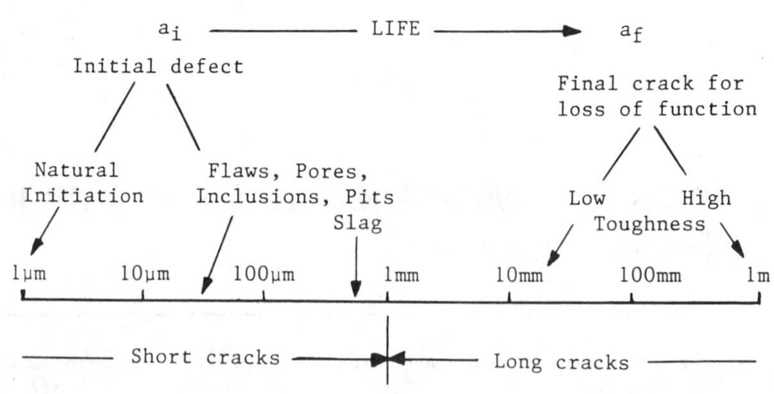

FIG. 1—*Various initial and final crack sizes defining crack propagation life.*

in correlating fatigue crack growth rate. During the past two decades, the mechanical treatment of fatigue crack propagation has been advanced by the developments in the field of fracture mechanics, by the introduction of servohydraulic testing machines, and by the availability of scanning electron microscopy (SEM) and other novel techniques for microscopic observations.

Most of the earlier works done between 1966 to 1975 dealt with the propagation of long cracks which were longer than about 5 mm. Around 1975, short cracks appeared as research subjects. Several principles established in the studies of long cracks have been applied to short crack growth, hoping to bridge the gap between the classical *S-N* approach and the fracture mechanics approach.

The present paper reviews major developments in the mechanical treatment of fatigue crack propagation over the past two decades and suggests directions of future developments. We deal exclusively with fatigue crack growth at an ambient temperature. Mechanics of plastic deformation and closure of fatigue cracks under cyclic loading will first be presented, then the propagation behavior of long and short cracks will be described. Since the propagation behavior of fatigue cracks is complex and there is no general theory for crack propagation, emphasis will also be placed on the experimental aspect as well as the theoretical aspect of the mechanical treatment of fatigue crack propagation.

Crack-Tip Deformation and Crack Closure Under Cyclic Loading

Plastic Deformation

Rice [*6*] analyzed the plastic deformation ahead of the crack tip under cyclic loading on the assumption of proportional plastic flow. This permits the plastic superposition of loading and unloading as shown in Fig. 2. By using the Dugdale type approximation [*7*] for the plastic zone ahead of the crack tip, the plastic zone size at the maximum applied stress σ_{max} is given by

$$\omega = a\left[\sec\left(\frac{\pi\sigma_{max}}{2\sigma_Y}\right) - 1\right] \quad (3)$$

and the crack-tip opening displacement (CTOD) is

$$\text{CTOD} = \frac{8\sigma_Y}{\pi E}\, a \ln\left[\sec\left(\frac{\pi\sigma_{max}}{2\sigma_Y}\right)\right] \quad (4)$$

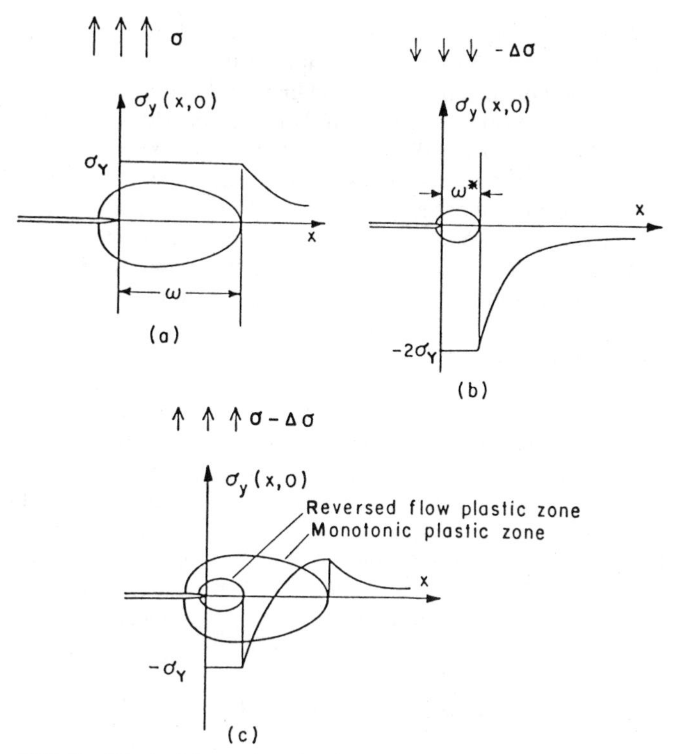

FIG. 2—*Plastic deformation ahead of crack tip under cyclic loading.*

where σ_Y is the yield strength. For the small-scale yielding (SSY) cases, ω and CTOD are functions of K_{max} $(=\sigma_{max}\sqrt{\pi a})$ as

$$\omega = \frac{\pi}{8}\left(\frac{K_{max}}{\sigma_Y}\right)^2 \tag{5}$$

$$\text{CTOD} = K_{max}{}^2/(E\sigma_Y) \tag{6}$$

When there is no crack closure during unloading, the changes in stresses, strains and displacements are obtained by replacing σ_{max} by $\Delta\sigma$ and σ_Y by $2\sigma_Y$ in the above equations $(\Delta\sigma = \sigma_{max} - \sigma_{min})$. The superposition yields the state at the minimum stress σ_{min} as shown in Fig. 2*c*. For SSY cases, this substitution into Eqs 5 and 6 yields

$$\omega^* = \frac{\pi}{8}\left(\frac{\Delta K}{2\sigma_Y}\right)^2 \tag{7}$$

$$\Delta\text{CTOD} = \Delta K^2/(2E\sigma_Y) \tag{8}$$

where

$$\Delta K = K_{max} - K_{min} \tag{9}$$

(K_{max} is the maximum stress intensity factor and K_{min} is the minimum stress intensity factor).

Two types of the plastic zone near a fatigue crack were clearly observed in 3% silicon iron by Hahn et al. [8]. Figure 3 shows an example of the etched micrograph of a fatigue crack near the threshold obtained in a load-shedding test [9]. The dark-etched region corresponds to the monotonic plastic zone, while the white region within the dark-etched region indicates the reversed plastic zone. While they showed the maximum plastic zone was proportional to $(K_{max}/\sigma_Y)^2$, the reversed plastic zone was not exactly a fourfold reduction of the monotonic plastic zone under zero tension loading. They ascribed a deviation from fourfold reduction to the cyclic work hardening. Several other experimental techniques have also shown similar results [10]. The study of the reversed plastic zone for a fatigue crack near the threshold indicated the significance of crack closure in determining the reversed plastic zone [9].

Crack Closure

According to elastic-plastic analysis of a stationary crack, the crack will touch at the center of the crack under a slight compression during reversed loading [6]. For a propagating fatigue crack, this is not always true; the crack can close even under the tensile-applied stress. Elber [11] first found the crack closure at a tensile load in 2024-T3 aluminum alloy. He ascribed the crack closure to the residual stretch left on the fatigue crack wake. This mechanism is now called plasticity-induced closure. Elber also suggested that the crack-tip deformation and crack propagation rate were controlled by the effective stress intensity range ΔK_{eff} defined by

$$\Delta K_{eff} = K_{max} - K_{op} = U\Delta K \tag{10}$$

where K_{op} is the SIF value at crack-tip opening and U is the effective fraction. When K_{op} is less than K_{min}, $\Delta K_{eff} = \Delta K$ and $U = 1$.

Later, several other mechanisms for crack closure have been found [12,13]. Figure 4 shows three principal forms of crack closure [13]. Oxide-induced closure is caused by oxide debris due to fretting of crack faces under cyclic loading in moist environment. Asperity contact of rough fracture surfaces is enhanced by an additional shear mode deformation (roughness-induced closure).

Budiansky and Hutchinson [14] analyzed the plasticity-induced closure of a semi-infinite crack propagating under a constant value of ΔK at various stress ratios $R = \sigma_{min}/\sigma_{max}$. Uniform residual extension is assumed to act on the face of the crack which is accompanied by a Dugdale plastic zone. They showed that K_{op}/K_{max} was 0.557 at $R = 0$ and increased with increasing R. The value of ΔCTOD is a unique function of ΔK_{eff} without respect to R value. The relation is

$$\Delta \text{CTOD} = 0.73 \frac{(\Delta K_{eff})^2}{E\sigma_Y} = \frac{(\Delta K_{eff})^2}{1.37E\sigma_Y} \tag{11}$$

It is interesting to compare Eq 11 with Eq 8. When crack closure takes place, $1.37\sigma_Y$ is used for reversed yielding instead of $2\sigma_Y$.

Tanaka and Nakai [15] extended the Budiansky-Hutchinson model to a short crack in the large-scale yielding situation. Figure 5 presents their result calculated for $R = -1$. The relation between the reversed plastic zone size (normalized by the half crack length, ω^*/a) and the applied stress amplitude (normalized by the yield strength, σ_a/σ_Y) is shown with the solid line, where the reciprocal of ω^*/a is taken as the abscissa. The effective stress range $\Delta\sigma_{eff} = \sigma_{max} - \sigma_{op}$ (σ_{max} equals the maximum applied stress; σ_{op} equals the applied stress at crack opening) normalized by σ_Y is shown with the dashed-dotted line as a function

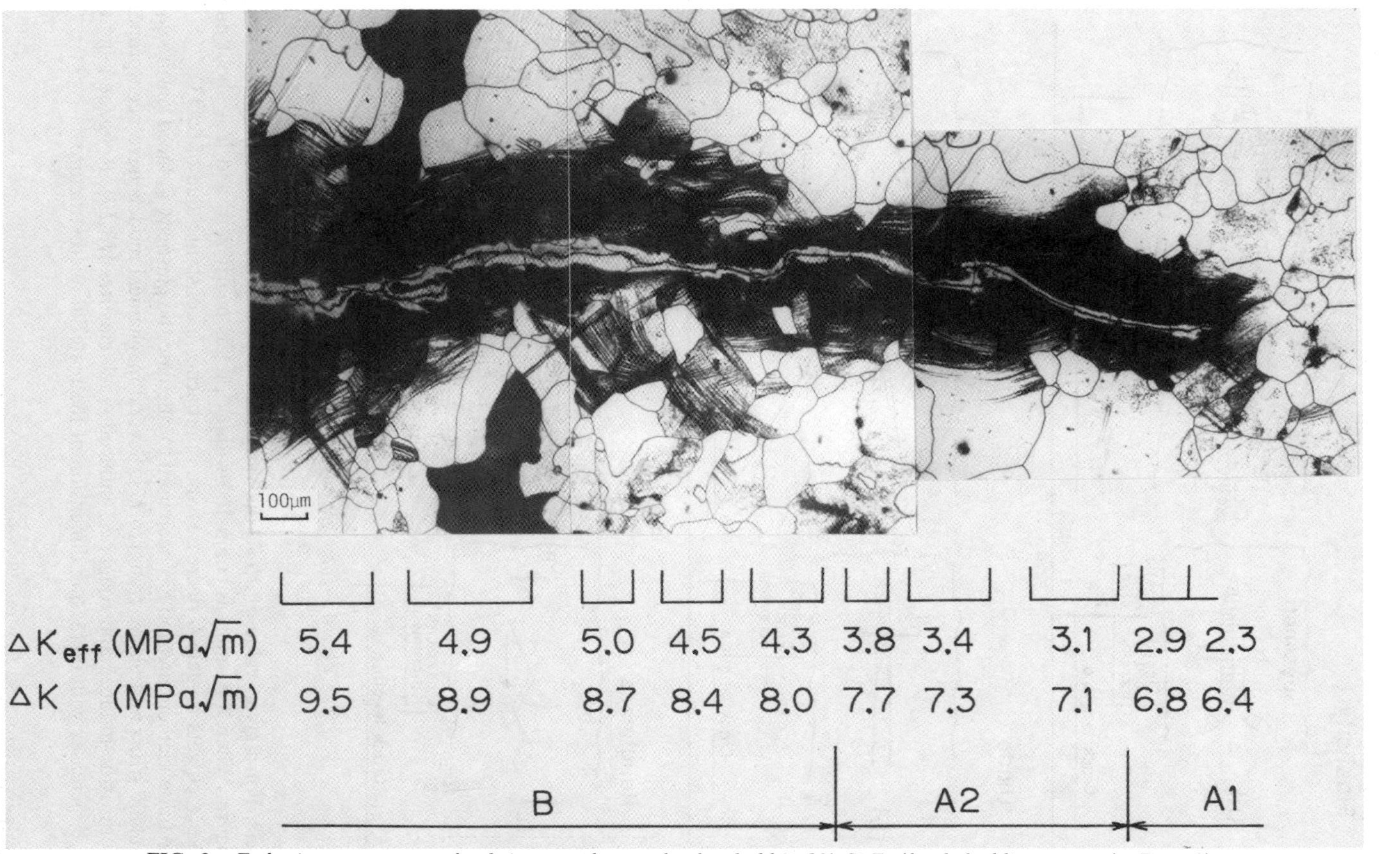

FIG. 3—*Etch pit patterns around a fatigue crack near the threshold in 3% Si-Fe (load-shedding test under* R = 0*).*

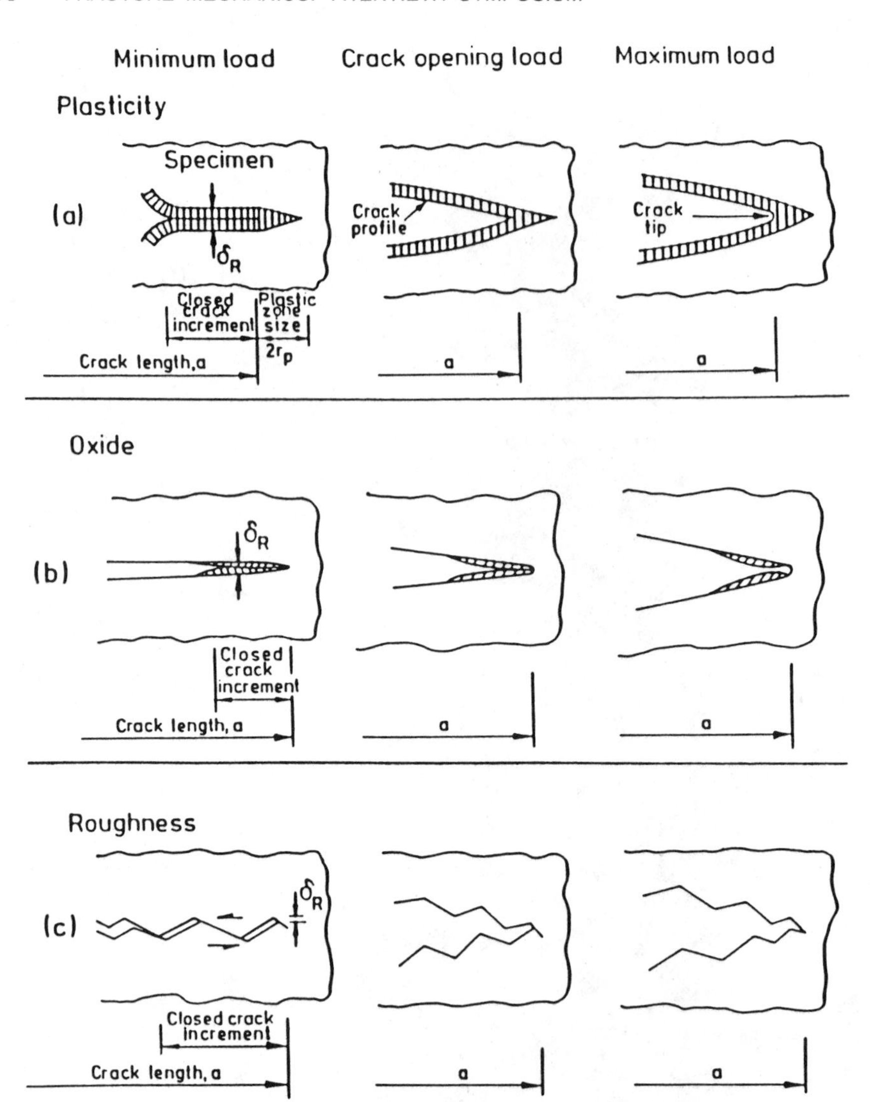

FIG. 4—*Three mechanisms of fatigue crack closure.*

of a/ω^*. For a given value of a/ω^*, the values of σ_a/σ_Y and $\Delta\sigma_{\text{eff}}/\sigma_Y$ are determined from the figure. When a is small as compared with ω^*, σ_{op} is nearly zero and U is about unity. As the crack gets longer, U decreases and approaches a constant value of 0.443.

The finite-element method has been used to calculate the plasticity-induced closure under both plane stress and plane strain [*16,17*]. Several mechanical models have been proposed for both oxide-induced and roughness-induced crack closures [*18,19*]. Roughness-induced closure increases with increasing dimension of the material microstructure.

Elastic-Plastic Parameters

Under large-scale yielding, gross and general yielding conditions, CTOD and J-integral are usable as a characterizing parameter of the crack-tip filed. Under Masing's assumption

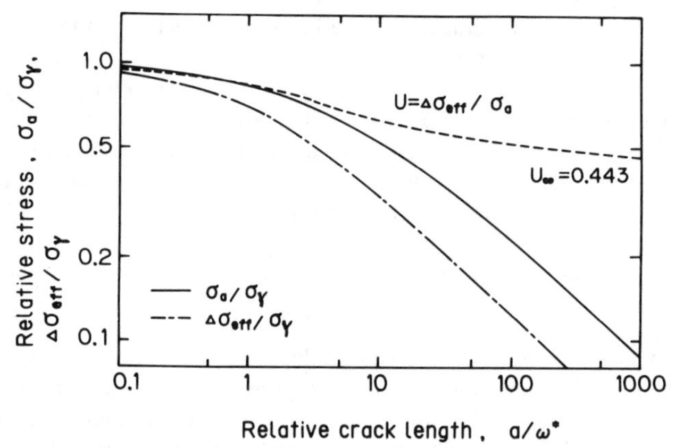

FIG. 5—*Changes of stress amplitude and effective stress with crack length relative to reversed plastic zone size.*

[*20*] that either branch of the stress-strain hysteresis is geometrically similar to the monotonic stress strain curve with a scale factor of two, the range of J-integral ΔJ is proved to be path-independent and characterize the range of change in stress, strain and displacement during one loading cycle [*21,22*].

The Similitude Concept

A unique correlation between the fatigue crack propagation rate and the stress intensity factor is based on the concept of similitude. Equal stress intensity (or J) will have equal consequences. This similitude requirement is not always satisfied, especially when we deal with small cracks.

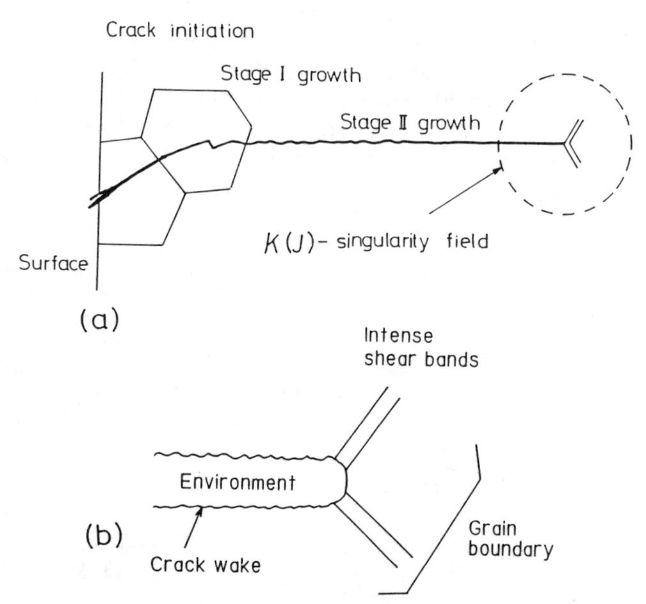

FIG. 6—*A fatigue crack nucleated from the surface:* (a) *Stage I and Stage II growth;* (b) *Details of fracture-related zone at the crack tip.*

Figure 6*a* illustrates a fatigue crack nucleated from the specimen surface. The spread of the K-singularity field near the crack tip where the singular term is predominant is about one-tenth of the crack length [*23*]. When the crack is large and when the physical (chemical) process of fatigue cracking occurs within the K-singularity field as shown in Fig. 6*a*, the rate of crack growth will be unique. On the other hand, when the crack length is small, the K similitude breaks down; the crack growth rate is no longer a unique function of K-values. For elastic-plastic cracks, the J-integral is to be used instead of K as a near-tip field parameter.

Referring to the fracture-related zone for the fatigue crack shown in Fig. 6*b*, the breakdown of the K similitude due to the shortness of the crack is possibly induced through several processes:

1. Breakdown of the microstructural similitude. When the crack length is on the order of the material microstructure, for example, the grain size, plastic deformation near the crack tip is very much influenced by the material microstructure, and the grain boundaries will block crack growth. The assumption of a macroscopic continuum in fracture mechanics is violated; the similitude is broken down because of microstructural inhomogeneity. These short cracks are *microstructurally short cracks* [*2*].
2. Breakdown of the mechanical similitude. The ΔK value loses its meaning as a crack driving force for the following two cases. The plastic zone size is large compared with the crack length, that is, the crack-tip yielding is of large scale. The crack closure is not fully developed because of the short crack wake. These cracks are *mechanically short cracks* [*3*].
3. Breakdown of the environmental similitude. The crack-tip chemical environment controls the crack growth rate and is often different from the bulk environment. Short cracks may have a different crack-tip environment from long cracks, so that the propagation rate will be different. Such cracks are *chemically short cracks* [*24*].

Table 1 presents the classification of the crack size according to mechanical and microstructural influences [*25*]. When the crack length is large compared with the microstructural dimension (microstructurally large), the material resistance is expected to be homogeneous. The subject is to find out an appropriate driving force. For a small crack whose length is comparable to the microstructural dimension (microstructurally small), the material resistance is influenced by crystal orientations and grain boundaries. Therefore, even when the mechanical driving force is properly expressed in terms of SIF (like in the case of Type 3 crack), a crack shows irregular, anomalous growth behavior.

The limitation of K for determining crack-tip plasticity due to large-scale yielding can be estimated on the basis of Dugdale model analysis [*7*]. Ten percent deviation of the plastic

TABLE 1—*Classification of crack size according to mechanical and chemical influences.*

	Mechanical Size	
Microstructural Size	Large: $a/\omega > 4 - 20$ (SSY) Large Crack Wake	Small: $a/\omega < 4 - 20$ (LSY) Small Crack Wake
Large: $a/M > 5 - 10$ $\omega/M \gg 1$	Type 1: Mechanically and microstructurally large (LEFM valid)	Type 2: Mechanically small/ microstructurally large
Small: $a/M < 5 - 10$ $\omega/M \sim 1$	Type 3: Mechanically large/ microstructurally small	Type 4: Mechanically and microstructurally small

zone size from Eq 5 occurs at the stress level $\sigma_{max}/\sigma_Y = 0.30$, and the corresponding plastic zone size relative to crack length is 0.12. Ten percent deviation of CTOD from Eq 6 occurs at $\sigma_{max}/\sigma_Y = 0.45$, and the corresponding ω/a is 0.32. When a similar 10% deviation criterion for the SSY limit is applied to the cyclic components, the SSY limit based on the reversed plastic zone size is at $\Delta\sigma/2\sigma_Y = 0.30$ and $\omega^*/a = 0.12$, and that based on ΔCTOD is at the $\Delta\sigma/2\sigma_Y = 0.45$ and $\omega^*/a = 0.32$, provided that there is no crack closure. Tanaka and Nakai [*15*] showed that the ΔK_{eff}-approach was applicable to even larger stress levels than the ΔK-approach. While crack closure is reduced due to the small amount of plasticity, ω^* and ΔCTOD are determined as a unique function of ΔK_{eff}, but not of ΔK.

Even for a long crack, the similitude will break down when the load is suddenly changed. Fatigue crack propagation is temporarily retarded immediately after overloading or load reduction. This violation comes from the history effect; the crack-tip deformation is different from that under constant-amplitude loading. Crack closure consideration will show that ΔK_{eff} is a proper local parameter.

Propagation of Long Fatigue Cracks

Relation between Crack Propagation Rate and Stress Intensity Factor

For a long crack, the steady-state fatigue crack growth rate is a function of the stress intensity factor. Paris and Erdogan [*5*] first obtained the fourth-power law between da/dN and ΔK

$$da/dN = C(\Delta K)^4 \tag{12}$$

A large number of works done in the following two decades have shown that the exponent of the power relation between da/dN and ΔK is not necessarily four, but varies between two to seven. Thus, a generalized Paris law is

$$da/dN = C(\Delta K)^m \tag{13}$$

where C is a weak function of R. A wide range of data of da/dN versus ΔK shows a sigmoidal variation, as shown in Fig. 7. The power law only represents the relation in the intermediate-rate regime, Regime B, from 10^{-9} to 10^{-6} m/cycle. In Regime A, there is a threshold stress intensity range ΔK_{th} below which the rate is practically zero. At high ΔK in Regime C, the acceleration of crack growth takes place, and unstable fracture starts at the maximum stress-intensity factor equal to the fracture toughness. The characteristics of each regime are described in Fig. 7 [*26*].

Many formulae have been proposed for expressing a sigmoidal variation in the relation between da/dN and ΔK. Several equations, Eqs 14 to 18, are listed in Table 2 [*27–32*]. Equations 14, 16, and 17 are empirical, and others are derived on the basis of some physical consideration. Even today, we do not have universal equations, and the values of C and m in the above equations need to be determined experimentally.

Mean-Stress and Variable-Amplitude Effects

The mean stress superposed on the fluctuating stress will influence the amount of static-mode growth and crack closure. The former contribution in Regime C is taken into account as a denominator in Eqs 14, 15, and 18. For some cases, the mean stress has a large influence on the growth rate even in Regime B. Several equations, Eqs 19 to 22, which include the

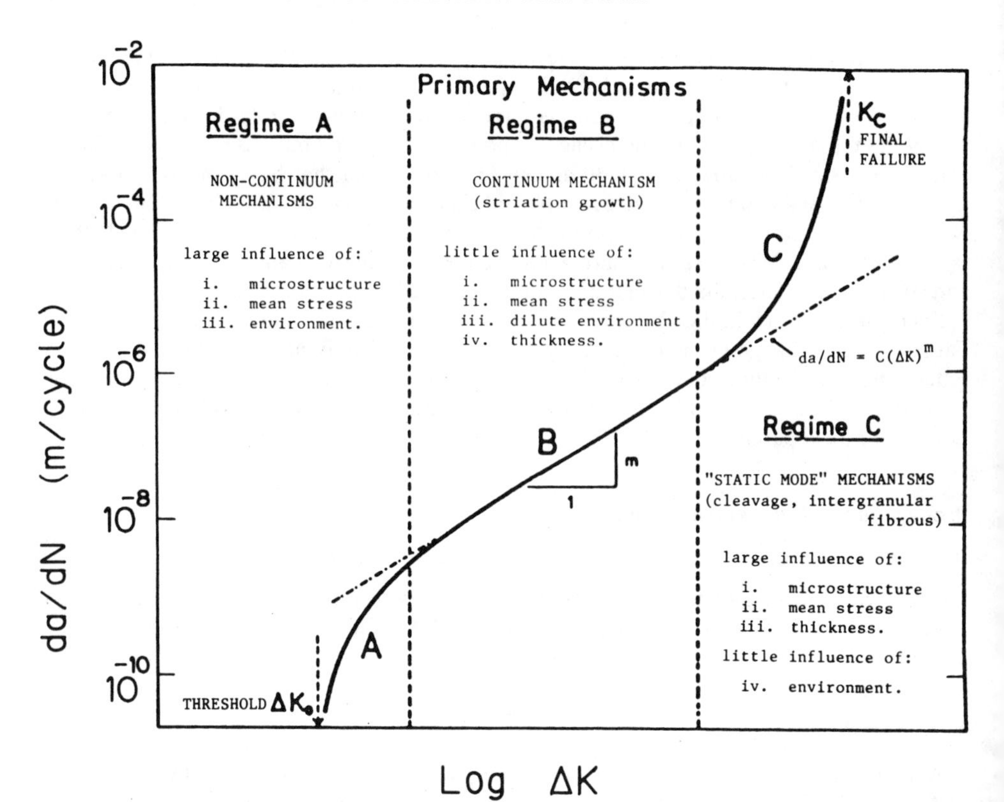

FIG. 7—*Relation between crack propagation rate and stress intensity range.*

mean-stress effect are given in Table 2 [*33–37*]. ΔG is the range of the energy release rate and has been used for cracking of adhesive joints. ΔS is the range of the strain energy density. Under Mode I loading, ΔG and ΔS are equivalent to $K_{max}{}^2 - K_{min}{}^2$ except for coefficients.

For brittle materials with low toughness, such as PMMA, epoxy, and graphite, the static-mode crack growth takes place even in the intermediate-rate regime. The value of K_{max} as well as ΔK contribute the total amount of crack growth. In Eq 19, $\gamma = q/(p + q)$ indicates the relative contribution of the maximum stress.

For ductile materials, the effect of the mean stress on the crack propagation rate in low- and intermediate-rate regimes has been ascribed to crack closure. The effect of R diminishes in the relation between da/dN and ΔK_{eff}, as shown in Fig. 8. The crack propagation rate is expressed by Elber equation [*11*] for any R value.

$$da/dN = C(\Delta K_{eff})^m \tag{23}$$

The mean-stress effect is pronounced near the threshold. The ΔK_{th} value normally decreases with increasing R. Phenomenological relations between ΔK_{th} and R proposed by Klesnil and Lukas [*34*] are

$$\Delta K_{th} = \Delta K_{tho}(1 - R)^\gamma \tag{24}$$

TABLE 2—*Empirical and semi-empirical equations for fatigue crack propagation.*

Equation Number	Equation	Author	Year
(14)	$\frac{da}{dN} = C\frac{\Delta K^m}{(1 - R)K_c - \Delta K}$	Forman [27]	1967
(15)	$\frac{da}{dN} = C\frac{\Delta K^4}{K_c^2 - K_{max}^2}$	Weertman [28,29]	1966, 1969
(16)	$\frac{da}{dN} = C(\Delta K^m - \Delta K_{th}^m)$	Klesnil [30]	1972
(17)	$\frac{1}{da/dN} = \frac{C_1}{(\Delta K)^{n_1}} + \frac{C_2}{(\Delta K)^{n_2}} - \frac{C_2}{[K_c(1 - R)]^{n_2}}$	Saxena [31]	1979
(18)	$\frac{da}{dN} = C(\Delta K - \Delta K_{th})^2\left(1 + \frac{\Delta K}{K_c - K_{max}}\right)$	McEvily [32]	1983
(19)	$\frac{da}{dN} = C\,\Delta K^p\,K_{max}{}^q$	Roberts [33] Klesnil [34]	1965 1972
(20)	$\frac{da}{dN} = C(K_{max}{}^2 - K_{min}{}^2)^m$	Arad [35]	1971
(21)	$\frac{da}{dN} = C(\Delta G)^m$	Mostovoy [36]	1975
(22)	$\frac{da}{dN} = C(\Delta S)^m$	Badaliance [37]	1980

where ΔK_{tho} is the ΔK_{th} value at $R = 0$. Schmidt and Paris [38] proposed

$$\begin{aligned} \Delta K_{th} &= \Delta K_{tho}(1 - \kappa R) && \text{for } R < R_c \\ \Delta K_{th} &= \Delta K_{tho}(1 - \kappa R_c) = \text{const} && \text{for } R > R_c \end{aligned} \tag{25}$$

Figure 9 shows the relation between ΔK_{th} and R for steels [39]. The contribution of crack closure has been found to be large near the threshold and the effective threshold stress intensity range has been found to be constant. The ΔK_{th} value consists of

$$\Delta K_{th} = \Delta K_{effth} + \Delta K_{clth} \tag{26}$$

where the first term is the intrinsic material resistance and ΔK_{clth} $(= K_{opth} - K_{min})$ is the extrinsic one. ΔK_{effth} equals to ΔK_{th} at stress ratios larger than R_c. It is between 2 to 3 $\text{MPa}\sqrt{\text{m}}$ for steels. It is now known that crack closure also accounts for the major part of the microstructural effect on the threshold.

Crack closure plays a significant role in the fatigue crack growth behavior under variable-amplitude spectrum loading. Once the value of ΔK_{eff} is evaluated experimentally or theoretically, the subsequent crack growth rate can be predicted from the $da/dN - \Delta K_{eff}$ relation obtained in the test under constant-amplitude loading. Therefore, the prediction of crack closure is the topics of current researches. For variable-amplitude loading including the stress level below the threshold, ΔK_{effth} may disappear and the power relation between da/dN and ΔK_{eff} should be extended below ΔK_{effth} for predicting crack growth rate [40]. Although crack closure is primarily significant in crack growth under single or multiple

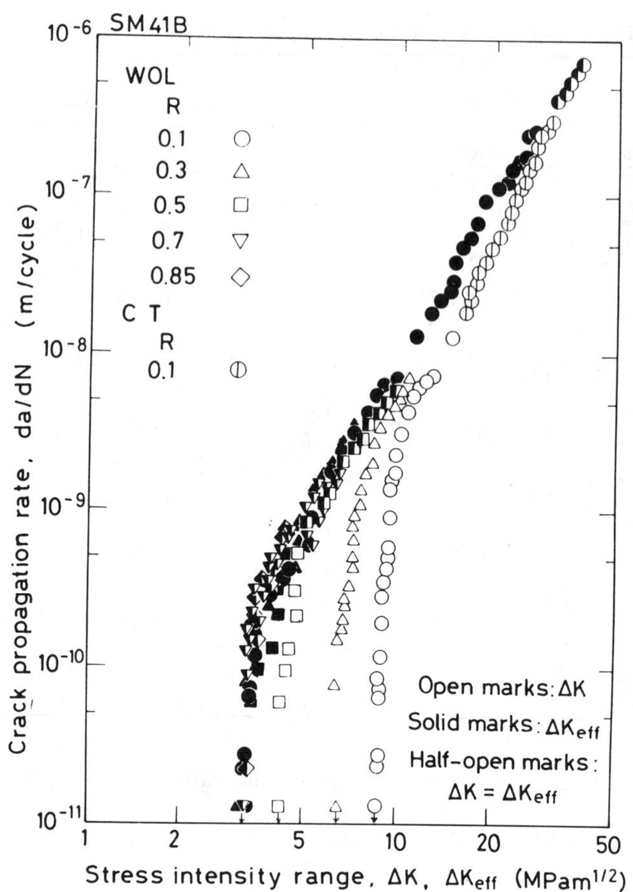

FIG. 8—*Relation between crack propagation rate and stress intensity range for mild steel under various* R *values.*

over loads, there may be other factors such as crack branching contributing crack growth retardation [*41*].

Mixed-Mode and Biaxial Loading Effects

The growth direction of fatigue a crack under mixed-mode loading is not coplanar in most cases because of crack-tip stress asymmetry, and is much influenced by material anisotropy [*42–44*]. The threshold condition for fatigue crack growth under a mixed mode of I and II was examined for a mild steel [*42*]. The shear-mode growth took place at low stress intensity, and the tensile-mode growth took over at higher stress intensities. The growth direction of tensile crack growth is roughly along the direction of the maximum tangential stress or the minimum strain energy density. A crack grows as to reduce the K_{II} component in isotropic materials. The ΔS parameter was used to predict the mixed-mode crack growth [*45*].

Fatigue cracks in anisotropic materials such as carbon-fiber-reinforced plastics (CFRP) grow coplanarly even under mixed-mode loading [*44*]. Figure 10 shows the rate of delam-

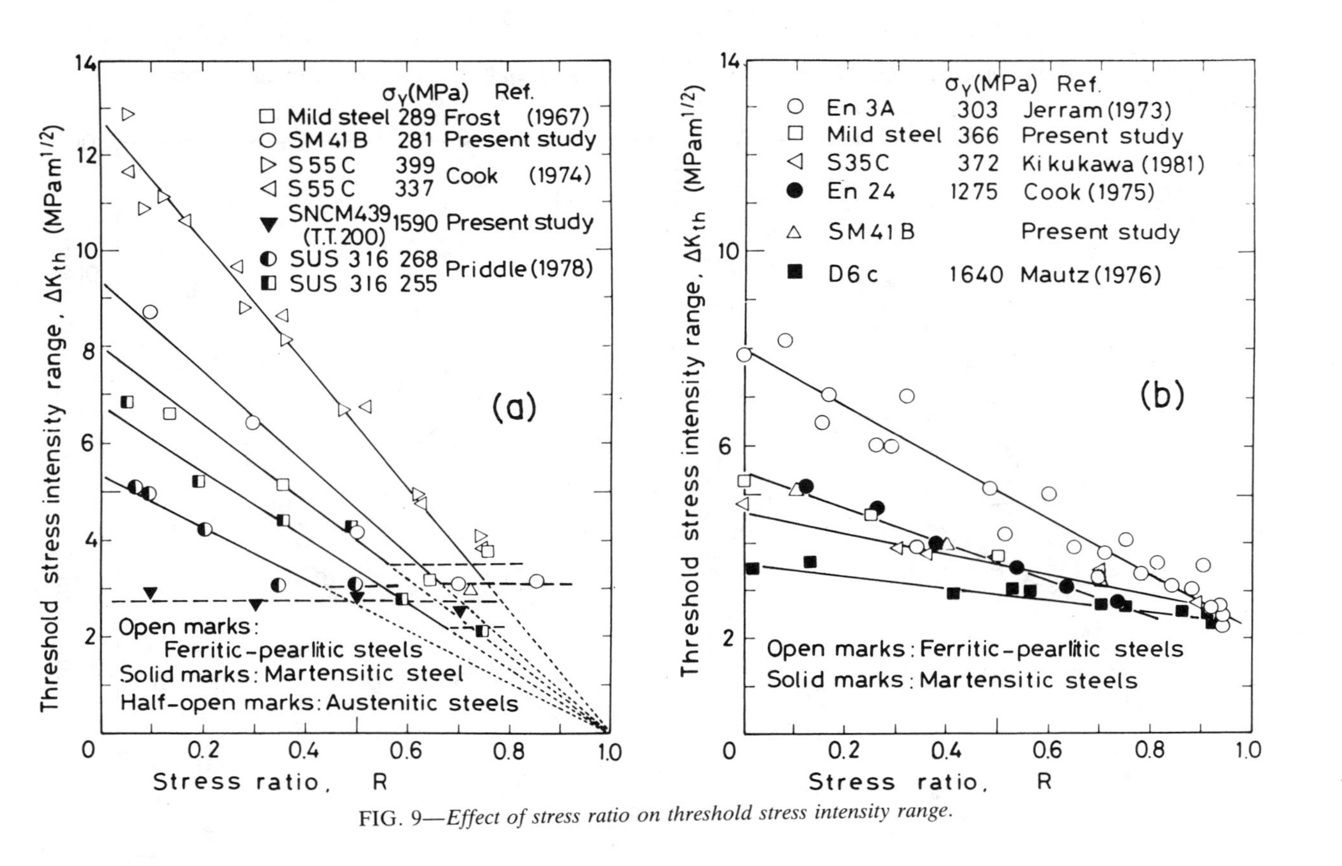

FIG. 9—*Effect of stress ratio on threshold stress intensity range.*

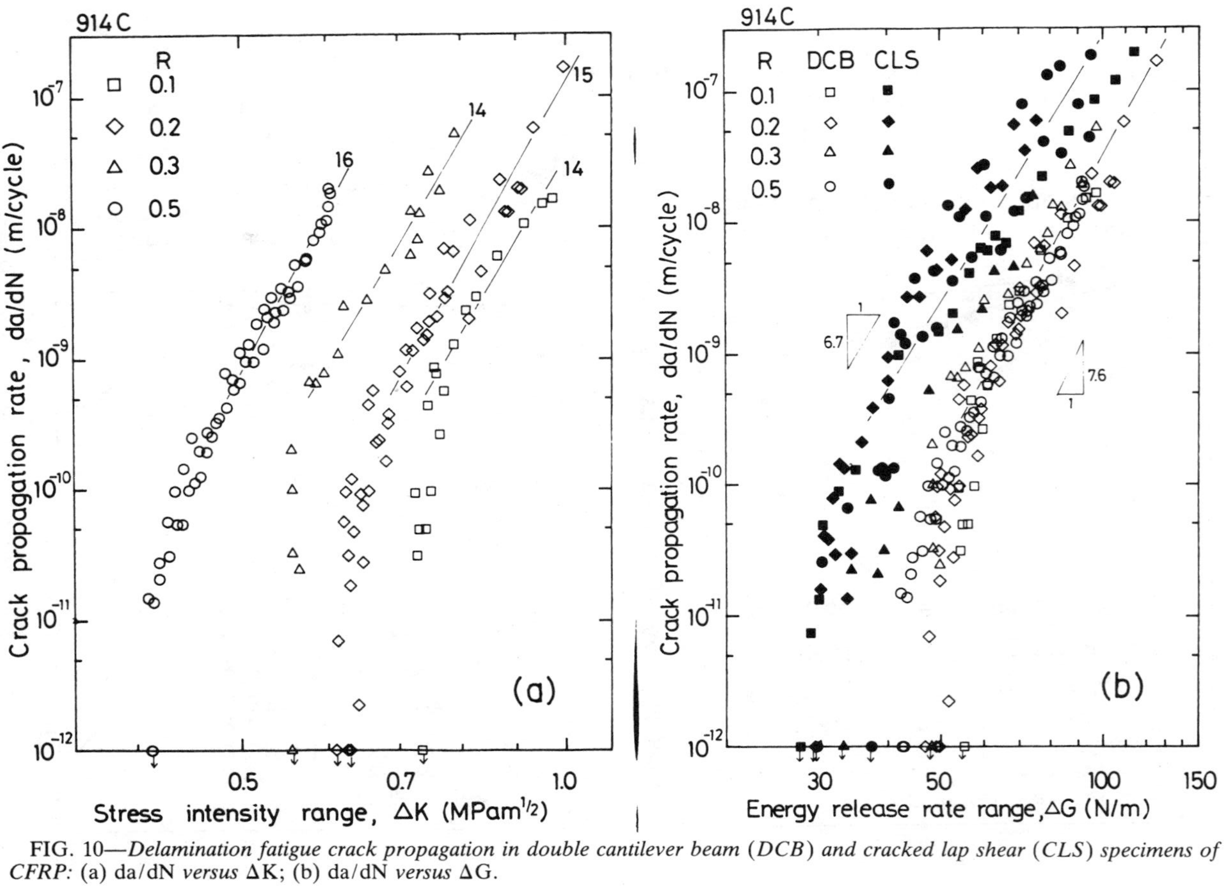

FIG. 10—*Delamination fatigue crack propagation in double cantilever beam (DCB) and cracked lap shear (CLS) specimens of CFRP:* (a) da/dN *versus* ΔK; (b) da/dN *versus* ΔG.

ination fatigue cracks under Mode I and a mixed mode of I and II at various R-values. Mixed mode data were obtained by using cracked lap shear specimens (CLS) where the ratio of Mode II component of the strain energy release rate ΔG_{II} to the total value ΔG is 0.70. The crack growth rate is much higher at higher R ratios and under mixed mode loading when it is correlated to ΔK_I. When da/dN is correlated to ΔG (= $\Delta G_I + \Delta G_{II}$), the effect of R disappears. However, ΔG is not enough to account for the contribution of Mode II component, and a new correlating parameter is necessary. The effect of anisotropy on crack growth under mixed-mode loading will be different depending on the types of composites. Further study is necessary.

Nonsingular stresses applied parallel to the crack plane often influence the fatigue crack growth rate in the biaxial stress field [*46–48*]. The plastic zone ahead of the crack tip tends to increase under pure shear (lateral compression) loading, while decreasing in biaxial tensile loading. Thus, a fatigue crack grows faster in pure shear than in biaxial tension in the large-scale yielding situation. There is no nonsingular stress effect in the SSY situation [*48*].

Elastic-Plastic Loading Effect

The rate of fatigue crack growth under elastic-plastic and gross plasticity conditions is higher than that predicted from the relation between da/dN and ΔK. Among various elastic-plastic parameters, the J-integral range seems to be most successful. Dowling and Begley [*49*] first correlated the rate to ΔJ and obtained the power-law relation

$$da/dN = C(\Delta J)^m \tag{27}$$

Under SSY conditions, Eq 27 becomes equivalent to Eq 23 because $\Delta K_{eff} = \sqrt{E\Delta J}$. Figure 11 shows the growth rate plotted against ΔK_{eff} and ΔJ [*50*]. The data obtained under strain-controlled condition show the acceleration in Fig. 11*a*, while all the data fall on the single line in Fig. 11*b*.

Other parameters which have been proposed are ΔCTOD [*51,52*] and the plastic zone size (PZS) [*53,54*]. The crack growth rate is given as a power function as

$$da/dN = C(\Delta \text{CTOD})^m \tag{28}$$

$$da/dN = C(\text{PZS})^m \tag{29}$$

Mechanisms of Fatigue Crack Propagation

Crack-Tip Blunting Model

Previous models for fatigue crack propagation are classified into two categories, one based on crack-tip blunting and the other based on damage accumulation [*32*]. In the intermediate-rate regime where no static mode fracture is involved, the crack-tip blunting model has been supported by various experimental evidences.

Crack-tip blunting model was first proposed by Laird and Smith [*55*] on the basis of the direct observation of crack opening profile. This type of model has been advanced by several investigators. Neumann [*56*] and Kikukawa et al. [*57*] made quantitative observation of the opening and closing of a crack and showed that the amount of crack growth per cycle was half the ΔCTOD, as derived from a simple geometrical relation. This geometrical relation is not satisfied for a wide range of fatigue crack growth data [*58–60*].

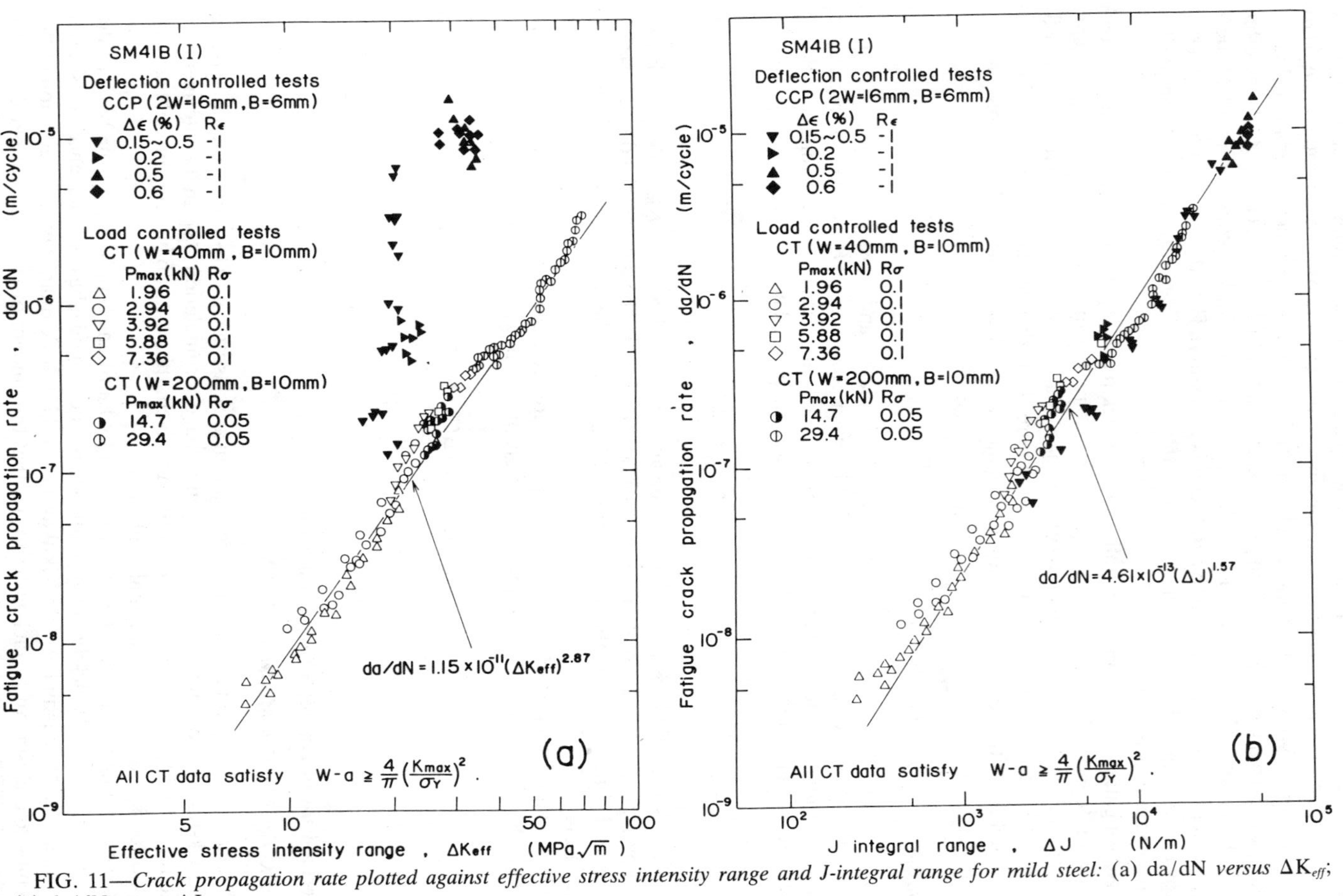

FIG. 11—*Crack propagation rate plotted against effective stress intensity range and J-integral range for mild steel:* (a) da/dN *versus* ΔK_{eff}; (b) da/dN *versus* ΔJ.

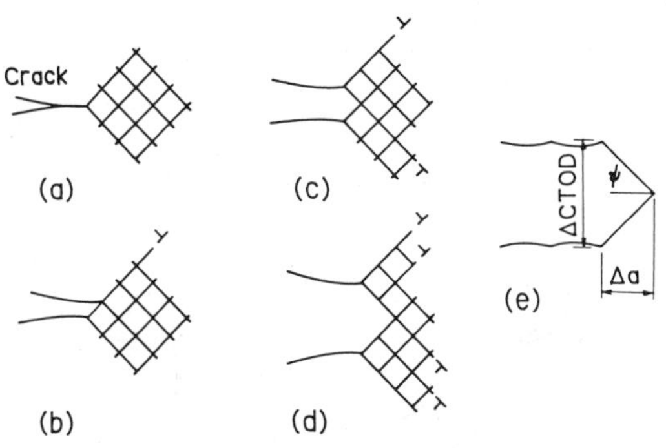

FIG. 12—*Crack-tip blunting by alternate shear.*

Figure 12 shows crack-tip blunting model depicted by Kuo and Liu [*58*]. The amount of crack extension per cycle Δa $(=da/dN)$ is proportional to ΔCTOD as

$$da/dN = \delta \Delta \mathrm{CTOD} \tag{30}$$

where δ is

$$\delta = 0.5 \cot \psi \tag{31}$$

and ψ is half the crack-tip opening angle. For the small-scale yielding case, ΔCTOD is given by

$$\Delta \mathrm{CTOD} = \beta' \Delta K^2/(E\sigma_Y') \tag{32}$$

where σ_Y' is the cyclic yield strength. From Eqs 30 and 32, we have the crack growth law

$$da/dN = \lambda \Delta K^2/(E\sigma_Y') \tag{33}$$

where $\lambda = \beta'\delta$. In the above derivation, ΔK is replaced by ΔK_{eff} when there is crack closure. Under elastic-plastic loading, ΔJ is used for ΔK, and the above equation becomes

$$da/dN = \lambda \Delta J/\sigma_Y' \tag{34}$$

The crack opening angle ψ is usually assumed to be 45 deg, which gives $\delta = 0.5$. According to the elastic-plastic analysis of a stationary crack, β' is between 0.15 and 0.73 [*61*]. Therefore, λ is between 0.08 and 0.37. Kuo and Liu [*58*] calculated $\lambda = 0.019$ by using the unzipping model, in which only a fraction of ΔCTOD is assumed to be effective for crack advance. Figure 13*a* shows the da/dN-ΔJ relation for three metals, and in Fig. 13*b* the rate is plotted against $\Delta J/\sigma_Y'$ (σ_Y' = the cyclic yield stress) [*60*]. Lines of several λ values are drawn in the figure. Electron fractography indicated complete covering of the fracture surface by striations at rates from 2×10^{-7} to 5×10^{-5} m/cycle. The value of λ decreases with decreasing rate. Careful observation of crack-tip opening profiles showed that the opening profile was not geometrically similar, as shown in Fig. 14. This gives rise to the exponent in Paris law larger than two even if striation formation is an operating mode. For several

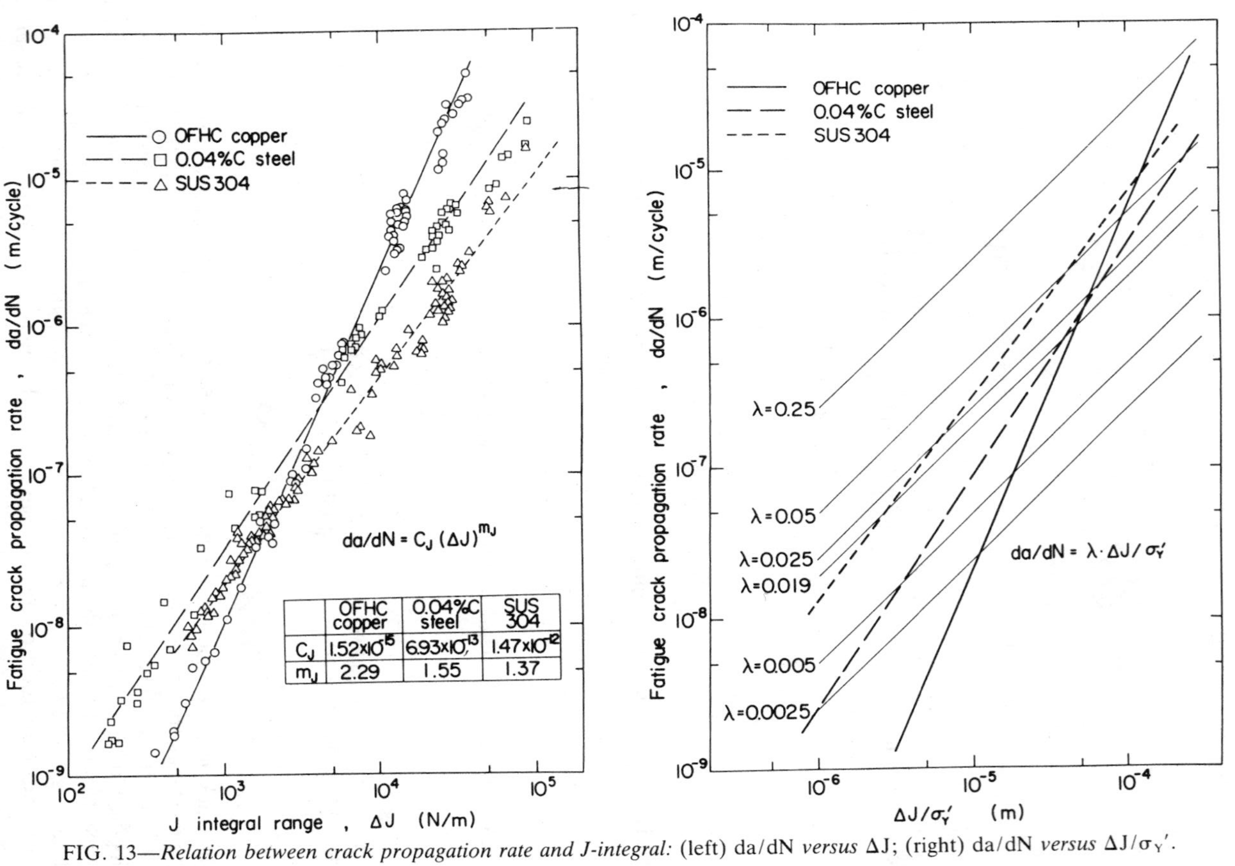

FIG. 13—*Relation between crack propagation rate and J-integral:* (left) da/dN *versus* ΔJ; (right) da/dN *versus* $\Delta J/\sigma_Y'$.

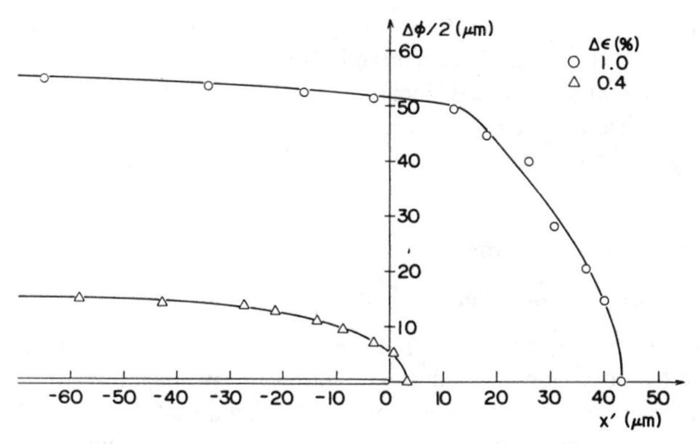

FIG. 14—*Crack-tip opening profile in pure copper.*

metals, geometrical similarity is satisfied and the striation spacing is proportional to ΔK^2 or to the crack length [*62*].

$$da/dN = Aa \tag{35}$$

Crack Propagation Through Multiple Mechanisms

When several mechanisms for crack growth are operating at the same time, the resultant crack growth rate will be a sum of the rate due to each mechanism. Consider the case in which only two mechanisms are operating. When two mechanisms operate in one cycle, the rate will be

$$da/dN = (da/dN)_1 + (da/dN)_2 \tag{36}$$

where $(da/dN)_1$ is the rate of crack growth due to Mechanism 1 and $(da/dN)_2$ is that due to Mechanism 2. When each mechanism occurs in only a certain area fraction of the fracture surface, the fracture surface consists of the area made by Mechanism 1 and that by Mechanism 2. The resultant rate will be

$$da/dN = f_1(da/dN)_2 + f_2(da/dN)_2 \tag{37}$$

where f_1 and f_2 are area fractions. In sequential model for crack growth, we have

$$da/dN = 1/[f_1/(da/dN)_1 + f_2/(da/dN)_2] \tag{38}$$

The resultant growth law is obtained by substituting the da/dN-ΔK relation for each mechanism.

Mechanisms of Crack Propagation Threshold

The threshold stress-intensity factor ΔK_{th} can be decomposed into the intrinsic (effective) component and the extrinsic (closure) component as given by Eq 26. Several models have been proposed for the threshold effective stress-intensity range ΔK_{effth}. The lower limiting value for ΔK_{effth} will be the stress-intensity factor required for dislocation generation at the crack tip. Yokobori et al. [*63*] and Weertman [*64*] have derived the threshold stress intensity factor for dislocation emission. The dislocation structure ahead of the crack tip also influences the threshold value. Tanaka et al. [*65*] used the propagation of the blocked slipband across the grain boundary as a threshold criterion for fatigue crack growth.

Propagation of Short Fatigue Cracks

Propagation Behavior of Short Cracks

Fatigue cracks nucleated in smooth specimens first grow in an irregular manner because of microstructural inhomogeneity. Figure 15 shows the propagation of a fatigue crack nucleated at an inclusion in a smooth specimen of aluminum alloy (2024-T3) [*66*]. The propagation rate is plotted against the crack length measured from the inclusion center. (The crack length is the length of the projection of a crack on the plane perpendicular to the applied stress axis.) The crack growing along slip bands is often decelerated when it hits the grain boundary and inclusions or when it makes a sharp bend.

The relation between the crack growth rate and the crack length for three cracks is shown in Fig. 16*a*. Crack F6 continues to propagate to become the main crack, while Crack F20 becomes nonpropagating. Figure 16*b* shows the da/dN-a relation for 40 cracks. The dotted lines are the upper and lower bounds of data. Most of crack deceleration down to a rate of 5×10^{-11} m/cycle occurs at crack lengths less than 0.1 mm. The solid line in the figure was derived from the da/dN-K_{max} relation for large cracks by assuming that the crack is semicircular. The dashed line is the da/dN-ΔK_{eff} relation for large cracks. For small Stage I cracks, the crack growth rate can be even higher than that predicted from the da/dN-ΔK_{eff} relation for large cracks. The scatter of the data in the da/dN-a relation diminishes, and the data converge to the long-crack growth law at a crack length of 0.6 mm and a growth rate of 10^{-8} m/cycle.

The smallest length of cracks to which the long crack law of da/dN-ΔK is applicable is significant in engineering applications. Lankford [*67*] examined many published data and concluded that the merger of growth rates of short and long cracks coincided with the state where the maximum plastic zone size was equal to the relevant microstructural dimension. On the other hand, the transition from Stage I to Stage II growth is controlled by the microstructural dimension rather than the stress intensity range [*66,68,69*].

Fast, irregular growth of small fatigue cracks in smooth specimens was first observed in an aluminum alloy by Pearson [*70*]. Later, similar behaviors of small cracks have been reported for various metals and alloys. Three principal reasons have been proposed for the high growth rates of short fatigue cracks [*2,3*]: microstructural effect; premature crack closure; and macro-plasticity effect. Most of the short crack data which show anomalous growth (including the data shown in Fig. 16) have been obtained under nominally (macroscopically) elastic condition. Under these conditions, faster growth of short fatigue cracks has been ascribed to microplasticity or to anomalous crack-tip plasticity which takes place under less-constrained conditions (as in a single crystal) in surface grains, and to premature crack closure. Mechanics for microstructurally short cracks and for mechanically short cracks will be described next.

Microstructurally Short Cracks

For microstructurally short cracks, the resistance to crack growth is much influenced by the material microstructure such as the grain orientation and grain boundary. The crack

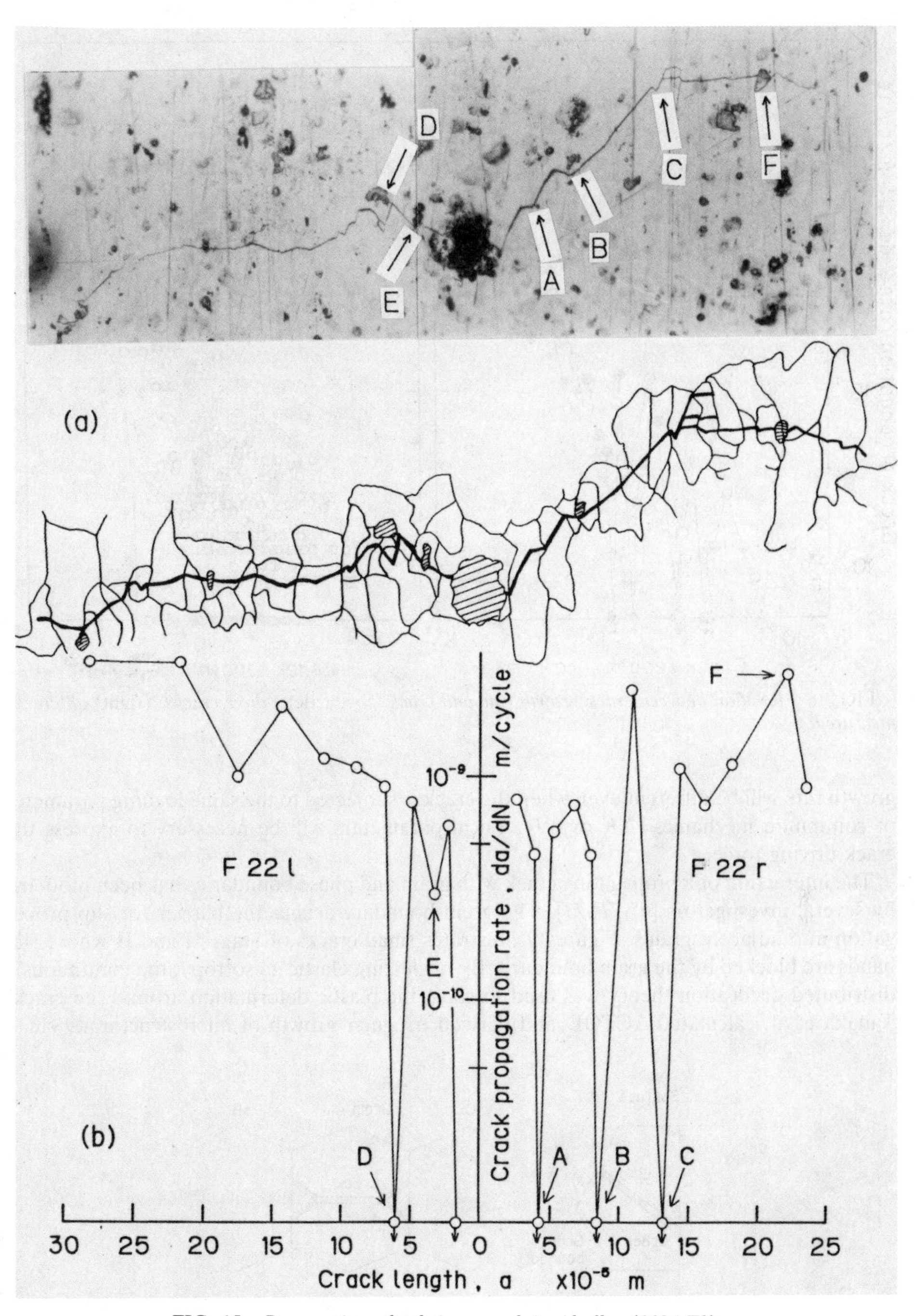

FIG. 15—*Propagation of a fatigue crack in Al alloy (2024-T3).*

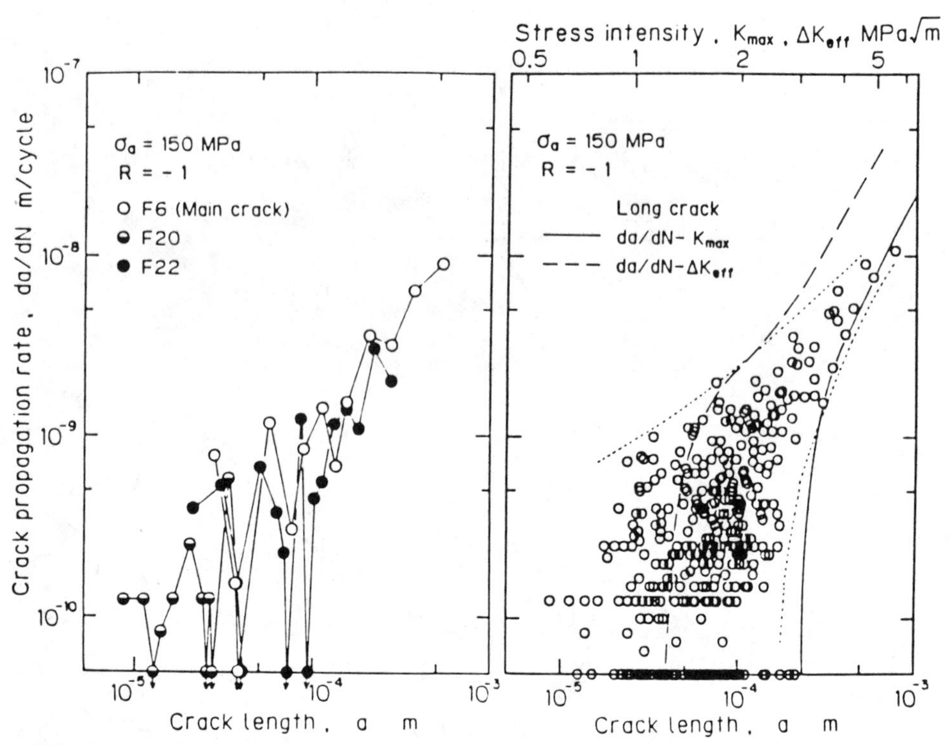

FIG. 16—*Relation between crack growth rate and crack length:* (left) *three cracks;* (right) *all cracks measured.*

growth rate will be different even when the crack is subjected to the same loading parameter of continuum mechanics (ΔK or ΔJ). Local parameters will be necessary to express the crack driving force.

The interaction of a propagating crack with grain and phase boundaries has been modeled by several investigators [*65,71–73*]. The grain boundary acts as the barrier for slip propagation into adjacent grains. Figure 17 illustrates small cracks of Stages I and II whose slip bands are blocked by the grain boundary. By neglecting elastic anisotropy, the continuously distributed dislocation theory was used to solve the plastic deformation around the crack. Tanaka et al. calculated ΔCTOD and derived irregular growth of microstructurally small

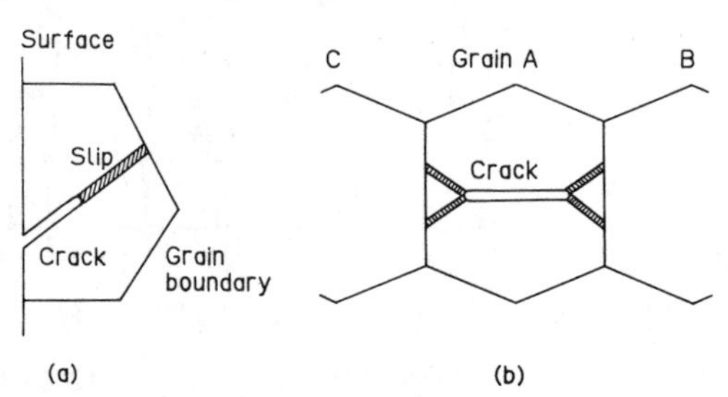

FIG. 17—*Micromechanical models for cracks interacting with grain boundary:* (a) *Stage I crack;* (b) *Stage II crack.*

cracks by assuming the crack growth range as a power function of ΔCTOD (Eq 28) [*73*]. Cracks decelerate as they approach the grain boundary.

The mechanical equation for Stage I crack growth is not well established. The contribution of crack closure is not clear. A mixed-mode loading may enhance the crack growth. Quantitative observation of the growth of Stage I crack in single crystals is in its infancy [*74,75*]. Any predictive method for the crack growth life of microstructurally small cracks must include the statistical nature of the interaction of cracks with the material microstructure.

Mechanically Short Cracks

When the crack length is larger than several times the grain size, the resistance of the material is regarded as homogeneous. For mechanically short (microstructurally long) cracks (Type 2 crack in Table 1), an appropriate choice of fracture mechanics parameters such as ΔK_{eff} or ΔJ will make crack-growth prediction possible.

All of the crack closure mechanisms indicate that the crack wake made by fatigue is responsible for crack closure. Since a short crack has a short crack wake, the amount of crack closure is expected to be smaller and, therefore, the growth rate is higher for shorter cracks. Premature crack closure plays a dominant role in anomalous, fast growth behavior of mechanically short cracks.

Jono et al. [*76*] measured the development of crack closure of a short crack (0.03 to 1 mm in length) in a specially designed specimen with a triangular cross section, using an improved unloading compliance method. Figure 18 shows the result for an aluminum alloy

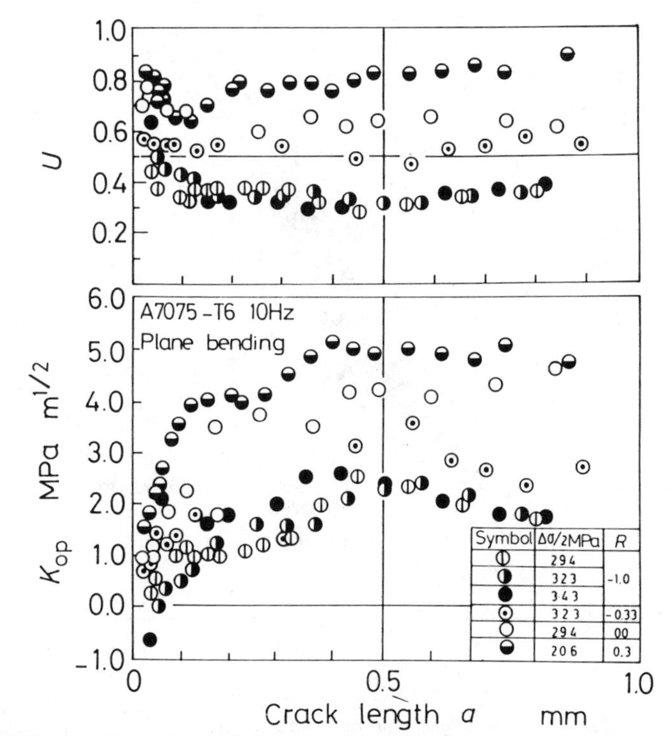

FIG. 18—*Change in crack closure with crack length in Al alloy (7075-T6).*

(7075-T6). For very short cracks, the opening stress is zero, as in an elastic crack, and increases with increasing crack length. U tends to decrease with crack length for R less than zero. The growth rate of short cracks is faster than that predicted from the da/dN-ΔK relation for long cracks; short and long crack data coincide in the da/dN-ΔK_{eff} relation.

Morris et al. [77] proposed a model for roughness-induced closure of short cracks, which utilized the measured value of surface roughness. The analysis of the plasticity-induced closure of short cracks by using Dugdale model indicates that the opening stress is lower under higher amplitude of stress cycling [*15*]. The acceleration of crack growth due to a small amount of macro-plasticity is predictable from an increase of ΔK_{eff} value.

When the amount of cyclic plasticity is large, crack closure adjustment is not enough; some elastic-plastic mechanics parameters are necessary to evaluate the crack growth rate. Among several parameters proposed, the J-integral seems to be most successful. The J-integral range, ΔJ, can be estimated from measurements of the cyclic hysteresis loop and crack closure [*78,79*]. The amount of crack closure is smaller for shorter cracks under larger plastic strain, and a crack is open even at a minimum (compressive) stress under large cyclic straining [*79*].

Figure 19 shows the relation between the crack propagation rate and ΔJ in the low-cycle

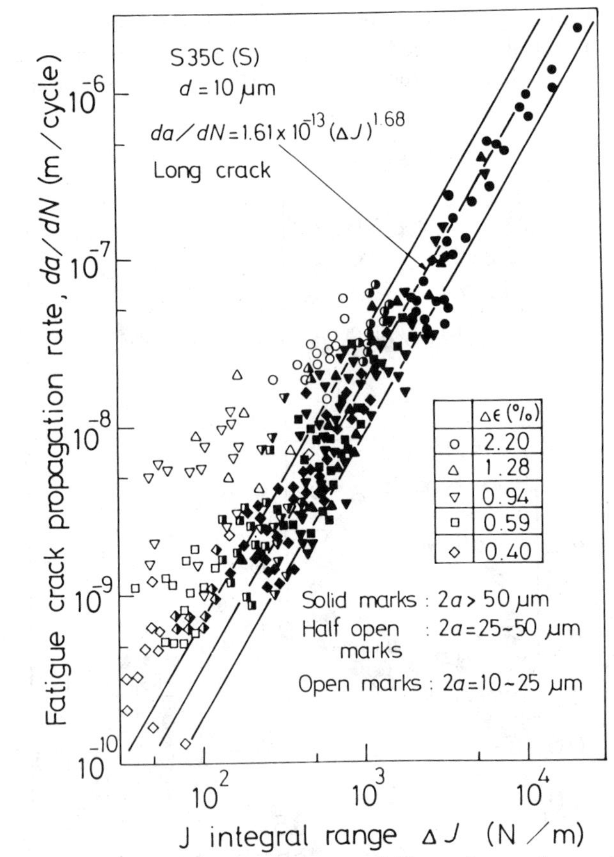

FIG. 19—*Relation between crack propagation rate and J-integral range for small cracks in medium-carbon steel.*

fatigue of smooth specimens of medium-carbon steel with a grain size of 10 μm [79]. The solid line indicates the da/dN-ΔJ relation for long cracks (with lengths greater than 5 mm) which is given by Eq 27 where $C = 1.61 \times 10^{-13}$ and $m = 1.68$. The solid marks show the data for cracks whose lengths are longer than three to five times the grain size. The data fall within the band (twice to half the rate given by the long crack relation). On the other hand, the data for cracks shorter than about twice the grain size, shown with the open marks, lie above the solid line. Hoshide et al. [79] attributed the growth acceleration of microstructurally small cracks to the difference in crack growth micromechanisms. A dominant mechanism is the intergranular fracture for smaller cracks, while the transgranular fracture with striations for larger cracks.

Fatigue Crack Propagation from Notches and Defects

Crack Propagation Behavior from Notches

The growth rate of a fatigue crack nucleated at the tip of a sharp notch first decreases as the crack propagates. The crack then becomes nonpropagating under low stresses, while it begins to accelerate after reaching a minimum growth rate under high stresses. Linear elastic fracture mechanics (LEFM) is applicable to a crack which has grown far away from the notch. Since ΔK is an increasing function of crack length under a constant applied stress, a dip in the growth rate of short cracks near the notch can not be predicted on the basis of the ΔK-based fracture mechanics approach.

Hammouda et al. [80] ascribed the crack acceleration and deceleration near the notch tip to the notch plasticity effect. Haddad et al. [81] explained the decreasing growth rate with crack length by combining the anomalous crack growth behavior due to the shortness of a crack with the notch plasticity effect. Tanaka and Nakai [82,83], Akiniwa and Tanaka [84], and Nishikawa et al. [85] have shown that crack closure is primarily responsible for the acceleration and deceleration of short crack growth near the notch.

Figure 20 shows the relations of dc/dN against ΔK and ΔK_{eff} obtained from a center-notched plate of low-carbon steel with various notch-tip radii between 0.16 and 0.83 mm. The solid and dashed lines are the relations for long cracks. A crack can propagate below ΔK_{th} and the rate decelerates with increasing crack length. In the dc/dN-ΔK_{eff} diagram, all the data follow a unique relation which agrees fairly well with the dashed line. The anomalous, decreasing behavior of short crack growth near the notch tip can be explained by the decreasing driving force due to the development of crack closure with crack growth.

When the applied stress is very high and when the notch plasticity is no longer constrained, an elastic-plastic fracture mechanics parameter, such as the J-integral, is necessary to predict the growth rate [85–87]. In the majority of notch fatigue cases, however, ΔK_{eff} is sufficient for explaining the growth behavior of short cracks. Several analytical [88] and finite-element [89,90] models have been proposed to estimate the crack closure of the notch-tip crack. Notch plasticity and the shortness of cracks are two reasons for a reduced amount of crack closure for short cracks at the notch root.

Crack Propagation Threshold of Notched and Cracked Components

For a nonpropagating short crack at the notch root, the threshold SIF range, ΔK_{th}, decrease as the crack becomes shorter, while the effective SIF range, ΔK_{effth}, is nearly constant [82–85]. The value of K_{opth} increases with crack length. Figure 21 shows the change of the crack opening stress intensity factor with the crack length at the threshold under $R = -1$ for low-

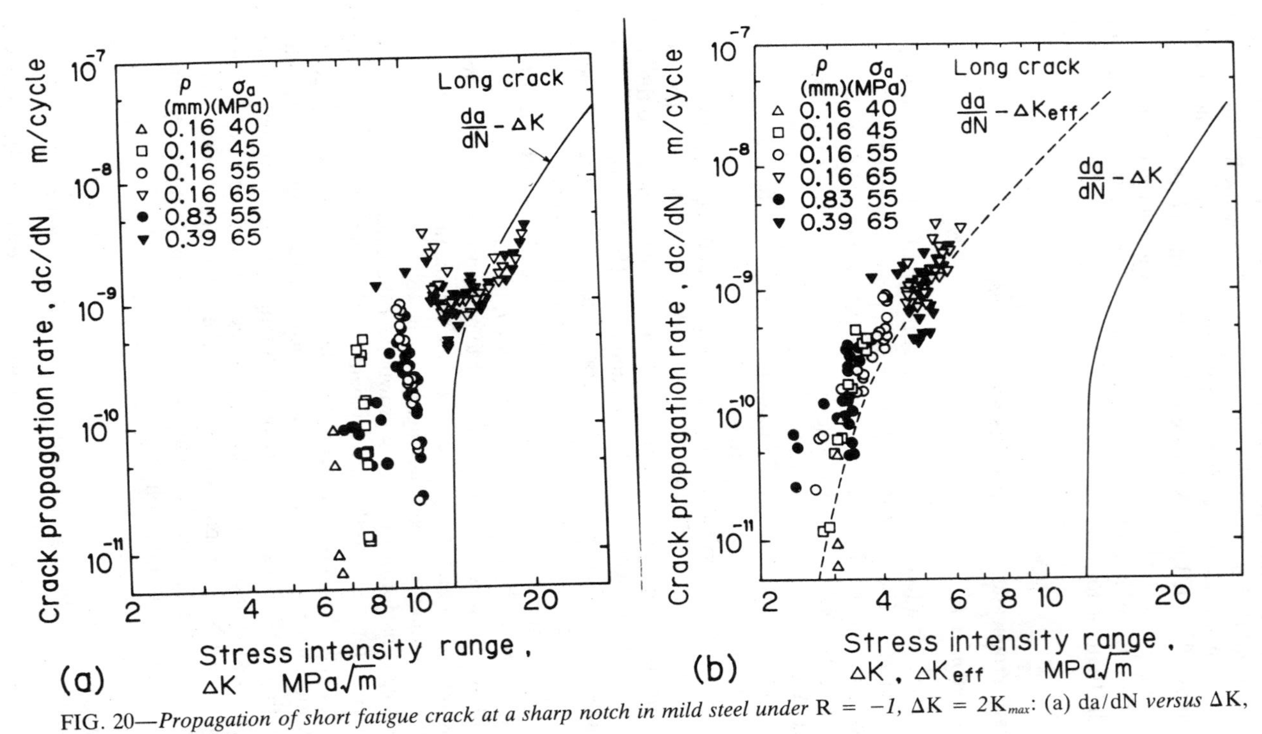

FIG. 20—*Propagation of short fatigue crack at a sharp notch in mild steel under* R = *−1*, $\Delta K = 2K_{max}$: (a) da/dN *versus* ΔK, (b) da/dN *versus* ΔK_{eff}.

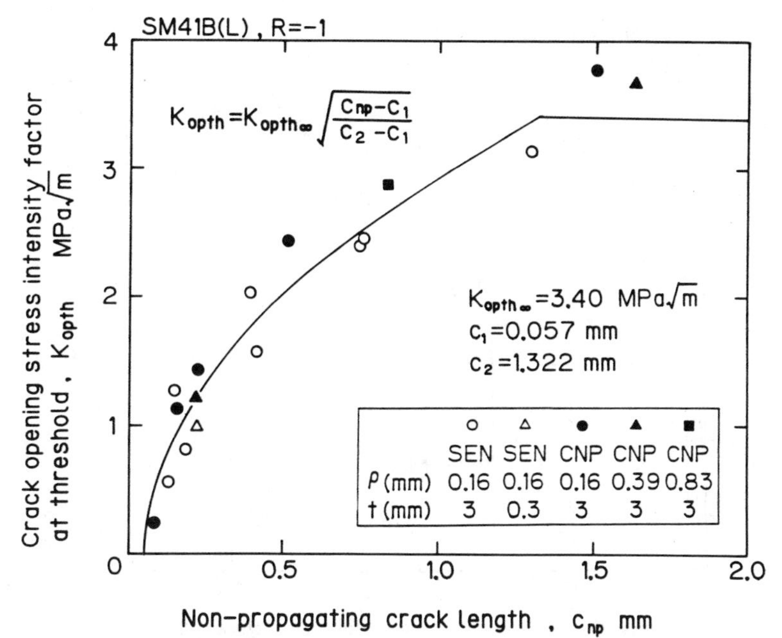

FIG. 21—*Change in crack closure with crack length at the threshold in mild steel.*

carbon steel specimens having various notch-tip radii and depths [*91*]. The change of K_{opth} with c_{np} is approximated by the following equation for all notches examined.

$$K_{\text{opth}} = K_{\text{opth}\infty} \left[\frac{c_{np} - c_1}{c_2 - c_1} \right]^{1/2} \tag{37}$$

where c_1 is the length of a Stage I crack measured on the fracture surface and c_2 is the limiting crack length above which $K_{\text{opth}} = K_{\text{opth}\infty}$ ($K_{\text{opth}\infty}$ is the value obtained for a long crack).

The resistance curve (R-curve) method can be applied to predict the threshold of crack growth of notched components, as shown in Fig. 22. The R curve is constructed in terms of the maximum stress intensity factor, which is given by

$$K_{\text{maxth}} = \Delta K_{\text{effth}} + K_{\text{opth}} \tag{38}$$

The change of the applied K_{max} under a given value of the applied nominal stress is shown with the dashed line in Fig. 22. A Stage II crack can be made at the notch root when the stress is larger than σ_{w1} and it continues to propagate when the stress is larger than σ_{w2}. A nonpropagating crack is formed for stresses between σ_{w1} and σ_{w2}. The effects of loading mode and specimen geometry on σ_{w1} and σ_{w2} are taken into account by the applied K_{max} value. A similar technique can be applied for predicting the fatigue threshold of pre-cracked components.

Kitagawa and Takahashi [*92*] first showed that the value of ΔK_{th} decreased as the crack became shorter and that the threshold stress approached the fatigue limit of smooth specimens, $\Delta\sigma_{w0}$. The crack length at the intersection of the SIF-constant criterion and the stress-

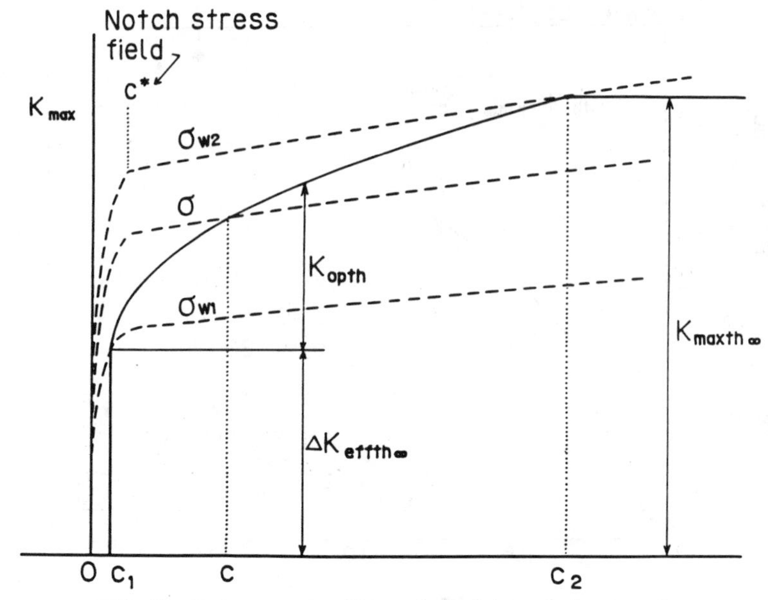

FIG. 22—*Resistance curve (R-curve) for fatigue cracks at notches.*

range-constant criterion is given by

$$a_0 = (\Delta K_{th\infty}/\Delta\sigma_{w0})^2/\pi \tag{39}$$

where $\Delta K_{th\infty}$ is the threshold SIF range for long cracks (SIF is assumed to be $K = \sigma(\pi a)^{1/2}$ as for an isolated crack in an infinite plate).

Figure 23 shows the change in the threshold stress range $\Delta\sigma_{th}$ with crack length a for various metals, where $\Delta\sigma_{th}$ is normalized by $\Delta\sigma_{w0}$ and a by a_0 [*65*]. The crack length is converted to the length of an isolated crack by using the SIF-equality equation. The horizontal dashed line corresponds to $\Delta\sigma_{th} = \Delta\sigma_{w0}$, and the dashed line with a slope of $-1/2$ to $\Delta K_{th} = \Delta K_{th\infty}$. The data are scattered along or below these dashed lines.

Haddad et al. [*93*] proposed the model of fictitious crack length which gives the solid-line relation in Fig. 23. They assumed that the threshold condition for crack growth was determined by the constant value of the SIF range ($=\Delta K_{th\infty}$) for a fictitious crack whose length was the actual length plus the intrinsic crack length a_0 where

$$\Delta K_{th\infty} = \Delta\sigma_{th}[\pi(a + a_0)]^{1/2} \tag{40}$$

Using Eq 39, Eq 40 can be rewritten as

$$\Delta\sigma_{th} = \Delta\sigma_{w0}[a_0/(a + a_0)]^{1/2} \tag{41}$$

In Fig. 23, the data lie on or above the solid line.

Although Haddad's model is successful in explaining the experimental data on $\Delta\sigma_{th} - a$, it does not include crack closure. Tanaka and Nakai [*83*] and Tanaka et al. [*65*] gave a different interpretation to Eq 41 on the basis of crack closure and micromechanical threshold

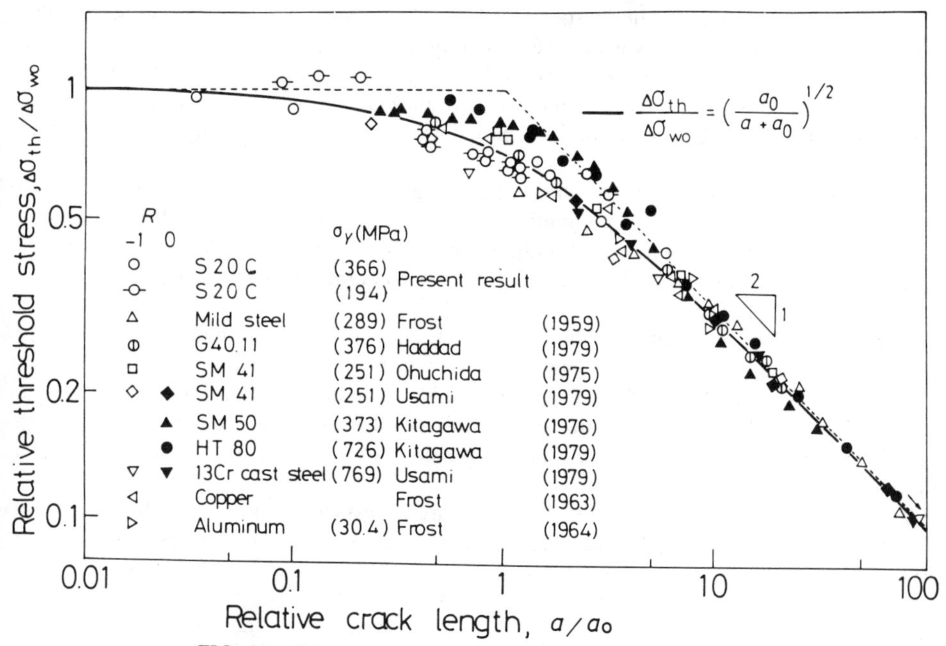

FIG. 23—*Relation between threshold stress and crack length.*

model. The effective stress intensity range ΔK_{effth} is nearly constant when the crack length is longer than the characteristic length given by

$$a_0' = (\Delta K_{effth\infty}/\Delta\sigma_{w0})^2/\pi \tag{42}$$

where $\Delta K_{effth\infty}$ is the effective stress intensity range for long cracks. For cracks with lengths between a_0' and a_0, the increase in ΔK_{th} with crack length is caused by crack closure. For cracks shorter than a_0', micromechanical analysis is necessary.

Conclusions

Current researches on the mechanics of fatigue crack propagation cover short cracks as well as long cracks, aiming to establish unified, quantitative description of crack propagation from the very early to the last stages of the fatigue fracture process. The subjects of future studies will be summarized as follows:

1. Mechanical modeling of mechanisms of fatigue crack propagation need to be advanced. Non-similarity of the geometry should be included in the crack-tip blunting mechanisms. Mechanics for other cracking mechanisms need to be developed.
2. The law of fatigue crack propagation under mixed-mode loading is not yet well established to give the rate and direction of crack growth in anisotropic as well as isotropic materials.
3. Most of the influences of mean stress and variable amplitude loading on crack propagation result from crack closure. ΔK_{eff} yields a unique correlation. The prediction method of crack closure under various loading conditions needs to be developed.

4. Under macroscopic elastic-plastic conditions, ΔJ seems to be the best parameter. The estimation method of ΔJ for various cracks is required.

5. For mechanically short (microstructurally long) cracks, an appropriate choice of fracture mechanics parameters (ΔK_{eff} or ΔJ) will give the same growth equation as long cracks. The values of ΔK_{eff} or ΔJ need to be determined as a function of crack length under various loading conditions.

6. For microstructurally short cracks, the crack growth rate will be different even when the crack is subjected to the same loading parameter. Local parameters will be necessary to express the crack driving force. ΔCTOD is a candidate. Mixed-mode loading complicates crack growth mechanics. Reliable experimental data on crack growth kinetics are essential for model development.

7. Because of microstructural inhomogeneities, the statistical treatment is necessary to derive any productive method for small crack growth life.

8. With respect to the nucleation stage of microstructurally small cracks, that is, the region of cracks shorter than about 10 μm, only preliminary results on the crack growth kinetics have been obtained. More quantitative data are required to advance mechanical modeling.

9. Both experimental and theoretical studies are necessary to understand the mechanics of fatigue crack propagation in new advanced structural materials, such as ceramics and composites.

References

[1] Tomkins, B. in *Fatigue Crack Growth,* R. A. Smith, Ed., Pergamon Press, Oxford, 1986, pp. 133–146.
[2] Suresh, S. and Ritchie, R. O., *International Metals Review,* Vol. 29, 1984, pp. 445–476.
[3] Tanaka, K. *JSME International Journal,* Japan Society of Mechanical Engineers, Vol. 35, 1987, pp. 1–13.
[4] Paris, P. C., Gomez, M. P., and Anderson, W. E., *The Trend in Engineering,* Vol. 13, 1961, pp. 9–14.
[5] Paris, P. C. and Erdogan, F., *Transactions,* American Society of Mechanical Engineers, *Journal of Basic Engineering,* Vol. 85, 1963, pp. 528–534.
[6] Rice, J. R. in *Fatigue Crack Propagation, ASTM STP 415,* American Society for Testing and Materials, Philadelphia, 1967, pp. 247–311.
[7] Dugdale, D. S., *Journal of the Mechanics and Physics of Solids,* Vol. 8, 1960, pp. 100–104.
[8] Hahn, G. T., Hoagland, R. G., and Rosenfield, A. R., *Metallurgical Transactions A,* Vol. 3A, 1972, pp. 1189–1202.
[9] Tanaka, K., Hojo, M., and Nakai, Y., *Materials Science and Engineering,* Vol. 55, 1982, pp. 85–96.
[10] Lankford, J., Davidson, D. L., and Cook, T. S. in *Cyclic Stress-Strain and Plastic Deformation Aspects of Fatigue Crack Growth, ASTM STP 637,* American Society for Testing and Materials, Philadelphia, 1977, pp. 36–55.
[11] Elber, W. in *Damage Tolerance in Aircraft Structures, ASTM STP 486,* American Society for Testing and Materials, Philadelphia, 1971, pp. 230–242.
[12] Suresh, S. and Ritchie, R. O. in *Fatigue Crack Growth Threshold Concepts,* D. Davidson and S. Suresh, Eds., The Metallurgical Society of the American Institute for Mining, Metallurgical, and Petroleum Engineers, 1984, pp. 227–262.
[13] Fleck, N. A. in *Fatigue Crack Growth,* R. A. Smith, Ed., Pergamon Press, Oxford, 1986, pp. 75–88.
[14] Budiansky, B. and Hutchinson, J. W., *Transactions,* American Society of Mechanical Engineers, *Journal of Applied Mechanics,* Vol. 45, 1978, pp. 267–276.
[15] Tanaka, K. and Nakai, Y. in *Fatigue Crack Growth Threshold Concepts,* D. L. Davidson and S. Suresh, Eds., The Metallurgical Society of the American Institute for Mining, Metallurgical, and Petroleum Engineers, 1984, pp. 497–516.

[16] Ohji, K., Ogura, K., and Ohkubo, Y., *Engineering Fracture Mechanics,* Vol. 7, 1975, pp. 457–464.
[17] Newman, J. C. in *Cyclic Stress-Strain and Plastic Deformation Aspects of Fatigue Crack Growth, ASTM STP 637,* American Society for Testing and Materials, Philadelphia, 1977, pp. 56–80.
[18] Suresh, S. and Ritchie, R. O. in *Fatigue Crack Growth Threshold Concepts,* D. L. Davidson and S. Suresh, Eds., The Metallurgical Society of the American Institute for Mining, Metallurgical, and Petroleum Engineers, 1984, pp. 227–261.
[19] Beevers, C. J., Bell, K., Carlson, R. L., and Starke, E. A., *Engineering Fracture Mechanics,* Vol. 19, 1984, pp. 93–100.
[20] Masing, G. in *Proceedings of 2nd International Congress on Applied Mechanics,* Zurich, 1926.
[21] Lamba, H. S., *Engineering Fracture Mechanics,* Vol. 7, 1975, pp. 693–703.
[22] Wüthrich, C., *International Journal of Fracture,* Vol. 20, 1982, pp. R35–R37.
[23] Smith, R. A., *International Journal of Fracture,* Vol. 13, 1977, pp. 717–720.
[24] Gangloff, R. P., *Res Mechanica Letters,* Vol. 1, 1981, pp. 299–306.
[25] Hudak, S. J. and Chan, K. S. in *Small Fatigue Cracks,* R. O. Ritchie and J. Lankford, Eds., The Metallurgical Society of the American Institute for Mining, Metallurgical, and Petroleum Engineers, 1986, pp. 379–405.
[26] Ritchie, R. O., *Metal Science,* Vol. 11, 1977, pp. 368–381.
[27] Forman, R. G., Kearing, V. E., and Engle, R. M., *Transactions,* American Society of Mechanical Engineers, *Journal of Basic Engineering,* Vol. 89, 1967, pp. 459–464.
[28] Weertman, J., *International Journal of Fracture Mechanics,* Vol. 2, 1966, pp. 460–467.
[29] Weertman, J., *International Journal of Fracture Mechanics,* Vol. 5, 1969, pp. 13–15.
[30] Klesnil, M. and Lukas, P., *Engineering Fracture Mechanics,* Vol. 4, 1972, pp. 77–92.
[31] Saxena, A., Hudak, S. J., and Jouris, G. M., *Engineering Fracture Mechanics,* Vol. 12, 1979, pp. 103–115.
[32] McEvily, A. J. in *Fatigue Mechanisms: Advances in Quantitative Measurement of Physical Damage, ASTM STP 811,* American Society for Testing and Materials, Philadelphia, 1983, pp. 283–312.
[33] Roberts, R. and Erdogan, F., *Transactions,* American Society of Mechanical Engineers, *Journal of Basic Engineering,* Vol. 89, 1967, pp. 885–892.
[34] Klesnil, M. and Lukas, P., *Materials Science and Engineering,* Vol. 9, 1972, pp. 231–240.
[35] Arad, S., Radon, J. C., and Culver, L. E., *Journal of Mechanical Engineering Science,* Vol. 13, 1971, pp. 75–81.
[36] Mostovoy, S. and Ripling, E. J., *Polymer Science Technology B,* Vol. 9B, 1975, p. 513.
[37] Badaliance, R., *Engineering Fracture Mechanics,* Vol. 13, 1986, pp. 657–666.
[38] Schmidt, R. A. and Paris, P. C. in *Progress in Flaw Growth and Fracture Toughness Testing, ASTM STP 576,* American Society for Testing and Materials, Philadelphia, 1973, pp. 79–94.
[39] Nakai, Y., Tanaka, K., and Kawashima, R., *Journal of the Society of Materials Science* (Japan), Vol. 33, 1984, pp. 1045–1051.
[40] Kikukawa, M., Jono, M., Tanaka, K., and Kondo, Y., *International Journal of Fracture,* Vol. 13, 1977, pp. 702–704.
[41] Suresh, S., *Engineering Fracture Mechanics,* Vol. 18, 1983, pp. 577–593.
[42] Otsuka, A., Mori, K., and Miyata, T., *Engineering Fracture Mechanics,* Vol. 7, 1975, pp. 429–439.
[43] Tanaka, K., *Engineering Fracture Mechanics,* Vol. 6, 1974, pp. 493–507.
[44] Hojo, M., Gustafson, C-G., Tanaka, K., and Hayashi, R., *Journal of the Society of Materials Science* (Japan), Vol. 36, 1987, pp. 222–228.
[45] Sih, G. C. and Barthelemy, B. M., *Engineering Fracture Mechanics,* Vol. 13, 1980, pp. 439–451.
[46] Kibler, J. J. and Roberts, R., *Transactions,* American Society of Mechanical Engineers, *Journal of Engineering for Industry,* Vol. 92, 1970, pp. 727–734.
[47] Miller, K. J., *Metal Science,* Vol. 11, 1977, pp. 432–438.
[48] Hoshide, T., Tanaka, K., and Yamada, A., *Fatigue of Engineering Materials and Structures,* Vol. 4, 1981, pp. 355–366.
[49] Dowling, N. E. and Begley, J. A. in *Mechanics of Crack Growth, ASTM STP 590,* American Society for Testing and Materials, Philadelphia, 1976, pp. 82–103.
[50] Hoshide, T., Tanaka, K., and Nakata, M., *Journal of the Society of Materials Science* (Japan), Vol. 31, 1982, pp. 566–572.
[51] McEvily, A. J., Beukelmann, D., and Tanaka, K. in *Proceedings of 1974 Symposium on Mechanical Behavior of Materials,* Kyoto, Vol. 1, 1974, pp. 269–281.
[52] Taira, S., Tanaka, K., and Ogawa, S., *Journal of the Society of Materials Science* (Japan), Vol. 26, 1977, pp. 93–98.

[53] Erdogan, F. and Roberts, R. in *Proceedings of International Conference on Fracture,* Sendai, Japan, Vol. 1, 1965, pp. 341–362.
[54] Taira, S. and Tanaka, K. in *Proceedings of International Conference on Mechanical Behavior of Materials,* Kyoto, Vol. 2, 1972, pp. 48–58.
[55] Laird, C. and Smith, G. C., *Philosophical Magazine,* Vol. 7, 1962, pp. 847–857.
[56] Neumann, P., *Acta Metallurgica,* Vol. 22, 1974, pp. 1155–1165.
[57] Kikukawa, M., Jono, M., and Adachi, M. in *Fatigue Mechanisms, ASTM STP 675,* American Society for Testing and Materials, Philadelphia, 1979, pp. 234–253.
[58] Kuo, A. S. and Liu, H. W., *Scripta Metallurgica,* Vol. 14, 1980, pp. 525–530.
[59] Tomkins, B., *Metal Science,* Vol. 14, 1980, pp. 408–417.
[60] Tanaka, K., Hoshide, T., and Sakai, N., *Engineering Fracture Mechanics,* Vol. 19, 1984, pp. 805–825.
[61] Shih, C. F., *Journal of the Mechanics and Physics of Solids,* Vol. 29, 1981, pp. 305–326.
[62] Bates, R. C. and Clark, W. G., Jr., *Transactions,* American Society of Mechanical Engineers, Vol. 62, 1969, pp. 380–389.
[63] Yokobori, T., Yokobori, A.T., Jr., and Kamei, A., *International Journal of Fracture,* Vol. 12, 1976, pp. 519–520.
[64] Weertman, J. in *Mechanics of Fatigue, ASME AMD-Vol. 7,* American Society of Mechanical Engineers, 1981, pp. 11–19.
[65] Tanaka, K., Nakai, Y., and Yamashita, M., *International Journal of Fracture,* Vol. 17, 1981, pp. 519–533.
[66] Takana, K., Akiniwa, Y., and Matsui, E., *Materials Science and Engineering,* Vol. A104, 1988, pp. 105–115.
[67] Lankford, J., *Fatigue of Engineering Materials and Structures,* Vol. 8, 1985, pp. 161–175.
[68] Tanaka, K., Hojo, M., and Nakai, Y. in *Fatigue Mechanisms, Advances in Quantitative Measurement of Fatigue Damage, ASTM STP 811,* American Society for Testing and Materials, Philadelphia, 1983, pp. 207–232.
[69] Tokaji, K., Ogawa, T., Harada, Y., and Ando, Z., *Fatigue of Engineering Materials and Structures,* Vol. 9, 1986, pp. 1–14.
[70] Pearson, S., *Engineering Fracture Mechanics,* Vol. 7, 1975, pp. 235–247.
[71] Morris, W. L., *Metallurgical Transactions A,* Vol. 11A, 1980, pp. 1117–1123.
[72] Rios, E. R., Mohamed, H. J., and Miller, K. J., *Fatigue of Engineering Materials and Structures,* Vol. 8, 1985, pp. 49–63.
[73] Tanaka, K., Akiniwa, Y., Nakai, Y., and Wei, R. P., *Engineering Fracture Mechanics,* Vol. 24, 1986, pp. 803–819.
[74] Basinski, Z. S. and Basinski, S. J., *Acta Metallurgica,* Vol. 33, 1985, pp. 1319–1327.
[75] Neumann, P. and Tonnessen, A. in *Small Fatigue Cracks,* R. O. Ritchie and J. Lankford, Eds., The Metallurgical Society of the American Institute for Mining, Metallurgical, and Petroleum Engineers, 1986, pp. 41–47.
[76] Jono, M., Song, J., Yama, Y., Nishigaichi, N., Okabe, N., and Kikukawa, M., *Transactions A,* Japan Society of Mechanical Engineers, Vol. 51, 1985, pp. 1677–1686.
[77] Morris, W. L., James, M. R., and Buck, O., *Engineering Fracture Mechanics,* Vol. 18, 1983, pp. 871–877.
[78] Dowling, N. E., in *Cyclic Stress-Strain and Plastic Deformation Aspects of Fatigue Crack Growth, ASTM STP 637,* American Society for Testing and Materials, Philadelphia, 1977, pp. 97–121.
[79] Hoshide, T., Yamada, T., Fujimura, S., and Hayashi, T., *Engineering Fracture Mechanics,* Vol. 21, 1985, pp. 85–101.
[80] Hammouda, M. M., Smith, R. A., and Miller, K. J., *Fatigue of Engineering Materials and Structures,* Vol. 2, 1979, pp. 139–154.
[81] El Haddad, M. H., Smith, K. N., and Topper, T. H. in *Fracture Mechanics (11th Conference), ASTM STP 677,* American Society for Testing and Materials, Philadelphia, 1979, pp. 274–289.
[82] Tanaka, K. and Nakai, Y., *Fatigue of Engineering Materials and Structures,* Vol. 6, 1983, pp. 315–327.
[83] Tanaka, K. and Nakai, Y., *Transactions,* American Society of Mechanical Engineers, *Journal of Engineering Materials and Technology,* Vol. 106, 1984, pp. 194–327.
[84] Akiniwa, Y. and Tanaka, K., *Transactions A,* Japan Society of Mechanical Engineers, Vol. A53, 1987, pp. 393–400.
[85] Nishikawa, I., Konishi, M., Miyoshi, Y., and Ogura, K., *Journal of the Society of Materials Science* (Japan), Vol. 35, 1986, pp. 904–910.

[86] El Haddad, M. H., Dowling, N. E., Topper, T. H., and Smith, K. N., *International Journal of Fracture,* Vol. 16, 1980, pp. 15–30.
[87] Ohji, K., Nakai, Y., Ochi, T., and Mura, M., *Transactions A,* Japan Society of Mechanical Engineers, Vol. A51, 1985, pp. 2067–2075.
[88] Tanaka, K. and Akiniwa, Y., *Transactions A,* Japan Society of Mechanical Engineers, Vol. A52, 1986, pp. 1741–1748.
[89] Ohji, K., Ogura, K., and Ohkubo, K., *Transactions,* Japan Society of Mechanical Engineers, Vol. 42, 1976, pp. 643–648.
[90] Newman, J. C., Jr. in *Proceedings of AGARD Conference,* North Atlantic Treaty Organization, No. 328, 1983, pp. 6.1–6.26.
[91] Tanaka, K. and Akiniwa, Y., *Engineering Fracture Mechanics,* Vol. 30, 1988, pp. 863–876.
[92] Kitagawa, H. and Takahashi, S. in *Proceedings of 2nd International Conference on Mechanical Behavior of Materials,* Boston, American Society for Metals, 1976, pp. 627–631.
[93] El Haddad, M. H., Smith, K. N., and Topper, T. H., *Transactions,* American Society of Mechanical Engineers, *Journal of Engineering Materials and Technology,* Vol. 101, 1979, pp. 42–46.

E. A. Starke, Jr., [1] *and J. C. Williams*[2]

Microstructure and the Fracture Mechanics of Fatigue Crack Propagation

REFERENCE: Starke, E. A., Jr., and Williams, J. C., **"Microstructure and the Fracture Mechanics of Fatigue Crack Propagation,"** *Fracture Mechanics: Perspectives and Directions (Twentieth Symposium), ASTM STP 1020,* R. P. Wei and R. P. Gangloff, Eds., American Society for Testing and Materials, Philadelphia, 1989, pp. 184–205.

ABSTRACT: Microstructure is the principal independent variable which can be used to control the fatigue crack growth rate (FCGR) once loading conditions and environment are established. Microstructure affects the FCGR through its influence on strain distribution (slip character), slip length, plastic zone size, and crack path. These factors, in turn, can influence both crack closure and the materials sensitivity to aggressive environments. This paper reviews the authors' perception of the current understanding of the effect of microstructure on FCGRs.

KEY WORDS: age-hardened, alloy, coherent, continuous ferrite (CF), continuous martensite (CM), crack path, dispersoids, environment, fatigue crack propagation (FCP), fatigue crack growth rate (FCGR), linear elastic fracture mechanics (LEFM), microstructure, plastic zone size, single phase, slip, strain, stress-intensity factor, two phase

Some order was put in the study of fatigue crack propagation (FCP) of metals and alloys by the introduction of linear elastic fracture mechanics (LEFM) concepts by Paris and coworkers in 1961 [*1*]. Later, Paris and Erdogan [*2*] showed the functional relationship between the fatigue crack growth rate (FCGR) expressed as increment of crack extension per cycle (da/dN) and the stress-intensity range ΔK. Following the establishment of the Paris-Erdogan relationship, a large volume of information was generated which related fatigue crack growth rates to other parameters, for example, temperature, frequency, environment, specimen geometry, and load ratio (R). The initial discovery of the crack closure phenomenon by Elber [*3*] illustrated some of the many complexities of the fatigue crack propagation process and helped to explain the effect of the load ratio R on da/dN. In addition to these external influences on fatigue crack propagation, the material parameters such as E, G, yield strength, cyclic work hardening behavior, slip mode, and grain orientation have also been shown to affect fatigue crack growth behavior. Recently, there has been an attempt to separate the influence of these "intrinsic" parameters from the influence of "extrinsic" effects associated with closure. However, these material parameters depend on an alloy's composition and microstructure, which in turn, control deformation and fracture behavior. Since the deformation and fracture characteristics may well determine the nature and extent to which crack

[1] Dean, School of Engineering and Applied Science, Thornton Hall, University of Virginia, Charlottesville, VA 22901.

[2] General manager, Engineering Materials Technology Laboratories, General Electric Company, P.O. Box 156301, Cincinnati, OH 45215-6301.

closure and environment influence FCGRs, there is no clear separation of intrinsic and extrinsic effects.

Plastic deformation is a prerequisite for fatigue failure and, therefore, the reversed plastic zone at the tip of the crack is an important parameter in the consideration of fatigue crack growth rates. Using LEFM concepts, the plastic zone at the tip of a crack is defined by the contour ahead of the crack tip at which the local stress is equal to the yield stress of the alloy. In the case of cyclic loading, the reversed plastic zone replaces this plastic zone. The reversed plastic zone is that volume which undergoes yielding due to both tensile and compressive stresses. Its maximum dimension can be expressed by [*4*]

$$r_{RP} = 0.033(\Delta K/\sigma'_{ys})^2 \tag{1}$$

where ΔK is the stress intensity range and σ'_{ys} is the cyclic yield stress. The corresponding tensile crack-opening displacement range can be expressed by [*4*]

$$\Delta\delta = [(1 + R)/(1 - R)]\Delta K^2/4\sigma'_{ys}E \tag{2}$$

where R is the load ratio ($R \propto K_{min}/K_{max}$). The LEFM approach assumes that the strain within the plastic zone is homogeneous. However, for real materials, this assumption is seldom fulfilled and further deviations occur in the presence of strain localization, which is fairly common in high strength materials. Strain localization occurs when a limited number of slip systems are activated, when the material deforms by planar slip, or when both occur. The strain distribution within the plastic zone may also be inhomogeneous if the plastic zone encompasses more than one grain or constituent since the slip character or the number of activated slip systems may vary between these microstructural entities.

Although the importance of the reversed plastic zone and crack opening displacement in fatigue crack propagation is widely accepted, there is some disagreement as to the detailed mechanism which causes each increment of fatigue crack growth during cyclic loading. Two different viewpoints were developed in the early 1960s [*5*] and both still have their advocates [*6*]. One view is that fracture occurs after "damage" accumulation, and the other is that fracture occurs by a plastic sliding-off process at the crack tip. An early observation of Forsyth and Ryder [*7*] that a fatigue crack advances by an increment Δa in each cycle, has been used to support the "plastic sliding off" mechanism. However, recent studies by Lankford and Davidson [*8*] have clearly shown that under cyclic loading a crack tip will tend to open and close for a number of cycles and then extend a small distance on the next cycle. Repetition of this process causes incremental crack advance requiring multiple cycles, the number of cycles decreasing with increasing cyclic stress intensity ΔK. Consequently, the relative importance of each mechanism is dependent on the magnitude of ΔK. Resolution of this point is complicated because the macroscopically measured crack growth rate represents an average value along the crack front, whereas the proposed mechanisms tend toward a localized, two-dimensional view. Lankford and Davidson's observations are interesting; however, they are of necessity made at a free surface. Thus, the broad applicability of these findings is not clear.

Although the fundamental mechanisms associated with fatigue extension are still unresolved and the interactions and effects of the various parameters mentioned previously are not clearly understood, enough information has been accumulated over the past 20 years to predict qualitatively the effect of a variety of microstructural features, and deformation behavior, on fatigue crack growth rates of metals and alloys.

Microstructural Effects on Fatigue Crack Propagation

There are five major factors which should be considered when attempting to establish the relationship between microstructure and fatigue crack propagation [*9*]:

1. Strain distribution (slip character).
2. Slip length and plastic zone size.
3. Crack path and crack extension forces.
4. Morphology and properties of constituents in multiphase alloys.
5. Environment.

Effect of Strain Distribution (Slip Character)

Starke and Lütjering [*10*] reviewed the effect of strain distribution on fatigue crack initiation and clearly showed that strain localization promotes crack initiation. However, strain localization almost always reduces the rate of crack propagation. The reasons for this reduction will be discussed in the following section.

Single-Phase Alloys—Descriptions of the slip character in single-phase alloys have been discussed elsewhere [*11*]. Slip may be considered as wavy or planar and coarse or fine. Since multiple slip planes with a common slip direction are involved in wavy slip, this type of plastic deformation may be considered to be the closest approximation to homogeneous. However, in materials that exhibit planar slip, the situation is not as simple: planar slip can be either fine or coarse. Fine planar slip usually denotes homogeneous deformation, whereas coarse planar slip implies strain localization. Hornbogen and Zum Gahr [*12*] have suggested that planar slip leads to slower fatigue crack growth since slip reversibility is enhanced. During loading of a cracked sample, slip is activated in the plastic zone and the dislocations that are generated glide until they come in contact with some barrier, resulting in a pileup with an associated back stress. During the unloading part of the cycle, the back stress within the pileup forces the dislocations to move in the opposite direction. In homogeneously deforming materials, where cross-slip has occurred dislocation debris forms, making reversed slip more difficult. However, reversed slip is more feasible in materials deforming by planar slip. A certain number of dislocations may egress from the crack tip during reversed slip, and crack advance will be correspondingly reduced. Consequently, single-phase alloys with low stacking fault energy (SFE) or long- or short-range order will generally have a higher resistance to fatigue crack growth than similar alloys without these characteristics. It is well known that alloying can affect SFE and the degree of order.

The effect of SFE is clearly seen in the comparison of FCP rates of various copper-aluminum alloys (Fig. 1) taken from the work of Ishii and Weertman [*13*]. As aluminum is added to copper, the stacking fault energy is reduced and slip becomes more planar, thus enhancing the possibility of slip reversibility during cyclic deformation and increasing the resistance to FCP. Similar effects of short-range, order-induced coarse planar slip have been demonstrated in titanium-aluminum alloys by Allison and Williams [*14*]. Here also coarse planar slip results in lower FCGR at a fixed ΔK. This effect is illustrated in Fig. 2, which shows an increase in the threshold stress intensity of a solution heat treated and quenched Ti-8Al alloy, which has short-range order, over that for a Ti-4Al alloy, which deforms by wavy slip [*14*]. Since a decrease in slip-barrier spacing (a decrease in grain size for single-phase alloys containing no substructure) enhances multiple slip and dislocation-dislocation interactions, it also reduces slip reversibility and thus fatigue crack growth resistance. When the slip character is kept constant, increasing grain size has a similar effect [*15*]. Grain size

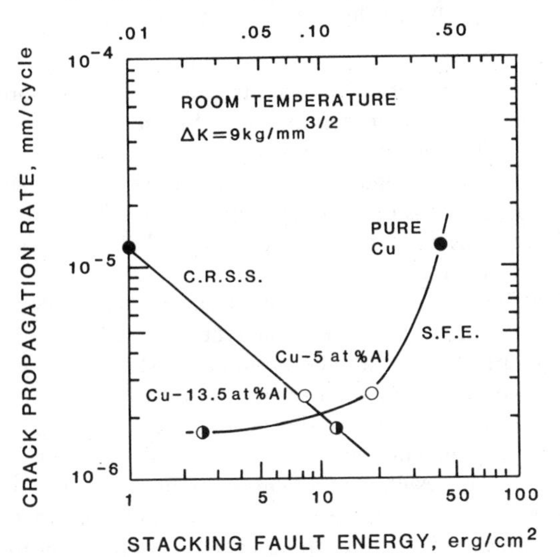

FIG. 1—*FCGRs as function of SFE and of CRSS for single crystals of approximately the same orientation* [13].

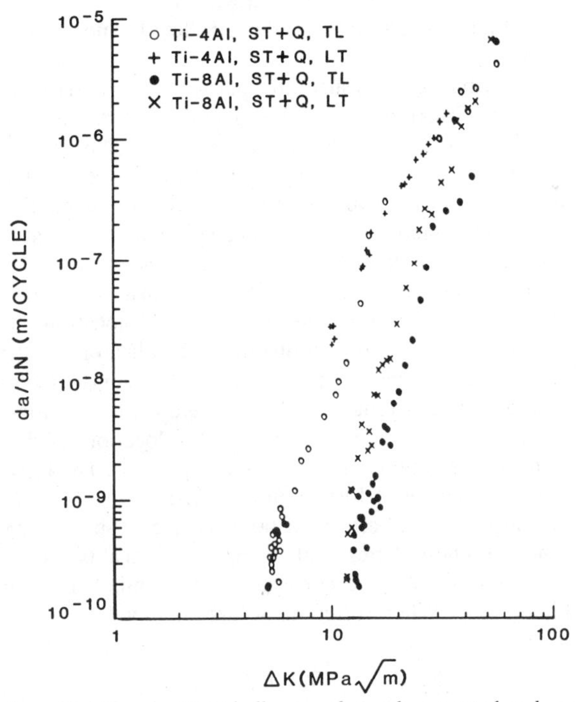

FIG. 2—*FCGRs for a Ti-4Al and a Ti-8Al alloy in solution heat-treated and quenched condition.*

effects will be discussed in more detail later on in this paper but are mentioned here to establish the interconnection between slip character and slip length.

Two-Phase Alloys—Most alloys of commercial interest consist of more than one phase and FCP resistance is a major consideration in many of these. Commercial alloys typically have higher strength than the single-phase model alloys discussed earlier. Many of these alloys derive their strength from second-phase particles, either precipitates or dispersoids. Effective strengthening of this class of materials depends on a fine distribution of second-phase particles. Therefore, consideration of precipitation and dispersion strengthened alloys may provide a useful basis for discussing FCP.

In precipitation-hardening alloys, three types of interfaces may develop during nucleation and growth of the strengthening precipitates: coherent, semi-coherent, and incoherent interfaces. No significant difference in atomic positions exist across a coherent interface, although a slight difference in lattice parameter may result in local elastic strains required to maintain coherency. If the coherency strains become excessive, which usually happens as aging proceeds, the coherent interface may be replaced by a periodic array of edge dislocations or structural ledges resulting in a semi-coherent interface. The third type is the noncoherent interface which has a degree of disorder comparable to a high-angle grain boundary. Dispersions, which usually refer to nonmetallic particles, normally have noncoherent interfaces. They are often produced by precipitation from the melt, eutectoid decomposition, internal oxidation, or mechanical mixing of particulate followed by powder metallurgy consolidation. Different modes of deformation are normally observed for precipitation-strengthened and dispersion-strengthened alloys. Coherent and semi-coherent particles present in precipitation-strengthened alloys are normally sheared by moving dislocations, the difficulty of shearing increasing with decreasing coherency. Noncoherent particles like those present in dispersion-strengthened alloys are normally looped or bypassed by moving dislocations.

The strengthening contribution from shearable particles is related to the particle's intrinsic properties and the volume fraction of particles. The overall strengthening contribution is reduced with increasing strain for particles that are sheared. This results in a local relative decrease in resistance to further dislocation motion in the active slip plane, which in turn leads to concentration of slip with a further reduction in local strength. The tendency for strain localization in age-hardened alloys containing coherent precipitates thus depends not only on the degree of coherency of the precipitates, but also on their volume fraction and size. The factors mentioned above which result in slip localization are partially offset by the back stress which develops as a result of the formation of dislocation pileups.

Duva et al. [*16*] have recently derived a quantitative indicator for strain localization in age-hardened alloys containing coherent precipitates. They consider both the degradation of strength with slip and also the strengthening associated with dislocation pileups, and they take as an indicator of slip localization the number of dislocations N that pass on a typical slip plane from the time deformation begins until local slip ends. The larger the slip intensity as measured by N, the coarser the deformation expected.

By considering the strengthening effects of both the second-phase particles and the dislocation pileups at the grain boundaries, and assuming spherical particles of equal size, one can derive two expressions for N, one corresponding to initial softening and one corresponding to initial hardening. When softening occurs initially

$$N = f^{1/2} r_p^{\ 1/2} r_G C_p / C_B b \tag{3}$$

where

f = volume fraction of the particles,
r_p = particle radius,
r_G = grain radius,
b = Burger's vector,
C_p = a constant that depends on the intrinsic properties of the particles, and
C_B = constant that depends on the elastic properties of the bulk material.

The slip intensity N increases with the square root of both the particle volume fraction and the particle radius and increases linearly with the grain radius. When hardening occurs initially

$$N = (2/b)r_p + (2/b)fr_G[C_p/4C_B]^2 \quad (4)$$

For this case, the slip intensity increases linearly with particle radius and volume fraction and increases as the square of the grain radius. These expressions predict that the propensity for strain localization increases during aging, up to the point at which the deformation mode changes from shearing to looping and bypassing of the particles. They also predict that strain localization would increase with increasing grain size.

The effect of strain localization, as enhanced by shearable precipitates and increasing grain size, is shown in Fig. 3, which compares the fatigue crack growth behavior in vacuum of 7475 in the underaged (shearable precipitates and coarse planar slip) condition with that of the overaged (nonshearable precipitates and homogeneous slip) for two different grain sizes [*17*]. Lin and Starke [*18*] examined a series of aluminum-zinc-magnesium-copper alloys

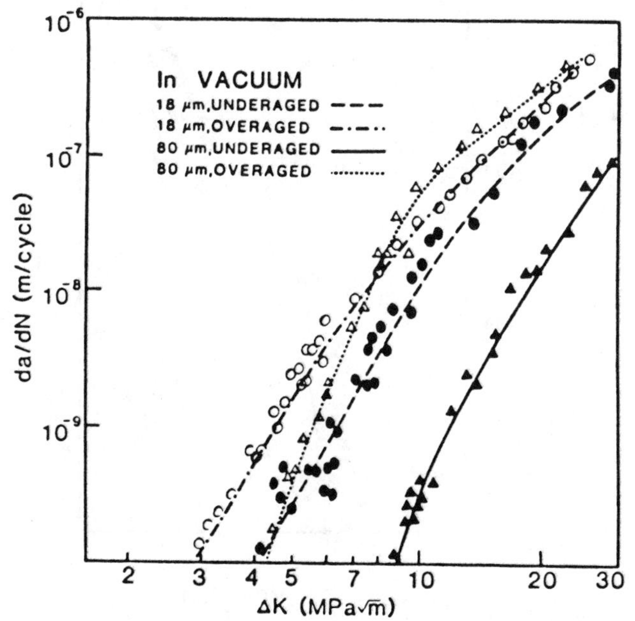

FIG. 3—*FCGRs for several microstructural variants of 7475 tested in vacuum.*

that had identical grain structures but contained strengthening precipitates of similar size but different degrees of coherency. The degree of coherency was varied by altering the copper content of the alloy for the peak-aged, T651, temper. The degree of coherency decreased as the copper content was increased and the corresponding deformation mode varied from coarse planar slip to fine homogeneous slip [*19,20*]. The results of this study (Fig. 4) clearly show that the fatigue crack growth resistance decreases as the coherency of the precipitates decreases and strain becomes more homogeneous. This figure also contains the fatigue crack growth results for all alloys in the overaged, T7351, temper. Since this temper results in homogeneous deformation for all of the alloys, no differences in FCP behavior occur.

The most effective type of second-phase particles for homogenizing slip are dispersoids, such as oxide or carbide particles, which are finely spaced and typically have a volume fraction of <0.04. The recent advent of rapid solidification processing technology has broadened the range of alloy systems which can be produced containing dispersoids. The effects of dispersoids on fatigue properties have been shown to be twofold [*21,22*]. First, the homogenous slip tends to increase the resistance to crack initiation, thereby increasing the fatigue life of uncracked specimens. This leads to improvements in fatigue strength. Second, because dispersoids tend to homogenize slip, the propensity for faceted or irregular crack paths is reduced; further, dispersoid-containing materials typically have finer grain sizes, which also reduces the roughness. Both of these factors increase the FCGR by reducing

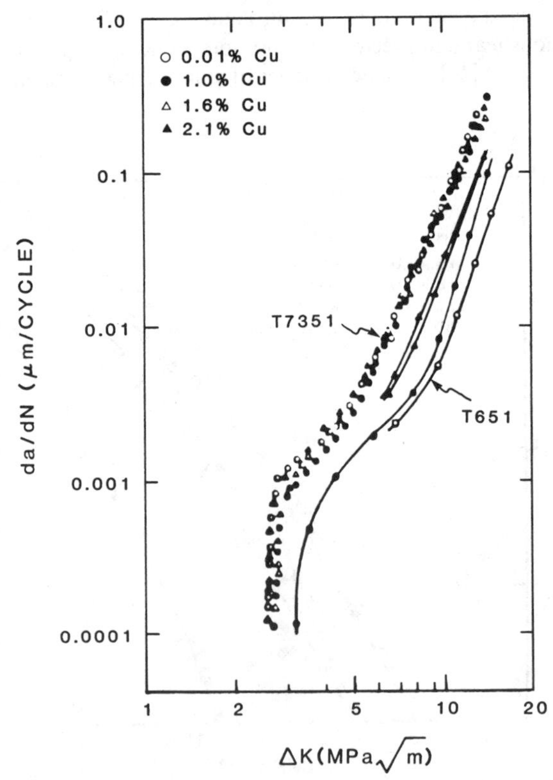

FIG. 4—*Effect of aging treatment and Cu content on the FCGR of Al-Zn-Mg alloys tested in dry air.*

fracture roughness and minimizing the role of roughness-induced closure in reducing the crack propagation process. Thus, for damage-tolerant applications, where FCGR is of paramount importance, dispersion-strengthened alloys are often less attractive than those hardened by precipitates. This is, of course, moderated by consideration of other properties such as toughness and environmental sensitivity. These latter factors also are affected by slip character so that the overall performance trade-off becomes relatively complex.

Slip Length and Plastic Zone Size

It was mentioned in the subsection on Single-Phase Alloys that a decrease in slip length reduces slip reversibility by reducing the number of dislocations that can move in the reverse direction during unloading. Also, the extent of strain localization in age-hardened materials depends on the grain size (slip length), Eqs 3 and 4. The most pronounced effects of slip length (grain size) on FCG are normally observed in the lower ΔK region. Robinson and Beevers [*23*] pointed out, after observing a grain-size effect on the FCP behavior of alpha-titanium alloys, that the ratio of the reversed plastic zone size to grain size could be the rate-controlling factor. This basic idea was extended by Hornbogen and Zum Gahr [*12*] to explain the differences in crack propagation rates in an iron-nickel-aluminum alloy with different age-hardening conditions.

Figure 5 illustrates schematically the relationship between slip length and the plastic zone size in planar-slip materials when the grain boundary is the major barrier to slip [*15*]. For the case in which grain size is smaller than the reverse plastic zone multiple slip is activated, even in planar slip materials, since the high local stress concentrations at the head of the pileup will induce slip on secondary slip systems in the same grain as well as slip in neighboring grains. In addition, the von Mises criterion [*24*] requires that multiple slip occur in order

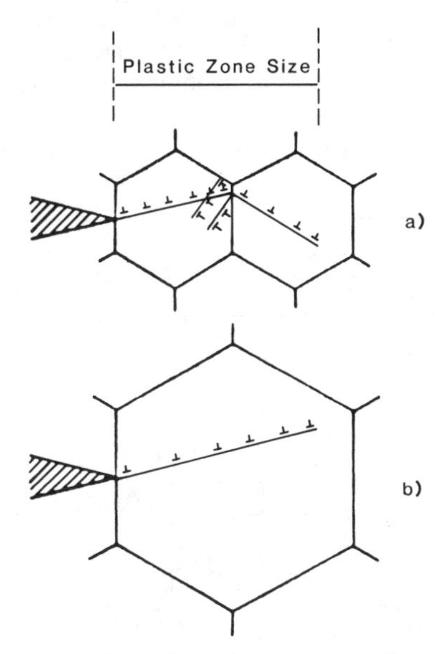

FIG. 5—*Schematic representation of the effect of grain size on dislocation structure at the crack tip:* (a) *small grains;* (b) *large grains* [15].

to maintain continuity across the grain boundaries during large-scale deformation. Upon unloading, dislocations in the neighboring grains become trapped within the grains and dislocations on secondary slip systems in the grain directly ahead of the crack interact with the dislocations on the primary slip system [*15*]. Both of these effects reduce slip reversibility during unloading. However, when the plastic zone is smaller than or equal to the grain size, slip may occur on only one slip system and slip reversibility may not be impeded. Of course, as the stress intensity and the associated reversed plastic zone increases, multiple slip is enhanced and slip reversibility is reduced even in coarse-grained materials.

Chakrabortty [*25*] has suggested an alternative to the reversibility concept to explain the effect of slip length (grain size) on fatigue crack growth rates. LEFM concepts predict that the stress (and thus the strain) at the tip of a Griffith-type crack is a continuously decreasing function of distance from the crack tip [*26*]. However, Chakrabortty suggests that for a real material, the plastic strain ahead of the crack tip will not decrease monotonically with distance but should be more uniform within a zone which is dependent on microstructure, as shown schematically in Fig. 6. For most materials, the zone will be equivalent to the distance between major deformation barriers, which is often equal to the grain size for single phase and age-hardenable alloys. Using this concept, the strain within each microdeformation zone will decrease as the slip length (grain size) increases. Consequently, this model predicts an increased resistance to fatigue crack growth as the slip length increases whether or not the "damage accumulation" or the "plastic sliding off" process occurs, since both require a critical strain for fatigue fracture.

Grain size and strength have been shown by numerous investigators to affect the threshold value of $\Delta K(\Delta K_{th})$, but its dependence on these variables is complex and somewhat controversial [*27*]. Many observers have noted an increase in ΔK_{th} with grain size, while others have noted an opposite effect [*28*]. Beevers [*29*] showed that ΔK_{th} increased with strength for a range of medium to high strength metals, while Minakawa and McEvily [*30*] showed an inverse dependence for a similar series of alloys. Models which consider both slip length (grain size) and cyclic yield strength [*31,32*] predict an increase in resistance to FCG when each parameter is increased. (For a review of the existing models, see Ref *33*). Figure 7 shows a schematic representation of the predicted change in ΔK_{th} with grain size and yield stress using the Chakrabortty model [*25*], as modified by Starke et al. [*31*] when the grain size is varied and the yield strength is held constant (Fig. 7*a*) and when the yield strength is varied and the grain size is held constant (Fig. 7*b*). In many cases, and especially for

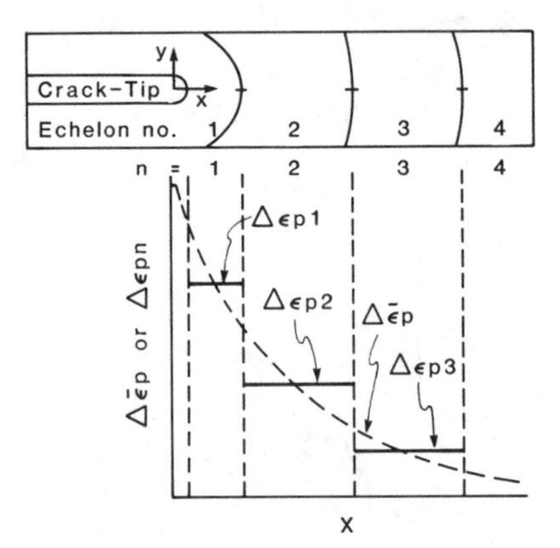

FIG. 6—*Schematic of strain distribution within microdeformation zones as a function of distance from the crack tip. Estimated cyclic plastic strain for a continuum is indicated by dashed line* [25].

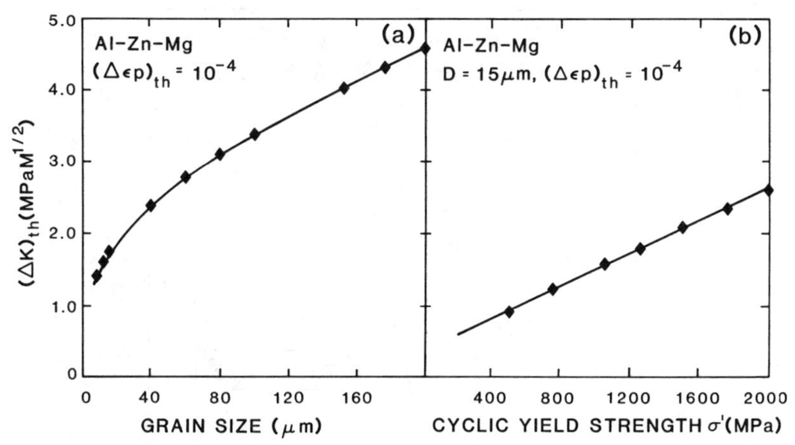

FIG. 7—*Schematic representation of predicted change in delta* Kth *with* (a) *grain and* (b) *cyclic yield strength.*

steels, an increase in grain size is accompanied by a concurrent decrease in strength, as described by the Hall-Petch relationship [*34,35*]. Since such a simultaneous change would affect the threshold in opposite directions, the confusion in the literature may be associated with failure to control each parameter individually and with different grain size-strength relationships for different alloys [*31*].

Recently, Yoder et al. [*36*] critically analyzed the grain size and yield strength dependence of near-threshold fatigue crack growth in steels and titanium alloys. This analysis covered a wide variety of alloys, microstructures, yield strengths and grain sizes, although the fatigue data were generated at roughly the same value of stress ratio, namely $0.0 \leq R \leq 0.1$. Yoder et al. utilize the basic concept previously presented by Hornbogen and Zum Gahr [*12*] of fatigue crack propagation being microstructurally sensitive when the grain size is larger than the reversed plastic zone and microstructurally insensitive when the grain size is smaller than the reversed plastic zone. They relate the transition T, that is, when the reversed plastic zone given by Eq 1 is equal to the mean grain size ($\bar{\ell}$), to a significant slope change in log da/dN versus log ΔK plots which occurs as the threshold is approached. It then follows that

$$\Delta K_T = 5.5\sigma_{ys}(\bar{\ell})^{1/2} \tag{5}$$

which results in a shift of the da/dN-ΔK plots, as represented schematically in Fig. 8. A plot of the observed transition values versus the predicted transition values using Eq 5 for a wide variety of steels and titanium alloys is given in Fig. 9. The level of agreement is remarkable, particularly in view of uncertainties involved in the measurements made by different investigators.

The microstructural parameter used in Eq 5 should be the slip length, that is, the maximum dimension to which a slip-band can develop without encountering a major obstacle. For many alloys, for example, those having planar slip, this may be equivalent to the grain size since the grain boundaries may be the major obstacle. However, the major barrier to slip may be intermetallic particles, dispersoids, or other types of precipitates. Davidson and Lankford [*37*] have shown that the dispersoid mean free path determined the critical slip

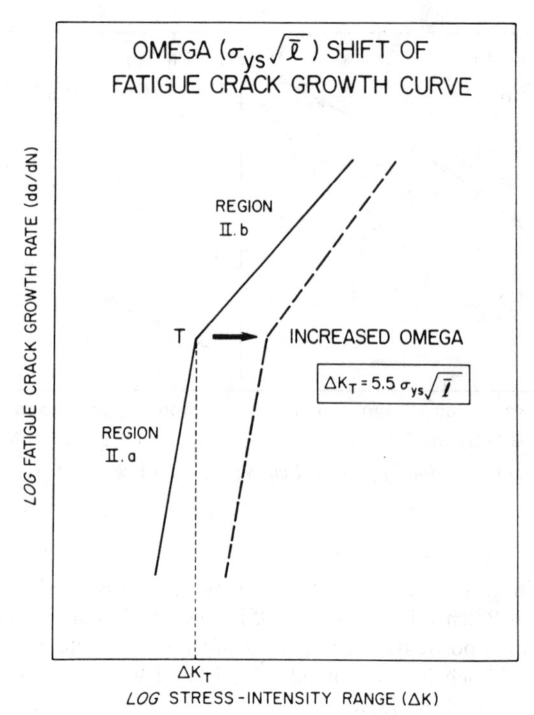

FIG. 8—*Schematic representation of shift in FCGR curve that is predicted quantitatively from the synergetic interaction of yield strength and grain size for a number of steels* [36].

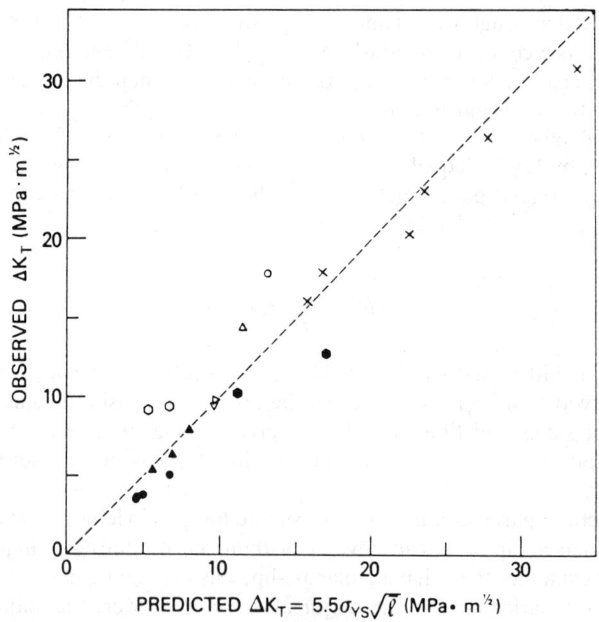

FIG. 9—*Comparison of observed values of transitional stress-intensity range versus predicted values for various steels and titanium alloys using Eq 5* [36].

length for two high-strength aluminum alloys. Since dispersoids are frequently used to control the grain structure in aluminum alloys, this parameter may often relate closely to the grain size. Kuo and Starke [*38*] observed that the effectiveness of grain boundaries as barriers to slip decreased in a strongly textured aluminum alloy when compared to a similar material having a weak texture. The differences in FCGRs observed by Kuo and Starke were related to the difference in the slip length which was influenced by the texture of the material.

Crack Path and Crack Extension Forces

Although slip reversibility should play a role in improving the fatigue resistance when strain localization occurs, it is by no means the only factor that should be considered. Lin and Starke [*18*] observed that the crack path features changed drastically as the degree of strain localization changed, varying from a relatively flat and straight crack when deformation was homogeneous to a zigzag crack with multiple branches when strain localization was extensive, as shown schematically in Fig. 10. Homogeneous deformation, and thus Type I fracture behavior, was also enhanced by increases in the stress-intensity factor and the aggressiveness of the environment. Strain localization favors crystallographic fracture and homogeneous deformation favors noncrystallographic fracture normal to the stress axis. The faster crack growth behavior observed for Type I behavior compared with Types II and III result from several factors: (*a*) the slip being more irreversible; (*b*) the projected crack length being more closely related to the true length; and (*c*) the calculated stress intensity factor being closer to the true value [*18*]. Similar observations and suggestions have been made by other researchers.

When a crack deviates from its main path, it undergoes both tensile opening and sliding displacements locally at the crack tip, even though the far-field loading may be purely tensile (Fig. 11) [*18*]. Mode I and Mode II stress intensity factors k_1 and k_2 at the deflected crack tip can be expressed as functions of the corresponding stress intensity factors for a linear crack of the same length, K_I and K_{II}, respectively, such that [*39*]

$$k_1 = a_{11}(\theta)K_I + a_{12}(\theta)K_{II} \tag{6}$$

$$k_2 = a_{21}(\theta)K_I + a_{22}(\theta)K_{II} \tag{7}$$

where θ is the angle denoting the extent of branching and $a_{ij}(\theta)$ are the angular functions associated with the deflected crack. Suresh has used this concept to develop simple elastic

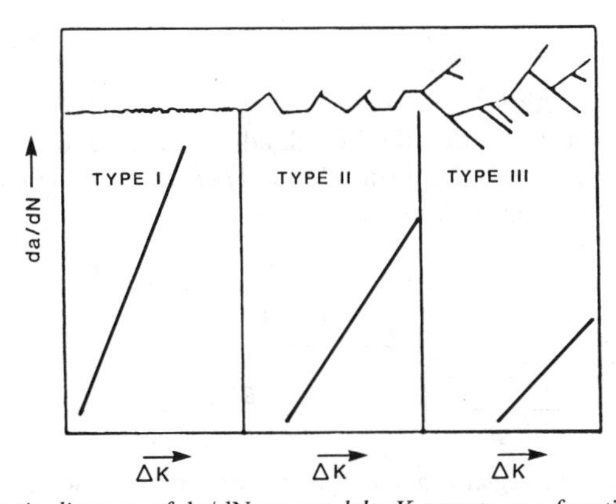

FIG. 10—*Schematic diagram of* da/dN *versus delta* K *curves as a function of crack path.*

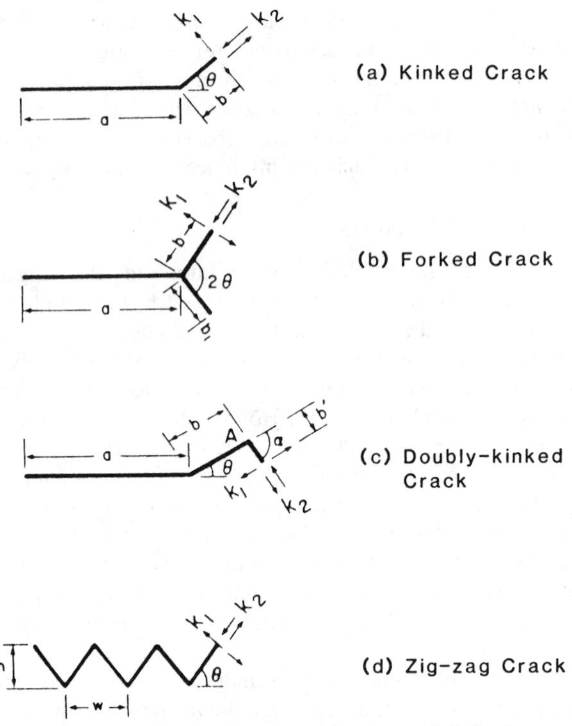

FIG. 11—*Schematic representation of possible types of fatigue crack deflections and the corresponding nomenclature to describe stress-intensity factors* [40].

deflection models which can be used to estimate the growth rates of nonplanar fatigue cracks subjected to various degrees of deflection [*40*]. Suresh has shown that the growth rate of an undeflected crack, da/dN, will be reduced when the crack is deflected. His model predicts that the crack growth rate of the deflected crack, $\overline{da}/dN$, may be represented by the relationship

$$\overline{da}/dN = (D \cos \theta + S/D + S)(da/dN) \tag{8}$$

where

θ = angle of deflection,
D = distance over which the tilted crack advances along the kink,
S = distance over which the linear (Mode I) crack grows (Fig. 12), and
da/dN = the linear crack growth ratio.

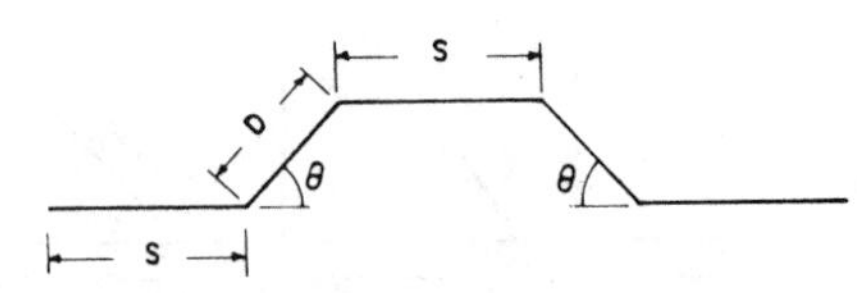

FIG. 12—*Model profile of a segment of a deflected crack with the associated nomenclature used in the text* [40].

In other words, the measured growth rates of a deflected crack are always apparently lower than those for an undeflected crack, subjected to the same effective stress intensity, if the deflections in crack path are not taken into consideration [*40*]. Figure 13 illustrates the magnitude of the effect that may be observed due to crack deflection alone.

A number of researchers [*41–43*] have shown that if slip is not completely reversible, the Mode II displacement may result in premature contact between the fracture surfaces, thus reducing the effective stress-intensity range, that is, producing a roughness-induced closure. Contact of this type is shown in Fig. 14. If slip is partially reversible, some sliding may occur after contact has been made, resulting in wear debris, as can be seen in Fig. 14. Suresh [*44*] has recently extended his earlier model [*40*] to include the Mode II displacement and thus roughness-induced closure. Using χ to represent the ratio of in-plane displacement to normal displacement, Suresh modifies the right-hand side of Eq 8 by the additional factor

$$1 - [(\chi \tan \theta)/(1 + \tan \theta)]^{1/2} \quad (8)$$

to account for roughness-induced closure. A schematic illustrating the magnitude of this effect is shown in Fig. 15. These results clearly show the importance of crack path on the crack extension forces.

Morphology and Properties of Phases in Multiphase Alloys

Many alloys are composed of two or more phases with different mechanical properties and a wide variety of morphologies. Age-hardenable and dispersion strengthened alloys

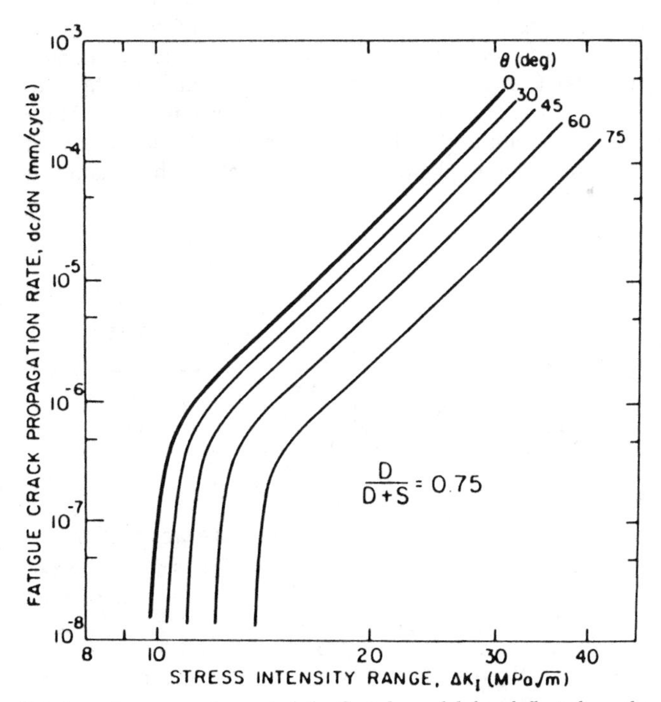

FIG. 13—*Predicted crack propagation using the Suresh model for deflected cracks as functions of angle of deflections,* θ, *and the extent of deflection* D/(D + S) [40].

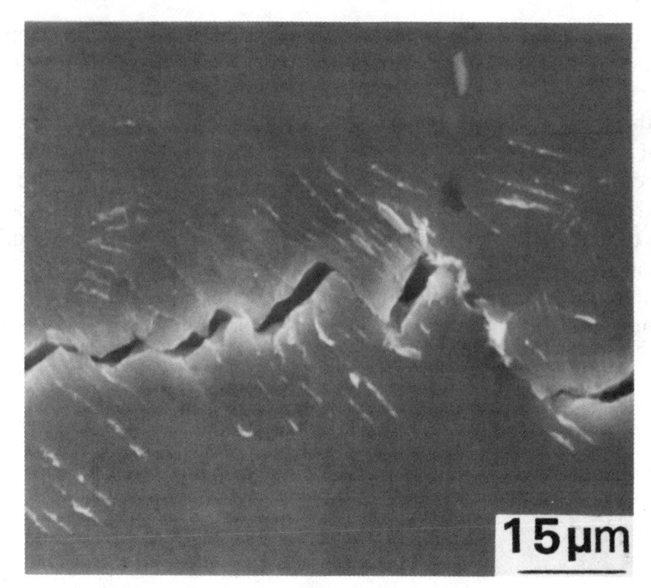

FIG. 14—*SEM showing zigzag crack growth, roughness induced closure, and wear debris of a peak-aged Al-Li-Mg alloy tested in vacuum.*

represent one class, as discussed previously. Another important class of microstructures may be described as dual-phase structures and these are often present in, but are not limited to, a wide variety of steels and titanium alloys. For example, by suitable selection of working and annealing temperatures, microstructures can be developed in alpha-beta titanium alloys having equiaxed, platelet, or grain boundary alpha in a retained beta matrix. Both phases can be fine, medium, or coarse, and continuous or noncontinuous. Martensite may also form in quenched alloys with a plate-like or lath morphology. Similar variations may be produced in steels. The different phases may both be deformable, that is, the yield strength

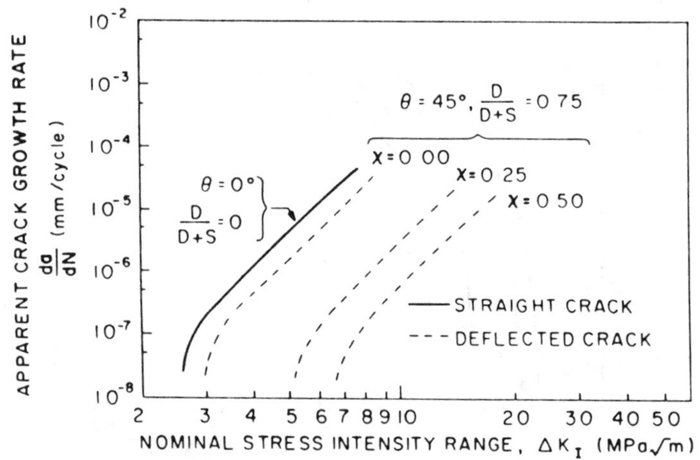

FIG. 15—*Predicted crack propagation rates for deflected cracks as functions of the mismatch factor* χ *using the Suresh model* [44].

of the strongest may be below the ultimate tensile strength of the weakest, or only one phase may be deformable. The strong phase may be the continuous phase, or the discontinuous phase. FCP behavior and preferred fracture path become a complex function of the microscopic stress and strain distribution and the fatigue and fracture resistance of each phase.

Because of the many microstructural possibilities in dual-phase structures, it is difficult to generalize. Consequently, we will use as an example the comparison of the FCP behavior of dual-phase steel having a continuous soft phase with a similar steel having a continuous hard phase. Suzuki and McEvily [*45*] have shown that this inversion of phase morphology results in an increase in strength and an increase in fatigue crack growth resistance. A continuous martensite network results in higher strengths and threshold stress intensities by restricting the plastic deformation to the ferrite at the crack tip [*46*]. The higher thresholds in a continuous martensite microstructure have been attributed to higher closure levels caused by a combination of Mode I and Mode II crack growth [*46*], crack deflections [*46,48*], and roughness-induced crack closure [*48*]. As will be described below, the magnitude of the different effects is very sensitive to the morphology of the strong phase.

Ramage et al. [*49*] recently studied the FCP behavior of two dual-phase steels similar to those examined by Suzuki and McEvily [*45*], designated CF for continuous ferrite and CM for continuous martensite. The mean intercept grain size for the different phases in the two structures were essentially the same. Figure 16 shows the da/dN-ΔK plots before and after correcting for closure and crack deflections. Before the closure and deflection corrections were applied, the FCGR for CF is considerably faster than the FCGR for CM, for comparable stress-intensity ranges. After accounting for the crack closure and crack deflections, the intrinsic or effective threshold stress intensities and the crack growth rates are observed to be the same. However, at similar stress intensities, CF had a more tortuous crack path than CM and a larger roughness. Consequently, when considering these parameters alone, one would have expected the CF material to have had larger closure and deflection corrections. Consequently, there must be some other closure contribution for the CM material.

There are two effects leading to the higher closure level in CM [*49*]. The first is caused by the martensite being the continuous phase. A crack tip in the ferrite will be surrounded by a martensite envelope. Upon applying a load, the martensite will be the load-bearing phase, thus shielding the crack tip from a part of the applied load and resulting in a reduced

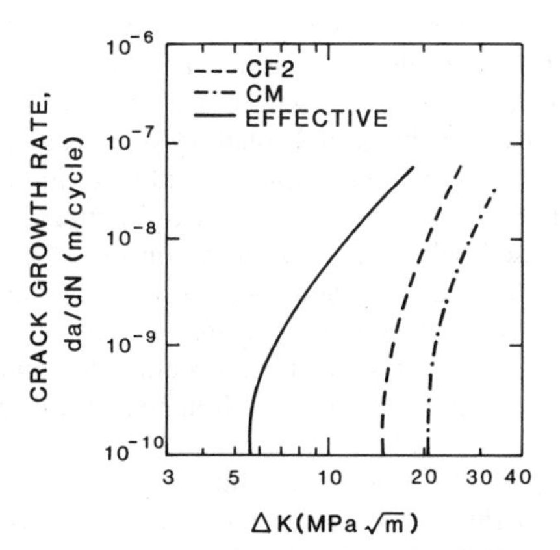

FIG. 16—*Observed and effective crack growth curves for CF and CM microstructures. The effective crack growth curves for the two microstructures are identical.*

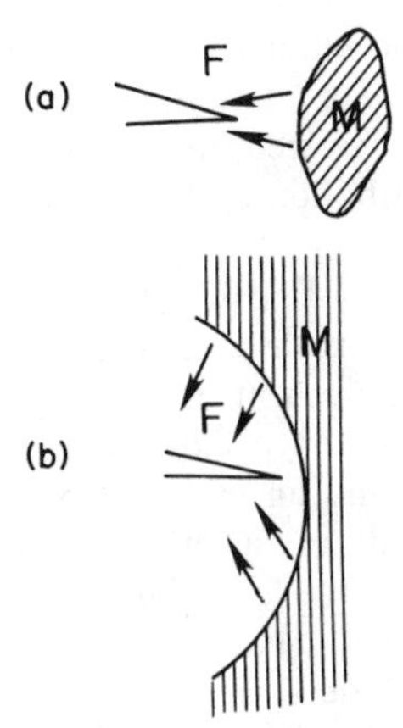

FIG. 17—*Schematic illustration of the constraint of the plastic deformation in the ferrite* (*F*) *by the martensite* (*M*): (a) *in CF;* (b) *in CM, the martensite envelope more effectively constrains ferrite deformation.*

cyclic plastic strain and effective stress-intensity range at the crack tip. The second effect is due to the interaction of the martensite and the ferrite. In CM, a martensite envelope surrounds a crack growing in ferrite. It has been shown that the martensite in a dual-phase structure will constrain the plastic deformation in the ferrite [*46,50*]. This constraining effect is primarily a function of the distance from the crack tip to the martensite and will cause a crack growing in ferrite to decelerate as it approaches a martensite particle [*50*]. The constraint of the plastic deformation in the ferrite will increase the crack tip opening displacement necessary for damage to accumulate and for crack extension to occur [*46*]. This effect will be more apparent to the CM microstructure than the CF microstructure due to the continuity of the martensite which will more effectively constrain plastic deformation in the ferrite, Fig. 17.

Environment

Although the effect of environment on FCP is covered in detail in other papers in this volume, a few comments will be made here since environmental effects may negate some of the microstructural advantages discussed earlier. For example, it is pointed out that strain localization improved the fatigue crack growth resistance since it enhances slip reversibility, crack path tortuosity, crack branching, and roughness-induced closure. Figure 5 illustrates that in age-hardened alloys, shearable precipitates and large grains (long slip length) which enhance strain localization improve the fatigue crack growth resistance in an inert environment. However, as discussed below, strain localization increases environmental sensitivity compared with materials that deform homogeneously, and thus these advantages often disappear when the tests are conducted in an aggressive environment. Figure 18 shows the results of tests conducted in laboratory air for the same alloy and treatments used for Fig. 3. Most of the advantages of strain localization disappear in this mildly aggressive environment.

In addition to studying the effect of deformation behavior, as influenced by both copper additions and aging treatments, as described previously (Fig. 4), Lin and Starke also examined the effect of these parameters on the fatigue crack propagation in corrosive envi-

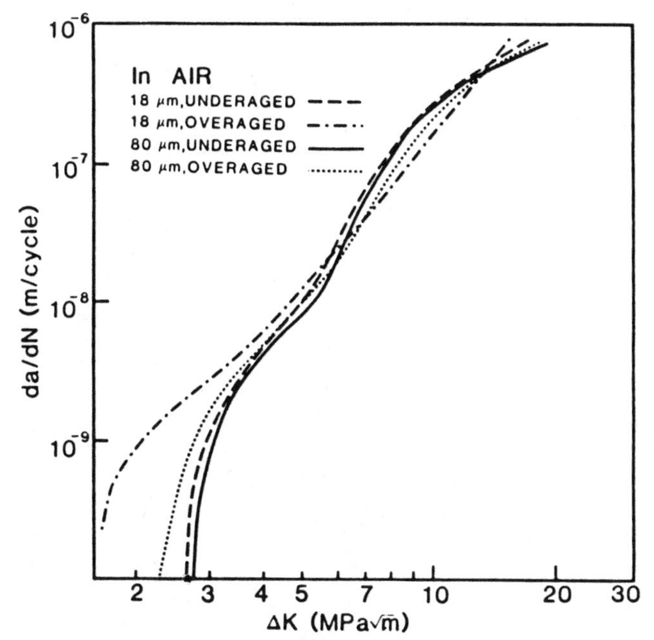

FIG. 18—*FCGR's for several microstructural variants of 7475 tested in air.*

ronments [*51*]. The ratios of the FCGR in distilled water to that in dry air, as a function of ΔK, for the various conditions described previously for Fig. 4, are shown in Fig. 19. This figure shows that the environmental sensitivity decreases when the copper content increases or when the alloys are overaged. Thus, the principal conclusion of these curves is that the environmental sensitivity is a function of deformation. This effect is largest for the highest degree of planar strain localization, such as the 0% copper alloy in the T651 condition, and decreases with increasing copper content. The smallest effect is for the overaged specimens regardless of the copper content since they all exhibit uniform deformation. The aggressive environment always enhanced Type I FCG behavior (Fig. 10), as shown in Fig. 20. The exact corrosion fatigue mechanism was not established in the Lin-Starke study; however, the results are consistent with a hydrogen embrittlement mechanism for corrosion fatigue, which relates the degree of susceptibility to slip planarity [*52–54*].

The interaction of aggressive environments and strain localization has also been observed in alloy systems other than aluminum. As mentioned earlier, Allison and Williams [*14*] studied the effect of deformation character on the FCP behavior of binary titanium-aluminum alloys. They varied the deformation mode from homogeneous wavy slip, Ti-4Al, to coarse planar slip, unaged Ti-8Al, to very intense coarse planar slip, aged Ti-8Al. The tests were conducted in laboratory air which should be considered as mildly aggressive for alpha-titanium alloys. Although the unaged Ti-8Al alloy exhibited growth rates 300 to 400 times lower than the Ti-4Al alloy (Fig. 2), the aged Ti-8Al alloy that had intense planar slip exhibited growth rates much faster than the unaged alloy—rates that approached the Ti-4Al which deformed homogeneously (Fig. 21). The differences between the aged and unaged Ti-8Al must be considered in light of the influence of environment, although the effect of α_2 precipitates on the propensity for basal and prism slip is not understood at

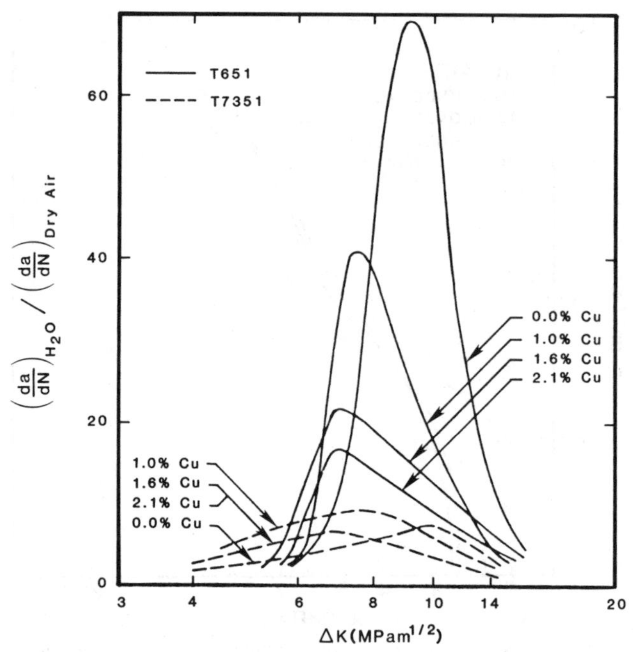

FIG. 19—*Environmental sensitivity of four Al-Zn-Mg-Cu alloys in the T651 and T7351 conditions.*

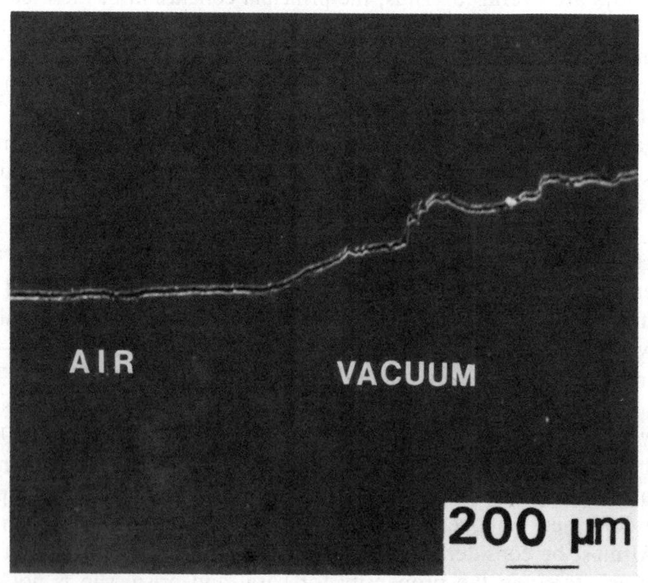

FIG. 20—*SEM showing different FCG profiles obtained when Al-Li-Mg specimen is tested in air and in vacuum.*

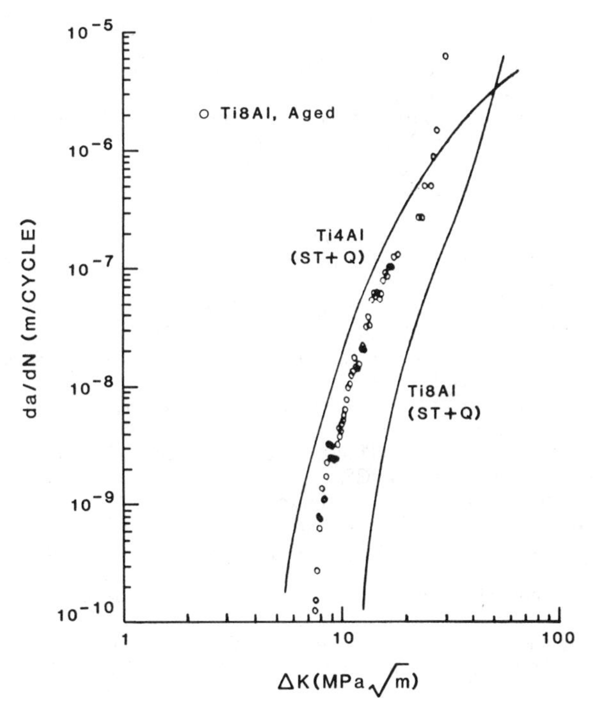

FIG. 21—*Comparison of FCGR of a Ti-8Al alloy in the aged condition with the growth rates of the specimens described in Fig. 3.*

present. Thus, the difference between aged and unaged Ti-8Al crack growth behavior could be due to more comparable amounts of basal and prism slip, or it may be another example of increased environmental sensitivity with an increase in strain localization. Further work is required to differentiate between these two effects.

Conclusions

From the above discussion, the following conclusions have been drawn:

1. Microstructure is the principal independent variable which can be used to control the fatigue crack growth rate (FCGR) once loading conditions and environment are established.
2. The principal factors which affect intrinsic FCGR are slip character, slip length, and slip reversibility. Each of these factors is affected by both alloy composition and microstructure.
3. Strain localization typically reduces FCGR and increases ΔK_{th} by enhancing crack path tortuosity and crack closure.
4. Wavy (homogeneous) slip usually corresponds to higher FCGR.
5. Increasing slip length tends to reduce FCGR, especially in alloys which exhibit strain localization.
6. Increased slip reversibility typically results in lower FCGR.
7. Environmentally related acceleration of FCGR is usually greater in the presence of strain localization.

References

[1] Paris, P. C., Gomez, M. P., and Anderson, W. E., *The Trend in Engineering at the University of Washington,* Vol. 13, No. 1, 1961, p. 9.
[2] Paris, P. C. and Erdogan, F., *Journal of Basic Engineering,* Vol. 85, 1963, pp. 528–534.
[3] Elber, W., *Engineering Fracture Mechanics,* Vol. 2, 1970, pp. 37–45.
[4] Knott, J. F., *Fundamentals of Fracture Mechanics,* Butterworths, London, 1973.
[5] *Fatigue Crack Propagation, ASTM STP 415,* American Society for Testing and Materials, Philadelphia, 1967.
[6] *Fatigue Mechanisms: Advances in Quantitative Measurement of Physical Damage, ASTM STP 811,* J. Lankford, D. L. Davidson, W. L. Morris, and R. P. Wei, Eds., American Society for Testing and Materials, Philadelphia, 1983.
[7] Forsyth, P. J. E. and Ryder, D. A., *Metallurgica,* Vol. 63, March 1961, p. 117.
[8] Lankford, J. and Davidson, D. L., *Acta Metallurgica,* Vol. 31, 1983, p. 1273.
[9] Hornbogen, E. in *Proceedings of the Sixth International Conference on Strength of Metals and Alloys,* Vol. 3, R. C. Gifkins, Ed., 1983, p. 1059.
[10] Starke, E. A., Jr., and Lütjering, G., *Fatigue and Microstructure,* American Society for Metals, Metals Park, OH, 1978, pp. 205–243.
[11] Williams, J. C., Thompson, A. W., and Baggerly, R. G., "Accurate Description of Slip Character," *Scripta Metallurgica,* Vol. 8, 1974, p. 625.
[12] Hornbogen, E. and Zum Gahr, K. H., *Acta Metallurgica,* Vol. 24, 1976, p. 581.
[13] Ishii, H. and Weertman, J., *Metallurgical Transactions,* Vol. 2, 1971, pp. 3441–52.
[14] Allison, J. E. and Williams, J. C. in *Proceedings Strength of Metals and Alloys,* Vol. 3, R. C. Gifkins, Ed., 1983, pp. 1219–1224.
[15] Lindigkeit, J., Terlinde, G., Gysler, A., and Lütjering, G., *Acta Metallurgica,* Vol. 27, 1979, pp. 1717–1726.
[16] Duva, J. M., Daeubler, M. A., Starke, E. A., Jr., and Lütjering, G., "Large Shearable Particles Lead to Coarse Slip in Particle Reinforced Alloys," *Acta Metallurgica,* Vol. 36, 1988, pp. 585–589.
[17] Carter, R. D., Lee, E. W., Starke, E. A., Jr., and Beevers, C. J., *Metallurgical Transactions,* Vol. 15A, 1984, pp. 555–563.
[18] Lin, F. S. and Starke, E. A., Jr., *Materials Science and Engineering,* Vol. 43, 1980, pp. 65–76.
[19] Lin, F. S. and Starke, E. A., Jr., *Materials Science and Engineering,* Vol. 39, 1979, pp. 27–41.
[20] Heikkenen, H. C., Lin, F. S., and Starke, E. A., Jr., in *Proceedings,* 2nd International Conference on Environmental Degradation of Engineering Materials in Aggressive Environments, M. R. Louthan, Jr., R. P. McNitt, and R. D. Sesson, Eds., Virginia Polytechnic Institute and State University, Blacksburg, VA, 1981, pp. 459–469.
[21] Terlinde, G., Peters, M., Lütjering, G., and Williams, J. C., "Microstructure and Mechanical Properties of a Dispersion-Hardened P/M Al-10MN and a P/M Al-10MN-2.5Si Alloy," *Zeitschrift für Metallkunde,* Vol. 78, 1987, pp. 607–617.
[22] Jata, K. V., Walsh, J. A., and Starke, E. A., Jr., in *Fatigue '87,* R. O. Richie and E. A. Starke, Jr., Eds., Engineering Materials Advisory Services, Ltd., West Midlands, UK, 1987, Vol. 1, pp. 517–526.
[23] Robinson, J. L. and Beevers, C. J., *Metallurgical Science Journal,* Vol. 7, 1973, p. 153
[24] Von Mises, R., *Zeitscrift für Angewandte Mathematik und Mechanik,* Vol. 8, 1928, p. 161.
[25] Chakrabortty, S. B., *Fatigue in Engineering Materials and Structures,* Vol. 2, 1979, p. 331.
[26] Majumdar, S. and Morrow, J. D. in *Fracture Toughness and Slow-Stable Cracking, (Eighth Conference), ASTM STP 559,* American Society for Testing and Materials, Philadelphia, 1974, p. 159.
[27] McKittrick, J., Liaw, P. K., Kwun, S. J., and Fine, M. E., *Metallurgical Transactions A,* Vol. 12A, 1981, p. 1535.
[28] Ritchie, R. O., *International Metals Reviews,* Vol. 24, 1979, p. 205.
[29] Beevers, C. J. in *Fatigue Thresholds: Fundamentals and Engineering Applications,* J. Backlund, A. Blom, and C. J. Beevers, Eds., Vol. 1, 1982, p. 257.
[30] Minakawa, K. and McEvily, A. J. in *Fatigue Thresholds: Fundamentals and Engineering Applications,* J. Backlund, A. Blom, and C. J. Beevers, Eds., Vol. 1, 1982, p. 373.
[31] Starke, E. A., Jr., Lin, F. S., Chen, R. T., and Heikkenen, H. C. in *Fatigue Crack Growth Threshold Concepts,* D. Davidson and S. Suresh, Eds., American Institute for Mining, Metallurgical, and Petroleum Engineers, 1984, pp. 43–62.
[32] Davidson, D. L., *Acta Metallurgica,* Vol. 32, 1984, pp. 707–714.
[33] Bailon, J. P. and Antolovich, S. D. in *Fatigue Mechanisms: Advances in Quantitative Measurement*

of Physical Damage, ASTM STP 811, J. Lankford, D. L. Davidson, W. L. Morris, and R. P. Wei, Eds., American Society for Testing and Materials, Philadelphia, 1983, pp. 313–349.
[34] Hall, E. O., *Proceedings of the Physical Society,* Vol. 64, 1951, p. 747.
[35] Petch, N. J., *Journal of the Iron and Steel Institute,* Vol. 197, 1953, p. 25.
[36] Yoder, G. R., Cooley, L. A., and Crooker, T. W. in *Fracture Mechanics: Fourteenth Symposium, ASTM STP 791,* Vol. 1, J. C. Lewis, and G. Sines, Eds., American Society for Testing and Materials, Philadelphia, 1983, pp. 348–365.
[37] Davidson, D. L. and Lankford, J., *Materials Science Engineering,* Vol. 74, 1985, pp. 189–199.
[38] Kuo, V. W. C. and Starke, E. A., Jr., *Metallurgical Transactions A,* Vol. 16A, 1985, pp. 1089–1103.
[39] Bilby, B. A., Cardew, G. E., and Howard, I. C. in *Fracture 1977,* Vol. 3, D. M. R. Toplin, Ed., University of Waterloo Press, 1977, p. 197.
[40] Suresh, S., *Metallurgical Transactions A,* Vol. 14A, 1983, pp. 2375–2385.
[41] Walker, N. and Beevers, C. J., *Fatigue of Engineering Materials and Structures,* Vol. 1, 1979, p. 135.
[42] Ritchie, R. O., Suresh, S., and Moss, C. M., *Journal of Engineering Materials and Technical Transactions,* American Society of Mechanical Engineers, Series H, Vol. 102, 1979, p. 293.
[43] Minakawa, K. and McEvily, A. J., *Scripta Metallurgica,* Vol. 15, 1981, p. 633.
[44] Suresh, S., *Metallurgical Transactions A,* Vol. 16A, 1985, pp. 249–260.
[45] Suzuki, H. and McEvily, A. J., *Metallurgical Transactions A,* Vol. 10A, 1979, pp. 475–481.
[46] Minakawa, K., Matsuo, Y., and McEvily, A. J., *Metallurgical Transactions A,* Vol. 13A, 1982, pp. 439–445.
[47] Wasynczuk, J. A., Ritchie, R. O., and Thomas, G., *Materials Science Engineering,* Vol. 62, 1984, pp. 79–92.
[48] Dutta, V. B., Suresh, S., and Ritchie, R. O., *Metallurgical Transactions A,* Vol. 15A, 1984, pp. 1193–1207.
[49] Ramage, R. M., Jata, K. V., Shiflet, G. J., and Starke, E. A., Jr., "The Effect of Phase Continuity on the Fatigue Crack Closure Behavior of a Dual-Phase Steel," *Metallurgical Transactions A,* Vol. 18A, 1987, pp. 1291–1298.
[50] Yang, C. Y. and Liu, H. W., *Fatigue in Engineering Materials and Structures,* Vol. 1, 1979, pp. 483–493.
[51] Lin, F. S. and Starke, E. A., Jr., in *Hydrogen Effects in Metals,* I. M. Bernstein, and A. W. Thompson, Eds., American Institute for Mining, Metallurgical, and Petroleum Engineers, 1980, pp. 485–492.
[52] Donovan, J. A., *Metallurgical Transactions A,* Vol. 7A, 1976, p. 1677.
[53] Louthan, M. R., Jr., Caskey, F. R., Jr., Donovan, J. A., and Rawl, R. E., Jr., *Material Science and Engineering,* Vol. 10, 1972, p. 357.
[54] Wei, R. P. in *Fatigue '87,* R. O. Ritchie and E. A. Starke, Jr., Eds., West Midlands, U.K., Vol. 3, pp. 1541–1560.

Environmentally Assisted Cracking

Russell H. Jones,[1] Michael J. Danielson,[1] and Donald R. Baer[1]

Microchemistry and Mechanics Issues in Stress Corrosion Cracking

REFERENCE: Jones, R. H., Danielson, M. J., and Baer, D. R., **"Microchemistry and Mechanics Issues in Stress Corrosion Cracking,"** *Fracture Mechanics: Perspectives and Directions (Twentieth Symposium), ASTM STP 1020,* R. P. Wei and R. P. Gangloff, Eds., American Society for Testing and Materials, Philadelphia, 1989, pp. 209–232.

ABSTRACT: Environmentally induced subcritical crack growth occurs in most metallic materials given the appropriate material, environment, and stress conditions. Much is known about the conditions which cause stress corrosion cracking (SCC) in metallic materials, but much less is known about the processes which control environment-induced crack initiation and propagation. There is presently no description of the flaw length, flaw shape, and local chemistry conditions that coincide with the transition from a localized corrosion phenomenon and the initiation of a crack. Although there have been considerable attempts to describe the rate-limiting steps in crack propagation, a number of issues are unresolved, including specifics about the crack-tip conditions, crack length effects, processes controlling the Stage I SCC region, and details about the crack advance processes in both transgranular stress corrosion cracking (TGSCC) and intergranular stress corrosion cracking (IGSCC). A summary of these unresolved issues and some analysis of their effect on SCC is given in this paper.

KEY WORDS: stress corrosion, issues, microstructure, chemistry, mechanics

Both the phenomenological and mechanistic aspects of stress corrosion cracking (SCC) have received considerable attention over the past 20 years. The level of research has been sufficient to allow numerous major topical conferences on the subject to be held during this time period. Also, a great deal is now known about the phenomenological aspects of SCC, such as the environmental, microstructural, and mechanics factors controlling SCC. Similarly, considerable effort has focused on identifying the mechanisms controlling the crack advance process for both intergranular stress corrosion cracking (IGSCC) and transgranular stress corrosion cracking (TGSCC), including both anodic dissolution and mechanical fracture processes.

A number of significant issues remain unresolved regarding the process of stress corrosion crack growth. Some of these issues include the details of crack initiation, several aspects of crack-tip conditions, crack length effects, an analytical model describing Stage I crack growth regime, and details about the crack extension process in both TGSCC and IGSCC.

The phenomenological conditions controlling crack initiation are well known. For instance, it is well established that cracks may initiate at crevices and pits which allow local concentration cells to form. Also, intergranular corrosion may lead to the initiation of an intergranular stress corrosion crack. However, details about the specific conditions that cause a localized corrosion reaction to initiate a crack are not well established. In particular, the

[1] Staff scientist, senior scientist, and senior scientist, respectively, Pacific Northwest Laboratory, Richland, WA 99352.

specific details about the crack length and shape and crack-tip electrochemistry that distinguish a growing crack from a nonpropagating, localized corrosion reaction are unknown. An example in which localized corrosion led to a propagating crack in one case and a nonpropagating crack in another was reported by Jones et al. [1] for nickel with grain boundaries enriched in phosphorus and sulfur.

Some specific crack-tip conditions that are significant to SCC and that need further evaluation include the effect of strain rate on the crack-tip surface chemistry and corrosion rate, crack wall chemistry effects, crack geometry factors, and salt film effects on crack-tip chemistry and corrosion rates. Crack length can affect the crack-tip chemistry and the crack-tip mechanics. Therefore, a strong dependence of crack growth rate on crack length should exist; however, there have been relatively few studies of this effect.

There are several factors regarding the crack extension process in SCC which remain unsettled. One of these issues is the lack of an analytical crack growth model which is consistent with the observed stress intensity dependence of the crack growth rate in the Stage I regime which is characterized by a strong stress-intensity dependence. Stress corrosion crack growth models either do not include a stress-intensity dependence or provide for much too small a dependence. Likewise, the rate-controlling process in the stress intensity-independent Stage II regime remains an open issue. Some possible rate-limiting steps include cation and anion transport in the crack tip solution, film rupture and repassivation rates, and surface reaction rates. Also, details about the crack extension process for both TGSCC and IGSCC are yet unsettled. There is considerable evidence that TGSCC proceeds in many systems by a brittle crack jump process [2–6]; however, much is unknown about the brittle films which induce the cleavage crack growth. The same is true about the role of mechanical fracture during crack growth in IGSCC.

The purpose of this paper is to review some of the key microchemistry and mechanics issues in SCC. A mixture of data from the literature and from research conducted at Pacific Northwest Laboratory (PNL)[2] will be used in this assessment.

Stress Corrosion Cracking Issues

Crack Initiation Processes

Stress corrosion cracking frequently initiates at preexisting or corrosion-induced surface features. These features may include grooves, laps, or burrs resulting from fabrication processes. Stress corrosion cracks can also initiate at pits that form during exposure to the service environment or by prior cleaning operations, such as pickling of Type 304SS before fabrication. Pits can form at inclusions that intersect the free surface or by a breakdown in the protective film. In electrochemical terms, pits may form when the potential exceeds the pitting potential. Parkins [7] showed that the SCC potential and pitting potential were identical for steel in nitrite solutions.

The transition between pitting and cracking is dependent on the same parameters that control SCC, that is, the electrochemistry at the base of the pit, pit geometry, chemistry of the material, and stress or strain rate at the base of the pit. A detailed description of the relationship between these parameters and crack initiation has not been developed because of the difficulty in measuring crack initiation. Methods for measuring short, surface cracks are under development but are limited to detecting cracks that are beyond the initiation stage. The pit geometry is important in determining the stress and strain rate at the base

[2] Operated for the U.S. Department of Energy by Battelle Memorial Institute under Contract DE-AC06-76RLO 1830.

of the pit; the aspect ratio between penetration and lateral corrosion of a pit must be much greater than 1 for a crack to initiate from a pit.

As in the case of a growing crack, the pit walls must exhibit some passive film-forming capability in order for the corrosion ratio to exceed 1. A change in the corrosive environment and potential within a pit may also be necessary for the pit to act as a crack initiator. Pits can act as occluded cells similar to cracks and crevices, although in general their volume is not as restricted. There are a number of examples in which stress corrosion cracks initiated at the base of a pit by intergranular corrosion. In these circumstances, the grain boundary chemistry and pit chemistry were such that intergranular corrosion was favored. Crack propagation was also by IGSCC in these cases.

While the local stresses and strain rates at the base of the pit play a role in SCC initiation, examples can be found of preexisting pits that did not initiate stress corrosion cracks. This observation led Parkins [7] to conclude that the electrochemistry of the pit is more important than the local stress or strain rate. A preexisting pit may not develop the same local electrochemistry as one grown during service because the development of a concentration cell depends on the presence of an actively corroding tip that establishes the anion and cation current flows. Similarly, an inability to reinitiate crack growth in specimens in which active crack growth was occurring before the specimens were removed from solution, rinsed, dried, and reinserted into solution also suggests that the local chemistry is very important.

Stress corrosion crack initiation can also occur in the absence of pitting by intergranular corrosion or slip-dissolution processes. Intergranular corrosion-initiated SCC requires that the local grain boundary chemistry differ from the bulk chemistry. This condition occurs in sensitized austenitic stainless steels or with segregation of impurities such as phosphorus, sulfur, or silicon in a variety of materials.

The specific pit geometry, localized chemistry, and mechanics conditions that correspond to the transition to an active crack are relatively unknown. An example where localized corrosion led to a propagating intergranular stress corrosion crack in one case but not the other is shown in Figs. 1 and 2 for nickel with sulfur- and phosphorus-enriched grain bound-

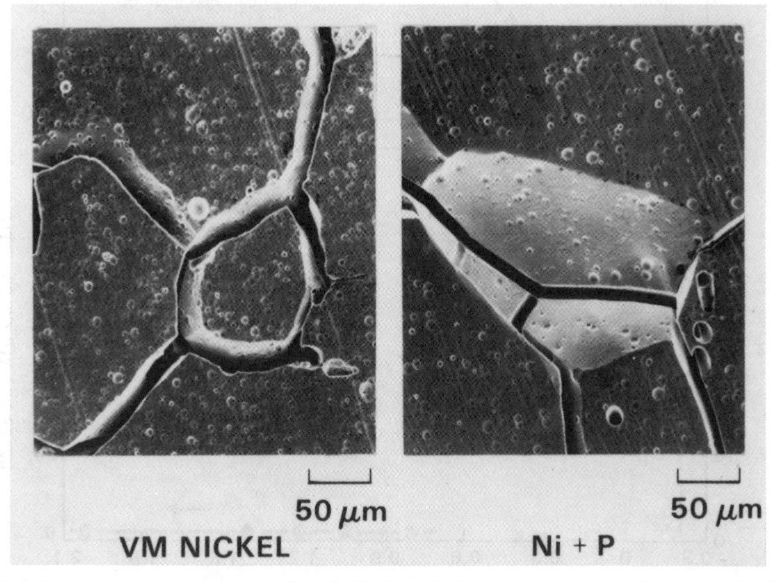

FIG. 1—*Intergranular corrosion in Ni and Ni + P polarized in the transpassive regime in 1N H_2SO_4 at 25°C.*

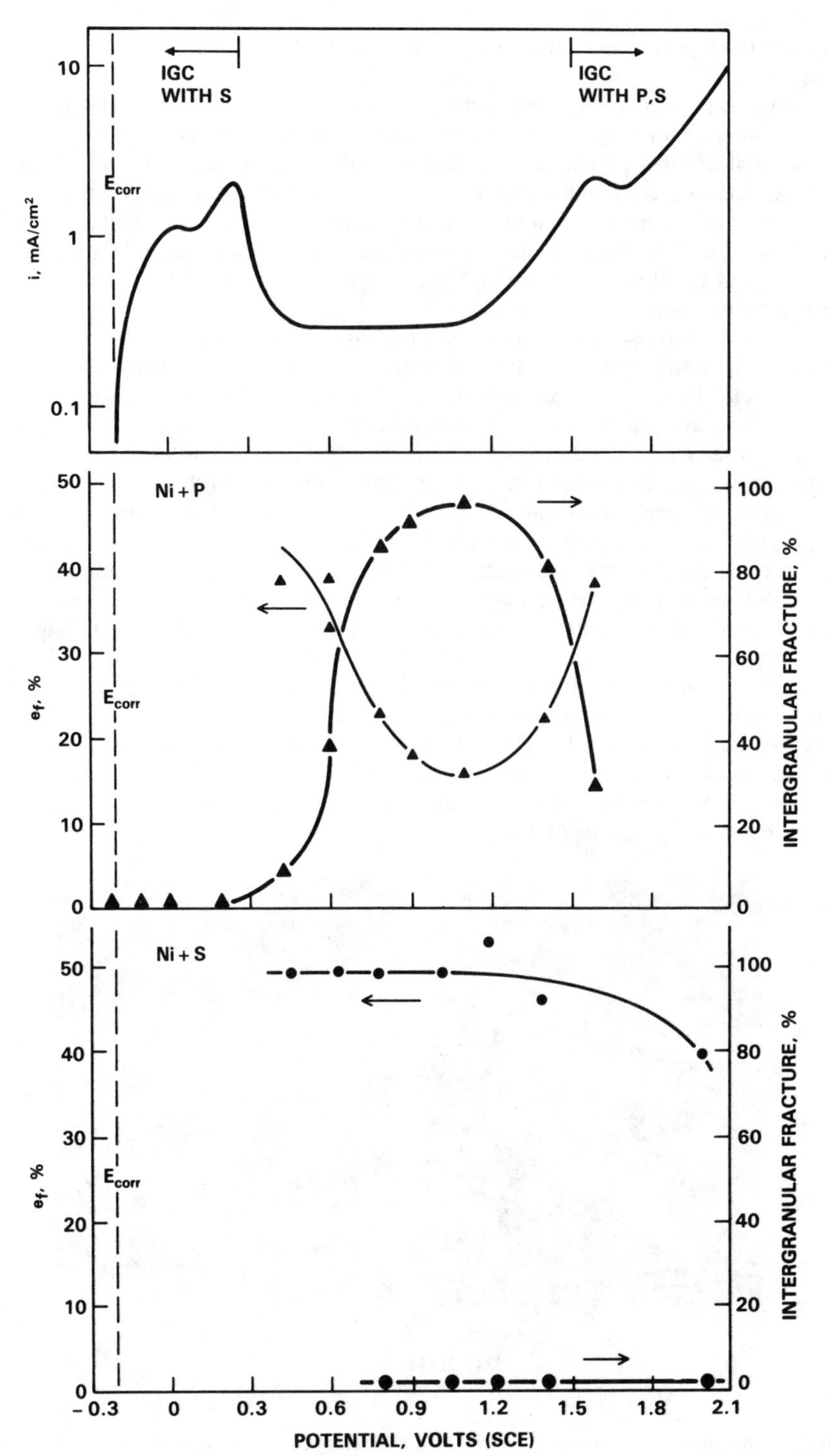

FIG. 2—*Polarization curve and slow strain rate results for Ni + P and Ni + S tested in 1N H_2SO_4 at 25°C.*

aries [1]. In this case, intergranular corrosion was apparent in both materials when exposed to 1N sulfuric acid (H_2SO_4) at a potential of 900 mV; however, SCC occurred only in the nickel with phosphorus segregation, as shown in Fig. 2. Upon further evaluation it became apparent that intergranular corrosion in the nickel with sulfur segregation occurred only to a depth of about 0.1 mm or about 1 grain diameter. Intergranular corrosion was arrested in the sulfur-enriched material when one of two conditions existed: a passive film did not form on the walls of the intergranular penetration or it did not form where the walls and the grain boundary at the base of the penetration passivated. It was shown, in subsequent corrosion studies with sulfur-implanted nickel, that both conditions could exist for sulfur in or on nickel depending on the electrochemical potential. Crack-tip chemistry or corrosion rate calculations using a model developed at PNL [8] which considers the effect of the corrosion rate of the walls substantiated that intergranular corrosion with nonpassive walls would cease corroding, as illustrated in Fig. 3. In contrast, when the tip remains active and the walls become passive, as in the case of nickel with segregated phosphorus, intergranular corrosion proceeded at a rate of 10^{-5} mm/s and IGSCC at a rate of 4×10^{-4} mm/s.

A somewhat opposite observation was made for iron with segregated sulfur and phosphorus tested in calcium nitrate solutions at 60°C (140°F). In this case, intergranular corrosion was slow to initiate and nonuniform, while crack growth was very rapid, as shown in Figs. 4 and 5. Intergranular corrosion is shown for three iron alloys in Fig. 4 where the grain boundary chemistries were (in monolayers) 0.23 sulfur, 0.0 phosphorus for iron, 0.11 sulfur, and 0.02 phosphorus for the iron + 0.1 atomic percent phosphorus, and 0.01 sulfur and 0.21 phosphorus for the iron + 0.1 atomic percent phosphorus + 0.1 atomic percent manganese alloy. Intergranular corrosion was not observed in any of the alloys after 25 h in the passive regime but was apparent after 50 h at 750 mV and 4 h at 1.5 V, which is in the transpassive regime. Crack growth rates of 10^{-3} mm/s and greater were observed in all the alloys at a potential of +750 mV for stress intensities of around 15 MPa$\sqrt{m}$. The stress-intensity thresholds for the cracking were similar in all the alloys as were the crack growth rates.

The examples given for nickel and iron illustrate just one aspect of the uncertainty regarding the conditions between localized corrosion and crack initiation. In the case of nickel, localized corrosion appeared to occur without difficulty, but sulphur and phosphorus segregation produced vastly different results; in the case of iron, localized corrosion was difficult, but crack growth was rapid and sulfur and phosphorus segregation produced essentially identical results. Therefore, intergranular corrosion cannot be used as a simple indicator of IGSCC; a more detailed understanding of the localized chemistry and mechanics is needed to properly predict IGSCC.

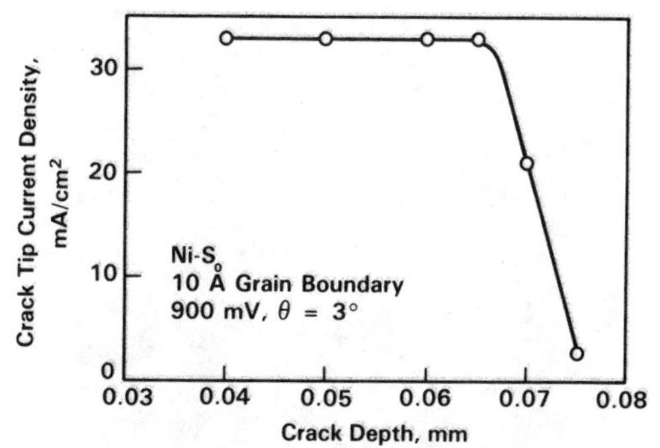

FIG. 3—*Calculated crack tip corrosion rate versus crack depth for Ni+S in 1N H_2SO_4 at 25°C.*

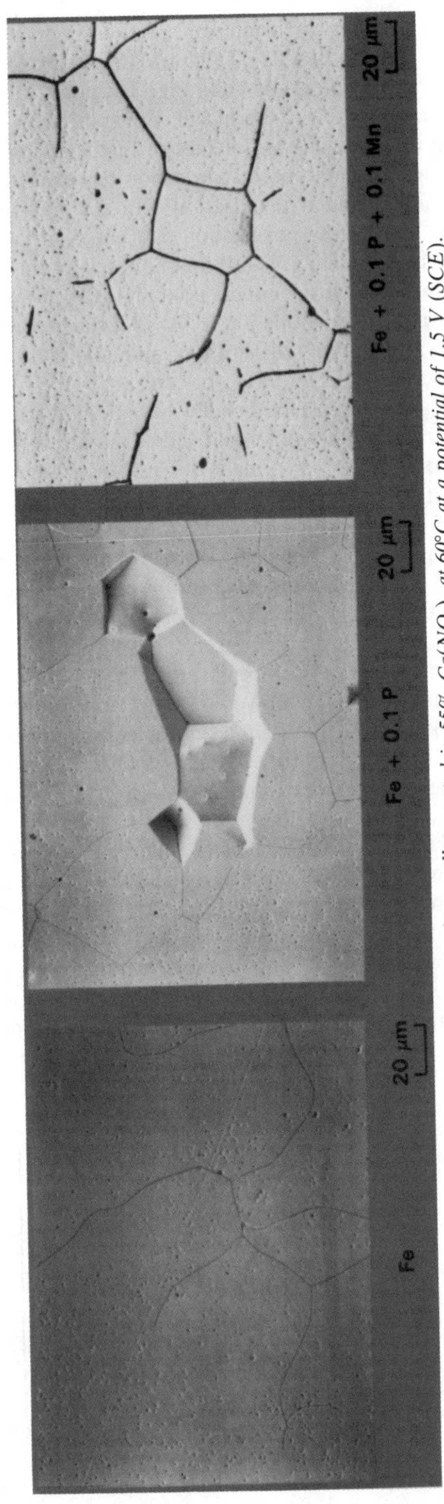

FIG. 4—*Intergranular corrosion of iron alloys tested in 55% $Ca(NO_3)_2$ at 60°C at a potential of 1.5 V (SCE).*

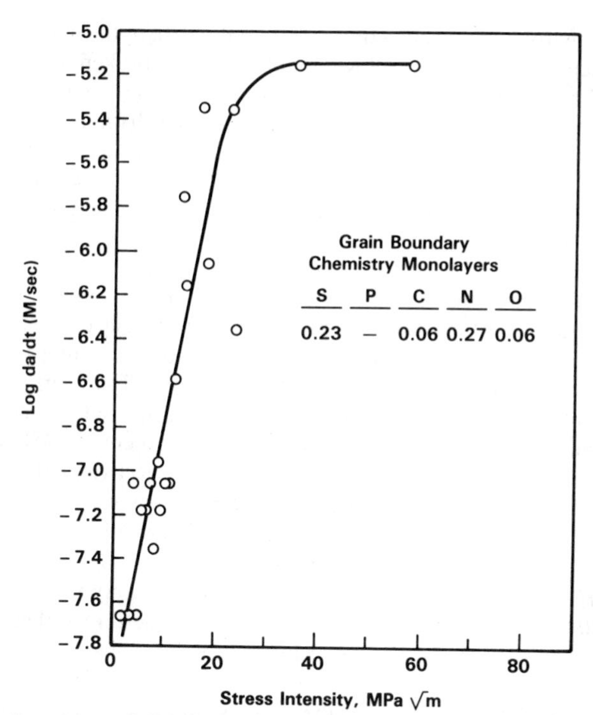

FIG. 5—*Subcritical crack growth (IGSCC) of iron in 55% $Ca(NO_3)_2$ at 60°C and +750 mV (SCE).*

Crack Propagation Issues

Crack-Tip Strain and Strain Rate Effects

The role of crack-tip strain and strain rate on the stress corrosion crack growth rate of passive-film-forming materials has been well established by Vermilyea [*9*] and Ford [*10*], among others. The models presented by these two researchers give the crack growth rate as a function of the crack-tip strain rate, passive film rupture rate, and repassivation rates. Strain rate effects in a carbon-manganese steel tested in a carbonate/bicarbonate solution have also been evaluated by Parkins [*11*], who showed a sharp strain rate threshold for cracking. Parkins also explained these results in terms of the strain maintaining an active crack tip while the walls form a protective film.

In general, the models of Vermilyea, Ford, and Parkins deal with the effect of crack-tip strain or strain rate increasing the crack-tip corrosion rate by exposing clean, unoxidized metal. However, direct surface chemistry measurements of the interaction of the environment with the clean, exposed metal and more specifically with a plastically deformed clean metal surface are relatively scarce. This reaction or adsorption rate could increase or decrease the crack growth rate, depending on the nature of the reaction. For a reaction with hydrogen sulfide in the crack-tip environment, strain could enhance the adsorption rate and hence the crack growth rate, while in the case of water reacting with a passive-film-forming material, the repassivation rate could be a function of strain or strain rate.

Influence of Deformation and Contamination on Surface Adsorption

Environmentally enhanced fracture in gaseous environments must involve either the interaction of one component of the environment with a material surface or the transport of some part of the environment through the surface to the bulk of the material. In some cases, a dissociation reaction at the surface allows a molecule to be broken into components which may react with the material very differently from the way the initial molecule reacts with it. Gaseous hydrogen [*12,13*], for example, is thought to adsorb on the surface in a molecular precursor state and to then dissociate to two hydrogen atoms. It is only the dissociated hydrogen atoms that cause embrittlement. The types of surface reactions that occur and their rate are, therefore, important components of any environmentally enhanced cracking process.

Influence of Deformation on Adsorption—It is well established that most types of defects that occur on a surface alter the adsorption process. Stepped surfaces on single-crystal silicon are up to ten times more reactive to oxygen than are smooth, unstepped surfaces [*14*]. In other cases, surface species are observed to bind preferentially to surface defects [*15,16*]. In addition to the role of defects, calculations suggest that an applied stress itself can influence adsorption [*17*] and studies of thin layers of copper on ruthenium [*18*], which are highly strained relative to normal copper lattice spacings, have a reactivity enhanced by almost an order of magnitude compared to pure copper for some reactions.

Although the above studies suggest that the highly deformed region around a crack tip might be more reactive than nearby regions, not all reactions are observed to be enhanced by defects and it is not obvious that studies on stepped single crystals directly relate to the complicated surfaces that are present during cracking. It is desirable, therefore, to examine the effect of strain on surface chemistry for environments and materials in which SCC is an issue.

Investigations at PNL have examined the effect of surface deformation on the adsorption of hydrogen sulfide (H_2S) and water (H_2O) on iron [*19,20*]. An enhanced embrittlement of many engineering alloys in the presence of H_2S provides reason for examination of its surface properties. Water is present in many environments and can provide hydrogen for embrittlement, oxygen for oxidation, and OH^- species often found in passive films. Two different approaches have been used. First, the adsorption of each gas on annealed and ion-bombarded iron was studied. Ion sputtering is a common method of cleaning specimens, and in addition to showing that surface defects can influence adsorption, the studies established annealing conditions necessary after sputter cleaning before the start of the second type of test. The second approach involved the use of an ultra-high vacuum stressing stage that was designed to fit into a surface analysis system [*19*]. In this study, the adsorption of gas is studied as a function of applied stress on the specimen.

The influence of ion beam damage on the adsorption of H_2O on an Fe(110) crystal is shown in Fig. 6. The oxygen on the surface was measured as a function of time using Auger electron spectroscopy. The flow of water molecules to the surface was the same for each set of data shown. As the amount of sputtering increased, the rate of adsorption increased. For both the H_2O and H_2S adsorption, surface damage was found to influence primarily the rate of adsorption. No systematic changes in the saturation coverage was found.

In order to determine the relative effect of the ion beam damage on the rate of adsorption, the slope of the adsorption for coverages below 0.3 monolayer ($\alpha = d\theta/dt$, where θ is oxygen coverage) is divided by the slope obtained for the annealed specimen (α_0) and this ratio (α/α_0) is plotted as a function of ion dose in ions per surface atom. As shown in Fig. 7, the rate of H_2O adsorption on iron increases by over a factor of 10 due to sputter damage. For comparison, the effect of sputter damage on the rate of H_2S adsorption on polycrystalline iron is also shown. It should be noted that H_2S is very reactive and that the adsorption rate did not increase very much.

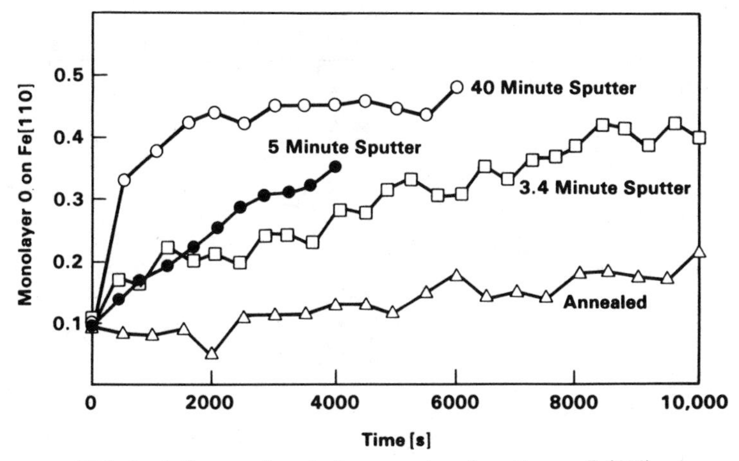

FIG. 6—*Influence of sputtering on water adsorption on Fe(110).*

Water, on the other hand, adsorbs slowly on iron, and the surface damage can markedly increase the rate of adsorption. The yield strength of the iron samples was 60 MPa, which exceeds the value at which enhanced adsorption was observed. This can be explained by near surface microstrain, which can occur at stresses less than the bulk yield strength.

The influence of applied load on the rate of adsorption on H_2S on iron was found to be comparable to that caused by sputter damage, as shown in Fig. 8. Similar studies for H_2O are in progress.

Both H_2O and H_2S are thought to adsorb on iron through precursor states such as those

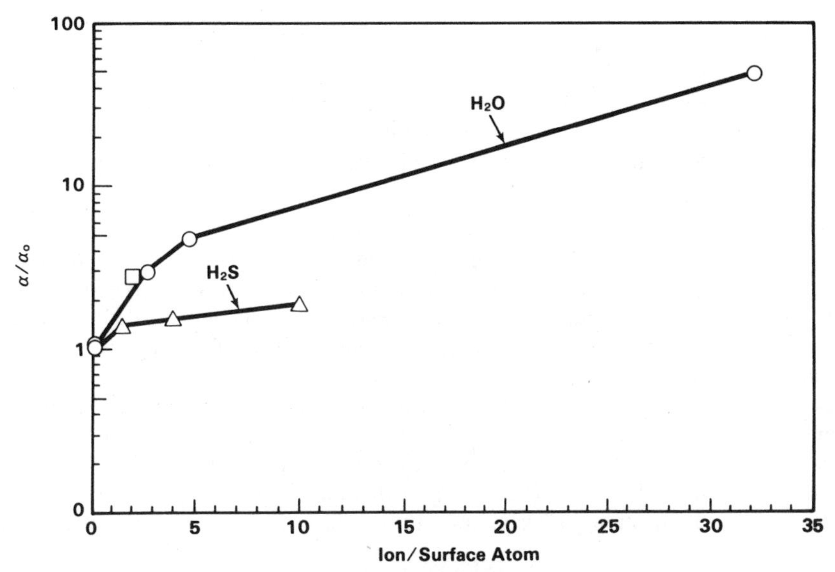

FIG. 7—*Influence of sputtering on water and H_2S adsorption on iron.*

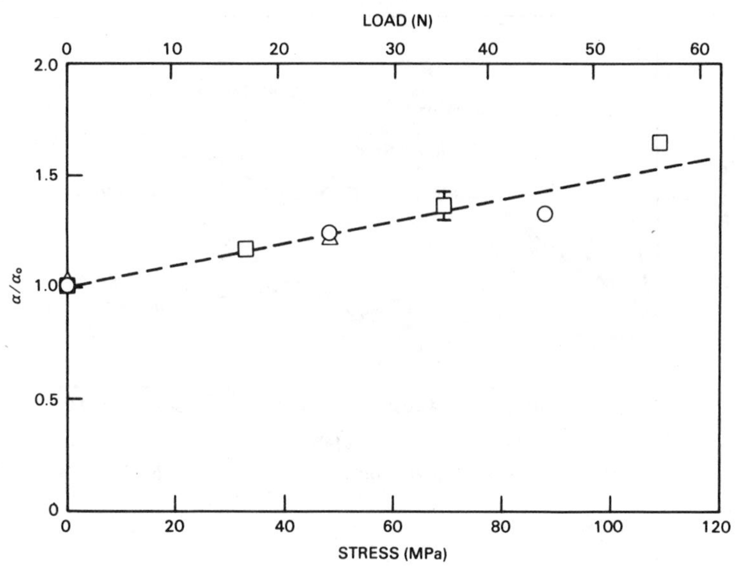

FIG. 8—*Relative adsorption rates for sulfur on iron as a function stress. Adsorption rates* α *are normalized by the adsorption rates for specimens in the annealed conditions* α_0.

listed below [*21,22*]

$$[H_2O]_g \underset{k_d}{\overset{k_aP}{\rightleftharpoons}} [H_2O]_p \xrightarrow{k_c} Fe - 0 + H_2 \tag{1}$$

$$[H_2S]_g \underset{k_d}{\overset{k_aP}{\rightleftharpoons}} [H_2S]_p \xrightarrow{k_c} \text{Fe-H} + \text{Fe-HS} \tag{2}$$

where

k_aP = rate of adsorption from the gas to the precursor state,
k_d = rate of desorption from the precursor state back to the gas phase, and
k_c = rate constant for adsorption into a dissociated chemisorbed state from the precursor.

For Reaction 2, surface damage appears to increase k_a [*19*]. Earlier work [*10*] on H_2O adsorption on Fe(100) suggested that increasing temperature increased k_c (rate of adsorption into the dissociated state). The measurements of H_2O adsorption on annealed and damaged Fe(110) may be consistent with the increased k_c, but more detailed analysis is needed to verify this conclusion.

These and other studies indicate that surface deformation caused by material deformation or iron beam damage can alter the rates of adsorption of components from the environment onto the surface. In some stages, crack growth can be controlled by the kinetics of adsorption from the environment [*12,23*]. When the growth rate is limited by adsorption kinetics, alterations in the adsorption pathway and rate may have a significant influence on crack growth.

Influence of Contamination on Hydrogen Adsorption

Clean metal surfaces can react with species in an environment in a manner different from that by surfaces covered or contaminated with other elements. Reaction rates may be decreased due to surface blocking or adsorption rates may be increased due to altered reaction pathways and more favorable energetics. As indicated earlier, H_2S increases the chances of hydrogen embrittlement. This may be due to a hydrogen-producing reaction such as discussed above, or it might be due to an influence of sulfur on H_2 dissociation. Pradier et al. [*24*] have shown that small coverages of sulfur on Pt(111) enhance H_2-D_2 equilibrium, presumably by speeding the dissociation recombination reactions for hydrogen on the platinum surface. Recent work by Rondulic et al. [*25*] has shown that half a monolayer of oxygen on Ni(111) increases the adsorption rate of hydrogen and changes the process from activated to nonactivated adsorption.

The influence of sulfur on the adsorption of hydrogen on Type 4340 steel was undertaken as part of a study of the influence of surface conditions on hydrogen embrittlement [*26,27*]. Electron-stimulated desorption measurements suggest that the amount of hydrogen on the Type 4340 surface at 30°C (86°F) varies with the amount of sulfur on the surface in a nonlinear fashion. Small amounts of sulfur enhance the hydrogen coverage, while large amounts of sulfur block hydrogen from the surface. Similar trends are observed in the fracture stress of the steel during hydrogen dosing (Fig. 9). Small amounts of sulfur encourage hydrogen embrittlement, while a surface saturated with sulfur partially blocks hydrogen entry onto and into the material.

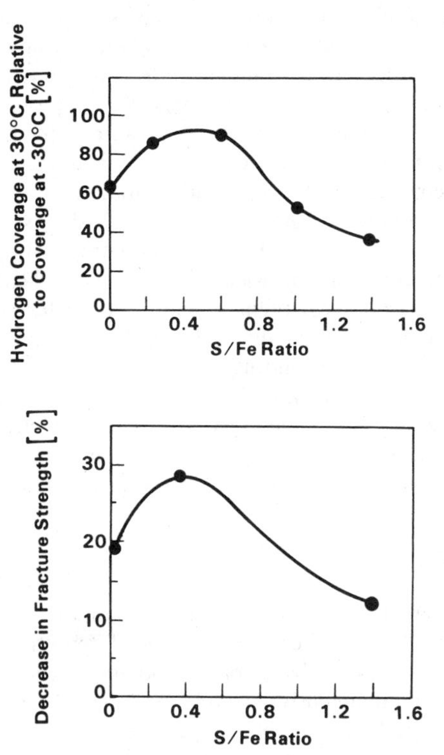

FIG. 9—(Top) *Hydrogen surface coverage at 30°C and* (bottom) *decrease in fracture stress of 4340 steel as a function of sulfur coverage at* $-30°C$ *(in percent).*

A metal surface such as a growing crack can be contaminated both by the environment surrounding it and by impurities from within the metal. The rate of reaction at such a surface may depend upon both the details of the surface composition and surface deformation created by the fracture process. It is clear that the proper combination of environment, surface deformation, and surface contamination can markedly alter specific surface reaction rates in a localized region.

Crack Microchemistry Effects—Stress corrosion can occur when the local crack tip chemistry supports a corrosion reaction at the crack tip. The local crack-tip chemistry is affected by several variables, including the bulk solution chemistry far from the crack or crack tip. Transport processes within the crack are controlled by diffusion, migration, and convection, and these processes in turn determine the rate at which ions are transported away from the corroding crack tip, hence the crack-tip corrosion rate, and also the rate at which anions are transported to the crack tip. The migration of ions to the crack tip that degrade passive film formation is one manner in which the bulk solution chemistry can affect the crack-tip chemistry. Other factors such as crack length, crack width and angle, convection from cyclic loading or corrosion currents, and the location of cathodes for anodically dominated SCC or anodes for cathodically dominated SCC are all important in determining the crack-tip corrosion rate.

Direct measurement of crack-tip chemistry, pH, and potential has been attempted using simulated cracks with electrodes inserted normal to the crack plane or with microelectrodes inserted from the crack mouth to a fraction of the crack length [*28–30*]. These evaluations are useful for illustrating the trends expected in real cracks; however, measurements are limited to crack geometries that are substantially different from real cracks. Crack-tip chemistry modeling has been used by several researchers [*8,31,32*] to predict the local corrosion rates, pH, and potentials in real crack geometries because of the restrictions of probing the small dimensions within real cracks. Turnbull and Thomas [*32*] developed a steady-state mass transport model considering diffusion and migration for steels with crevices in 3.5 percent sodium chloride solutions. Anodic dissolution, hydrolysis of the ferrous ions, and cathodic reduction of hydrogen ions and H_2O were considered in the model and were assumed to take place at both the crack tip and walls. Theoretical predictions of the potential and pH were found to be in reasonable agreement with values determined experimentally with electrodes in simulated cracks. The methodology of Turnbull and Thomas was applied to cracks in nickel with segregated phosphorus or sulfur by Danielson et al. [*8*]. The emphasis of this evaluation was the role of crack wall chemistry, crack length, and crack angle on the crack-tip corrosion rates and hence the crack growth rates. Further evaluation has since been conducted on the effect of crack geometry and salt film formation on the crack-tip corrosion rates. These effects will be reviewed in the subsequent sections of this paper.

Crack Wall Chemistry Effects

The concept that the crack wall corrosion rate must differ from the crack-tip corrosion rate for SCC to occur has been well established. This differential has generally been associated with SCC occurring in the active-to-passive potential regimes, where these two conditions could be satisfied simultaneously. The relationship between this corrosion differential and SCC has generally been linked merely to the aspect ratio of the crack where high wall corrosion rates result in a low aspect ratio and hence a blunted crack or shallow pit and low wall corrosion rates with a high aspect ratio crack.

Danielson et al. [*8*] evaluated the effect of variable wall chemistries on the crack wall corrosion rates and also considered the wall corrosion rate effect in terms of the crack-tip chemistry rather than the crack aspect ratio. Variable crack wall chemistries resulted from intergranular stress corrosion cracks propagating along either phosphorus- or sulfur-enriched grain boundaries in nickel. Jones et al. [*1*] showed that for the conditions of these tests in 1N H_2SO_4, phosphorus is oxidized from the grain boundary at the crack tip leaving the crack walls with the matrix nickel composition and thus allowing passivation to occur. Sulfur is very surface active and remains on the crack walls, which keeps them in the actively corroding condition at potentials in the active-to-passive transition, while sulfur is converted to a sulfate at transpassive potentials which allows nickel to passivate. Therefore, sulfur-enriched grain boundaries can either be active walls/active tip or passive walls/passive tip. Using the crack tip chemistry model adapted from Turnbull and Thomas [*32*], Danielson et al. [*8*] calculated the effect of the crack wall corrosion rate on the crack-tip corrosion rate. Similar concentrations of phosphorus and sulfur in the grain boundary would produce similar crack-tip corrosion rates based on tests with flat samples doped with phosphorus or sulfur. The results of these calculations, shown in Figs. 10 and 11, as a function of crack angle clearly illustrates the suppression of the crack-tip corrosion rate resulting from the high wall corrosion rate of the sulfur-enriched crack walls. For a 2-mm-long crack at low crack angles, the potential in the crack of the sulfur-enriched intergranular cracks is such that the crack walls are active, which causes nickel sulfate salt precipitation, while as the crack opens to an angle greater than 0.32 deg, the crack walls and crack tip are passive. The result of these crack wall conditions limit the crack tip corrosion rate to 0.065 mA/cm² for the sulfur-enriched cracks; a crack-tip corrosion rate of 175 mA/cm² resulted for a 2-mm-long phosphorus-enriched crack at an angle of about 0.5 deg.

There are other SCC examples in which crack wall chemistry could play a significant role in the crack growth behavior. These examples include sensitized stainless steel, for which

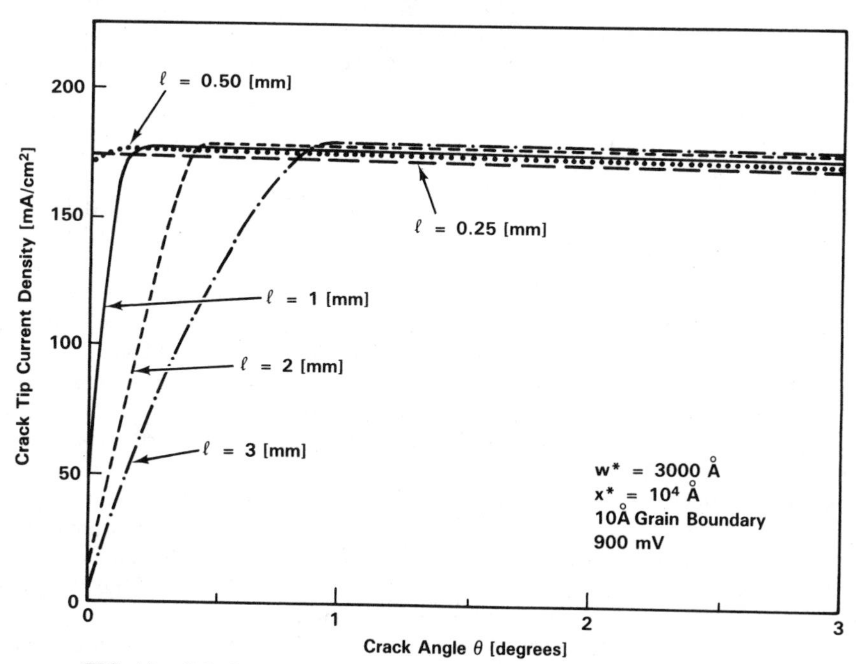

FIG. 10—*Calculated crack-tip corrosion rates versus crack angle for Ni + P.*

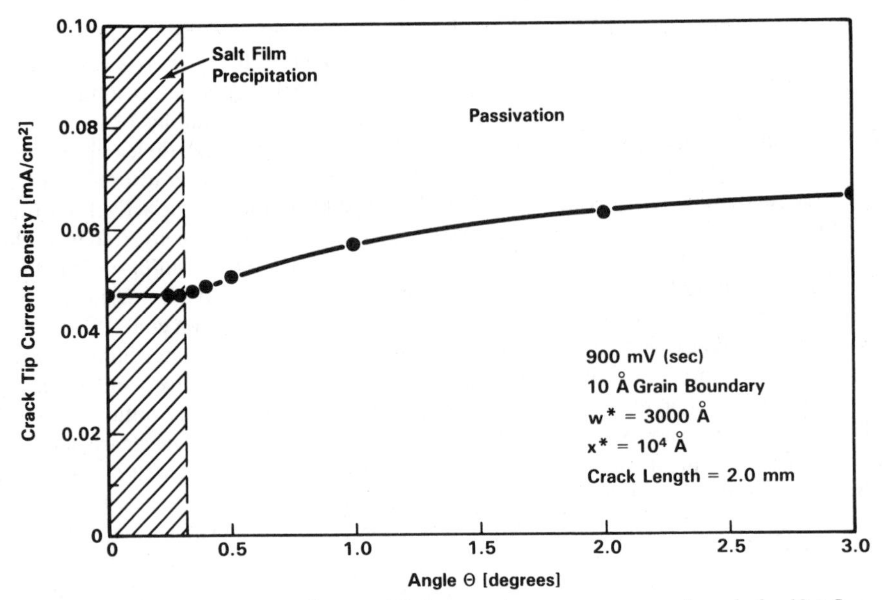

FIG. 11—*Effect of crack angle on crack-tip corrosion rates versus crack angle for Ni+S.*

the chromium concentration along the crack wall is dependent on the minimum and width of the chromium profile; alloys with active precipitates at their grain boundaries such as gamma prime hardened nickel-based alloys or aluminum alloys; transgranular cracks that intersect active inclusions, such as sulfides; or the migration of surface-active species, such as sodium thiosulfate or hydrogen sulfide into the crack. Therefore, it would appear that crack wall chemistry and corrosion rates could be important in the SCC of numerous materials.

Crack Length and Geometry Effects

The length and geometry of a stress corrosion crack can potentially affect the crack tip chemistry and mechanics of a growing crack. Crack length effects have been demonstrated for fatigue crack growth and are generally understood as resulting from crack closure [*33*] or, at crack lengths approaching the scale of the microstructure, the interaction of the stress field with the microstructure. Crack closure considerations are not relevant under static load stress corrosion, while the interaction of the stress field with the microstructure at very short crack lengths is potentially a factor. Consideration of the stress and stress intensity at very short crack lengths is an important factor in the development of models describing crack initiation conditions. A clear treatment of crack mechanics at very short crack lengths in statically loaded conditions has not been rigorously treated.

Crack length effects on crack-tip chemistry have been evaluated experimentally and theoretically for hydrogen-induced crack growth of a steel by Gangloff and Turnbull [*34*]. In this evaluation they found that the stress intensity threshold, K_{th}, for hydrogen-induced crack growth in a 3% sodium chloride solution at −600 mV was independent of the crack length for cracks of 1 to 25 mm, but that the threshold stress intensity decreased at crack lengths less than 1.0 mm. Theoretical calculations based on the relationship between pH and hydrogen solubility, solubility and K_{th} and calculated crack tip pH versus crack length gave values in fair agreement with the experimental measurements. The crack length dependence results from the predicted hydrogen production rate, which increases with decreasing crack length.

Crack length effects on the crack-tip corrosion rate have also been evaluated by Danielson et al. [*8*] for IGSCC in nickel with segregated phosphorus, as shown in Fig. 10. These results are shown as a function of crack angle, which is the angle between the crack tip and crack mouth, which can be a function of load or stress intensity. The effect of varying the crack length from 0.25 to 3 mm is shown in Fig. 10, where it can be seen that the crack-tip corrosion rate decreases at a critical crack angle for cracks of 1 to 3 mm but not for 0.25 and 0.5-mm-long cracks. At crack angles greater than 1 deg, there is no crack length dependence on the crack-tip corrosion rate or, equivalently, the crack growth rate. This suggests that at high stress intensities such as in the Stage II regime, there may be little crack length dependence; at lower stress intensities, a significant crack length dependence is suggested. If K_{th} is a function of the angle at which the current decreases, a plot of this angle θc versus crack length should indicate the relationship between K_{th} and crack length for IGSCC in nickel with segregated phosphorus. This relationship is shown in Fig. 12*a*, where a linear decrease in θc with crack length is evident for crack lengths of 0.5 to 3 mm. A crack-length-independent regime, as seen by Gangloff and Turnbull [*34*], Fig. 12*b*, did not occur in nickel up to a crack length of 3 mm, possibly because the cathode was exterior to the crack in the nickel while anodes and cathodes were within the crack in the steel. Of course, the similarity in Figs. 12*a* and 12*b* is probably only fortuitous, since one case is for anodic dissolution-controlled intergranular crack growth in nickel and the other is for hydrogen-induced crack growth of a steel.

Crack-tip geometry will vary significantly with yield strength and applied stress intensity where the crack tip opening displacement (CTOD) and crack angle is a function of $(K/\sigma y)^2$. Small CTOD and crack angle values are associated with small values of $K/\sigma y$ with increasing values for increasing $K/\sigma y$. Since the crack-tip electrochemical conditions are crucial to the crack tip corrosion rate and crack growth rate, anything that affects the crack tip electrochemistry should affect stress corrosion crack growth rate. Saturation of the crack-tip solution with metal ions from the crack tip region is a factor that is expected to be a function of the crack-tip geometry and that can alter the crack-tip corrosion rate. An assessment for IGSCC in nickel with segregated phosphorus for changing CTOD and crack angle was conducted using the crack-tip chemistry model by Danielson et al. [*8*]; the results are shown in Fig. 13. This analysis considered a range of crack-tip geometries from a wedge with a 10 Å width at the tip of an angle of 0.005 deg up to a crack tip plastically strained to an opening of 0.3 μm and an angle of 0.2 deg. The plastically deformed crack tip was taken as parabolic to a distance x^*, beyond which the crack walls were taken as linear with an angle θ. A detailed description of the crack geometry parameters is given in Ref *8*. Both the crack angle and the CTOD had a significant effect on the crack-tip corrosion rate. A change in the crack angle from 0.005 to 0.2 deg at a CTOD of 10 Å was sufficient to increase the crack-tip current density from 2 mA/cm^2 to 71 mA/cm^2; at a CTOD of 0.3 μm, this increase in the angle resulted in an increase in the crack-tip current density from 15 to 100 mA/cm^2. These calculations suggest a clear relationship between the crack-tip mechanics and the crack-tip corrosion rate for IGSCC. Some features unique to this analysis include the active grain boundary and passive crack wall relationship and the separation between the anode and cathode where the anode is at the crack tip and the cathode is at the crack mouth. Further analysis for TGSCC and for TGSCC/IGSCC with cathodes along the crack wall is needed to determine the more general role of crack geometry in SCC.

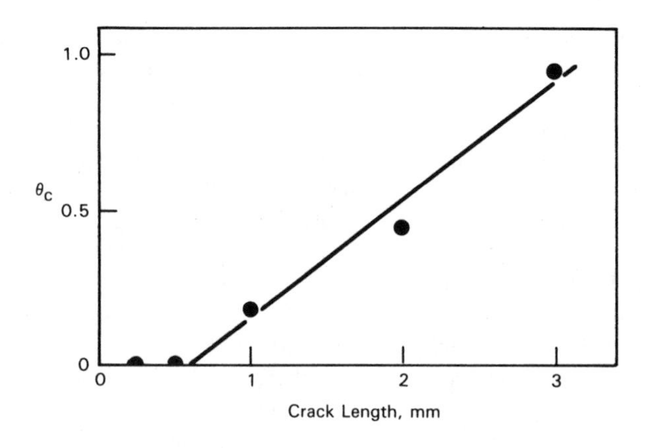

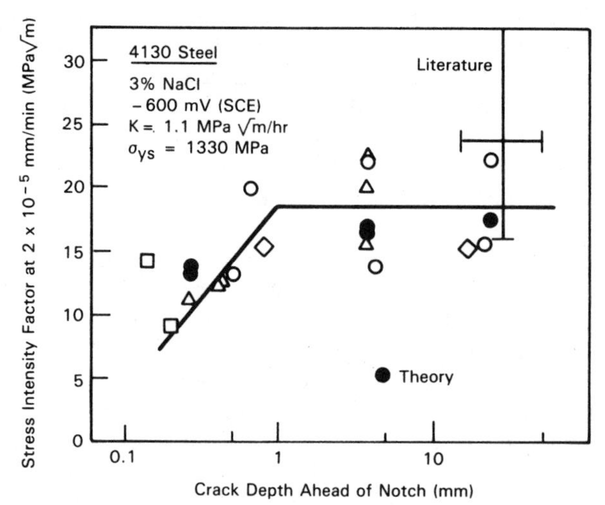

FIG. 12—*Effect of crack depth on threshold stress corrosion cracking parameters:* (top) *calculated critical angle for Ni + P and* (bottom) *calculated and measured* K_{th} *for Type 4130 Steel in 3% NaCl.*

Crack Growth Processes in Stage I

The crack velocity-stress intensity relationship during SCC of a wide variety of materials is characterized by a Stage I regime in which the crack velocity has a strong stress-intensity dependence and a Stage II regime in which the crack velocity has a weak stress-intensity dependence. The following relationship is frequently used to express the Stage I crack velocity-stress intensity relationship

$$da/dt = A\ K^m \quad (3)$$

where

A = a constant,
K = stress intensity, and
m = stress-intensity exponent.

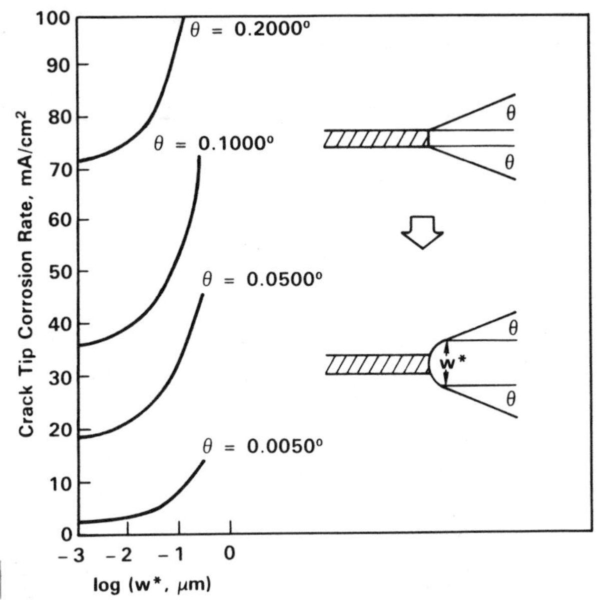

FIG. 13—*Calculated crack-tip corrosion rate as a function of crack geometry for Ni+P.*

A summary of *m* values for a variety of alloys and test environments is given in Table 1, where it is seen that *m* is 7 to 24 for Type 304SS [*9,35–38*], 11 for Ni+P [*1*], 19 for an aluminum alloy [*39*], and 8 for iron. The value of *m* will clearly depend on the geometry and compliance of the specimen and method of loading (increasing, constant, or decreasing load), which could explain the variation in these data. For the Type 304SS data, which were obtained with a variety of specimen types (DCB, SEN, CT), the value for *m* ranged from 7 to 24. In any event, the data summarized in Table 1 illustrate the strong dependence of the crack growth rate on *K* in Stage I.

A number of SCC models give predicted crack velocity-stress intensity relationships (Table 2). Andresen [*40*] has recently expanded on Ford's [*10*] stress corrosion crack growth rate-strain rate model and shown that the crack growth rate can be expressed by the following relationship (Table 2)

$$da/dt = [1.5 \times 10^{-8} K^{2.8}(A)^{1/n}]^{2n/2-n} \tag{4}$$

where *A* and *n* come from the following equation

$$da/dt = A\, e_{ct}^{n} \tag{5}$$

A value of *n* equal to 0.76 was given by Andresen [*40*], while Ford [*10*] has assumed *n* to be 0.5. The crack growth rate is proportional to *K* to the 1.9 to 3.4 power based on *n* of 0.5 to 0.76.

These values of *m* are substantially less than those observed experimentally (Table 1) and are more consistent with the stress-intensity dependence observed in the Stage II regime. However, the rate-controlling step in the Stage II regime is thought to be either ion transport in the crack electrolyte or the environment/crack-tip reaction rate, and not the passive film rupture rate as assumed by the Ford/Andresen analysis.

TABLE 1—*Summary of experimental Stage I crack growth parameters* m *and* B.

Material	Test Condition	m	B	Reference
304SS	sensitized 97°C, 1.5 ppm O_2	7	0.14	[35]
304SS	sensitized 99°C, 22% NaCl	24	1.2	[36]
304SS	25°C, $Na_2S_2O_3$	11	0.26	[37]
304SS	sensitized 90°C, $Na_2S_2O_3$	17	0.25	Fig. 14*b*
304SS	sensitized, 250°C 10^{-2} m, Na_2SO_4	8	0.18	[38]
304SS	sensitized, 289°C 10^{-2} m, Na_2SO_4	14	0.8	[9]
Ni + P	25°C, 1N H_2SO_4 900 mV (SCE)	11	1.1	[1]
7079-T651 Al	25°C, saturated NaCl	19	1.3	[39]
Fe	60°C, 55% Ca $(NO_3)_2$ + 750 mV	8	0.25	Fig. 5*a*
Mild steel	92°C, 33% NaOH E_{corr}	7	0.16	[42]
Brass	23°C, ammoniacal copper sulfate (pH = 7)	1.8	0.098	[43]

Newman [41] presented a film rupture stress corrosion model for steel in sodium hydroxide (Table 2). Newman assumed that crack extension was limited to the thickness of the corrosion product film, L_c, which fractured in a time period, t_c, determined by the stress intensity and hence the crack tip strain rate. The crack velocity was expressed very similarly to that given by Vermilyea [9]

$$da/dt = L_c/t_c \tag{6}$$

The parameters L_c and t_c are not independent because the value of L_c depends on the strain rate and hence on t_c. The expression for L_c was given as

$$L_c = e_c\, k/s\, f(K) \tag{7}$$

TABLE 2—*Summary of predicted Stage I crack growth parameter* m.

Material	Crack Growth Model	m	Reference
304SS	$da/dt = [1.5 \times 10^{-8}\, K^{2.8}(A)^{1/n}]^{2n/2-n}$	1.9–3.4	[40]
Steel	$da/dt = -e_c\, k\beta/f(K)\, \ell n\, (1 - G)$	25	[41]
304SS	$da/dt = k''\, K^{p/\eta}$	p/η	[9]
Brass	$da/dt = A \exp\,(-Q + f(K)/RT)$	...	[43]
Ni + P	CTOD, CTOA, crack-tip corrosion rate	13	[44]

where

e_c = film rupture strain,
k = stress relaxation constant,
s = a constant which relates the strain rate to the stress, and
$f(K)$ = s stress intensity function.

The expression for the critical strain, e_c, was given as

$$e_c = s\, f(K)\, v\, [1 - \exp(-\beta t_c)]/K\, \beta \tag{8}$$

where v is the initial corrosion rate and β is the passivation rate constant. Solving Eq 8 for t_c and substituting Eqs 7 and 8 into Eq 6 gives the following relationship for the crack velocity

$$da/dt = -\, e_c\, k\beta/sf(K)\, \ln\,(1 - G) \tag{9}$$

where G is equal to $e_c\, K\beta/sf(K)v$. Equation 9 is valid for the following values: $0 < G < 1$. Newman [*41*] solved this relationship numerically and obtained curves for the crack velocity versus K for $f(K)$ proportional to K and K^2. The two functions for K result in similar-shaped curves, with K^2 dependence giving a slightly higher threshold value; however, the predicted slope of the Stage I regime was 25 for both cases. Newman concluded that the stress corrosion model was consistent with experimental data, but no comparison was made between calculated and experimental crack velocity-stress intensity relationships. The predicted slope of the Stage I regime is only slightly larger than those observed for sensitized stainless steel, but is much larger than that for iron in calcium nitrate ($Ca(NO_3)_2$) or mild steel in sodium hydroxide [*42*] as given in Table 1.

Vermilyea [*9*] considered the effect of the stress intensity and the solution chemistry by proposing that the relationship between crack velocity and stress intensity has the form

$$da/dt = k'CK^p \tag{10}$$

where k' is a constant and C is the concentration of some species in the crack tip solution. Then, assuming that the difference between the crack tip and mouth concentration is a function of crack velocity, he obtained the following relationship for crack velocity

$$da/dt = k''K^{p/\eta} \tag{11}$$

where η is a small constant in the crack-tip concentration/crack velocity relationship and considered to have a value of about 0.1 by Vermilyea. The suggestion was made that small changes in η and hence in the crack-tip chemistry could have large effects on the stress-intensity dependence of the crack velocity, which is consistent with the values of m ($m = p/\eta$) given in Table 1. A systematic study of crack-tip chemistry effects on the parameter m has not been carried out, but the data in Table 1 suggest that for sensitized Type 304SS, m is less than 10 for high-temperature water, 8 to 14 for sodium sulfate (Na_2SO_4), 11 to 17 for sodium thiosulfate ($Na_2S_2O_3$), and 24 for 22% sodium chloride. Increasing values of m suggest decreasing values of η in Vermilyea's model, which results from increasing concentrations of a crucial species at the crack tip for a given crack velocity. Further work under more controlled conditions is needed to determine if Vermilyea's concept is valid, and if so, to then determine the relationship between crack tip chemistry and crack velocity-stress intensity.

While the crack velocity-stress intensity relationship is frequently assumed to have the form given in Eq 2, a considerable amount of data fit the following form

$$\log da/dt = \log A + B\,K \tag{12}$$

or

$$da/dt = A \exp\,(B\,K) \tag{13}$$

Lee and Tromans [43] found that the crack velocity-stress-intensity relationship for a cold-worked brass tested in an ammoniacal copper sulfate solution could be described by the following relationship, which is of the form of Eqs 12 and 13

$$da/dt = A \exp\,(-Q + f(K)/RT) \tag{14}$$

where Q was 48.9 kJ/mole at $K = 0$ and $f(K)$ was linear with K. The experimental data for crack velocity-stress intensity when fitted to Eq 2 is consistent with an m value of 1.8, which is the smallest value given in Table 1. Values for B in Eq 12 are given in Table 1, from which it can be seen that a large value of B corresponds to a large value of m. Accuracy in the determination of B and m from published figures limits the accuracy of this comparison, so that the largest value of m does not correspond to the largest value of B.

Three of the stress corrosion models presented in Table 2 are expressed in the form of Eq 2; the Lee and Tromans model is in the form of Eq 13. Comparison between experimental data and the stress corrosion models of Andresen [40], Newman [41], and Vermilyea [9] is difficult because the experimental values of m were obtained by assuming that the data are of the form of Eq 2, while in reality they are not. Even with this assumption, the experimental values for m suggest a strong dependence of the crack velocity on the stress intensity, while the model by Andresen gives a value of m between 1.9 and 3.4. On the other hand, the model by Newman results in a large value of m and that by Vermilyea in a value of m which is dependent on the chemistry at the crack tip. For small values of η, the model of Vermilyea predicts large values for m. The model presented by Lee and Tromans is of the same form as most experimental data; however, their model was not compared to their experimental results and a value of B was not given, so comparison with experimental crack growth-stress intensity relationship is not possible.

Using the relationship between the CTOD and crack-tip opening angle (CTOA) and crack-tip corrosion rate, Jones, Danielson, and Oster calculated the crack velocity-stress intensity relationship for Ni+P [44]. The CTOD and CTOA were related to the stress intensity using the elastic-plastic analysis of Kumar, German, and Shih [45]. The calculated crack growth rate-stress intensity relationship was of the form of Eq 2 with a linear (log da/dt) versus (log K) relationship. A value of 13 for the exponent m was obtained for the case where the CTOD was uncorrected relative to the measured CTOD. This compares to an experimental value of 11. The calculated and experimental crack growth rate-stress intensity curves differ in the stress intensity threshold, which indicates that Stage I in Ni+P is only partially described by the crack-tip chemistry model. Evaluation of the effect of salt film formation and crack-tip strain on the passive wall corrosion rate on the crack growth rate-stress intensity relationship is in progress. Both of these factors are expected to cause a shift in Stage I to higher stress intensities.

Crack Extension Processes

Transgranular SCC by film-induced cleavage has been proposed by Newman and Sieradzki [6] to explain crack extension processes in a variety of materials. In the case of brass, they propose that a dealloyed layer is the brittle film that induces cleavage in normally ductile brass. In a later analysis, Sieradzki [*46*] concluded that the elastic modulus, film-substrate lattice parameter mismatch, film thickness, and interfacial bond strength are the important parameters that determine whether film-induced cleavage will occur. There is a clear need to determine the relationship between film chemistry and properties and the onset of film-induced cleavage if this process is a key factor in TGSCC. Therefore, the microchemistry and mechanics of TGSCC are key issues in SCC which needs considerably more attention before a complete picture emerges. Also, the potential to control SCC is dependent on knowing more about the properties and chemistry of these films.

Intergranular SCC involves the propagation of cracks along grain boundaries which are either embrittled with segregated impurities or boundaries in which the properties are likely altered by the presence of precipitates, such as carbides in sensitized stainless steel or intermetallic compounds in aluminum alloys. Therefore, it is possible that IGSCC occurs by a combination of anodic dissolution and mechanical fracture. A series of experiments have been conducted at PNL using acoustic emission (AE) as a sensor to determine whether mechanical fracture accompanies IGSCC. Tests have been conducted on Ni + P, Type 304SS and iron with segregated sulfur or phosphorus at anodic potentials. The Ni + P was held at 900 mV in 1N H_2SO_4, the Type 304SS at open circuit in $Na_2S_2O_3$ at 90°C (194°F), and the iron at 750 mV in $Ca(NO_3)_2$. Examples of the crack growth rate-stress intensity curves correlated to the sum of the AE events for Ni + P and Type 304SS are shown in Fig. 14. In the examples shown, the sum of the AE events begins to increase significantly at the same stress intensity at which measurable IGSCC occurs. The behavior for iron in $Ca(NO_3)_2$ was similar to Ni + P and Type 304SS. In many cases, there was evidence for AE before cracking, and this AE is thought to be associated with the development of a plastic zone at the crack tip. In all three materials, plastic zone sizes exceeding 1 mm would exist at the threshold stress intensity. The results of tests on Ni + P, Type 304SS, and iron indicates that AE accompanies IGSCC; however, the source of the AE is still being determined. It has been noted in preliminary analysis that the number of AE events per unit of crack extension increases with the amount of transgranular fracture accompanying the IGSCC. This result suggests that the IGSCC proceeds by an anodic process and that there is a "process zone" with transgranular ligaments which fracture and produce the measured AE during crack extension. Further evaluation is in progress, but these results indicate the need for further mechanics analysis of IGSCC.

Summary

A review of some outstanding unresolved issues in SCC has been presented. Many of these issues relate to the microchemistry or mechanics of the crack tip region, although there are macroscopic chemistry or mechanics issues that have not been considered. Issues concerning crack initiation, crack length, crack tip chemistry, analytical description of Stage I and crack extension processes in Stages I and II have been reviewed.

Most laboratory tests of SCC have been conducted with either smooth samples tested to failure or precracked samples containing relatively long cracks. Because neither of these tests reveal much information about the initiation and early propagation of short cracks by a SCC mechanism, virtually no SCC data exist for this regime. Models describing the transition from localized corrosion to actively growing cracks are only rudimentary and need further development before a clear understanding of SCC initiation can be obtained. The behavior of short stress corrosion cracks is currently being evaluated experimentally and in conjunction with crack-tip chemistry modeling in a few materials, but further evaluation is needed for a comprehensive picture of crack initiation and growth.

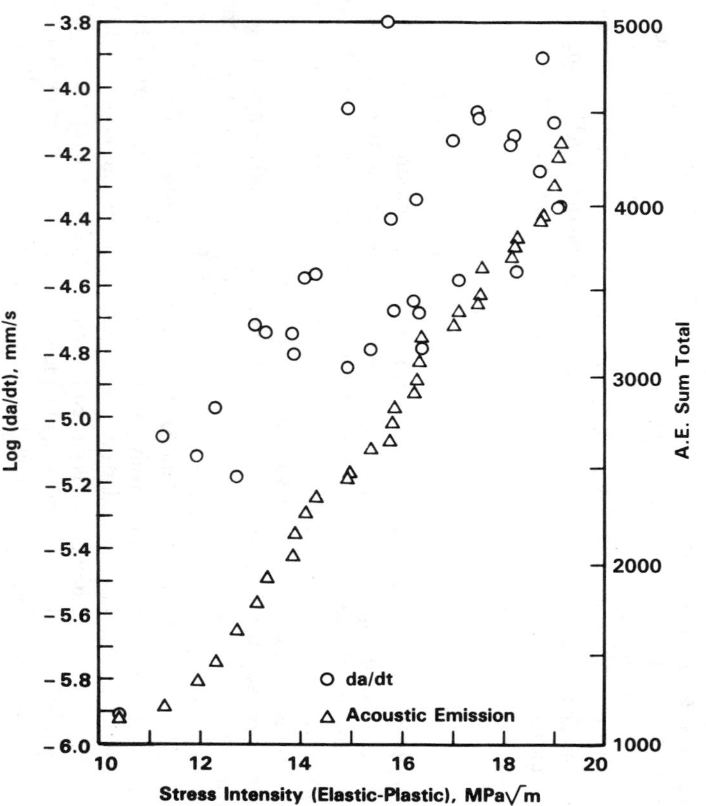

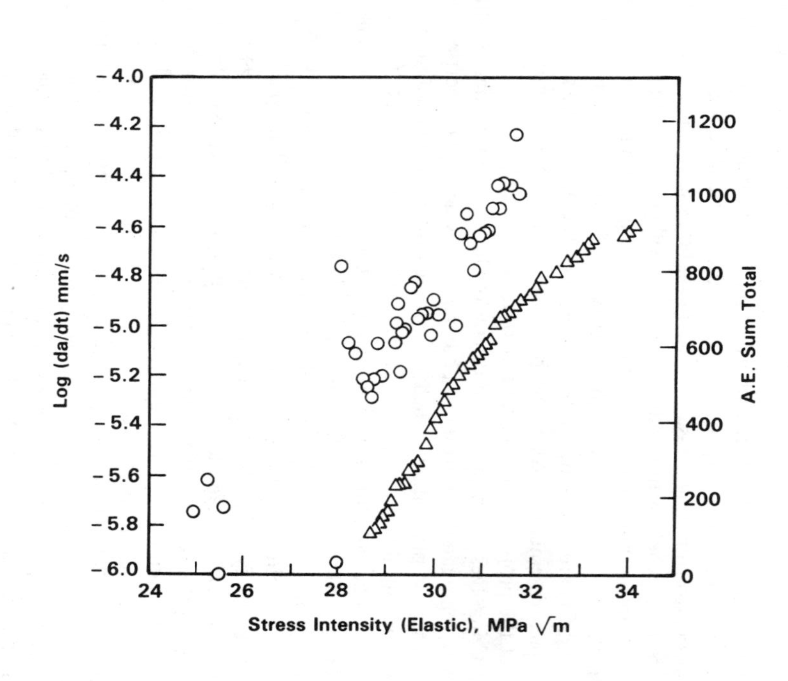

FIG. 14—*Crack growth rate and total number of acoustic emission events versus stress-intensity for IGSCC in* (left) *Ni + P tested in 1N H_2SO_4 at 900 mV (SCE) and* (right) *Type 304SS tested in $Na_2S_2O_3$ at 90°C.*

The phenomenology of stress corrosion crack propagation has received considerable attention over the years such that the conditions that cause SCC in many materials are known. Also, considerable attention has been given to the mechanisms of crack propagation; however, a number of key issues remain unsettled. Some of these issues include some of the details of the role of crack-tip strain and strain rate, crack-tip chemistry, crack length and geometry, the rate-controlling process in Stage I crack growth, and some of the details in the crack extension process. In summary, most of the SCC models describe the Stage I crack growth regime in a form and with a stress-intensity dependence that does not agree with the experimental data. Much of the experimental data for SCC in Stage I fit a log a versus K relationship, while most of the SCC models are linear in log a versus log K. Also, in regard to the crack extension process, details surrounding the initiation of a brittle crack in the TGSCC of materials and the source of AE during IGSCC need further elucidation.

References

[1] Jones, R. H., Danielson, M. J., and Baer, D. R., "Corrosion of Nickel-Base Alloys," *Journal of Materials for Energy Systems,* Vol. 18, 1986, p. 185.
[2] Edeleneau, C. and Forty, A. J., *Philosophical Magazine,* Vol. 46, 1960, p. 521.
[3] Beavers, J. A. and Pugh, E. N., *Metallurgical Transactions,* Vol. 11A, 1980, p. 809.
[4] Hahn, M. T. and Pugh, E. N., *Corrosion Journal,* Vol. 36, 1980, p. 380.
[5] Pugh, E. N. in *Atomistics of Fracture,* R. M. Latanision and J. R. Pickens, Eds., Plenum Press, New York, 1983, pp. 997–1010.
[6] Newman, R. C. and Sieradzki, K. in *Chemistry and Physics of Fracture,* R. M. Latanision and R. H. Jones, Eds., Martinus Nijhoff Publishers, The Netherlands, 1987, p. 597.
[7] Parkins, R. N., *Materials Science Technology,* Vol. 1, 1985, p. 480.
[8] Danielson, M. J., Oster, C. A., and Jones, R. H., *Corrosion Science,* 1988, in press.
[9] Vermilyea, D. A. in *Stress Corrosion Cracking and Hydrogen Embrittlement of Iron Base Alloys,* R. W. Staehle, J. Hochmann, R. D. McCright, and J. E. Slater, Eds., 1973, National Association of Corrosion Engineers, Houston, TX, pp. 208–217.
[10] Ford, F. P., EPRI NP-2589, Electric Power Research Institute, Palo Alto, CA, 1982.
[11] Parkins, R. N. in *Atomistics of Fracture,* R. M. Latanision and J. R. Pickens, Eds., Plenum Press, New York, 1983, pp. 969–990.
[12] Pasco, R. W., Sieradzki, K., and Ficalora, P. J., *Scripta Metallurgica,* Vol. 16, 1982, p. 881.
[13] Shanabarger, M. R., *Surface Science,* Vol. 150, 1985, p. 451.
[14] Kasupke, N. and Herzler, M., *Surface Science,* Vol. 92, 1980, p. 407.
[15] Fujiwara, K., *Physics,* Rev. B, Vol. 24B, 1981, p. 2240.
[16] Holloway, P. J., *Journal of Vacuum Science and Technology,* Vol. 18, 1981, p. 653.
[17] Ciftan, M. and Ruck, V., *Physics Status Solidi,* Vol. 99b, 1979, p. 237.
[18] Peden, C. H. F., Blair, D. S., and Goodman, D. W., *Journal of Vacuum Science and Technology,* Vol. A4, 1986, p. 1354.
[19] Baer, D. R., Thomas, M. T., and Jones, R. H., *Metallurgical Transactions,* Vol. 15A, 1984, pp. 853–860.
[20] Baer, D. R. and Thomas, M. T., *Applied Surface Analysis,* Vol. 26, 1986, p. 150.
[21] Dwyer, D. J., Simmons, G. W., and Wei, R. P., *Surface Science,* Vol. 64, 1977, p. 617.
[22] Shanabarger, M. R. in *Hydrogen Effects in Metals,* I. M. Bernstein and A. W. Thompson, Eds., The Metallurgical Society of the American Institute of Mining, Metallurgical, and Petroleum Engineers, Warrandale, PA, 1981, pp. 135–141.
[23] Williams, D. P. and Nelson, H. G., *Metallurgical Transactions,* Vol. 1, 1970, p. 63.
[24] Pradier, C. M., Bertheir, Y., and Oudar, J., *Surface Science,* Vol. 130, 1983, p. 229.
[25] Rondulic, K. D., Winkler, A., and Karner, H., *Journal of Vacuum Science and Technology,* Vol. A5, No. 4, 1987, p. 488.
[26] Jones, R. H. and Baer, D. R., *Scripta Metallurgica,* Vol. 20, 1986, p. 927.
[27] Baer, D. R. and Jones, R. H. in *Chemistry and Physics of Fracture,* R. M. Latanision and R. H. Jones, Eds., Martinus Nijhoff Publishers, The Netherlands, 1987, p. 552.
[28] Turnbull, A. in *Embrittlement by the Localized Crack Environment,* R. P. Gangloff, Ed., The Metallurgical Society of the American Institute for Mining, Metallurgical, and Petroleum Engineers, Warrendale, PA, 1984, pp. 3–32.

[29] Landles, K., Congleton, J. and Parkins, R. N. in *Embrittlement by the Localized Crack Environment,* R. P. Gangloff, Ed., The Metallurgical Society of the American Institute for Mining, Metallurgical, and Petroleum Engineers, Warrendale, PA, 1984, pp. 59–74.
[30] Alavi, A. and Cottis, R. A. in *Embrittlement by the Localized Crack Environment,* R. P. Gangloff, Ed., The Metallurgical Society of the American Institute for Mining, Metallurgical, and Petroleum Engineers, Warrendale, PA, 1984, pp. 75–88.
[31] Shuck, R. R. and Swedlow, J. L. in *Localized Corrosion,* B. F. Brown, R. W. Staehle, J. Kruger, and A. Agrawal, Eds., National Association of Corrosion Engineers, Houston, TX, pp. 190–220.
[32] Turnbull, A. and Thomas, J. G. N., *Journal of the Electrochemical Society,* Vol. 129, 1982, pp. 1412–1422.
[33] McEvily, A. J., Minakawa, K., and Nakamura, H. in *Fracture: Interactions of Microstructure, Mechanisms and Mechanics,* J. M. Wells and J. D. Landes, Eds., The Metallurgical Society of the American Institute for Mining, Metallurgical, and Petroleum Engineers, Warrendale, PA, 1984, pp. 215–233.
[34] Gangloff, R. P. and Turnbull, A. in *Modeling Environmental Effects on Crack Growth Processes,* R. H. Jones, Ed., The Metallurgical Society of the American Institute for Mining, Metallurgical, and Petroleum Engineers, Warrendale, PA, 1986, pp. 55–81.
[35] Ford, R. P. in *Embrittlement by the Localized Crack Environment,* R. P. Gangloff, Ed., The Metallurgical Society of the American Institute for Mining, Metallurgical, and Petroleum Engineers, Warrendale, PA, 1984, pp. 117–148.
[36] Speidel, M. O., EPRI NP-2531, Electric Power Research Institute, Palo Alto, CA, 1982.
[37] Newman, R. C., Sieradzki, K., and Isaacs, H. S., *Metallurgical Transactions,* Vol. 13A, 1982, p. 2015.
[38] Chung, P., Yoshitake, A., Cragnolino, G., and Macdonald, D. D., Paper No. 166, Corrosion/84, New Orleans, 1984.
[39] Speidel, M. O. in "Predictive Capabilities in Environmentally Assisted Cracking," R. Rungta, Ed., 1985, American Society of Mechanical Engineers, New York, p. 443.
[40] Andresen, P. L., *Corrosion Journal,* Vol. 44, 1988, p. 450.
[41] Newman, J. F., *Corrosion Science,* Vol. 21, 1981, p. 487.
[42] Singbeil, D. and Troman, D., *Metallurgical Transactions,* Vol. 13A, 1982, p. 1091.
[43] Lee, L. D. and Tromans, D. in *Environment-Sensitive Fracture of Engineering Materials,* Z. A. Foroulis, Ed., The Metallurgical Society of the American Institute for Mining, Metallurgical, and Petroleum Engineers, Warrendale, PA, 1979, pp. 232–240.
[44] Jones, R. H., Danielson, M. J., and Oster, C. A. in *Proceedings, Modeling Environmental Effects on Crack Growth Processes,* R. H. Jones and W. W. Gerberich, Eds., The Metallurgical Society of the American Institute of Mining, Metallurgical, and Petroleum Engineers, Warrendale, PA, 1986, pp. 41–54.
[45] Kumar, V., German, M. D., and Shih, C. F., EPRI NP-1931, Electric Power Research Institute, Palo Alto, CA, 1981.
[46] Sieradzki, K. in *Proceedings, Modeling Environmental Effects on Crack Growth Processes,* R. H. Jones and W. W. Gerberich, Eds., The Metallurgical Society of the American Institute for Mining, Metallurgical, and Petroleum Engineers, Warrendale, PA, 1986, pp. 187–196.

Robert P. Wei[1] *and Richard P. Gangloff*[2]

Environmentally Assisted Crack Growth in Structural Alloys: Perspectives and New Directions

REFERENCE: Wei, R. P. and Gangloff, R. P., **"Environmentally Assisted Crack Growth in Structural Alloys: Perspectives and New Directions,"** *Fracture Mechanics: Perspectives and Directions* (*Twentieth Symposium*), *ASTM STP 1020,* R. P. Wei, and R. P. Gangloff, Eds., American Society for Testing and Materials, Philadelphia, 1989, pp. 233–264.

ABSTRACT: Environmentally assisted crack growth (namely, stress corrosion cracking and corrosion fatigue) in alloys is one of the principal determining factors for durability and reliability of engineering structures. Over the past 20 years, activities in this area have transformed from screening and qualitative characterizations of the phenomena to quantitative assessment and scientific understanding. This work has enabled the recent development of life-prediction procedures.

In this paper, the contributions of fracture mechanics in this transformation are reviewed. Current mechanistic understanding of environmentally assisted crack growth by hydrogen embrittlement is summarized, and is placed in perspective. Applications to mitigate stress corrosion and corrosion fatigue cracking in marine environments are summarized. Outstanding issues and new directions for research are discussed.

KEY WORDS: environmentally assisted crack growth, stress corrosion cracking, aluminum alloys, titanium alloys, steels, corrosion fatigue, surface chemistry, electrochemistry, fracture mechanics

Environmentally assisted cracking of structural alloys (incorporating the well-known phenomena of stress corrosion cracking and corrosion fatigue) is well recognized as an important cause for the failure or early retirement of engineered structures. Stress corrosion cracking (SCC) was first recognized as a technological problem in the last half of the nineteenth century as "season cracking" in cold-drawn brass [*1*], with corrosion fatigue (CF) being recognized in the early 1900s [*2,3*]. Brown, tracing the historical background through 1972, concluded that "SCC, once thought confined to a few systems (combinations of metals and environments), must now be regarded as a general phenomenon which any alloy family may experience, given the wrong combination of heat treatment and environment" [*1*].

Bolstered by defense-related interests and by safety issues in the energy industry, there was a decade of unusually high research activity from the early 1960s to the early 1970s [*1*]. Because of these needs and of the impact of the "energy crisis," work continued through the 1970s in support of offshore oil exploration and alternative energy systems, such as coal gasification and liquefaction and solar energy. Coincidental to these activities, fracture

[1] Professor, Department of Mechanical Engineering and Mechanics, Sinclair Laboratory #7, Lehigh University, Bethlehem, PA 18015.

[2] Associate professor, Department of Materials Science, Thomas Hall, University of Virginia, Charlottesville, VA 22901.

mechanics was undergoing considerable development and maturation and became increasingly accepted as an important tool in structural analysis and materials research.

In this paper, a heuristic summary of the key developments in the understanding of environmentally assisted crack growth in structural alloys is given, and key issues and new directions for research are discussed. The intent of this summary is to highlight the significant milestones and the contributions of fracture mechanics; as such, it does not include a complete chronology of all of the developments and contributions to this field. A complete view of these developments over the past 20 years may be gleaned from several monographs and from the proceedings of many international conferences [*1,4–23*].

Chronologically, it is convenient to think of three periods of development, 1966 to 1972, 1972 to 1978, and 1978 to the present. Much of the groundwork in the United States was established during the first of these periods under two major programs, one sponsored by the Advanced Research Projects Agency of the Department of Defense (ARPA Coupling Program on Stress Corrosion Cracking) and the other by the Air Force Materials Laboratory [*1,24*]. It was during this period that the fracture-mechanics methodology was first introduced and applied to quantify the environmental cracking resistance of high-strength alloys.

Activities in the 1972 to 1978 period were devoted principally to in-depth phenomenological characterizations of cracking response and to the development of empirically based design and failure-analysis methods. Low-strength, fracture-resistant materials were increasingly studied, particularly in view of the demonstrated deleterious effects of cyclic loading. Efforts toward quantitative mechanistic understanding were launched, and science-based approaches were initiated.

The past decade, beginning with 1978, was a period of transition. Scientific understanding of environmentally assisted crack growth and engineering application of this understanding were placed on a quantitative footing. Significant advances have been made, in part because of the development of sophisticated analytical instrumentation and experimental techniques. More emphasis is now needed to translate improved scientific understanding into methods for quantitative design and to improve structural reliability.

Initiation Versus Propagation (Dawn of an Era)

A key turning point in the study of environmentally assisted crack growth and in the approach to design occurred in 1965. Brown and coworkers [*25,26*] at the Naval Research Laboratory were encouraged to investigate the SCC susceptibility of titanium alloys by using specimens which were deliberately precracked in fatigue. Due to this initial crack, the alloy was highly susceptible and fractured in a matter of minutes, even though it appeared immune to SCC when stressed in the smooth (uncracked) state in the same electrolyte.

As a result of these findings, a major shift in emphasis was made from testing of smooth and mildly notched specimens to that of cracked bodies. Fracture mechanics was introduced as a basis for analyzing environmentally assisted cracking in a paper by Johnson and Paris [*27*] in the First National Symposium on Fracture Mechanics in 1967. Experimental support for the use of the crack-tip stress-intensity factor (K) to describe the mechanical driving force was provided by Smith, Piper, and Downey [*28*] for stress corrosion crack growth (Fig. 1) and by Feeney, McMillan, and Wei [*29*] for CF (Fig. 2). A formal discussion of the use of crack-tip stress intensity to describe the mechanical driving force for crack growth was given by Wei [*30*].

Early users of this emerging technology for environmental cracking investigations included Steigerwald [*31*], Hanna, Troiano, and Steigerwald [*32*], Johnson and Willner [*33*], and Hancock and Johnson [*34*] for SCC (or crack growth under sustained loading), and Bradshaw

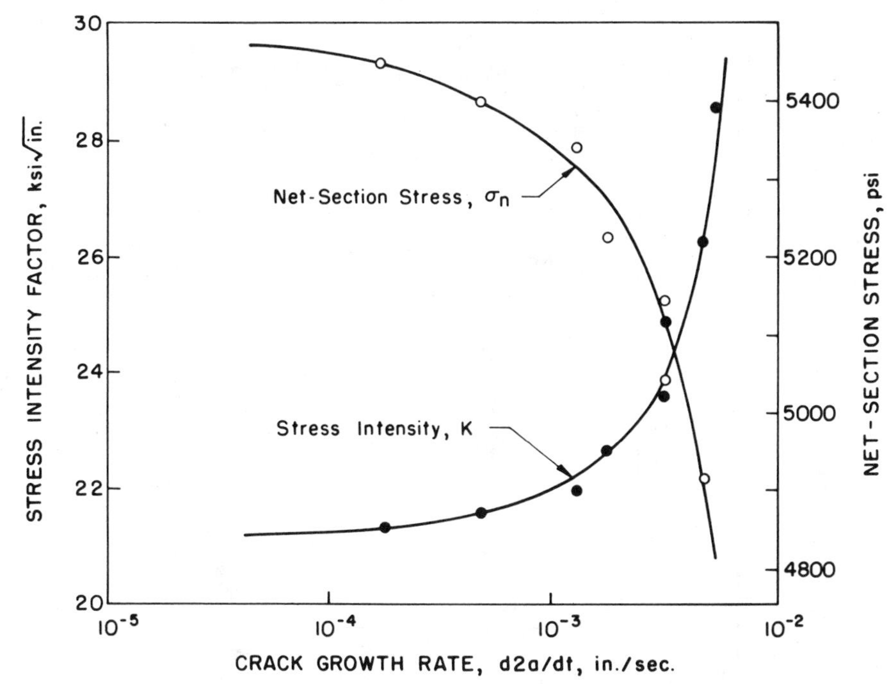

FIG. 1—*Stress-intensity factor and net-section stress versus crack growth rate for a wedge-force-loaded specimen* [28].

and Wheeler [*35*], Hartman [*36*], and Wei, Talda, and Li [*37*] for CF. Some early work predated the usage of the K concept. The effort associated with the ARPA Coupling Program made extensive use of fracture mechanics and contributed to the development of fracture-mechanics-based technology for materials evaluation and for design [*1*].

Phenomenological Characterization (1966–1972)

Two important approaches emerged from the early activities on SCC using precracked specimens [*30,38*]: the threshold and the kinetics approaches. The choice of a particular approach was determined in part by tradition and design philosophy and in part by practical considerations of experimentation and cost.

The simpler and less expensive approach involves the measurement of time-to-failure for precracked specimens under different applied loads, and the determination of a so-called threshold stress-intensity factor (designated as K_{Iscc}), below which it is presumed that no failure occurs by SCC [*25,30,38,39*]. The level of K_{Iscc} in relation to K_{Ic}, the plane-strain fracture toughness of a material, provided a measure of SCC susceptibility. The use of the threshold approach was favored for material selection and for safe-life design.

The second approach was more complex and involved the determination of crack-growth kinetics [*30,38*]. It required the measurement of crack growth rate (da/dt) under controlled environmental conditions and as a function of the mechanical crack driving force, which is characterized by the stress-intensity factor K. This approach required greater effort and more sophisticated instrumentation and was favored for mechanistic studies and for fail-safe design. A similar distinction in approach existed for corrosion fatigue [*40*].

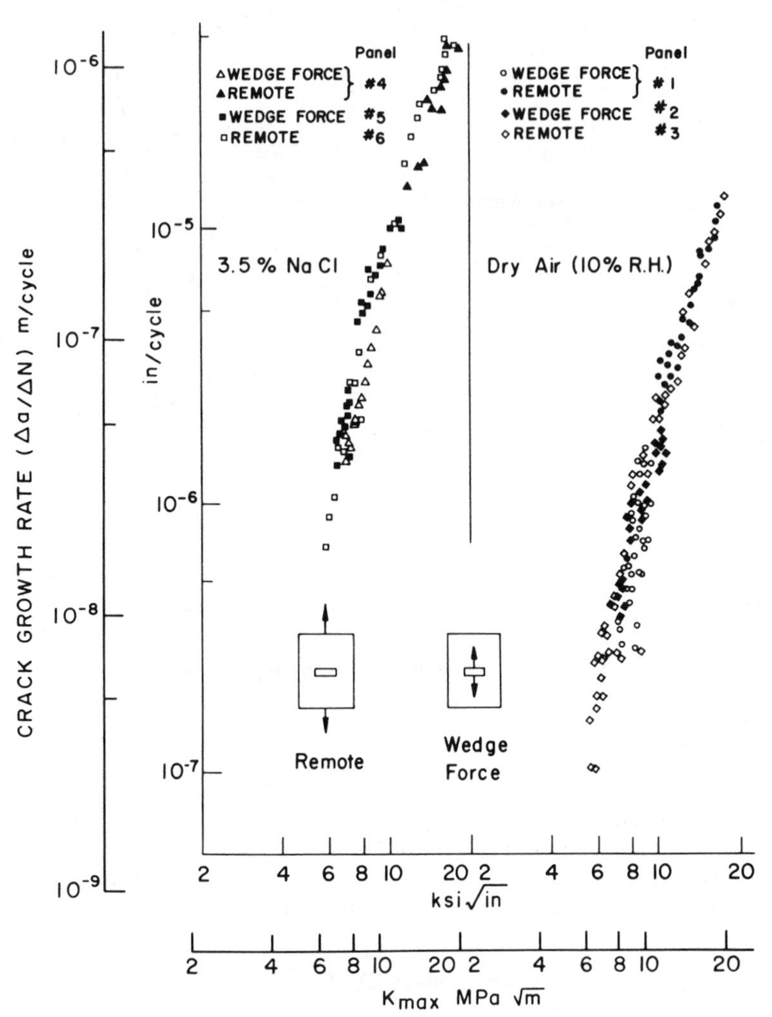

FIG. 2—*Fatigue crack growth rate correlation with stress-intensity range for aluminum alloy 7075-T6 in dry air and 3.5% sodium chloride* [29].

Both of these approaches were widely used from 1966 to 1972 to characterize material response and to develop empirically based design approaches. The final report for the ARPA Coupling Program, published as a monograph [*1*], reflects the typical efforts during this period.

A number of key studies took place which helped to set the stage for the development of quantitative understanding of environmentally assisted crack growth over the next decade to 1982 [*30,33–37,41–47*]. These studies showed the importance of the kinetics approach and served to establish its use in subsequent investigations. Some of the important findings are as follows:

1. There was an increasing awareness of the importance of the kinetics approach and a recognition of the fact that stress corrosion crack growth progresses in three stages (Fig. 3)

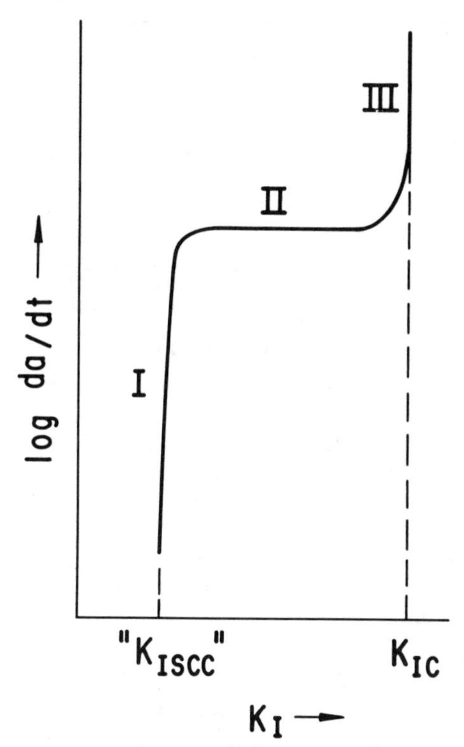

FIG. 3—*Schematic of monotonic load environment-enhanced crack growth showing classical three-stage response.*

[*31,41*]. The particular significance of the K-independent Stage II growth rate, in terms of mechanistic understanding of chemically limited crack growth response, was also recognized. The approach was used by Speidel [*42*] to study the influence of halide ions on stress corrosion crack growth in high-strength aluminum alloys, and by Wei and co-workers [*43,44*] to examine hydrogen embrittlement of high-strength steels.

2. The effectiveness of using crack growth rate as a means for understanding the mechanisms for environmental enhancement of crack growth under sustained loading was demonstrated by Johnson and co-workers [*33,34*], and of fatigue crack growth by Bradshaw and Wheeler [*35*], Hartman [*36*], and Wei et al. [*37*]. Through studies of the influences of different environments and of the inhibiting effect of trace amounts of oxygen, these researchers demonstrated the importance of surface reactions as a part of the embrittlement sequence (Fig. 4).

3. Pressure and temperature were recognized as important probes for identifying the processes that control environmentally assisted crack growth under sustained loading [*31,33,34,45,46*] and in fatigue [*31,35,37,47*], Fig. 5.

4. The relationship between stress corrosion and corrosion fatigue crack growth was recognized, with corrosion fatigue crack growth modelled as a superposition of fatigue and stress corrosion cracking [*48*].

As an illustration of the phenomenological characterization effort, typical Stage II crack growth rate data for a high-strength (AISI 4340) steel, stressed in various environments,

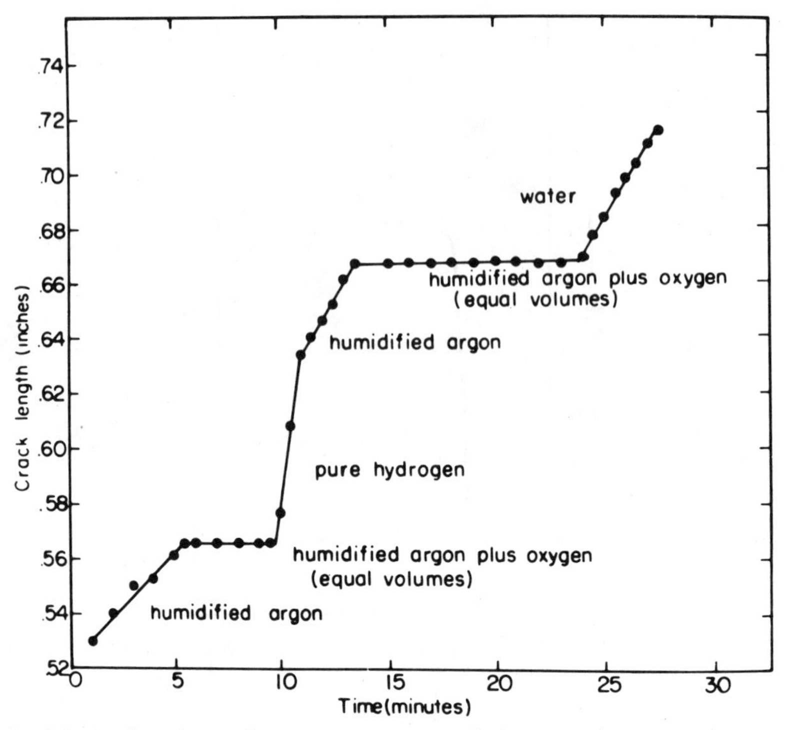

FIG. 4—*Subcritical crack growth in water, water vapor, hydrogen, and oxygen environments. H-11 Steel, 1585 MPa yield strength* [34].

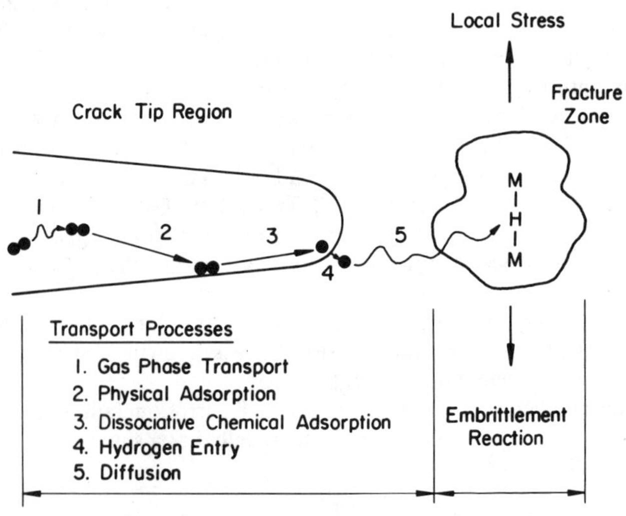

FIG. 5—*Schematic of sequential processes in environmental enhancement of crack growth by hydrogenous gases. Embrittlement is depicted by the metal-hydrogen bond* [50].

are shown as a function of temperature in Fig. 6 [*44,49*]. These data demonstrate that the crack growth response varied widely with environmental conditions and depended uniquely on temperature. Fractographic evidence suggested that the micromechanism for crack growth in these diverse, hydrogen-producing environments was the same. Thus, the different responses had to be attributed to one of several chemical processes in the overall chain, Fig. 5.

Based on these initial findings, it was recognized that further progress in understanding environmentally assisted crack growth was not possible without integrating, at least, the fracture-mechanics-based kinetic measurements with independent measurements of the kinetics of participating chemical processes.

Development of a Scientific Approach (1972–1978)

Beginning in 1972, a scientific approach to the study of environmentally assisted crack growth in high-strength alloys began to be developed, and evolved over the following years [*40,43,44,50,51*]. The approach was grounded in linear fracture mechanics and was predicated on the recognition that environmentally assisted crack growth is the result of a sequence of processes and is controlled by the slowest process in the sequence [*40,50*].

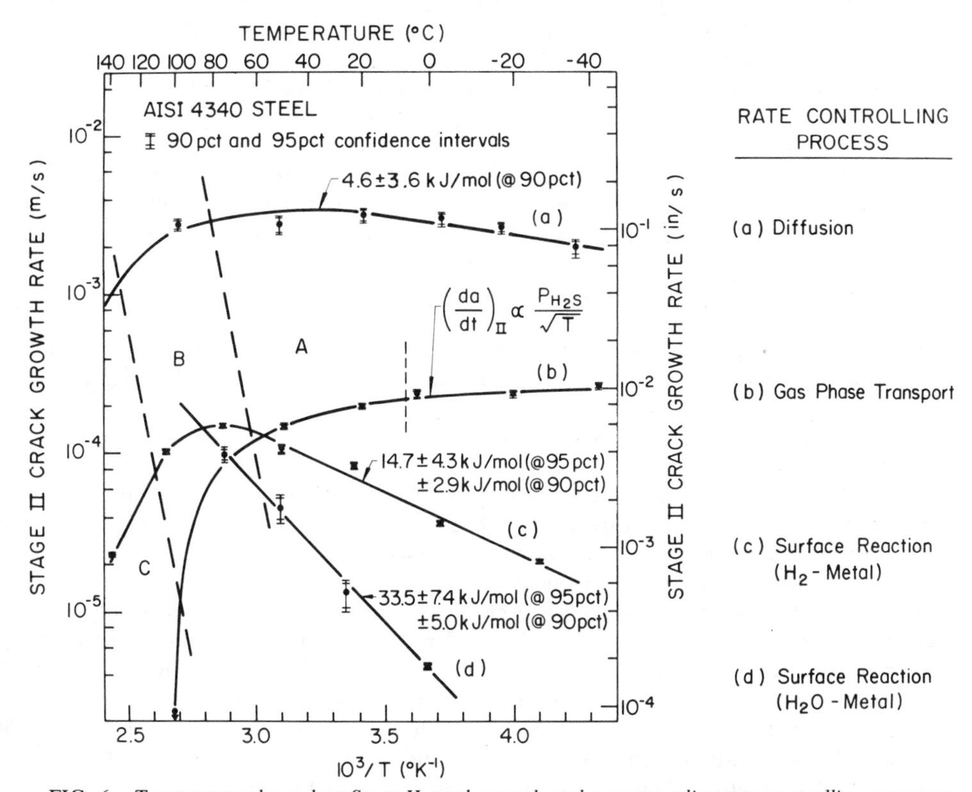

FIG. 6—*Temperature-dependent Stage II crack growth and corresponding rate-controlling processes at low temperatures (Region A) for AISI 4340 Steel in various hydrogenous environments:* (a) *hydrogen sulfide at 2.66 kPa,* (b) *hydrogen sulfide at 0.13 kPa,* (c) *hydrogen at 133 kPa, and* (d) *water* [44,49,51,60,61].

The processes that are involved in the environmental enhancement of crack growth in high-strength alloys by hydrogen and hydrogenous gases (such as water and hydrogen sulfide), and by aqueous environments are as follows and are illustrated schematically in Fig. 5 for the case of gaseous environments [*40,50*]:

1. Transport of the deleterious environment to the crack tip.
2. Reactions of the deleterious environment with newly produced crack surfaces to effect localized dissolution and to produce hydrogen.
3. Hydrogen entry (or absorption).
4. Diffusion of hydrogen to the fracture (or embrittlement) site.
5. Partitioning or distribution of hydrogen among various microstructural sites.
6. Hydrogen-metal interactions leading to embrittlement at the microstructural sites.

The actual processes depend on the mechanism of crack growth enhancement: namely, active path dissolution or hydrogen embrittlement. For a dissolution mechanism, only the first two steps in the sequence need to be considered, and the anodic (dissolution) reactions in the second step are directly responsible for crack growth enhancement. On the other hand, if hydrogen embrittlement is the responsible mechanism, then the reaction step serves only as the source for hydrogen; the remaining processes (last four processes, above) must be considered. These steps are identical for aqueous and gaseous environments.

The overall crack growth response is governed by the rate-controlling process in conjunction with the mechanical driving force, which is characterized here by either the local stress or the crack-tip stress-intensity factor, K [*27,31,40*].

Embrittlement, or the final step in the sequence, is a function of microstructure. The extent of embrittlement, or the rate of cracking along each microstructural path, depends on the local hydrogen concentration, which depends in turn upon the external environmental conditions (that is, pressure, pH, electrode potential, temperature). Cracking along the various microstructural paths can take place concurrently (in parallel) with the overall crack growth rate given as the weighted average of the individual rates.

Clearly, the understanding of crack growth required identification and quantification of the rate-controlling process. The evolving scientific approach, therefore, focused upon well-defined and coordinated chemical/electrochemical, mechanical, and metallurgical experiments.

It was recognized from the outset that the concept of rate-controlling processes applied to both stress corrosion (or sustained-load) and corrosion fatigue crack growth and that considerations of both modes of crack growth would afford a synergism in exploring the fundamental issues. Because of the ready linkage of a K-independent regime (Stage II) of crack growth to the underlying rate-controlling process, early efforts were directed at sustained-load crack growth. Detailed studies of the kinetics and mechanisms of reactions of water with iron and steel and of Stage II crack growth were carried out as a function of temperature [*51–55*]. By 1977, these efforts led to an unambiguous identification of the iron-water oxidation reaction as the rate-controlling process for crack growth in high-strength steels and to a realization of the very limited extent of these reactions (see Fig. 6).

With additional work on surface reactions and on crack growth, a broader based understanding of rate-controlling processes emerged during the remainder of the 1970s, and modeling of crack growth was begun. In 1977, the integrated fracture-mechanics-reaction chemistry approach was used to examine the role of water vapor in enhancing fatigue crack growth in high-strength aluminum alloys [*56*] and the role of hydrogen sulfide in enhancing fatigue crack growth in steels [*57*]. These efforts led to a model of corrosion fatigue crack growth [*58,59*], which in turn led to the recognition and modeling of transport-controlled crack growth under sustained loading [*60*].

Chemical and Microstructural Modeling

The modeling effort begun in the late 1970s emerged as an important adjunct of the scientific approach over the past ten years. It has provided guidance and a formalized framework for examining the fundamental issues and has served as a basis for the utilization of data in design. Chemical and microstructural modeling of stress corrosion and corrosion fatigue crack growth are briefly summarized to provide perspective and to serve as a basis for discussing directions for future research.

Modeling of crack growth may be subdivided in terms of loading conditions and in addition, according to the various processes that affect crack growth (see Fig. 5). If hydrogen embrittlement is assumed to be the responsible mechanism, modeling may be approached in terms of hydrogen supply (transport, reactions, and diffusion), hydrogen distribution within the material (partition), and embrittlement reactions (which determine the rates of cracking along the various microstructural paths) and can incorporate suitable methods for predicting an average or macroscopic crack growth rate. Most of the effort during this period has concentrated on chemical modeling and reflects developments in understanding the chemical and fracture-mechanics aspects of environmentally assisted crack growth. More recently, this effort has been extended to include the influence of microstructure.

Sustained-Load Crack Growth

Initial modeling of sustained-load crack growth was largely phenomenological and limited to the case of hydrogen-supply-controlled Stage II crack growth in the lower temperature region (Region A in Fig. 6). The principal thrust was directed at obtaining chemical reaction and crack growth rate data to confirm the concept of rate-controlled crack growth and to identify the controlling process. Obvious deviations of the crack growth response curves (Fig. 6) from single-process behavior and the observed changes in fracture paths (or micromechanisms) led to considerations of rate-controlling process transfer and of the role of microstructure.

Models Based on Hydrogen Supply—Models for Stage II crack growth were proposed on the basis of extensive data on the kinetics of surface reactions and crack growth for high-strength steels in water/water vapor, hydrogen, and hydrogen sulfide [*45,46,50,52–62*]. When the reactions are slow (as in hydrogen and in water), Stage II crack growth rate is controlled by the rate of surface reaction. On the other hand, for very rapid reactions (as in hydrogen sulfide), the growth rate is determined either by the rate of transport of the gases to the crack tip or by the rate of diffusion of hydrogen to the embrittlement site. The models of Stage II crack growth rate, expressing the specific dependence on gas pressure (p_0) and temperature (T), are [*44,63*]

$$\text{Transport Control:} \quad (da/dt)_{\text{II}} = C_t p_0 / T^{1/2} \quad (1)$$

(for Knudsen flow [*64*])

$$\text{Surface Reaction Control:} \quad (da/dt)_{\text{II}} = C_s p_0^m \exp(-E_s/RT) \quad (2)$$

$$\text{Diffusion Control:} \quad (da/dt)_{\text{II}} = C_d p_0^{1/2} \exp(-E_d/2RT) \quad (3)$$

The constants C_i contain chemical and physical quantities that relate to gas transport, surface reaction, and crack geometry and reflect the susceptibility of specific alloys to embrittlement by specific environments (namely, the embrittlement reaction term). E_s and E_d are the activation energies for surface reaction and hydrogen diffusion respectively. Good agreement with experimental observations in the low-temperature region (Region A) is indicated in Fig. 6.

In these models, a single process is assumed to be in control, and the terms C_i are assumed to be sensibly constant. It is recognized that transfer of control from one process to another may occur as the environmental conditions are changed. The consequences of this transfer have been discussed [*63*]. In the formulation of these phenomenological models, single-step reactions were implicitly assumed. Because the reactions tend to be more complex, these models are viewed as starting points for developing further understanding of environmentally assisted crack growth.

Partitioning of Hydrogen and Surface Phase Transformation—The observed decrease in sustained-load crack growth rate with increasing temperature in the "high-temperature" region (Region C in Fig. 6) for carbon martensitic steels (such as AISI 4340 steel) has been analyzed based on surface chemistry [*46,65,66*]. Similar analyses account for the much steeper decrease in Region C crack growth rate for 18Ni maraging steels [*67*]. These analyses, however, have ignored the important role of microstructure and micromechanism in hydrogen embrittlement or have made unrealistic assumptions regarding surface coverage by hydrogen. The models proposed on the basis of these analyses, therefore, cannot explain the observed changes in fracture mode with temperature [*49*].

To account quantitatively for the role of hydrogen-microstructure interactions, a "hydrogen partitioning" model was developed [*49*]. The model suggests that the rate of hydrogen-assisted crack growth is determined by two factors: the rate of hydrogen supply to the fracture process zone and the partitioning of hydrogen amongst different microstructural elements or traps (principally between prior-austenite grain boundaries and martensitic matrix/interfaces). The partitioning of hydrogen is controlled by hydrogen-trap interactions and determines the contribution by each structural feature to the overall crack growth rate, which is the weighted average of rates of cracking along the different microstructural paths. This model is illustrated schematically in Fig. 7 for hydrogen-assisted crack growth in a high-strength steel. Detailed considerations and a derivation of the model are given in Ref *49*.

At low temperatures (in Region A of Fig. 6), hydrogen resides primarily at prior-austenite grain boundaries and slip planes. Crack growth tends to be predominantly intergranular

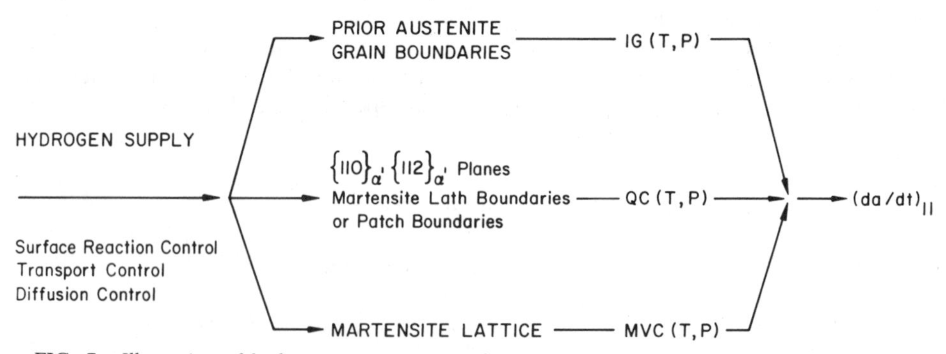

FIG. 7—*Illustration of hydrogen partitioning and its relationship to hydrogen supply and resulting Stage II crack growth rate* [49].

(IG), and would conform to Eqs 1 to 3. With increasing temperature into Region C, the hydrogen supply processes remain in control, but hydrogen concentration at the grain boundaries and in the slip planes decreases, and more hydrogen would reside in the martensitic structure. This temperature-induced partitioning of hydrogen leads to increasing amounts of microvoid coalescence (MVC) and to slower crack growth rates. The change in crack growth rate with temperature reflects a transfer of cracking mechanisms, rather than (or in addition to) a change in the process of hydrogen supply. Based on these considerations, the predicted temperature and pressure dependences for Stage II crack growth in high-strength steels are in good agreement with crack growth data for an AISI 4340 steel in hydrogen and in hydrogen sulfide (see, for example, Fig. 8) [*49,61*].

For the 18Ni maraging steels, a phase transition model was proposed to account for the abrupt decrease in crack growth rate that had been observed at the higher temperatures [*50,67,68*]. The model was based on the suggestion by Hart [*69*] that adsorbed atoms on a metal surface can undergo a phase transformation at critical temperatures and pressures and it is consistent with experimental observations.

Clearly, a number of factors influence the kinetics of environmental crack growth. The hydrogen partitioning and surface phase transition models have provided some insight into and a clear indication of the need for a broadly based understanding, including that of the embrittlement mechanisms.

Fatigue Crack Growth

Based on the understanding developed for sustained-load crack growth, models for surface reaction and transport-controlled fatigue crack growth were developed [*58,59*] and have

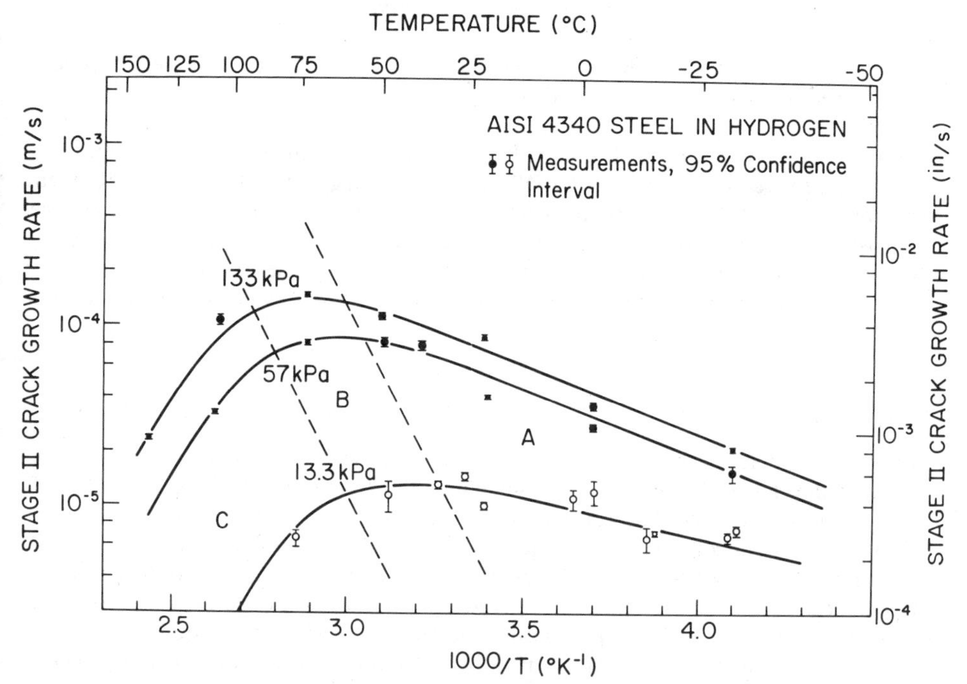

FIG. 8—*Comparison between model predictions and sustained-load crack growth date for AISI 4340 steel tested in hydrogen at different pressures and temperatures* [49,61].

been used successfully to explain the observed dependence of fatigue crack growth rates on cyclic load frequency and pressure in gaseous environments [56–59]. Insight obtained from the gaseous case has been applied to the consideration of corrosion fatigue in aqueous environments.

Superposition Model—Modeling was based on the proposition that the mechanical and environmental contributions are decoupled such that the rate of crack growth in a deleterious environment, $(da/dN)_e$, may be written as the sum of three components [40,70,71]

$$(da/dN)_e = (da/dN)_r(1 - \phi) + (da/dN)_c\phi + (da/dN)_{SCC} \quad (4)$$

where

$(da/dN)_r$ = mechanical fatigue rate,
$(da/dN)_c$ = "pure" corrosion fatigue rate,
ϕ = fractional area of crack that is undergoing pure corrosion fatigue, and
$(da/dN)_{SCC}$ = contribution of sustained-load crack growth.

These rates may be composed of contributions from several concurrent micromechanisms. For simplicity, the sustained-load growth term is not included in the following discussions. Equation 4 is rewritten as [66]

$$(da/dN)_e = (da/dN)_r + [(da/dN)_c - (da/dN)_r]\phi \quad (5)$$

or

$$(da/dN)_{cf} = [(da/dN)_c - (da/dN)_r]\phi \quad (6)$$

where $(da/dN)_{cf}$ denotes the incremental increase in growth rate above the reference level resulting from the embrittling environment.

In the limit for $\phi = 0$, or for a test in an inert environment, $(da/dN)_e = (da/dN)_r$, corresponding to pure fatigue. For $\phi = 1$, corresponding to chemical reaction saturation [53,54], $(da/dN)_e = (da/dN)_{e,s} = (da/dN)_c$, and measured growth rates correspond to pure corrosion fatigue rates. In essence, the parameter ϕ represents material response to changes in environmental conditions. It is directly related to its counterpart, the fractional surface coverage (θ), in chemical modeling; that is, $\phi = \theta$ [58,59]. The maximum in corrosion fatigue crack growth rate, therefore, corresponds to the maximum extent of chemical reaction ($\theta = 1$).

Chemical Modeling

Important understanding of corrosion fatigue crack growth response in gaseous environments was developed through chemical modeling [58,59] and through experimental verification of the role of gas transport and surface reactions on $(da/dN)_{cf}$ [56,57,72]. Similar understanding is being developed for aqueous environments [16,23,73–75].

Assuming that the environmental enhancement of fatigue crack growth results from embrittlement by hydrogen produced by the reactions of hydrogenous gases with freshly produced crack surfaces, models for transport and surface reaction controlled crack growth were developed [58,59]. An analogous model for electrochemical reaction controlled crack growth was proposed for steels in aqueous environments, where the kinetics of reaction are assumed to be slow [73–75]. In these models, the environmental contribution is assumed to be proportional to the extent of chemical or electrochemical reaction per cycle, which is given by the fractional surface coverage θ. The crack growth rate, $(da/dN)_{cf}$, is given by Eq 6 with $\phi = \theta$.

Models for diffusion-controlled growth [76] and strain-induced hydride formation [77–79] have also been suggested. The latter model relates to metallurgical changes and the consequent effect on crack growth rates, and is considered later. Diffusion-controlled crack growth occurs when the preceding transport and surface reaction processes are rapid, and must be considered outside of the context of limited surface coverage per cycle.

Transport-Controlled Growth—For highly reactive gas-metal systems, crack growth is controlled by the rate of transport of gases to the crack tip [58,59]. The surface coverage (θ) is linearly proportional to pressure (p_0) and inversely proportional to frequency (f). The environmental contribution to fatigue crack growth is given by the relationships

$$(da/dN)_{cf} = [(da/dN)_c - (da/dN)_r] \cdot [(p_0/f)/(p_0/f)_s] \tag{7a}$$

$$\text{for } (p_0/f) < (p_0/f)_s$$

$$(da/dN)_{cf} = [(da/dN)_c - (da/dN)_r] = \text{constant} \tag{7b}$$

$$(da/dN)_{e,s} \text{ for } (p_0/f) \geqq (p_0/f)_s$$

The term $[(da/dN)_c - (da/dN)_r]$ is the maximum enhancement in the rate of cycle-dependent corrosion fatigue crack growth and is a consequence of the finite extent of surface reaction (that is, $\theta \rightarrow 1$) [56,57]. The saturation exposure, $(p_0/f)_s$, is a function of pressure, temperature, and molecular weight of the gas, and of stress-intensity level, load ratio, yield strength, and crack length and opening geometry [58,59,72].

Surface and Electrochemical-Reaction-Controlled Growth—With less reactive systems, crack growth is controlled by the rate of surface or electrochemical reactions at the crack tip. For first-order reactions, the crack growth rate in gaseous environments is given by Eq 8 in terms of pressure, frequency, and the reaction rate constant k_c [58,59]

$$(da/dN)_{cf} = [(da/dN)_c - (da/dN)_r] \cdot [1 - \exp(-k_c p_0/f)] \tag{8}$$

A more general interpretation of surface coverage can be made to accommodate multi-step reactions, with the actual response reflecting the nature and kinetics of the individual reaction steps.

For aqueous environments, $(da/dN)_{cf}$ may be expressed as an analog to Eq 8 [75]

$$(da/dN)_{cf} = [(da/dN)_c - (da/dN)_r] \cdot [q/q_s] \tag{9}$$

where q is the amount of electrochemical charge transferred per cycle, q_s is the "saturation" amount or that charge required to complete the reactions, and the ratio q/q_s is identified with θ.

Diffusion-Controlled Growth—When transport and surface reaction processes are sufficiently rapid, the crack growth rate is determined by the rate of diffusion of hydrogen in the metal and from the crack tip to the "fracture process zone." According to Kim et al.

[76], $(da/dN)_{cf}$ is given by

$$(da/dN)_{cf} = A_0 \exp(-H_B/RT) \cdot (p_0 D/f)^{1/2} \Delta K^2 \quad (10)$$

where

A_0 = empirical constant,
H_B = binding enthalpy of hydrogen to dislocations,
R = universal gas constant,
D = hydrogen diffusivity, and
f = frequency.

In both the transport- and reaction-controlled models, a growth rate dependence upon ΔK^2 is implicitly assumed to reflect the expected proportionality between the sizes of the "hydrogen-damaged" zone and the crack-tip plastic zone [59,72]. This dependence is explicitly incorporated in the diffusion-controlled model [76]. The temperature dependence is reflected through thermal influences on reaction rates, the inert environment fatigue process, and on gas transport [58,59,73,74]. If the reaction mechanisms remain unchanged with temperature, then the maximum enhancement in rate [or $(da/dN)_c$] is expected to remain constant. The temperature dependence for corrosion fatigue would be reflected principally through its frequency dependence below the saturation condition and as illustrated by Eq 7.

Experimental Support—The transport and surface chemical/electrochemical reaction-controlled models have been examined by coordinated studies of the kinetics and mechanisms of gas or electrolyte-metal reactions, and of corrosion fatigue crack growth response as functions of pressure, temperature, electrode potential, ion concentration, and loading time (or frequency). Good agreement between these models and the experimental data on crack growth and surface and electrochemical reactions has been obtained (Figs. 9 to 11, for example). The transport-controlled case is represented by an aluminum alloy in water vapor (Fig. 9), and steel in hydrogen sulfide (Fig. 10) at low pressures. The reaction controlled case is represented by high-strength steel in aqueous electrolytes (Fig. 11, for example [80]).

The form of the crack growth rate-frequency response depends on the kinetics and on the mechanism or mechanisms of the surface reactions, and may reflect both transport and reaction control. For the case of reaction-controlled crack growth, the response may reflect both the fact that the reactions do not follow first-order kinetics and the presence of more than one reaction step. For example, for the case of 7075-T651 aluminum alloy (Fig. 9), the additional enhancement at the higher pressures is surface reaction controlled and is attributed to a slow step in the reactions of water with segregated magnesium [81]. Similarly, the increase in growth rate observed for the 2¼Cr-1Mo steel in hydrogen sulfide at the higher pressures (Fig. 10) is surface reaction controlled, and is identified with the slower second step in the reactions of hydrogen sulfide with iron [57,60]. A similar situation exists for crack growth in high-strength steels in water vapor and in aqueous solutions. The situation in water vapor may be further complicated by capillary condensation at the crack tip [53,73,74].

Evidence for diffusion-controlled crack growth is provided by data on titanium alloys (Fig. 12, for example). At higher frequencies, $(da/dN)_{cf}$ is inversely proportional to the square root of frequency [82]. This dependence, coupled with the surface reactivity of titanium, is consistent with diffusion control. The abrupt decrease in growth rates at the lower frequencies is attributed to a hydride mechanism that depends on both strain and

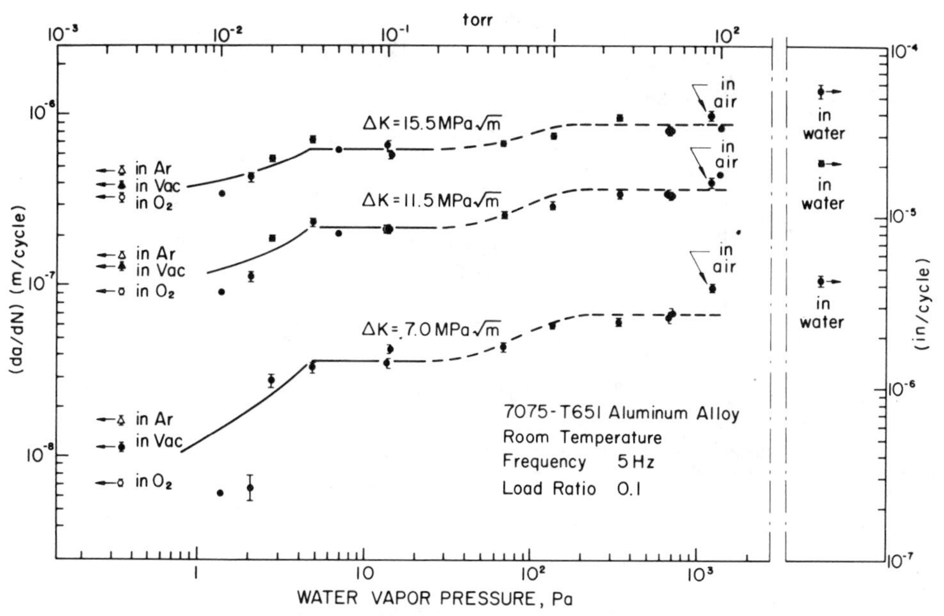

FIG. 9—*Influence of water-vapor pressure on fatigue crack growth in 7075-T651 alloy at room temperature. Solid lines represent predictions of Eq 7. Dashed lines indicate surface-reaction-controlled growth and reflect influence of segregated magnesium* [81,85].

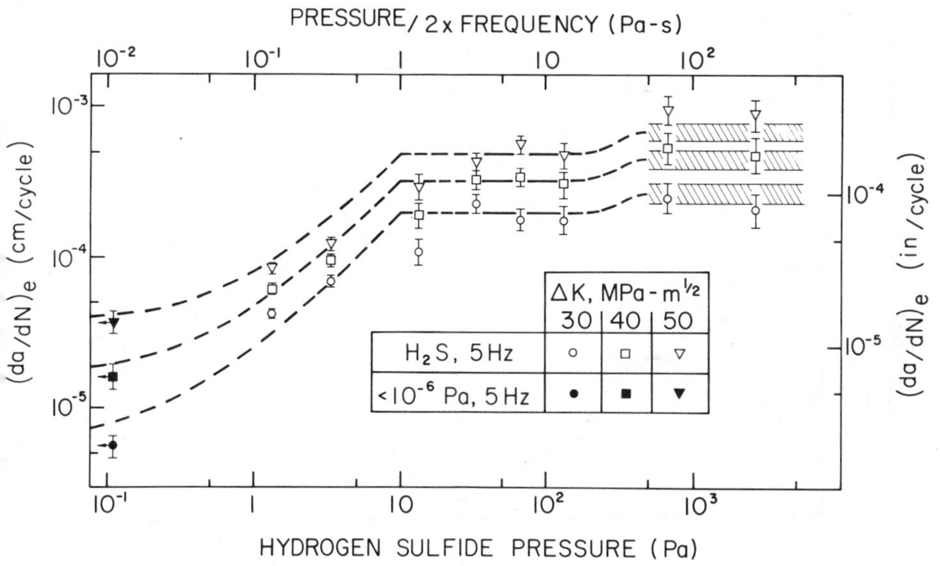

FIG. 10—*Influence of hydrogen sulfide pressure on fatigue crack growth in a 2¼Cr-1Mo (A542 Class 2) steel at room temperature. Dashed lines represent predictions of Eq 7. Solid lines indicate surface-reaction-controlled growth and reflect the second step of hydrogen sulfide-iron surface reactions* [57].

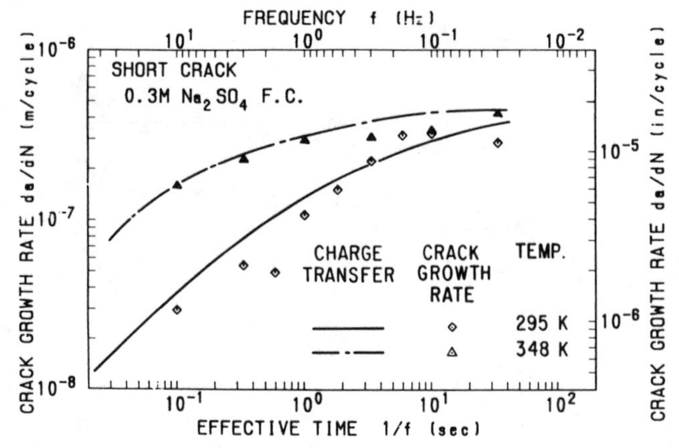

FIG. 11—*Influence of frequency and temperature on fatigue crack growth for NiCrMoV steel in 0.3N sodium sulfate solution* [80].

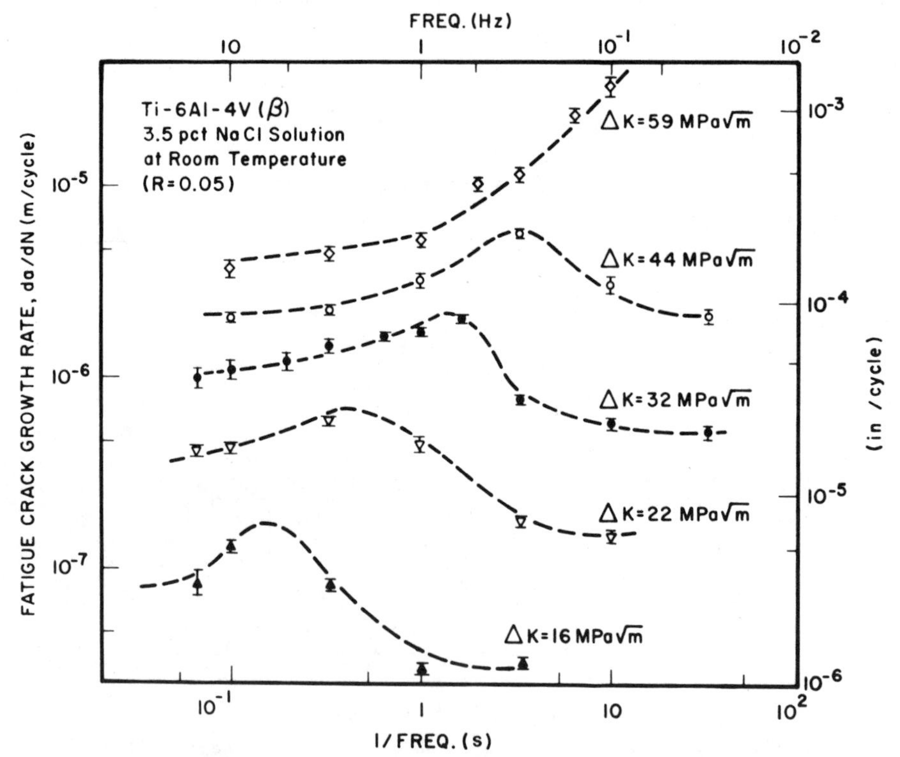

FIG. 12—*Influence of frequency on fatigue crack growth in a Ti-6Al-4V alloy exposed to 0.6M sodium chloride solution at room temperature and* R = *0.1* [82].

strain rate [77–79,82]. There is, however, no quantitative model for hydride-induced crack growth.

Microstructural Modeling

The important role of micromechanisms was discussed by Gerberich and Peterson [83]. The role of micromechanism (or of microstructure) is explicitly incorporated in Eqs 4 and 6. The implications of the model are as follows: the partitioning of hydrogen to microstructural sites need not be uniform, and the fractional area of fracture surface (ϕ) produced by pure corrosion fatigue is equal to the fractional surface coverage (θ) for chemical reactions. The relationship between the microstructural and environmental parameters (ϕ and θ) was examined by Ressler [84] and by Gao et al. [85].

For an AISI 4340 steel in water vapor (585 Pa) at room temperature, Ressler [84] determined the corrosion fatigue crack growth rate as a function of frequency (Fig. 13). Fractographic data show a change in fracture surface morphology with decreasing frequency from a predominantly transgranular mode (relative to the prior-austenite grains) to predominantly intergranular cracking, Fig. 14. By the identification of the intergranular failure mode with pure corrosion fatigue and the transgranular mode with mechanical fatigue, the fraction of pure corrosion fatigue (ϕ) was estimated fractographically. A comparison was made between ϕ and θ, based on independent surface reaction measurements [51] and an adjustment of exposure to account for capillary condensation in the fatigue crack (Fig. 15). Agreement is excellent. A similar good correlation was reported by Gao et al. [85] for a 7075-T651 aluminum alloy.

These results indicate the important role of microstructure and of the interactions between the environmental and microstructural variables. More work is needed to broaden the scope of this understanding and to provide statistically reliable support. The framework for understanding is in place.

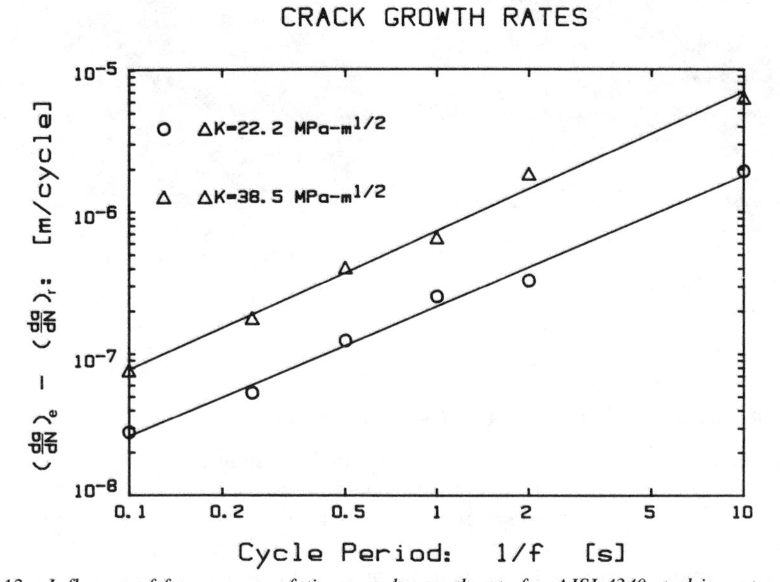

FIG. 13—*Influence of frequency on fatigue crack growth rate for AISI 4340 steel in water vapor at 585 Pa and room temperature* [84].

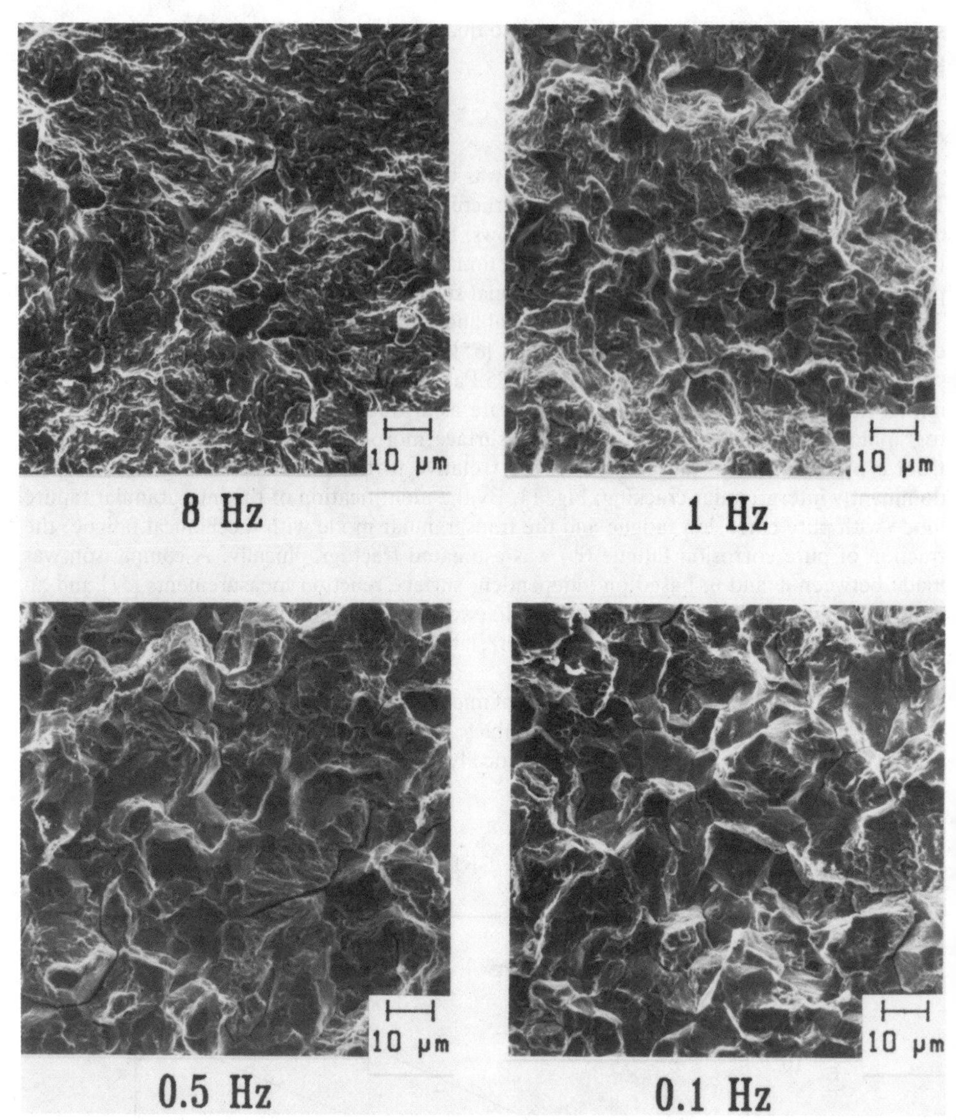

FIG. 14—*Fractographs of AISI 4340 steel stressed in water vapor, showing changes in fracture surface morphology with frequency* [84].

Implementation of the Fracture-Mechanics Approach (1980s)

Significant complexities must be overcome in implementing the fracture-mechanics approach for the quantitative prediction of component life to control environmentally assisted crack growth. During the 1980s, two critical issues emerged. First, the principle of fracture-mechanics similitude (that is, equal subcritical crack growth rates are produced by equal applied stress intensities) may be violated because of the effects of crack closure and geometry-dependent variations in environment chemistry within the crack. Second, the large

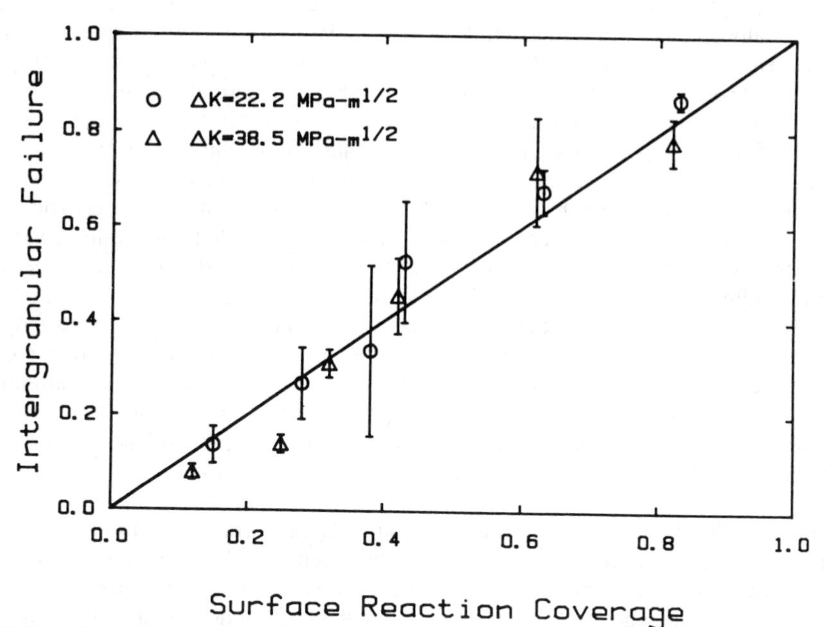

FIG. 15—*Correlation between surface coverage and fraction of intergranular failure (or ϕ) for corrosion fatigue in AISI 4340 steel-water vapor* [84].

number of relevant variables and their time-dependent interactions greatly complicate life predictions, particularly for lower-strength materials which were not extensively characterized during the 1970s and for which linear fracture mechanics may not suffice.

Mechanistic understanding and scientific modeling provide the means for characterizing subcritical cracking in terms of a scalable crack-tip driving force, for extrapolating short-term data to predict long-term component behavior, and for defining the effects of mechanical, chemical, and metallurgical variables.

Fracture-Mechanics Similitude

The use of fracture-mechanics similitude to scale environmentally assisted crack growth rates, for different crack sizes and loadings, is permitted only if the following two conditions are met: the applied driving force parameter (for example, stress-intensity factor) uniquely defines the stresses, strains, and strain rates near the crack tip, and the composition and conditions of the environment at the crack tip are constant for a given applied driving force, irrespective of crack size and opening shape. Investigations of environmentally assisted crack growth at low stress intensities, in low-strength or anisotropic materials, for small crack geometries and in complex embrittling environments demonstrate that the validity of similitude cannot always be assumed [*16,17,86–88*]. The applied stress-intensity basis shown in Figs. 1 and 2 may need to be modified to reflect the actual crack driving force, as described in the following subsections.

Crack Closure Problem—Premature contact of fatigue crack surfaces during unloading, or namely crack closure, reduces the effective crack driving force relative to the applied value (for example, ΔK). For corrosion fatigue, likely crack closure mechanisms include environmentally assisted crack deflection, intergranular cracking and enhanced surface roughness, enhanced plasticity, fluid pressure, and corrosion product wedging [*89,90*].

The effect of corrosion product wedging is shown in Fig. 16. Here, the reduction in fatigue growth rates at low mean stress is attributed to wedging by a thin surface oxide produced by fretting of the fracture surfaces in moist or oxygenated environments [*91*]. This closure mechanism is relevant when the crack-tip opening displacements are small and may become more pronounced because of enhanced (crevice) corrosion within fatigue cracks. The effect was documented for cathodically polarized steels in seawater [*92,93*]. Generally, corrosion debris does not induce crack closure at high mean stress as suggested in Fig. 16, because fretting is minimal and crack-tip openings are large.

For laboratory specimens, crack closure effects are accounted for through measurements of specimen compliance [*89*]. The implications of environment-induced closure for components subjected to complex loading spectra, however, are unclear and require attention. Even though elementary micromechanical models have been proposed for specific closure mechanisms, the environment-sensitive processes are poorly understood [*90*]. The problem is exacerbated by the time-dependent nature of corrosion processes.

Small Crack Problem—Because of the importance of early crack formation and growth to component fatigue life, considerable emphasis has been placed through the 1980s on the fracture mechanics of small cracks [*87*]. Generally, stress corrosion and corrosion fatigue cracks sized below 1 to 5 mm have been found to grow unexpectedly rapidly relative to long cracks at the same K level and to grow below the threshold K level for long cracks [*87,90,94,95*]. This behavior must be understood in order to apply the fracture-mechanics approach to predict environmental cracking life in components.

The observed crack size effect may be caused by inappropriate formulation of the mechanical driving force and by differences in crack-tip environment. A variety of mechanisms and corresponding bounding crack sizes were identified [*96*]. For fatigue in benign environments, increased crack-tip plasticity, underdeveloped crack wake closure, three-dimensional crack shape, and large-scale yielding can increase growth rates of small cracks. Interactions with grain boundaries can arrest growth. The same mechanisms may affect environmentally assisted cracking through changes in crack-tip environment, environmental modification of crack closure, or strain-enhanced film rupture [*90,95,97–99*].

The occluded environments within short cracks can differ from the bulk and from those within long cracks. For the cases examined to date, the short crack environments appear to be more deleterious. Sustained-load cracking data for a high-strength steel in aqueous chloride solution (Fig. 17) show that stress intensity similitude is not obeyed, with K_{Iscc} decreasing with crack size below about 1 mm [*100*]. Values for K_{Iscc} at the smallest crack size are about one-third those obtained with specimens containing 15 to 30-mm-long cracks. This crack size effect is chemical in origin and has been predicted successfully through calculations of solution pH, electrode potential and total rate of hydrogen production within cracks of varying length, and the use of an empirical relationship between K_{Iscc} and adsorbed hydrogen concentration [*100–102*].

Gangloff showed that small CF cracks (below about 3 mm) grew at rates up to 10^3 times faster than long cracks at equivalent applied ΔK [*103,104*]. This effect is illustrated in Fig. 18 by comparing data for 25 to 40-mm-long cracks with those of 0.1 to 3-mm-long elliptical surface and through-thickness edge cracks for a high-strength steel in aqueous 3% sodium chloride solutions. The concomitant absence of a crack size effect in vacuum and moist air,

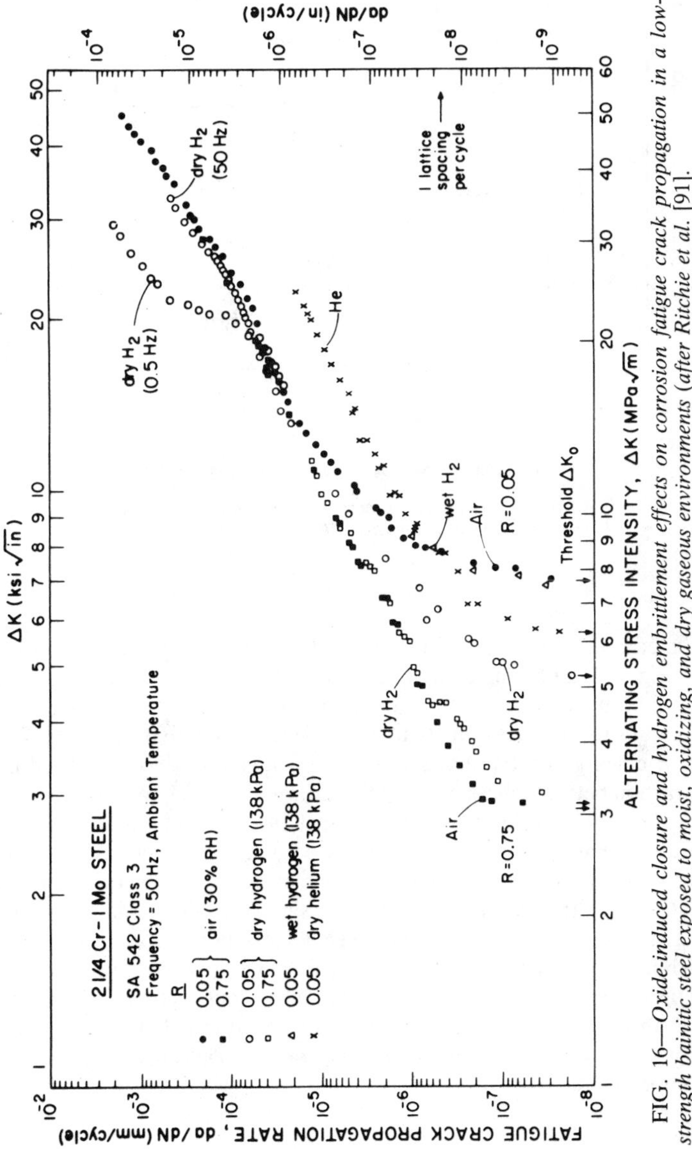

FIG. 16—*Oxide-induced closure and hydrogen embrittlement effects on corrosion fatigue crack propagation in a low-strength bainitic steel exposed to moist, oxidizing, and dry gaseous environments (after Ritchie et al.* [91].

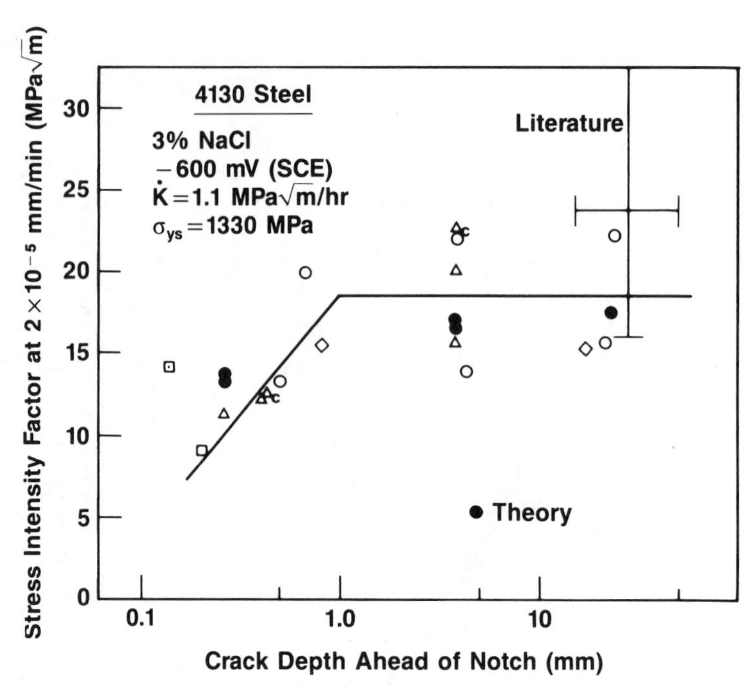

FIG. 17—*Measured and predicted effect of precrack depth on the threshold stress intensity for stress corrosion cracking of high-strength steel exposed to 3% sodium chloride (after Gangloff and Turnbull* [100].

predicted by mechanical mechanisms for this high-strength steel [*96*], further demonstrates the chemical origin of the breakdown in similitude. The chemical crack size effect has been confirmed by literature data and by experiments with lower-strength steels. The effect, however, decreases from a factor of 10^3 to 2 as the yield strength of the steel decreases [*95*].

The effect of crack geometry shown in Fig. 18 is qualitatively understood in terms of a complex interaction between mass transport by diffusion, ion migration and convection, and electrochemical reactions within the occluded crack [*94,104,105*]. An initial effort at modeling this effect has been made, but is not supported by critical experiments [*104–106*].

In essence, it is necessary to examine the coupled reaction and transport problem as a whole in dealing with the apparent violation of similitude. Turnbull has approached this problem by modeling steady-state reactions and mass transport in trapezoidally shaped cracks [*107,108*]. While an important contribution, this work must be modified to include the important transient reaction kinetics discussed in a previous section and must be related to the micromechanics of hydrogen and dissolution/film rupture-enhanced crack advance. Such an approach will provide the foundation for developing quantitative predictive methods for long-term component service.

Life Prediction for Environmental Cracking

The time dependence of environmental crack growth, the many relevant variables, and the complexities affecting similitude hinder fracture-mechanics based life prediction. Nonetheless, the phenomenological and scientific foundations have been developed during the

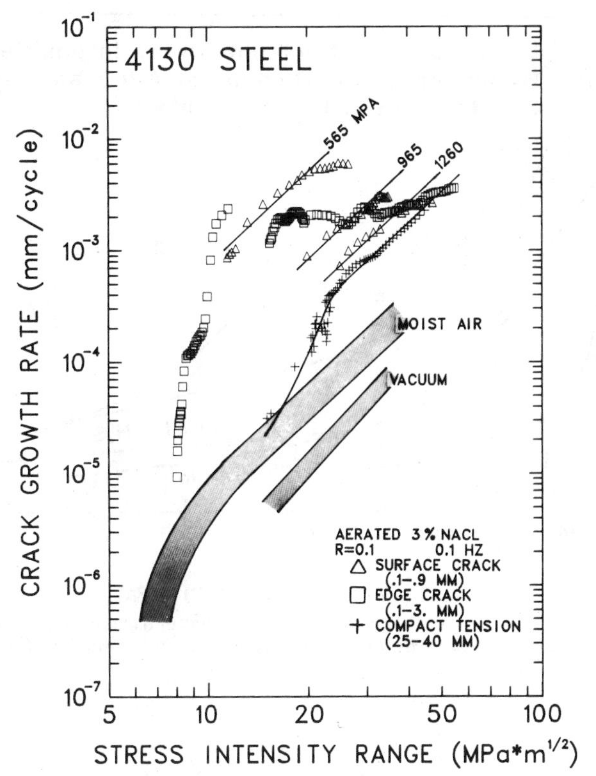

FIG. 18—*Effect of short crack size on corrosion fatigue propagation rates in high-strength* (σ_{ys} = *1340 MPa) steel in 3% sodium chloride. Applied stress ranges are shown for surface cracks, and the shaded bands describe data for each indicated crack size and geometry (after Gangloff* [104]*).*

past 15 years. The challenge is to refine and integrate this understanding to produce life-prediction methods which overcome the weaknesses of smooth specimen, time-indifferent design rules. Significant progress has been reported in this regard for stress corrosion and CF in nuclear systems, based on film rupture mechanistic modeling [*109*].

Recent advances and future directions for life prediction of steels, cracking by hydrogen embrittlement, are reviewed for static and cyclic loading.

Stress Corrosion Cracking—Stress corrosion cracking is sensibly predicted based on the threshold stress-intensity (K_{Iscc}) concept, within the bounds summarized elsewhere [*110*]. Two variables critically affect K_{Iscc}: steel yield strength and hydrogen uptake.

Lower-bound relationships between K_{Iscc} and yield strength are presented in Fig. 19, based on over 500 measurements reported during the past 20 years [*111*]. The beneficial effect of decreasing strength is shown for five specific environments. For constant strength, K_{Iscc} correlates with the steady-state concentration of dissolved hydrogen produced by gaseous chemical or aqueous electrochemical reactions on the input surface of a steel foil and measured by a permeation experiment, as illustrated in Fig. 20 [*111,112*]. Such data, when employed with permeation based sensors of hydrogen uptake in plant components [*113*], enable conservative life predictions for environmental cracking under monotonic loading.

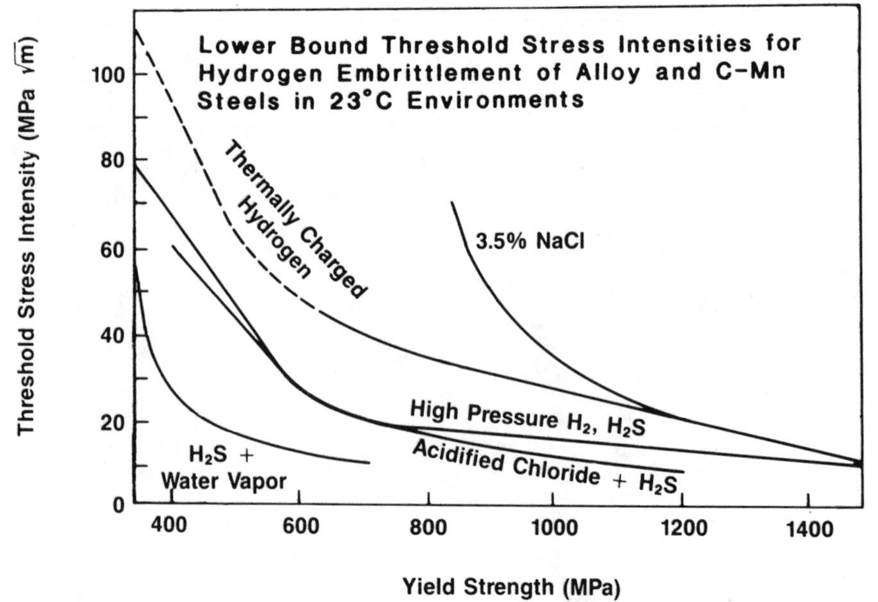

FIG. 19—*Lower bounds on* K_{Iscc}-σ_{ys} *for ferrite-pearlite, bainitic, and tempered martensitic steels stressed in five hydrogen-producing environments. Over 500 measurements are represented (after Gangloff* [111]).

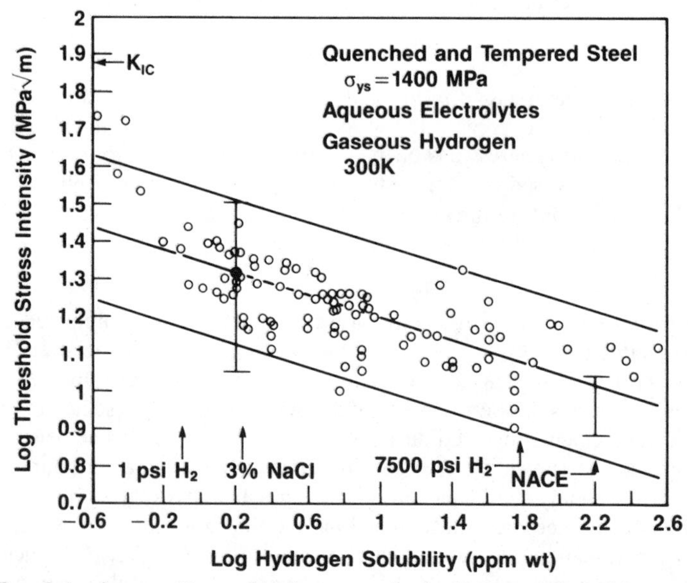

FIG. 20—*Correlation between* K_{Iscc} *and hydrogen uptake for high-strength steel fractured in gaseous hydrogen and electrolytes (after Gangloff* [111]).

This approach may be compromised by cyclic loading, unique microstructures, and new environment chemistries [*110,111*].

Corrosion Fatigue—Significant progress was achieved in the 1980s for fracture-mechanics based predictions of long-term corrosion fatigue in offshore structural applications, particularly large welded joints between low-strength carbon steel tubulars [*109,114*]. The elements of this approach are reviewed.

1. *Data base:* CF cracking in the low alloy steel/aqueous environment system occurs at stress intensities well below K_{Iscc} and within the regime relevant to tubular joint performance, as illustrated by the data and arrow in Fig. 21 [*93,115–119*]. Crack growth is enhanced by reduced loading frequency and increased cathodic polarization; such effects must be understood to predict the long-term (10^8 cycles at 0.1 Hz) life of cathodically protected tubulars.
2. *Hydrogen embrittlement models:* The aim of mechanistic modeling is to predict cor-

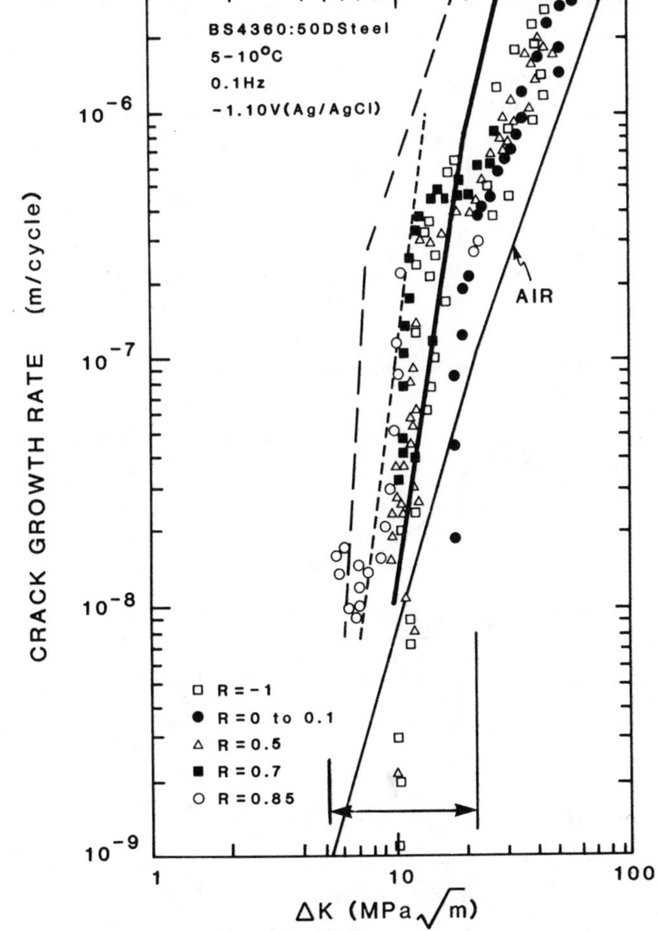

FIG. 21—*Corrosion fatigue crack growth rate versus* ΔK *for low-strength BS4360:50D C-Mn steel in seawater after Scott et al.* [93]; *seawater + hydrogen sulfide* ——— [115]; *oil + hydrogen sulfide* —— [117]; *and high presure hydrogen* ------ [116].

rosion fatigue crack growth rate as a function of stress intensity, stress ratio, environment composition, microstructure, electrode potential, and loading frequency. No single, quantitative model exists. The foundation has been established, however, by research on crack-tip reactions and the frequency dependence of CF [*44,59,75*], on crack closure [*89*], on steady-state crack chemistry modeling [*100,102,105–108*], and on hydrogen embrittlement [*115,120*]. Mechanisms for near-threshold CF have been proposed also [*90,91,93, 94,115,121*]; however, quantitative formulations and experimental evaluations are lacking.

3. *Stress-intensity and life predictions:* To date, only approximate K solutions for complex tubular joints have been developed [*122*]. Such equations were integrated with laboratory crack growth rate data (Fig. 21) to yield the prediction shown in Fig. 22 [*122*].

4. *Full-scale component testing:* Sophisticated capabilities exist in several countries to conduct fatigue and corrosion fatigue experiments with large welded tubular joints and to monitor crack growth continuously by electrical potential techniques [*123–125*]. As shown in Fig. 22, fatigue life data obtained from full-scale joints are in excellent agreement with fracture-mechanics predictions for cycle-dependent cracking in moist air. Measurements of surface crack shape development are also consistent with model predictions.

Limited CF experiments on full-scale joints demonstrate the deleterious effect of the seawater environment and the inability of simple cycle-based fatigue design rules to adequately describe tubular life [*109*]. Modeling of these effects requires improved crack growth rate data, mechanistic models and understanding of similitude, particularly for near-threshold corrosion fatigue.

Outstanding Issues and New Directions

The foregoing sections provide a perspective summary of the modeling effort to connect quantitatively chemical and physical processes with environmental crack growth response.

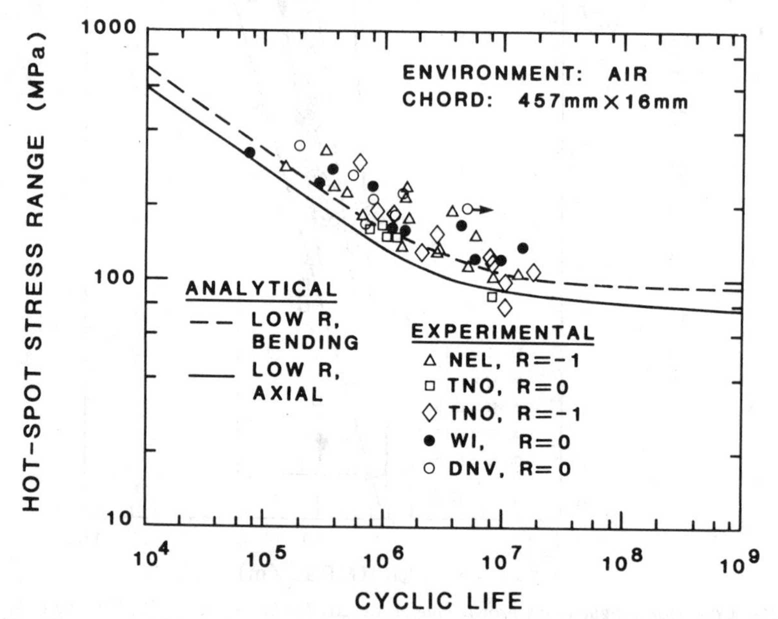

FIG. 22—*Fracture-mechanics based life prediction for fatigue crack propagation in welded tubular joints exposed to moist air and compared to full-scale component measurements* [122].

From an initial effort that was narrowly focused on the rate-controlling processes for crack growth in steels exposed to gaseous environments, ensuing studies broadened the understanding to cracking in high-strength aluminum and titanium alloys and extended the approach to the complex problem of crack growth in aqueous environments. The simplified models serve as a framework for research and design, as illustrated for offshore structures. They serve also as a basis for furthering the understanding of environmentally assisted crack growth. Results from these studies will aid in the development of alloys and of methods to minimize sensitivity to environmentally assisted cracking, and of procedures for making reliable predictions of long-term service. Fracture mechanics has played a major role in the development of this understanding. Significant issues remain unresolved. Several topics that require attention are as follows:

1. Detailed understanding of the kinetics and mechanisms of surface/electrochemical reactions with clean metal surfaces is needed, over a broad range of environmental conditions: to establish the form and quantity of hydrogen that is produced, and the fraction that enters the material to effect embrittlement; to explore and confirm the transfer of rate-controlling processes with changes in environmental conditions; and to determine the relationship between the kinetics of surface reactions and crack growth response to improve the predictive capability of crack-growth models.
2. For environmentally assisted crack growth in aqueous environments, the traditional electrochemical measurement of polarization response is inadequate. A new technique for measuring the kinetics of equilibration reactions at the crack tip has been developed. These measurements, however, must be coupled with a detailed understanding of the reaction mechanisms and modeling of the crack-tip chemistry.
3. Greater effort is needed to understand the factors and processes that control or inhibit electrochemical reactions of bare, straining metal surfaces with electrolytes.
4. Quantitative understanding of the physical-chemical interactions between hydrogen and metal (that is, the embrittlement mechanisms) is needed to establish the roles of microstructure and of other metallurgical variables in determining the rate of crack growth or the degree of susceptibility.
5. Better understanding of the influences of alloying and impurity elements and of microstructure is needed. It is essential to determine whether such influences on environmental cracking result from alterations of the reaction kinetics (chemical effects), from their influence on the mechanical properties of alloys (physical effects), or from both (physical-chemical effects).
6. Better understanding of the processes that control environmental crack initiation and early growth (namely, threshold and Stage I crack growth) is needed.
7. Mechanistic descriptions of crack chemistry, transient reactions, and micromechanical embrittlement must be integrated to produce a predictive model of environmentally assisted crack growth rate. The model must quantitatively predict both specimen and component cracking performance and must be amenable to experimental confirmations and mechanistic refinements.
8. Experimental research and modeling must be exploited to develop *in situ* monitors of environmental crack growth in complex components. Sensors for hydrogen uptake from service environments must be further developed and implemented.
9. New experimental methods must be developed for direct measurement of environmental damage processes at the crack tip.
10. Research on environmentally assisted cracking must be extended to include novel monolithic and composite materials.

Summary

Fracture mechanics-based experimental and analytical work in the past two decades has contributed significantly to the phenomenological and mechanistic understanding of environmentally assisted crack growth. Crack-growth response reflects the complex interplay among chemical, mechanical, and metallurgical factors and is dependent on the rate-controlling processes and on the micromechanisms for crack growth and the mechanisms of the relevant chemical reactions. On the basis of this understanding, modeling of environmentally assisted crack growth, under sustained-load and in fatigue, has advanced. This modeling effort has placed the study of this technologically important problem on a sound footing and provides a framework for new understanding and for the development and utilization of data in design. To make significant further advances in understanding, continued emphasis on multidisciplinary approaches which incorporate chemistry, physics, materials science, and fracture mechanics, and long-term support are essential.

Acknowledgment

This work was supported by the Office of Naval Research under Contract N00014-83-K-0107 NR 036-097 (RPW) and by the National Aeronautics and Space Administration, Langley Research Center, Research Grant NAG-1-745 (RPG).

References

[1] Brown, B. F. in *Stress-Corrosion Cracking in High Strength Steels and in Titanium and Aluminum Alloys,* B. F. Brown, Ed., U.S. Naval Research Laboratory, Washington, DC, 1972, pp. 2–16.
[2] Haigh, B. P., *Journal of the Institute of Metals,* Vol. 18, 1917, p. 55.
[3] Gough, H. J. and Sopwith, D. G., *Journal of the Institute of Metals,* Vol. 49, 1932, p. 93.
[4] *The Theory of Stress Corrosion Cracking in Alloys,* J. C. Scully, Ed., North Atlantic Treaty Organization, Brussels, Belgium, 1971.
[5] *Fundamental Aspects of Stress Corrosion Cracking,* NACE-1, R. W. Staehle, A. J. Forty, and D. van Rooyen, Eds., National Association of Corrosion Engineers, Houston, TX, 1969.
[6] *Corrosion Fatigue: Chemistry, Mechanics and Microstructure,* NACE-2, O. F. Devereux, A. J. McEvily, and R. W. Staehle, Eds., National Association of Corrosion Engineers, Houston, TX, 1972.
[7] *Stress Corrosion Cracking and Hydrogen Embrittlement of Iron Base Alloys,* NACE-5, R. W. Staehle, J. Hochmann, R. D. McCright, and J. E. Slater, Eds., National Association of Corrosion Engineers, Houston, TX, 1978.
[8] *L'Hydrogene Dans Les Metaux,* J. P. Fidelle and M. Rapin, Eds., Commissariat Energie Atomique, Paris, France, 1967.
[9] *Hydrogen in Metals,* I. M. Bernstein and A. W. Thompson, Eds., American Society for Metals, Metals Park, OH, 1974.
[10] *Effect of Hydrogen on Behavior of Materials,* A. W. Thompson and I. M. Bernstein, Eds., American Society for Metals, Metals Park, OH, 1976.
[11] *Hydrogen in Metals, Proceedings of Second Japan Institute of Metals International Symposium (JIMS-2),* Supplement to *Transactions of the Japan Institute of Metals,* Vol. 21, 1980.
[12] *Environment-Sensitive Fracture of Engineering Materials,* Z. A. Foroulis, Ed., The Metallurgical Society of the American Institute of Mining, Metallurgical, and Petroleum Engineers, New York, 1979.
[13] *Hydrogen Embrittlement and Stress Corrosion Cracking,* R. Gibala and R. F. Hochmann, Eds., American Society for Metals, Metals Park, OH, 1984.
[14] Wei, R. P. in *Environmental Degradation of Engineering Materials in Aggressive Environments,* Vol. 2, M. R. Louthan, Jr., R. P. McNitt, and R. D. Sisson, Jr., Eds., Virginia Polytechnic Institute and State University, Blacksburg, VA, 1981, pp. 73–81.
[15] *Environment-Sensitive Fracture: Evaluation and Comparison of Test Methods, ASTM STP 821,* S. W. Dean, E. N. Pugh, and G. M. Ugiansky, Eds., American Society for Testing and Materials, Philadelphia, 1984.

[16] *Embrittlement by the Localized Crack Environment,* R P. Gangloff, Ed., The Metallurgical Society of the American Institute of Mining, Metallurgical, and Petroleum Engineers, New York, 1984.
[17] *Modeling Environmental Effects on Crack Initiation and Propagation,* R. H. Jones and W. W. Gerberich, Eds., The Metallurgical Society of the American Institute of Mining, Metallurgical, and Petroleum Engineers, Warrendale, PA, 1986.
[18] *Corrosion Fatigue Technology, ASTM STP 642,* H. L. Craig, Jr., T. W. Crooker, and D. W. Hoeppner, Eds., American Society for Testing and Materials, Philadelphia, 1978.
[19] *Hydrogen Effects in Metals,* I. M. Bernstein and A. W. Thompson, Eds., The Metallurgical Society of the American Institute of Mining, Metallurgical, and Petroleum Engineers, Warrendale, PA, 1981.
[20] *Basic Questions in Fatigue,* Vols. I and II, *ASTM STP 924,* R. P. Wei, J. T. Fong, R. J. Fields, and R. R. Gangloff, Eds., American Society for Testing and Materials, Philadelphia, 1988.
[21] *Environmental Degradation of Engineering Materials-III,* M. R. Louthan, R. P. McNitt, and R. D. Sisson, Eds., Pennsylvania State University, 1987.
[22] *Mechanisms of Environment Sensitive Cracking of Materials,* P. R. Swann, F. P. Ford, and A. R. C. Westwood, Eds., The Metals Society, London, 1977.
[23] *Corrosion Fatigue, ASTM STP 801,* T. W. Crooker and B. N. Leis, Eds., American Society for Testing and Materials, Philadelphia, 1983.
[24] Fontana, M. G. and Staehle, R. W., "Stress Corrosion Cracking of Metallic Materials, Part III," *Hydrogen Entry and Embrittlement in Steel,* Final Report, AFML-TR-72-102, Air Force Materials Laboratory, Wright Patterson AFB, Ohio, April 1975.
[25] Brown, B. F. and Beachem, C. D., *Corrosion Science,* Vol. 5, 1965, pp. 745–750.
[26] Brown, B. F., *Materials Research and Standards,* Vol. 6, No. 3, 1966, p. 129.
[27] Johnson, H. H. and Paris, P. C., *Engineering Fracture Mechanics,* Vol. 1, 1968, pp. 3–45.
[28] Smith, H. R., Piper, D. E., and Downey, F. K., *Engineering Fracture Mechanics,* Vol. 1, 1968, pp. 123–128.
[29] Feeney, J. A., McMillan, J. C., and Wei, R. P., *Metallurgical Transactions,* Vol. 1, 1970, pp. 1741–1757.
[30] Wei, R. P. in *Fundamental Aspects of Stress Corrosion Cracking,* NACE-1, R. W. Staehle, A. J. Forty, and D. van Rooyen, Eds., National Association of Corrosion Engineers, Houston, TX, 1969, pp. 104–112.
[31] Steigerwald, E. A., "Delayed Failure of High-Strength Steel in Liquid Environments," *Proceedings,* American Society for Testing and Materials, Philadelphia, Vol. 60, 1960, pp. 750–760.
[32] Hanna, G. L., Troiano, A. R., and Steigerwald, E. A., *American Society for Metals Transactions Quarterly,* Vol. 57, 1964, p. 658.
[33] Johnson, H. H. and Willner, A. M., *Applied Materials Research,* Vol. 1, 1965, p. 34.
[34] Hancock, G. G. and Johnson, H. H., *Transactions,* The Metallurgical Society of the American Institute of Mining, Metallurgical, and Petroleum Engineers, Vol. 236, 1966, p. 513.
[35] Bradshaw, F. J. and Wheeler, C., *Applied Materials Research,* Vol. 5, 1966, p. 112.
[36] Hartman, A., *International Journal of Fracture Mecahnics,* Vol. 1, 1965, p. 167.
[37] Wei, R. P., Talda, P. M., and Li, C.-Y. in *Fatigue Crack Propagation, ASTM STP 415,* American Society for Testing and Materials, Philadelphia, 1967, pp. 460–485.
[38] Wei, R. P., Novak, S. R., and Williams, D. P., *Materials Research and Standards,* Vol. 12, 1972, p. 25.
[39] Novak, S. R. and Rolfe, S. T., *Corrosion,* Vol. 26, No. 4, 1970, pp. 121–130.
[40] Wei, R. P. in *Fatigue Mechanisms, ASTM STP 675,* J. T. Fong, Ed., American Society for Testing and Materials, Philadelphia, 1979, pp. 816–840.
[41] Wei, R. P. in "Steels, ARPA Coupling Program on Stress Corrosion Cracking: (Seventh Quarterly Report)," NRL Memorandum Report 1941, Naval Research Laboratory, Washington, DC, Oct. 1968, p. 49.
[42] Speidel, M. O., "Current Understanding of Stress Corrosion Crack Growth in Aluminum Alloys," *The Theory of Stress Corrosion Cracking in Alloys,* J. C. Scully, Ed., NATO Scientific Affairs Division, Brussels, Belgium, 1971, pp. 289–344.
[43] Wei, R. P., Klier, K., Simmons, G. W., and Chou, Y. T. in *Hydrogen Embrittlement and Stress Corrosion Cracking,* R. Gibala and R. F. Hehemann, Eds., American Society for Metals, Metals Park, OH, 1984, pp. 103–133.
[44] Wei, R. P. and Gao, M. in *Hydrogen Degradation of Ferrous Alloys,* R. A. Oriani, J. P. Hirth, and M. Smialowski, Eds., Noyes Publications, Park Ridge, NJ, 1985, pp. 579–607.
[45] Williams, D. P. and Nelson, H. G., *Metallurgical Transactions,* Vol. 1, 1970, p. 63.
[46] Nelson, H. G., Williams, D. P., and Tetelman, A. S., *Metallurgical Transactions,* Vol. 2, 1971, p. 953.

[47] Wei, R. P., *International Journal of Fracture Mechanics,* Vol. 4, 1968, pp. 159–170.
[48] Wei, R. P. and Landes, J. D., *Materials Research and Standards,* Vol. 9, No. 7, 1969, pp. 25–28.
[49] Gao, M. and Wei, R. P., *Metallurgical Transactions A,* Vol. 16A, 1985, pp. 2039–2050.
[50] Gangloff, R. P. and Wei, R. P., *Metallurgical Transactions A,* Vol. 8A, 1977, pp. 1043–1053.
[51] Simmons, G. W., Pao, P. S., and Wei, R. P., *Metallurgical Transactions A,* Vol. 9A, 1978, p. 1147.
[52] Simmons, G. W. and Dwyer, D. J., *Surface Science,* Vol. 48, 1975, p. 373.
[53] Dwyer, D. J., Simmons, G. W., and Wei, R. P., *Surface Science,* Vol. 64, 1977, p. 617.
[54] Dwyer, D. J., Ph.D. dissertation, Lehigh University, Bethlehem, PA, 1977.
[55] Wei, R. P. and Simmons, G. W. in *Stress Corrosion Cracking and Hydrogen Embrittlement of Iron Base Alloys,* R. W. Staehle, J. Hochmann, R. D. McCright, and J. E. Slater, Eds., NACE-5, National Association of Corrosion Engineers, Houston, TX, 1979, pp. 751–765.
[56] Wei, R. P., Pao, P. S., Hart, R. G., Weir, T. W., and Simmons, G. W., *Metallurgical Transactions A,* Vol. 11A, 1980, pp. 151–158.
[57] Brazill, R., Simmons, G. W., and Wei, R. P., *Journal of Engineering Materials and Technology,* Transactions, American Society of Mechanical Engineers, Vol. 101, July 1979, pp. 199–204.
[58] Weir, T. W., Simmons, G. W., Hart, R. G., and Wei, R. P., *Scripta Metallurgica,* Vol. 14, 1980, pp. 357–364.
[59] Wei, R. P. and Simmons, G. W. in *FATIGUE: Environment and Temperature Effects,* John J. Burke and Volker Weiss, Eds., Sagamore Army Materials Research Conference Proceedings, Vol. 27, 1983, pp. 59–70.
[60] Lu, M., Pao, P. S., Weir, T. W., Simmons, G. W., and Wei, R. P., *Metallurgical Transactions A,* Vol. 12A, 1981, pp. 805–811.
[61] Lu, M., Pao, P. S., Chan, N. H., Klier, K., and Wei, R. P. in *Hydrogen in Metals,* supplement to *Transactions of the Japan Institute of Metals,* Vol. 21, 1980, p. 449.
[62] Chan, N. H., Klier, K., and Wei, R. P. in *Hydrogen in Metals,* supplement to *Transactions of the Japan Institute of Metals,* Vol. 21, 1980, p. 305.
[63] Wei, R. P. in *Hydrogen Effects in Metals,* I. M. Bernstein and A. W. Thompson, Eds., The Metallurgical Society of the American Institute of Mining, Metallurgical and Petroleum Engineers, Warrendale, PA, 1981, p. 677.
[64] Dushman, S., "Molecular Flow" in *Scientific Foundations of Vacuum Technique,* 2nd ed., J. M. Lafferty, Ed., Wiley, New York, 1962, pp. 87–104.
[65] Pasco, R. W. and Ficolora, P. J., *Scripta Metallurgica,* Vol. 15, 1980, p. 1019.
[66] Pasco, R. W., Sieradzki, K., and Ficalora, J. P., *Scripta Metallurgica,* Vol. 16, 1982, p. 881.
[67] Chan, N. H., Klier, K., and Wei, R. P., *Scripta Metallurgica,* Vol. 12, 1978, pp. 1043–1046.
[68] Pao, P. S. and Wei, R. P., *Scripta Metallurgica,* Vol. 11, 1977, pp. 515–520.
[69] Hart, E. W. in *The Nature and Behavior of Grain Boundaries,* Hsun Hu, Ed., Plenum, New York, 1970, p. 155.
[70] Wei, R. P. and Landes, J. D., *Materials Research and Standards,* Vol. 9, No. 7, 1969, pp. 25–28.
[71] Wei, R. P. and Gao, M., *Scripta Metallurgica,* Vol. 17, 1983, pp. 959–962.
[72] Shih, T.-H. and Wei, R. P., *Engineering Fracture Mechanics,* Vol. 18, No. 4, 1983, pp. 827–837.
[73] Wei, R. P. and Shim, G. in *Corrosion Fatigue, ASTM STP 801,* T. W. Crooker and B. N. Leis, Eds., American Society for Testing and Materials, Philadelphia, 1983, pp. 5–25.
[74] Shim, G. and Wei, R. P., *Materials Science and Engineering,* Vol. 86, 1986, pp. 121–135.
[75] Wei, R. P. in "Microstructural and Mechanical Behavior of Materials," *Proceedings,* International Symposium on Microstructure and Mechanical Behaviour of Materials, Engineering Materials Advisory Services Ltd., West Midlands, U.K., 1986, pp. 507–526.
[76] Kim, Y. H., Speaker, S. M., Gordon, D. E., Manning, S. D., and Wei, R. P., Report No. NADC-83126-60, Vol. 1, Naval Air Development Center (604), Warminster, PA, March 1983.
[77] Pao, P. S. and Wei, R. P. in *Titanium: Science and Technology,* G. Lutjering, U. Zwicker, and W. Bunk, Eds., Deutsche Gesellschaft Fur Metall-kunde e.V., Oberursel, Federal Republic of Germany, 1985, p. 2503.
[78] Birnbaum, H. K., "On the Mechanisms of Hydrogen Related Fracture in Metals," *Environment-Sensitive Fracture of Metals and Alloys,* Office of Naval Research, Arlington, VA, 1987, pp. 105–113.
[79] Peterson, K. P., Schwanebeck, J. C., and Gerberich, W. W., *Metallurgical Transactions A,* Vol. 9A, 1978, p. 1169.
[80] Nakai, Y., Alavi, A., and Wei, R. P., "Effects of Frequency and Temperature on Short Fatigue Crack Growth in Aqueous Environments," *Metallurgical Transactions A,* Vol. 19A, 1988, pp. 543–548.

[81] Wei, R. P., Gao, M., and Pao, P. S., *Scripta Metallurgica,* Vol. 18, 1984, pp. 1195–1198.
[82] Chiou, S. and Wei, R. P., "Corrosion Fatigue Cracking Response of Beta Annealed Ti-6Al-4V Alloy in 3.5% NaCl Solution," Report No. NADC-83126-60 (Vol. 5), U.S. Naval Air Development Center, Warminster, PA, 30 June 1984.
[83] Gerberich, W. W. and Peterson, K. A. in *Micro and Macro Mechanics of Crack Growth,* K. Sadanada, B. B. Rath, and D. J. Michel, Eds., The Metallurgical Society of the American Institute of Mining, Metallurgical, and Petroleum Engineers, Warrendale, PA, 1982, pp. 1–17.
[84] Ressler, D., "An Examination of Fatigue Crack Growth in AISI 4340 Steel in Respect to Two Corrosion Fatigue Models," M.S. thesis, Department of Mechanical Engineering and Mechanics, Lehigh University, Bethlehem, PA, 1984.
[85] Gao, M., Pao, P. S., and Wei, R. P., "Chemical and Metallurgical Aspects of Environmentally Assisted Fatigue Crack Growth in 7075-T651 Aluminum Alloy," *Metallurgical Transactions A,* Vol. 19A, 1988, p. 1739.
[86] *Fatigue Crack Growth Threshold Concepts,* D. Davidson and S. Suresh, Eds., The Metallurgical Society of the American Institute of Mining, Metallurgical, and Petroleum Engineers, Warrendale, PA, 1984.
[87] *Small Fatigue Cracks,* R. O. Ritchie and J. Lankford, Eds., The Metallurgical Society of the American Institute of Mining, Metallurgical, and Petroleum Engineers, Warrendale, PA, 1986.
[88] *Aluminum-Lithium Alloys II,* T. H. Sanders, Jr., and E. A. Starke, Jr., Eds., The Metallurgical Society of the American Institute of Mining, Metallurgical, and Petroleum Engineers, Warrendale, PA, 1984.
[89] Suresh, S. and Ritchie, R. O. in *Fatigue Crack Growth Threshold Concepts,* D. Davidson and S. Suresh, Eds., The Metallurgical Society of the American Institute of Mining, Metallurgical, and Petroleum Engineers, Warrendale, PA, 1984, pp. 227–261.
[90] Gangloff, R. P. and Ritchie, R. O. in *Fundamentals of Deformation and Fracture,* B. A. Bilby, K. J. Miller, and J. R. Willis, Eds., Cambridge University Press, Cambridge, U.K., 1985, pp. 529–558.
[91] Suresh, S., Zamiski, G. F., and Ritchie, R. O., *Metallurgical Transactions A,* Vol. 12A, 1981, pp. 1435–1443.
[92] Hartt, W. H., Culberson, C. H., and Smith, S. W., *Corrosion,* Vol. 40, 1984, pp. 609–618.
[93] Thorpe, T. W., Scott, P. M., Rance, A., and Silvester, D., *International Journal of Fatigue,* Vol. 5, 1983, pp. 123–133.
[94] Gangloff, R. P. and Duquette, D. J. in *Chemistry and Physics of Fracture,* R.M. Latanision and R. H. Jones, Eds., Martinus Nijhoff Publishers BV, Amsterdam, Netherlands, 1987.
[95] Gangloff, R. P. and Wei, R. P. in *Small Fatigue Cracks,* R. O. Ritchie and J. Lankford, Eds., The Metallurgical Society of the American Institute of Mining, Metallurgical, and Petroleum Engineers, Warrendale, PA, 1986, pp. 239–264.
[96] Ritchie, R. O. and Lankford, J., *Materials Science and Engineering,* Vol. 84, 1986.
[97] Tanaka, K. and Wei, R. P., *Engineering Fracture Mechanics,* Vol. 21, 1985, p. 293–305.
[98] Zeghloul, A. and Petit, J. in *Small Fatigue Cracks,* R. O. Ritchie and J. Lankford, Eds., The Metallurgical Society of the American Institute of Mining, Metallurgical, and Petroleum Engineers, Warrendale, PA, 1986, pp. 225–235.
[99] Ford, F. P. and Hudak, S. J., Jr., in *Small Fatigue Cracks,* R. O. Ritchie and J. Lankford, Eds., The Metallurgical Society of the American Institute of Mining, Metallurgical, and Petroleum Engineers, Warrendale, PA, 1986, pp. 289–308.
[100] Gangloff, R. P. and Turnbull, A. in *Modeling Environmental Effects on Crack Initiation and Propagation,* R. H. Jones and W. W. Gerberich, Eds., The Metallurgical Society of the American Institute of Mining, Metallurgical, and Petroleum Engineers, Warrendale, PA, 1986, pp. 55–81.
[101] Clark, W. G., Jr., in *Cracks and Fracture, ASTM STP 601,* American Society for Testing and Materials, Philadelphia, 1976, pp. 138–153.
[102] Brown, B. F. in *Stress Corrosion Cracking and Hydrogen Embrittlement of Iron Based Alloys,* J. Hochmann, J. Slater, R. D. McCright, and R. W. Staehle, Eds., National Association of Corrosion Engineers, Houston, TX, 1976, pp. 747–751.
[103] Gangloff, R. P., *Res Mechanica Letters,* Vol. 1, 1981, pp. 299–306.
[104] Gangloff, R. P., *Metallurgical Transactions A,* Vol. 16A, 1985, pp. 953–969.
[105] Gangloff, R. P. in *Embrittlement by the Localized Crack Environment,* R. P. Gangloff, Ed., The Metallurgical Society of the American Institute of Mining, Metallurgical, and Petroleum Engineers, Warrendale, PA, 1984, pp. 265–290.
[106] Gangloff, R. P. in *Critical Issues in Reducing the Corrosion of Steel,* H. Leidheiser, Jr., and S. Haruyama, Eds., National Science Foundation/Japan Society for the Promotion of Science, Tokyo, 1985, pp. 28–50.

[*107*] Turnbull, A. and Ferriss, D. H. in *Proceedings of Conference on Corrosion Chemistry within Pits, Crevices and Cracks,* A. Turnbull, Ed., National Physical Laboratory, Teddington, U.K., 1987.
[*108*] Turnbull, A. in *Embrittlement by the Localized Crack Environment,* R. P. Gangloff, Ed., The Metallurgical Society of the American Institute of Mining, Metallurgical, and Petroleum Engineers, Warrendale, PA, 1984, pp. 3–48.
[*109*] Andresen, P. L., Gangloff, R. P., Coffin, L. F., and Ford, F. P. in *Fatigue 87,* R. O. Ritchie and E. A. Starke, Jr., Eds., Engineering Materials Advisory Services, Ltd., West Midlands, U.K., 1987.
[*110*] "Characterization of Environmentally Assisted Cracking for Design," Report NMAB-386, National Materials Advisory Board, 1982.
[*111*] Gangloff, R. P. in *Corrosion Prevention and Control, Proceedings of 33rd Sagamore Army Materials Research Conference,* S. Isserow, Ed., U.S. Army Materials Technology Laboratory, Watertown, MA, 1987.
[*112*] Yamakawa, K., Tsubakino, H., and Yoshizawa, S. in *Critical Issues in Reducing the Corrosion of Steel,* H. Leidheiser, Jr., and S. Haruyama, Eds., National Science Foundation/Japan Society for the Promotion of Science, Tokyo, 1985, pp. 348–358.
[*113*] Yamakawa, K., Harushige, T., and Yoshizawa, S. in *Corrosion Monitoring in Industrial Plants Using Nondestructive Testing and Electrochemical Methods, ASTM STP 908,* G. C. Moran and P. Labine, Eds., American Society for Testing and Materials, Philadelphia, 1986, pp. 221–236.
[*114*] Dover, W. D. and Dharmavasan, S. in *Fatigue 84,* C. J. Beevers, Ed., Engineering Materials Advisory Services, Ltd., West Midlands, U.K., 1984, pp. 1417–1434.
[*115*] Austen, I. M. and Walker, E. F., in *Fatigue 84,* C. J. Beevers, Ed., Engineering Materials Advisory Services, Ltd., West Midlands, U.K., 1984, pp. 1457–1469.
[*116*] Cialone, H. J. and Holbrook, H. J., *Metallurgical Transactions A,* Vol. 16A, 1985, pp. 115–122.
[*117*] Vosikovsky, O. and Cooke, R. J., *International Journal of Pressure Vessels and Piping,* Vol. 6, 1978, pp. 113–129.
[*118*] Vosikovsky, O., *Journal of Testing and Evaluation,* 1978, pp. 175–182.
[*119*] Vosikovsky, O., *Journal of Testing and Evaluation,* Vol. 8, 1980, pp. 68–73.
[*120*] Scott, P. M., Thorpe, T. W., and Silvester, D. R. V., *Corrosion Science,* Vol. 23, No. 6, 1983, pp. 559–575.
[*121*] Duquette, D. J. and Uhlig, H. H., *Transactions,* American Society for Metals, Vol. 61, 1968, pp. 449–456.
[*122*] Hudak, S. J., Burnside, O. H., and Chan, K. S., *Journal of Energy Resources Technology,* Transactions, American Society of Mechanical Engineers, Vol. 107, 1985, pp. 212–219.
[*123*] *Fatigue in Offshore Structural Steels,* Thomas Telford, Ltd., London, 1981.
[*124*] *Proceedings of the International Conference on Steel in Marine Structures,* Comptoir des produits Siderurgiques, Paris, 1981.
[*125*] *Proceedings Institute of Mechanical Engineers Conference on Fatigue and Crack Growth in Offshore Structures,* Institute of Mechanical Engineers, London, 1986.

Fracture Mechanics of Nonmetals and New Frontiers

A. G. Evans[1]

The New High-Toughness Ceramics

REFERENCE: Evans, A. G., **"The New High-Toughness Ceramics,"** *Fracture Mechanics: Perspectives and Directions (Twentieth Symposium), ASTM STP 1020,* R. P. Wei and R. P. Gangloff, Eds., American Society for Testing and Materials, Philadelphia, 1989, pp. 267–291.

ABSTRACT: The principal microstructural sources of toughening are reviewed, with emphasis on comparisons between theory and experiment. Process zone mechanisms, such as transformation and microcracking, as well as bridging mechanisms induced by either ductile networks or fiber/whiskers are afforded primary consideration. Microstructural variables that allow toughness optimization are established.

KEY WORDS: composites, ceramics, toughness, micromechanics, fibers, ductile phases, bridging, microcracking, transformation, process zone

The past decade has witnessed major advances in the development of ceramics having enhanced toughness. All of the mechanisms that provide appreciable toughening have the common feature that material elements at, or near, the crack surfaces exhibit nonlinear behavior, with hysteresis, as schematically illustrated in Fig. 1. Indeed, the toughening can be explicitly related to the hysteresis, as will be elaborated for each of the important mechanisms. Furthermore, in most cases, the mode of toughening results in resistance-curve characteristics (Fig. 2), wherein the fracture resistance systematically increases with crack extension. The individual mechanisms that have been established include displacive transformations, microcracking, ductile networks, and fiber/whisker reinforcement. The more significant materials and the peak verifiable toughnesses measured on these materials are summarized in Table 1.

The known mechanisms can be conveniently considered to involve either a process zone or a bridging zone (Fig. 3). The former category exhibits a toughening fundamentally governed by a critical stress for the onset of nonlinearity (σ_0 in Fig. 1) and by the permanent strain induced by the nonlinear mechanism (ϵ_0 in Fig. 1). The hysteresis is dictated by the stress-strain behavior of composite elements within the process zone, such that integration over the zone gives

$$\Delta G_c \approx 2 f \sigma_0 E h \tag{1}$$

where

ΔG_c = increase in toughness,
f = volume fraction of toughening agent, and
h = width of the process zone (Fig. 3).

Transformation and microcrack toughening are mechanisms of this type.

[1] Chairman, Materials Department, College of Engineering, University of California, Santa Barbara, CA 93106.

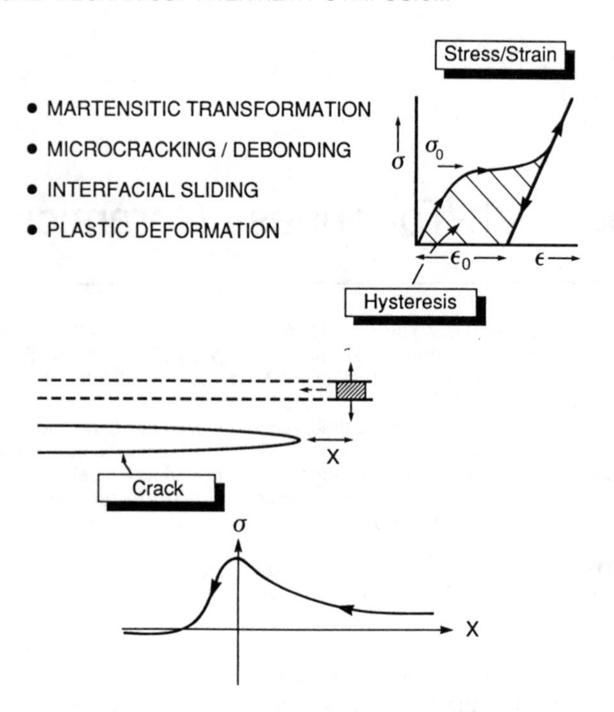

FIG. 1—*Nonlinear, hysteretic elemental response and associations with enhanced toughness.*

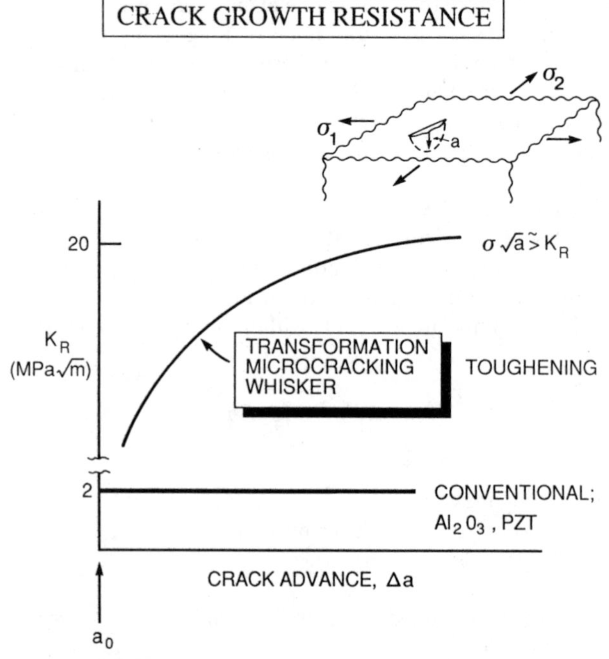

FIG. 2—*Resistance curve behavior characteristically encountered in tough ceramics:* K_R *is the fracture resistance and* Δa *the crack advance.*

TABLE 1—*Ceramics with enhanced toughness.*

Toughening Mechanism	Material	Maximum Toughness, MPa $\sqrt{m}$	Comments
Fiber reinforced	LAS/SiC	>20	steady-state cracking
	Glass/C	>20	
	SiC/SiC	>20	
Whisker reinforced	Al_2O_3/SiC(0.2)	10	amorphous interphase
	Si_3N_4/SiC(0.2)	14	
Ductile network	Al_2O_3/Al(0.2)	>12	steady-state cracking
	B_4C/Al(0.2)	>14	
	WC/Co(0.2)	20	
Transformation toughened	PSZ	18	nonlinear
	TZP	16	
	ZTA	10	
Microcrack toughened	ZTA	7	
	Si_3N_4/SiC	7	

The latter category exhibits toughening governed by hysteresis along the crack surface,

$$\Delta G_c = f \int_0^{u_*} \sigma(u)\, du \tag{2}$$

where

u = crack opening,
u_* = opening at the edge of the bridging zone, and
σ = the tractions on the crack surfaces exerted by the intact toughening agent (Fig. 3).

Ductile networks as well as whiskers and fibers toughen by means of bridging tractions.

In the toughest materials, a steady-state cracking phenomenon has been identified, wherein the crack extension stress becomes independent of crack length. For such materials, the toughness is usually non-unique and not, therefore, a useful material parameter. Conditions that provide steady-state cracking and the associated material behavior are described first, using a mechanism map approach.

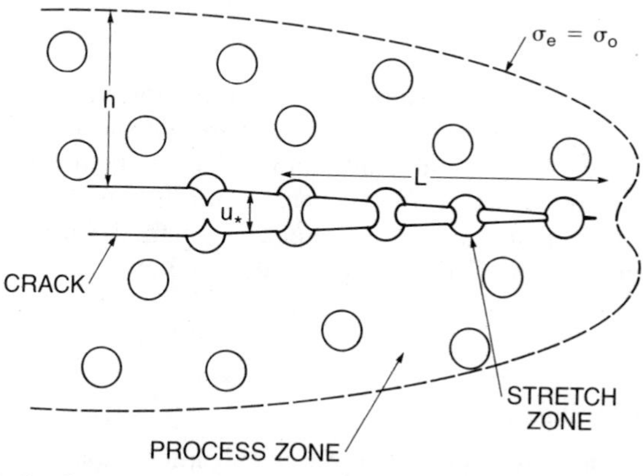

FIG. 3—*Process zone and bridging zone mechanisms of toughening.*

Steady-State Cracking

In certain composite ceramics, matrix cracking can occur at constant stress [1–3]. This phenomenon is referred to as steady-state cracking. Such behavior is most widely known in fiber-reinforced ceramics, but is also possible in ductile phase reinforced ceramics. Steady-state conditions exist when a crack is bridged by strong reinforcements (Fig. 4). An equilibrium crack opening then develops, with the load fully supported by the intact reinforcements. For such cases, the stress intensity associated with matrix cracking becomes independent of crack length. The steady-state stress depends on the response of the interface and matrix to the load on the reinforcements between the crack surfaces. For fibers that resist opening by friction, the cracking stress is [1–3]

$$\sigma_S = [6\tau K_0^2 f^2 E_f E^2/(1 - f)E_m^3 R]^{1/3} + \sigma_R E/E_m \quad (3)$$

where

τ = shear resistance of the interface,
K_0 = fracture resistance of the matrix,
f = volume of fibers,
E = Young's modulus,
R = fiber radius,
σ_R = residual stress in the matrix, and
subscripts f,m = fiber and matrix, respectively.

This prediction agrees well with various experimental results [2,3] obtained on fiber reinforced ceramics and glasses.

The equivalent result for partially debonded fibers without friction is [3]

$$\sigma_S = (K_0 E/E_m)[6f^2 E_f/E(1 - f)^2 R(1 + \upsilon_m)]^{1/4} \times [1 + 4f\ell\theta/R(1 - f)]^{1/2}[1 + \rho\ell/R]^{-1} + \sigma_R E/E_m \quad (4)$$

where

ℓ = debonded length,
θ = ratio of the debonded to matrix toughness,
υ = Poisson's ratio, and
$\rho = [0.8/(1 - f)][6E/E_f(1 + \upsilon_m)]^{1/2}$.

Steady state crack growth conditions exist when the reinforcement strength, $\overline{S}$, satisfies the inequality, $\overline{S} > \sigma_S/f$. Such materials have very desirable engineering properties, by virtue of the associated damage tolerance and significant nonlinearity before ultimate failure [4,5]. The transition from matrix cracking to toughness controlled fracture can sometimes be conveniently represented as a map. For example, when frictional sliding dominates the composite fracture process, imposing $\overline{S} = \sigma_S/f$ onto Eq 3 gives a transition condition, as illustrated in Fig. 5.

Transformation Toughening

The stress-induced transformations that can cause significant toughening include martensitic [6,7] and ferroelastic [8] transformations, as well as twinning. The former involves both

FIG. 4—*Steady-state cracking in ceramic matrix fiber composite.*

dilatational and shear components of the transformation strain, while the latter typically have only a shear component. Martensitic transformation toughening in zirconium dioxide (ZrO_2) has been most extensively investigated and will be given primary emphasis.

At the simplest level, transformation toughening can be regarded as a process dominated by the dilatational component of the stress free strain, e^T_{ii}, as expressed by the associated

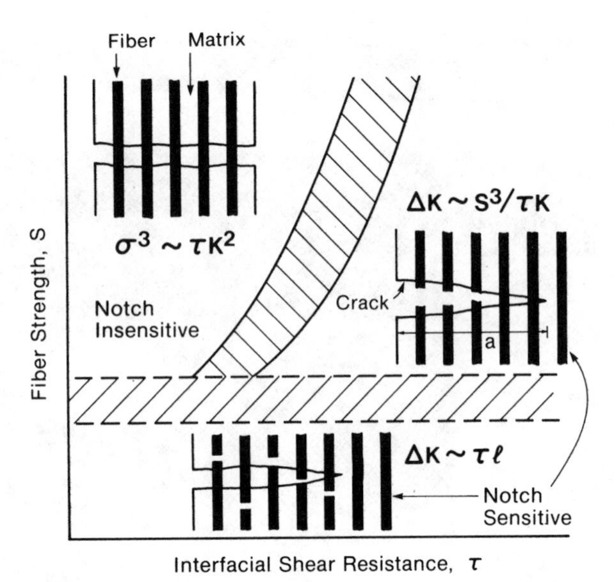

FIG. 5—*Transition from steady-state cracking to toughness controlled fracture in a composite dominated by interfacial sliding resistance.*

supercritical[2] stress-strain curve. Then, based on the path independence of the J-integral, initial crack growth occurs without toughening (Fig. 2). However, upon crack extension, J becomes path dependent, and toughening develops. For steady-state conditions, the supercritical toughening is readily derived as [*6,9–11*]

$$\Delta G_c = 2hf \int \sigma \, d\epsilon \tag{5}$$

where the integral represents the hysteresis area depicted in Fig. 1, such that (see Eq 1)

$$\Delta G_c = 2hf\sigma_0 e_{ii}{}^T \tag{6}$$

This steady-state level is attained after crack extensions of ~5h [*9*]. A directly equivalent result for the increase in critical stress intensity factor, ΔK_c, can be derived by considering the residual stress field created by transformation, giving [*9*]

$$\Delta K_c = \lambda E e_{ii}{}^T f\sqrt{h}/(1 - \upsilon) \tag{7}$$

where E is the composite modulus and λ is a coefficient equal to 0.22.

Comparison with experimental data [*6*] (Fig. 6) has revealed that Eq 6 consistently underestimates the toughness, because shear effects and zone widening have not been incorporated. One hypothesis regarding the shear strain that seemingly coincides with existing observations and measurements involves nonassociated flow [*6*]. Specifically, it is presumed that the shear stress dominates the nucleation of the transformation, because of the large transformation shear strain associated with the nucleus, but that the residual particle strain

[2] *Supercritical* refers to the condition wherein all particles within the process zone fully transform.

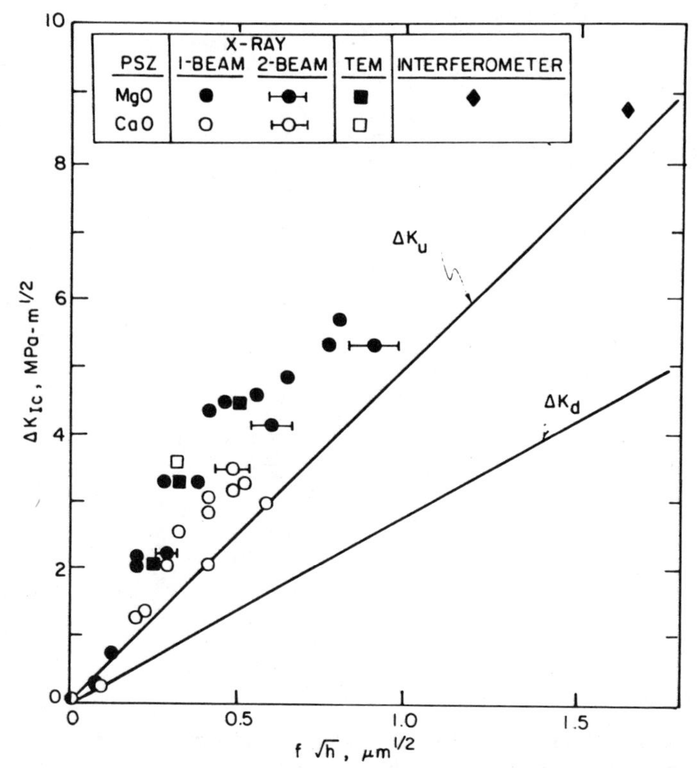

FIG. 6—*Comparison between theory and experiment for various PSZ materials:* ΔK_u *refers to a zone shape dictated by the equivalent stress, whereas* ΔK_d *refers to a zone shape governed by the mean stress.*

is predominantly dilatational, because extensive twinning eliminates long range shear strains [*12*]. This premise results in the zone profile, in plane strain, depicted in Fig. 7 which, by virtue of the absence of the transformed material ahead of the crack, predicts a toughening that exceeds Eq 6 and, furthermore, agrees well with experimental data (Fig. 6). Such zone profiles are also consistent with available observations [*13,14*]. While the above notions offer a self-consistent explanation of the transformation toughening in some cases, it is emphasized that alternative models of the influence of shear strains must be used to address other transformation problems, especially twin and ferroelastic toughening.

The zone size h represents the major microstructural influence on toughness. Clearly, h is governed by a nucleation law [*15*]. However, a fully validated law for nucleation does not yet exist, because the nucleation sequence has not been established. Consequently, connections between h and the microstructure still cannot be specified. Nevertheless, certain trends are apparent, based on the free energy of the fully transformed product [*16,17*]. Specifically, h invariably decreases with increase in temperature and decrease in particle size. *A temperature and particle size dependent toughness is thus inevitable for this mechanism.*

Finally, it should be noted that h is not usually unique, but varies with crack extension, especially during initial growth [*9,18,19*]. This behavior leads to resistance curves that increase over larger crack extension lengths than expected from the constant h analysis [*20*]. The full simulation of this process has yet to be performed.

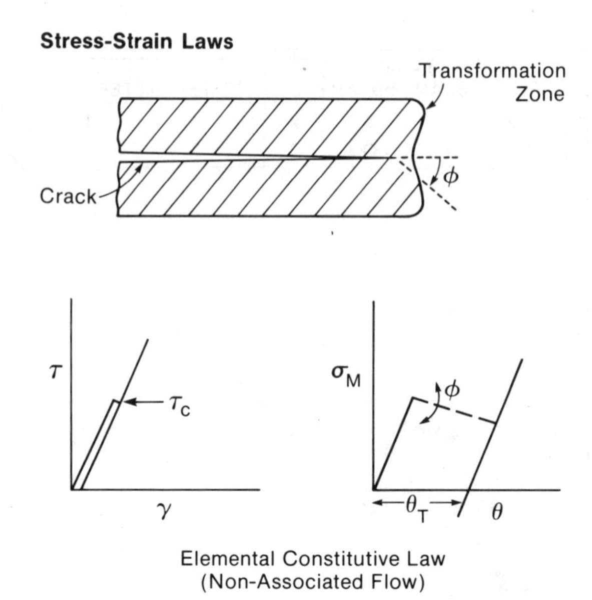

FIG. 7—*Zone profile influenced by shear and nonassociated flow characteristics.*

Microcrack Toughening

The phenomenon of microcrack toughening has been postulated for over a decade [*21–23*], and indeed, a range of materials exhibits trends in toughness with particle size, temperature, etc., qualitatively consistent with this mechanism. However, as yet there is only one fully validated example of this mechanism: aluminum oxide (Al_2O_3) toughened with monoclinic zirconium dioxide (ZrO_2) [*24*]. The fundamental premise concerning the mechanism is depicted in Fig. 8. Microcracks occur within regions of local residual tension, caused by thermal expansion mismatch or by transformation [*25,26*]. The microcracks locally relieve the residual tension and thus cause a dilatation governed by the volume displaced by the microcrack. Furthermore, the microcracks reduce the elastic modulus within the microcrack process zone. Consequently, the elemental stress-strain curve for a microcracking solid has the form depicted in Fig. 8. The hysteresis dictated by this curve, when the microcracks are activated by the passage of a macrocrack, contributes to the change in toughness, as elaborated below. However, this contribution is partially counteracted by a degradation of the material ahead of the microcrack. The extent of the degradation is presently unknown.

The crack shielding can be conveniently separated into dilatational and modulus contributions. The former depends on the process zone size and shape, while the latter depends only on the zone shape. The *dilatational* contribution ΔK_d has essentially the same form as Eq 7. Notably for steady-state supercritical conditions [*24,26*],

$$\Delta K_d = -\lambda E f \theta^T \sqrt{h} \tag{8}$$

where θ^T is the volume strain, as governed by the microcrack shape and the prior residual tension (Table 2). The steady-state *modulus* contribution ΔK_m is [*26*]

$$(1 - \upsilon)\Delta K_m / K_c = (k_1 - 5/8)(\mu/\bar{\mu} - 1) + (k_2 + 3/4)(\bar{\upsilon}\mu/\bar{\mu} - \upsilon) \tag{9}$$

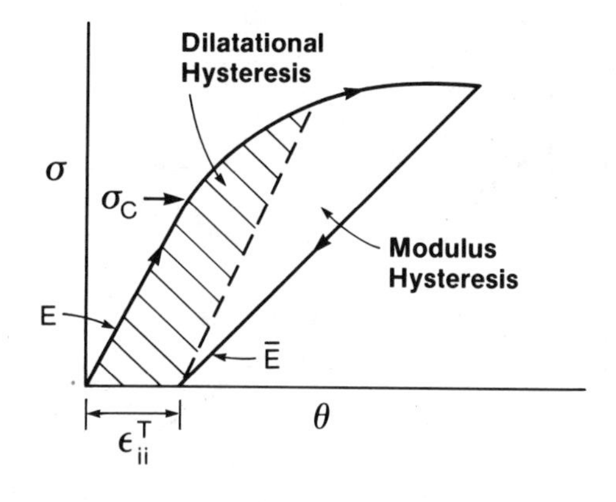

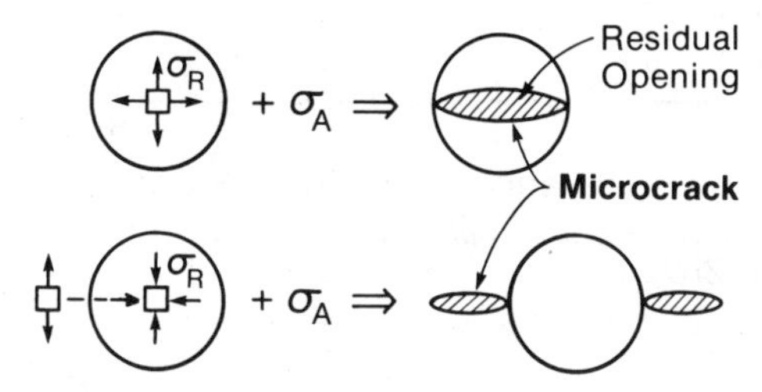

FIG. 8—*Basic concepts of microcrack toughening.*

where the bar refers to microcracked material,

$k_1 \approx 0.017$,
$k_2 \approx -0.043$,
μ = shear modulus, and
K_c = *composite* toughness.

Some results for $\mu/\bar{\mu}$ are presented in Table 2. The contributions ΔK_d and ΔK_m are additive when both are small compared with K_c. Otherwise, interaction effects occur and numerical procedures are then needed to determine the change in toughness [27].

Comparison between theory and experiment has been made for the Al_2O_3/ZrO_2 system

TABLE 2—*Some typical microcrack parameters.*[a]

Crack Morphology	Volume per Microcrack ΔV	Modulus Ratio $\mu/\bar{\mu}$
Circular microcrack	$(16/3)b^3(1 - \upsilon^2)\sigma/E$	$1 + (32/45)(1 - \upsilon)(5 - \upsilon)\epsilon/(2 - \upsilon)$
Annular microcrack: intact interface	$(\pi^2/2)(1 - \upsilon^2)ab^2(1 - b/a)^2\sigma/E$	$1 + (32/45)(1 - \upsilon)(5 - \upsilon)\epsilon/(2 - \upsilon)$
Annular microcrack: failed interface	$\sim 3.6cb\delta$	$[1 - \epsilon(5.8 - 8c/b)]^{-1}$

[a] b is the particle radius, a the crack radius, $c = a - b$, σ the residual stress in the particle, ϵ the microcrack density, and δ the microcrack opening at the particle interface.

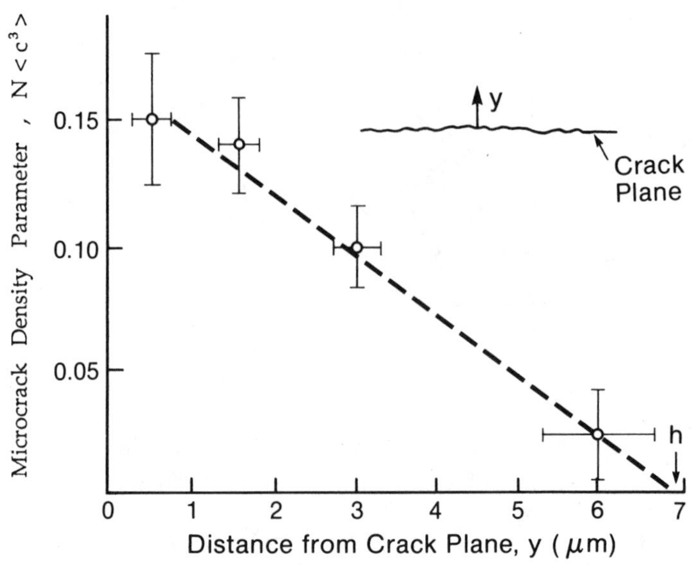

FIG. 9—*Trends in microcrack density with distance from crack surface for a ZTA material with predominantly monoclinic* ZrO_2.

[*24*], based on the microcrack density measurements presented in Fig. 9, (obtained using electron microscopy), and the θ^T results for annular microcracks with interface opening, noted in Table 2. For this case, the contributions to toughening from the dilatation and the modulus are evaluated as ~2.5 and 5 MPa$\sqrt{m}$ respectively, compared with a measured composite toughness of ~6 MPa$\sqrt{m}$. While this comparison is reasonable and validates microcracking as the prevalent toughening mechanism, present understanding of microcrack toughening is, nevertheless, incomplete. In particular, interaction effects between the modulus and dilatational contribution, as well as near-tip degradation effects require appreciable further investigation.

Because of the experimental difficulty involved in detecting microcracks,[3] there are no other validated examples of microcrack toughening. However, it is noted that several particulate reinforced systems exist wherein enhanced toughness only develops when the particulates exceed a critical size. Microcracking is a likely contribution to the toughening in several of these systems.

One potentially detrimental feature of microcrack toughening is the incidence of thermal microcracks at the largest particles in the distribution [*24*]. Such cracks can be strength limiting, resulting in material that is relatively tough but has only moderate strength [*28*]. Avoidance of such strength-limiting cracks requires stringent control of the size distribution of the reinforcing particles, just beneath the critical size for thermal microcracking.

Finally, it is noted that while the dilatational contribution to microcrack toughening would usually be temperature dependent because of the reduction in residual stress with increase in temperature, the modulus contribution is temperature invariant, subject to the process zone being relatively large compared with the microcrack spacing. Microcrack toughening is, typically, less potent than the transformation toughening.

[3] Typical residual crack openings are <2 nm.

Ductile Phase Toughening

Two sources of toughening are governed by the plasticity of ductile reinforcements. Particles intercepted by the crack, when bonded to the matrix, could exhibit extensive plastic stretching in the crack wake [*29,30*] (Fig. 3) and contribute to the toughness by inhibiting crack opening. Simultaneously, plastic straining of particles in a process zone causes crack shielding (Fig. 3). The former mechanism has the general characteristics that high toughness is encouraged by a large value of the product of the work of rupture for the particles and particle size [*29,30*]. By contrast, the crack-shielding process tends to become larger for ductile particles having small size and low yield strength [*31*].

An estimate of the shielding has been derived by considering the residual strain in the crack wake caused by plastic distortion of ductile enclaves in an elastic matrix, giving [*30,31*]

$$\Delta K/K_c = -(5/6\pi)f[\ln(r_p/y_0) - 2] \tag{10}$$

where r_p is the width of the plastic zone and y_0 is the distance to the first particles. Significant values of $\Delta K/K_c$ only obtain when $r_p/y_0 < 10^{-6}$. Consequently, since r_p increases as the yield strength Y decreases and y_0 decreases as the particle size R decreases, a combination of low yield strength and small particle size is needed to develop an appreciable shielding contribution to the toughness. Typically, Y must be less than ~10 MPa and R less than ~1 μm to achieve significant toughening.

The contribution to the toughness from the stretching of particles between crack surfaces is explicitly governed by the stress/opening curve (Eq 2), as depicted in Fig. 10, such that [*30*]

$$\Delta G_c = \chi(n)fYr^* + \alpha\sigma_R fu^* \tag{11}$$

where r^* is either the reinforcement radius (when the particle necks to a point, Fig. 11*a*) or the half spacing between voids (when rupture is terminated by hole formation, Fig. 11*b*),

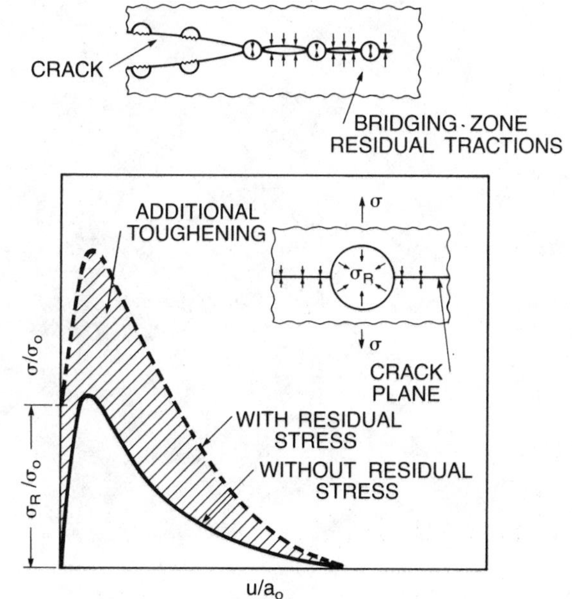

FIG. 10—*Stress, crack opening relation for a crack bridged by a ductile particle.*

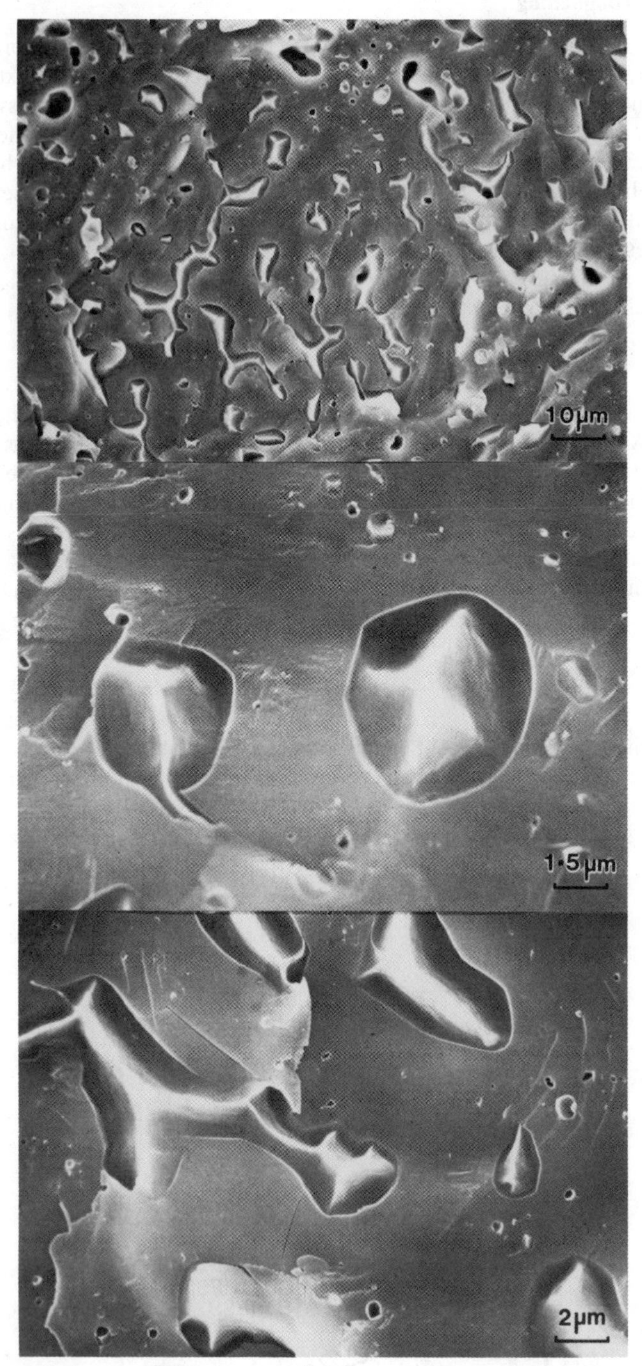

FIG. 11a—*Ductile rupture in ceramic/metal composites: Al_2O_3/Al revealing that the aluminum fails by necking to a point.*

FIG. 11*b*—*Ductile rupture in ceramic/metal composite: WC/Co with cobalt failure occurring by hole nucleation and growth.*

σ_R is the residual tension in the metal phase, χ is the function plotted in Fig. 12, with n being the work hardening rate.

Comparison between theory and experiment is presently limited to two composite systems: aluminum oxide/aluminum and tungsten carbide/cobalt (Al_2O_3/Al and WC/Co). For these materials, appreciable bridging zones are observed (Fig. 13). However, comparison with Eq 11 reveals that the measured toughness is underestimated when interface debonding is neglected.

It is useful to note that zone size measurements provide a self-consistent check regarding the role of bridging in toughness [*32*], as demonstrated by data for several composite systems [*30*] (Fig. 14).

Finally, it is evident that ductile reinforcement toughening should be temperature sensitive, because of the temperature dependence of both the yield strength and the residual stress.

Fiber and Whisker Toughening

The toughening of ceramics by brittle fibers or whiskers or both occurs subject to debonding at the interface, as depicted in Fig. 15. In the absence of debonding, because the fiber and matrix typically have comparable toughness, the composite is brittle and satisfies a rule of mixtures (Fig. 15). Debonding reduces the amplitude of the singularity at the fiber and, when sufficiently extensive, allows the crack to circumvent the fiber, leaving the fiber intact in the crack wake. The intact fiber inhibits crack opening and allows a composite toughness exceeding that of either constituent (Fig. 15). The overwhelmingly important issue in fiber/whiskers toughening thus concerns the extent of debonding, its dependence on interface properties, and its effect on crack opening and fiber failure.

The full debonding and fiber fracture problem has several different aspects that eventually require resolution. These include the characteristics of interface fracture properties and associated requirements for the initiation and extent of debonding at the matrix crack tip, as well as the influence of debond length on the crack opening force and the relationship

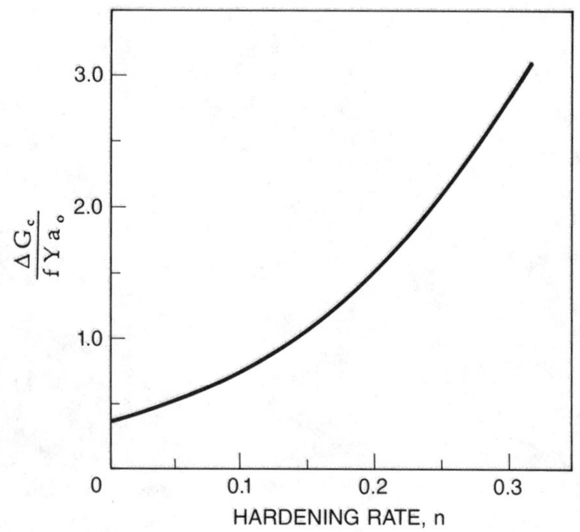

FIG. 12—*Trends in toughness with hardening rate in the* absence *of residual stress: with* $\Delta G_c/fYa_0$ *being equivalent to* χ(n) *in Eq 11.*

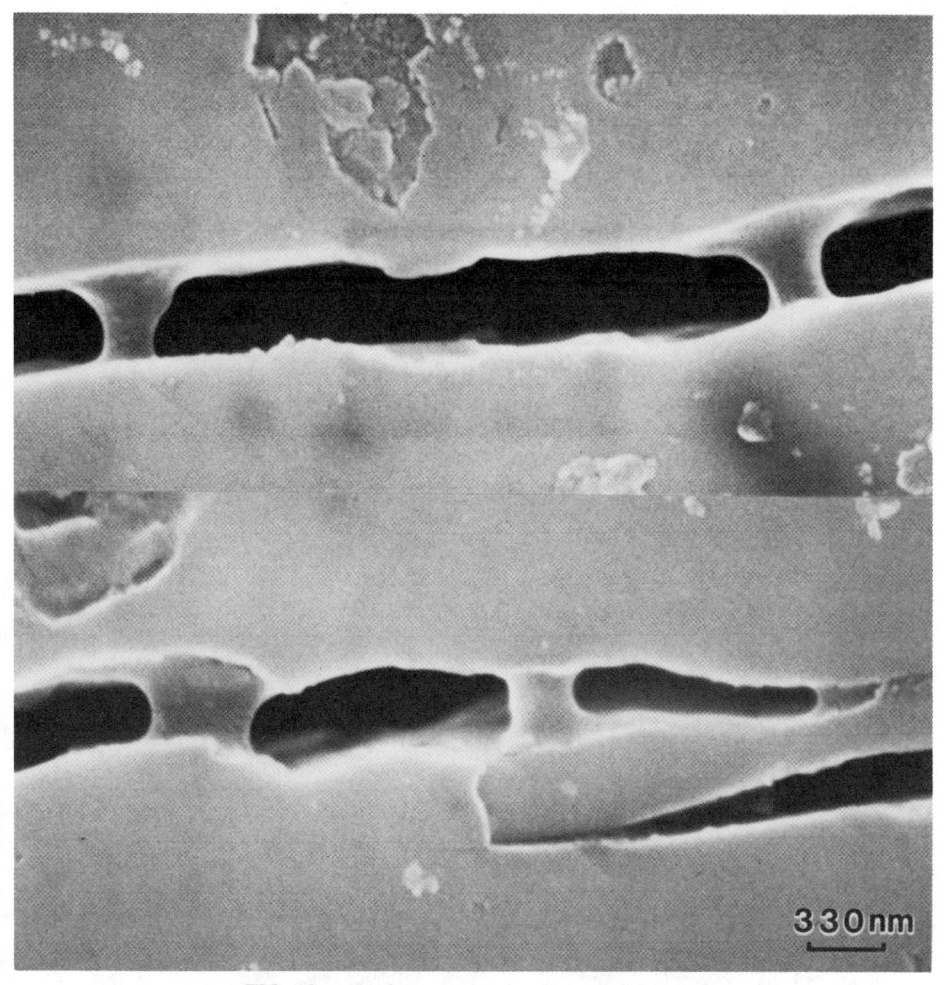

FIG. 13*a*—*Bridging zones observed in* Al_2O_3/Al.

between debond length and the fiber fracture force. While some progress has been made toward understanding these problems, several important questions remain. The present status is briefly reviewed.

Initial Debonding

Even though debonding at the matrix crack tip is a prerequisite to fiber toughening, there is minimal quantitative understanding of the phenomenon. Two mechanics solutions have relevance: one for forked or kinked cracks [*33*] and the other for debonding ahead of a crack tip [*3*]. The solution to the former clearly indicates that debonding is a mixed-mode problem (Fig. 16). For very short debonds, $K_I \approx K_{II} \approx 0.4K_\infty$, implying that debonding would initiate when the interface fracture energy is about a quarter that of the fibers. However, further debonding becomes increasingly difficult. The extent of the difficulty is

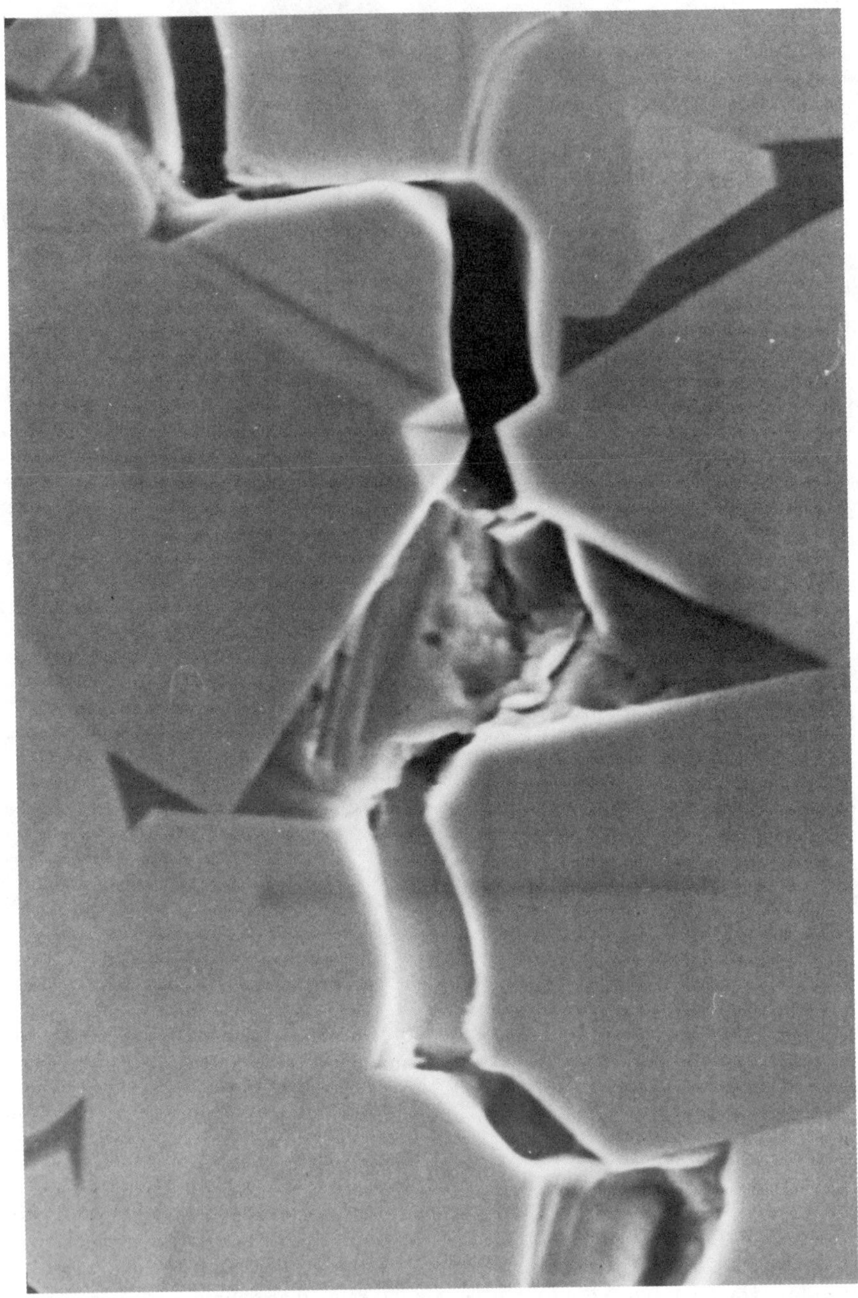

FIG. 13*b*—*Bridging zones observed in WC/Co.*

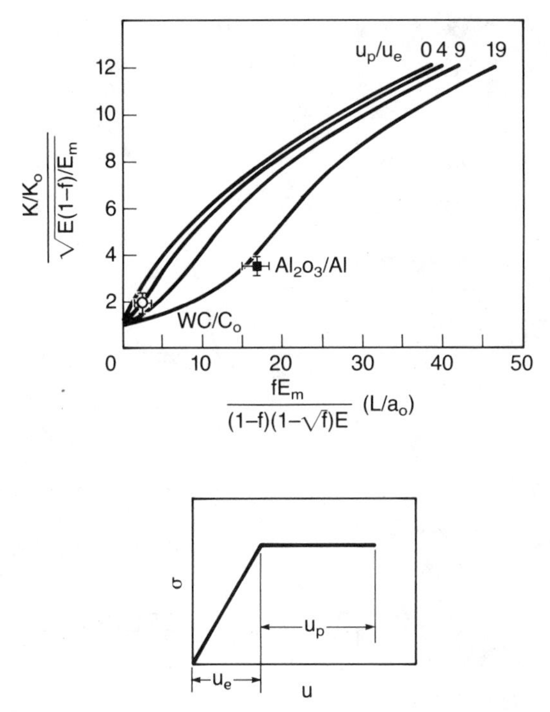

FIG. 14—*Relations between toughness and zone size:* u_e *is elastic displacement of particle and* u_p *the plastic displacement before failure.*

estimated from the solution for debonding in advance of the crack tip. This solution reveals that, in order to achieve appreciable debonding, the interface fracture resistance should be about an order of magnitude less than that of the matrix (Fig. 17).

Further experimental and theoretical study of this important phenomenon would greatly facilitate the design of interfaces that allow optimal debonding.

Crack Opening

Preliminary results have been obtained for the crack opening forces associated with intact fibers normal to the crack plane. For the extreme condition of extensive debonding, with opening inhibited by friction, a shear lag analysis [*2,3,34*] indicates that the initial segment is linear and represents behavior wherein the interfacial shear stress is less than the sliding stress, τ, such that the stress on the fiber end is

$$\sigma = [\ell f E_f(1 + \xi)/R]u \tag{12}$$

where, $\xi = E_f f/(1 - f)E_m$. At larger stresses, $\sigma > [2\tau f E_f(1 + \xi)/\ell]$, the behavior is dominated by sliding and is given by

$$\sigma = 2f(1 + \xi)[E_f\tau/R]^{1/2}[u - \tau R/E_f]^{1/2} \tag{13}$$

For debonds without friction, it is noted that a steady-state condition exists when the debond length $\ell > 2R$, wherein the stress intensities and strain energy release rate for the

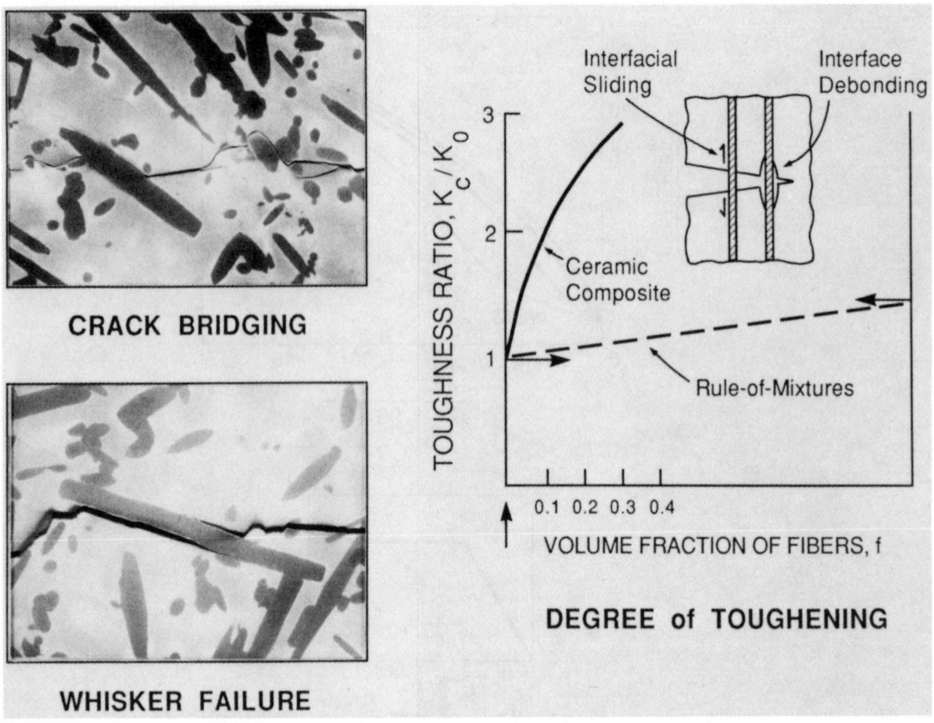

FIG. 15—*Role of debonding in toughening: SiC whisker-toughened alumina.*

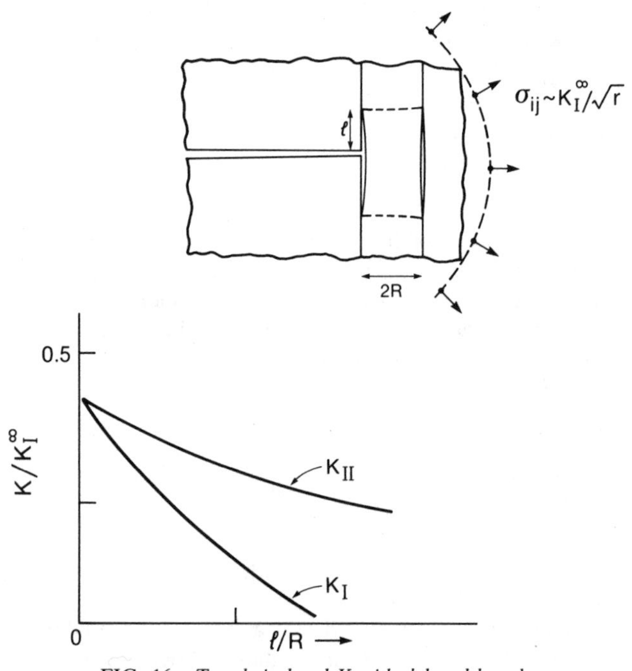

FIG. 16—*Trends in local K with debond length.*

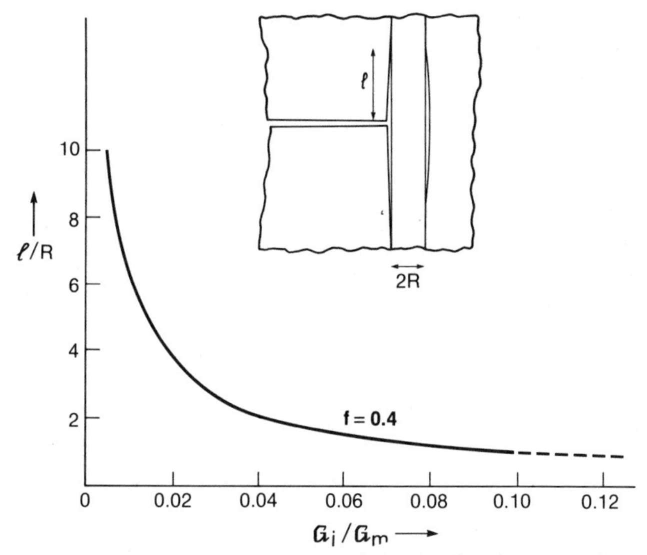

FIG. 17—*Trends in debond length with relative interface fracture resistance.*

debond become independent of ℓ [35] (Fig. 18). For fibers and matrix having the same elastic properties, the steady-state quantities are

$$K_{I} \approx 0.11\bar{\sigma}\sqrt{R}$$

$$K_{II} \approx 0.46\bar{\sigma}\sqrt{R}$$

$$G \approx 0.21\bar{\sigma}^{2}R/E \tag{14}$$

where $\bar{\sigma}$ is the stress on the fiber between the crack surfaces. The coefficients vary somewhat with modulus ratio. Steady-state behavior dictates that the crack opening, u, varies linearly with debond length, such that [36]

$$u = (\sigma/E_{f})\ell\beta(f) \tag{15}$$

where β is a function of the volume fraction of fibers, but is insensitive to the modulus ratio. A solution by finite elements for $f = 0.15$ gives $\beta = 0.95$.

When fibers fail, the above functions are clearly modified and the further opening of the crack often occurs subject to a declining stress, as elaborated below.

Fiber Failure

The process of fiber failure is not yet understood in detail. Available analyses [36,37] have focused on the problem of fracture behind the matrix crack front, neglecting fractures that might occur in the crack-tip field. This approach unquestionably has merit for tough composites. Inclined fibers fail in the wake from the end of the debond zone [38] (Fig. 15) subject to the debond length being small. Conversely, large debond lengths encourage a statistical mode of fracture within the debond zone, as governed by the flaw size distribution in the fiber. Some useful results have been obtained for two extreme possibilities. *Failure at the end of the debond* in the absence of friction is found to be independent of debond length, as evident from the existence of steady-state (Fig. 18). The stress at failure on fibers *normal* to the crack plane has the form

length, as evident from the existence of steady-state (Fig. 18). The stress at failure on fibers *normal* to the crack plane has the form

$$S = (K_f/\sqrt{R})\omega(f, \Sigma) \tag{16}$$

where ω is a function of order 2–6, and K_f is the Mode I fiber toughness. When the stress reaches S, there is an abrupt drop in stress on the fiber.

The statistical mode of fiber failure leads to a fiber failure probability that increases with increase in distance behind the crack tip [37], while the most probable fiber failure location displaces further from the crack surface. The associated stress/displacement curves have sliding and pullout contributions that can be derived in terms of the fracture properties of the fibers and the shear resistance of the interface. For fibers normal to the crack having a flaw strength distribution

$$\int g(S)\, dS = (S/S_0)^m$$

where S_0 is a scale parameter and m a shape parameter, weakest link statistics give the crack opening stresses, plotted in Fig. 19, in terms of the nondimensional qualities σ/Σ and u/υ where, in this case, Σ and υ are defined by

$$\Sigma = [S_0^m \tau(m + 1)/2\pi R^2]^{1/(m+1)} : \upsilon = \Sigma^2 R/4E_f\tau(1 + \xi) \tag{17}$$

When the fibers or whiskers are *inclined* to the crack plane, crack opening interference may persist because of the trajectory of the associated fiber crack (Fig. 20). An enhanced declining segment of the $\sigma(u)$ function may then obtain.

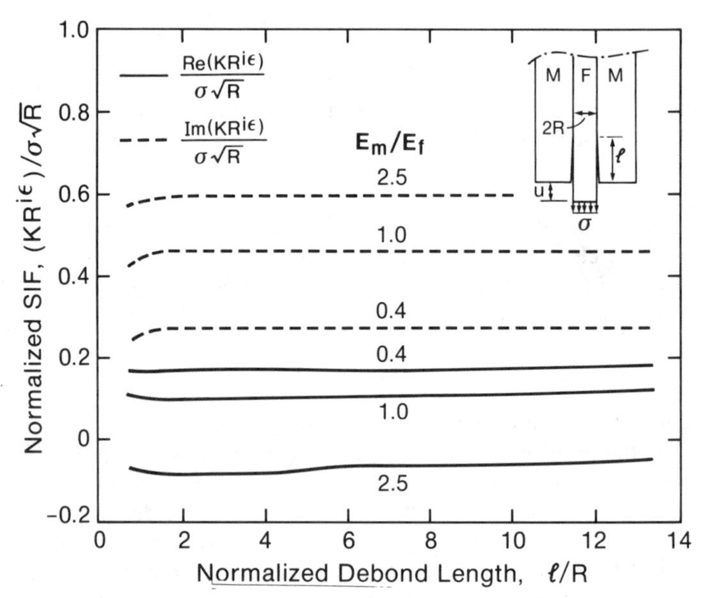

FIG. 18—*Effects of debond length on nondimensional stress intensities at debond tip.*

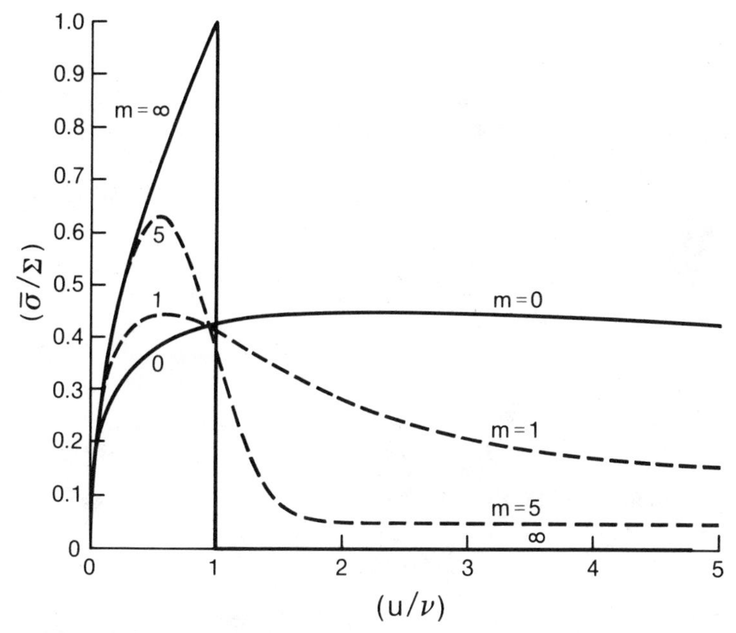

FIG. 19—*Stress, crack opening functions for statistical fiber failure with friction.*

The Toughening

Trends in toughening can be deduced from the $S(u)$ function, using Eq 2. However, resistance curve and steady-state cracking effects must be considered as important related effects. Perhaps the most straightforward case involves the whisker reinforced systems, Al_2O_3/SiC and Si_3N_4/SiC. For these materials, limited debonding along an amorphous interphase, without friction, seems to contribute to the toughness. Consequently, for aligned whiskers normal to the crack plane, Eqs 2, 15, and 18 would predict a toughening

$$\Delta G_c E_f / K_f^2 \approx \tfrac{1}{2}\omega^2 \Sigma f(\ell/R)$$

$$\approx 8 f(\ell/R) \tag{18}$$

The toughening is thus extremely sensitive to debonding. Residual stress must also be involved, because bridging contributes importantly to the toughness. However, the fiber fracture condition, as well as the initial crack opening force, is influenced by the residual field. Analysis of this problem has not yet been conducted, and hence, trends in toughness with residual stress are presently unknown.

Composites that exhibit extensive debonding and frictional sliding typically have pull-out contributions to the toughness [*37,39*]. The associated $\sigma(u)$ functions (Fig. 19) may simply be integrated to predict the *asymptotic* toughness, using Eq 2. Frequently, however, the mean pullout length exceeds the crack mouth opening (except for very long cracks), indicating that asymptotic behavior does not occur and instead, fracture is governed by a rising resistance curve. Simulation of trends in toughness with crack extension has yet to be conducted.

Comparisons between theory and experiment for fiber toughening are limited to a zone

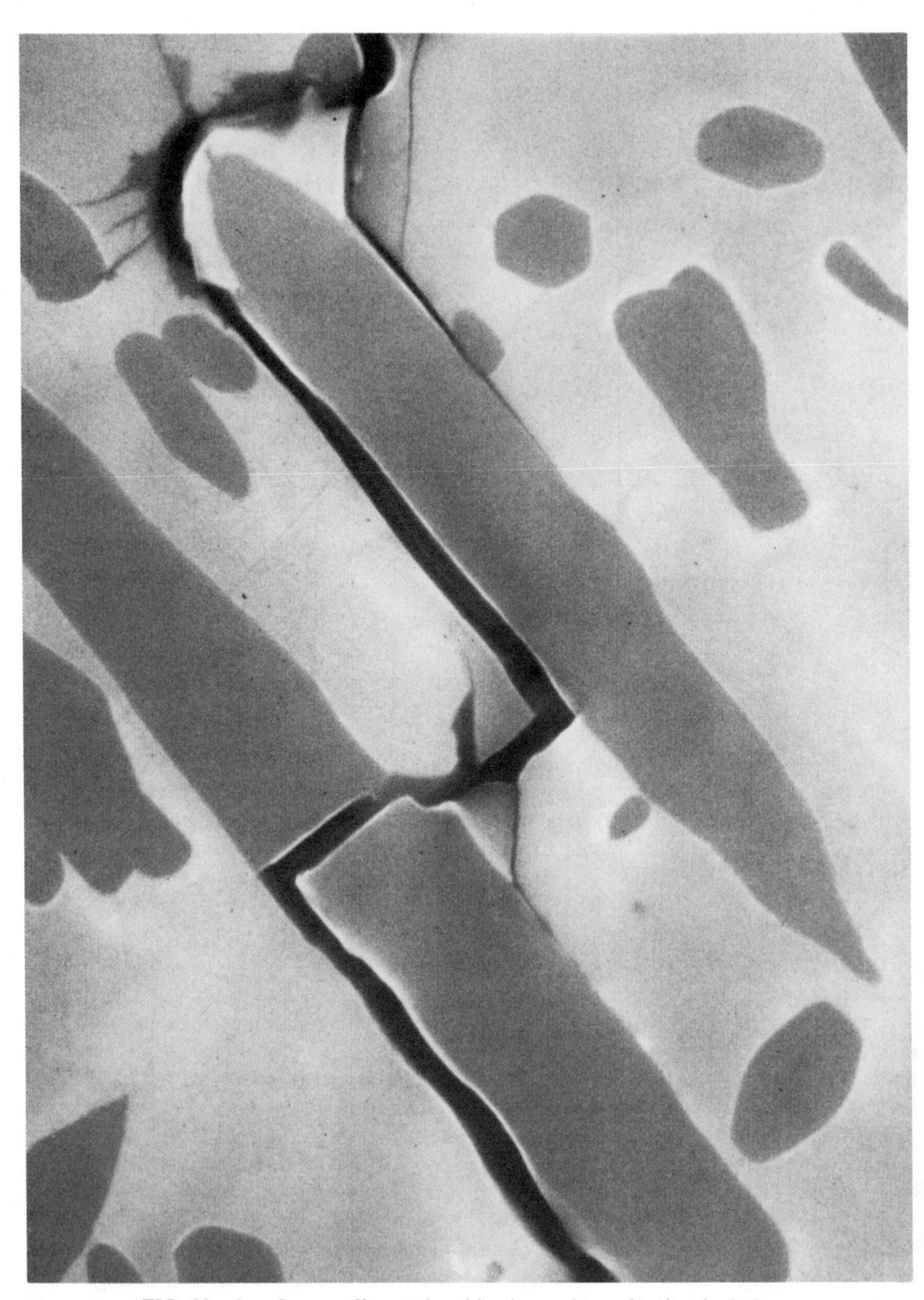

FIG. 20—*Interference effects induced by the cracking of inclined whiskers.*

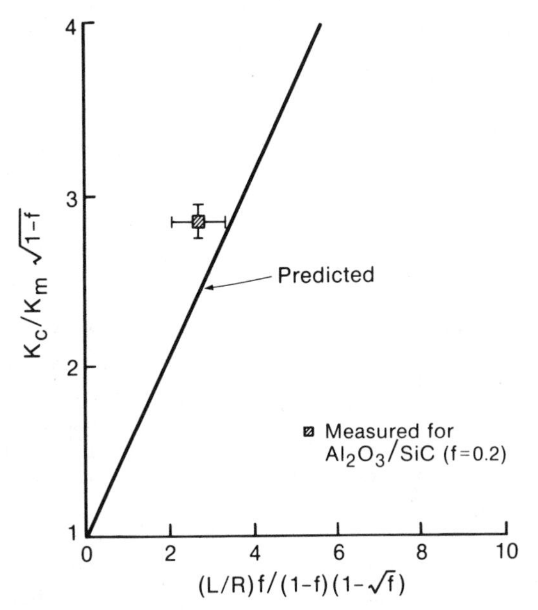

FIG. 21—*Relation between zone size and toughness for elastic reinforcements, with data for* Al_2O_3 *reinforced using SiC whiskers.*

size comparison for the Al_2O_3/SiC system [*36*] (Fig. 21). Recalling that zone size comparisons are insensitive to material properties, it is emphasized that the relatively good correspondence indicates only that bridging contributes importantly to the enhanced toughness. Toughness comparisons based on Eq 19 have yet to be conducted, although inspection indicates that the magnitude is reasonable, given that $\ell/R \sim 8$ for this material (see Fig. 21).

One important feature of fiber and whisker toughening is the absence of an obvious temperature dependent parameter, except for possible temperature effects on the debond resistance. This toughening approach thus seems to have the greatest potential for toughness at elevated temperature.

Other Mechanisms

A number of ceramic materials have a contribution to toughness from intact grains along the crack surface [*40*]. The phenomenon is evident in several large grained polycrystals, most notably Al_2O_3, and in certain glass ceramics. The contribution of these intact grains to toughness is formally similar to whisker toughening. However, the corresponding $\sigma(u)$ law and the analogous criteria that govern grain intactness are not yet well understood. Nevertheless, by direct analogy with whiskers toughening, grain boundaries weakened by additives have been demonstrated [*41*] to encourage bridging effects and, thus, enhance toughness. The ability of certain grains to remain intact probably involves residual stresses, with associated grain size effects, perhaps coupled with statistical variability of the grain boundary fracture resistance.

Synergism

When bridging and process zone mechanisms occur in conjunction, synergistic toughening becomes possible [*42,43*]. The basic concept is illustrated in Fig. 3. Specifically, the tractions exerted on the crack surface by the intact particles can locally widen the process zone. This zone widening, in turn, enhances the tip shielding above the level expected in the absence of bridging. The net effect of both processes on toughness is larger than the sum of the individual mechanisms. Indeed, cases exist wherein the toughening is multiplicative [*42,43*]. This intriguing topic is expected to be the subject of appreciable future research.

Future Research

A number of substantial issues remain to be addressed, particularly for fiber- and whisker-toughened materials. Most importantly, since the toughening is strongly influenced by the debond length, the development of relationships between this length and the relative properties of the interface is critical to the prediction of microstructures that provide optimum toughness. The complexity of this problem would suggest a coupled experimental/mechanics study. A related issue concerns the identity of interphases that have low (mixed mode) debond resistances that also remain stable at elevated temperature in oxidizing environments. Presently, boron nitride, silicon oxide, and carbon are used as interphases, but each has limitations at elevated temperatures in air.

In randomly oriented whisker composites, trends in properties with whisker inclination require analysis. Specifically, the effects of inclination on debonding, compliance, and crack surface interference are presently unknown. Again, a coupled experimental/mechanics study that elucidates orientation effects is envisaged.

Finally, potential synergistic effects between process and bridging zone mechanisms merit detailed investigation, from both analytical and experimental perspectives.

References

[*1*] Aveston, J., Cooper, G. A., and Kelly, A. in *Proceedings,* National Physical Laboratory Conference, IPC Science and Technology Press, 1971, p. 15.
[*2*] Marshall, D. B., Cox, B. N., and Evans, A. G., *Acta Metallurgica,* Vol. 33, 1985, p. 2013.
[*3*] Budiansky, B., Hutchinson, J. W., and Evans, A. G., *Journal of the Mechanics and Physics of Solids,* Vol. 2, 1986, p. 167.
[*4*] Prewo, K. and Brennan, J. J., *Journal of Materials Science,* Vol. 15, 1980, p. 463.
[*5*] Marshall, D. B. and Evans, A. G., *Journal of the American Ceramic Society,* Vol. 68, 1985, p. 225.
[*6*] Evans, A. G. and Cannon, R. M., *Acta Metallurgica,* Vol. 34, 1986, p. 761.
[*7*] *Advances in Ceramics,* N. Claussen, M. Rühle, and A. H. Heuer, Eds., Vol. 12, 1985.
[*8*] Virkar, A. and Matsumoto, R., *Journal of the American Ceramic Society,* Vol. 69, 1986, p. 224.
[*9*] McMeeking, R. and Evans, A. G., *Journal of the American Ceramic Society,* Vol. 65, 1982, p. 242.
[*10*] Marshall, D. B., Drory, M. D., and Evans, A. G., *Fracture Mechanics of Ceramics,* Vol. 5, 1983, p. 289.
[*11*] Budiansky, B., Hutchinson, J. W., and Lambropolous, J., *International Journal of Solids and Structures,* Vol. 19, 1983, p. 337.
[*12*] Rühle, M. and Kriven, W. M., *Solid-State Phase Transformations,* H. Aaronson, D. Laughlin, R. F. Sekerka, and C. M. Wayman, Eds., American Institute for Mining, Metallurgical, and Petroleum Engineers, 1982.
[*13*] Swain, M. V. and Rose, L. R. F., *Journal of the American Ceramic Society,* Vol. 69, 1986, p. 511.
[*14*] Rühle, M., Kraus, B., Strecker, A., and Waidelich, D., *Advances in Ceramics,* Vol. 12, 1985, p. 256.
[*15*] Rühle, M. and Heuber, A. H., *Advances in Ceramics,* Vol. 12, 1985, p. 14.
[*16*] Evans, A. G., Drory, M. D., Burlingame, N., and Kriven, W., *Acta Metallurgica,* Vol. 29, 1981, p. 447.
[*17*] Lange, F. F., *Journal of Materials Science,* Vol. 17, 1982, p. 225.
[*18*] Marshall, D. B., *Journal of the American Ceramic Society,* Vol. 69, 1986, p. 173.

[19] Heuer, A. H., *Journal of the American Ceramic Society,* Vol. 70, 1987, p. 689.
[20] Swain, M. V. and Hannick, R. J. H., *Advances in Ceramics,* Vol. 12, 1985, p. 225.
[21] Hoagland, R. G., Embury, J. D., and Green, D. J., *Scripta Metallurgica,* Vol. 9, 1975, p. 907.
[22] Claussen, N., Steeb, J., and Pabst, R. F., *American Ceramic Society Bulletin,* Vol. 56, 1977, p. 559.
[23] Evans, A. G., *Scripta Metallurgica,* Vol. 10, 1976, p. 93.
[24] Rühle, M., Evans, A. G., Charalambides, P. G., McMeeking, R. M., and Hutchinson, J. W., *Acta Metallurgica,* Vol. 35, 1987, p. 2701.
[25] Evans, A. G. and Faber, K. T., *Journal of the American Ceramic Society,* Vol. 67, 1984, p. 255.
[26] Hutchinson, J. W., *Acta Metallurgica,* Vol. 35, 1987, p. 1605.
[27] Charalambides, P. G. and McMeeking, R. M., *Mechanics of Materials,* Vol. 6, 1987, p. 71.
[28] Rühle, M., Claussen, N., and Heuer, A. H., *Journal of the American Ceramic Society,* Vol. 69, 1986, p. 195.
[29] Sigl, L. and Exner, E., *Metallurgical Transactions A,* in press.
[30] Sigl, L., Mataga, P., Dalgleish, B. J., McMeeking, R. M., and Evans, A. G., *Acta Metallurgica,* Vol. 36, 1988, p. 945.
[31] Budiansky, B., unpublished work.
[32] Budiansky, B., Amazigo, J., and Evans, A. G., *Journal of the Mechanics and Physics of Solids,* Vol. 36, 1988, p. 167.
[33] Cotterell, B. and Rice, J. R., *International Journal of Fracture Mechanics,* Vol. 16, 1980, p. 155.
[34] Marshall, D. B., unpublished analysis.
[35] Charalambides, P. G. and Evans, A. G., *Journal of the American Ceramic Society,* in press.
[36] Rühle, M., Dalgleish, B. J., and Evans, A. G., *Scripta Metallurgica,* Vol. 21, 1987, p. 681.
[37] Thouless, M. D. and Evans, A. G., *Acta Metallurgica,* Vol. 36, 1988, p. 517.
[38] Evans, A. G., Rühle, M., and Thouless, M. D., *Proceedings,* Materials Research Society, Vol. 78, 1987, p. 259.
[39] Phillips, D. C., *Journal of Materials Science,* Vol. 7, 1972, p. 1175.
[40] Swanson, P. L., Fairbanks, C. J., Lawn, B. R., Mai, Y. W., and Hockey, B. J., *Journal of the American Ceramic Society,* Vol. 70, 1987, p. 289.
[41] Fu, Y. and Evans, A. G., *Acta Metallurgica,* Vol. 30, 1982, p. 1619.
[42] Budiansky, B., *Micromechanics II, Proceedings,* 10th U.S. Congress of Applied Mechanics, 1986.
[43] Evans, A. G., Beaumont, P. W. R., Ahmad, Z., and Gilbert, D. G., *Acta Metallurgica,* Vol. 34, 1986, p. 79.

PART II
Contributed Papers

Analytical Fracture Mechanics

Ivatury S. Raju,[1] *Satya N. Atluri,*[2] *and James C. Newman, Jr.*[3]

Stress-Intensity Factors for Small Surface and Corner Cracks in Plates

REFERENCE: Raju, I. S., Atluri, S. N., and Newman, J. C., Jr., **"Stress-Intensity Factors for Small Surface and Corner Cracks in Plates,"** *Fracture Mechanics: Perspectives and Directions (Twentieth Symposium), ASTM STP 1020,* R. P. Wei and R. P. Gangloff, Eds., American Society for Testing and Materials, Philadelphia, 1989, pp. 297–316.

ABSTRACT: Three-dimensional finite-element and finite-element-alternating methods were used to obtain the stress-intensity factors for small surface and corner cracked plates subjected to remote tension and bending loads. The crack-depth-to-crack-length ratios (a/c) ranged from 0.2 to 1, the crack-depth-to-plate-thickness ratios (a/t) from 0.05 to 0.2. The performance of the finite-element alternating method was studied on these crack configurations. A study of the computational effort involved in the finite-element-alternating method showed that several crack configurations can be analyzed with a single rectangular mesh idealization, whereas the conventional finite-element method requires a different mesh for each configuration. The stress-intensity factors obtained with the finite-element-alternating method agreed well (within 5%) with those calculated from the finite-element method with singularity elements.

The stress-intensity factors calculated from the empirical equations proposed by Newman and Raju were generally within 5% of those calculated by the finite-element method. The stress-intensity factors given herein should be useful in predicting crack-growth rates and fracture strengths of surface- and corner-cracked components.

KEY WORDS: crack, elastic analysis, stress-intensity factor, finite-element method, finite-element-alternating method, surface crack, corner crack, tension and bending loads

Surface and corner cracks may occur in many structural components. These cracks initiate near regions of stress concentrations and may cause premature failure of aircraft landing gears, spars, stiffeners, and other components [*1*]. Accurate stress-intensity factor solutions for these components are needed for reliable prediction of crack-growth rates and fracture strengths.

Most of the life of these cracked components is spent when the cracks are small. Also, many applications of damage tolerance or durability analyses require the computation of stress-intensity factors for small cracks. Previous analyses of surface- and corner-crack configurations, using three-dimensional finite-element analyses [*2–4*], boundary-integral equation methods [*5*], and alternating methods [*6–8*] have considered crack-depth-to-plate-thickness ratios greater than or equal to 0.2. Engineering judgment or extrapolations were used to estimate stress-intensity factors for small surface and corner cracks [*9,10*]. Therefore, more analyses are needed to verify these extrapolations for small cracks. The purpose of this paper is to present stress-intensity factors for a wide range of semi-elliptical surface

[1] Senior scientist, Analytical Services and Materials, Inc., Hampton, VA 23666.

[2] Regents' professor and director, Center for the Advancement of Computational Mechanics, Georgia Institute of Technology, Atlanta, GA 30332.

[3] Senior scientist, Materials Division, NASA Langley Research Center, Hampton, VA 23665-5225.

cracks and quarter-elliptical corner cracks in plates with crack-depth-to-plate-thickness ratios less than 0.2 and to obtain asymptotic values as crack-depth-to-plate thickness ratios approach zero.

Two popular methods to obtain the stress-intensity factors for the surface- and corner-crack configurations are the finite-element method with singularity elements [*2–4*] and the finite-element-alternating method [*11–13*]. In the finite-element method, large numbers of elements are needed with customized modeling near the crack front with singularity elements. Once such models are developed, accurate stress-intensity factors can be obtained [*2–4*]. In contrast, the finite-element-alternating method does not need customized modeling near the crack front. This is because the uncracked solid is analyzed by the finite-element part of the method. The second objective of this paper is to study various types of modeling that could be used and to study the computational efficiency of the method. The stress-intensity factors obtained with the finite-element-alternating method were compared with those from the finite-element method with singularity elements for surface- and corner-crack configurations. The stress-intensity factors obtained by these methods were also compared with values calculated from empirical equations for surface- and corner-cracked plates with crack-depth-to-plate-thickness ratios less than 0.2.

Analysis

Two types of crack configurations, a surface- and corner-cracked plate, as shown in Fig. 1, were analyzed. The three-dimensional finite-element method and finite-element-alternating method were used to obtain the Mode I stress-intensity factors. In these analyses, Poisson's ratio (ν) was assumed to be 0.3.

Loading

Two types of loading were applied to the crack configurations: remote uniform tension and remote out-of-plane bending (bending about the x-axis). The remote uniform tensile stress is S_t in the z-direction, and the remote outer-fiber bending stress is S_b. The bending stress S_b is the outer fiber stress calculated at the origin ($x = y = z = 0$ in Fig. 1) without the crack present.

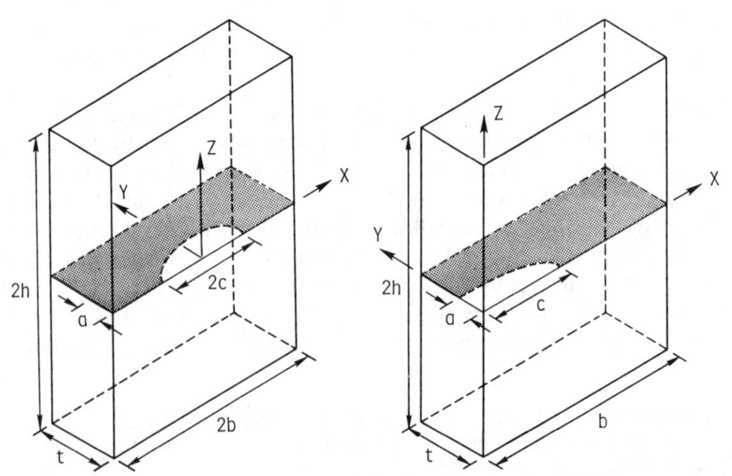

FIG. 1—*Crack configurations analyzed:* (left) *surface crack;* (right) *corner crack.*

Stress-Intensity Factor

The tensile and bending loads cause only Mode I deformations. The Mode I stress-intensity factor K for any point along the crack front was taken to be

$$K = S_i(\pi a/Q)^{1/2} \cdot F(a/t, a/c, \phi) \quad (1)$$

where

subscript i = tension load ($i = t$) or bending load ($i = b$),
a = crack depth,
c = surface length,
t = thickness of the plate,
ϕ = parametric angle of the ellipse, and
Q = shape factor of the ellipse (which is equal to the square of the complete elliptic integral of the second kind).

The half-length of the bar, h, and the width, b (Fig. 1), were chosen large enough ($h/b > 2$ and $b/c > 5$) to have negligible effects on the stress-intensity factors. Values of F, the boundary-correction factor, were calculated along the crack front for various crack shapes (a/c = 0.2 to 1) with a/t values of 0.05, 0.1, and 0.2. The crack dimensions and the parametric angle are defined in Figs. 1 and 2.

Three-Dimensional Finite-Element Method

Figure 3 shows a typical finite-element model for a surface or corner crack in a rectangular plate. The same finite-element model was used to obtain the stress-intensity factors for the surface- and corner-crack configurations. For the surface-crack configuration, symmetric boundary conditions were imposed on the $z = 0$ and $x = 0$ planes, whereas for the corner-crack configuration, symmetric boundary conditions were imposed only on the $z = 0$ plane. The finite-element models employed six-noded, pentahedron, singularity elements at the crack front and eight-noded, hexahedral elements elsewhere. Stress-intensity factors were evaluated using the nodal-force method [2]. Details of the formulation of these types of elements and development of the finite-element models are given in Ref *2* through *4* and are not repeated here.

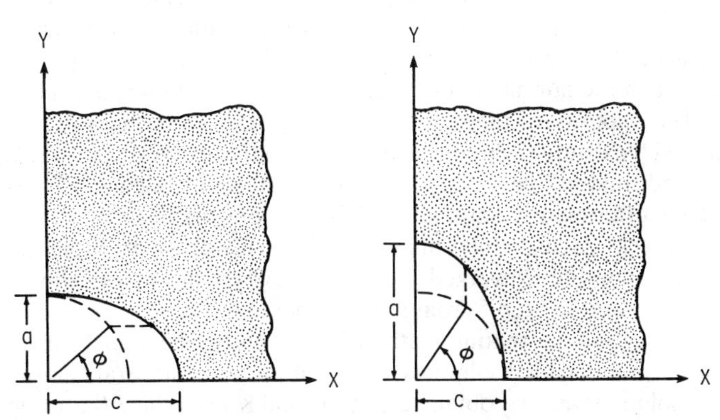

FIG. 2—*Definition of parametric angle:* (left) $a/c \leq 1$; (right) $a/c > 1$.

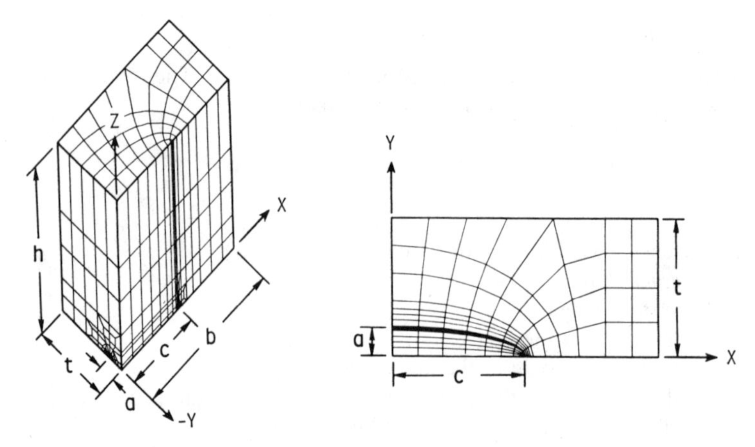

FIG. 3—*Finite-element model for surface- and corner-crack configuration:* (left) *specimen model;* (right) *element pattern on* z = *0 plane.*

Finite-Element-Alternating Method

This method is based on the Schwartz-Neumann alternating technique. The alternating method uses two basic solutions of elasticity and alternates between these two solutions to satisfy the required boundary conditions of the cracked body [*6–8*]. One of the solutions is for the stresses in an uncracked finite solid, and the other is for the stresses in an infinite solid with a crack subjected to arbitrary normal and shear tractions. The solution for an uncracked body may be obtained in several ways, such as the finite-element method or the boundary-element method. In this paper, the three-dimensional finite-element method was used.

The procedure that is followed in the alternating method is summarized in the flow chart in Fig. 4 and is briefly explained here for Mode I problems. First, solve the uncracked solid subjected to the given external loading using the three-dimensional finite-element method (Step 1 in Fig. 4). The finite-element solution gives the stresses everywhere in the solid including the region over which the crack is present (Step 2). The normal stresses acting on the region of the crack need to be erased to satisfy the crack-boundary conditions. The opposite of the stresses calculated in Step 2 are fit to an nth degree polynomial in terms of x- and z-coordinates (Step 4). Due to the polynomial stress distributions obtained in Step 4, calculate the stress-intensity factor [*11*] for the current iteration (Step 5). Use the analytical solution of an embedded elliptic crack in a infinite solid subjected to the polynomial normal traction [*11*] to obtain the normal and tangential stresses on all the external surfaces of the solid (Step 6). The opposite of these stresses on the external surfaces obtained in Step 6 are then considered the externally prescribed stresses on the uncracked solid (Step 7). Again, solve the uncracked solid problem due to the surface tractions calculated in Step 7. This is the start of the next iteration. Continue this iteration process until the normal stresses in the region of the crack are negligibly small or lower than a prescribed tolerance level. The stress-intensity factors in the converged solution are simply the sum of the stress-intensity factors that are computed in Step 5 from all iterations.

The key element in the alternating method is, obviously, the analytical solution for an infinite solid with an embedded elliptical crack subjected to arbitrary normal and shear tractions. Such a solution was first obtained by Shah and Kobayashi [*14*] for tractions normal to the crack surface. However, this solution was limited to a third-degree polynomial function

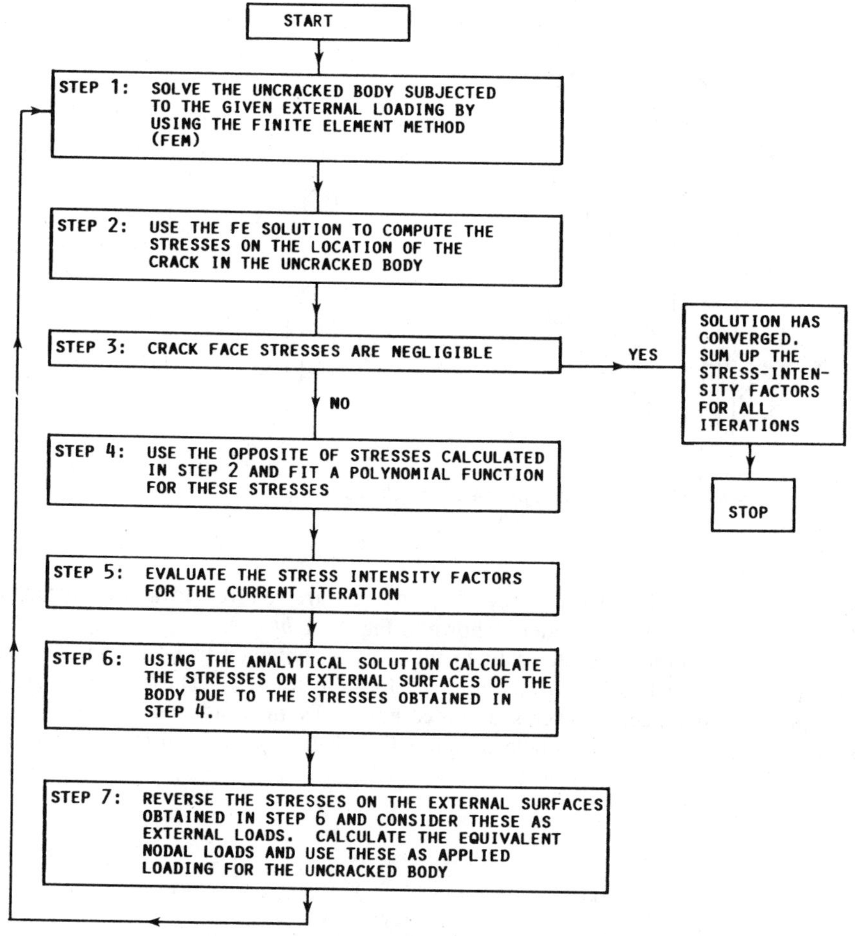

FIG. 4—*Flow chart for finite-element-alternating method.*

in each of the Cartesian coordinates describing the ellipse. Vijayakumar and Atluri [*15*] overcame this limitation and obtained a general solution of arbitrary polynomial order. Nishioka and Atluri [*11–13*] improved and implemented this general solution in a finite-element-alternating method and analyzed surface- and corner-cracked plates. The details of the method are well documented in Refs *11* through *13*, but they are briefly described herein.

In the three-dimensional finite-element solution, 20-noded isoparametric parabolic elements were used to model the uncracked solid. Two types of idealizations were used. In the first type, the idealization was such that the elements on the $z = 0$ plane conform to the shape of the crack in the cracked solid (Fig. 5, *left*). Although the finite-element solution is for the uncracked body, such an idealization is convenient to perform the polynomial fit using the finite-element stresses from the elements that are contained in the region of the crack. The three-dimensional mesh is then generated by simply translating in the z- direction the mesh on the $z = 0$ plane. These models will be referred to as the mapped models. A typical mapped model is shown in Fig. 5, *left*. In the second type, simple rectangular ideal-

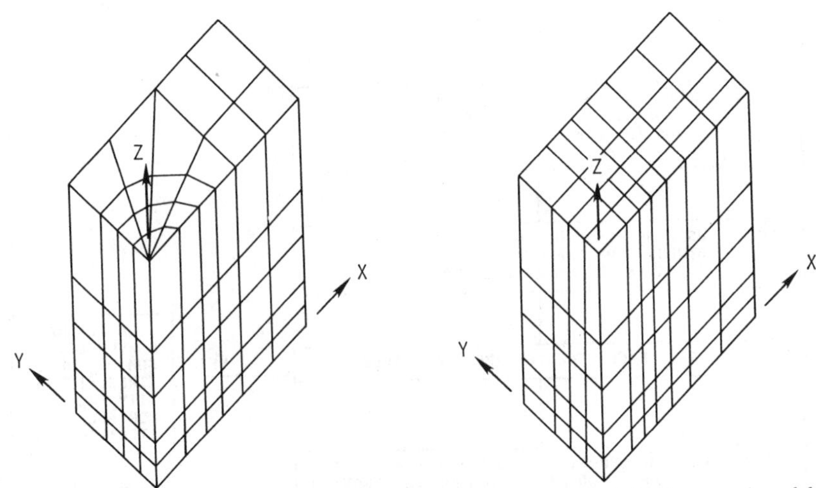

FIG. 5—*Finite-element models used in finite-element-alternating method:* (left) *mapped model;* (right) *rectangular model.*

izations were used to model the solid. These models are referred to as the rectangular models. A typical rectangular model is shown in Fig. 5, *right*.

The alternating method requires a fit to the stresses, obtained from the finite-element solution (of the uncracked body), at the crack location (Step 4). These stresses are the residual pressures that need to be erased. For corner cracks, the residual crack-face pressure distribution, $\sigma^R{}_z$, was assumed to be a complete fifth-degree polynomial in x and y with 21 terms as shown in the Pascal's triangle

$$\begin{array}{ccccccccccc} & & & & & 1 & & & & & \\ & & & & x & & y & & & & \\ & & & x^2 & & xy & & y^2 & & & \\ & & x^3 & & x^2y & & xy^2 & & y^3 & & \\ & x^4 & & x^3y & & x^2y^2 & & xy^3 & & y^4 & \\ x^5 & & x^4y & & x^3y^2 & & x^2y^3 & & xy^4 & & y^5 \end{array}$$

For surface cracks, the residual pressure $\sigma^R{}_z$ had only twelve terms because of symmetry about the y-axis. These twelve terms were obtained by neglecting the terms involving odd powers of x in the fifth-degree polynomial shown in the Pascal's triangle. For mapped models, the residual pressure was fit over the complete region of the crack. For rectangular models, the residual pressure was fit over a rectangular region bounded by the semi-minor and semi-major axes of the crack (see shaded region in Fig. 6).

Because the continuum solution corresponds to that of an embedded elliptic crack in an infinite solid, it is necessary to define the residual stresses not only on the region of the crack but also on the "fictitious" portion of the crack which lies outside of the finite solid. Nishioka and Atluri [*11–13*] suggested the residual-pressure distribution, through numerical experimentation, to be

$$\sigma^R{}_z = \begin{cases} \sigma^R{}_z\,(0, y) & \text{for } x \le 0,\ y \ge 0 \\ \sigma^R{}_z\,(0, 0) & \text{for } x \le 0,\ y \le 0 \\ \sigma^R{}_z\,(x, 0) & \text{for } x \ge 0,\ y \le 0 \end{cases}$$

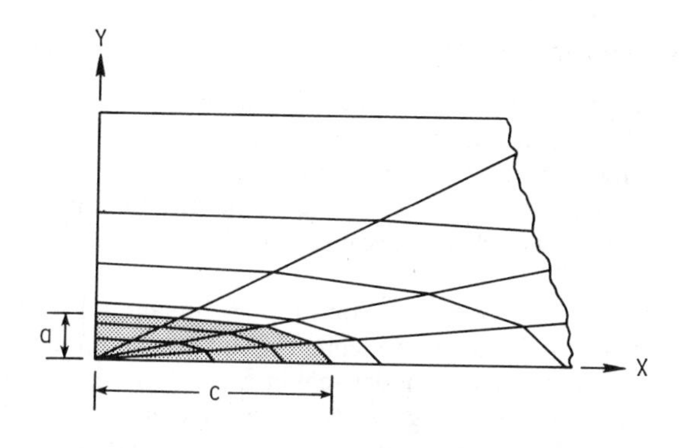

(a) Mapped model.

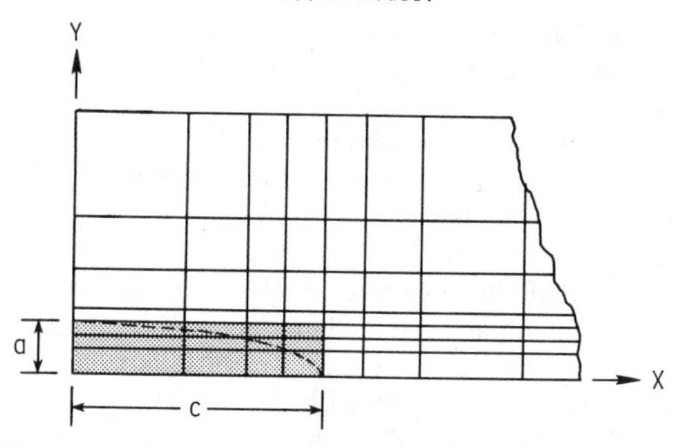

(b) Rectangular model.

FIG. 6—*Crack-surface area used in residual pressure* ($\sigma^R{}_z$) *fit.*

for corner cracks, and

$$\sigma^R{}_z = \sigma^R{}_z\,(x, 0) \quad \text{for } x \geq 0, y \leq 0$$

for surface cracks.

The stresses computed at the nodal points of a 20-node element in a finite-element analysis can be inaccurate [*16*]. Therefore, the stresses were evaluated at the 2 × 2 × 2 Gaussian points of an element and then were extrapolated to the element nodes as suggested by Hinton et al. [*16,17*].

Results and Discussion

In this section, the convergence of the finite-element-alternating method is studied. Then, the stress-intensity factors obtained from this method are compared to those calculated by the three-dimensional finite-element method. Next, the stress-intensity factors for various crack configurations are compared to those calculated from empirical stress-intensity factor equations.

Convergence of the Finite-Element-Alternating Method

To study the convergence of the finite-element-alternating method, an oblong corner crack subjected to remote uniform tension with an a/c ratio of 0.2 was considered. The corner-crack configuration was chosen because the configuration is more severe than the surface-crack configuration because of the existence of an additional free surface ($x = 0$ plane). The a/c ratio of 0.2 was chosen because larger areas of the external surfaces need to be made stress free.

Figure 7*a* shows a typical mapped model on the $z = 0$ plane for a shallow corner crack ($a/t = 0.2$) with 20-noded isoparametric elements. This coarse model had 982 nodes and 162 elements and uses 4 elements to model region corresponding to the crack face. Two other models, medium and fine, using eight- and twelve-elements to model the region corresponding to the crack face were developed; see Figs. 7*b* and 7*c*, respectively. All three models had nine unequal layers of elements in the height (z) direction. For all three models, the stress-intensity factors converged to within 1% accuracy in five iterations. The average residual pressure on the crack face normalized by the remote tension stress showed excellent convergence, as shown in Table 1.

Figure 8 presents the normalized stress-intensity factors all along the crack front for the three models. The stress-intensity factors from the three models agreed well with one another and indicated that even coarse models give accurate results.

Figure 9 shows the three rectangular models (on the $z = 0$ plane) that were used in the analyses: coarse, medium, and fine. The three models were developed such that the coarse model is a subset of the medium and the medium model is a subset of the fine model. All models had the same refinement in the height (z) direction. The coarse and medium models had only four elements, while the fine model had nine elements in the crack region. The coarse model had only five elements in the y-direction while the medium model had seven elements in the y-direction. In both cases, the x-refinement was held constant. The fine

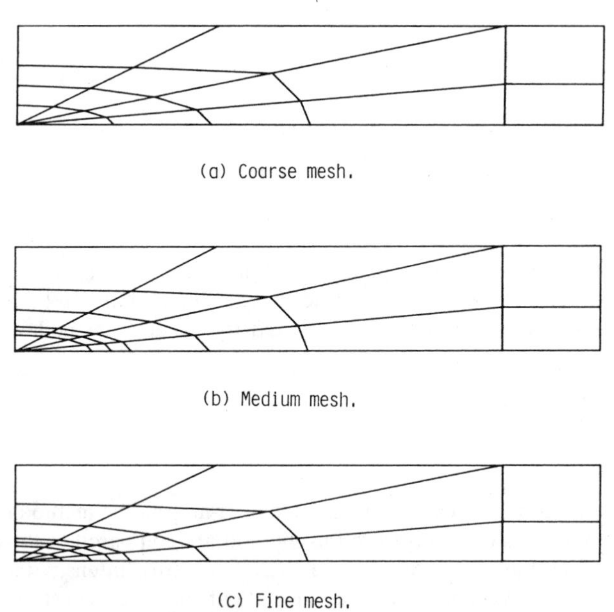

(a) Coarse mesh.

(b) Medium mesh.

(c) Fine mesh.

FIG. 7—*Mapped models used in finite-element-alternating method.*

TABLE 1—*Average normalized residual pressure on the crack face from mapped models.*

Iteration No.	Finite-Element Model		
	Coarse	Medium	Fine
1	0.975	0.975	0.975
2	0.215	0.201	0.199
3	0.034	0.044	0.043
4	0.013	0.017	0.017
5	0.002	0.003	0.003

mesh, on the other hand, had nine elements in the y-direction and seven elements in the x-direction. Therefore, the fine model had better refinement near the crack front. The plate was idealized with 175, 245, and 441 elements for the coarse, medium, and fine models, respectively. For these models, the stress-intensity factors converged to within 1% accuracy in four iterations. The average residual pressure on the crack face normalized by the remote uniform tension stress, again, showed excellent convergence, as shown in Table 2.

Figure 10 presents the normalized stress-intensity factors obtained from the three rectangular models for a slightly different corner-crack configuration ($a/c = 0.2$ and $a/t = 0.1$) than used for the mapped models. Small differences in stress-intensity factors were found between the medium and fine models (about 0.5%). However, for larger values of ϕ, considerable differences were observed between the coarse and medium models. This behavior was caused by inadequate refinement in the y-direction for the coarse model. These results suggest that accurate stress-intensity factors can be obtained from rectangular models with as little as four elements in the region of the crack, provided adequate refinement is used in modeling the free surfaces.

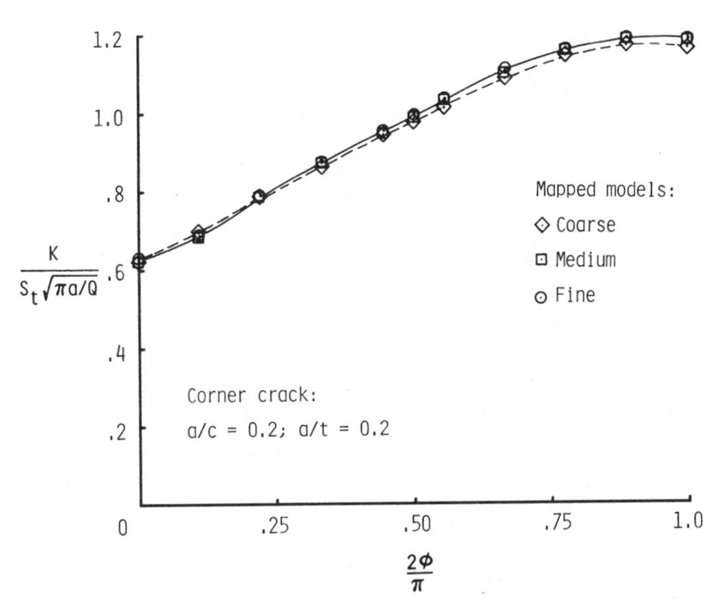

FIG. 8—*Convergence of normalized stress-intensity factors from finite-element-alternating method using mapped models for quarter-elliptic corner crack in a plate under remote tension.*

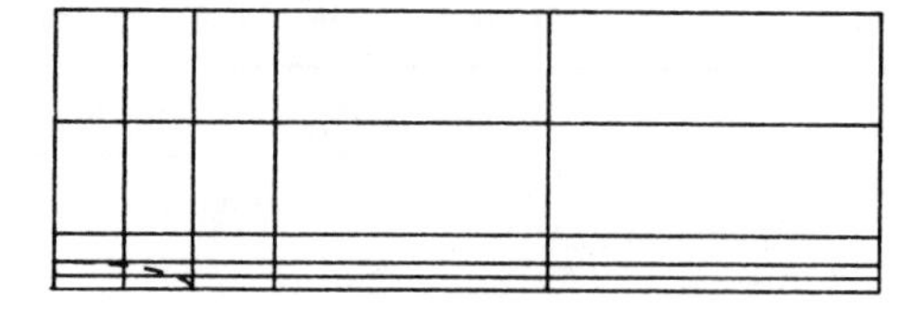

(a) Coarse mesh.

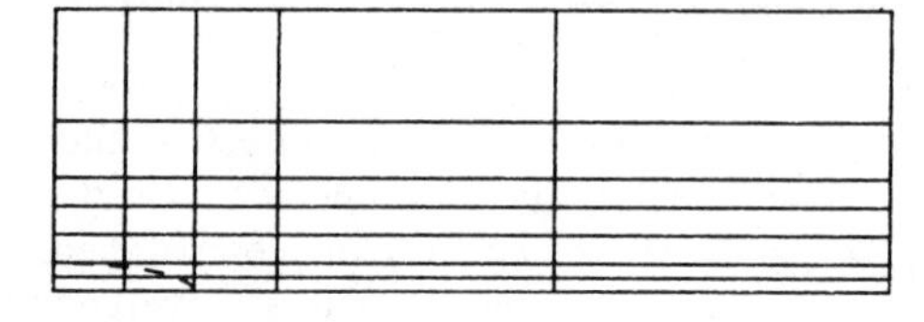

(b) Medium mesh.

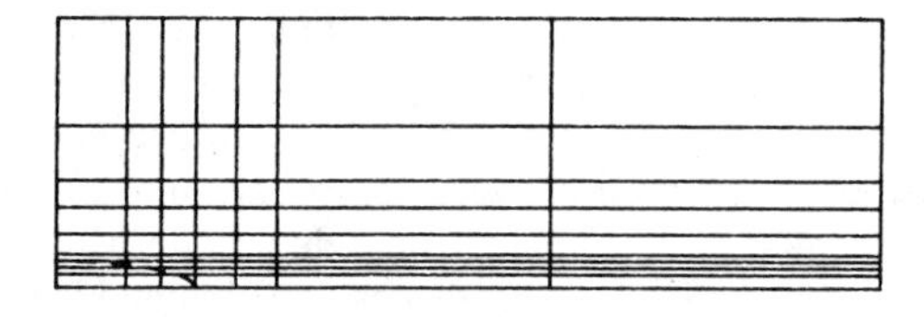

(c) Fine mesh.

FIG. 9—*Rectangular models used in finite-element-alternating method.*

Comparison of Stress-Intensity Factors

Figures 11 and 12 compare the stress-intensity factors for shallow corner cracks (a/c = 0.2) obtained with the finite-element method and the finite-element-alternating method for a/t = 0.2 and 0.1, respectively. The results from the mapped and rectangular (medium) models are shown in Fig. 11, while the results from a rectangular (medium) model are shown in Fig. 12. The results shown from both analyses in Fig. 11 agreed well. The maximum difference was near $\phi = 0$ and was about 4%. (Herein, "percent difference" is defined as the difference between the two solutions normalized by the largest value for that configuration.) Figure 12 shows that the results from the rectangular model agreed well with those obtained from the finite-element method, except near $\phi = 0$. The maximum difference, however, was about 6%.

TABLE 2—*Average normalized residual pressure on the crack face from the rectangular models.*

Iteration No.	Finite-Element Model		
	Coarse	Medium	Fine
1	1.273	1.273	1.273
2	0.144	0.171	0.171
3	0.020	0.027	0.027
4	0.006	0.007	0.008

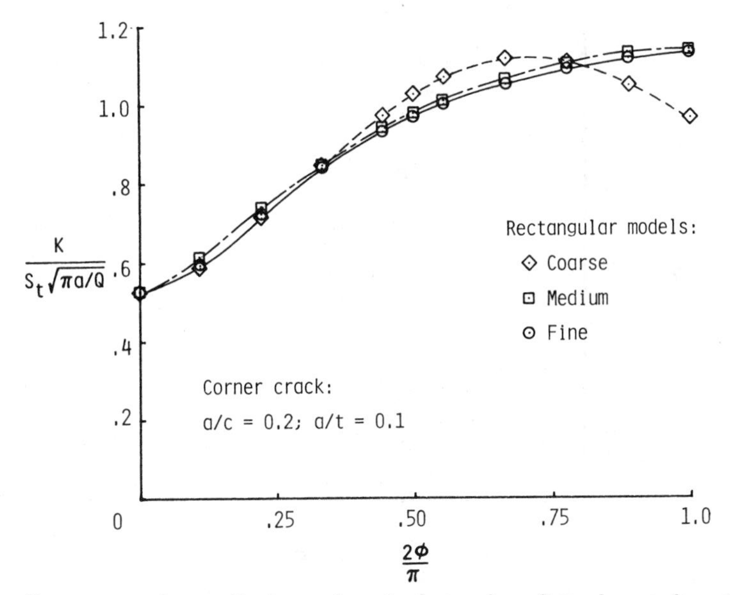

FIG. 10—*Convergence of normalized stress-intensity factors from finite-element-alternating method using rectangular models for a quarter-elliptical corner crack in a plate under remote tension.*

Figures 13 and 14 present comparisons of stress-intensity factors for a nearly semicircular surface crack and nearly quarter-circular corner crack, respectively, obtained with the finite-element method and the finite-element-alternating method. Note that the finite-element-alternating method cannot be used for cracks with an a/c ratio of unity because the elliptic functions have indefinite forms. From numerical experimentation, the limiting values of the

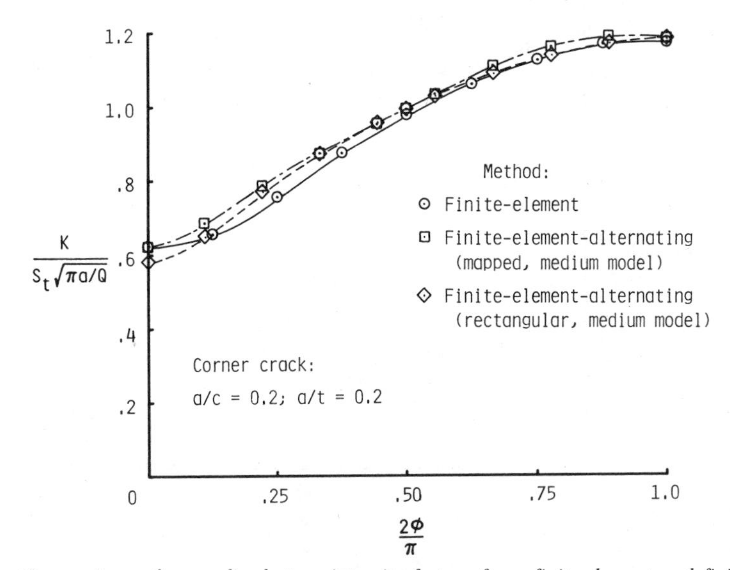

FIG. 11—*Comparison of normalized stress-intensity factors from finite-element and finite-element-alternating methods for a quarter-elliptical corner crack in a plate under remote tension.*

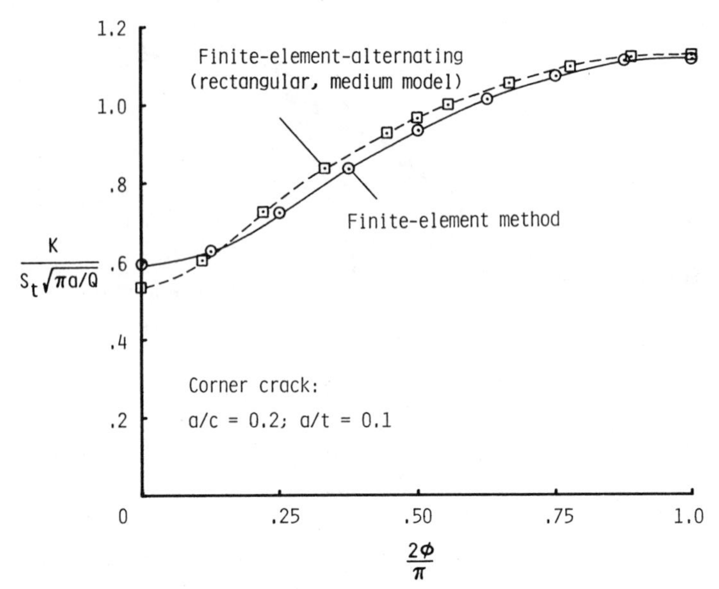

FIG. 12—*Comparison of normalized stress-intensity factors from finite-element and finite-element-alternating methods for a quarter-elliptical corner crack under remote tension.*

a/*c* ratio appear to be 0.98 for embedded cracks, 0.92 for surface cracks, and 0.85 for corner cracks. Thus, an *a*/*c* value of 0.85 was chosen for both crack configurations. These figures show reasonable agreement between the two methods for both the surface- and corner-crack configurations. The largest discrepancy occurred where the crack front intersects a free surface (about 5%).

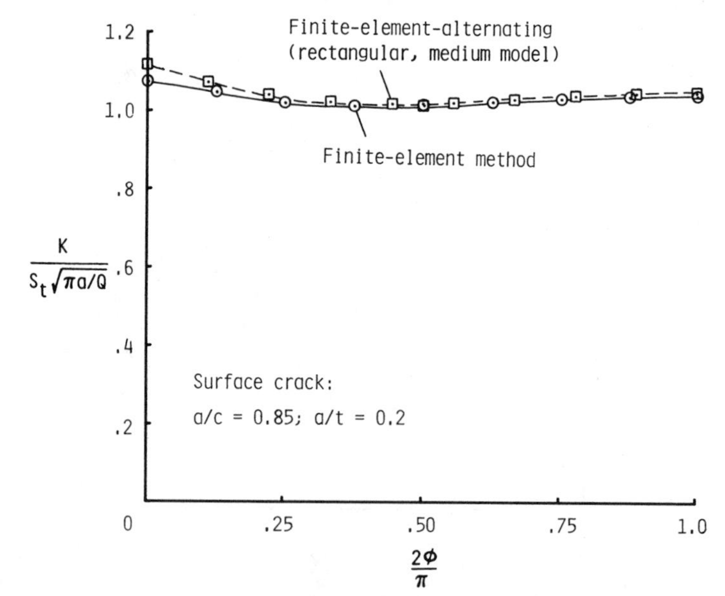

FIG. 13—*Comparison of normalized stress-intensity factors from finite-element and finite-element-alternating methods for a near semi-circular surface crack under remote tension.*

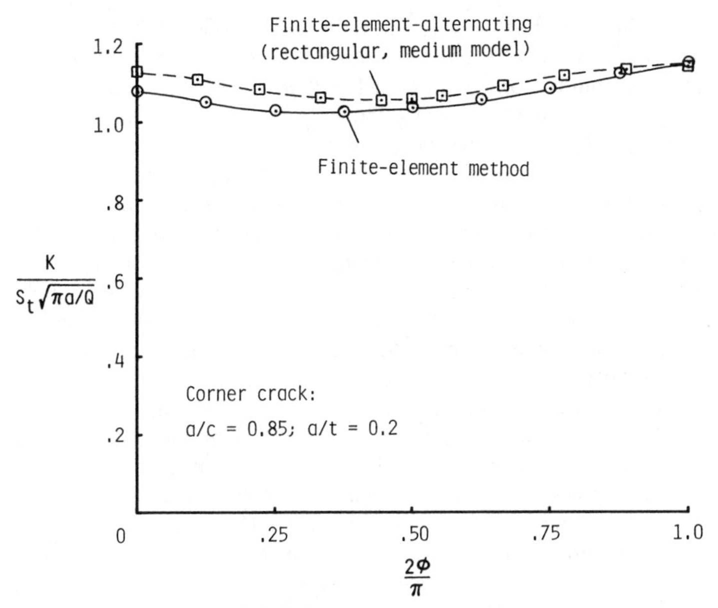

FIG. 14—*Comparison of normalized stress-intensity factors from finite-element and finite-element-alternating methods for a near quarter-circular corner crack under remote tension.*

Computational Effort of the Finite-Element-Alternating Method

A study of the computational time for the alternating method indicated that assembling and decomposing the finite-element stiffness matrix was the most dominant computational effort. Each iteration was approximately one percent of the time required to assemble and decompose the stiffness matrix. For the configurations studied, convergence to less than 1% error bound in the stress-intensity factors was achieved in four or five iterations. The results shown in Fig. 12 suggest that rectangular models provide accurate solutions and these models are easier to generate than the mapped models. The rectangular models give accurate results provided that adequate refinement is made along each coordinate axis. Most importantly, a single rectangular fine mesh can be used to analyze a wide range of crack shapes and sizes without repeated assembly and decomposition of the stiffness matrix. For example, the computational time required to analyze three crack shapes (a/c) and three crack sizes (a/t), or nine crack configurations, using the finite-element-alternating method was 630 CPU s (VPS-32 Computer). The conventional finite-element method requires nine separate computer runs. The computational time for one run was about 400 CPU s (VPS-32 Computer). Therefore, 3600 CPU are required for the finite-element method.

Stress-Intensity Factors for Small Cracks

Stress-intensity factor equations [*9,10*] have been developed by using the stress-intensity factors obtained from the finite-element method, engineering judgement, and extrapolations. To evaluate the equations for $a/t < 0.2$, therefore, it is logical that the values from the equation be compared with those from the finite-element method. Furthermore, the differences between the results from the finite-element-alternating method and the finite-element method with singularity elements were about 3% for most of the crack front.

Therefore, stress-intensity factors were calculated for various crack shapes ($a/c = 0.2$ to 1) with a/t ratios ranging from 0.05 to 0.2 by using the finite-element method. A typical stress-intensity factor distribution for a corner crack with $a/c = 0.4$, subjected to remote uniform tension loading, for various a/t ratios are shown in Fig. 15. For remote tensile loading and all a/c ratios considered, smaller a/t values always gave slightly lower stress-intensity factors all along the crack front. However, the difference in stress-intensity factors from an a/t value of 0.2 to 0.05 was less than 3%. For remote bending loading and all a/c ratios, smaller values of a/t gave higher stress-intensity factors all along the crack front. This is expected because the crack experiences a more uniform stress gradient as a/t approaches zero. At $a/t = 0$, the bending stress-intensity correction factors (F) are exactly equal to those due to remote tension. At $a/t = 0.05$, the maximum differences between the stress-intensity correction factors at the deepest point of the crack due to remote tension and remote bending loading are about 10%.

The present results were also compared to the empirical stress-intensity factor equations proposed by Newman and Raju [*9,10*]. As previously mentioned, the empirical equations were obtained by a curve fitting procedure to the finite-element results in the range $0.2 \leq a/t \leq 0.8$ for various crack shapes. In developing the empirical equations, some engineering judgment and extrapolations were used for the limiting solution for $a/t = 0$. The present finite-element results are compared with calculations from the empirical equations in Figs. 16 through 19 for surface or corner cracks. For surface cracks, the comparisons were made at the maximum depth point ($2\phi/\pi = 1$) and near the free surface ($2\phi/\pi = 0.125$) for the surface crack. At the free surface ($2\phi/\pi = 0$), the finite-element results are influenced by the boundary-layer effect and the results are mesh dependent [2]. For corner cracks, the comparisons are also made near the two free surfaces ($2\phi/\pi = 0.125$ and 0.875). The results from the empirical equations (solid curves) are generally within about 5% of the finite-element results for the range of a/c ratios considered.

From the finite element results, asymptotic stress-intensity correction values at $a/t = 0$

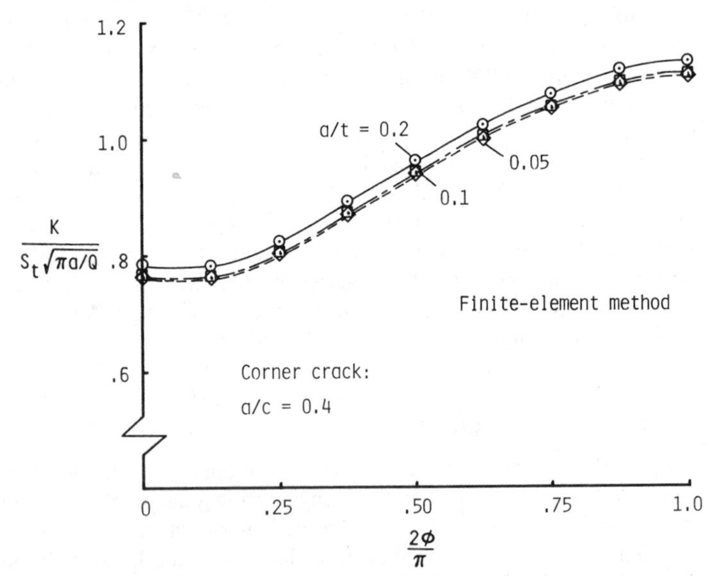

FIG. 15—*Normalized stress-intensity factors for a quarter-elliptical corner crack* (a/c = *0.4*) *in a plate under remote tension.*

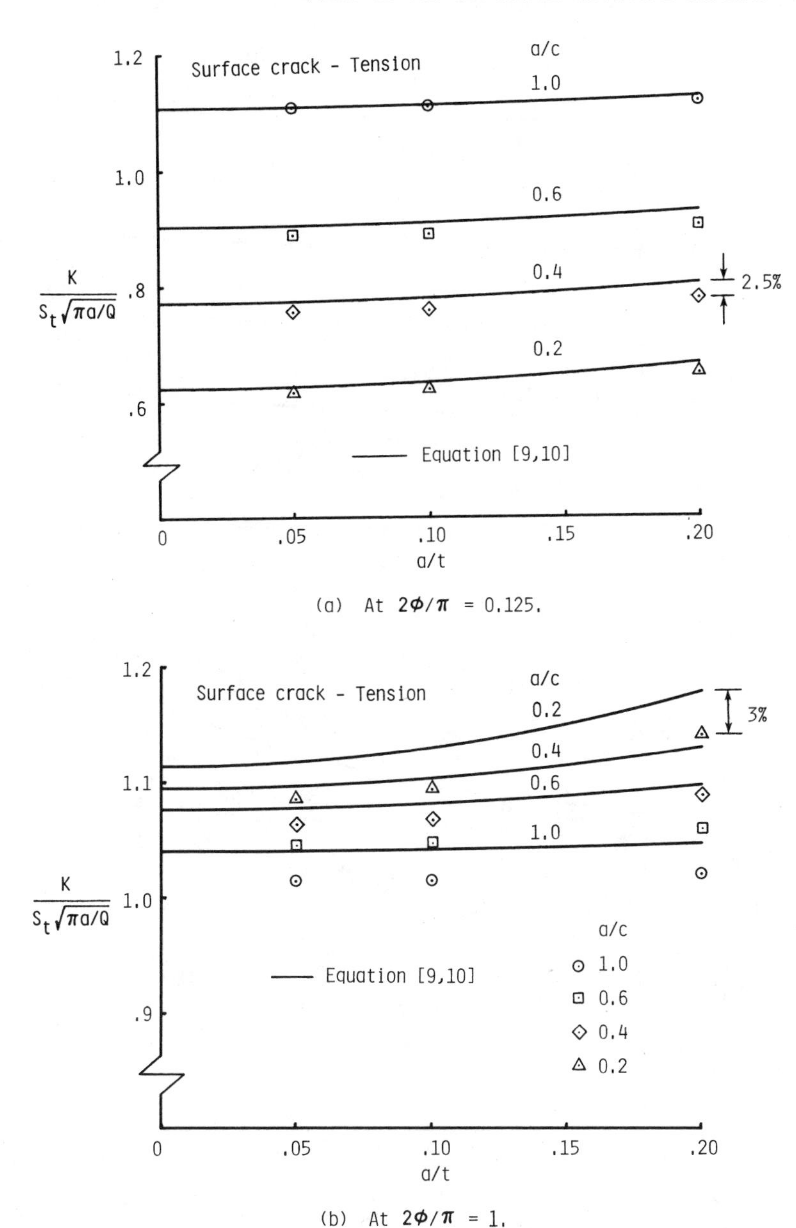

FIG. 16—*Comparisons of normalized stress-intensity factors from finite-element method and empirical equations for surface cracks in a plate under remote tension.*

were computed by fitting a quadratic equation in terms of a/t to the results for a/t values of 0.2, 0.1, 0.05. These asymptotic values (average of the tension and bending loads) are shown in Table 3. These values are also compared with those obtained from the empirical equations. The asymptotic values from the empirical equation and the extrapolated finite-element results at $a/t = 0$ agreed well (within 3%). Thus, the empirical equations have an accurate limit as a/t approaches zero.

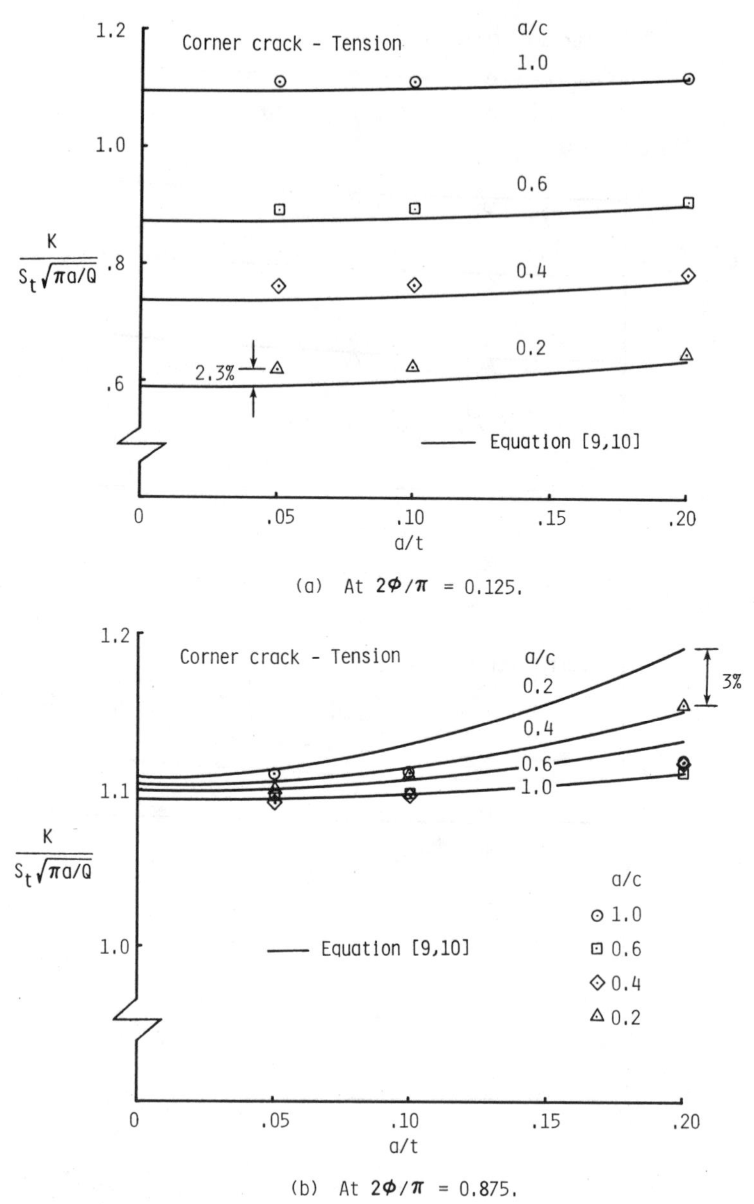

FIG. 17—*Comparisons of normalized stress-intensity factors from finite-element method and empirical equations for corner cracks in a plate under remote tension.*

Concluding Remarks

Stress-intensity factors for shallow surface and corner cracks in rectangular plates were obtained using the three-dimensional finite-element and finite-element-alternating methods. The plates were subjected to remote tension and remote out-of-plane bending loads. A wide range of crack shapes were considered (a/c = 0.2 to 1). The crack-depth-to-plate-thickness (a/t) ratios ranged from 0.05 to 0.2.

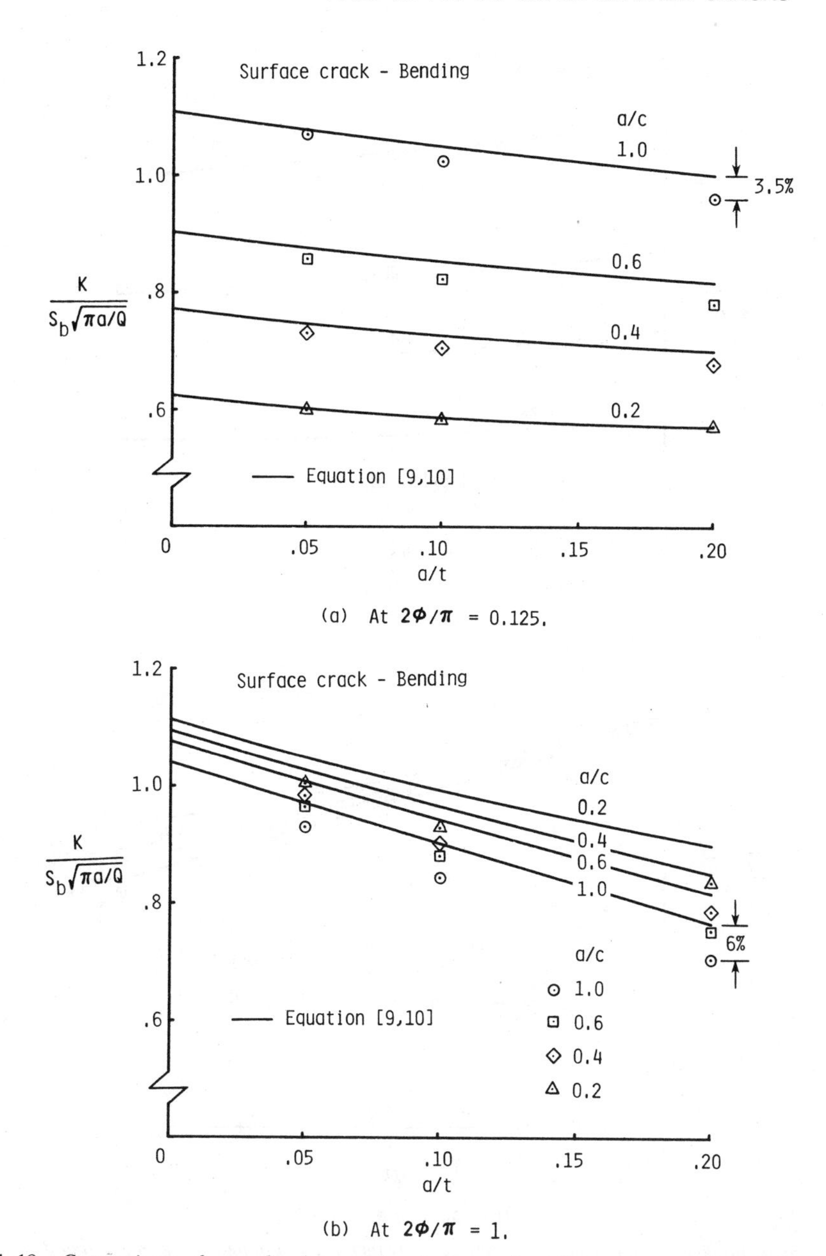

FIG. 18—*Comparisons of normalized stress-intensity factors from finite-element method and empirical equations for surface cracks in a plate under remote bending.*

The performance of the finite-element-alternating method was studied by considering two types of models: mapped and rectangular models. The mapped models used idealizations that conform to the shape of the crack while the rectangular models used a rectangular idealization throughout the solid. The stress-intensity factors obtained by either model showed excellent convergence and showed that about four to eight elements are sufficient to model the crack region. The stress-intensity factors obtained from the finite-element-

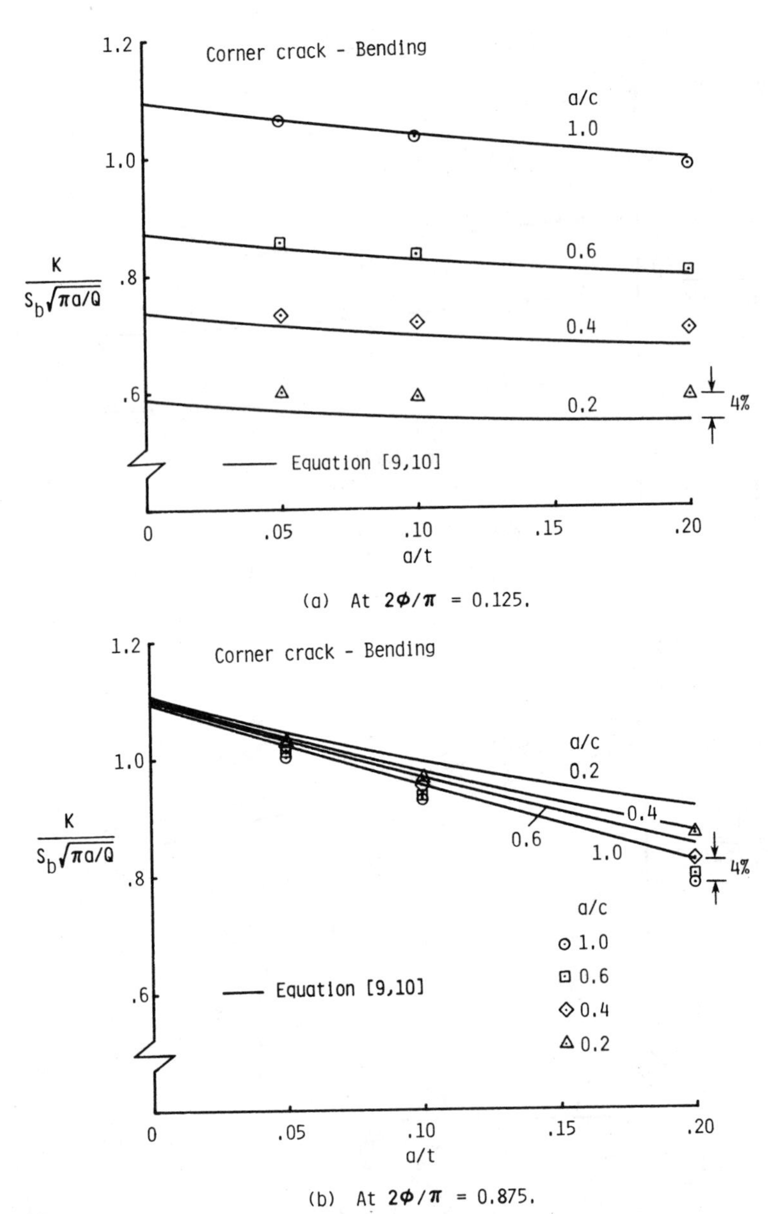

FIG. 19—*Comparisons of normalized stress-intensity factors from finite-element method and empirical equations for corner cracks in a plate under remote bending.*

alternating method agreed well with those obtained from the finite-element method with singularity elements (maximum difference was about 5%). The study of the computational effort involved in the finite-element-alternating method showed that a single rectangular idealization could be used to analyze several crack configurations. The method produced accurate stress-intensity factors at a lower cost compared to the conventional finite-element method.

For remote tensile loading and all crack shapes (crack depth-to-surface-length ratios,

TABLE 3—*Comparison of asymptotic limits of the normalized stress-intensity factors,* F, *as* a/t *approaches zero, obtained by the finite-element analysis (average value of tension and bending loading) and the empirical equation* $K = S_i (\pi a/Q)^{1/2} F$.

	Surface Cracks		Corner Cracks	
a/c	Finite-Element Analysis	Empirical Equation	Finite-Element Analysis	Empirical Equation
	$2\phi/\pi = 0.125$		$2\phi/\pi = 0.125$	
0.2	0.621	0.623	0.618	0.588
0.4	0.764	0.771	0.752	0.736
0.6	0.896	0.902	0.886	0.871
1.0	1.119	1.107	1.104	1.094
	$2\phi/\pi = 1.0$		$2\phi/\pi = 0.875$	
0.2	1.094	1.112	1.095	1.108
0.4	1.073	1.094	1.085	1.104
0.6	1.055	1.076	1.090	1.100
1.0	1.022	1.040	1.104	1.094

a/c) considered, lower values of the crack-depth-to-plate-thickness (a/t) ratios gave lower stress-intensity factors for surface- and corner-cracked plates. However, the largest difference between the stress-intensity factors for a/t values ranging from 0.05 to 0.2 was only about 5%. The results of an a/t ratio of 0.05 were very nearly equal to the asymptotic values at $a/t = 0$. For remote bending loading and all crack shapes (a/c), lower values of a/t gave higher stress-intensity factors all along the crack front. This was expected because the crack experiences a more uniform stress gradient as a/t approaches zero. The asymptotic limits of the stress-intensity factors (as a/t approaches zero) given the empirical equations proposed by Newman and Raju were within 3% of the limits obtained by the finite-element method.

The stress-intensity factors given in this paper should be useful in predicting crack-growth rates and fracture strengths, in designing structural components, and in establishing inspection intervals for surface- and corner-cracked components.

Acknowledgment

The first author's contribution to this work was performed under NASA contract NAS1-18256 at the NASA Langley Research Center, Hampton, Virginia.

References

[1] Gran, R. J., Orazio, F. D., Paris, P. C., Irwin, G. R., and Hertzberg, R. H., "Investigation and Analysis Development of Early Life Aircraft Structural Failures," AFFL-TR-70-149, Air Force Flight Laboratory, Wright-Patterson Air Force Base, Ohio, 1971.

[2] Raju, I. S. and Newman, J. C., Jr., "Stress-Intensity Factors for a Wide Range of Semi-Elliptical Surface Cracks in Finite-Thickness Plates," *Engineering Fracture Mechanics*, Vol. 11, No. 4, 1979, pp. 817–829.

[3] Newman, J. C., Jr., and Raju, I. S., "Analyses of Surface Cracks in Finite Plates under Tension or Bending Loads," NASA TP-1578, National Aeronautics and Space Administration, 1979.

[4] Raju, I. S. and Newman, J. C., Jr., "Finite-Element Analysis of Corner Cracks in Rectangular Bars," NASA TM-89070, National Aeronautics and Space Administration, 1987.

[5] Heliot, J., Labbens, R. C., and Pellissier-Tanon, A., "Semi-elliptical Surface Cracks Subjected to Stress Gradients," *Fracture Mechanics (11th Conference), ASTM STP 677,* C. W. Smith, Ed., American Society for Testing of Materials, Philadelphia, 1979, pp. 341–364.

[6] Shah, R. C. and Kobayashi, A. S., "On the Surface Flaw Problem," *The Surface Crack: Physical Problems and Computational Solutions,* J. L. Swedlow, Ed., American Society of Mechanical Engineers, 1972, pp. 79–142.

[7] Smith, F. W., "The Elastic Analysis of the Part-circular Surface Flaw Problem by the Alternating Method," *The Surface Crack: Physical Problems and Computational Solutions,* J. L. Swedlow, Ed., American Society of Mechanical Engineers, 1972, pp. 125–152.

[8] Smith, F. W. and Kullgren, T. E., "Theoretical and Experimental Analysis of Surface Cracks Emanating from Fastener Holes," AFFDL-TR-76-104, Air Force Flight Dynamics Laboratory, Wright-Patterson Air Force Base, Ohio, 1977.

[9] Newman, J. C., Jr., and Raju, I. S., "Stress-Intensity Factor Equations for Cracks in Three-Dimensional Finite Bodies," *Fracture Mechanics: Fourteenth Symposium. Volume I: Theory and Analysis, ASTM STP 791,* J. C. Lewis and G. Sines, Eds., American Society for Testing and Materials, Philadelphia, 1983, pp. I-238–I-265.

[10] Newman, J. C., Jr., and Raju, I. S., "Stress-intensity Factor Equations for Cracks in Three-dimensional Finite Bodies Subjected to Tension and Bending Loads," *Computational Methods in the Mechanics of Fracture,* S. N. Atluri, Ed., North Holland, Amsterdam, 1986, pp. 312–334.

[11] Nishioka, T. and Atluri, S. N., "Analytical Solution for Embedded Elliptical Cracks, and Finite Element-Alternating Method for Elliptical Surface Cracks, Subjected to Arbitrary Loadings," *Engineering Fracture Mechanics,* Vol. 17, 1983, pp. 247–268.

[12] Nishioka, T. and Atluri, S. N., "An Alternating Method for Analysis of Surface Flawed Aircraft Structural Components," *American Institute of Aeronautics and Astronautics Journal,* Vol. 21, 1983, pp. 749–757.

[13] Atluri, S. N. and Nishioka, T., "Computational Methods for Three-dimensional Problems of Fracture," *Computational Methods in Mechanics of Fracture,* S. N. Atluri, Ed., North Holland, Amsterdam, 1986, pp. 230–287.

[14] Shah, R. C. and Kobayashi, A. S., "Stress-Intensity Factors for an Elliptic Crack under Arbitrary Loading," *Engineering Fracture Mechanics,* Vol. 3, 1971, pp. 71–96.

[15] Vijayakumar, K. and Atluri, S. N., "An Embedded Elliptical Flaw in an Infinite Solid, Subject to Arbitrary Crack-Face Tractions," *Transactions,* American Society of Mechanical Engineers, Series E, *Journal of Applied Mechanics,* Vol. 48, 1981, pp. 88–96.

[16] Hinton, E. and Campbell, J. S., "Local and Global Smoothing Discontinuous Finite Element Functions Using Least Squares Method," *International Journal for Numerical Methods in Engineering,* Vol. 8, 1974, pp. 461–480.

[17] Hinton, E., Scott, F. C., and Rickets, R. E., "Local Least Squares Stress Smoothing for Parabolic Isoparametric Elements," *International Journal for Numerical Methods in Engineering,* Vol. 9, 1975, pp. 235–256.

C. W. Smith,[1] T. J. Theiss,[1] and M. Rezvani[1]

Intersection of Surface Flaws with Free Surfaces: An Experimental Study

REFERENCE: Smith, C. W., Theiss, T. J., and Rezvani, M., **"Intersection of Surface Flaws with Free Surfaces: An Experimental Study,"** *Fracture Mechanics: Perspectives and Directions (Twentieth Symposium), ASTM STP 1020,* R. P. Wei and R. P. Gangloff, Eds., American Society for Testing and Materials, Philadelphia, 1989, pp. 317–326.

ABSTRACT: A series of frozen stress experiments was conducted on surface flaws of varying aspect ratios in finite-thickness plates in order to study the practical aspects of the well-known loss of the inverse square-root singularity when the flaw borders intersect the free surface. Optical data are converted into classical stress-intensity factors ignoring and accounting for the above-noted effect and compared with a well-established numerical result. Results are discussed regarding their usefulness to both the research and practicing engineer.

KEY WORDS: linear elastic fracture mechanics (LEFM), stress singularity, stress intensity factor (SIF), frozen stress photoelasticity, free-surface effects, dominant eigenvalues

Nomenclature

σ_{ij}	Components of stress
σ^0_{ij}	Components of nonsingular stress
$f_{ij}(\theta)$	Functional form of θ in singular stresses for Mode I loading
$f_{ij}(\theta,\phi),h_i(\theta,\phi)$	Functional forms for σ_{ij}, u_i in spherical coordinates
σ^0	Component of nonsingular normal stress parallel to crack plane
$\overline{\sigma}$	Remote uniform stress not affected by presence of crack
τ_{max}	Maximum shear stress in plane perpendicular to crack front direction
τ_0	Nonsingular maximum in-plane shearing stress near crack tip
x_1,x_2,r,θ,ζ	Crack tip coordinates (Fig. 5)
a	Crack depth
K	Mode I stress-intensity factor (SIF)
K_P	Mode I photoelastic SIF
K_{NR}	Mode I SIF from Newman-Raju Theory
K_{cor}	"Corresponding" Mode I SIF (see Eq 1)
$(K)_{Ap}$	Apparent SIF (see Eq 7)
$(K_{\lambda_\sigma})_{Ap}$	Apparent stress eigenfactor (see Eq 14)
λ_σ, λ_u	Lowest eigenvalues for eigenfunction expansions of stress, displacement components, respectively
$Z(\zeta)$	Quasi two-dimensional stress function

There are a number of practical situations where the inverse square-root stress singularity of linear elastic fracture mechanics (LEFM) may be lost or altered. The alteration at a bi-

[1] Alumni professor, graduate research assistant, and graduate research assistant, respectively, Department of Engineering Science and Mechanics, Virginia Polytechnic Institute and State University, Blacksburg, VA 24061.

material interface was noted in 1962 by Zak and Williams [1], and such effects are now being recognized in the composite material literature. One such effect, which has significance in both heterogeneous and homogeneous materials, concerns the loss of the inverse square-root singularity when a crack intersects a free surface. Around 1980, Benthem [2,3] provided a very accurate value of the lowest dominant eigenvalue when a quarter infinite crack intersects the free surface of a half space at right angles. For his analysis, Benthem expressed the components of stress and displacement in the form of the eigenfunction series

$$\sigma_{ij} = \sum_{k=1}^{\infty} r^{\lambda_\sigma^k} f_{ij}(\theta,\phi)_k$$

$$u_i = \sum_{k=1}^{\infty} r^{\lambda_\mu^k} h_i(\theta,\phi)_k$$

where (r,θ,ϕ) are the usual spherical coordinates and λ_σ^k and λ_u^k are kth order eigenvalues for the stresses and displacements, respectively. Benthem's result was verified experimentally [4,5] using compact bending specimens with straight front cracks which intersected the side surfaces of the beam at right angles. The free-surface value of the lowest dominant eigenvalue in the displacement λ_u ($k = 1$) was obtained by high-density moiré interferometry. However, high-density moiré interferometry is a rather specialized and expensive method, and several attempts [6,7] have been made to develop the frozen stress photoelastic method for achieving the same purpose for surface flaws which also intersect the surfaces at right angles. However, results from these studies suggest the need for an improved algorithm for converting optical data into appropriate fracture parameters.

By conducting frozen stress and moiré experiments in tandem on compact bending specimens [7], it has been possible to develop an approximate, but improved algorithm for this purpose. It is approximate in the sense that it utilizes a two-dimensional algorithm to characterize the near tip behavior in the interior of the body and then utilizes a modification based upon experimental data to account for the complex three-dimensional stress state existing at the free surface-crack front intersection region. However, since it is our intent to focus upon the experimental results obtained by applying the refined algorithm to experiments conducted on surface flaws in finite thickness plates rather than the algorithm per se, the latter has been placed in the Appendix.

The Experiments

Test specimens were large, stress-free plates made from stress freezing photoelastic material known as PSM-9 and cast by Photoelastic Inc. The plates were nominally 12.7 mm thick. Small surface cracks (less than two tenths of the plate thickness in depth) were inserted into the plates by striking a sharp blade held normal to the specimen surface with a hammer, causing a small crack to emanate from the blade tip. This crack was enlarged above the critical temperature in the stress freezing oven under a deadweight Mode I loading. When the cracks reached the desired size, the load was reduced and the specimens were cooled under the reduced load to room temperature. Upon load removal, thin slices were removed from the near tip region which were mutually orthogonal to the crack surface and the crack border. These slices were then analyzed in a crossed circular polariscope using both partial mirror fringe multiplication and the Tardy Method, and the data were then fed into a computer program for extracting the desired fracture parameters.

A plot of raw data from one of the interior slices of one of the surface flaws showing the

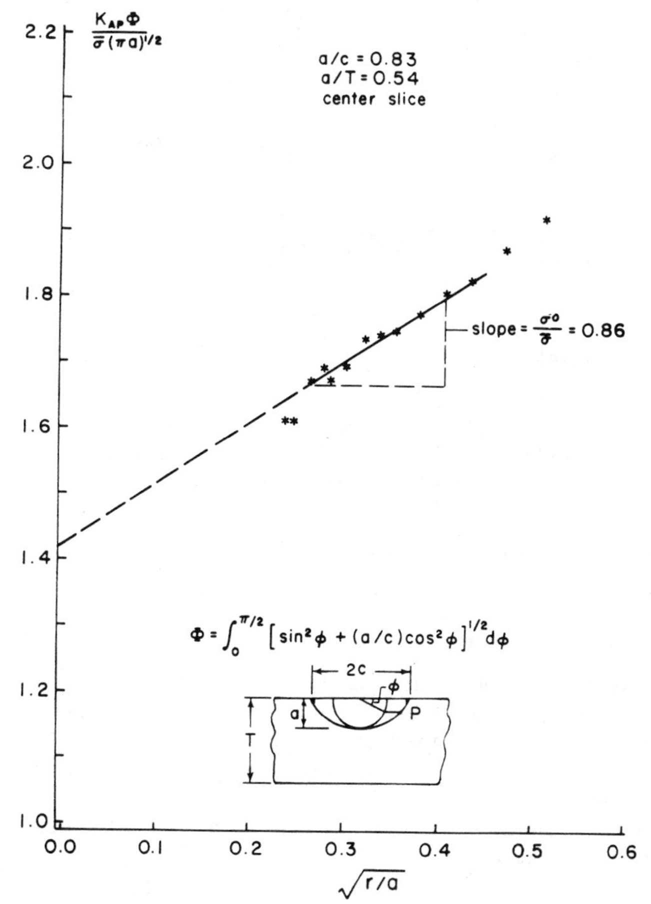

FIG. 1—*Determination of* $\sigma^0\sqrt{\sigma}$ *from raw data from a slice from a surface flaw.*

data zone of interest is presented in Fig. 1. From the slope of this curve (Eq 8), τ_0 may be computed (Eq 15).

The distribution of τ_0 values for the tests for which $a/c = 0.85$, $a/t = 0.56$, as determined from LEFM using the procedure described in the Appendix below Eq 14, are given in Fig. 2 for all slices except the surface slice. From these data, it was concluded that the trend toward varying τ_0 values as one moves toward the free surface had not begun to manifest itself in the inner slices. Consequently, only the surface slice required adjustment for this trend, noted in the Appendix. For the surface slice, τ_0 was assigned a value of zero. The same procedure was used for the deeper flaw. Clearly, this adjustment is approximate in nature and would not be expected to provide an accurate description of the loss of singularity effect in the transition zone near the free surface, across which a continuous variation in λ_u may occur. However, it does allow us to describe approximately conditions near the free surface in such a way as to obtain reasonable agreement with Benthem's result at the free surface and Moiré results obtained from compact bending specimens containing straight front cracks as reported in Ref *4*.

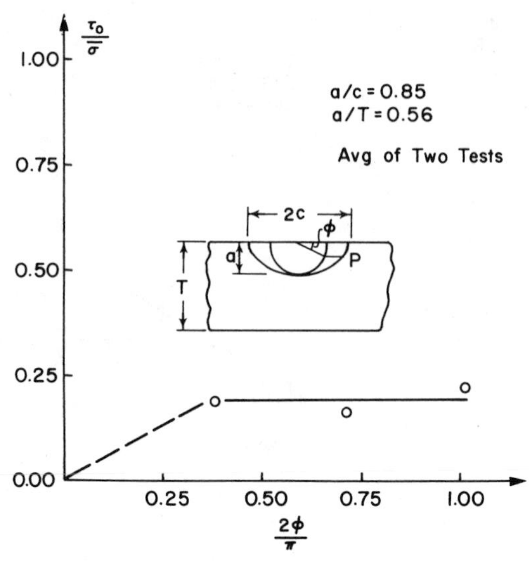

FIG. 2—*Distribution of* τ_0 *for surface flaw.*

Application to Fracture Mechanics

Since the value of the lowest dominant eigenvalue differs from that associated with classical fracture mechanics, one is led to address the issue of how the quantified knowledge of this phenomena can be introduced into the framework of classical fracture mechanics so it can be utilized in engineering analysis. A simple way in which to achieve this is to introduce a stress-intensity factor (SIF) which would be required to produce the same value of τ_{max} as one measures in the actual test where λ_σ is not equal to the classical value. Calling this the corresponding SIF, one may write

$$\frac{(K_{cor})_{Ap}}{r^{1/2}} = \frac{(K_{\lambda_\sigma})_{Ap}}{r^{\lambda_\sigma}} \tag{1}$$

Now, substituting Relation 1 into Eq 14, one has

$$\tau_{max} = \frac{\lambda_\sigma (K_{cor})_{Ap}}{\sqrt{2\pi}\, r^{1/2}} \tag{2}$$

Now, combining Eqs 7 and 2, we have

$$\frac{\lambda_\sigma (K_{cor})_{Ap}}{\sqrt{2\pi}\, r^{1/2}} = \frac{K_{Ap}}{\sqrt{8\pi}\, r^{1/2}}$$

where

$$(K_{cor})_{Ap} = \frac{K_{Ap}}{2\lambda_\sigma} \tag{3}$$

Since values of K and K_{cor} are obtained by extrapolating plots of K_{Ap} versus $r^{1/2}$ across a nonlinear zone to the origin, it follows that

$$K_{cor} = \frac{K}{2\lambda_\sigma} \tag{4}$$

It should be emphasized that the introduction of Relation 1 is quite arbitrary and is based upon purely dimensional grounds. Moreoever, since the analysis here is quasi-two dimensional, one depends on moiré results using Benthem's analysis to validate our stress algorithm at the free surface.

Experimental Results

The flaw geometries involved moderate aT values (≈ 0.5) to deep ($a/T \approx 0.75$) flaws for which a/c values for one group were less than unity and for the deeper flaw, $a/c \approx 1.12$. Symmetrical slices were taken on both sides of the flaws and averaged. The photoelastic results for Mode I for the set for $a/c \approx 0.85$ are shown in Fig. 3, together with results computed from the Newman-Raju theory [8] (adjusted to $\nu = 0.5$) and also the normalized K_{cor} values. The adjustments made to the Newman-Raju Theory, which was originally presented only for $\nu = 0.3$, involved additional computer runs at $\nu = 0.45$, which those authors graciously conducted at our request for a series of aspect ratios and relative depths, and these results were used together with small extrapolations to reach the results plotted herein. These results suggest the test data may be affected by the singularity loss at the free surface to some extent near the free surface, yielding lower values than predicted by Ref *8*. However, when the surface slice data are converted to K_{cor}, a significant increase above both LEFM results is noted. This may suggest the presence of a strong boundary layer effect which becomes diffused across the thickness of the surface slice. The photoelastic and Newman-Raju values should coalesce near the center of the crack border. Differences in that region may be due partly to error in adjustment of the Newman-Raju theory to $\nu = 0.5$ and partly due to experimental error.

The same information is provided for the deeper flaws in Fig. 4 and the same trends are noted. K_{cor} may be regarded as the value of the stress intensity factor which would have resulted at the free surface had the dominant surface value of λ_σ been one half as in LEFM. As such, it may be regarded as a conservative quantity which might be used to fit the results into a LEFM framework. One would expect this effect to be greatest on surface measurements in incompressible materials.

Summary

The loss of the inverse square root singularity in nearly incompressible materials when surface flaws intersect free surfaces has been studied experimentally using frozen stress photoelasticity. The data are interpreted through an algorithm which is modified near the free surface (Eqs 15 through 17) to improve its accuracy. Based upon the results included here and other moiré results [7], the following points may be made:

1. The loss in singularity effect at the free surface will be significant in incompressible materials and will be strongest in surface measurements on such materials.
2. When the concept of the corresponding SIF to correct surface slice data from photo-

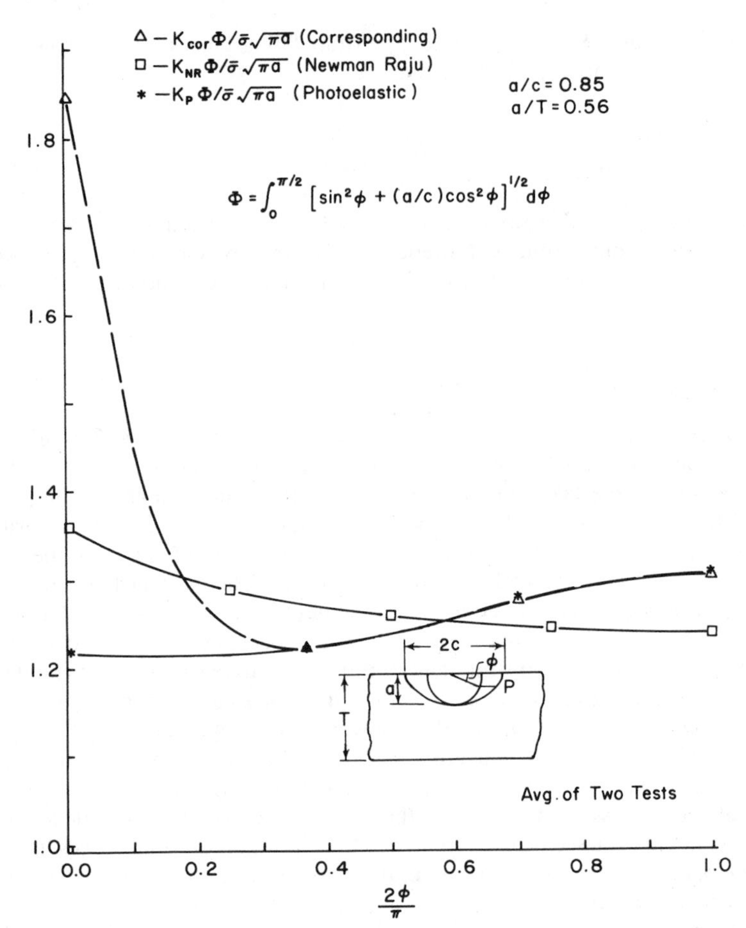

FIG. 3—*Comparison of numerical LEFM, photoelastic, and* K_{cor} *stress intensity factor distributions along the border of a surface flaw of moderate depth.*

elastic experiments is used, an overcorrection (on the conservative side) results in surface flaw data. This is conjectured to be due to the fact that the effect for which K_{cor} corrects is concentrated very near to the free surface and the thickness averaging through the surface slice of the photoelastic data may substantially mitigate its influence on the slice data.

One may conclude, then, that corrections for the loss in singularity at the free surface using K_{cor} will be significant and accurate only for surface measurements (that is, moiré) on nearly incompressible materials, and that, when the K_{cor} method is used on thickness averaged data near the free surface, a substantial overcorrection will likely result.

However, from the point of view of the designer, who desires to accomodate this effect into the framework of LEFM in a conservative manner, use of K_{cor} on slice data still provides a simple, safe approach.

Acknowledgments

The authors wish to acknowledge the contributions of colleagues as noted in the references together with W. R. Lloyd and the support of the National Science Foundation under Grant No. MEA 832-0252.

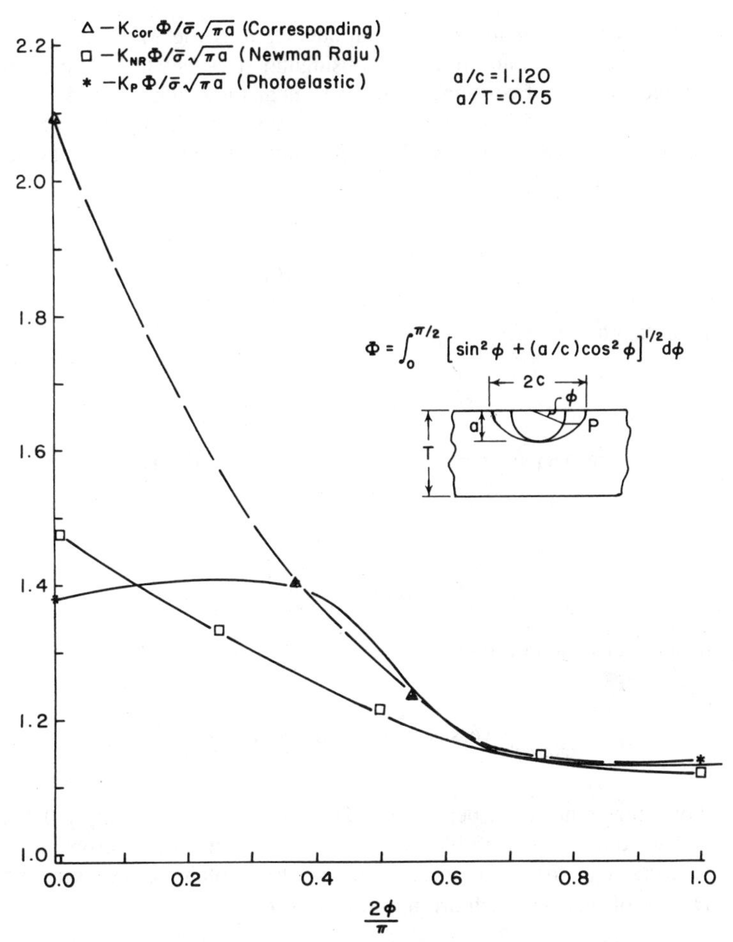

FIG. 4—*Comparison of numerical LEFM, photoelastic, and* K_{cor} *stress intensity factor distributions along the border of deep surface flaws.*

APPENDIX

Algorithms for Converting Optical Data into Fracture Parameters

In developing algorithms for converting free-surface data into fracture parameters, the authors used both LEFM concepts and the results of Benthem's analysis [*2,3*]. For benefit of the reader and preservation of continuity, both will be briefly reviewed here.

Mode I LEFM Algorithm for Photoelastic Data

We may describe the near tip stresses in LEFM for two-dimensional problems following Ref *9* as

$$\sigma_{ij} = \frac{K}{(2\pi r)^{1/2}} f_{ij}(\theta) - \sigma^0_{ij}(\theta) \; i,j = 1,2 \tag{5}$$

where σ_{ij} are the near-tip in-plane stress components, K is the Mode I SIF, σ_{ij}^0 may be regarded as the near-tip contribution of the nonsingular stresses which, experience shows, can be represented as a set of constant values very near the crack tip, and (r,θ) are polar coordinates as shown in Fig. 5. Along $\theta = \pi/2$, we may compute from Eqs 5 [when $(\sigma^0)^2$ is small relative to $8\tau_{max}^2$ (Eq 5)], where $\sigma^0/\sqrt{8}$ represents the contribution of the nonsingular stress state to τ_{max} locally as [9]

$$\tau_{max} = \frac{K}{(8\pi r)^{1/2}} + \frac{\sigma^0}{\sqrt{8}} \tag{6}$$

Now defining an "apparent" SIF

$$(K)_{Ap} = \tau_{max}(8\pi r)^{1/2} \tag{7}$$

and normalizing with respect to $\bar{\sigma}(\pi a)^{1/2}$ where $\bar{\sigma}$ represents the remote stress and a the crack length, we have

$$\frac{(K)_{Ap}}{\bar{\sigma}(\pi a)^{1/2}} = \frac{K}{\bar{\sigma}(\pi a)^{1/2}} + \frac{\sigma^0}{\bar{\sigma}}\left(\frac{r}{a}\right)^{1/2} \tag{8}$$

which suggests a linear zone in a plot of

$$\frac{(K)_{Ap}}{\bar{\sigma}(\pi a)^{1/2}} \text{ versus } \left(\frac{r}{a}\right)^{1/2} \text{ with a slope of } \frac{\sigma^0}{\bar{\sigma}}$$

Experience shows this zone to lie between $(r/a)^{1/2}$ values of approximately 0.2 to 0.4 (or above) in most two-dimensional problems. By extracting optical data from this zone and extrapolating across a near-tip nonlinear zone, we can obtain an accurate estimate of $K_1/\bar{\sigma}(\pi a)^{1/2}$. Details of this approach are found in Ref 9.

Photoelastic Algorithms for Accounting for Free-Boundary Effect on the Crack Tip Singularity Near the Free Surface

Although the above noted effect involves a three-dimensional state of stress, one may consider measurements on the free surface or from internal slices as being associated with a quasi-two dimensional stress state. Thus one may begin, as in LEFM, with a two-dimensional stress function $f(\zeta)$ of the form

$$Z = \frac{f(\zeta)}{\zeta^{\lambda_\sigma}} \tag{9}$$

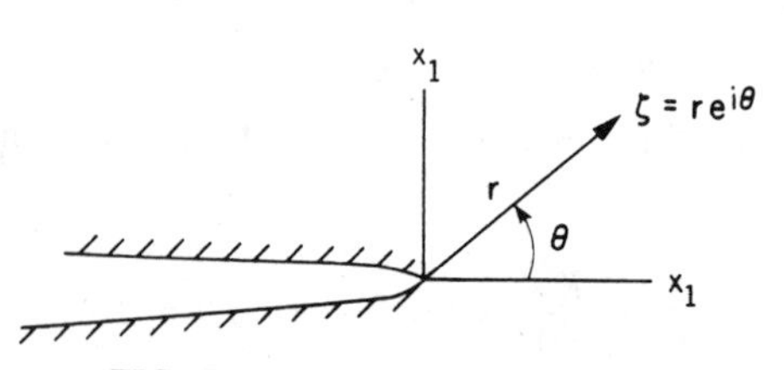

FIG. 5—*Near-tip coordinate system.*

where $\zeta = re^{i\theta}$ and is measured from an origin at the crack tip. Expanding $f(\zeta)$ in a MacLaurin series about the crack tip and letting $\zeta \to 0$, we recover

$$Z \downarrow_{\zeta \to 0} = \frac{K_{\lambda_\sigma}}{\sqrt{2\pi}\zeta^{\lambda_\sigma}} \tag{10}$$

where K_{λ_σ} may be designated as a stress eigenfactor. Now substituting Eq 10 into

$$\sigma_{11} = \text{Re}Z - x_2\text{Im}Z$$

$$\sigma_{22} = \text{Re}Z + x_2\text{Im}Z$$

$$\sigma_{12} = -x_2\text{Re}Z'$$

we obtain [4][2]

$$\sigma_{11} = \frac{K_{\lambda_\sigma}}{\sqrt{2\pi}\, r^{\lambda_\sigma}}\{\cos\lambda_\sigma\theta - \lambda_\sigma\sin\theta\,\sin(\lambda_\sigma + 1)\theta\} - \sigma^0$$

$$\sigma_{22} = \frac{K_{\lambda_\sigma}}{\sqrt{2\pi}\, r^{\lambda_\sigma}}\{\cos\lambda_\sigma\theta + \lambda_\sigma\sin\theta\,\sin(\lambda_\sigma + 1)\theta\}$$

$$\sigma_{12} = \frac{K_{\lambda_\sigma}}{\sqrt{2\pi}\, r^{\lambda_\sigma}}\{\lambda_\sigma\sin\theta\,\cos(\lambda_\sigma + 1)\theta\} \tag{11}$$

where σ^0 plays the same role as σ_{ij}^0 in LEFM. From Eqs 11, one may compute

$$\tau_{\max} = \left[\left(\frac{\sigma_{22} - \sigma_{11}}{2}\right)^2 + \sigma_{12}^2\right]^{1/2} \tag{12}$$

Now, noting that

$$\frac{\sigma^0(2\pi)^{1/2}r^{\lambda_\sigma}}{K_{\lambda_\sigma}\lambda_\sigma\sin\theta} < 1$$

and truncating the resulting expression, one obtains

$$\tau_{\max} = \frac{\lambda_\sigma K_{\lambda_\sigma}\sin\theta}{\sqrt{2\pi}\, r^{\lambda_\sigma}} + \frac{\sigma^0}{2}\sin(\lambda_\sigma + 1)\theta \tag{13}$$

Along $\theta = \pi/2$, Eq 13 reduces to

$$\tau_{\max} = \frac{\lambda_\sigma K_{\lambda_\sigma}}{\sqrt{2\pi}\, r^{\lambda_\sigma}} + \frac{\sigma^0}{2}\sin(\lambda_\sigma + 1)\frac{\pi}{2} = \lambda_\sigma\frac{(K_{\lambda_\sigma})_{\text{Ap}}}{\sqrt{2\pi}\, r^{\lambda_\sigma}} \tag{14}$$

and for $\lambda_\sigma = 1/2$, the LEFM case, we recover Eq 6.

[2] The third equation in Eq 11 corrects an error in Eq 10*c* of Ref *4*.

Since one expects LEFM to prevail away from the free surface, one can, by assuming $\lambda_\sigma = ½$, obtain σ^0 from the slope of a plot of

$$\frac{(K)_{Ap}}{\bar{\sigma}(\pi a)^{1/2}} \text{ versus } \left(\frac{r}{a}\right)^{1/2} \text{ (see Eq 8)}$$

This, in effect determines the value of the nonsingular term on the right-hand side of Eq 14.

If we define this term as τ_0, that is

$$\tau_0 = \frac{\sigma^0}{2}\sin(\lambda_\sigma + 1)\frac{\pi}{2} \tag{15}$$

then Eq 15 suggests that τ_0 may vary if λ_σ varies. Thus, as one approaches the free surface, one should experience a variation in τ_0. Experimental LEFM determinations of σ^0 show that it begins to change significantly a short distance from the free surface. If one then imposes a linear decrease in τ_0 from this point to zero for the surface slice, one finds reasonable correlation between the photoelastic and moiré data within the transition zone. Using this approach and Eq 14 in the form

$$\tau_{max} = \frac{D(\lambda_\sigma)}{r^{\lambda_\sigma}} + \tau_0 \tag{16}$$

whence

$$\log(\tau_{max} - \tau_0) = \log D(\lambda_\sigma) - \lambda_\sigma \log r \tag{17}$$

we may extract values of λ_σ in the transition region from a plot at $\log(\tau_{max} - \tau_0)$ versus $\log r$.

References

[1] Zak, A. R. and Williams M. L., "Crack Point Stress Singularities at a Bimaterial Interface," GALCIT SM-42-1, California Institute of Technology, Jan. 1962.

[2] Benthem, J. P., "On an Inversion Theorem for Conical Regions in Elasticity Theory," *Journal of Elasticity,* Vol. 9, No. 2, 1979, pp. 159–169.

[3] Benthem, J. P., "The Quarter Infinite Crack in a Half-Space: Alternative and Additional Solutions," *International Journal of Solids and Structures,* Vol. 16, 1980, pp. 119–130.

[4] Smith, C. W. and Epstein, J. S., "Measurement of Three Dimensional Effects in Cracked Bodies," *Proceedings, 5th International Congress on Experimental Mechanics,* 1984, pp. 102–107.

[5] Smith, C. W. and Epstein, J. S., "Boundary Layer Measurements in Cracked Body Problems," *Proceedings, 12th Southeastern Conference on Theoretical and Applied Mechanics,* Vol. 1, 1984, pp. 355–360.

[6] Smith, C. W., Olaosebikan, O., and Epstein, J. S., "Boundary Effects in Fracture Mechanics," *Proceedings, Annual Conference on the Society for Experimental Mechanics,* 1985, pp. 113–118.

[7] Ruiz, C. and Epstein, J. S., "On the Variation of the Stress Intensity Factor Along the Front of a Surface Flaw," *International Journal of Fracture,* Vol. 28, 1985, pp. 231–238.

[8] Newman, J. C., Jr., and Raju, I. S., "Analyses of Surface Cracks in Finite Plates Under Tension on Bending Loads," NASA T.R. 1578, National Aeronautics and Space Administration, Dec. 1979.

[9] Smith, C. W., "Use of Three Dimensional Photoelasticity and Progress in Related Areas," *Experimental Techniques in Fracture Mechanics,* Vol. 2, A. S. Kobayashi, Ed., Society for Experimental Stress Analysis Monograph No. 2, 1975, pp. 3–58.

George T. Sha,[1] *Chien-Tung Yang,*[2] *and James S. Ong*[1]

An Efficient Finite-Element Evaluation of Explicit Weight Functions for Mixed-Mode Cracks in an Orthotropic Material

REFERENCE: Sha, G. T., Yang, C.-T., and Ong, J. S., **"An Efficient Finite-Element Evaluation of Explicit Weight Functions for Mixed-Mode Cracks in an Orthotropic Material,"** *Fracture Mechanics: Perspectives and Directions, (Twentieth Symposium), ASTM STP 1020,* R. P. Wei and R. P. Gangloff, Eds., American Society for Testing and Materials, Philadelphia, 1989, pp. 327–350.

ABSTRACT: An efficient finite-element methodology, which uses the energy perturbation concept for evaluating the decoupled weight functions ($h_{I(II)}$) of a mixed-mode crack in an orthotropic material, is presented. The methodology is achieved by applying a single colinear virtual crack extension technique to a symmetric mesh in the crack-tip neighborhood with the crack lying on one of the elastic symmetry planes. The use of symmetric mesh permits analytical separation of all field parameters of stresses, strains, displacements, tractions, and strain energy release rates into Mode I and Mode II components within the symmetric mesh zone. Once the decoupled explicit weight functions are predetermined explicitly for a given crack geometry, the decoupled stress-intensity factors, $K_{I(II)}$, under any loading conditions can be evaluated efficiently and accurately by a sum of worklike products between the applied tractions and the explicit weight functions at their application locations. Comparing the decoupled stress-intensity factors ($K_{I(II)}$) obtained from the predetermined explicit weight functions for the orthotropic crack with those of well-established literature data shows good agreement with identical crack geometry, loading, and orthotropic properties but with different approaches.

KEY WORDS: orthotropic crack, explicit weight functions, stress intensity factors, biaxial loading

Nomenclature

a	Crack length
δa	Infinitesimally virtual crack extension in direction colinear to crack
H	Effective modulus
E	Young's modulus of isotropic material
ν	Poisson's ratio of isotropic material
E_{ii}	Moduli of principal axes of material orthotropy with $i = 1,2$
ν_{xy}, ν_{yx}	In-plane Poisson's ratio of orthotropic material
G_{xy}	In-plane shear modulus
$h_{I(II)}$	Decoupled explicit weight functions with h_I for Mode I and h_{II} for Mode II
$K_{I(II)}$	Decoupled stress-intensity factors with K_I for Mode I and K_{II} for Mode II

[1] Staff research scientist/engineer and analytical methods engineer, respectively, Allison Gas Turbine Division, General Motors Corporation, P. O. Box 420, Indianapolis, IN 46206.

[2] Associate professor, Department of Naval Architecture, National College of Marine Science and Technology, Keelung, Taiwan, Republic of China.

$\frac{\partial u_{I(II)}}{\partial a}$ Decoupled partial displacement derivatives with respect to crack length
γ Angle between remote loading axis and an inclined straight crack
$H_{I(II)}$ Finite correction factors with H_I for Mode I and H_{II} for Mode II
S_i Complex roots of characteristic equation
α_i Real constants of real part of S_i roots
β_i Real constants of imaginary part of S_i roots
s Traction application surfaces
f_i Consistent nodal forces at i's location
f_i^b Body force at i's location per unit volume
V Volume
ϕ_1 Angle of rotation between x', y' coordinates and materials' orthotropic axes
ϕ_2 Angle of rotation from x', y' coordinate system to x, y coordinate system (see Fig. 1)
$A_n^{T(B)}$ Least-square fitted coefficients for Mode I crack-face weight functions with superscripts T and B referring to top and bottom crack-faces, respectively
$B_n^{T(B)}$ Least-square fitted coefficients for Mode II crack-face weight functions
r_s Radial distance from crack tip along crack faces

Fracture of anisotropic media such as fiber-reinforced composite materials is of great economic and technical importance. The intent of this paper is to demonstrate an efficient finite-element evaluation of the explicit weight functions, which permit accurate and economical calculation of stress-intensity factors of orthotropic materials.

The use of Bueckner's weight functions [1] can obviate repeated finite-element calculation of stress-intensity factors for a given crack geometry under different loading conditions. This is accomplished due to the load-independent characteristics of the explicit weight functions, which serve well as the universal function for determining stress-intensity factors under different loading conditions for a given crack geometry. The virtual crack extension (VCE) technique as suggested by Parks [2] and Hellen [3] has been successfully combined with the singular elements [4,5] for efficient finite-element evaluation of explicit Mode I weight functions (h_I) [6] according to Rice's displacement derivative representation [7] of Bueckner's weight functions. Recently, Rice's displacement derivative representation of Mode I cracks has been extended to mixed-mode cracks in isotropic material by Sha and Yang [8] in terms of decoupled stress intensity factors and decoupled displacement derivatives under the identical * loading system at i's location within the structure of interest as

$$h_{I(II)i}\,(x_i, y_i, a, \gamma) = \frac{H}{2K^*_{I(II)}} \frac{\partial U^*_{I(II)i}\,(x_i, y_i, a, \gamma)}{\partial a} \tag{1}$$

where

H = effective modulus; for an isotropic material $H = E$ (Young's modulus) for plane stress and $H = E/(1 - \nu^2)$ for plane strain with ν = Poisson ratio; for anisotropic materials the effective modulus can be chosen from the work of Sih et al. [9],
$K^*_{I(II)}$ = decoupled stress-intensity factors with K_I for Mode I and K_{II} for Mode II under * loading system,
$\frac{\partial U^*_{I(II)i}}{\partial a}$ = decoupled partial displacement derivatives with respect to crack length (a) under * load system with $\partial u_I/\partial a$ for Mode I or $\partial u_{II}/\partial a$ for Mode II, and
γ = oblique angle between loading axis and straight crack $\gamma = \pi/2 - \phi_2$ with $\phi_2 = 0$ deg for Mode I crack geometry.

Application of the collinear virtual crack extension technique to the symmetric mesh with singular elements in the immediate crack-tip neighborhood can provide efficient finite element evaluation of the decoupled explicit weight functions with Eq 1. This is accomplished through efficient evaluation of (1) decoupled strain energy release rates, $G^*_{I(II)}$, which permit accurate calculation of the decoupled stress intensity factors, $K^*_{I(II)}$, and (2) the decoupled partial displacement derivatives with respect to the infinitesimal virtual crack extension, $\partial u^*_{I(II)i}/\partial a$, which are obtained efficiently for the entire structure with the two-step process as shown by Sha and Yang [*8,10*]. The use of symmetric mesh in the crack-tip vicinity permits analytical separation of all field parameters of stresses, strains, displacements, tractions, and strain energy release rates within the symmetric mesh zone into Mode I and Mode II components. The extension from isotropic material to orthotropic material for the weight function evaluation with the symmetric mesh approach is shown in this paper for a crack lying along the symmetry plane for both orthotropic material properties and symmetric mesh in the crack-tip neighborhood. Further extension of the efficient finite-element evaluation of explicit weight functions with analytical near-tip displacement solutions will be presented shortly for orthotropic bodies containing cracks that do not lie on a plane of elastic symmetry.

Few relatively simple orthotropic crack solutions are available in either closed or numerical forms. Sih and Liebowitz [*11*] presented an analytical solution to the problem with the complex variable approach. Bowie and Freese [*12*] developed a modified mapping collocation technique and numerically analyzed a number of orthotropic rectangular sheets containing central cracks. Using essentially the same technique, Gandhi [*13*] analyzed the presence of an inclined crack centrally located in an orthotropic plate. The collocation technique [*12,13*] is primarily used for simple crack configurations.

Lin and Tong [*14*] developed the hybrid finite-element technique for stress intensity factor calculation of Mode I and mixed-mode cracks in an orthotropic composite material with good agreement with collocation results of Bowie and Freese [*12*]. The finite-element technique with stress-intensity factors as part of global unknowns for direct calculation of $K_{I(II)}$ values has been used by Foschi and Barrett [*15*] and Gifford and Hilton [*16*]. The boundary integral equation (BIE) method has also been used to determine Mode I and Mode II stress intensity factors of a cracked anisotropic plate by Snyder and Cruse [*17*]. The finite-element approach as shown in this paper for efficient calculations of both stress-intensity factors and explicit weight functions of an orthotropic crack is obtained by applying the virtual crack extension technique to the symmetric mesh with singular elements in the immediate crack-tip vicinity. Once the decoupled explicit weight functions are predetermined for a given orthotropic crack geometry, the accurate decoupled stress-intensity factors under any loading conditions can be obtained economically by performing the worklike products between the applied tractions and the explicit weight functions as

$$K_{I(II)} = \int_s f_i \cdot h_{I(II)i} ds + \int_v f_i^b \cdot h_{I(II)i} dv \tag{2}$$

where

f_i = consistent nodal force at i's location for the applied pressure at surface s,
f_i^b = body force at i's location within volume v,
$h_{I(II)i}$ = decoupled explicit weight function vectors for Mode I (h_I) or Mode II (h_{II}) at i's location, and
$K_{I(II)}$ = decoupled stress intensity factors for Mode I (K_I) or Mode II (K_{II}).

The finite-element results agree remarkably well with Gandhi's results [*13*] for a 45-deg inclined crack in a square orthotropic plate with $0.05 \leq 2a/W \leq 0.7$. The weight function characteristics along the crack faces and the other key traction application boundaries are also presented in this paper for the centrally inclined crack in a two-dimensional orthotropic plate. By applying the superposition principle to the weight function concept, the stress-intensity factors under complex loading conditions are illustrated in terms of flexibility and versatility aspects of the weight function concept. In addition, by using the physical interpretation, these predetermined explicit weight functions can be used uniquely for guiding the damage-tolerant design applications, as shown recently by Sha et al. [*19*].

Numerical Results

One of the examples in Gandhi's work [*13*] was used extensively for assessing the mixed-mode capability and accuracy of the finite-element approach presented here. The problem of a centrally inclined crack in a square orthotropic plate is shown in Fig. 1 with both crack and principle axes of symmetry at an angle of 45 deg from the X' axis. The orthotropic elastic constants used in the numerical assessment were identical to Gandhi's work [*13*] with $E_{xx} = 7.0\ 10^6$ psi, $E_{yy} = 2.5\ 10^6$ psi, $G_{xy} = 1.0\ 10^6$ psi, and $\nu_{xy} = 0.29$ to facilitate direct numerical comparisons (1 psi = 6.895 kN/m²). A wide crack range of $0.05 \leq 2a/W \leq 0.7$ with $\phi_1 = \phi_2 = 45$ deg was studied for assessment purposes with the incorporation of a finite-element virtual crack extension technique with the symmetric mesh of singular elements in the crack-tip neighborhood. The typical finite element mesh used in the numerical

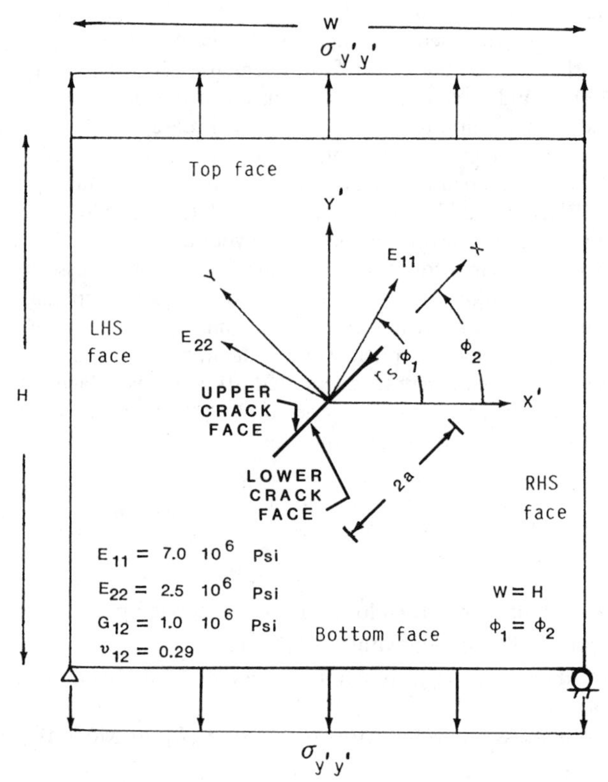

FIG. 1—*Centrally located oblique crack in a square plate with* E_{11} *and* E_{22} *moduli in the principal axes of material orthotropy.*

studies is shown in Fig. 2. The stress-intensity factors of the present study for the mixed-mode orthotropic cracks agree very well with Gandhi's, as shown in Fig. 3. The symmetric mesh in the immediate vicinity of the crack-tip node consists of singular quarter-point elements [4,5], which were surrounded by the regular eight-noded elements for the remainder of the asymmetric two-dimensional orthotropic crack geometry. The virtual crack extension operation of an asymmetric orthotropic crack is simulated with infinitesimal advancement of the crack-tip node in the collinear direction by Δa. With this proposed virtual crack extension mode, only a few elements are subjected to the elemental stiffness changes involved in calculating decoupled stress-intensity factors and decoupled displacement derivatives for the decoupled weight function evaluation for the entire structure of interest, as shown in Eq 1. This is one reason that the proposed finite-element methodology for explicit weight function evaluation is computationally efficient.

The symmetric mesh in the crack-tip neighborhood in isotropic and orthotropic materials

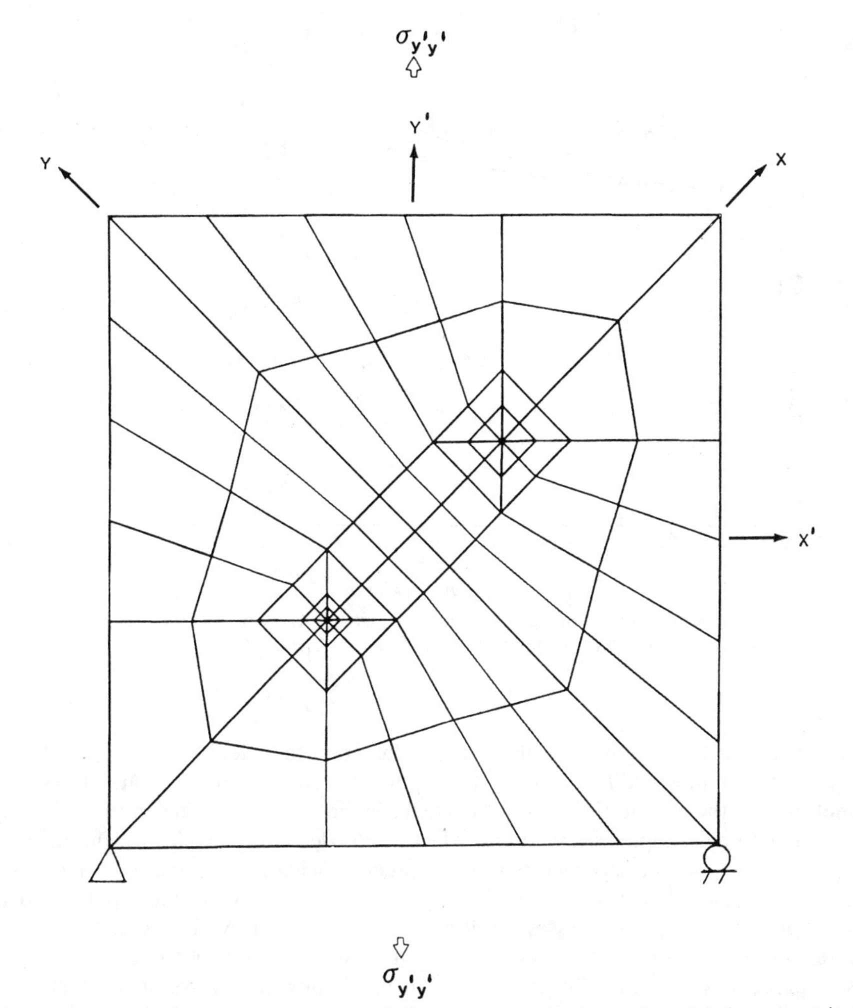

FIG. 2—*Typical finite-element mesh for modeling an orthotropic square plate with the symmetric mesh in the crack-tip neighborhood of a centrally located oblique crack.*

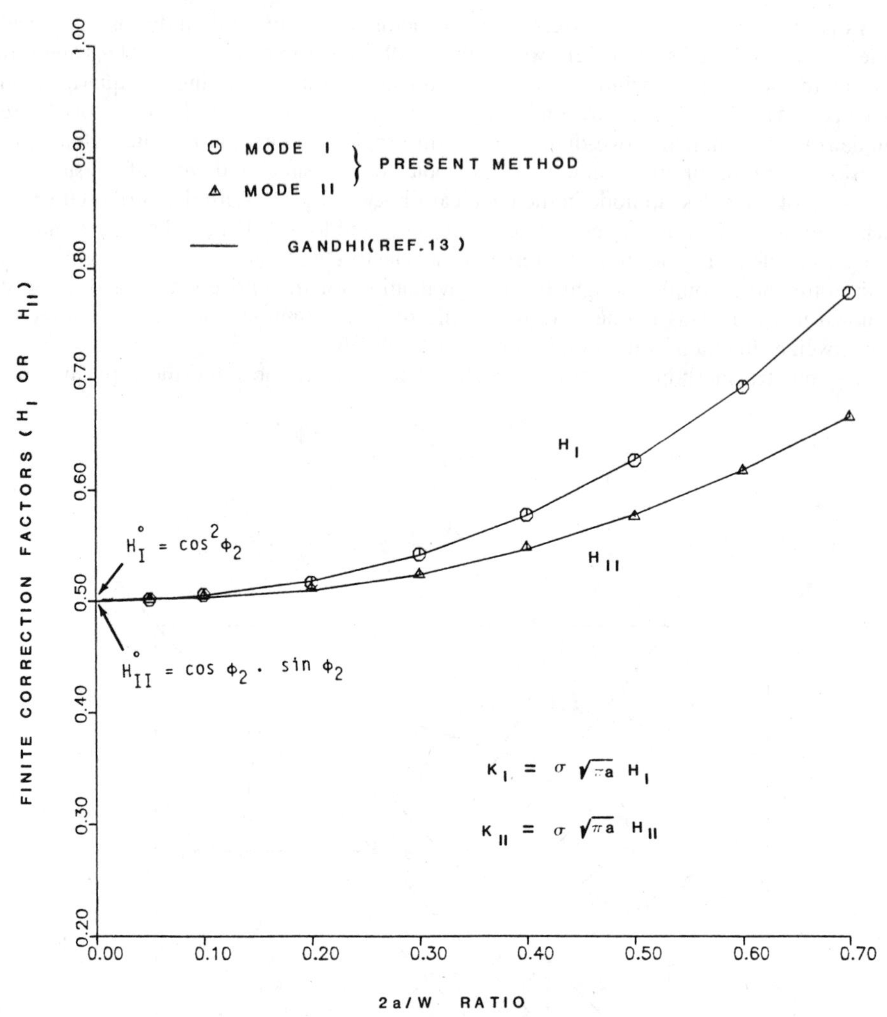

FIG. 3—*Finite correction factors of the stress-intensity factors* ($K_{I(II)}$) *as a function of* 2a/W *ratios for an oblique crack in a square orthotropic plate with* $\phi_1 = \phi_2 = 45$ *deg.*

permits analytical separation of displacement fields and other field parameters within the symmetric mesh into Mode I and Mode II components. This analytical separability of field parameters with the symmetric mesh in the crack-tip vicinity allows efficient finite-element evaluation of both decoupled stress-intensity factors and decoupled explicit weight functions. The decoupled stress-intensity factors were obtained efficiently from the decoupled strain energy release rates that were obtained from the decoupled displacement fields and the perturbation in elemental stiffnesses resulting from the colinear VCE operation. The displacement derivatives for the entire structure of interest can be obtained efficiently with the two-step process as shown by Sha et al. [*8,10*] by combining the inverse of global stiffness, $[K]^{-1}$, and the elemental stiffness perturbation of a few elements coupled with the decoupled displacement fields. For brevity, the technical details for finite-element evaluation of decoupled stress-intensity factors and decoupled displacement derivatives with the virtual crack extension technique will not be repeated here.

The present extension from the mixed-mode isotropic crack to the orthotropic crack with the symmetric mesh approach depends largely on the analytical separation of all field parameters into Mode I and Mode II components for the orthotropic crack when the crack lies along one of the material's principal orthotropic axes. The finite-element techniques for efficient explicit weight function evaluation as presented in this paper are confined to the orthotropic body containing cracks that lie on one of the elastic symmetry planes. Weight functions for orthotropic cracks that do not lie on the orthotropic symmetry plane will be the subject of a forthcoming publication.

The numerical results of the present study for the centrally located crack in an orthotropic plate with a 45-deg oblique angle are compared directly with Gandhi's results in terms of the finite correction factors ($H_{\rm I}$, $H_{\rm II}$), which are defined as

$$K_{\rm I} = \sigma \sqrt{\pi a}\, H_{\rm I} \tag{3a}$$

$$K_{\rm II} = \sigma \sqrt{\pi a}\, H_{\rm II} \tag{3b}$$

If the square plate boundaries of an isotropic medium with infinite extent, $H_{\rm I}$ and $H_{\rm II}$, equal $H_{\rm I}^{\circ} = \cos^2\phi_2$ and $H_{\rm II}^{\circ} = \cos\phi_2 \cdot \sin\phi_2$, respectively, the amount of deviation from $H_{\rm I(II)}$ to $H_{\rm I(II)}^{\circ}$ represents a measure of the material anisotropy and the boundary effects at the crack tip. The anisotropic effect on $H_{\rm I(II)}^{\circ}$ diminishes when the orthotropic plate boundaries are infinite in extent [9].

If $2a/W = 0.05$ is a fair representation of infinite anisotropic medium simulation, the error in $K_{\rm I(II)}$ values is less than 0.3%, as indicated in the numerical results with the symmetric mesh approach. The finite-element results with the symmetric mesh approach for $K_{\rm I(II)}$ determination of two-dimensional orthotropic cracks agree well with Gandhi's results, as shown in Fig. 3.

In the practical application of fracture mechanics for life prediction purposes, determination of the explicit weight function is far more advantageous than calculation of stress-intensity factors alone. In addition to the decoupled stress-intensity factors, finite-element evaluation of the decoupled explicit weight functions needs supplemental calculation of the decoupled displacement derivatives under the same reference loading condition, as shown in Eq 1. As shown by Sha and Yang [8], this added effort for obtaining displacement derivatives with a two-step process is computationally efficient with the application of the collinear virtual crack extension technique to the symmetric mesh in the crack-tip neighborhood. Although both the decoupled stress intensity factors ($K_{\rm I(II)}$) and the decoupled displacement derivatives depend strongly on the reference loading system, the decoupled explicit weight functions obtained according to Eq 1 are indeed invariant with respect to the loading conditions for a given crack geometry of preselected constraint conditions under different loading conditions. In principle, the use of a simple loading such as a concentrated point load is sufficient for explicit finite-element determination of decoupled weight functions, which depend on geometry, composition, and constraint conditions but are independent of loading conditions.

Crack-face weight function components in the x, y coordinate system can be grouped into primary weight function components, which consist of $h_{\rm Iy}$ and $h_{\rm IIx}$ components, and secondary weight function components, which consist of $h_{\rm Ix}$ and $h_{\rm IIy}$ components. Note that the primary weight function components demonstrate $1/\sqrt{r}$ singular behavior in the crack-tip neighborhood, as shown in Figs. 4 and 5 for $h_{\rm Iy}$ and $h_{\rm IIx}$ components, respectively, with different

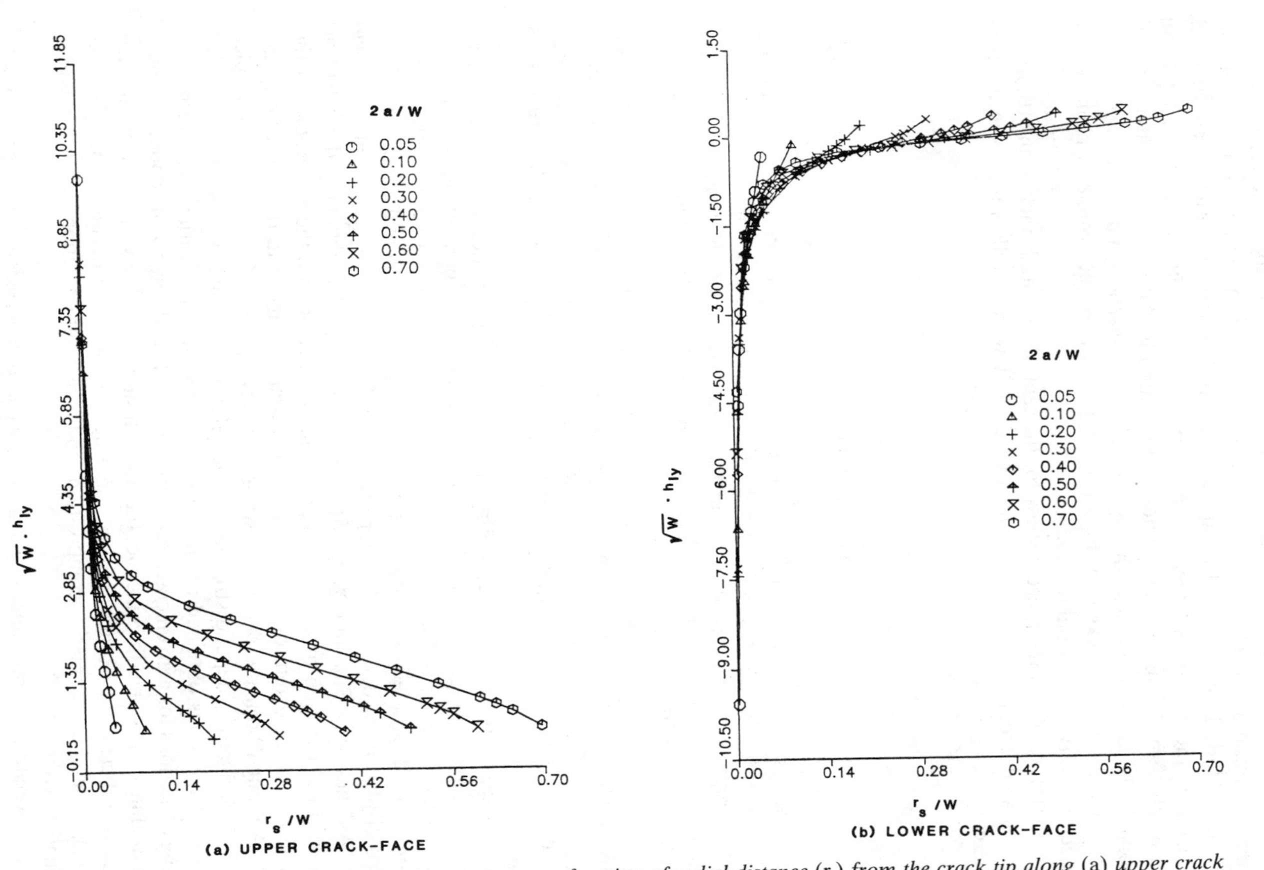

FIG. 4—$\sqrt{W}h_{Iy}$ *explicit weight function component as a function of radial distance* (r_s) *from the crack tip along* (a) *upper crack face and* (b) *lower crack face for a centrally oblique crack in an orthotropic square plate with* $0.05 \leq 2a/W \leq 0.7$ *and* $\phi_1 = \phi_2 =$ *45 deg.*

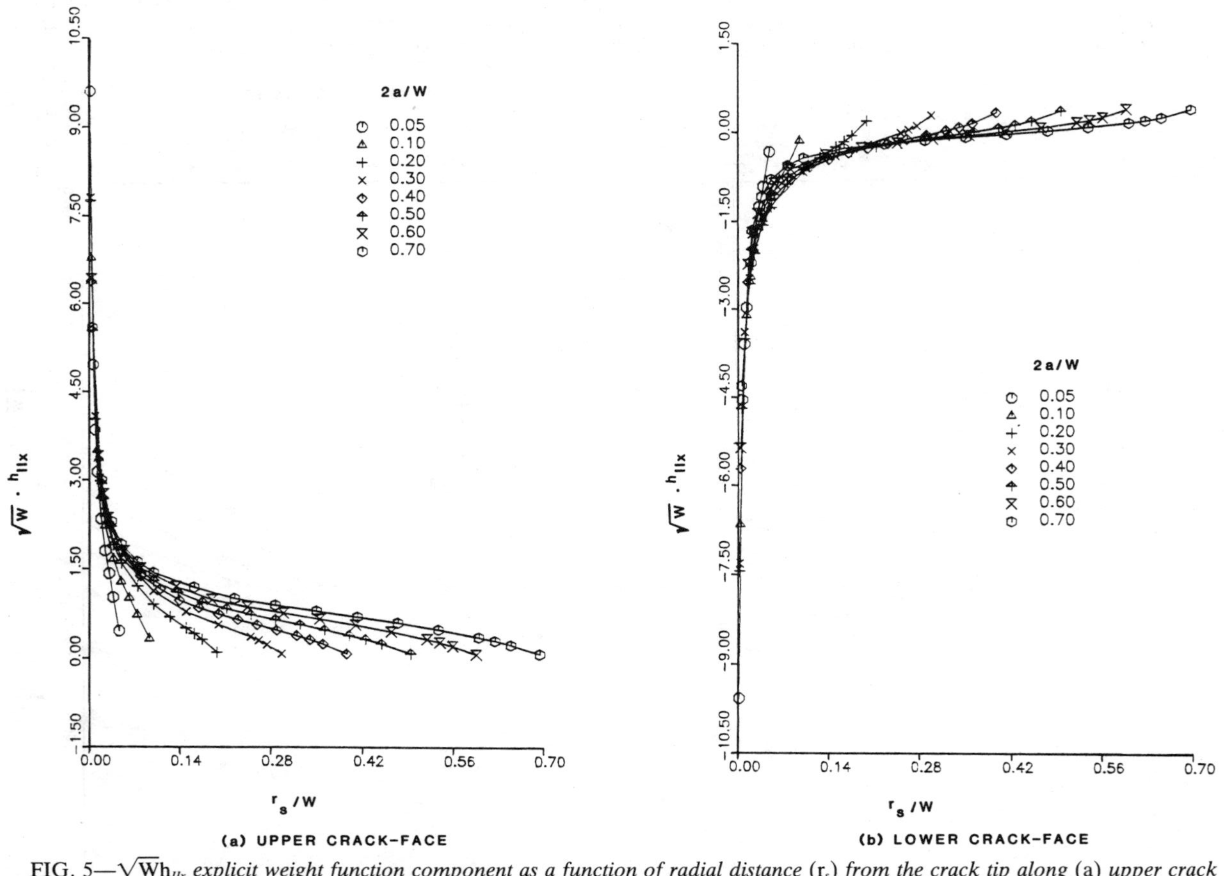

FIG. 5—$\sqrt{W}h_{IIx}$ *explicit weight function component as a function of radial distance* (r_s) *from the crack tip along* (a) *upper crack face and* (b) *lower crack face for a centrally oblique crack in an orthotropic square plate with* $0.05 \leq 2a/W \leq 0.7$ *and* $\phi_1 = \phi_2 = 45$ *deg.*

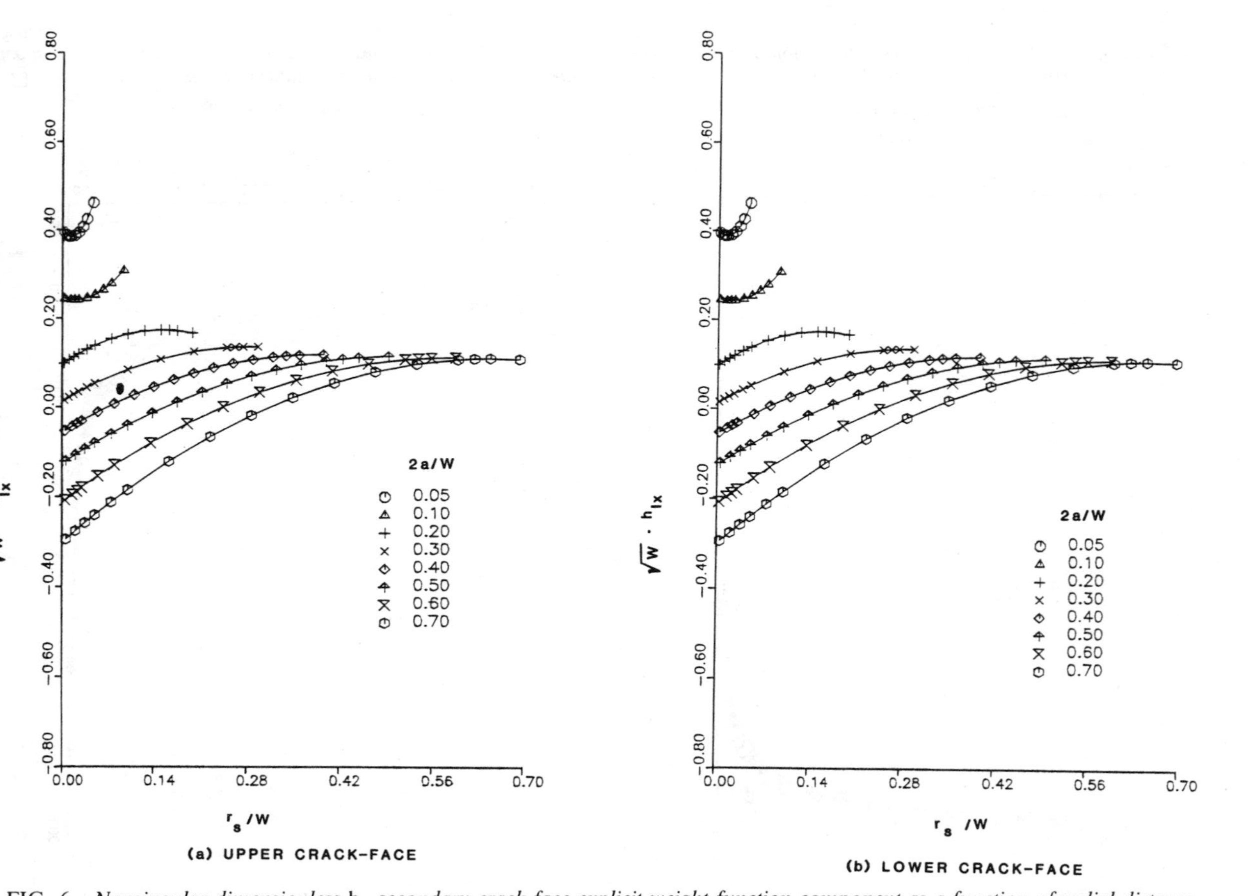

FIG. 6—*Nonsingular dimensionless* h_{Ix} *secondary crack-face explicit weight function component as a function of radial distance* (r_s) *from the crack tip along* (a) *upper crack face and* (b) *lower crack face of a centrally oblique crack in an orthotropic square plate with* $0.05 \leq 2a/W \leq 0.7$ *and* $\phi_1 = \phi_2 = 45$ *deg.*

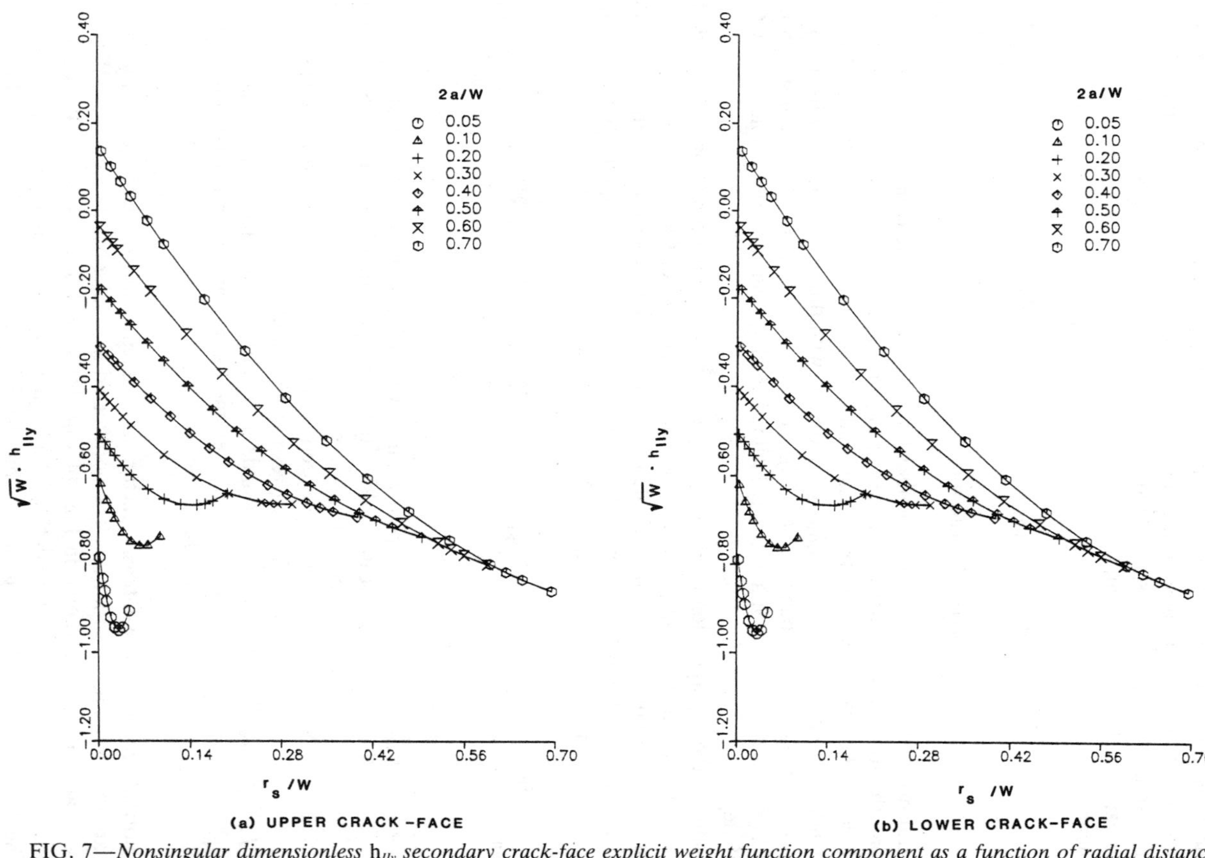

FIG. 7—*Nonsingular dimensionless* h_{IIy} *secondary crack-face explicit weight function component as a function of radial distance* (r_s) *from the crack tip along the* (a) *upper crack face and* (b) *lower crack face of a centrally oblique crack in an orthotropic square plate with* $0.05 \leq 2a/W \leq 0.7$ *and* $\phi_1 = \phi_2 = 45$ *deg.*

crack lengths. The secondary crack-face weight function components such as h_{Ix} and h_{IIy} are nonsingular in the crack tip neighborhood, as shown in Figs. 6 and 7, respectively.

In general, to calculate mixed-mode stress intensity factors with the weight function concept for traction application locations other than crack-face loading, the weight function components with respect to (x', y') coordinates ($h_{Ix'}$, $h_{Iy'}$, $h_{IIx'}$, and $h_{IIy'}$ components) are more convenient to use than those of (x, y) coordinates (h_{Ix}, h_{Iy}, h_{IIx}, and h_{IIy} components). Since the explicit weight functions are vectors, the simple coordinate transformation can be made to express $H_{Ix'}$, $H_{Iy'}$, $h_{IIx'}$, and $h_{IIy'}$ components as a function of h_{Ix}, h_{Iy}, h_{IIx}, and h_{IIy} components as

$$h_{I(II)x'} = h_{I(II)x} \cos \phi_2 - h_{I(II)y} \sin \phi_2 \tag{4a}$$

$$h_{I(II)y'} = h_{I(II)x} \sin \phi_2 + h_{I(II)y} \cos \phi_2 \tag{4b}$$

The transformed $h_{Iy'}$ and $h_{IIy'}$ components with different crack lengths ($0.05 \leq 2a/W \leq 0.7$) along the top and the bottom faces are shown in Figs. 8 and 9, respectively, for the centrally located crack in the orthotropic material with oblique angle $\phi_2 = 45$ deg. On the other hand, the transformed $h_{Ix'}$ and $h_{IIx'}$ components with different crack lengths ($0.05 \leq 2a/W \leq 0.7$) along the right-hand side (RHS) and left-hand side (LHS) faces are shown in Figs. 10 and 11, respectively, for a centrally inclined crack in a square orthotropic plate with $\phi_2 = 45$ deg.

By applying the superposition principle to biaxial remote loading conditions, the Mode I and Mode II stress-intensity factors for Case 12*a* are the sum of Cases 12*b* and 12*c* (Fig. 12) with biaxiality ratio $\eta = \sigma_{x'x'}/\sigma_{y'y'}$. Evaluation of the mixed-mode stress intensity factors under biaxial loading conditions requires the explicit weight function components $h_{I(II)y'}$ along the top and bottom faces and $h_{I(II)x'}$ along RHS and LHS faces, respectively, as

$$K_{I(II)} = \sum_{i=1}^{N_{TOP}} f_{y_i'}^{TOP} \cdot h_{I(II)y_i'}^{TOP} + \sum_{i=1}^{N_{BOT}} f_{y_i'}^{BOT} \cdot h_{I(II)y_i'}^{BOT} + \sum_{i=1}^{N_{RHS}} f_{x_i'}^{RHS} \cdot h_{I(II)x_i'}^{RHS} + \sum_{i=1}^{N_{LHS}} f_{x_i'}^{LHS} \cdot h_{I(II)x_i'}^{LHS} \tag{5}$$

where

$h_{I(II)y_i'}^{TOP}, h_{I(II)y_i'}^{BOT}, h_{I(II)x_i'}^{RHS}, h_{I(II)x_i'}^{LHS}$ = decoupled explicit weight function components at *i*'s nodal location along the top, bottom, RHS, and LHS traction application faces,

$N_{TOP}, N_{BOT}, N_{RHS}, N_{LHS}$ = total discretized nodes along the top, bottom, RHS, and LHS faces, respectively, and

$f_{y_i'}^{TOP}, f_{y_i'}^{BOT}, f_{x_i'}^{RHS}, f_{x_i'}^{LHS}$ = consistent nodal forces for $\sigma_{y'y'}$ applied stress along the top and bottom faces and for $\sigma_{x'x'}$ applied stress along RHS and LHS faces, respectively.

The normalized $K_{I(II)}$ stress-intensity factors as a function of a wide range of η ($-2 \leq \eta \leq 2$) are shown in Fig. 13 with Eq 5. The normalized $K_{I(II)}$ values as a function $2a/W$ ratio for $\eta = 0$ and $\eta = 1$ are shown in Fig. 14 with $H/W = 1.0$, $\phi_1 = \phi_2 = 45$ deg, and $0.05 \leq 2a/W \leq 0.7$. With these predetermined explicit weight function components ($h_{I(II)y'}$ along the top and bottom faces and $h_{I(II)x'}$ along RHS and LHS faces, as presented in Figs. 8 to 11), the $K_{I(II)}$ values at any biaxiality ratio η value as shown in Fig. 12 can be obtained efficiently and accurately by coupling the weight function concept with the superposition principles.

In order to collate the present finite element study of $K_{I(II)}$ calculations in orthotropic

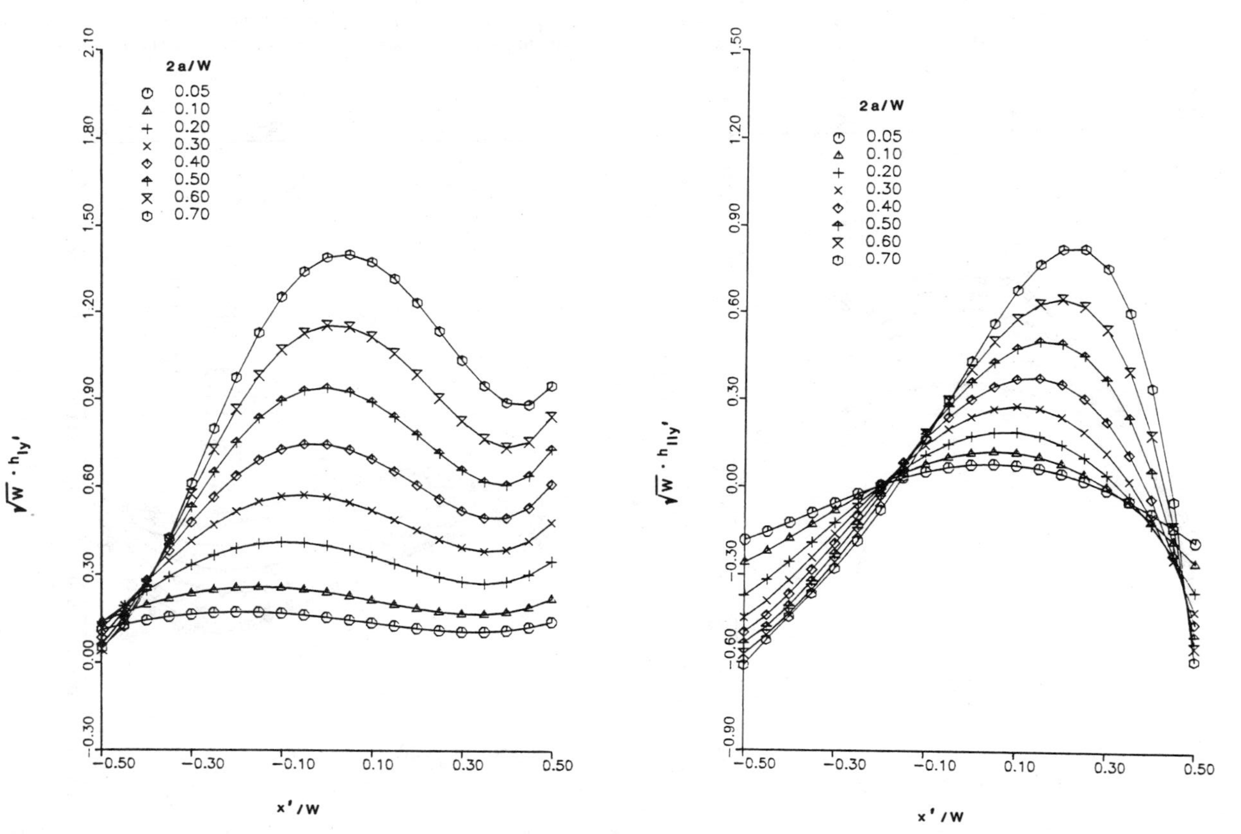

FIG. 8—*Dimensionless* $h_{Iy'}$ *and* $h_{IIy'}$ *explicit weight function components as a function of x' along the top face of a centrally oblique crack in an orthotropic square plate with* $0.05 \leq 2a/W \leq 0.7$ *and* $\phi_1 = \phi_2 = 45$ *deg.*

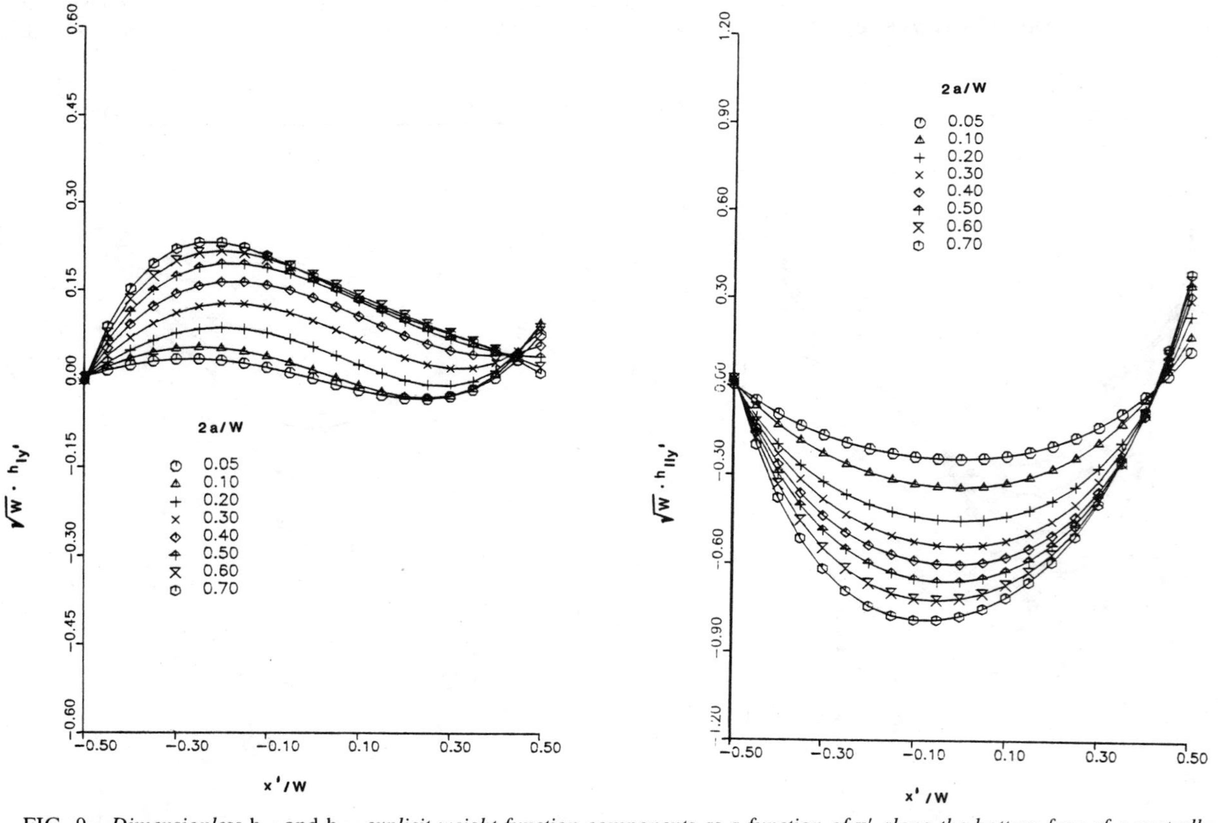

FIG. 9—*Dimensionless* $h_{Iy'}$ and $h_{IIy'}$ *explicit weight function components as a function of* x' *along the bottom face of a centrally oblique crack in an orthotropic square plate with* $0.05 \leq 2a/W \leq 0.7$ *and* $\phi_1 = \phi_2 = 45$ *deg.*

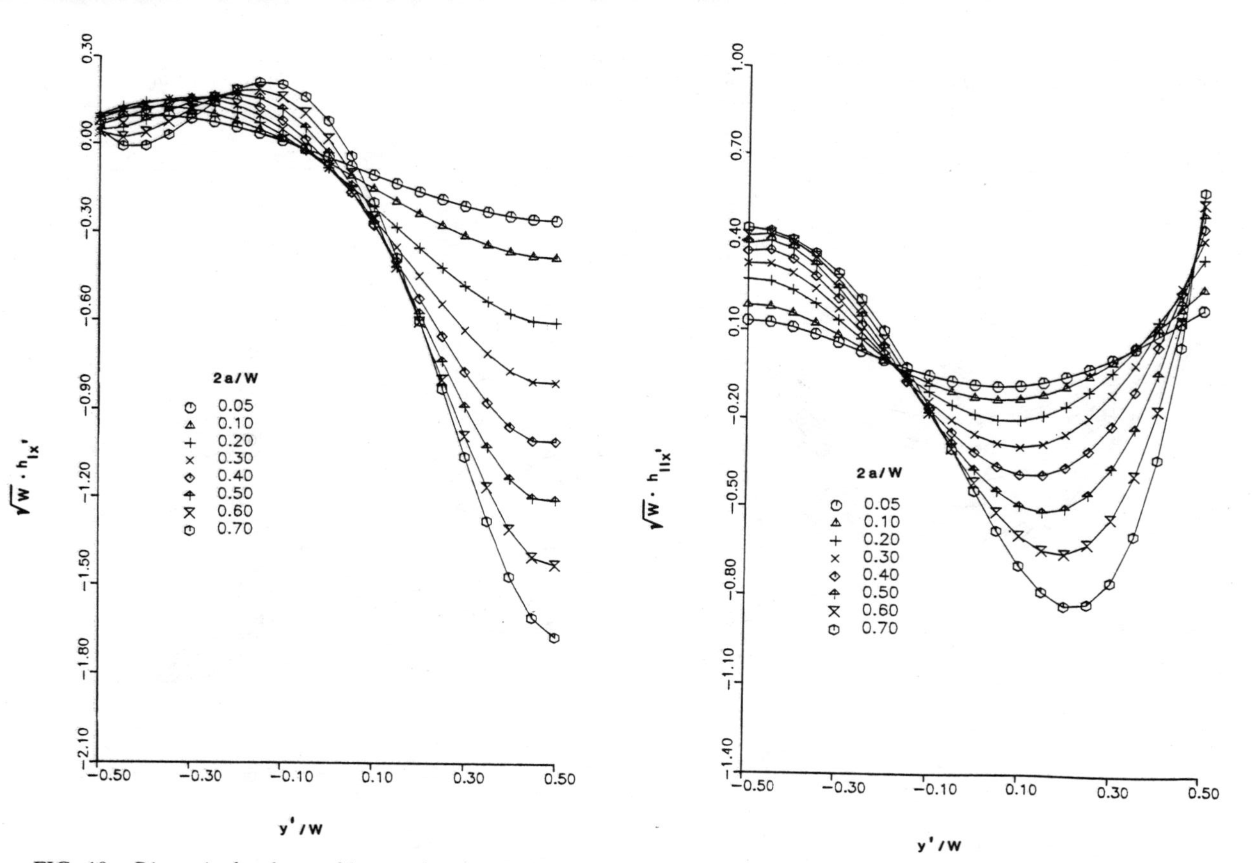

FIG. 10—*Dimensionless* $h_{Ix'}$ *and* $h_{IIx'}$ *explicit weight function components as a function of* y' *along the RHS face of a centrally oblique crack in an orthotropic square plate with* $0.05 \leq 2a/W \leq 0.7$ *and* $\phi_1 = \phi_2 = 45$ *deg.*

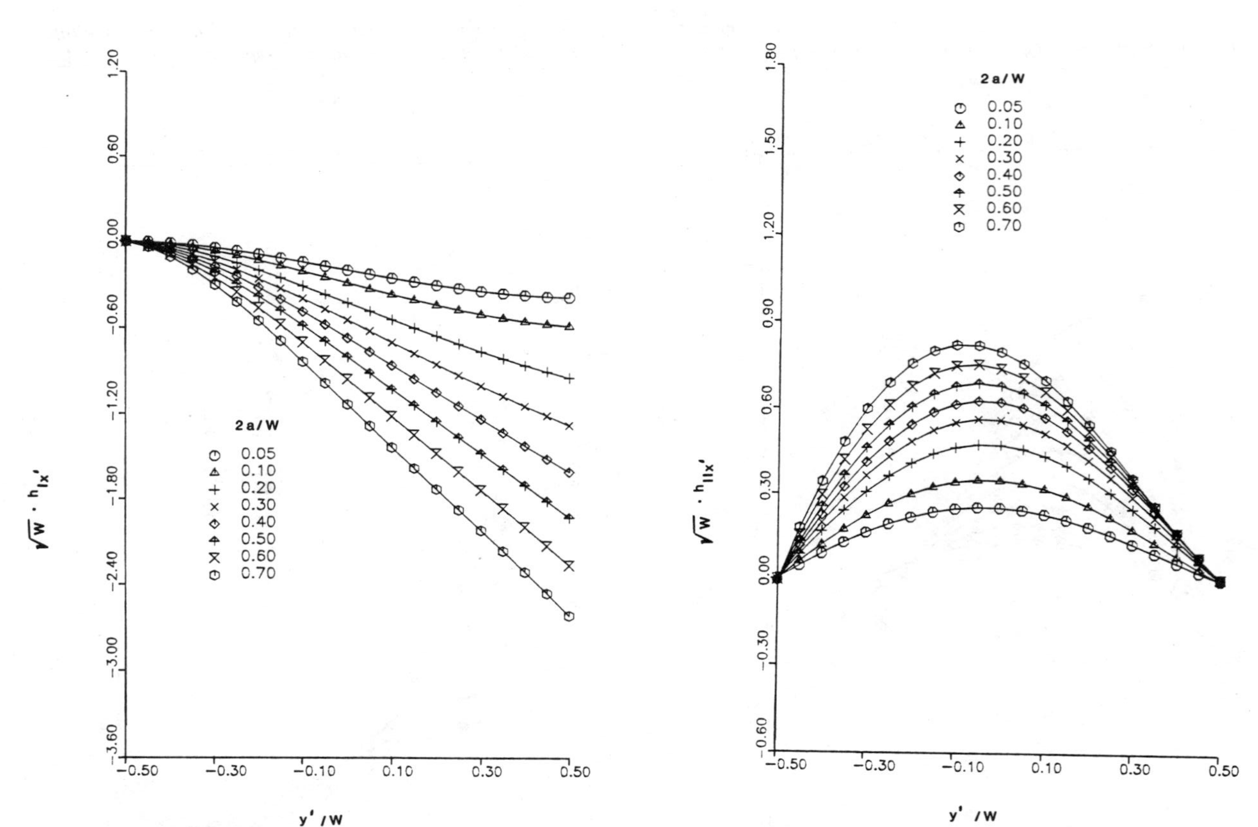

FIG. 11—*Dimensionless* $h_{Ix'}$ *and* $h_{IIx'}$ *explicit weight function components as a function of* y' *along the LHS face of a centrally oblique crack in an orthotropic square plate with* $0.05 \leq 2a/W \leq 0.7$ *and* $\phi_1 = \phi_2 = 45$ *deg.*

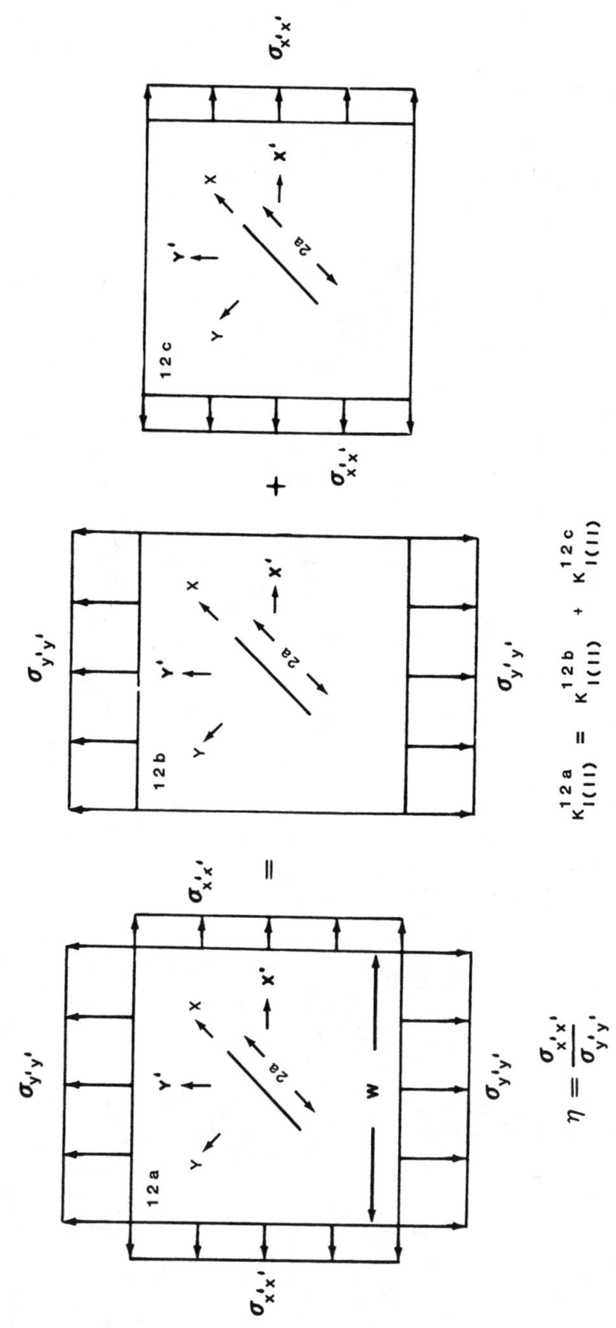

FIG. 12—*Evaluating the decoupled stress-intensity factors under biaxial loading conditions with the linear superposition principle coupled with the weight function concept.*

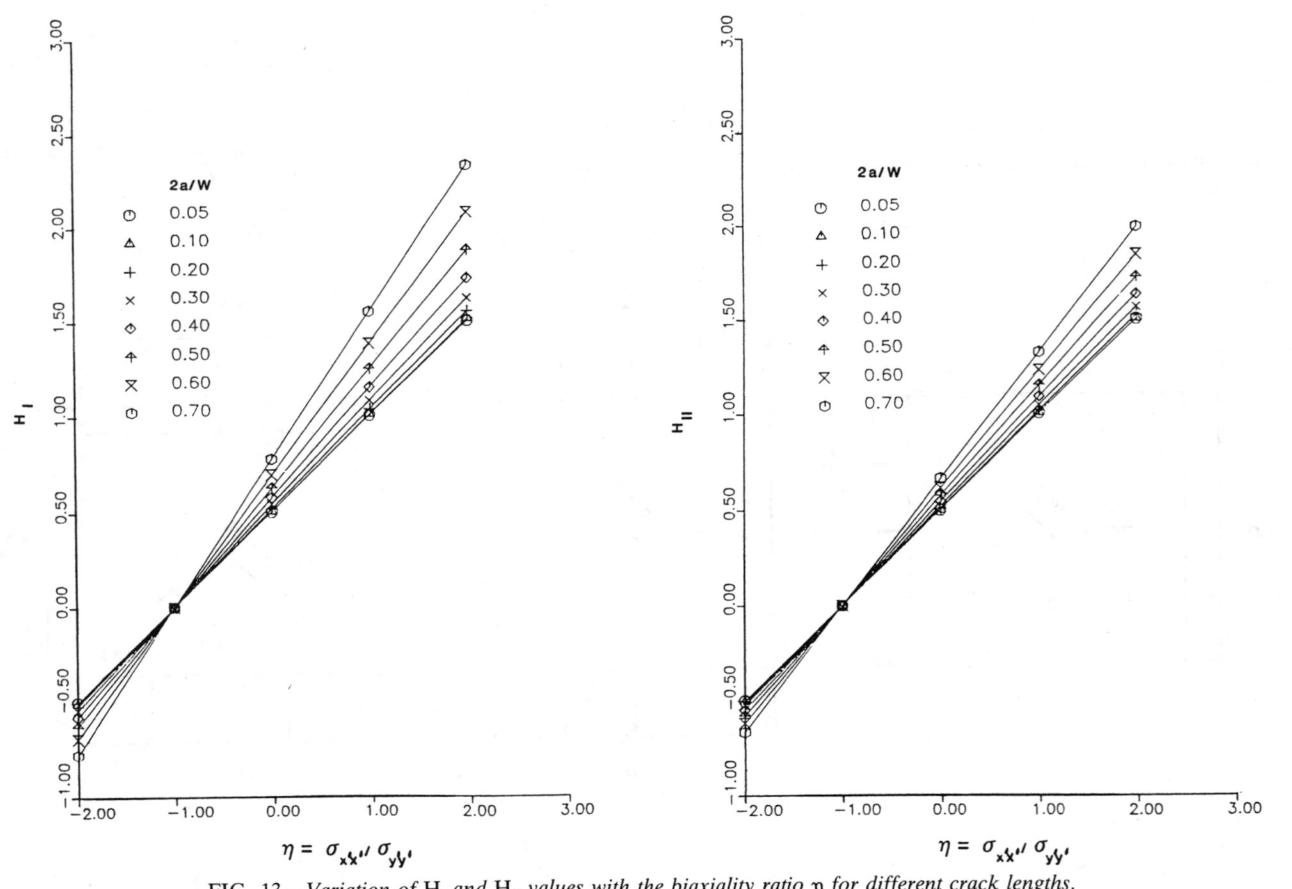

FIG. 13—*Variation of* H_I *and* H_{II} *values with the biaxiality ratio* η *for different crack lengths.*

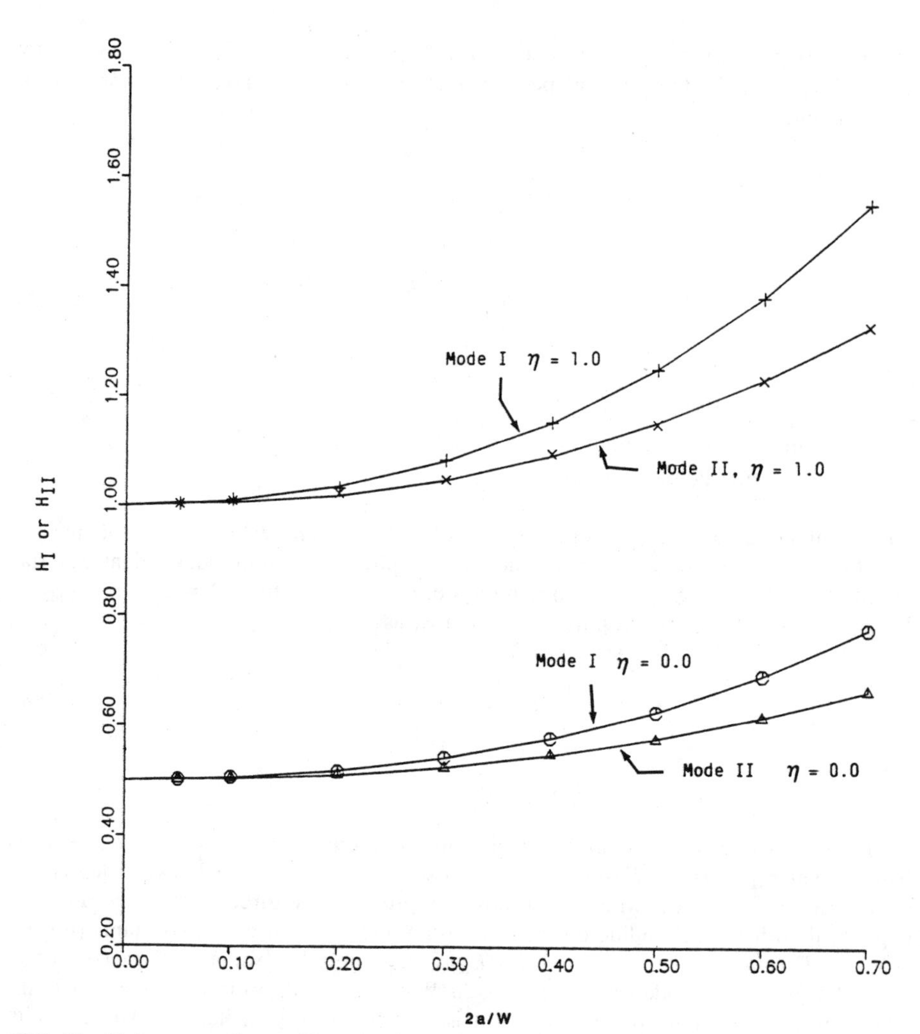

FIG. 14—*Finite correct factors* (H_I *and* H_{II}) *as a function of* 2a/W *ratios for* $\eta = 0.0$ *and* $\eta = 1.0$.

material with boundary collocation results, the effect of the E_{11}/E_{22} ratio on finite width correction factor (H) is studied in terms of the β_i values, which represent the imaginary part of the complex roots of the characteristic equation (as shown by Sih and Liebowitz [*11*]). By introducing the Airy stress function to the rectilinear anisotropic body, the governing equation is converted to the characteristic equation with four roots that must be either complex or imaginary, as given by Lekhnitskii [*18*] with energy consideration. These four roots are

$$
\begin{aligned}
s_1 &= \alpha_1 + i\beta_1 \\
s_2 &= \alpha_2 + i\beta_2 \\
s_3 &= \bar{s}_1 \\
s_4 &= \bar{s}_2
\end{aligned}
\qquad (6)
$$

where α_i, β_i (i = 1, 2) are real constants with $\beta_1 > 0$, $\beta_2 > 0$, and $\beta_1 \neq \beta_2$. As given by Bowie and Freese [12], the four independent material properties have been expressed in terms of β_1 and β_2 as

$$\beta_1\beta_2 = \left(\frac{E_{xx}}{E_{yy}}\right)^{1/2} \tag{7a}$$

$$\beta_1 + \beta_2 = \sqrt{2}\left\{\frac{E_{xx}}{E_{yy}} + \frac{E_{xx}}{2G_{xy}} - \nu_{xy}\right\}^{1/2} \tag{7b}$$

where

E_{xx}, E_{yy} = principal Young's moduli,
G_{xy} = in-plane shear modulus, and
ν_{xy} = in-plane Poisson's ratio.

For the intended purposes, use of the validated finite element technique for studying the effect of E_{xx}/E_{yy} ratios on $K_{I(II)}$ values was not made explicitly with four independent material constants but with the superficially chosen ones of $\alpha_1 = \alpha_2 = 0$ fixed $\beta = 1$, varying $\beta_2 = \sqrt{E_{xx}/E_{yy}}$, $\nu_{12} = 0.3$, and constraints for G_{xy} and ν_{yx} as

$$\frac{1}{G_{xy}} = \frac{1}{E_{xx}} + \frac{1}{E_{yy}} + \frac{2\nu_{xy}}{E_{xx}} \tag{8a}$$

$$\nu_{yx}E_{xx} = \nu_{xy}E_{yy} \tag{8b}$$

With these two constraint equations, the isotropic requirement of $G = E/2(1 + \nu)$ is satisfied when $E_{11} = E_{22} = E$ of the isotropic assumption of having only two independent constants. In other words, with the proposed approach only three elastic constants are independent with G_{xy} depending on the other three independent elastic constants (for example, E_{xx}, E_{yy}, and ν_{xy}). Varying $\beta_2^2 = (E_{11}/E_{22})$ values affect $K_{I(II)}$ values or finite correction factors (H) for mixed-mode cracks ($0.05 \leq 2a/W \leq 0.8$) in an orthotropic plate or both, as shown in Fig. 15. Note that $\beta_1 = \beta_2 = 1$ is used for isotropic solutions. These results confirm that for a large value of $E_{x'x'}/E_{y'y'}$, the correction factor converges to the isotropic solution of infinite medium solutions.

Dimensionless explicit weight function components such as $\sqrt{W}\, h_{I(II)x}$, $\sqrt{W}\, h_{I(II)y}$ are invariant with similar centrally oblique crack geometries (same $H/W = 1, 2a/W$) regardless of their actual W size. Assuming two centrally inclined crack geometries with two different widths (W_1 and W_2) and with weight function components ($h^{W_1}_{I(II)x}$, $h^{W_1}_{I(II)x}$, $h^{W_1}_{I(II)x}$, and $h^{W_1}_{I(II)y}$) exist at a similar location, the following invariant relationship holds as

$$\sqrt{W_1}\, h^{W_1}_{I(II)x} = \sqrt{W_2}\, h^{W_2}_{I(II)x} \tag{9a}$$

$$\sqrt{W_1}\, h^{W_1}_{I(II)y} = \sqrt{W_2}\, h^{W_2}_{I(II)y} \tag{9b}$$

The primary crack-face weight function components (h_{Iy} and h_{IIx}) for a centrally oblique crack ($\phi_2 = 45$ deg) in a square orthotropic plate can be expressed analytically as a function

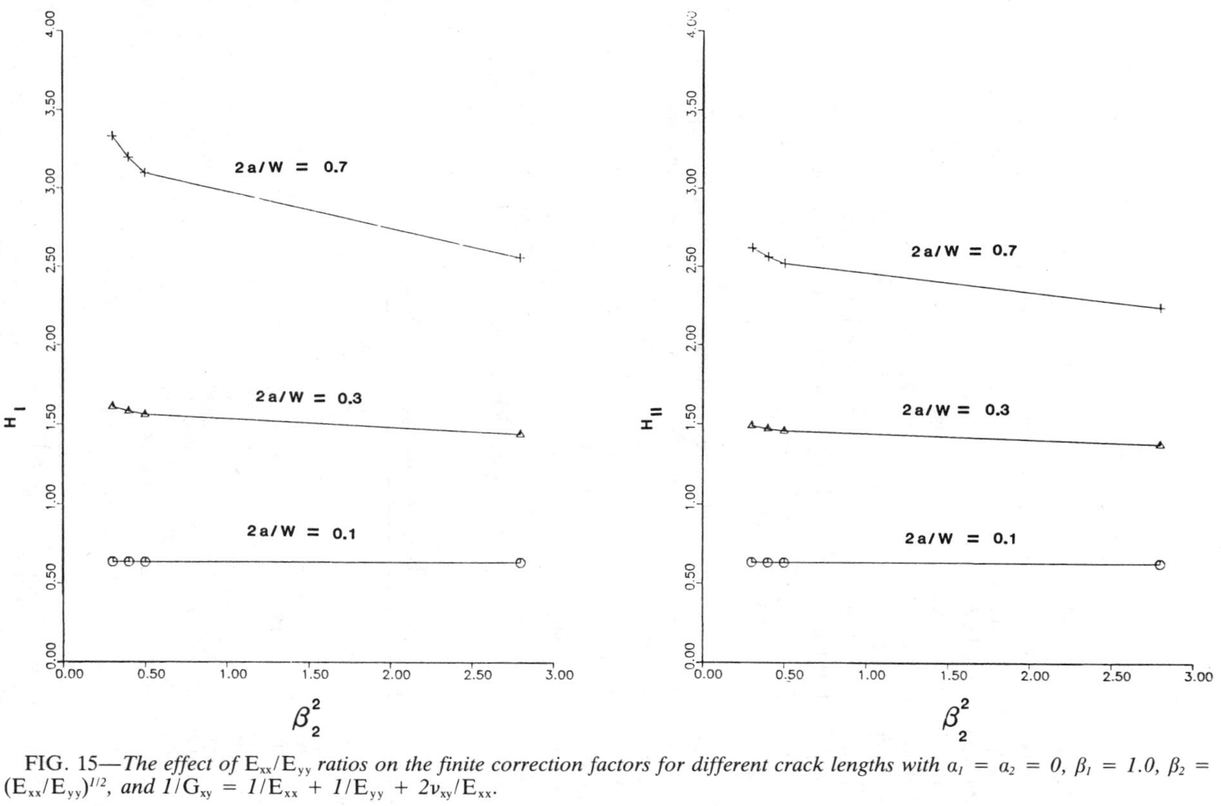

FIG. 15—*The effect of* E_{xx}/E_{yy} *ratios on the finite correction factors for different crack lengths with* $a_1 = a_2 = 0$, $\beta_1 = 1.0$, $\beta_2 = (E_{xx}/E_{yy})^{1/2}$, *and* $1/G_{xy} = 1/E_{xx} + 1/E_{yy} + 2\nu_{xy}/E_{xx}$.

of radial distance (r_s) along the crack face from the crack tip as

$$\sqrt{W}\, h_{\mathrm{Iy}}^{T(B)} = \sum_{n=1}^{N} A_n^{T(B)} \left(\frac{r_s}{2a}\right)^{(n/2-1)} \tag{10a}$$

$$\sqrt{W}\, h_{\mathrm{IIx}}^{T(B)} = \sum_{n=1}^{N} B_n^{T(B)} \left(\frac{r_s}{2a}\right)^{(n/2-1)} \tag{10b}$$

The least-square coefficient $A_n^{T(B)}$ and $B_n^{T(B)}$ is shown in Tables 1 and 2, respectively, with superscripts T and B referring to the top and bottom crack face, respectively.

The decoupled explicit weight function component at i's location can be interpreted quantitatively as the decoupled stress-intensity factors $K_{\mathrm{I(II)}}$ with unit traction application at i's nodal location with respect to the global axes. In other words, the explicit decoupled weight function is, in fact, the Green's function for determining the decoupled stress intensity factors. Since the explicit weight functions can be efficiently made for the entire structure of interest as shown in this paper with the energy perturbation approach, they can be logically applied to guide the damage-tolerant design concept. According to the physical interpretation of the explicit weight functions, the damage-tolerant design from a stress analyst's viewpoint can be applied in the following manner:

1. The designer can determine where to avoid the boundary traction application locations that can produce elevated $K_{\mathrm{I(II)}}$ values due to high $h_{\mathrm{I(II)}}$ values of an orthotropic structure containing a crack under the preselected constraint conditions.
2. Constraint conditions that can reduce $K_{\mathrm{I(II)}}$ values for a given crack geometry with prefixed boundary traction applications can be adjusted.
3. By applying the superposition principle, the life-impacting stress intensity factor of a flawed structure can be determined by the designer with the "uncracked" stress field coupled with the predetermined explicit weight function for fatigue life evaluation

TABLE 1—*Least-square fitted coefficients of Eq 10*a *of* $\sqrt{w}\, h_{Iy}$ *versus* $r_s/2a$.

2a/W	A_1^T	A_2^T	A_3^T	A_4^T	A_5^T	A_6^T
			A_i^T COEFFICIENTS FOR THE UPPER CRACK FACE			
0.05	0.21373E+01	−0.50128E+01	0.23191E+02	−0.50120E+02	0.48875E+02	−0.18687E+02
0.1	0.15197E+01	−0.35221E+01	0.16906E+02	−0.36442E+02	0.35519E+02	−0.13599E+02
0.2	0.95344E+00	−0.99985E+00	0.83153E+01	−0.22161E+02	0.24992E+02	−0.10735E+02
0.3	0.76224E+00	−0.39215E+00	0.66353E+01	−0.19112E+02	0.22777E+02	−0.10239E+02
0.4	0.67065E+00	−0.25520E+00	0.67783E+01	−0.18121E+02	0.20553E+02	−0.91440E+01
0.5	0.60388E+00	−0.18445E−01	0.67188E+01	−0.17253E+02	0.18993E+02	−0.85271E+01
0.6	0.53571E+00	0.50408E+00	0.57013E+01	−0.14808E+02	0.16245E+02	−0.76367E+01
0.7	0.50155E+00	0.70855E+00	0.61954E+01	−0.14543E+02	0.14405E+02	−0.67180E+01
			A_i^B COEFFICIENTS FOR THE LOWER CRACK FACE			
2a/W	A_1^B	A_2^B	A_3^B	A_4^B	A_5^B	A_6^B
0.05	−0.21418E+01	0.53556E+01	−0.23413E+02	0.50523E+02	−0.49234E+02	0.18800E+02
0.1	−0.15212E+01	0.39954E+01	−0.16980E+02	0.36533E+02	−0.35650E+02	0.13642E+02
0.2	−0.95363E+00	0.17728E+01	−0.83278E+01	0.21995E+02	−0.25015E+02	0.10743E+02
0.3	−0.76220E+00	0.14957E+01	−0.66336E+01	0.18721E+02	−0.22772E+02	0.10237E+02
0.4	−0.67078E+00	0.17246E+01	−0.67912E+01	0.17503E+02	−0.20583E+02	0.91546E+01
0.5	−0.60372E+00	0.18697E+01	−0.67050E+01	0.16255E+02	−0.18965E+02	0.85169E+01
0.6	−0.53574E+00	0.17842E+01	−0.57019E+01	0.13445E+02	−0.16245E+02	0.76368E+01
0.7	−0.50160E+00	0.20646E+01	−0.62014E+01	0.12720E+02	−0.14418E+02	0.67223E+01

TABLE 2—*Least-square fitted coefficients of Eq 10b of $\sqrt{w}\, h_{IIx}$ versus $r_s/2a$.*

$2a/W$	B_1^T	B_2^T	B_3^T	B_4^T	B_5^T	B_6^T
		B_i^T COEFFICIENTS FOR THE UPPER CRACK FACE				
0.05	0.42979E+00	0.13628E+02	−0.48030E+02	0.73687E+02	−0.52541E+02	0.13087E+02
0.1	0.30794E+00	0.95816E+01	−0.33690E+02	0.51593E+02	−0.36652E+02	0.90483E+01
0.2	0.65704E+00	0.31367E+01	−0.13766E+02	0.24874E+02	−0.20619E+02	0.58117E+01
0.3	0.58669E+00	0.21645E+01	−0.10272E+02	0.19328E+02	−0.15896E+02	0.41782E+01
0.4	0.53462E+00	0.13026E+01	−0.53873E+01	0.89706E+01	−0.61948E+01	0.85497E+00
0.5	0.46487E+00	0.13282E+01	−0.51461E+01	0.83590E+01	−0.52683E+01	0.34234E+00
0.6	0.46278E+00	0.76558E+00	−0.27181E+01	0.40354E+01	−0.12295E+01	−0.12344E+01
0.7	0.41798E+00	0.85015E+00	−0.27698E+01	0.43831E+01	−0.16036E+01	−0.11922E+01

$2a/W$	B_1^B	B_2^B	B_3^B	B_4^B	B_5^B	B_6^B
		B_i^B COEFFICIENTS FOR THE LOWER CRACK FACE				
0.05	−0.43270E+00	−0.13573E+02	0.47885E+02	−0.73456E+02	0.52326E+02	−0.13013E+02
0.1	−0.31238E+00	−0.95155E+01	0.33460E+02	−0.51174E+02	0.36265E+02	−0.89140E+01
0.2	−0.65703E+00	−0.31368E+01	0.13767E+02	−0.24876E+02	0.20620E+02	−0.58119E+01
0.3	−0.58693E+00	−0.21602E+01	0.10249E+02	−0.19275E+02	0.15843E+02	−0.41587E+01
0.4	−0.53462E+00	−0.13027E+01	0.53874E+01	−0.89709E+01	0.61951E+01	−0.85511E+00
0.5	−0.46487E+00	−0.13282E+01	0.51457E+01	−0.83577E+01	0.52668E+01	−0.34178E+00
0.6	−0.46279E+00	−0.76542E+00	0.27167E+01	−0.40312E+01	0.12243E+01	0.12365E+01
0.7	−0.41797E+00	−0.85027E+00	0.27707E+01	−0.43861E+01	0.16074E+01	0.11906E+01

without physically involving the stress analyses of stress intensity factor evaluations. Additional technical details on using the energy perturbation concept for two-dimensional cracks and notches in isotropic materials for damage-tolerant design applications can be found in a recent paper by Sha, Chen, and Yang [*19*].

Conclusions

An efficient finite-element methodology for accurate evaluation of decoupled stress-intensity factors and decoupled explicit weight functions for the mixed-mode orthotropic crack has been developed by applying the virtual crack extension technique to the symmetric mesh in the crack-tip neighborhood. The proposed finite-element technique for the mixed-mode orthotropic crack has been validated by comparing the computed $K_{I(II)}$ values numerically and analytically with the existing $K_{I(II)}$ solutions. This proposed finite-element technique is limited to the crack face lying along both the principal axes of orthotropy and the symmetric axis of symmetric mesh in the crack-tip neighborhood. Further extension of the weight function evaluation for the two-dimensional crack in orthotropic material with a crack that does not lie on the principal axis of orthotropy will be the subject of a forthcoming publication.

The additional finite-element efforts needed to obtain the explicit weight functions for the entire structure of interest with an asymmetric orthotropic crack are not much greater than the normal finite-element evaluation of the stress intensity factors for given loading and constraint conditions with a single collinear VCE technique coupled with symmetric mesh in a crack-tip neighborhood. From the applied fracture mechanics practitioner's viewpoint, the payoffs far exceed the additional efforts for these reasons:

1. Explicit weight function can serve as the universal function for $K_{I(II)}$ determinations according to Eq 2 for a given orthotropic crack geometry under any loading condition.
2. Explicit weight function can provide a unique capability in guiding damage-tolerant design applications from the stress analyst's viewpoint. This is accomplished with the

physical interpretation of the explicit weight functions as the Green's function of stress-intensity factor calculation.

This finite-element approach with symmetric mesh in the crack-tip neighborhood represents the first effort in explicit weight function evaluation of an orthotropic material with a crack lying along one of the symmetry planes.

References

[1] Bueckner, H. F., "Field Singularities and Related Integral Representations," *Mechanics of Fracture I. Method of Analysis and Solutions of Crack Problems*, G. C. Sih, Ed., Noordhoff, the Netherlands, 1972.

[2] Parks, D. M., "A Stiffness Derivative Finite Element Technique for Determination of Crack Tip Stress Intensity Factors," *International Journal of Fracture*, Vol. 10, 1974, pp. 487–501.

[3] Hellen, T. K., "On the Method of Virtual Crack Extensions," *International Journal of Numerical Methods in Engineering*, Vol. 9, 1975, pp. 187–207.

[4] Henshell, R. D. and Shaw, K. G., "Crack Tip Elements Are Unnecessary," *International Journal of Numerical Methods in Engineering*, Vol. 9, 1975, pp. 495–507.

[5] Barsoum, R. S., "On the Use of Isoparametric Finite Elements in Linear Fracture Mechanics," *International Journal of Numerical Methods in Engineering*, Vol. 10, 1976, pp. 25–27.

[6] Sha, G. T., "Stiffness Derivative Finite Element Technique to Determine Nodal Weight Functions with Singularity Elements," *Engineering Fracture Mechanics*, Vol. 19, 1984, pp. 685–699.

[7] Rice, J. R., "Some Remarks on Elastic Crack-Tip Stress Fields," *International Journal of Solids Structures*, Vol. 8, 1972, pp. 751–758.

[8] Sha, G. T. and Yang, C-T, "Weight Function Calculation for Mixed-Mode Fracture Problems with the Virtual Crack Extension Technique," *Journal of Engineering Fracture Mechanics*, Vol. 21, 1985, pp. 1119–1149.

[9] Sih, G. C., Paris, P. C., and Irwin, G. R., "On Cracks in Rectilinearly Anisotropic Bodies," *International Journal of Fracture Mechanics*, Vol. 1, 1965, pp. 189–203.

[10] Sha, G. T. and Yang, C-T, "Determination of Mixed Mode Stress Intensity Factors Using Explicit Weight Functions," *Fracture Mechanics: Eighteenth Symposium, ASTM STP 945*, D. T. Read and R. P. Reed, Eds., American Society for Testing and Materials, Philadelphia, 1987, pp. 301–330.

[11] Sih, G. C. and Liebowitz, H., "Mathematical Theories of Brittle Fracture," *Fracture*, Vol. 2, H. Liebowitz, Ed., Academic Press, New York, 1968

[12] Bowie, O. L. and Freese, C. E., "Central Crack in Plane Orthotropic Rectangular Sheet, *International Journal of Fracture*, Vol. 8, 1972, pp. 49–58.

[13] Gandhi, K. R., "Analysis of an Inclined Crack Centrally Placed in an Orthotropic Rectangular Plate," *Journal of Strain Analysis*, Vol. 7, 1972, pp. 157–163.

[14] Lin, K. Y. and Tong, P., "A Hybrid Crack Element for the Fracture Mechanics Analyses of Composite Materials," *Numerical Methods in Fracture Mechanics, Proceedings*, 1st International Conference, University College of Wales, Swansea, U.K., 9–13 Jan. 1978, A. R. Luxmoore and D. R. J. Own, Eds., pp. 733–746.

[15] Foschi, R. O. and Barrett, J. D., "Stress Intensity Factors in Anisotropic Plates Using Singular Isoparametric Elements," *International Journal of Numerical Methods in Engineering*, Vol. 10, 1976, pp. 1281–1287.

[16] Gifford, L. N., Jr., and Hilton, P. D., "Stress Intensity Factors by Enriched Finite Elements," *Engineering Fracture Mechanics*, Vol. 10, 1978, pp. 485–496.

[17] Snyder, M. D. and Cruse, T. A., "Crack Tip Stress Intensity Factors in Finite Anisotropic Plates," AFML-TR-73-209, Air Force Materials Laboratory, Aug. 1973.

[18] Lekhnitskii, S. G., *Anisotropic Plates*, translated by S. W. Tsai and T. Cheron, Gordon and Breach, New York, 1968.

[19] Sha, G. T., Chen, J. K., and Yang, C-T, "Energy Perturbation Finite Element Technique for Damage Tolerant Design Applications," *Journal of Engineering Fracture Mechanics*, Vol. 29, 1988, pp. 197–218.

Robert A. Sire,[1] David O. Harris,[1] and Ernest D. Eason[2]

Automated Generation of Influence Functions for Planar Crack Problems

REFERENCE: Sire, R. A., Harris, D. O., and Eason, E. D., **"Automated Generation of Influence Functions for Planar Crack Problems,"** *Fracture Mechanics: Perspectives and Directions (Twentieth Symposium), ASTM STP 1020,* R. P. Wei and R. P. Gangloff, Eds., American Society for Testing and Materials, Philadelphia, 1989, pp. 351–365.

ABSTRACT: A numerical procedure for the generation of influence functions for Mode I planar problems is described. The resulting influence functions are in a form for convenient evaluation of stress-intensity factors for complex stress distributions. Crack surface displacements are obtained by a least-squares solution of the Williams eigenfunction expansion for displacements in a cracked body. Discrete values of the influence function, evaluated using the crack surface displacements, are curve fit using an assumed functional form. The assumed functional form includes appropriate limit-behavior terms for very deep and very shallow cracks. Continuous representation of the influence function provides a convenient means for evaluating stress-intensity factors for arbitrary stress distributions by numerical integration. The procedure is demonstrated for an edge-cracked strip and a radially cracked disk. Comparisons with available published results demonstrate the accuracy of the procedure.

KEY WORDS: influence function, stress-intensity factor, planar crack, least-squares

Nomenclature

- A_i Coefficients in Williams stress function
- a Crack length
- $C_{i,n}$ Coefficients in influence function
- E Young's modulus
- h Mode I influence function
- K Mode I stress-intensity factor
- u_r, u_θ Displacements in the r and θ directions, respectively
- W Body width at crack
- α Dimensionless crack length, a/W
- ξ Dimensionless crack surface position, x/a
- ν Poisson's ratio

Stress-intensity factor solutions exist for numerous planar crack problems [*1–3*]. However, they are generally for relatively simple stress systems, such as uniform or linearly varying stress. In many instances, more complex stress distributions are of interest, such as steep nonlinear stress gradients encountered in thermal and residual stress problems and in the vicinity of geometric stress concentrations. Crack configurations are often encountered for which the stress-intensity factor solutions are not available, even for uniform or linear stress

[1] Senior engineer and managing engineer, respectively, Failure Analysis Associates, Inc., Palo Alto, CA 94303.

[2] President, Modeling and Computing Services, Sunnyvale, CA 94089.

distributions. Consequently, there is a need for an efficient numerical procedure for evaluation of stress-intensity factors in a wide variety of crack geometries and stress distributions. This paper describes such a procedure.

Several well-developed numerical procedures for generation of stress intensity factor solutions are available. Examples include finite-element crack modeling, boundary integral equations, and boundary collocation. These procedures, however, are generally utilized only for the evaluation of stress-intensity factors for a given stress system. Advantage is often not taken of the other numerical information provided by these techniques that can be used to evaluate the influence function (IF). Crack surface displacements, as well as K, can be readily obtained using these same techniques. This paper describes how the crack surface displacement information for any reference stress system can be used to evaluate the influence function.

Influence Function Method

The development and application of the influence function method in fracture mechanics has been presented elsewhere [*4–6*]; therefore, only a brief review is given here. The influence function h represents the value of the crack-tip stress-intensity factor K resulting from a unit point load at a given position on the crack surface. Therefore, h is a function of geometry and crack surface position only, and is otherwise independent of loading.

Consider a planar crack problem in which the stress normal to the crack plane is $\sigma(x)$, where x is the position on the crack surface. The differential load $\sigma(x)\,dx$ at position x on the crack surface causes a differential increment of K given by

$$dK = h(a,x)\,\sigma(x)dx \tag{1}$$

The stress-intensity factor due to the stress $\sigma(x)$ is obtained by integration over the crack surface

$$K = \int_0^a h(a,x)\,\sigma(x)dx \tag{2}$$

Following the development of Cruse and Besuner [*5,7*], the influence function $h(a,x)$ can be obtained from crack surface displacements $u_\theta(a,x)$ using the equation

$$h(a,x) = \frac{H}{2K^*(a)}\,\frac{\partial u_\theta{}^*(a,x)}{\partial a} \tag{3}$$

where H is $E/(1-\nu^2)$ for plane strain or E for plane stress. The superscript * indicates K and u_θ values determined for the given geometry under an arbitrary reference loading. By utilizing crack surface displacement information available from many of the common numerical procedures for obtaining K, one can use Eq 3 to generate influence function data for discrete values of a and x. The influence function data can then be curve fit using an appropriate functional form to provide a continuous representation of h over the full range of a and x. This continuous influence function is then numerically integrated with any arbitrary stress distribution $\sigma(x)$ using Eq 2 to obtain values of K for any crack size.

The functional form assumed for the influence function must be capable of representing both the crack surface behavior near the crack tip as well as the short-crack and deep-crack behavior of K. For planar cracks, the general form of the influence function in terms of dimensionless parameters is assumed to be

$$h(a,\xi) = \frac{2}{\sqrt{\pi a}} \frac{1}{\sqrt{1-\alpha}} \left[\frac{\sqrt{1-\alpha} + \sum_{n=1}^{N} B_n(\alpha)(1-\xi)^n}{\sqrt{1-\xi^2}} + \frac{1.122\ \pi\phi(1-\xi)}{(1-\alpha)} \right] \tag{4}$$

The last term in Eq 4 represents the effect of bending induced by load eccentricity relative to the uncracked ligament. For cases with no bending constraint, such as edge cracks, the variable ϕ is given a value of 1. For cases in which bending displacement is constrained, such as center cracks and full-circumferential cracks in pipe walls, ϕ is given a value of 0. The assumed functional form of $B_n(\alpha)$ is

$$B_n(\alpha) = C_{0.n} + C_{1.n}\alpha + C_{2.n}\alpha^2 + C_{3.n}\alpha^3 + C_{4.n}\alpha^4 \tag{5}$$

Sufficient accuracy has been obtained using five terms ($N = 5$) in the summation in Eq 4. The 25 undetermined coefficients ($C_{i,n}$; $i = 0$ to 4 and $n = 1$ to 5) are used to fit Eq 4 to data obtained using Eq 3. By this method, the influence function for any planar crack is reduced to a set of 25 coefficients, which, along with Eqs 2, 4, and 5, can then be used to compute values of K for any crack size with minimal computational effort. As is shown in the following paragraphs, only 20 of the 25 coefficients need to be fitted; the other 5 are predetermined to obtain proper short-crack behavior.

The need for generating influence function data for very short cracks can be eliminated with no loss in accuracy by taking advantage of the limit behavior of K as the value of α approaches zero. From Eq 5 it can be seen that the terms $C_{0.n}$ dominate as α goes to zero. A short crack at a free surface may be treated as an edge crack in a semi-infinite half space. Tada [1] gives a solution

$$K = \frac{2}{\sqrt{\pi a}} \frac{P}{\sqrt{1-(b/a)^2}} [1.297 - 0.297\ (b/a)^{5/4}] \tag{6}$$

for a point-loaded crack in a semi-infinite half space, where the load is positioned b units from the free surface. If the load P is represented as a stress on an infinitesimal length of crack surface dx, then the stress distribution $\sigma(x)$ has a non-zero value only at $x = b$. The integral in Eq 2 reduces to

$$K = h(a,b)\ P \tag{7}$$

If Eqs 6 and 7 are combined and it is recognized that b is merely a crack surface position parameter that can be replaced with x, the short crack influence function can be written in terms of the dimensionless parameters α and ξ as

$$h(\alpha,\xi) = \frac{2}{\sqrt{\pi a}} \frac{1}{\sqrt{1-\xi^2}} [1.297 - 0.297(\xi)^{5/4}] \tag{8}$$

Letting α go to zero in Eq 4 and substituting for h using Eq 8, values of the coefficients $C_{0.n}$ can be obtained for edge-cracks, with $\phi = 0$ or 1, by a least-squares solution of the following equation for ξ between 0 and 1

$$\sum_{n=1}^{5} C_{0.n}(1-\xi)^n = [1.297 - 0.297(\xi)^{5/4} - 1.122\ \pi\phi(1-\xi)]\ \sqrt{1-\xi^2} - 1 \tag{9}$$

A similar development for short cracks with no free surface and $\phi = 0$ (for example, center cracks) requires the coefficients $C_{0,n}$ be zero. The unconstrained bending case ($\phi = 1$) for short cracks with no free surface is of little practical importance. Table 1 summarizes the values of the coefficients $C_{0,n}$. By imposing these restrictions on the $C_{0,n}$ terms, the proper short crack behavior of the influence function is ensured and the number of unknown coefficients in the assumed form of the influence function is reduced to 20.

The behavior of h for very long cracks approaching a free surface is obtained in a manner similar to that for very short cracks. However, α appears in all the terms of the influence function containing the remaining 20 undetermined coefficients. Consequently, long crack behavior cannot be isolated to a few terms, as was done for short crack behavior. Therefore, influence function data for long cracks is generated using functions derived from known K-solutions [1] for α approaching 1. These data are curve fit, along with influence function data generated from the solution of the elastic crack problem, to obtain the 20 remaining influence function coefficients.

Planar Pressurized Crack Problem

Generation of the influence function for a planar crack geometry requires solution of a planar elastic problem to obtain the stress intensity factor and crack surface displacements for a reference loading condition. The influence function depends only on geometry and displacement boundary conditions and is independent of the boundary tractions used for the reference loading condition. For computational convenience, the case of uniform crack surface pressure was chosen for the reference loading. If superposition theory is used, the pressurized crack problem can be decomposed into the sum of a stress-free crack and uniform tension with no crack, as shown schematically in Fig. 1. The Williams eigenfunction expansion [8,9] is a general solution for stresses and displacements in a planar body containing a stress-free crack under otherwise arbitrary boundary conditions. The Williams stress function is an Airy stress function which identically satisfies both the biharmonic equation and the traction-free boundary condition on the crack surface. A particular solution for a given set of boundary conditions is obtained by numerical evaluation of the coefficients of the Williams expansion. The solution for uniform tension without a crack is trivial. Crack surface displacements can be obtained directly from the planar displacement solution. The stress-intensity factor is obtained from the coefficient of the singular ($1/\sqrt{r}$) term in the stress expansion.

Details of the Williams eigenfunction expansion and the corresponding expressions for stresses and displacements are given in the Appendix. Errors which appeared in the original Williams papers [8,9] have been corrected.

TABLE 1—*Summary of coefficients* $C_{0,n}$ *for Eq 5.*

	$\phi = 0$		$\phi = 1$
Coefficient	Center Crack	Edge Crack	Edge Crack
$C_{0,1}$	0.0	0.209	−0.591
$C_{0,2}$	0.0	1.127	−7.335
$C_{0,3}$	0.0	−2.325	9.145
$C_{0,4}$	0.0	1.960	−6.597
$C_{0,5}$	0.0	−0.671	2.153

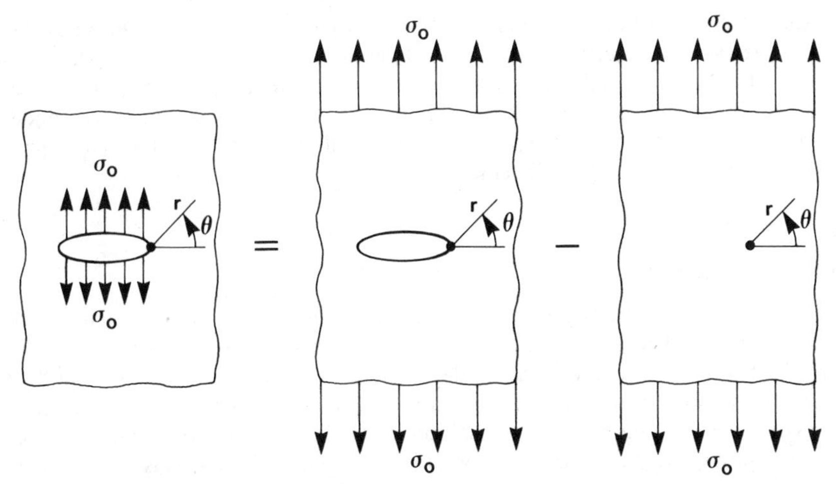

FIG. 1—*Planar pressurized crack.*

Least-Squares Solution

Least-squares (LSQ) methods for solving differential equation problems are entirely general and well-suited to singularity problems. The LSQ method is one of the oldest methods for solving differential equation problems (traceable to Gauss in 1794), but convenient and general numerical implementation is only as recent as 1977 [*10,11*]. The LSQ method is used in the current procedure for general curve-fitting of data (for example, fitting Eq 4 to data from Eq 3), as well as solution of the planar elastic crack problem.

The basic LSQ approach to the planar elastic crack problem is a generalization of boundary collocation, which has been widely used to generate K solutions [*12,13*]. The LSQ approach uses expansion functions, such as Eqs 17 and 18, which contain undetermined constants in each term. The expansion functions need not satisfy interior (equilibrium) or boundary conditions exactly, thus giving greater flexibility than boundary collocation in choice of expansion functions. If, however, the expansion function identically satisfies equilibrium and compatibility, as does the Williams expansion, it is only necessary to perform the least-squares approximation at points on the boundary. At discrete boundary points, an imposed traction or displacement can be represented by an analytic function in the undetermined coefficients. The coefficients are fit to the boundary conditions by minimizing the squared residuals at the discrete points, where a residual is the difference between the imposed boundary value of stress or displacement and the value obtained by evaluating the analytic function using the coordinates of the point and a given set of coefficients.

The complete solution is obtained by minimizing the sum of the squared residuals for all discrete boundary points. A suitable number and location of points along the boundary is dependent upon the number of unknown coefficients in the expansion and the severity of gradients in the boundary conditions. If enough points are used, the solution is much less sensitive to point location than collocation, and the accuracy between points is better. The computational accuracy of the solution depends upon the order at which the infinite series expansion is truncated for numerical evaluation. The order of the truncated series determines the number of unknown coefficients. The number of terms in the series is often limited by the numerical accuracy limits of the computer used for the solution. Successive terms in Eqs 17 and 18 contain increasing powers of the coordinate parameter r. Numerical evaluation of high-order terms for boundary points with very large or very small r-values can result in computer overflow or underflow errors. Coordinate normalization relative to the boundary point with the largest radial coordinate, r, is used to eliminate the numerical overflow problem. By requiring solutions for only a limited range of α, one minimizes the underflow problem by limiting the minimum value of the radial coordinate for the boundary point closest to the coordinate origin, the crack tip.

Automated Procedure

The procedure for generating planar influence functions has been automated in a computer program NEWINF. The program solves for the Williams expansion coefficients using the LSQ method, for crack lengths ranging between 20% and 60% of the body width. Since the loading condition of uniform, unit pressure on the crack surface is built into the modified Williams expansion being used, only displacement boundary conditions must be specified for a given geometry. Numerical differentiation by finite differences is used to generate influence function data using Eq 3. The influence function data are combined with the appropriate long-crack data and curve fit using the LSQ method to obtain the influence function coefficients.

Inputs to the NEWINF program consist of parameters identifying the type of crack geometry (that is, $\phi = 0$ or 1, with or without free surface), material properties E and ν, the coordinates of boundary nodes, and the displacement boundary conditions. The boundary must be divided into segments with continuous first derivative and constant boundary conditions. Boundary node locations are entered in cartesian coordinates and internally transformed into a radial coordinate system with its origin at the crack tip. Radial coordinates are normalized such that the boundary node furthest from the origin has a radial coordinate of 1. This normalization eliminates computational overflow problems in evaluation of the Williams series expansions.

A least-squares solution to the Williams expansion for given displacement boundary conditions is obtained at dimensionless crack lengths of α and 1.02α, for $\alpha = 0.2$ to 0.6 in increments of 0.1. Crack surface displacements u_θ at 20 uniformly spaced points ($\xi = 0$ to 0.95) are recovered by setting $\theta = \pi$ in Eq 18. Finite differencing is used to approximate the derivative $\partial u_\theta / \partial a$ in Eq 3 for each 2% change in crack length. Influence function results for discrete α and ξ are curve fit using Eq 4 and appropriate coefficients $C_{0,n}$ ($n = 1,5$) to obtain the remaining 20 influence function coefficients $C_{i,n}$ ($i = 1$ to 4).

Numerical Results

Two crack geometries were selected to evaluate the influence function generation program NEWINF: edge-cracked strip and edge-cracked disk. Alternative influence functions exist for the edge-cracked strip, as well as stress-intensity factor solutions for uniform and linearly varying tension. Stress-intensity factor solutions exist for the edge-cracked disk under uniform and parabolic tension. All of the stress intensity factor solutions can be found in Ref *1*. Influence functions were generated for both geometries using NEWINF, with the Williams series expansion truncated at 50 terms. The influence function for the edge-cracked strip is compared directly with the influence function of Orange [*14*]. Both new influence functions were integrated with the appropriate stress distributions to obtain values of K for α between 0.0 and 0.9. These results are compared with the existing solutions from Ref *1*.

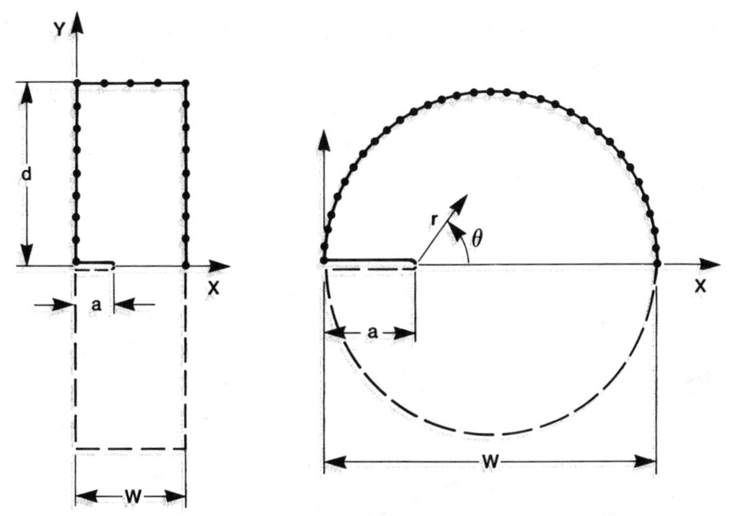

FIG. 2—*Planar boundary node models:* (left) *edge-cracked strip;* (right) *edge-cracked disk.*

Edge-Cracked Strip

A finite-width strip containing an edge crack was modeled as shown in Fig. 2*a*. Symmetry about the line of the crack is implicit in the assumed form of the Williams expansion. The boundary is described by three segments, each containing 20 nodes (fewer are shown in Fig. 2 for clarity). All boundary nodes are free to displace. Traction-free conditions for boundaries other than the crack surface are generated internally by the program for all boundary nodes that do not have a specified displacement. The resulting influence function coefficients for an edge-cracked strip are included in Table 2. The influence function is compared with that from Orange [*14*] in Fig. 3. The new influence function is in excellent agreement with Orange for the range of a/W tabulated in Ref *9*. The influence function was integrated to obtain stress-intensity factors for uniform, linear, and parabolic tension at values of a/W up to 0.9, as shown in Fig. 4. Results obtained using the K-solutions in Ref *1* for uniform and linear tension are plotted for comparison. The maximum differences are 6.2% for uniform tension at $a/W = 0.9$ and 3.8% for linear tension at $a/W = 0.8$. The K-solution for parabolic tension is not available for comparison.

TABLE 2—*Influence function coefficients for edge-cracked strip.*

	$C_{i,n}$				
n/i	0	1	2	3	4
1	−0.59	3.99	−10.88	41.87	−33.91
2	−7.34	−23.78	57.59	−124.90	76.10
3	9.14	35.61	−82.96	159.60	−82.74
4	−6.55	−26.37	52.58	−93.81	39.56
5	2.15	8.09	−13.64	22.21	−6.71

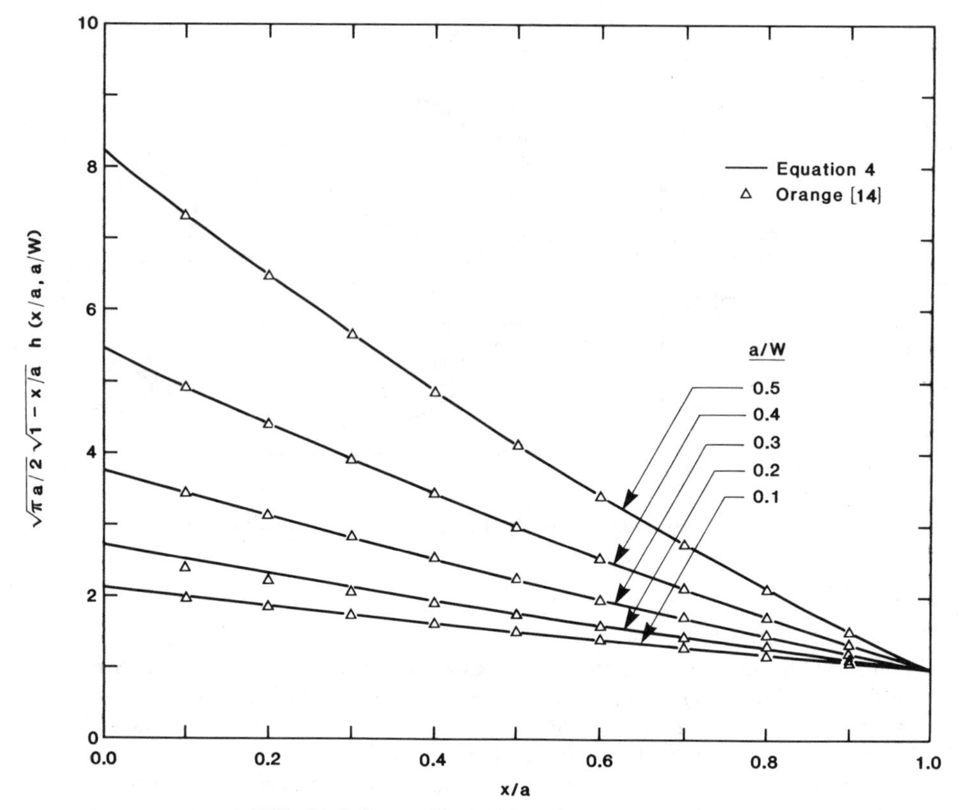

FIG. 3—*Influence function for edge-cracked strip.*

Edge-Cracked Disk

A circular disk containing an edge crack was modeled as shown in Fig. 2*b*. Again, only half of the symmetric geometry is modeled. The boundary is described by one segment containing 61 nodes. Boundary nodes are free to displace and are stress free. The coefficients in the edge-cracked disk influence function were generated using NEWINF and are included in Table 3. The influence function was also integrated to obtain stress-intensity factors for uniform and parabolic tension for comparison with the available solutions in Ref *1*. Results are shown in Fig. 5. The uniform tension case shows excellent agreement. However, normalized *K*-values for parabolic tension differ by as much as 30%.

In order to determine the cause for the discrepancy in results for parabolic tension, we modified the NEWINF program to compute the stress-intensity factors for parabolic stress directly. This was accomplished by first deriving the boundary traction distribution that produces a parabolic stress distribution on the line of the crack when the crack is not present. This boundary traction distribution was then used in the program to generate the traction boundary conditions at each node. The LSQ solution to the Williams expansion with these boundary conditions imposed directly yields the stress-intensity factor for a parabolic stress distribution on the line of the crack. The stress-intensity factors obtained from integration of the influence function and those from the direct LSQ solution of the parabolic loading case were renormalized to remove the a/W singularity and plotted in Fig. 6 for the full

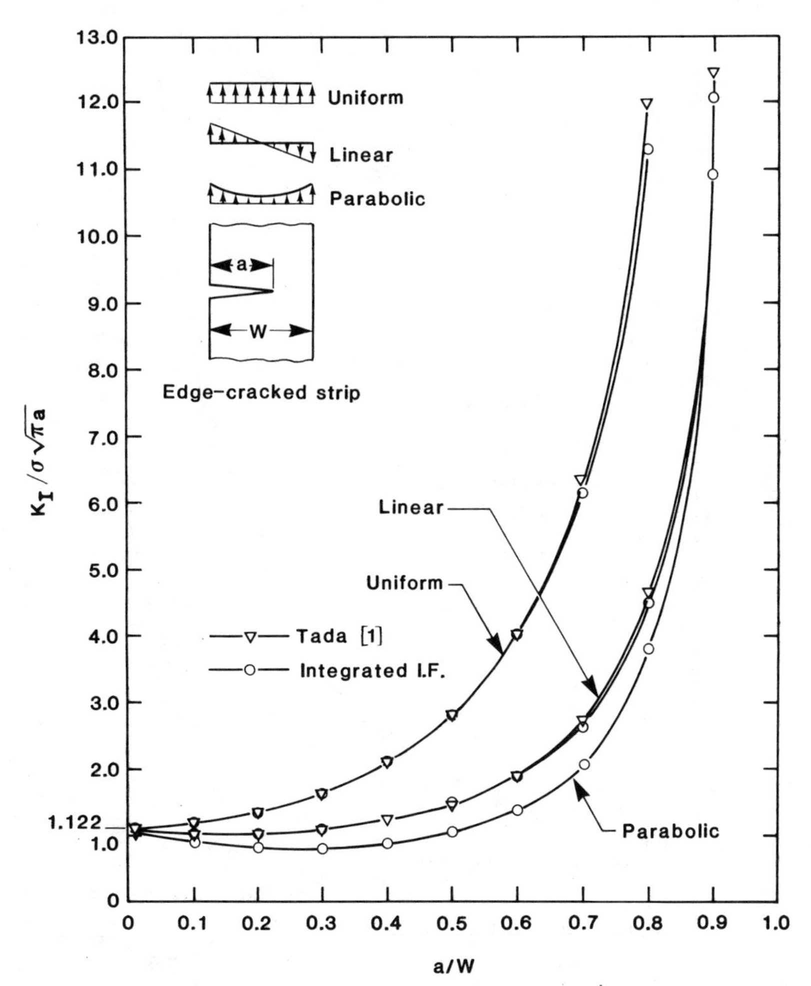

FIG. 4—*Stress-intensity factors for edge-cracked strip.*

TABLE 3—*Influence function coefficients for edge-cracked disk.*

	$C_{i,n}$				
n/i	0	1	2	3	4
1	−0.59	8.76	−32.92	82.44	−58.04
2	−7.34	−20.29	48.44	−108.14	62.01
3	9.14	29.64	−56.16	112.38	−49.76
4	−6.55	−20.08	17.12	−32.26	0.39
5	2.15	5.12	2.87	−5.47	10.06

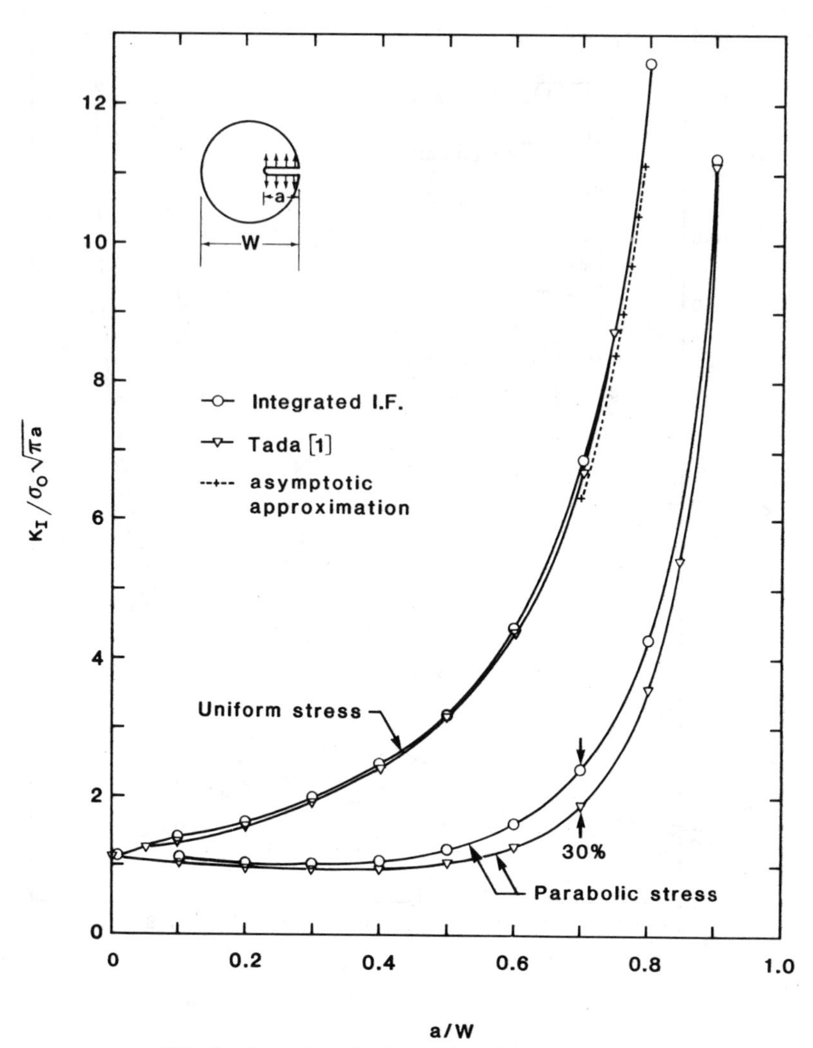

FIG. 5—*Stress-intensity factors for edge-cracked disk.*

range of a/W. Values from the direct LSQ solution closely follow the IF results. The Tada solution for $a/W > 0.5$ is an asymptotic estimate based on results by Rooke [*2*], which only cover a/W up to 0.5. Figure 6 shows that the Rooke solution follows more closely the new IF results at $a/W = 0.5$ than the Tada estimate. As an additional check, the results from Fig. 4 for the edge-cracked strip with parabolic stress are plotted on Fig. 6. This provides a lower bound on the disk solution; it can be argued that for a given crack size, body width, and stress, K for an edge-cracked strip will always be less than K for an edge-cracked disk due to the additional material providing greater stiffness. The authors have confirmed this for the uniform tension case. The comparisons made in Figs. 5 and 6 indicate that the new influence function for the edge-cracked disk obtained using NEWINF provides accurate K-values over the full range of a/W, and it provides significant improvement over previous estimates (Ref *1*, p. 11.14).

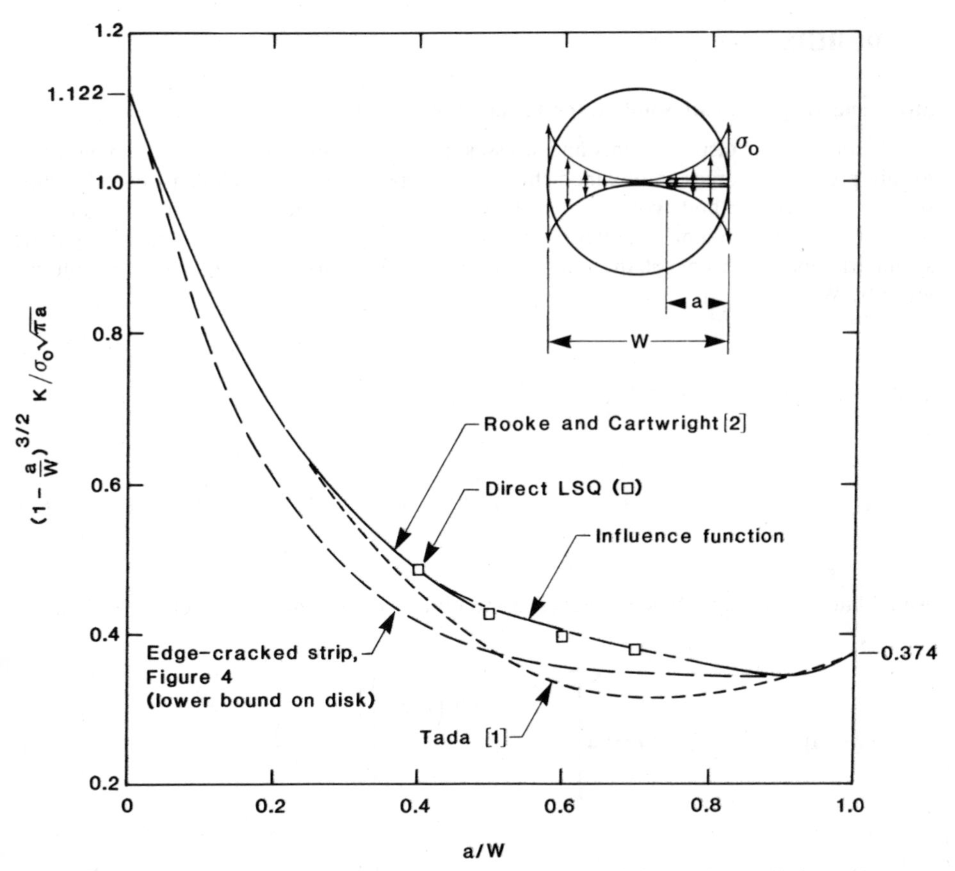

FIG. 6—*Renormalized stress-intensity factors for edge-cracked disk under parabolic tension.*

Conclusion

An efficient automated procedure has been developed for generation of influence functions for planar cracks. Stress-intensity factors calculated using generated influence functions for several planar crack geometries are in excellent agreement with published results. The continuous functional representation of the planar influence function provides a simple means for computation of stress-intensity factors for arbitrary stress distributions and crack lengths.

The planar influence function generation program NEWINF is being developed as part of the general purpose two- and three-dimensional fracture mechanics program NASCRAC [*15*]. The NASCRAC code currently contains a library of over 30 influence functions for two- and three-dimensional crack geometries. The NEWINF module will allow a NASCRAC user to expand the current library of influence functions to include geometries of particular importance to the user.

Acknowledgments

The authors would like to acknowledge the support of the Marshall Space Flight Center of the National Aeronautics and Space Administration through Contract NAS8-36171, and in particular, of Mr. Larry Salter, the contract manager at NASA.

Appendix

Stress and Displacement Solution for Planar Pressurized Cracks

The solution for displacements and stresses in a planar isotropic body containing a uniformly pressurized crack is obtained through superposition of the Williams eigenfunction solution for a planar body containing a stress-free crack and the solution for a planar body with no crack. If one limits the development to cases involving stress distributions that are symmetric about the line of the crack [that is, $\sigma(r,\theta) = \sigma(r,-\theta)$, Mode I], the solution given by Williams is

$$\chi(r,\theta) = \sum_{n=1}^{\infty} \left\{ (-1)^{n-1} A_{2n-1} r^{n+1/2} \left[-\cos\left(n - \frac{3}{2}\right)\theta + \frac{2n-3}{2n+1}\cos\left(n + \frac{1}{2}\right)\theta \right] + (-1)^n A_{2n} r^{n+1} \left[-\cos(n-1)\theta + \cos(n+1)\theta \right] \right\} \quad (10)$$

An additional harmonic function necessary to obtain displacements is given by Williams as

$$\psi(r,\theta) = \sum_{n=1}^{\infty} (-1)^{n+1} r^{n-1} 4 \left[-\frac{A_{2n-1}}{r^{1/2}} \frac{\sin\left(n - \frac{3}{2}\right)\theta}{n - \frac{3}{2}} + \frac{A_{2n}}{n-1} \sin(n-1)\theta \right] \quad (11)$$

The stresses and displacements can be represented in terms of derivatives of the Williams functions (Eqs 10 and 11) as

$$\sigma_{rr}(r,\theta) = \frac{1}{r^2}\frac{\partial^2 \chi}{\partial \theta^2} + \frac{1}{r}\frac{\partial \chi}{\partial r}$$

$$\sigma_{\theta\theta}(r,\theta) = \frac{\partial^2 \chi}{\partial r^2}$$

$$\sigma_{r\theta}(r,\theta) = -\frac{\partial}{\partial r}\left(\frac{1}{r}\frac{\partial \chi}{\partial \theta}\right) \quad (12)$$

$$u_r(r,\theta) = \frac{1}{2\mu}\left[-\frac{\partial \chi}{\partial r} + (1-\gamma) r \frac{\partial \psi}{\partial \theta} \right]$$

$$u_\theta(r,\theta) = \frac{1}{2\mu}\left[-\frac{1}{r}\frac{\partial \chi}{\partial \theta} + (1-\gamma) r^2 \frac{\partial \psi}{\partial r} \right] \quad (13)$$

where

$$\gamma = \begin{cases} \dfrac{\nu}{(1 + \nu)} & \text{plane stress} \\ \nu & \text{plane strain} \end{cases}$$

$$\mu = \frac{E}{2\,(1 + \nu)} \tag{14}$$

Stresses and displacements for the planar elastic problem of uniform, uniaxial tension with no crack are given by

$$\sigma_{rr}\,(r,\theta) = \sigma_0 \sin^2 \theta$$

$$\sigma_{\theta\theta}\,(r,\theta) = \sigma_0 \cos^2 \theta$$

$$\sigma_{r\theta}\,(r,\theta) = \sigma_0 \sin \theta \cos \theta \tag{15}$$

$$u_r\,(r,\theta) = \frac{\sigma_0 r}{2\mu}\,(\sin^2 \theta - \gamma)$$

$$u_\theta\,(r,\theta) = \frac{\sigma_0 r}{2\mu} \sin \theta \cos \theta \tag{16}$$

where μ and γ are defined in Eq 14.

Subtracting Eqs 15 and 16 from Eqs 12 and 13, respectively, and substituting with Eqs 10 and 11, the stresses and displacements for the uniformly pressurized crack are given by

$$\sigma_r = \sum_{n=1}^{\infty} \Bigg\{ (-1)^{n-1} A_{2n-1} \left(n - \frac{1}{2}\right) r^{n-3/2} \times \left[\left(n - \frac{7}{2}\right) \cos \left(n - \frac{3}{2}\right) \theta - \left(n - \frac{3}{2}\right) \cos \left(n + \frac{1}{2}\right) \theta \right] + (-1)^n A_{2n}\, n r^{n-1} \left[(n - 3) \cos\,(n - 1)\,\theta - (n + 1) \cos\,(n + 1)\,\theta\right] \Bigg\}$$

$$\sigma_\theta = \sum_{n=1}^{\infty} \Bigg\{ (-1)^{n-1} A_{2n-1} \left(n - \frac{1}{2}\right) r^{n-3/2} \times \left[-\left(n + \frac{1}{2}\right) \cos \left(n - \frac{3}{2}\right) \theta + \left(n - \frac{3}{2}\right) \cos \left(n + \frac{1}{2}\right) \theta \right] + (-1)^n A_{2n}\, n(n + 1)\, r^{n-1} \left[- \cos\,(n - 1)\,\theta + \cos\,(n + 1)\,\theta\right] \Bigg\}$$

$$\sigma_{r\theta} = \sum_{n=1}^{\infty} \Bigg\{ (-1)^{n-1} A_{2n-1} \left(n - \frac{1}{2}\right) \left(n - \frac{3}{2}\right) r^{n-3/2} \left[\sin \left(n - \frac{3}{2}\right) \theta + \sin \left(n + \frac{1}{2}\right) \theta \right] + (-1)^n A_{2n}\, n\, r^{n-1} \left[- (n - 1) \sin\,(n - 1)\,\theta + (n + 1) \sin\,(n + 1)\,\theta\right] \Bigg\} \tag{17}$$

$$u_r(r,\theta) = \frac{\sigma_0}{2\mu}\sum_{n=1}^{\infty}\Bigg\{(-1)^n A_{2n-1}\, r^{n-1/2}$$

$$\times\left[\left(\frac{7}{2} - n - 4\gamma\right)\cos\left(n - \frac{3}{2}\right)\theta + \left(n - \frac{3}{2}\right)\cos\left(n + \frac{1}{2}\right)\theta\right]$$

$$+ (-1)^{n-1} A_{2n}\, r^n\,[(3 - n - 4\gamma)\cos(n - 1)\,\theta + (n + 1)\cos(n + 1)\,\theta]\Bigg\}$$

$$u_\theta(r,\theta) = \frac{\sigma_0}{2\mu}\sum_{n=1}^{\infty}\Bigg\{(-1)^n A_{2n-1}\, r^{n-1/2}$$

$$\times\left[\left(\frac{5}{2} + n - 4\gamma\right)\sin\left(n - \frac{3}{2}\right)\theta - \left(n - \frac{3}{2}\right)\sin\left(n + \frac{1}{2}\right)\theta\right]$$

$$+ (-1)^{n-1} A_{2n}\, r^n\,[(3 + n - 4\gamma)\sin(n - 1)\,\theta - (n + 1)\sin(n + 1)\,\theta]\Bigg\} \quad (18)$$

Crack surface displacements are obtained by setting $\theta = \pi$ in Eq 18. The stress-intensity factor is obtained from the coefficient of the singular $(1/\sqrt{r})$ term in the stress expansions (Eq 17) as

$$K_I = -A_1\sqrt{2\pi} \quad (19)$$

References

[1] Tada, H., Paris, P., and Irwin, G., *The Stress Analysis of Cracks Handbook,* Del Research Corporation, Hellertown, PA, 1985.

[2] Rooke, D. P. and Cartwright, D. J., *Compendium of Stress Intensity Factors,* Her Majesty's Stationery Office, London, 1976.

[3] Sih, G. C., *Handbook of Stress Intensity Factors,* Institute of Fracture and Solid Mechanics, Lehigh University, Bethlehem, PA, 1973.

[4] Bueckner, H. F., *Zeitschrift för Angewandte Mathematik und Mechanik,* Vol. 50, No. 9, 1970, pp. 529–546.

[5] Besuner, P. M., *Nuclear Engineering and Design,* Vol. 43, 1977, pp. 115–154.

[6] Rice, J. R., *International Journal of Solids and Structures,* Vol. 8, 1972, pp. 751–758.

[7] Cruse, T. A. and Besuner, P. M., *Journal of Aircraft,* Vol. 12, No. 4, 1975, pp. 369–375.

[8] Williams, M. L., *Journal of Applied Mechanics,* Vol. 24, No. 1, 1957, pp. 109–114.

[9] Williams, M. L., *Journal of Applied Mechanics,* Vol. 28, No. 1, 1961, pp. 78–82.

[10] Eason, E. D. and Mote, C. D., Jr., *International Journal of Numerical Methods in Engineering,* Vol. 11, 1977, pp. 641–652.

[11] Eason, E. D. and Mote, C. D., Jr., *International Journal of Numerical Methods in Engineering,* Vol. 12, 1978, pp. 597–612.

[12] Gross, B., Srawley, J. E., and Brown, W. R., Jr., "Stress Intensity Factors for a Single-Edge-Notch Tension Specimen by Boundary Collocation of a Stress Function," NASA Technical Note NASA-TN-D-2395, Lewis Research Center, OH, Aug. 1964.

[13] Newman, J. C., Jr., "An Improved Method of Collocation for the Stress Analysis of Cracked

Plates with Various Shaped Boundaries," NASA Technical Note NASA-TN-D-6376, National Aeronautics and Space Administration, 1971.

[*14*] Orange, T. W., *Fracture Mechanics: Sixteenth Symposium, ASTM STP 868,* American Society for Testing and Materials, Philadelphia, 1985, pp. 95–105.

[*15*] Harris, D. O., Bianca, C. J., Eason, E. D., Salter, L. D., and Thomas, J. M., "NASCRAC: A Computer Code for Fracture Mechanics Analysis of Crack Growth," Paper No. AIAA-87-0847-CP, American Institute of Aeronautics and Astronautics, April 1987.

Nonlinear and Time-Dependent Fracture Mechanics

Terry Ingham,[1] Nigel Knee,[2] Ian Milne,[2] and Eddie Morland[1]

Fracture Toughness in the Transition Regime for A533B Steel: Prediction of Large Specimen Results from Small Specimen Tests

REFERENCE: Ingham, T., Knee, N., Milne, I., and Morland, E., **"Fracture Toughness in the Transition Regime for A533B Steel: Prediction of Large Specimen Results from Small Specimen Tests,"** *Fracture Mechanics: Perspectives and Directions* (*Twentieth Symposium*), *ASTM STP 1020*, R. P. Wei and R. P. Gangloff, Eds., American Society for Testing and Materials, Philadelphia, 1989, pp. 369–389.

ABSTRACT: The results of an experimental test program to investigate methods of predicting the influence of specimen size on the cleavage fracture toughness of A533B-1 steel are presented. Three-point bend specimens from 10 to 230-mm thick were tested; to minimize scatter in the results, these were all extracted from a single plate of A533B, such that their crack tips lay on the midplane of the plate. A total of 217 tests were performed so that enough data could be obtained to allow statistical analysis.

At any one temperature there was considerable scatter in the results, which did not diminish with increasing specimen size. However, at a particular temperature, the mean or lower-bound fracture toughness at cleavage decreased with increasing specimen size. The fracture toughness was analyzed as a function of size and temperature using exponential curve fitting procedures, multiple linear regression analysis, Weibull statistics and the β_{Ic} correction, and the significance of each of these methods is reviewed. The size effect can be quantified in terms of a shift in temperature for a given toughness level. The shift, however, is dependent on toughness, being smaller at the higher levels of toughness, and there is a tendency toward saturation as the fully ductile state is approached.

The implications of this work in respect to obtaining data for use in structural integrity analysis is discussed.

KEY WORDS: brittle fracture, cleavage, K_{Ic}, A533B, ductile-brittle transition, size effect, fracture toughness, elastic-plastic fracture

There are three fracture regimes of importance in ferritic steels: the lower transition, the upper transition, and the fully ductile upper shelf. These are identified schematically in Fig. 1. Within the lower transition region, the fracture mechanism is brittle (cleavage) and the fracture toughness increases slowly with temperature. As the temperature increases, the rate of increase in toughness also increases until cracking initiates by a ductile tearing mechanism. In this, the upper transition region, the ductile crack propagates a small distance

[1] Principal scientific officer and higher scientific officer, respectively, Risley Nuclear Power Development Laboratories, UKAEA, Risley, Warrington, WA3 6AT, United Kingdom. Mr. Ingham is now at HMNII, St. Peter's House, Bootle, L20 3LZ, United Kingdom.

[2] Research officer and section manager, respectively, Central Electricity Research Laboratories, Kelvin Ave., Leatherhead, Surrey, KT22 7SE, United Kingdom.

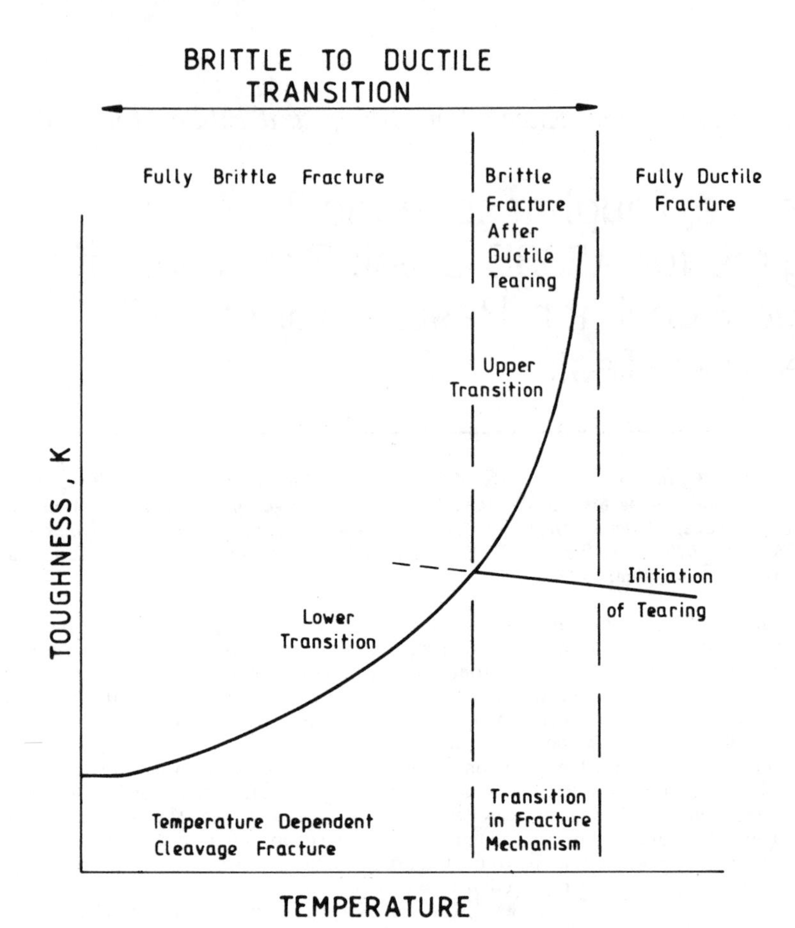

FIG. 1—*Schematic representation of fracture behavior in PWR pressure vessel steels.*

before it is interrupted by the brittle fracture mechanism. In this regime, although the toughness at which ductile tearing initiates is relatively temperature independent, that for brittle fracture is very sensitive to temperature, as indicated in Fig. 1 [*1*], so that the quantity of ductile tearing preceding brittle fracture also increases with temperature [*2*]. Eventually, the brittle fracture toughness curve rises so steeply that the conditions for brittle fracture cannot be achieved [*1*], and the material fails in a fully ductile manner.

Fracture in all of these regimes can be characterized by fracture mechanics parameters, although at high levels of toughness the relevant parameters can be obtained only by using techniques based on elastic-plastic *J* analysis. The results are often found to be size dependent, even when tests are performed under so-called *J*-valid conditions.

In the lower transition regime, the size effect appears as a tendency for small specimens to produce higher values of fracture toughness than large ones. This can generally be represented as a shift in the ductile-brittle transition temperature caused by the temperature-dependent brittle fracture regime appearing at higher temperatures in the larger specimens [*3*]. The effect has been explained in terms of the reduced constraint in the smaller specimens [*1*] and in terms of the higher probability of a region of low toughness material occurring close to the crack tip in the larger specimens [*4*]. There is evidence to support both explanations [*5,6*], and the effect has been observed in a variety of ferritic steels [*7,8*].

In the fully ductile regime, size effects are less apparent, and in the J-valid region they appear to be minor. Ingham et al. [*9*] observed that J-resistance curve values, obtained from small specimens under conditions well beyond those for J-validity, fell within the scatter band of J-valid data from large specimens. There are, however, some contradictions to these observations which indicate that ductile crack initiation begins at a higher level of J as the specimen size is increased [*6,8*].

The effect of specimen size at high toughness levels in the upper transition regime has not been studied as closely as in the other two regimes. In this regime, cracks growing initially by a ductile mechanism follow the J resistance curve until cleavage occurs [*2*], where the level of J at which the brittle fast fracture mechanism intervenes can be taken as a characterizing parameter. There is, therefore, a combination of fracture mechanisms which may complicate the interpretation of the size effects observed. It is the brittle fracture event which is important in this regime, and it might be expected that specimen size has the same effect on the brittle fracture toughness obtained in the transition region as it does in the fully brittle regime. However, evidence suggests that the small amount of tearing which precedes brittle fracture in the transition regime tends to reduce this size effect [*6,7*].

The nuclear power industry is presented with a conflict of requirements arising out of these size effects. First, to monitor material properties both during the manufacturing stage and after service exposure, it is necessary to use small specimens. Second, to perform a fracture analysis of a pressure vessel, conservative values of the fracture toughness are needed, necessitating the use of large test specimens. There is, therefore, an incompatibility of requirements which needs to be rationalized. The test program reported here was set up with this in mind.

Program Objectives

The primary objective was to quantify the effect of specimen size in the ductile-brittle transition regime of a pressurized water reactor (PWR) pressure vessel steel. For practical purposes, the brittle fracture toughness range of interest was taken from 150 to 400 $MPa\sqrt{m}$. It was anticipated that the quantification would be via a shift in temperature at a given mean level of toughness which either would saturate at a certain size of specimen or from which trend curves could be obtained. The test matrix was designed to accommodate this, with sufficient specimens for statistical analyses. A secondary objective was, therefore, to evaluate statistical methods of predicting size effects. Ductile cracking in these steels typically initiates at a fracture toughness of 200 to 250 $MPa\sqrt{m}$, so at the higher end of the fracture toughness range some tearing was expected to precede brittle fracture. In some cases this involved testing well beyond the limits for J-controlled growth [*10*] in order to achieve a cleavage event.

Experimental Procedures

Material

The test material was a 257-mm-thick A533B Class 1 steel plate which was produced by the basic electric process. The steel was vacuum degassed and aluminium grain refined, resulting in the chemical composition given in Table 1. The ingot was hot-worked, rolled

TABLE 1—*Chemical composition product analysis, weight percent.*

C	Mn	Mo	Ni	Si	Cr	S	P	Cu	Al
0.21	1.44	0.48	0.67	0.28	0.18	0.005	0.006	0.05	0.021

to size and heat-treated as follows:

Normalize at 880/940°C, 9 h
Temper at 640/685°C, 8.5 h/air cool
Austenitize at 844/928°C, 8.3 h/water quench
Temper at 635/670°C, 6.75 h/air cool

Microstructural examination of the center of the plate (the region sampled in the test program) revealed a banded bainitic structure, typical of A533B.

Test Specimens

To minimize scatter due to the inherent inhomogeneity of the plate, it was decided initially to align the notch of every test specimen along one of three lines, drawn along the surface of the plate parallel to the final rolling direction, with the centerplane of each specimen in the centerplane of the plate (Fig. 2). With crack propagation in the through-thickness direction (the T-S orientation) and a crack depth to width ratio of approximately 0.5, this meant that all crack tips sampled material at the same depth in the plate, regardless of specimen size. Square section three-point bend specimens were chosen with thicknesses, B, standardized initially at 25, 50, 100, and 230 mm. These specimens were carefully

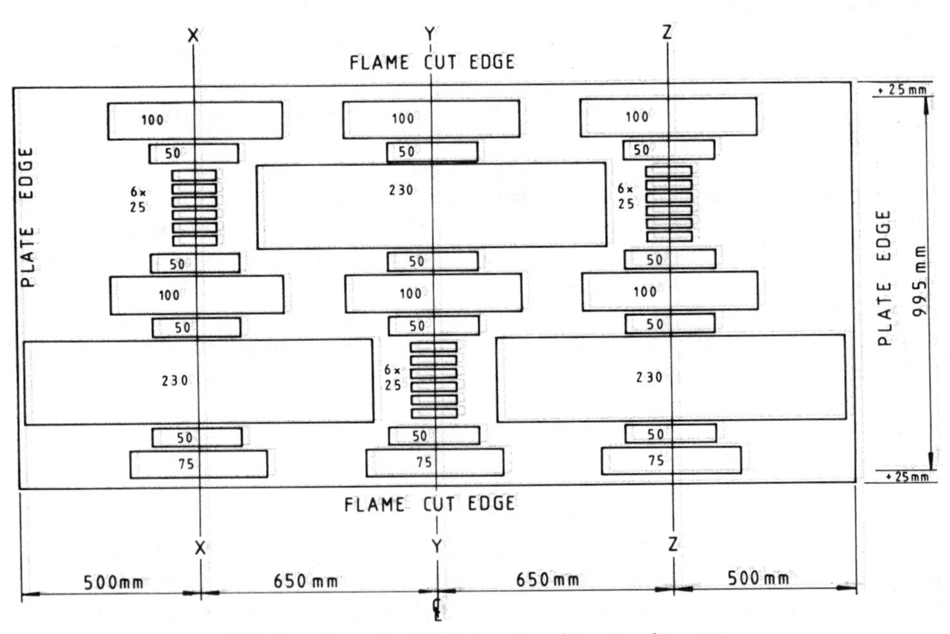

FIG. 2*a*—*Example of initial cutting plan: view from top.*

SECTION X-X AND SECTION Z-Z

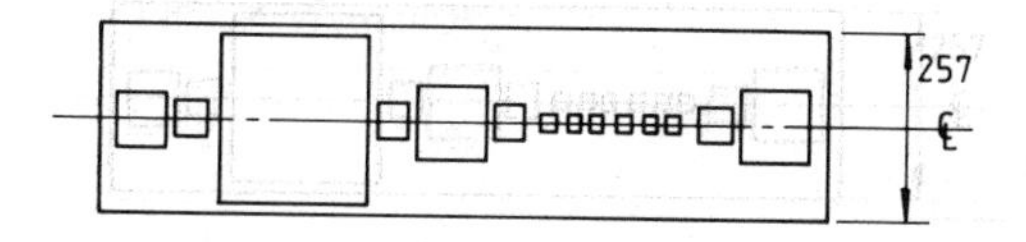

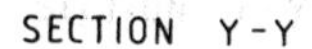

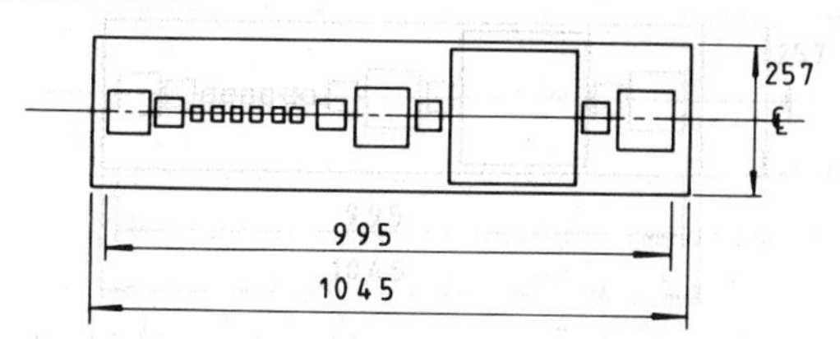

FIG. 2b—*Example of initial cutting plan: sections along* x-x, y-y *and* z-z *of Fig.* 2a.

distributed so that all regions of the plate were sampled by each specimen size. At an advanced stage in the program, contingency material was used to manufacture 10-mm-thick specimens. Once again specimens were located on one of three lines across the plate and with crack tips at middepth positions. However, sampling of areas along the plate was less extensive for the 10-mm three-point bend specimens than for larger specimens. The total numbers of specimens in the program were 44 by 10 mm, 70 by 25 mm, 61 by 50 mm, 30 by 100 mm, and 12 by 230 mm.

Mechanical Properties

Standard tension tests were performed over the temperature range −90° to +20°C using specimens extracted in the transverse orientation (T) at plate midthickness ($t/2$). Ambient temperature tests were also performed on transverse specimens extracted from the quarter-thickness ($t/4$) location for routine quality-control purposes. The results of these tests are summarized in Table 2.

The results of Charpy V-notch impact tests, which were performed for both T-S and T-L orientations at the $t/2$ location and for T-L specimens at $t/4$, are summarized in Table 3. A tanh fit to the C_v notch impact data was obtained for each set. The nil ductility temperature (NDT) from Pellini drop-weight tests was −15°C at $t/2$.

TABLE 2—*Tensile properties.*

Location in Plate	Test Temperature, °C	Yield Stress $\sigma_{0.2}$, MPa	Tensile Strength, σ_u, MPa	Elongation, %	Reduction of Area, %
$t/2$	20	470	620	25	66
$t/2$	−10	491	653	24	65
$t/2$	−50	522	693	24	64
$t/2$	−90	560	740	ND[a]	ND[a]
$t/4$	20	486	635	23	65

[a] Not determined.

TABLE 3—*Summary of impact properties.*

		Charpy V-Notch Data				
Orientation	Location	T_{41J}, °C	T_{68J}, °C	FATT, °C	Upper Shelf, *J*	Pellini NDT, °C
T-S	t/2	−53	−32	0	169	−15
T-L	t/2	−69	−45	0	176	−15
T-L	t/4	−59	−39	+5	196	−20

Fracture Toughness Tests

Testing was performed at temperatures ranging from −90° to +20°C.

The specimens were loaded in three-point bending with a span of $4W$, at displacement rates equivalent to an increase in K of 2.5 MPa$\sqrt{m}$/s^{-1} in the elastic range. In some cases, brittle fractures were not obtained, and these specimens were unloaded at displacements well beyond the maximum load condition. Toughness was quantified in terms of K, as calculated from the J obtained at failure. Inevitably many of the specimens were taken to toughness levels well beyond the limits for J-controlled growth. J was evaluated from the area, U, under each load-point displacement curve from

$$J = \frac{2U}{B(W - a)} [1 - 0.5\Delta a/(W - a)] \tag{1}$$

where Δa is the crack extension measured at the final failure or unloading point [11]. Ductile crack growth was relatively uniform over at least 75% of the crack front. The crack growth correction was generally small, except for a few results at the highest toughness levels, where it was up to 25% of the uncorrected value. This was then converted to units of toughness, K_{Jc}, assuming plain strain conditions prevail, using

$$K_{Jc} = EJ/(1 - \nu^2) \tag{2}$$

taking Poisson's ratio, ν, as 0.3 and Young's modulus, $E = 210 - 0.054T$, where E is in GPa and T in °C. The variation of E with temperature was derived from data given in British Standards Institution Specification for Unfired Fusion Welded Pressure Vessels (BS 5500:1985). The initial crack lengths and the amount of ductile crack extension were calculated using either a nine-point average (ASTM Test Method for J_{Ic}, a Measure of Fracture Toughness [E 813-81]) or by area measurements from photographs.

Analysis

The analysis of the fracture data in the transition regime was complicated by three factors: uncertainty about the functional form of the temperature dependence of fracture toughness, the considerable scatter in the data, and the fact that some 15% of the specimens did not fail by the brittle cleavage mechanism. The last factor is a consequence mainly of the physical limitations of a given specimen size, which prevents loading beyond a certain level of J. Specimens which attained the maximum J level for a given size have been designated as nonbrittle, although it is anticipated that cleavage fracture could have occurred, had they been capable of being loaded to higher levels of J. This led to a size dependent limit to the level for cleavage. Thus, any calculated temperature dependence is distorted by the existence of nonbrittle specimens. In consequence, a number of different approaches to data analysis were considered and are described below.

Exponential Curve Fits

Using only those data where the test was terminated by a cleavage event, the exponential curve fits for each specimen size were obtained from least squares regression analyses. K_{Jc}-temperature plots and the associated exponential functions are presented for each specimen size in Figs. 3*a* to 3*e*. The mean transition curves are compared directly in Fig. 4, which also shows the American Society of Mechanical Engineers (ASME) XI Appendix A K_{Ic} curve indexed to $RT_{NDT} = -15°C$.

The values of A and σ for expressions of the form

$$K_{Jc} = A \exp(\sigma T) \tag{4}$$

which describe the increase in toughness (K_{Jc}) with temperature (T) are presented in Table 4. This form implies that the log (K_{Jc}) values are normally distributed about the best fit line. Confidence limits calculated from the regression procedure were obtained and approximated using the exponential representation of Eq 4.

The mean K_{Jc}-T relationships for each specimen size were then used to develop a method for predicting the effect of thickness on cleavage fracture toughness in the transition region. The exponential relationships were used to define, as a function of specimen thickness, the temperatures corresponding to specified mean toughness levels. These have been plotted in Fig. 5, in terms of the temperature shift as a function of mean toughness, for a specimen size increase from B to 230 mm (that is, the maximum size tested). Figure 5 clearly shows that these temperature shifts decrease as the mean toughness level increases; that is, there is a tendency to saturate the size effects at high toughness values.

Multiple Linear Regression Analysis

As for the exponential curve fits, only those data where the test ended in a cleavage event were included in the multiple linear regression analysis, and again it was assumed that K_{Jc} increases exponentially with temperature. Terms were included that depend on the thickness and temperature, such that for a constant specimen thickness, the multiple regression reduces to equations of the form of Eq 4.

The response variable is log (K_{Jc}), and the following independent variables were found to be significant: T, ($T \times B$), and log (B). The goodness of fit, measured by the multiple correlation coefficient, was 0.8.

The resultant multiple linear regression is

$$\ell n(K_{Jc}) = 6.849 + 0.01467T - 0.188\ \ell n(B) + 66.0 \times 10^{-6} (B \times T) \tag{5}$$

where B is in mm and T is in °C.

Figure 6 shows the predicted mean fracture toughness for the five specimen thicknesses that have been tested. For all specimen sizes, the mean lines are very similar to those obtained by performing a separate regression on each specimen size, which can be seen by comparing Figs. 4 and 6. Considering all the data together in this way assumes that there are systematic trends which depend on specimen size and that there are no abrupt changes in behavior as the specimen size increases.

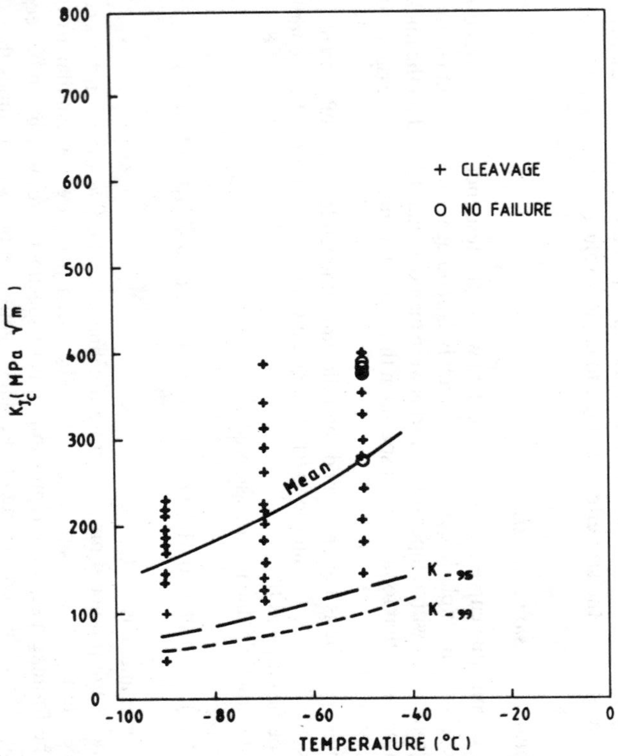

FIG. 3a—*Fracture transition data (exponential curve fits to cleavage data only): 10-mm-thick specimens.*

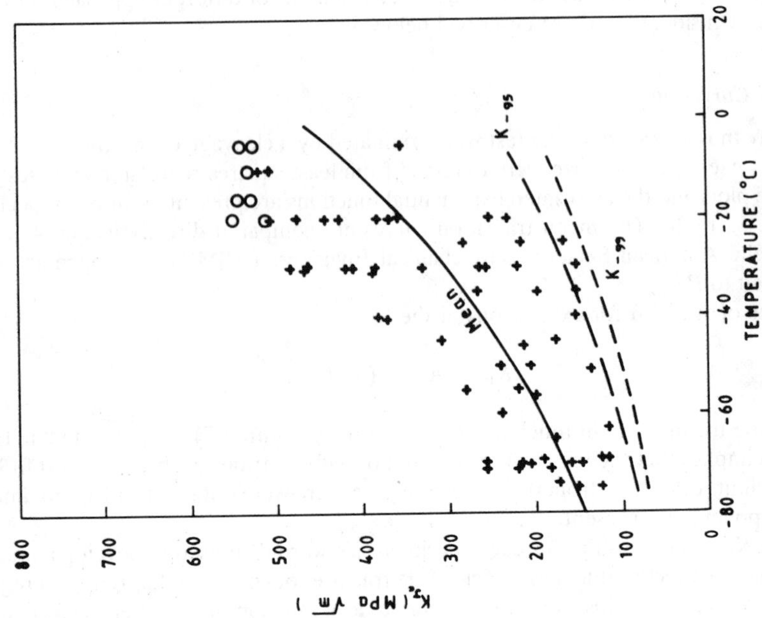

FIG. 3b—*Fracture transition data (exponential curve fits to cleavage data only): 25-mm-thick specimens.*

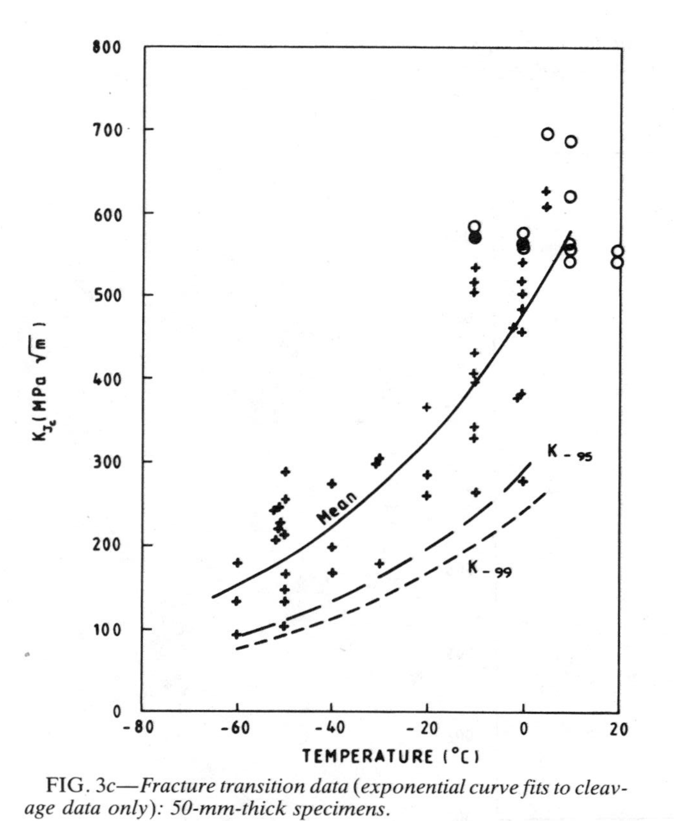

FIG. 3c—*Fracture transition data (exponential curve fits to cleavage data only): 50-mm-thick specimens.*

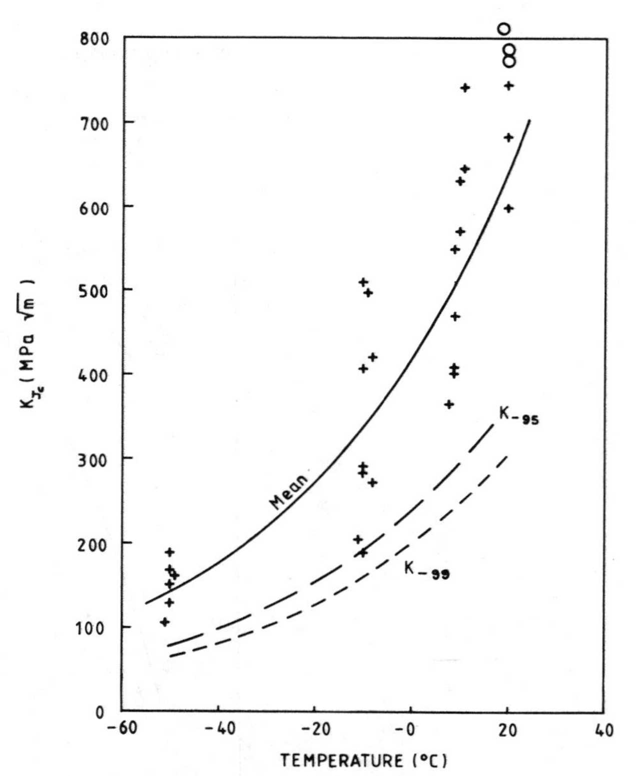

FIG. 3d—*Fracture transition data (exponential curve fits to cleavage data only): 100-mm-thick specimens.*

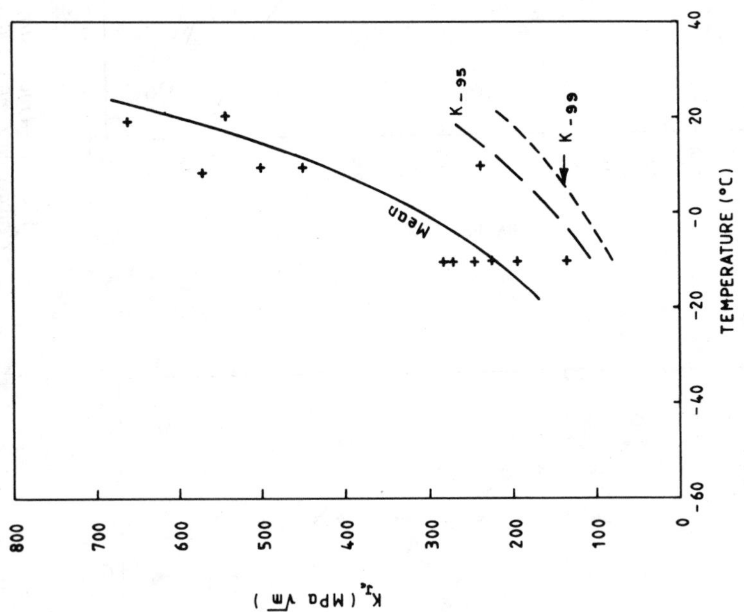

FIG. 3e—*Fracture transition data (exponential curve fits to cleavage data only): 230-mm-thick specimens.*

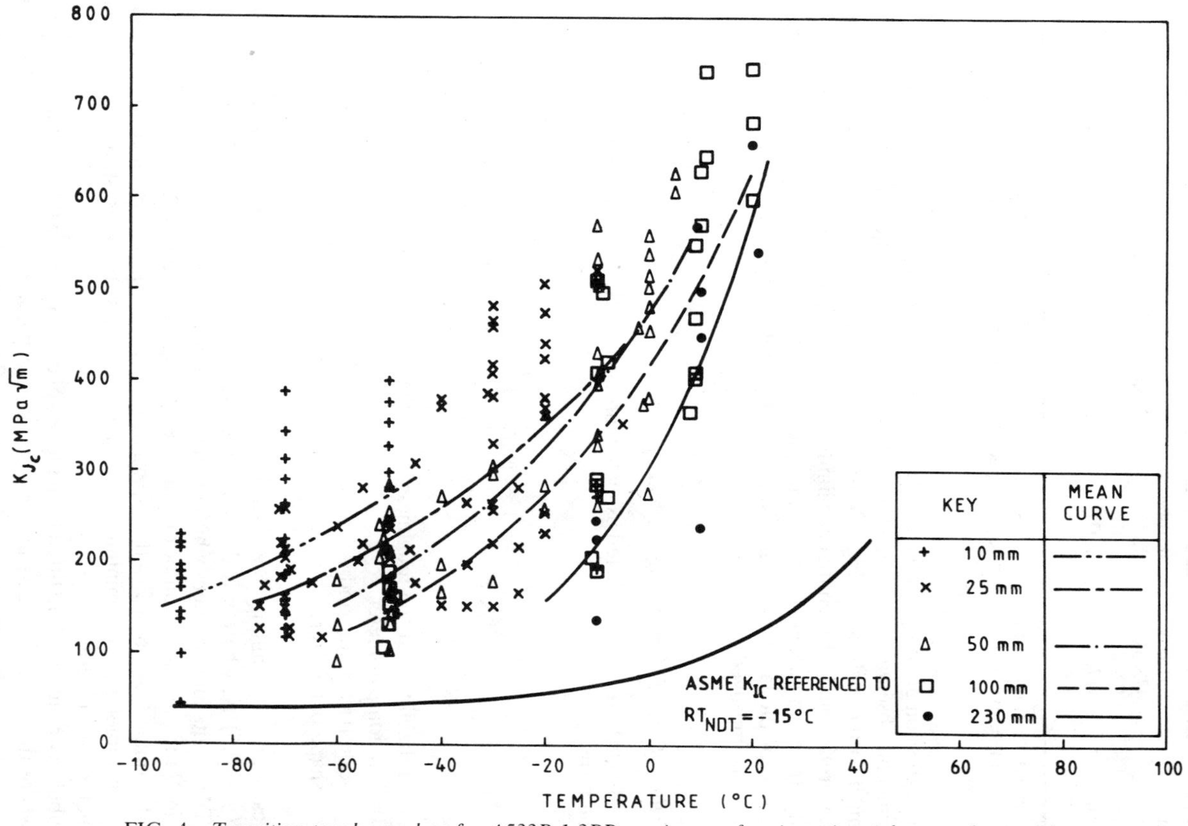

FIG. 4—*Transition toughness data for A533B-1 3PB specimens of various sizes (cleavage data only).*

TABLE 4—*Regression coefficients in the expression* $K_{Jc} = A \exp(\sigma T)$.[a]

Specimen Thickness	Mean Curve		−95% Level[b]		−99% Level[b]	
	A	σ	A	σ	A	σ
10 mm	551.6	0.014	257.3	0.014	202.4	0.014
25 mm	473.6	0.015	255.6	0.015	210.4	0.015
50 mm	478.4	0.019	285.4	0.019	242.5	0.019
100 mm	417.8	0.021	240.9	0.022	198.4	0.022
230 mm	304.4	0.033	152.3	0.031	113.7	0.031

[a] K_{Jc} in MPa$\sqrt{m}$ and T in °C.
[b] Approximate exponential expressions to describe the confidence levels.

The regression analysis can be used to calculate the confidence limits within which a new experimental observation will be expected to lie. The confidence limits were also found to be similar to those shown in Figs. 3*a* to 3*e*, but are somewhat narrower for $B = 230$ mm. The lower 95% and 99% confidence limits for this specimen thickness are shown in Fig. 6. This illustrates an advantage of the multiple linear regression, in that the confidence in the prediction for the largest specimen size, for which relatively few experimental results were obtained, is improved.

Prediction of Size Effects Using Weibull Statistics

Size effects have been predicted using the three-parameter Weibull [*12*] distribution which has the form

$$P = 1 - \exp\left\{-\left[\frac{K - K_e}{K_0}\right]^m\right\} \tag{6}$$

where

P = probability that the toughness is less than K,
K_e = location parameter,
K_0 = scale parameter, and
m = shape parameter.

P is calculated by ranking the results in order of increasing K and using the formula $P = (n - 0.5)/N$, where n is the rank order number and N is the total number of results [*13*]. The results for specimens which did not fail by cleavage were incorporated by including them in N for the calculation of probabilities, but ignoring them when best fits to the data were determined. A "best-fit" three-parameter distribution was obtained by adjusting the location parameter (K_e) to minimize the standard error, with the constraint that the location parameter must be greater than or equal to zero.

Weibull distributions were fitted to all the sets of data for a given temperature and specimen size where there were at least nine results, and to the large specimen ($B = 100$ and 230 mm) data sets where there were only six results. There were 13 sets of data available, and values of the parameters in the distribution for these sets are given in Table 5. The wide range of estimated values for K_e shows that in general the data sets were too small, as discussed later.

Assuming that specimen failure can be described using a "weakest link" theory, it is

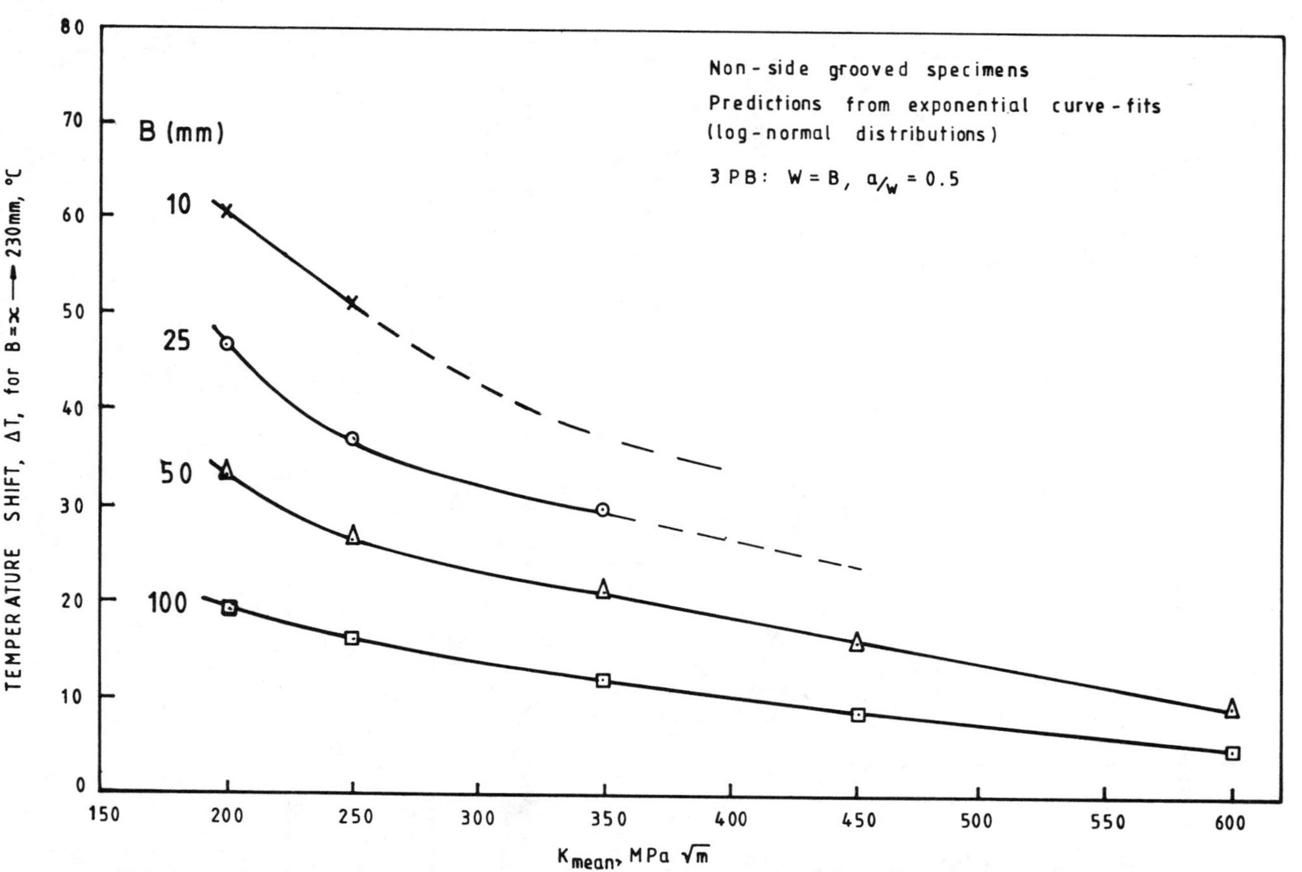

FIG. 5—*Predicted temperature shifts to achieve mean toughness levels when thickness increases from "B" to 230 mm.*

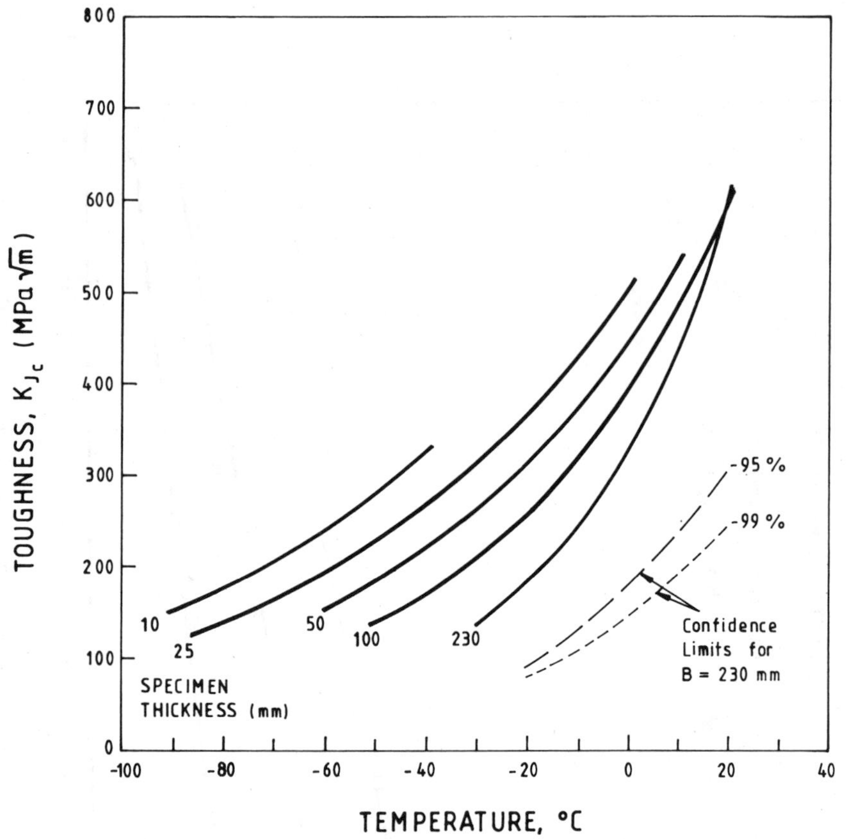

FIG. 6—*Mean toughness predicted using the multiple linear regression equation for various specimen thicknesses.*

TABLE 5—*Parameters for Weibull distributions.*

Thickness B, mm	Test Temperature, °C	Sample Size	n^a	Weibull Parameters: Median, K, MPa$\sqrt{m}$	Shape, m	Location, K_e, MPa$\sqrt{m}$	Scale, K_0, MPa$\sqrt{m}$
10	−90	12	0	184.5	2.48	0	196.5
10	−70	16	0	210.5	1.33	104.1	135.4
10	−50	16	4	311.0	2.44	73.2	287.1
25	−70	12	0	185.8	1.76	99.8	97.0
25	−30	14	0	384.3	3.62	0	395.9
25	−20	12	2	403.6	4.24	0	453.4
50	−50	12	0	214.6	3.94	0	223.9
50	−10	12	2	468.1	2.19	187.4	323.0
50	0	14	4	510.6	4.91	0	555.6
100	−50	6	0	156.9	5.72	0	163.0
100	−10	9	0	288.7	1.26	166.0	200.4
100	+10	9	0	510.1	1.30	340.2	217.2
100	+20	6	3	760.4	3.19	412.7	402.8
230	−10	6	0	235.1	4.33	0	248.0

[a] n = Number of specimens which did not fail.

possible to quantify the effects of specimen size at a given temperature, using the Weibull distribution [14]. Providing that the parameters of the distribution remain constant with specimen size, Eq 6 can be rewritten as

$$1 - P_N(K) = \exp\left\{-\left[\frac{K - K_e}{K_0}(N)^{1/m}\right]^m\right\} \quad (7)$$

where $P_N(K)$ is the probability that the toughness is less than K for a specimen N times thicker than the specimen thickness for which the distribution was derived. Equation 7 has been used to predict the median and the 99% lower bound toughness values at temperatures of −50° and −10°C, using the Weibull distribution parameters for 10- and 50-mm-thick specimens, respectively. These results are plotted in Figs. *7a* and *7b* and show that although this approach fits the experimental results at −50°, it is nonconservative for the −10°C results. This may indicate that a weakest-link theory is more appropriate to the lower shelf/lower transition regimes than to the upper transition.

Merkle's β_{Ic} *Adjustment*

The β_{Ic} method for correcting for the effect of specimen size in fracture toughness tests was first proposed by Irwin [15], and Merkle [16] has applied the method to estimate plane strain fracture toughness values. The K_{Ic} value is calculated from the expression

$$K_{Ic} = K_{Jc}\left\{\frac{\beta_{Ic}}{\beta_c}\right\}^2 \quad (8)$$

where

$$\beta_c = \beta_{Ic} + 1.4\beta_{Ic}^3 \text{ and } \beta_c = 1/B(K_{Jc}/\sigma_y)^2 \quad (9)$$

Merkle's suggestion to use β_{Ic} [16] has been recently criticized by Munz [17]. When the adjustment was applied to the present results, the effect was to reduce the scatter, lower the mean, and invert the size effect so that the smallest specimens define the lower bound. All of the results still fell above the ASME XI Appendix A reference K_{Ic} curve indexed to $RT_{NDT} = -15°C$.

Discussion

The present test program has generated a large amount of data on the effects of specimen size in the upper transition regime of a single plate of A533 Grade B Class 1 pressure vessel steel. In the planning of this program, precautions were taken to minimize the scatter in toughness results by ensuring that all of the specimens were centrally located within a single large plate of A533B-1. In spite of these precautions, the observed scatter in the upper transition regime is large at all sizes, Fig. 4. In order to cope with this scatter, a number of alternative analytical approaches were used, the merits, disadvantages and inherent assumptions of which are discussed below.

The two curve fitting methods are both empirical and result in similar relationships between the brittle fracture toughness, specimen size, and temperature. For the mean values of the fracture toughness the size effect is toughness dependent and shows signs of saturation as the toughness level increases (Fig. 5). This saturation effect could be real or it could be a consequence of the limited J capacity of the smaller specimens. The limit to the maximum J obtainable for a given specimen size tends to depress both the means and the lower

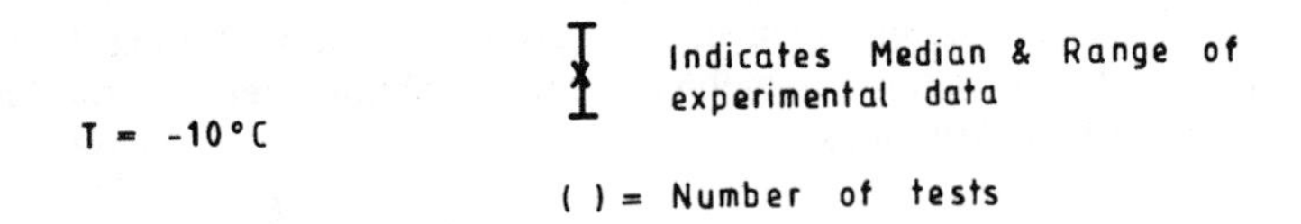

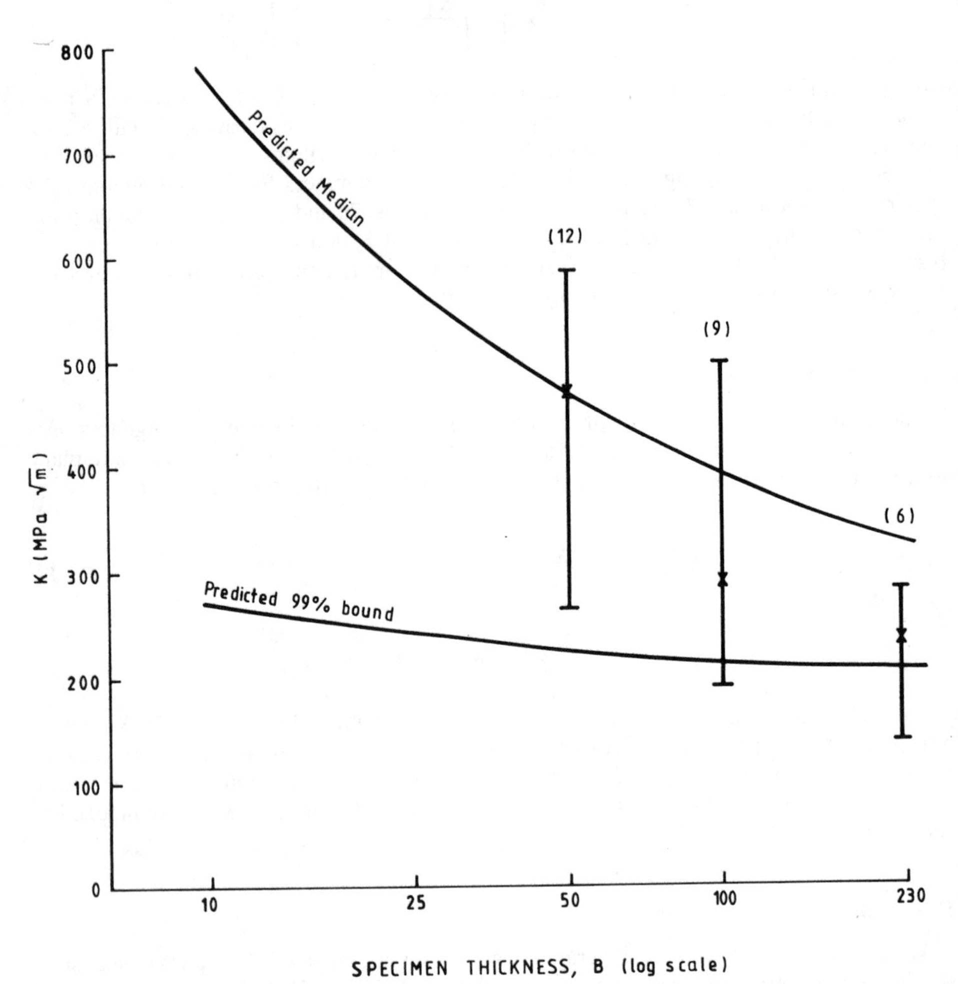

FIG. 7a—*Comparison of toughness predictions from estimated Weibull distributions with measured toughness values at the same temperature: prediction from 50-mm specimens at −10°C.*

confidence limits at the higher test temperatures and may give a misleading picture of the true trend in fracture toughness at this extreme. In reality, the true variation of the lower-bound toughness with temperature in the transition is unlikely to be a simple exponential increase, and further work is necessary to clarify the area of uncertainty as the failure mechanism becomes fully ductile.

For these regression methods, two assumptions are important. The first is the assumption that the values of the independent variables (temperature and thickness) are known exactly. Because the errors in the measurement of these variables are negligible compared with the

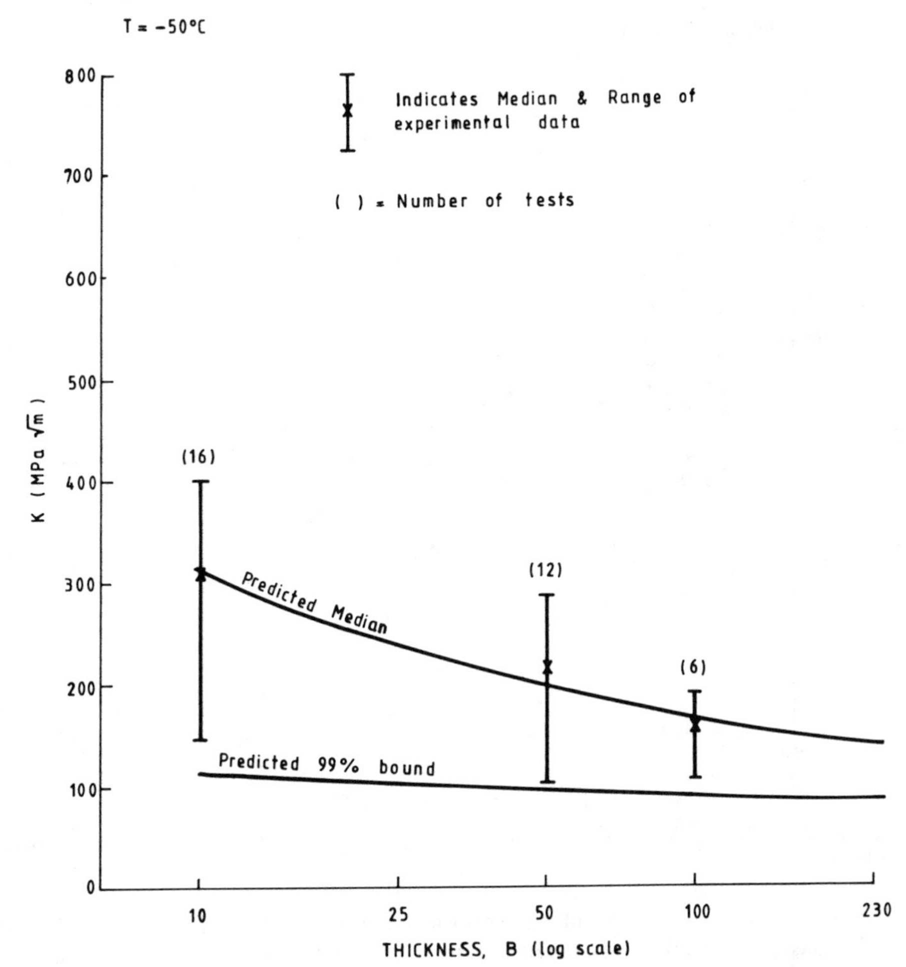

FIG. 7b—*Comparison of toughness predictions from estimated Weibull distributions with measured toughness values at the same temperature: prediction from 10-mm specimens at −50°C.*

uncertainty in the response variable (K_{Jc}), this assumption is justified. The second is the assumption that the variance in the response variable is constant. The validity of this is best tested by examining the residuals, that is, the difference between the measured and predicted values of the response variable. The residuals from the multiple linear regression are shown as a function of the predicted values in Fig. 8. If the assumption of constant variance is valid, then the residuals should fall within a parallel band, equally distributed about the line of zero residual. Divergence from this condition indicates that the variance is not constant. Figure 8 shows that the variance is approximately constant over the range of predicted values, although the data at predicted K_{Jc} values greater than about 500 Mpa$\sqrt{m}$ are rather scarce.

The Weibull analysis has been used previously [*5,13*] and is capable of a physical interpretation via the weakest-link model, although it does not account for effects due to loss of constraint. It allows prediction of specimen size effects directly, but the relationship with temperature can only follow from this empirically.

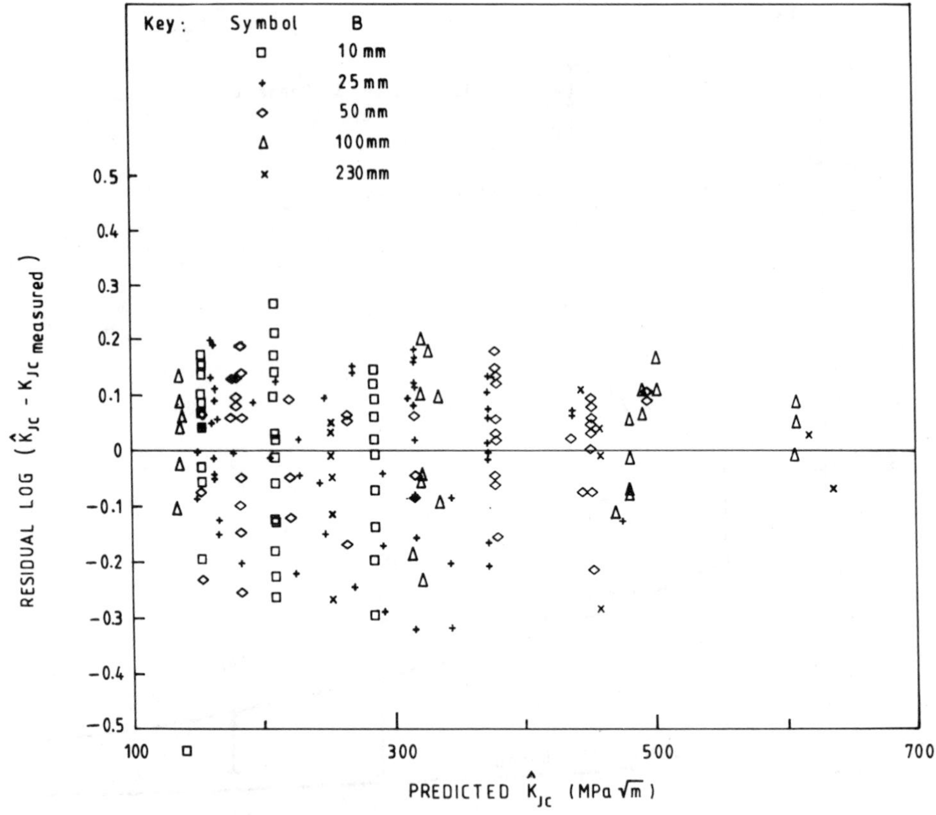

FIG. 8—*Plot of residuals from multiple regression against predicted values of cleavage fracture toughness* ($\hat{K}_{Jc}$).

Landes and McCabe [*14*] highlighted two assumptions implicit in the use of Weibull distributions: first, that the shape parameter, m, is independent of specimen size and second, that the location parameter, K_e, represents a lower bound for all specimen sizes. Neither of these assumptions were upheld by the present work, but since the distributions were fitted to small data sets there will be a large uncertainty in the estimated Weibull parameters. Nevertheless, the relevance of the Weibull approach must be questioned when the scatter in fracture toughness data has been shown to be largely independent of specimen size, Figs. 4 and 8. Also, the weak link model is not strictly applicable if some ductile tearing occurs prior to cleavage.

The logic behind the Merkle correction is somewhat obscure, since Irwin's original correction was applied to ductile instabilities at maximum load. There is no physical basis for applying the method to cleavage instabilities [*17*], nor is there any independent test of applicability possible, in the absence of valid K_{Ic} data. In fact, the method simply factors the calculated value of K_{Jc} by an amount which increases as K_{Jc} increases or as thickness decreases. The scatter is therefore inevitably reduced and the mean value is lowered.

The possibility of dividing the data up further, treating those results in which crack growth preceded brittle fracture separately from those in which there was no ductile tearing, was considered. However, it was not pursued because the final failure mechanism was still cleavage and because, given the large scatter, subdividing the results into two populations would have made the identification of systematic size effects more uncertain.

The objective of a practical test program is to obtain data for use in a structural integrity analysis. The normal limitations on material availability mean that both specimen size and numbers are often restricted. It is therefore an advantage to define ways of optimizing the test program by choosing the best combination of numbers of tests, specimen size, and temperature of tests. The optimum test program will of course depend upon the use to which data will be put, and this section addresses the main considerations.

The present work has shown that if scatter is a problem, this cannot be addressed by spreading a few tests over a wide temperature range. Nor can it be reduced by testing large specimens. The only way to address scatter is to test a large number of specimens at a single temperature. The number of specimens which should be tested depends upon how the results are to be analyzed. In regression analyses, errors are reduced as the number of data points increases. For linear regression techniques (such as those used to fit Eq 4), the confidence limits on y depend on the variance at x. Beyond 30 tests there is little benefit to be gained from testing more specimens. This criterion was met for transition curves fitted to each specimen size, apart from the largest (230 mm thick), where only twelve specimens were available. Wallin [*18*] has suggested that to define a Weibull distribution, 50 or more tests are needed to minimize errors in predicted parameters, while Iwadate et al. [5] claim that the number (N) of tests required is a function of specimen thickness (B), and the toughness level of interest (J_c). For the upper transition they suggested an empirical relationship of the form

$$N > (1000\, J_c)/(\sigma_y\, B) \qquad (10)$$

These requirements can be very difficult to satisfy. For example, Eq 10 requires that if $K_c = 300\ \text{MPa}\sqrt{\text{m}}$ ($J_c > 0.385\ \text{MJ/m}^2$) and 10-mm-thick specimens are used, N is approximately 70. A much smaller number of tests (between 6 and 16, depending on specimen size) were used in this program to determine the distribution.

In practice it will usually not be possible to test the ideal number of specimens. In these circumstances, assumptions can be made about the temperature dependence of toughness, or about the form of the distribution of results at a given temperature, based on published literature or existing databases. A limited number of tests of the relevant thickness can then be used to define the scatter and level of toughness at particular temperatures of interest. The limitations of any statistical fit to the data must be considered; for example, there may be a change in the failure mechanism which cannot be accommodated by the statistical model. Alternatively, the statistics can be improved, by testing smaller specimens and using trend curves obtained from the literature or from a specially developed program to infer the fracture toughness appropriate to the structure. In adopting this route, it should be recognized that the size effect trends may be material dependent and cannot be established without testing a representative number of large specimens as well as small ones.

Conclusions

1. A study of the effect of specimen size on the fracture toughness of A533B steel tested in the ductile-brittle transition regime, using 217 specimens from 10 to 230 mm thick, has

shown that at any one temperature there was considerable scatter in the data, which did not vary with specimen size.

2. Four methods were used to represent the size effect: exponential curve fitting; multiple linear regression analysis; Weibull statistics; and the β_{Ic} correction.

3. The exponential and multiple linear regression analyses for defining temperature shifts gave comparable results. These were capable of producing trend curves: for example, at a toughness of 200 MPa$\sqrt{m}$, the exponential fit predicts a shift of 47°C for an increase in specimen thickness from 25 to 230 mm.

4. Because results from specimens which did not fail were not incorporated in the regression analyses, this caused a downward bias in means and −99% confidence limits. This downward bias occurred at lower toughness levels (and temperatures) as the specimen size decreased.

5. Variation in estimated distribution parameters, between specimens of different sizes, precludes accurate prediction of size effects using Weibull statistics. In spite of the large numbers of specimens tested in this program, these variations arise partly because of an insufficient number of tests at individual temperatures.

6. The β_{Ic} adjustment proposed by Merkle produced anomalies in respect of the effect of specimen size and material variability.

Acknowledgments

The work was carried out at the Central Electricity Research Laboratories, Risley Nuclear Power Development Laboratories, and Springfields Nuclear Laboratories, and is published by permission of the Central Electricity Generating Board and the United Kingdom Atomic Energy Authority.

References

[1] Milne, I. and Chell, G. G., "Effect of Size on the *J* Fracture Criterion," *Elastic-Plastic Fracture, ASTM STP 668,* American Society for Testing and Materials, Philadelphia, 1979, pp. 358–377.

[2] Milne, I. and Curry, D. A., "Ductile Crack Growth Within the Ductile-Brittle Transition Regime: Predicting the Permissible Extent of Ductile Crack Growth," *Elastic Plastic Fracture: Second Symposium. Volume II: Fracture Curves and Engineering Applications, ASTM STP 803,* American Society for Testing and Materials, Philadelphia, 1981, pp. II278–II291.

[3] Sumpter, J. D. G., "The Prediction of K_{Ic} Using *J* and COD from Small Specimen Tests," *Metal Science,* Vol. 10, 1976, pp. 354–356.

[4] Landes, J. D. and Shaffer, D. H., "Statistical Characterisation of Fracture in the Transition Region," *Fracture Mechanics (12th Conference), ASTM STP 700,* American Society for Testing and Materials, Philadelphia, 1980, pp. 368–382.

[5] Iwadate, T., Tanaka, Y., Ono, S., and Watanabe, J., "An Analysis of Elastic-Plastic Fracture Behaviour for J_{Ic} Measurement in the Transition Region," *Elastic-Plastic Fracture: Second Symposium. Volume II: Fracture Curves and Engineering Applications, ASTM STP 803,* American Society for Testing and Materials, Philadelphia, 1983, pp. II531–II561.

[6] Milne, I. and Curry, D. A., "Fracture Toughness, Tearing Resistance and Specimen Size Effects in Submerged Arc and Manual Metal Arc Welds in A533B Plate," CEGB Report TPRD/L/2882/R85, Central Electricity Generating Board, U.K., 1985.

[7] Milne, I. and Curry, D. A., "The Effect of Triaxiality on Ductile-Cleavage Transitions in a Pressure Vessel Steel," *Fracture and Fatigue, Proceedings,* 3rd Colloquium on Fracture, London, 1980, pp. 39–47.

[8] De Roo, P., Marandet, B., Phillipeau, G., and Rousellier, G., "Effect of Specimen Dimensions on Critical *J*-Value at the Onset of Crack Extension," *Fracture Mechanics: Fifteenth Symposium, ASTM STP 833,* American Society for Testing and Materials, Philadelphia, 1984, pp. 606–621.

[9] Ingham, T., Bland, J. T., and Wardle, G., "The Influence of Specimen Size on the Upper Shelf

Toughness of SA 533B-1 Steel" in *Proceedings,* 7th International Conference on Structural Materials in Reactor Technology, Paper G2/3, Chicago, 1983.

[*10*] Hutchinson, J. W., "Singular Behaviour at the End of a Tensile Crack in a Hardening Material," *Journal of the Mechanics and Physics of Solids,* Vol. 16, 1968, pp. 13–31.

[*11*] Neale, B. K., Curry, D. A., Green, G., Haigh, J. R., and Akhurst, K. N., "A Procedure for the Determination of the Fracture Resistance of Ductile Steels," *International Journal of Pressure Vessels and Piping,* Vol. 20, 1985, pp. 155–179.

[*12*] Weibull, W., "A Statistical Theory of Strength of Materials," Royal Swedish Institute of Engineering Research, No. 15, 1939.

[*13*] Andrews, W. R., Kumar, V., and Little, M. M., "Small Specimen Brittle Fracture Toughness Testing," *Fracture Mechanics (13th Conference) ASTM STP 743,* American Society for Testing and Materials, Philadelphia, 1981, pp. 576–598.

[*14*] Landes, J. D. and McCabe, D. E., "The Effect of Section Size on the Transition Temperature Behaviour of Structural Steels," *Fracture Mechanics: Fifteenth Symposium, ASTM STP 833,* American Society for Testing and Materials, Philadelphia, 1984, pp. 378–392.

[*15*] Irwin, G. R., "Fracture Mode Transition for a Crack Traversing a Plate," *Journal of Basic Engineering,* Vol. 82, No. 2, 1960, pp. 417–425.

[*16*] Merkle, J. G., "An Examination of the Size Effects and Data Scatter Observed in Small Specimen Cleavage Fracture Toughness Testing," NUREG/CR-3672, Nuclear Regulatory Commission, Washington, DC.

[*17*] Munz, D., "The Size Effect of Fracture Toughness and the Irwin J_{Ic} Adjustment," *International Journal of Fracture,* Vol. 32, 1986, pp. R17–R19.

[*18*] Wallin, K. "The Scatter in K_{Ic} Results," *Engineering Fracture Mechanics,* Vol. 19, 1984, pp. 1085–1093.

Anthony A. Willoughby[1] *and Tim G. Davey*[1]

Plastic Collapse in Part-Wall Flaws in Plates

REFERENCE: Willoughby, A. A. and Davey, T. G., **"Plastic Collapse in Part-Wall Flaws in Plates,"** *Fracture Mechanics: Perspectives and Directions (Twentieth Symposium), ASTM STP 1020,* R. P. Wei and R. P. Gangloff, Eds., American Society for Testing and Materials, Philadelphia, 1989, pp. 390–409.

ABSTRACT: Experimental results on plastic collapse stresses in flat plates containing part-wall flaws have been collated. These are compared with two-dimensional limit load solutions, assuming both free rotation (pin jointing) and rigid restraint. Allowance for finite flaw lengths is made by calculating an effective flaw depth to thickness ratio, based on the area of the flaw divided by an effective sectional area. Few data on embedded flaws were found, and so an experimental program using diffusion bonded tension specimens containing internal slits was undertaken.

The results of this program, coupled with the analysis of experimental data on surface flaws, show that for surface flaws in sections experiencing some restraint, the assumption of rigid restraint is reasonable. For embedded flaws, pin jointing was assumed, rather than rigid restraint, since the equations are much simpler and the difference is smaller. The proposed correction for finite flaw length is shown to give good results, with overall safety factors under tensile loads in the range 1.08 to 1.64 being obtained. For highly work-hardening materials such as austenitic steels, the method overpredicts the collapse loads, due almost certainly to inappropriate estimates of flow strength.

KEY WORDS: plastic collapse, part-wall flaws, plates, fracture, surface flaws, embedded flaws, ferritic steel, stainless steel, flow strength, limit load

Nomenclature

- C — t/e
- C_{eff} — Effective value of t/e
- e — Thickness of cross section
- h — Width of cross section
- ℓ — Flaw length
- L — Tensile load
- M — Bending moment
- p — Ligament height between tip of buried flaw and surface
- P_m — Membrane (tensile) stress
- P_b — Bending stress (outer fibre)
- t — Depth of surface flaw or height of embedded flaw
- $\overline{\sigma}$ — Flow strength
- σ_Y — Yield strength
- σ_u — Tensile strength

[1] Principal research engineers, The Welding Institute, Abington Hall, Abington, Cambridge, CB1 6AL, U.K. Dr. Willoughby is at present with the Open University, Milton Keynes, U.K.

Introduction

A number of methods are now available for assessing the significance of flaws in welded structures under elastic plastic conditions. The most widespread are the PD6493 approach,[2] the Central Electricity Generating Board (CEGB) "R6" method [*1,2*], and the Electric Power Research Institute (EPRI) approach [*3*]. The PD6493 method is currently undergoing revision. When fracture due to crack-like flaws is considered, plastic collapse must be taken into account, because a high net section stress increases the driving force for crack propagation. Therefore, all the above methods of flaw assessment require an estimate of the stress for plastic collapse.

It should be noted that in flaw assessment, the plastic collapse calculation complements the fracture calculation. In high-toughness (strength-dominated) materials, failure will occur by collapse without flaw extension. In low-toughness (toughness-dominated) materials, flaw extension will occur because the fracture condition is attained before the plastic collapse condition. In many practical situations, failure will best be described by a combination of plastic collapse and fracture mechanisms.

In the simpler methods of flaw assessment (see PD6493:1980 and Ref *1*), plastic collapse is deemed to occur when the net section stress on the ligaments surrounding the flaw reaches the flow strength, $\bar{\sigma}$, of the material. The flow strength in a work-hardening material is equivalent to the yield strength in a non-work-hardening material. It is commonly taken as the average of the yield and tensile strengths of the material, a definition employed here; that is

$$\bar{\sigma} = \tfrac{1}{2}(\sigma_y + \sigma_u) \tag{1}$$

This definition has generally been found to apply to materials of low and moderate work-hardening capacity, such as ferritic steels. For high work-hardening materials, such as austenitic stainless steels, it may be an overestimate. Consequently, the more sophisticated methods [*2,3*] base the criterion on the yield strength and the strain-hardening characteristics of the material.

This paper arises from two related projects. A report by Willoughby [*4*] collated experimental data on collapse stresses from various geometries of flawed specimens. There was, however, a paucity of data on embedded flaws, which led Davey [5] to undertake experiments on tension specimens with embedded flaws. Both these reports are available only to research members of The Welding Institute, and therefore it was decided to combine them in the current paper and to present a new analysis of the data.

Methods of Analysis

Two-Dimensional Limit Load Analysis

In an uncracked rectangular section subjected to remotely applied tension loading, the neutral axis and the tensile axis will coincide, and there will be no bending moment on the cross section. The presence of a flaw, however, will cause the neutral axis to move away from the tensile axis, resulting in an additional bending moment. The appropriate limit load solution will depend on whether this bending moment is assumed to apply to the net section,

[2] British Standards Institution Guidance on Some Methods for Some Derivation of Acceptance Levels for Defects in Fusion Welded Joints (PD6493:1980).

or whether it is carried externally. The first case is analogous to "pin-jointing" (that is, the tensile load is applied through pin-jointed couplings), the second to rigid restraint.

For pin-jointing, it is shown in the appendix that a general equation for embedded flaws can be derived

$$\frac{2}{3}\frac{P_b}{\bar{\sigma}} + 2\frac{t}{e}\frac{P_m}{\bar{\sigma}} + \left(\frac{P_m}{\bar{\sigma}}\right)^2 = \left(1 - \frac{t}{e}\right)^2 + \frac{4pt}{e^2} \quad (2)$$

The general equation for surface breaking flaws is obtained by substituting $p = 0$.

Simplified equations for the cases of applied bending or tension alone can be derived as follows, by substituting $P_m = 0$, or $P_b = 0$:

1. Embedded flaws, pin jointed

$$P_m = 0, \quad \frac{P_b}{\bar{\sigma}} = 1.5\,(1 - C)^2 + 6\,C\frac{p}{e} \quad (3)$$

$$P_b = 0, \quad \frac{P_m}{\bar{\sigma}} = \sqrt{\left[2C^2 - 2C\left(1 - 2\frac{p}{e}\right) + 1\right]} - C \quad (4)$$

where C is equal to t/e

2. Surface flaws, pin jointed

$$P_m = 0, \quad \frac{P_b}{\bar{\sigma}} = 1.5\,(1 - C)^2 \quad (5)$$

$$P_b = 0, \quad \frac{P_m}{\bar{\sigma}} = \sqrt{[2C^2 - 2C + 1]} - C \quad (6)$$

The equivalent relations for rigid restraint are more complex, since three different sets of equations are required, depending on the position of the flaw in the cross section and its height. The appendix gives the derivation of the expression for a surface flaw as

$$\frac{2}{3}\frac{P_b}{\bar{\sigma}} + \left(\frac{P_m}{\bar{\sigma}}\right)^2 = (1 - C)^2 \quad (7)$$

For P_m or $P_b = 0$, this reduces to

$$P_m = 0, \quad \frac{P_b}{\bar{\sigma}} = 1.5\,(1 - C)^2 \quad \text{(same as Eq 5)} \quad (8)$$

$$P_b = 0, \quad \frac{P_m}{\bar{\sigma}} = 1 - C \quad (9)$$

PD6493 gives a graphical solution which is approximately described by Eq 2 for the case of embedded flaws (that is, it assumes pin jointing). For surface flaws, however, Eq 7 (rigid restraint) is used.

R6-Rev. 2 [*1*] bases its solutions on the assumption of pin jointing for surface flaws.

Embedded flaws are treated as surface flaws, with an imaginary surface passing through the center of the flaw. This is clearly unrealistic, since bending restraint will be provided by the other ligament.

In both PD6493 and R6-Rev. 2, the flow strength $\overline{\sigma}$ of the material is taken as the average of the yield and ultimate tensile strengths, to allow for work hardening.

Three-Dimensional Flaws

The equations presented in the previous section apply to infinitely long part wall flaws. In flaws of finite length they will be conservative, since part of the load can be carried by the unflawed section on either side of the flaw.

PD6493 makes no direct allowance for this effect. R6-Rev. 2 takes it into account by assuming that the flaw is semi-elliptical, of area $\pi \ell t/4$, and the load is distributed over a section of area $e(\ell + e)$ (Fig. 1). The effective flaw depth/thickness then becomes, for $t/\ell > 0.1$

$$C_{\text{eff}} = \frac{\pi \ell t}{4e(\ell + e)} \quad (10)$$

This method was shown empirically to be reasonable in Ref *4*. It is difficult, however, to apply it to embedded flaws. A recommendation was made in Ref *4* that Eq 10 could be rewritten, assuming an elliptical flaw in a cross section of $e(\ell + e)$.

Chell [*6*] proposes a transformation

$$C_{\text{eff}} = \left(1 - \frac{1}{f}\right) \Big/ \left(1 - \frac{t}{e} \cdot \frac{1}{f}\right) \quad (11)$$

where

$$f = \left(1 + \frac{\ell^2}{2e^2}\right)^{1/2}$$

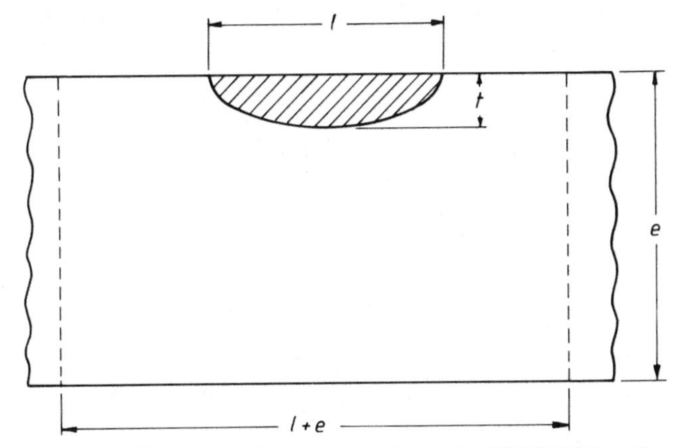

FIG. 1—*Effective flaw area and cross-sectional area in CEGB R6-Rev. 2 method.*

This method is analogous to that given by Keifner et al. [7] for axial flaws in pipelines. Its derivation is not made clear, however, and it is not obvious how to adapt it to apply to embedded flaws.

A new expression for C_{eff} is proposed here. The flaw is assumed to be rectangular, in an effective cross section of $e(\ell + 2e)$ (Fig. 2). The effective flaw height to thickness ratio is then given by

$$C_{eff} = \frac{\ell t}{e(\ell + 2e)} = \frac{t}{e}\frac{1}{(1 + 2e/\ell)} \quad \text{for } h > \ell + 2e \tag{12a}$$

or

$$C_{eff} = \frac{t}{e} \cdot \frac{\ell}{h} \quad \text{for } h \leqq \ell + 2e \tag{12b}$$

This expression has the following advantages: $C_{eff} = t/e$ in the limit as $\ell \to \infty$; it is relatively simple; and it is equally applicable to surface and embedded flaws.

Comparison with Experimental Data on Surface Flaws

Experimental data on limit loads for a variety of ferritic steel specimens containing surface flaws under either bending or tension loads were compiled in Ref *4*. There were also limited data on stainless steels, and on center-cracked tension panels, analogous to centrally placed embedded flaws. These data are summarized in Tables 1, 2, and 3.

When comparing the theoretical and experimental limit loads, it is assumed that the extension of the flaw by ductile tearing is negligible. This is a reasonable assumption for relatively tough materials, but for those of low toughness, such as the ferritic pipeline steel, it could result in an overprediction of the maximum load.

Long Surface Flaws in Tension

Data are given in Table 1 for those results for which $p = 0$. They were mostly obtained from large tensile panels ("wide plates") having plate widths in the range $h = 500$ to 900

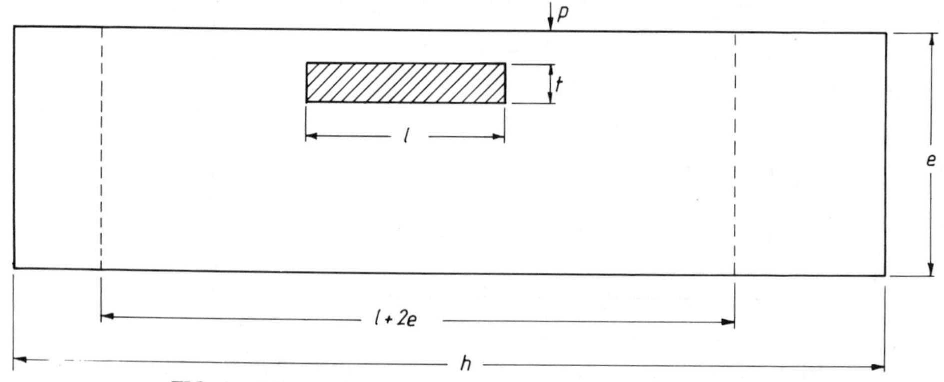

FIG. 2—*Effective flaw area and cross-sectional area proposed here.*

TABLE 1—*Summary of test results—flaws under tension loading.*

Steel	Specimen No.	h, mm	e, mm	ℓ, mm	t, mm	p, mm	σ_Y, N/mm^2	σ_u, N/mm^2	P_m (fail), N/mm^2
A533B	1009	508	110	∞	44.1	0	462	601	394
	1044	508	110	∞	57.6	0	462	601	325
	1045	508	110	∞	43.4	0	462	601	425
316 stainless	OCR1	150	49.4	∞	20.4	0	185	470	183
steel (HAZ)[a]	OCR4	150	49.2	∞	27.6	0	387	497	160
(WM)[b]	OCR5	147	49.8	∞	25.8	0	387	497	158
A533B	1006	508	110	152	22.8	0	462	532	608
	1007	510	113	240	38.7	0	462	532	502
	1046	508	113	328	56.2	0	462	532	416
Low alloy	2	38.7	19.7	18.0	11.7	0	446	603	433
	4	36.4	19.8	18.1	10.7	0	446	603	440
BS4360:50D (weld)	1060	900	50	97.3	12.9	0	466	614	504
	1064	900	50	108	14.1	0	466	614	490
316 stainless (weld)	OCR7	51.0	148.2	∞	78.6	34.8	387	497	176
	OCR8	51.4	147.1	∞	90.7	28.2	185	470	142
(HAZ)	OCR9	50.4	147.2	∞	72.8	37.2	387	497	199
A533B	1008	110	490	∞	210	140	462	601	372
	1042	110	490	∞	262	114	462	601	288
	1043	110	490	∞	134	178	462	601	472
X56 pipeline	1176	858	13	42	2.7	0	438	584	483
steel	1177	899	13	90	4.8	0	438	584	451
	1178	13	898	∞	86	406	438	584	403
	1179	13	892	∞	265	314	438	584	316
	(1180	885	13	∞	3.5 (max)	0	438	584	414)
	(1180	885	13	∞	2.13 (ave)	0	438	584	414)
	(1181	898	13	∞	5.5 (max)	0	438	584	351)
	(1181	898	13	∞	4.6 (ave)	0	438	584	351)
Low alloy steel	1	83	78.6	∞	44.9	16.85	446	603	258
	2	83	77.5	∞	38.1	19.7	446	603	272
	3	83	78.5	∞	41.6	18.45	446	603	275
Mild steel	1	50.8	6.35	25.4	5.62	0	300	410	219
	2	50.8	6.35	25.4	5.27	0	300	410	220
	3	50.8	6.35	25.4	4.91	0	300	410	243
	4	50.8	6.35	25.4	4.78	0	300	410	248
	5	50.8	6.35	25.4	4.50	0	300	410	264
	6	50.8	6.35	25.4	3.85	0	300	410	260
	7	50.8	6.35	25.4	2.72	0	300	410	330

[a] WM = weld metal.
[b] HAZ = heat affected zone.

mm. Some narrower tensile bars were also tested. Data from Table 4, from the experimental program described in the next section, are included. In all cases the flaw length extended the full width of the test specimen (that is, $\ell = h$, denoted as $\ell = \infty$ in Table 1). From Fig. 3 it can be seen that Eq 9 (rigid restraint) gives a conservative prediction for all the cases except for the stainless steels.

Short Surface Flaws in Tension

Data were mostly obtained from wide plates containing short surface flaws (that is, $p = 0$ and $\ell < h$ in Table 1). The data were analyzed using Eq 12 to calculate C_{eff}, and were then compared with the predictions of Eqs 6 and 9, as shown in Fig. 4. Equation 9 (rigid restraint) gives conservative predictions for all the cases. Equation 6 (pin jointing) gives very conservative predictions for the deeper flaws.

TABLE 2—*Summary of test results—long surface flaws in bending.*

Steel	Test No.	Specimen Dimensions, mm		Flaw depth t, mm	σ_Y, N/mm²	σ_u, N/mm²	P_b, at Failure, N/mm²
		h	e				
316 stainless	22551/1	50	50	29.8	250	590	144
	22551/2A	50	50	28.4	250	590	171
	22551/4A	50	50	27.3	250	590	160
	22551/1B	50	50	25.0	250	590	212
	22551/1C	50	50	24.1	250	590	221
	22551/5C	50	50	23.6	250	590	223
	OCR2	151	51	21.6	185	470	234
(weld)	OCR3	149	51	29.0	387	497	195
(weld)	OCR6	148	51	29.2	387	497	162
BS4360:50D	3571/1-1	52	105	53.3	393	567	250
	3571/4-2	52	104	52.0	393	567	269
	3571/1-2	51.5	104	51.9	413	587	260
MMA welds in pipeline steel	W10-3	15.9	17.7	3.6	465[a]	...[a]	809
	W10-6	15.9	16.3	2.9	465[a]	...[a]	945
	W10-9	15.9	17.6	3.7	465[a]	...[a]	743
	W10-12	15.9	17.6	4.5	465[a]	...[a]	709
X56 pipeline steel	9CQ 8	13.0	24.0	11.3	438	584	225
	9CQ 9	13.0	24.0	11.6	438	584	215
	9CQ12	13.0	24.0	11.6	438	584	225
	9CQ28	12.7	13.2	4.2	438	584	429
	9CQ29	12.7	13.2	3.3	438	584	419
	9CQ32	12.7	13.2	3.7	438	584	439
	9CQ33	12.7	13.2	3.9	438	584	419
	9CQ34	12.7	13.2	3.9	438	584	470
	9CQ35	12.7	13.2	3.6	438	584	480
	9CQ36	12.7	13.2	3.8	438	584	440

[a] Yield strength estimated from hardness and flow strength taken at 1.15 × yield.

Long Surface Flaws in Bending

The data were all taken from three-point bend fracture toughness specimens (Table 2). Figure 5 shows that Eq 5 gives conservative predictions, with the possible exception of some of the pipeline steel specimens.

Short Surface Flaws in Bending

Only three experimental results were available (Table 3). Equation 5, with $C = C_{eff}$ as defined by Eq 12, gave conservative predictions, as shown in Fig. 6.

TABLE 3—*Summary of test results—short surface flaws in bending.*

Steel	Specimen No.	h	e	ℓ	t	p	σ_Y, N/mm²	σ_u, N/mm²	P_b (Fail), N/mm²
Low alloy	1	37.4	20.3	17.8	10.7	0	446	603	>875
	3	39.2	20.3	19.2	10.8	0	446	603	859
C-Mn weld	W5	750	50	123.3	10.4	0	565	687	>853

TABLE 4—*Parent material data.*

CHEMICAL COMPOSITION[a]

Element Weight %																	
C	S	P	Si	Mn	Ni	Cr	Mo	V	Cu	Nb	Ti	Al	B	Sn	Co	N_2	O_2
0.16	0.019	0.019	0.44	1.41	0.03	0.01	<0.005	<0.002	0.04	0.034	<0.002	0.036	<0.0003	<0.005	0.005	0.0049	0.0016

TWI analysis reference numbers S/84/375 and O/N 84/156.
[a] By optical emission spectrography, oxygen and nitrogen by inert gas fusion.

TENSILE PROPERTIES AFTER DIFFUSION BONDING CYCLE

Specimen No.	Yield Strength, N/mm^2	Tensile Strength, N/mm^2	Elongation, %	Reduction of Area, %
11	302	508	27	64
12	305	506	29	64
13	309	506	25	60

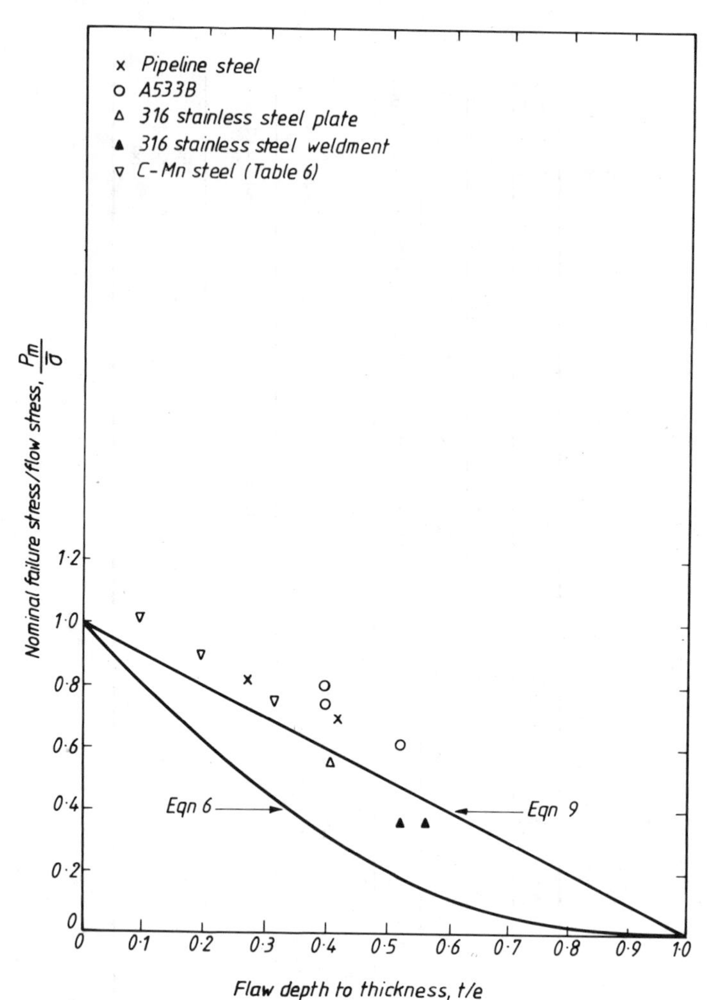

FIG. 3—*Variation of failure stress with flaw depth, for long surface flaws in tension.*

Long Embedded Flaws in Tension

The only data found were obtained on center-cracked tension panels (Table 1, with $p > 0$). Figure 7 shows a plot of the experimental $P_m/\bar{\sigma}$ against t/e. With regard to the analysis, since the flaw is centrally positioned, $p = (e - t)/2$ and Eq 4 reduces to Eq 9. It can be seen from Fig. 7 that the equation gives conservative predictions for the tough ferritic steels, but not for the low-toughness pipeline steel or the stainless steel.

Experimental Work

As noted in the previous section, the only available data for embedded flaws were obtained on center-cracked tension (CCT) panels. It was decided, therefore, to carry out an experimental investigation into the limit loads for a variety of embedded flaws. Some surface flaws were included for comparison.

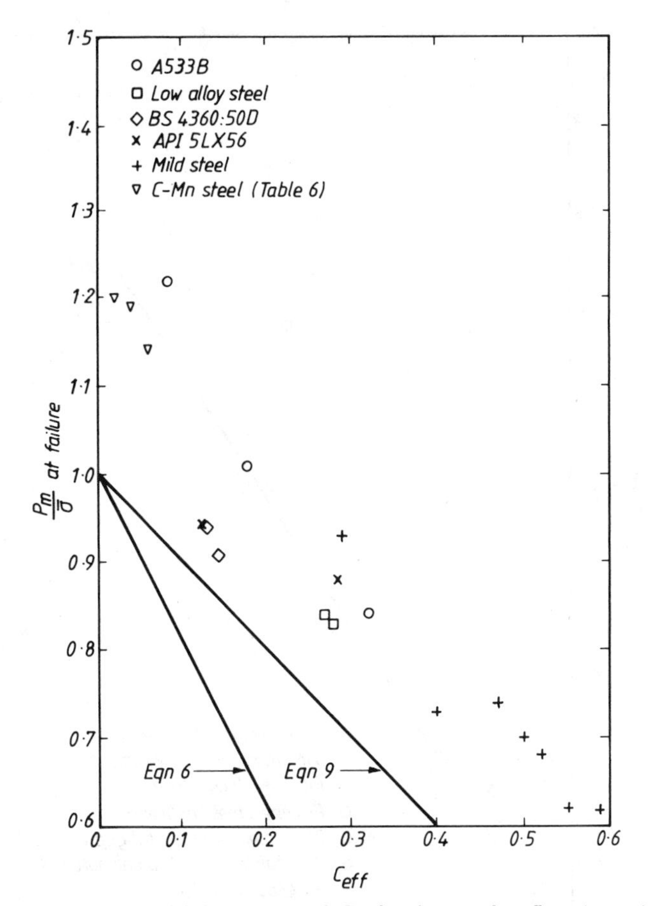

FIG. 4—*Variation of failure stress with* C_{eff} *for short surface flaws in tension.*

Outline of Experimental Method

To produce buried and surface breaking flaws of a controlled size in material with uniform mechanical properties, blocks of carbon-manganese steel with and without spark-eroded slots were diffusion-bonded together to produce composite specimens containing rectangular flaws, as shown in Fig. 8. Even where diffusion bonding was not needed, the blocks were placed in the diffusion bonding machine and subjected to the bonding thermal cycle. These composite specimens were tested in tension, and the ability of the various plastic collapse analyses to predict the maximum load was evaluated. These analyses require the flow strength of the material as an input so this had to be determined for the material in its diffusion bonded condition.

Test Material

This material used for the test program was extracted from a 25-mm-thick plate to British Standard 4360 (Weldable Structural Steels) Grade 50D BS4360:50D. The composition of the test material is given in Table 4. However, it was recognized that the properties of the

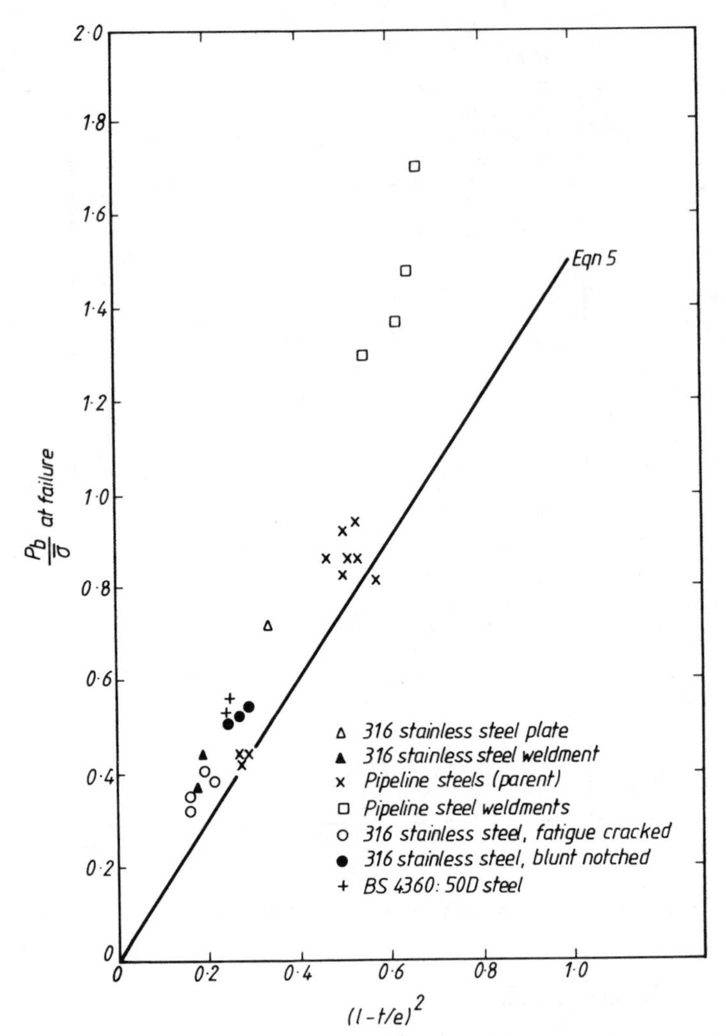

FIG. 5—*Relationship between failure stress and flaw depth for long surface flaws in bending.*

test material would be affected by the diffusion bonding cycle, so the tensile properties of the thermally affected material were determined for use in the plastic collapse analysis. Three tension specimens were extracted from a block of the test material which has been subjected to the same diffusion bonding cycle as the flawed specimens. These tension specimens were tested at room temperature, and the results are also given in Table 4. The resulting flow strength is 406 N/mm^2.

The diffusion bonding conditions are shown in Table 5. Although the bonding process considerably alters the material properties, the thermal cycle is similar to that used in a normalizing treatment. It is thus probable that the results obtained could be applied to most carbon, carbon-manganese or low-alloy ferritic steels. As noted earlier (in the first section), the results are inappropriate to stainless steels, for reasons not connected with the use of diffusion bonding.

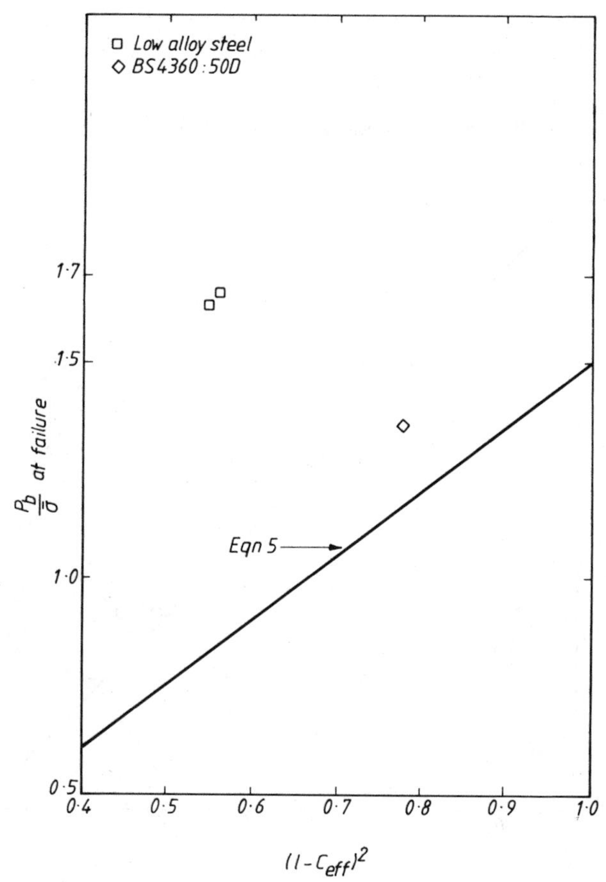

FIG. 6—*Variation of failure stress with* C_{eff} *for short surface flaws in bending.*

Manufacture and Testing of Flawed Specimens

In order for the flawed specimens to be produced, blocks 75- by 75-mm square and of varying thicknesses were extracted from the test material. Slots approximately 0.15 to 0.20 mm wide were produced in some of these blocks by spark erosion or abrasive wheel cutting, and the blocks were then diffusion-bonded together to produce a composite block 75 by 75 by 25 mm thick containing an embedded or surface breaking flaw. The plane of the diffusion bonds was parallel to the specimen axis and thus should not have influenced the failure behavior. The composite blocks produced were electron beam welded to extension bars for tension testing. The completed specimens were then waisted to give a parallel central portion 50 mm long and 50 by 25 mm in cross section, containing the flaw. Since notch tip acuity has only a second-order effect on plastic collapse behavior, the flaws were not sharpened by fatigue cracking. The specimen manufacturing process and the nominal specimen dimensions are given in Fig. 8.

Results

The sizes of the flaws inserted in the specimens are given in Table 6, using the nomenclature of PD6493. Also shown are the maximum loads sustained and the corresponding nominal

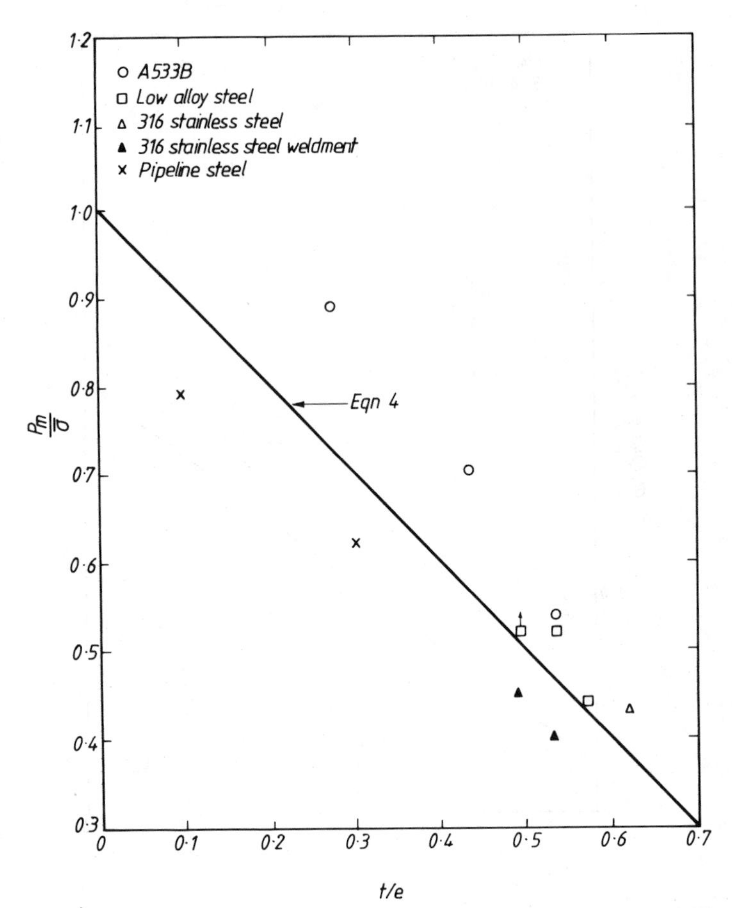

FIG. 7—*Variation in failure stress with* t/e *for long embedded flaws in tension* (*center-cracked tension panels*).

stress on the specimens. All specimens showed extensive necking, indicating that, as intended, they had failed by plastic collapse without intervention of any other failure mechanism.

Figure 9 shows the results of these tests in terms of C versus $P_m/\overline{\sigma}$ and compares the experimental data with the predictions of Eq 4 or Eq 6. The predictions are all conservative by a reasonable margin and show no tendency for the degree of conservatism to change with flaw depth (that is, with C).

One could also postulate that the reduction in limit load in the tension case is simply proportional to the reduction in cross section caused by the presence of the flaw, that is

$$P_m/\overline{\sigma} = 1 - \ell t/he \tag{13}$$

This is shown in Fig. 9 and also provides conservative predictions. However, the degree of conservatism of this approach decreases as t/e increases, and it is possible that it might become unconservative if t/e were to exceed about 0.4, particularly for surface flaws. This is probably due to the fact that the approach assumes perfectly rigid restraint, which is likely

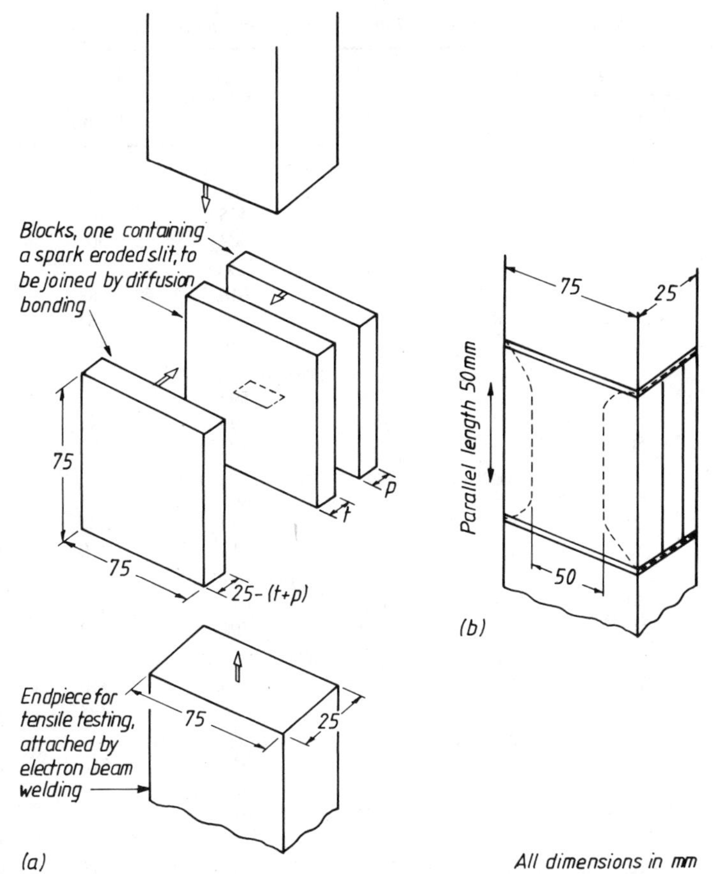

FIG. 8—*Method for production of defect-containing tension specimens:* (a) *component parts, containing spark-eroded slots;* (b) *waisted specimen cut from composite block.*

to be unrealistic for large surface flaws. This method is also not easily generalized to encompass combined tension and bending and so will not be considered further.

Discussion

For surface breaking flaws, the limit load equation based on rigid restraint (Eq 9) appears to predict all the experimental results conservatively, with the exception of the specimens made from 316 stainless steel and from the X56 pipeline steel. The use of the normal definition of flow strength has been shown to be inappropriate (causing overestimation of

TABLE 5—*Diffusion bonding conditions.*

Bonding temperature	1050°C
Bonding hold time	15 min
Applied pressure	5 N/mm^2
Cooling	natural

TABLE 6—*Results of tension tests on flaw-containing specimens.*

Specimen No.	Flaw dimensions, mm			Max Load, kN	Failure Stress,[a] N/mm²	t/e	p/e
	Length ℓ	Height t	Depth p				
			LONG SURFACE-BREAKING FLAWS				
1	50	2.5	0	564	451	0.1	0
4	50	5.0	0	455	364	0.2	0
20	50	7.5	0	384	307	0.3	0
			SHORT SURFACE-BREAKING FLAWS				
8	10	2.5	0	610	488	0.1	0
15	10	5.0	0	606	485	0.2	0
11	10	7.5	0	580	464	0.3	0
			LONG BURIED FLAWS, CLOSE TO SURFACE				
2	50	2.5	2.5	592	474	0.1	0.1
16	50	5.0	2.5	490	392	0.2	0.1
21	50	7.5	2.5	402	322	0.3	0.1
			LONG BURIED FLAWS, INTERMEDIATE DEPTH				
13	50	2.5	5.0	606	485	0.1	0.2
6	50	5.0	5.0	540	432	0.2	0.2
5	50	7.5	5.0	512	410	0.3	0.2
			LONG, DEEPLY BURIED FLAWS				
3	50	2.5	10.0	615	492	0.1	0.4
7	50	5.0	10.0	502	402	0.2	0.4
			SHORT, BURIED FLAWS NEAR TO SURFACE				
9	10	2.5	2.5	638	510	0.1	0.1
17	10	5.0	2.5	616	493	0.2	0.1
			SHORT BURIED FLAWS, INTERMEDIATE DEPTH				
14	10	2.5	5.0	645	516	0.1	0.2
18	10	5.0	5.0	627	502	0.2	0.2
12	10	7.5	5.0	615	492	0.3	0.2
			SHORT DEEPLY BURIED FLAWS				
10	10	2.5	10.0	640	512	0.1	0.4
19	10	5.0	10.0	507	486	0.2	0.4
22	10	7.5	7.5	623	498	0.3	0.3

[a] Failure stress = failure load/original nondefective cross section.

flow resistance) for austenitic stainless steels and other materials of very high work-hardening capacity [8], so the discrepancy in the results on the stainless steel is not surprising.

The X56 pipeline steel was known to have a particularly low resistance to ductile tearing [9], and it is highly likely that significant crack extension occurred prior to maximum load. This is particularly true for the CCT panels, which contained relatively long cracks and large ligaments, and which also gave collapse stresses which were below the predictions of Eq 4 by the greatest margin. The results on this steel can therefore be regarded as being controlled more by toughness than by plastic collapse, at least for some of the geometries.

Ignoring these results, therefore, it would appear that Eq 9 (rigid restraint), coupled with the allowance for finite flaw length of Eq 12, gives conservative predictions for all the surface flaws considered. Since all the test specimens had some degree of rotational restraint, it is possible that the limit load for an unrestrained section (that is, pin jointed) would be overpredicted. This situation is seldom encountered in a structure, but if it did occur, it would be prudent to revert to Eq 6 (pin jointed), at least for infinitely long surface flaws.

A rigid restraint solution for the embedded flaw should be equally applicable, but its

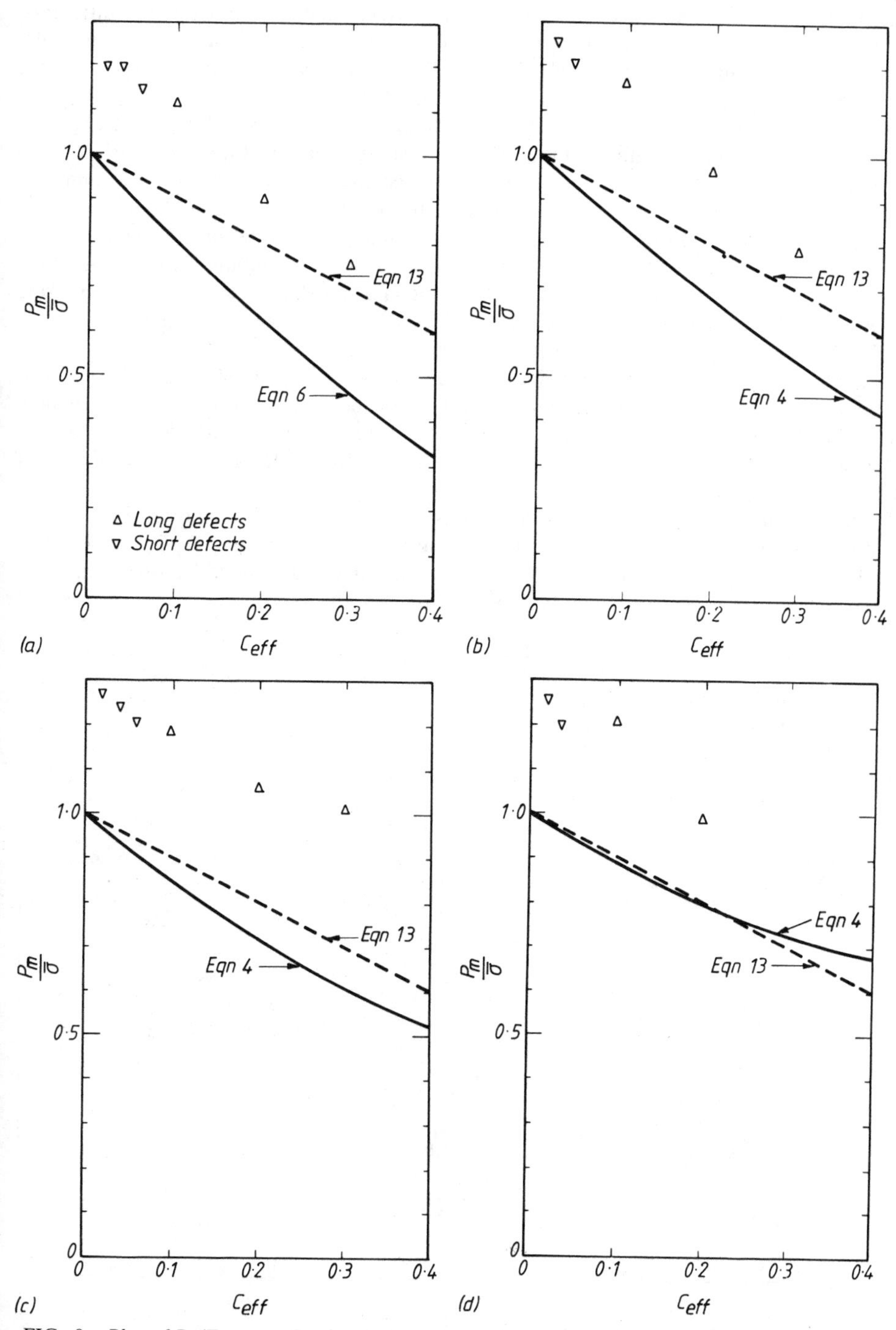

FIG. 9—*Plot of* $P_m/\bar{\sigma}$ *versus* C_{eff} *for diffusion-bonded specimens:* (a) *surface flaws,* p/e = *0;* (b) *embedded flaws,* p/e = *0.1;* (c) *embedded flaws,* p/e = *0.2; and* (d) *embedded flaws,* p/e = *0.4.*

derivation is by no means as simple as for the surface flaw. In any case, the difference between rigid restraint and pin jointing is much smaller for the embedded flaw, unless the flaw is large and just below one surface. When the flaw is located on the center-line, there is no induced bending, and the two solutions would be the same.

For pure tensile loading and ferritic steels, therefore, it is proposed that Eq 12 be used to allow for finite flaw length, and then the result substituted into Eq 9 for a surface flaw, or into Eq 4 for an embedded flaw, to obtain the collapse stress. The result of doing this for all the data presented here is shown in Fig. 10, which gives a plot of the safety factor obtained versus the t/e ratio. The safety factor is defined as the ratio of the actual collapse stress to that predicted. It is apparent that safe predictions are obtained for all the cases, except for some of the stainless steel and low-toughness pipeline steel specimens. Ignoring these data, safety factors in the range greater than 1.02 to 1.67 were obtained.

For pure bending stresses, Eq 5 gives conservative predictions for all long surface flaws, except for one result on the low-toughness pipeline steel (Fig. 5). Interestingly, the stainless steel specimens are also predicted safely by Eq 5. For the shorter flaws, there are few data available, but the allowance of Eq 12 appears to give a very safe prediction (Fig. 6). No data were obtained on embedded flaws in bending, but there is no reason to consider that Eq 3 will not apply.

There are no data available under conditions of combined bending and tension, but the conservatism of equations for either bending or tension suggest that Eq 7 can be safely applied to surface flaws with rotational restraint and Eq 2 to all embedded ones. Equation 12 could be used to correct for finite flaw length.

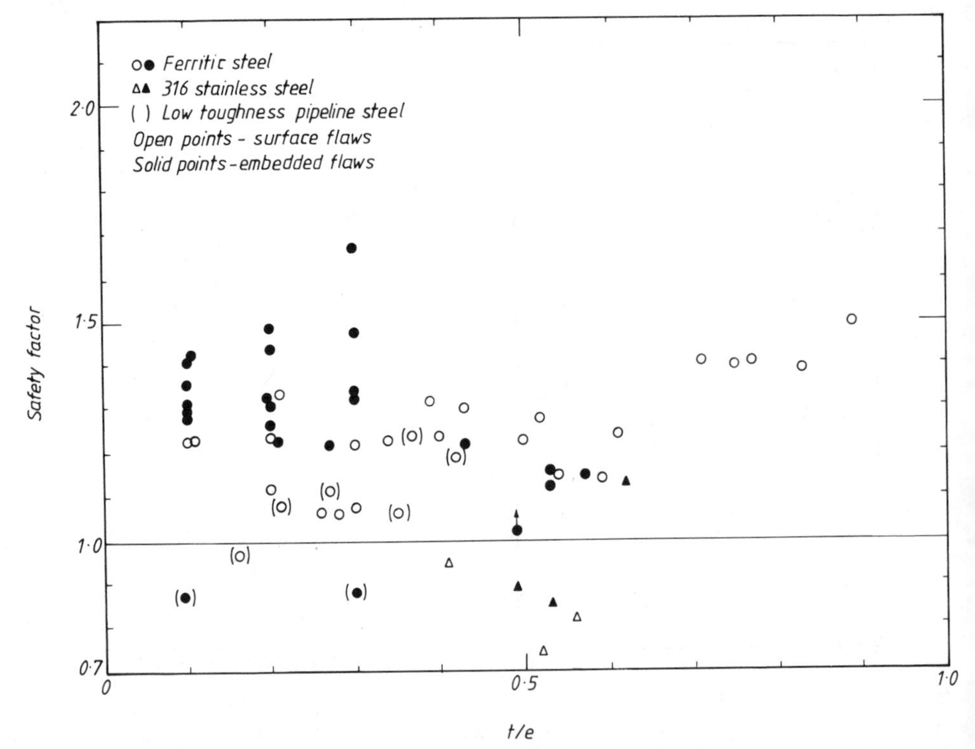

FIG. 10—*Variation of safety factor inherent in proposed analysis for all surface and buried flaws in tension.*

Conclusions

In this work comparing theoretical and experimental plastic collapse loads for flawed steel plates, the effective flow strength of the steels was taken as the mean of the yield (or proof) and tensile strengths. The following conclusions were reached:

1. For long surface-breaking flaws in plates of tough ferritic steels, loaded in tension with typical rotational restraint, the limit load solution for rigid restraint (Eq 9) gives conservative estimates of the plastic collapse stress. The assumption of pin jointing for the same geometry gives very conservative predictions.
2. For long embedded flaws in tension, the assumption of pin jointing is not excessively conservative and is preferred because of the simplicity of the resulting solution, compared to that for rigid restraint.
3. Limited data on long surface and embedded flaws in bending indicate that the limit load solution provides conservative estimates of collapse moments.
4. The correction proposed for finite flaw length (Eq 12), based on the area of the rectangular flaw and the area of an effective cross section, appears to give reasonably consistent factors of safety for all flaw aspect ratios considered, and is equally applicable to surface or embedded flaws.
5. Collapse loads for austenitic stainless steel plates under tension were not predicted conservatively, due probably to the inappropriate estimate of flow strength for materials of very high work-hardening capacity.

Acknowledgments

The authors would like to thank T. R. Gurney, R. H. Leggatt, and S. J. Garwood for helpful discussions. Financial support from the research members of The Welding Institute and from the Minerals and Metals Division of the U.K. Department of Trade and Industry is gratefully acknowledged.

APPENDIX

Two-Dimensional Limit Load Analysis

Embedded and Surface Breaking Flaw with no Bending Restraint

Consider the embedded flaw shown in Fig. 11, under applied tensile load L and bending moment M, in a rigid plastic nonwork-hardening material of yield strength σ_Y. The tensile load L contributes to the bending moment, due to the movement of the neutral axis away from the center, as a result of the presence of the flaw. The following equations can be written down:

Resolving horizontally

$$L = \sigma_Y p + \sigma_Y z - \sigma_Y x = \sigma_Y(p + z - x) \tag{14}$$

Taking moments about the line AB

$$M + L\frac{e}{2} = \sigma_Y p \left(e - \frac{p}{2}\right) + \sigma_Y z \left(x + \frac{z}{2}\right) - \sigma_Y x \frac{x}{2} \tag{15}$$

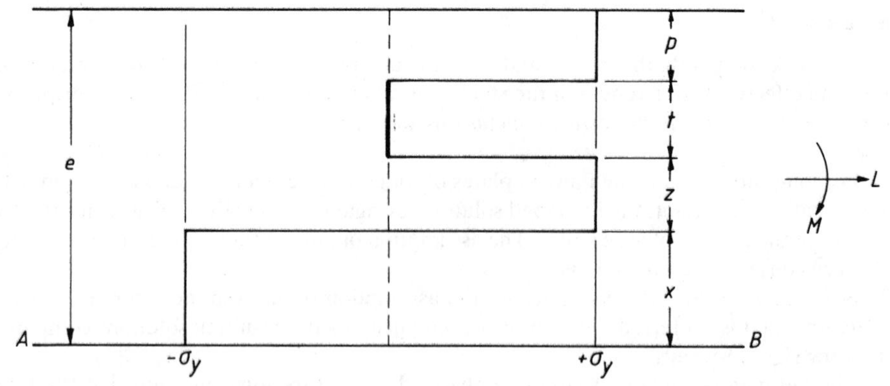

FIG. 11—*Limit load on an embedded flaw with no rotational restraint.*

From geometry

$$e = p + t + x + z \tag{16}$$

Additionally

$$L = P_m e \tag{17}$$

$$M = \frac{P_b e^2}{6} \tag{18}$$

Eliminating x and z from Eqs 14 to 16 and substituting Eqs 17 and 18 results in the general expression

$$\frac{2}{3}\frac{P_b}{\sigma_Y} + 2\frac{P_m}{\sigma_Y}\frac{t}{e} + \frac{P_m^2}{\sigma_Y^2} = \left(1 - \frac{t}{e}\right)^2 + \frac{4pt}{e^2} \tag{19}$$

Substituting $p = 0$ gives the equivalent relationship for a surface breaking flaw

$$\frac{2}{3}\frac{P_b}{\sigma_Y} + 2\frac{P_m}{\sigma_m}\frac{t}{e} + \frac{P_m^2}{\sigma_Y^2} = \left(1 - \frac{t}{e}\right)^2 \tag{20}$$

Surface Flaws with Bending Restraint

In this case it is assumed that the tensile load L does not contribute to the bending moment. A general expression equivalent to Eq 20 for an embedded flaw cannot be readily derived, since the initial conditions depend on the location and size of the flaw. The situation for the surface flaw is illustrated in Fig. 12.

Resolving horizontally

$$L = \sigma_Y (e - t - 2) \tag{21}$$

Taking moments about AB

$$M = \sigma_Y x \left(e - t - \frac{x}{2} - \frac{x}{2}\right) = \sigma_Y x (e - t - x) \tag{22}$$

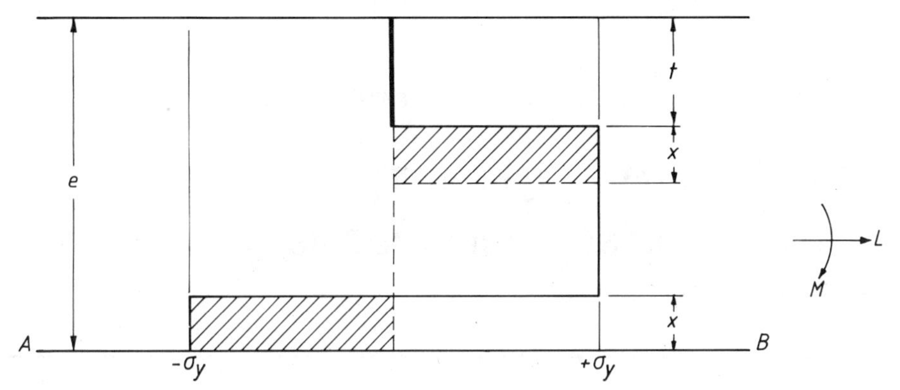

FIG. 12—*Limit load on a surface flaw with rotational restraint. Shaded region is the part which resists bending moment.*

Eliminating x and substituting Eqs 17 and 18

$$\frac{2}{3}\frac{P_b}{\sigma_Y} + \frac{P_m^2}{\sigma_Y^2} = \left(1 - \frac{t}{e}\right)^2 \tag{23}$$

References

[1] Harrison, R. P. et al., "Assessment of the Integrity of Structures Containing Defects," CEGB Report R/H/R6-Rev. 2, Central Electricity Generating Board, 1980.

[2] Milne, I. et al., "Assessment of the Integrity of Structures Containing Defects," CEGB Report R/H/R6-Rev. 3, Central Electricity Generating Board, 1986.

[3] Kumar, V., German, M. D., and Shih, C. F., "An Engineering Approach for Elastic-Plastic Fracture Analysis," EPRI Report NO-1931, Electric Power Research Institute, July 1981.

[4] Willoughby, A. A., "A Survey of Plastic Collapse Solutions Used in Failure Assessment of Part Wall Defects," Welding Institute Research Report 191/1982. (Available only to research members of The Welding Institute.)

[5] Davey, T. G., "An Investigation of Plastic Collapse Assessments for Ferritic Steels for Part Wall Defects in Tension," Welding Institute Research Report. (Available only to research members of The Welding Institute.)

[6] Chell, G. G., "Elastic-Plastic Fracture Mechanics," *Developments in Fracture Mechanics, Part I,* G. G. Chell, Ed., Applied Science Publishers, Ltd., London, 1979.

[7] Kiefner, J. F., Maxey, W. A., Eiber, R. J., and Duffy, A. R., "Failure Stress Levels of Flaws in Pressurized Cylinders," *Progress in Flaw Growth and Fracture Toughness Testing, ASTM STP 536,* The American Society for Testing and Materials, Philadelphia, 1973, pp. 461–481.

[8] Akhurst, K. N. and Milne, I., "Failure Assessment Diagrams and J Estimates: Validation for an Austenitic Steel," *Proceedings,* International Conference on Application of Fracture Mechanics to Materials and Structures (AFMMS), Freiburg, Federal Republic of Germany, June 1983.

[9] Garwood, S. J., Willoughby, A. A., and Rietjens, P., "The Application of CTOD Methods for Safety Assessment in Ductile Pipeline Steels," *Proceedings,* International Conference on Fitness for Purpose Validation of Welded Constructions, The Welding Institute, London, 1981.

J. Robin Gordon[1] *and Stephen J. Garwood*[2]

A Comparison of Crack-Tip Opening Displacement Ductile Instability Analyses

REFERENCE: Gordon, J. R. and Garwood, S. J., "**A Comparison of Crack-Tip Opening Displacement Ductile Instability Analyses,**" *Fracture Mechanics: Perspectives and Directions (Twentieth Symposium), ASTM STP 1020,* R. P. Wei and R. P. Gangloff, Eds., American Society for Testing and Materials, Philadelphia, 1989, pp. 410–430.

ABSTRACT: This paper presents the results of a large test program undertaken to study ductile fracture in aluminum weldments. The main body of the experimental program consisted of a series of wide-plate tests on 10-mm-thick parent material and welded aluminum alloy center-cracked panels. In addition to the wide-plate tests, tension, and single-specimen unloading compliance, R-curve tests were conducted on specimens extracted from the various parent materials and weld metals.

The wide-plate tests were analyzed using a crack-tip opening displacement (CTOD) strip yield model, a CTOD reference stress model, and the CTOD equivalent of the J estimation scheme published by the Electric Power Research Institute. It was found that all three models provided reasonably accurate predictions of maximum far-field stress if a representative CTOD R-curve was incorporated in the analysis.

KEY WORDS: fracture mechanics, fracture toughness, ductile fracture, crack tip opening displacement (CTOD), R-curves, aluminum, welded joints, stable crack growth

It is generally accepted that most welded engineering structures enter service containing defects. Moreover, depending on the operating conditions and the type of applied loading, these defects may extend to a size at which the overall integrity of the structure is impaired. It is therefore important that designers and engineers be able to assess the significance of defects which are detected during the manufacture of the structure or after it has entered its service life.

Over the past few years, The Welding Institute has been actively involved in developing assessment procedures which are less restrictive and more accurate than the current crack-tip opening displacement (CTOD) design curve approach [*1*]. This has resulted in the proposal of

1. A CTOD plastic collapse modified strip yield model [*2*].
2. A CTOD reference stress model [*3*].

Both these models can be used to assess ductile fracture by incorporating the material's crack growth resistance curve.

[1] Senior research engineer, Edison Welding Institute, 1100 Kinnear Road, Columbus, OH 43212.
[2] Head, Engineering Department, The Welding Institute, Abington Hall, Abington CB1 6AL, United Kingdom.

In a ductile instability analysis, the resistance curve for the material is compared against a series of driving force curves. These driving force curves, which represent the applied or predicted values of CTOD (δ_A) that exist in the structure for a given set of conditions, are generally expressed as a series of plots of CTOD against crack length for different values of applied load (or stress). Instability occurs when the rate of increase in the applied driving force equals or exceeds the gradient of the R-curve.

The principle behind this analysis is shown schematically in Fig. 1. In this diagram the material's CTOD R-curve is represented by the solid curve denoted δ_R. The R-curve is positioned so that it intersects the crack length axis at a value corresponding to the initial crack length in the structure, a_0.

The maximum load-carrying capacity of the structure is defined by P_2, where the driving force curve is tangential to the R-curve, that is, Point C. This condition can be expressed mathematically as

$$\delta_A = \delta_R \quad \text{and} \quad \frac{d\delta_A}{da} = \frac{d\delta_R}{da} \tag{1}$$

This paper presents the results of four wide-plate and associated small-scale fracture tests on 10-mm-thick parent material and welded aluminum panels. All test panels were supplied by Alcoa Technical Centre. The tests are analyzed using CTOD strip yield and reference stress models and the CTOD estimation scheme published by the Electric Power Research Institute (EPRI). The predicted and actual conditions at maximum load are compared.

Assessment Methods

General Considerations

The various fracture assessment equations compared in this paper are expressed as functions of effective primary stress σ_p, effective secondary stress σ_s, and effective net section stress terms σ_n [3]. The effective primary and secondary stress terms are given by the following expressions where $K_I{}^p$ and $K_I{}^s$ are the stress intensity factors due to primary and

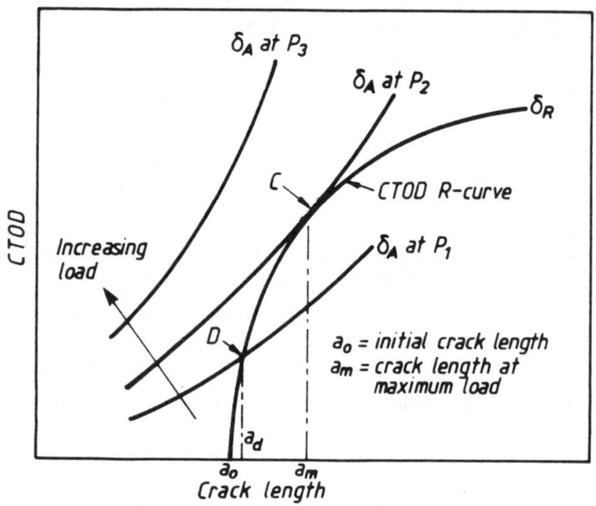

FIG. 1—*Principle of* R-*curve analysis.*

secondary stresses, respectively

$$\sigma_p = \frac{K_I^p}{\sqrt{\pi a}} \qquad \sigma_s = \frac{K_I^s}{\sqrt{\pi a}} \tag{2}$$

The dimension a corresponds to the flaw dimension of interest, that is, the half length of a through thickness flaw, the depth of a surface flaw, or the half height of an embedded flaw. The effective primary and secondary stresses have the units of stress but, in general, are not equal to the actual primary and secondary stresses. (The exception is an infinite plate in uniform tension with a through thickness flaw of length $2a$). The effective primary and secondary stresses are simply convenient parameters that contain the stress intensity and flaw size.

The effective net section stress, σ_n, characterizes the primary stress in the uncracked area of the section under consideration. Net section yield occurs when σ_n is equal to the yield stress.

In the case of a center-cracked panel subjected to a uniform membrane stress, σ_m, consideration of force equilibrium leads to the following simple expression for effective net section stress

$$\sigma_n = \sigma_m \frac{A_g}{A_n} \tag{3}$$

where A_g is the area of the gross section and A_n is the area of the net section.

Plastic Collapse Modified Strip Yield Model

The CTOD plastic collapse modified strip yield fracture assessment model employed in this investigation is based on the model originally proposed by Garwood [2]. However, to enable residual stresses to be incorporated in fracture assessments, the model has been reformulated to produce the expression

$$\delta = \frac{\sigma_{YS}\pi a}{E}\left\{\left(\frac{\sigma_p + \sigma_s}{\sigma_{YS}}\right)^2 + \frac{1}{2}\left(\frac{\sigma_s}{\sigma_{YS}}\right)^4 + \left(\frac{\sigma_p}{\sigma_{YS}}\right)^2\left[\left(\frac{\sigma_{flow}}{\sigma_n}\right)^2 \frac{8}{\pi^2} \ln \sec\left(\frac{\pi}{2}\frac{\sigma_n}{\sigma_{flow}}\right) - 1\right]\right\} \tag{4}$$

where

σ_{YS} = yield strength,
σ_{flow} = flow strength, and
E = Young's modulus.

In the above expression the total applied CTOD is made up of:

1. An elastic component based on primary and secondary stresses.
2. A first-order plastic zone correction term based on secondary stresses.
3. A plastic component based on primary stresses, which is assumed to include implicitly a first-order plastic zone correction for primary stresses.

Reference Stress Model

The CTOD reference stress fracture assessment equation employed in this work is based on the model originally proposed by Anderson et al. [3]. However, to enable residual stresses to be incorporated in fracture assessments, the model has been reformulated to produce the expression

$$\delta = \frac{\pi\sigma_{YS}a}{E}\left\{\left[\frac{\sigma_p + \sigma_s}{\sigma_{YS}}\right]^2\left(\frac{E}{E'}\frac{a_e}{m_{el}a}\right) + \left[\frac{\sigma_p}{\sigma_{YS}}\right]^2\frac{\mu}{m_{FP}}\left(\frac{E\epsilon_{ref}}{\sigma_{ref}} - 1\right)\right\} \tag{5}$$

where

$$a_e = a + \frac{1}{\beta\pi}\frac{n-1}{n+1}\frac{(\sigma_p + \sigma_s)^2\pi a}{\sigma_{YS}^2}\frac{1}{1 + (\sigma_{ref}/\sigma_{YS})^2} \tag{6}$$

and

m_{el} = 1 for plane stress and m_{el} = 2 for plane strain,
m_{FP} = 1.1 for tension and $m_{FP} \simeq$ 1.3–1.8 for bending,
μ = 0.75 for plane stress and μ = 1.0 for plane strain,
E' = E for plane stress and $E' = E/(1 - \nu^2)$ for plane strain, and
β = 2 for plane stress and β = 6 for plane strain.

The reference stress, (σ_{ref}) in Eqs 5 and 6 is equal to the effective net section stress. The reference strain, (ϵ_{ref}) is defined as the value of strain corresponding to σ_{ref} in a uniaxial tension test.

In the CTOD reference stress equation the total applied CTOD is made up of

1. An elastic component (which includes a first order plastic zone correction) based on primary and secondary stresses.
2. A fully plastic component based on primary stresses.

EPRI CTOD Estimation Scheme

The CTOD estimation equations published by EPRI [4] can be reformulated to give the fracture assessment equation

$$\delta = \frac{(\sigma_p + \sigma_s)^2\,\pi a_e}{m_{el}\sigma_{YS}E'} + \alpha\epsilon_0 c d_n h_1\left(\frac{\sigma_n}{\sigma_0}\right)^{n+1} \tag{7}$$

where

$$a_e = a + \frac{1}{\beta\pi}\frac{n-1}{n+1}\frac{(\sigma_p + \sigma_s)^2\pi a}{\sigma_0^2}\frac{1}{1 + (\sigma_n/\sigma_0)^2} \tag{8}$$

where

α and n = Ramberg-Osgood constants,
c = charactristic dimension (usually the remaining ligament),
d_n = material constant which relates the fully plastic J and CTOD's, and
h_1 = geometry-dependent function obtained using finite-element analysis.

In the CTOD estimation scheme, the total applied CTOD is made up of

1. An elastic component of CTOD (which includes a first order plastic zone correction) based on primary and secondary stresses.
2. A fully plastic component of CTOD based on primary stresses.

Details of the three fracture assessment models are given in the Appendix.

TABLE 1—*Details of aluminum alloy center-cracked panel wide-plate specimens.*

Test Panels	Parent Material	Welding Consumable
W11, W12	5456-H116	5556
W13, W14	6061-T651	5356
W15, W16	6061-T651	4043
MO2	6061-T651	...

Details of Aluminum Alloy Test Panels

A total of four wide plate tests were conducted in this investigation; the first three tests were performed on 10-mm-thick welded aluminum alloy panels, while the fourth test was conducted on 10-mm-thick 6061-T651 aluminum parent plate. In the case of the three welded panels, a second series of three panels of identical material/weld combinations were cut up to provide a series of small scale fracture toughness specimens. Details of the parent plate/weld combinations of the panels are given in Table 1, while the chemical compositions of both the plate materials and the welding consumables are summarized in Table 2.

The welded panels were approximately 915 mm square and comprised two rectangular plates each approximately 915 by 460 mm which were joined with a double V, four pass butt weld. All the panels were welded by Alcoa using the metal inert gas (MIG) process.

Surface residual stress measurements were made on each welded panel using a center hole rosette gage technique [*5*]. The internal residual stress distribution was determined using a two-stage sectioning technique similar to that proposed by Rosenthal and Norton [*6*]. The maximum transverse residual stresses measured in panels W12, W14, and W16 were 38, 20, and 25 N/mm^2, respectively.

Details of Test Program

Tension Tests

The true stress-true strain behavior of the three aluminum alloy weld metals was determined by testing flat cross weld tension specimens with high elongation strain gages attached to the weld (one on each side of the specimen). Before the specimens were strain gaged, the weld overfill was machined off flush with the surface of the parent plate. The tension tests on the 6061 T651 parent plate were conducted using a conventional extensometer to measure the strain. The true stress-true strain curves of the three weld metals and the 6061 parent plate are compared in Fig. 2, while the tensile properties are summarized in Table 3.

TABLE 2—*Chemical compositions of aluminum alloy parent materials and welding consumables in percent.*[a]

Description	Si	Fe	Cu	Mn	Mg	Cr	Zn	Ti	V
6061-T651 plate	0.64	0.37	0.21	0.06	0.86	0.16	0.09	0.02	0.01
5456-H116 plate	0.09	0.24	0.04	0.74	5.02	0.09	0.03	0.02	0.01
4043 electrode	5.69	0.32	0.24	0.02	0.02	0.00	0.03	0.01	0.01
5356 electrode	0.07	0.13	0.02	0.13	4.94	0.08	0.04	0.08	0.02
5556 electrode	0.08	0.13	0.03	0.72	5.00	0.07	0.03	0.13	0.02

[a] Chemical analyses supplied by Alcoa.

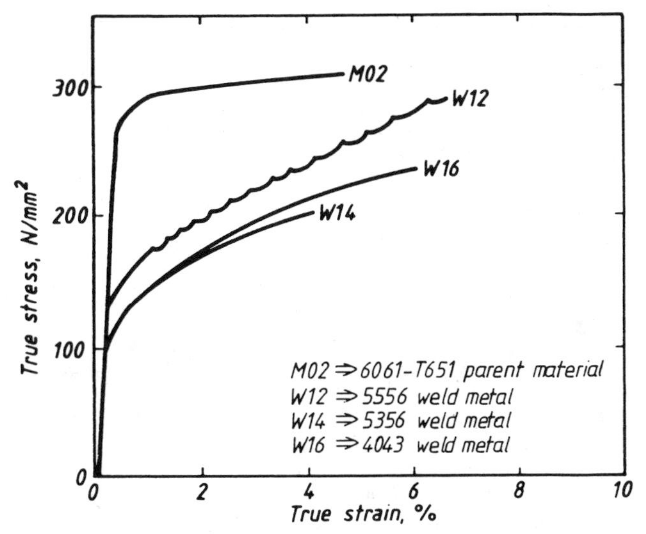

FIG. 2—*Comparison of true stress-true strain curves.*

Fracture Toughness Tests

CTOD R-curves were obtained for the three aluminum weld metals by performing small-scale fracture toughness tests on 10- by 20-mm (B by $2B$) single-edge notch bend (SENB) and $B = 10$ mm, $W = 120$ mm compact specimens. These specimens were notched and fatigue cracked in the weld metal, the through-thickness notches being located on the weld centerline. The specimens, which had initial fatigue crack length to specimen width (a/W) ratios of approximately 0.5, were tested at room temperature using the unloading compliance method [7].

Since the small-scale fracture toughness specimens were the same thickness as the wide-plate panels, it was decided to test the small-scale specimens in the plane-sided configuration. Testing plane-sided specimens generally results in crack tunnelling, which can cause the crack growth predicted by the unloading compliance method to underestimate the physical crack extension. However, all the single-specimen unloading compliance R-curves obtained in this investigation from both the SENB and compact specimens satisfied the requirement that the difference between the estimated and measured crack growths must not exceed

TABLE 3—*Tensile properties of aluminum alloy parent materials and weld metals.*

			Parent Plate, N/mm²		Weld Metal, N/mm²	
Test Panels	Parent Material	Welding Consumable	0.2% Proof Stress	Tensile Strength	0.2% Proof Stress	Tensile Strength
W11, W12	5456-H116	5556	228[a]	317[a]	144	308[b]
W13, W14	6061-T651	5356	292	316	121	212
W15, W16	6061-T651	4043	292	316	123	216
MO2	6061-T651	...	292	316	...	...

[a] Estimated from material specification.
[b] Failed under rising load deflection curve.

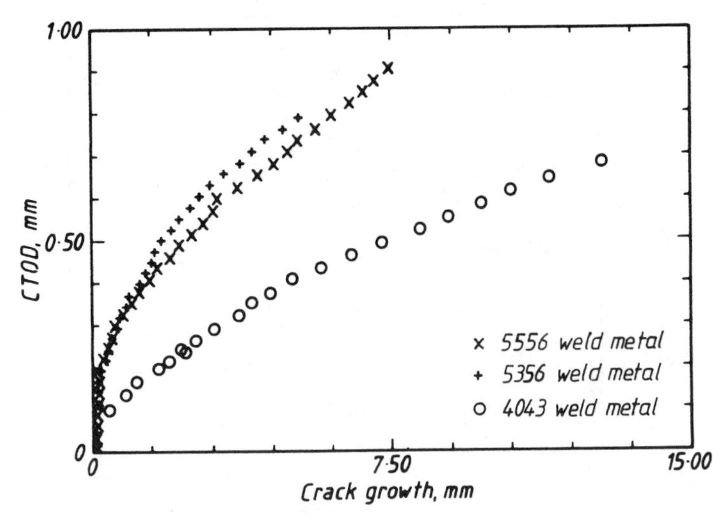

FIG. 3—*Typical compact single specimen* R-*curves.*

10% [7]. The CTOD R-curves obtained from the compact specimens are compared in Fig. 3.

Wide-Plate Tests

Details of the center-cracked panel (CCP) wide-plate specimens are summarized in Fig. 4. The welded specimens were notched in the weld metal with the weld axis at right angles to the loading direction of the plate. All the wide-plate specimens were fatigue precracked in cyclic tension prior to testing.

The wide-plate tests were conducted in displacement control. During the tests, the applied load and three clip gages were continually monitored by a computerized data acquisition system. The three clip gages were located at the center of the notch and at both ends of the original fatigue crack. A number of partial unloadings were performed during each test,

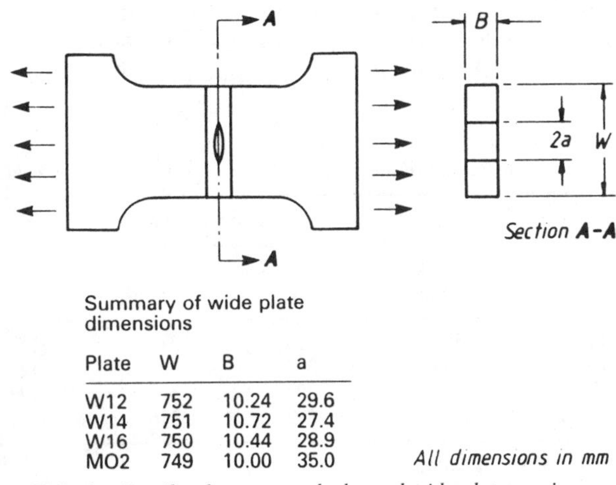

Summary of wide plate dimensions

Plate	W	B	a
W12	752	10.24	29.6
W14	751	10.72	27.4
W16	750	10.44	28.9
MO2	749	10.00	35.0

FIG. 4—*Details of center-cracked panel wide-plate specimens.*

and the crack length at each unloading (measured at the plate surface using a closed-circuit television system) was recorded. The CTOD's at each end of the crack immediately prior to each unloading were calculated using the formula

$$\delta = \frac{K^2(1 - \nu^2)}{\sigma_{YS} E} + V_p \tag{9}$$

where

$K = \sigma(\pi a)^{1/2}$,
σ = far-field stress,
a = half-length of current crack,
ν = Poisson's ratio, and
V_p = plastic component of clip gage displacement corresponding to the appropriate end of the crack.

The plastic component of clip gage displacement at each unloading was estimated by extrapolating the partial unloading line corresponding to the clip gage of interest (that is, either Clip Gage 1 or 3 depending on which end of the crack was being considered) back down to the clip gage displacement axis. The plastic component of clip gage displacement was assumed to equal the value of clip gage displacement at the intersection.

The CTOD calculation presented above is synonymous to the procedure presented in the British Standards Institution Methods for Crack Opening Displacement (COD) Testing (BS 5762-1979) for calculating CTOD for SENB specimens.

CTOD R-curves were obtained by plotting the CTODs for each end of the crack against the corresponding crack growth and drawing a best-fit curve through the data. The CTOD R-curves determined from the four wide-plate tests are compared in Fig. 5.

Comparison of Fracture Assessment Models

General

Driving force/R-curve assessments were conducted on both the welded and parent material wide-plate tests. Since the nominal thickness of the wide-plate specimens was only 10 mm,

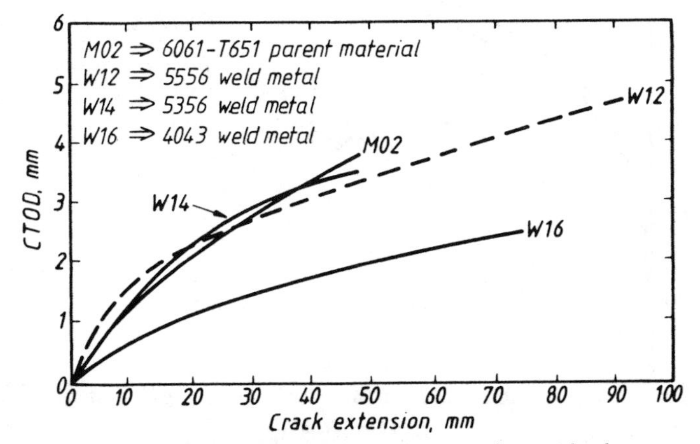

FIG. 5—*Crack tip opening displacement* R-*curves from wide-plate tests.*

TABLE 4—*Ramberg-Osgood true stress-true strain constants.*

Test Panels	Material/ Weld Metal	Young's Modulus, E N/mm²	σ_0, N/mm²	ϵ_0	α	n
W11, W12	5556 weld metal	70 000	152.2	2.17×10^{-3}	1.333	5.338
W13, W14	5356 weld metal	70 000	120.4	1.72×10^{-3}	1.658	4.983
W15, W16	4043 weld metal	70 000	129.6	1.85×10^{-3}	2.054	4.752
MO2	6061-T651 parent plate	70 000	294.0	4.20×10^{-3}	1.936	33.59

the analyses were performed assuming plane stress. The analyses performed on the welded wide-plate specimens were based on the tensile properties of the weld metals. Furthermore, the Young's modulus of both the parent material and weld metals was taken as 70×10^3 N/mm².

The reference stress model and the Electric Power Research Institute (EPRI) CTOD estimation scheme both require stress-strain information of the parent material/weld metal under consideration. In this investigation, the assessments based on the reference stress model were conducted using the actual true stress-true strain behavior of the appropriate parent plate/weld metal. In comparison, the assessments based on the EPRI estimation scheme were performed using the best-fit Ramberg-Osgood relationship to the actual true stress-true strain data determined, using the method proposed by Dorn and McCracken [8]. The Ramberg-Osgood constants for the three weld metals and the 6061-T651 parent plate are summarized in Table 4. Although the true stress-true strain behavior of the 6061-T651 plate material (that is, wide plate MO2) and the 5356 and 4043 weld metals in Panels W13, 14, 15, and 16 were satisfactorily represented by Ramberg-Osgood fits, this was not the case for the 5556 weld metal in Panels W11 and W12. This problem is demonstrated in Fig. 6, which shows the actual true stress-true strain behavior of 5556 weld metal and the corre-

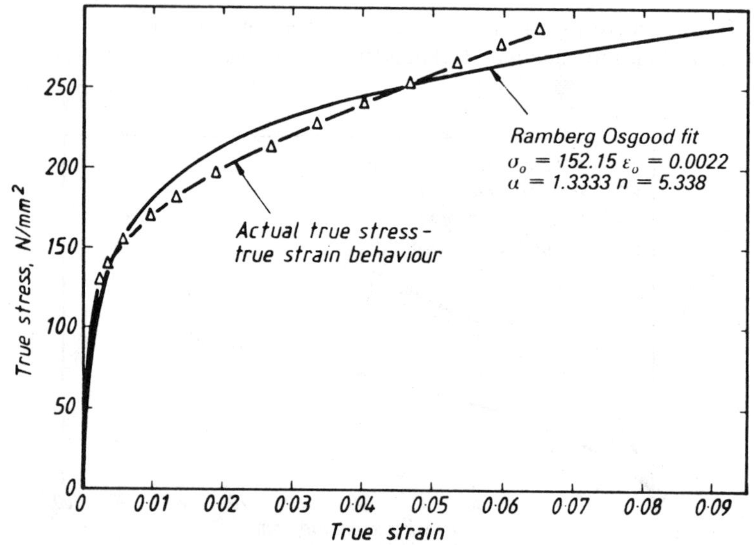

FIG. 6—*Comparison of true stress-true strain behavior of 5556 weld metal with Ramberg-Osgood fit.*

sponding Ramberg-Osgood relationship. This highlights a possible drawback of the EPRI estimation scheme when assessing high work-hardening materials.

Ductile Instability Assessments

Driving force/R-curve ductile instability analyses were conducted for each wide-plate specimen, by comparing the CTOD R-curve obtained from the wide-plate specimen against computer-generated driving force curves calculated using the collapse modified strip yield, reference stress, and EPRI models, that is, Eqs 4, 5, and 7. In the case of the welded wide-plate specimens, assessments were conducted both with and without residual stresses. The values of effective secondary stress employed in the assessments were 38, 20, and 25 N/mm^2 for wide plates W12, W14, and W16, respectively. These values correspond to the maximum transverse residual stresses measured in each of the welded panels. Note that since the wide-plate specimens were center-cracked panels, the residual stress values presented above are equivalent to effective secondary stresses. If the test specimens had been a different geometry it would have been necessary to calculate the appropriate value of effective secondary stress for each specimen using Eq 2.

It should be emphasized that, even if elastic-plastic plane-strain thickness requirements are satisfied, the slope of an R-curve is dependent on the geometry of the specimen, and in particular the mode of loading. The lowest R-curves are produced by specimens which experience a large bending component. Consequently, it is normal practice to employ lower bound R-curves obtained from laboratory bend type specimens (that is, SENB or compact) in ductile structural integrity assessments, to increase the degree of conservatism. In this investigation the R-curves employed in the assessments were those determined from the wide-plate specimens (that is, the structures being assessed). The reason for this deviation from normal practice was to produce a critical analysis. Consequently, the driving force/R-curve analyses performed on the wide-plate tests should have predicted the actual conditions at maximum load, thereby enabling the accuracy of the various fracture models to be assessed.

The results of the ductile instability predictions obtained ignoring residual stresses are compared with the recorded maximum stresses and crack extensions up to maximum load in Table 5, while Fig. 7 compares the agreement between the actual and predicted maximum stresses. It is evident that the maximum far-field stress predictions obtained from all three models are in reasonable agreement, with the reference stress and EPRI models producing virtually identical results (even for wide-plate W12 where the Ramberg-Osgood stress-strain fit of the high work-hardening 5556 weld metal was not totally representative). The three models are compared in Fig. 8, which shows plots of the predicted CTOD (estimated from the three models) against applied load, for wide-plate W12 assuming a constant crack size a of 30 mm.

In the case of the low work-hardening parent material wide-plate specimen, it is apparent from Table 4 that not only are all the predicted maximum far-field stresses slightly larger than the actual maximum stress, but that the predicted stresses are all similar in magnitude. This indicates that the three fracture models are in close agreement for the low work-hardening 6061 parent plate test panel. This is confirmed in Fig. 9, which shows plots of the predicted CTODs (estimated from the three models) against applied load assuming a constant crack size a of 30 mm. In this test, the net section stress at maximum load conditions was approximately 269 N/mm^2 compared to the material's yield strength of 292 N/mm^2. Consequently, the test was dominated by the R-curve characteristics of the material. In comparison, the net section stresses in the welded wide plates at maximum load conditions were larger than the flow stresses of the weld metals.

TABLE 5—*Ductile instability predictions (ignoring residual stresses).*

Wide Plate	Actual Maximum Stress, N/mm²	Actual Crack Extension at Maximum Stress, mm	Strip Yield Model			Reference Stress Model			EPRI Model		
			Predicted Stress, N/mm²	Predicted Crack Extension, mm	Actual Stress/Predicted Stress	Predicted Stress, N/mm²	Predicted Crack Extension, mm	Actual Stress/Predicted Stress	Predicted Stress, N/mm²	Predicted Crack Extension, mm	Actual Stress/Predicted Stress
W12	200	≃20	202	6	0.99	183	11	1.09	185	9	1.08
W14	181	≃30	150	8	1.21	148	19	1.22	152	19	1.19
W16	154	≃40	146	13	1.05	132	24	1.17	134	19	1.15
MO2	223	≃18	261	11	0.85	260	10	0.86	264	7	0.84

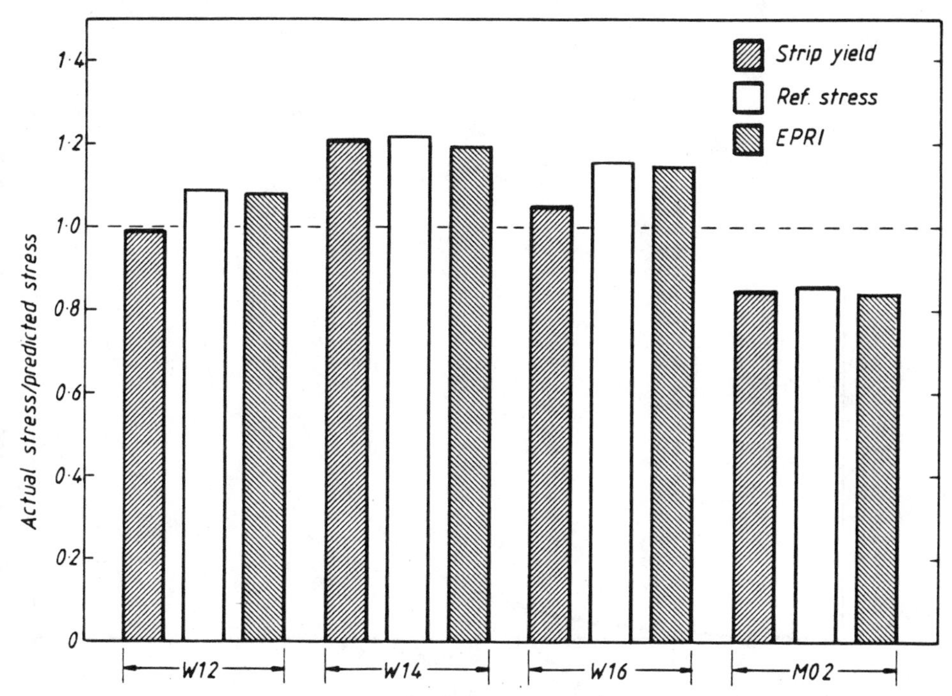

FIG. 7—*Comparison of predicted far-field stresses obtained from driving force*/R-*curve analyses.*

The general effect of including secondary stresses in the CTOD collapse-modified strip yield and reference stress fracture assessment models presented in this report is demonstrated in Fig. 10, which shows plots of CTOD against applied load for wide-plate W16 (4043 weld metal) both with and without residual stresses. These curves assume a constant crack size a of 30 mm and residual stresses equal to the yield strength of the weld metal. The most obvious effect is that the curve including residual stresses no longer starts from the origin; that is, the secondary stress produces an elastic CTOD which exists in the structure before any primary stress is applied. However, as the primary stress is increased it appears that the two curves gradually merge, resulting in almost exact agreement as collapse is approached. This however is not the case: in reality, the difference between the curves actually increases as the applied load gets larger. However, beyond net section yield, the rate of increase in the applied CTOD becomes so large that in relative terms the effect of residual stresses becomes less pronounced. It is therefore clear that including residual stresses in the models described in this report should not have a dramatic effect on instability load predictions that are above net section yield.

This is confirmed in Table 6, which compares the results of the ductile instability analyses performed on the welded wide-plate specimens, (including the influence of residual stresses) with the actual maximum far-field stresses and crack extensions up to maximum load. As expected, including residual stresses in the assessments has had little effect on either the maximum far-field stress or crack extension predictions.

It should be stressed that the significance of residual stresses with respect to crack growth in the transitional and upper shelf regimes of fracture behavior is not fully understood. Indeed, it is frequently assumed that residual stresses can be ignored in ductile fracture assessments, since the large applied strains will cause the residual stresses to relax. Whether

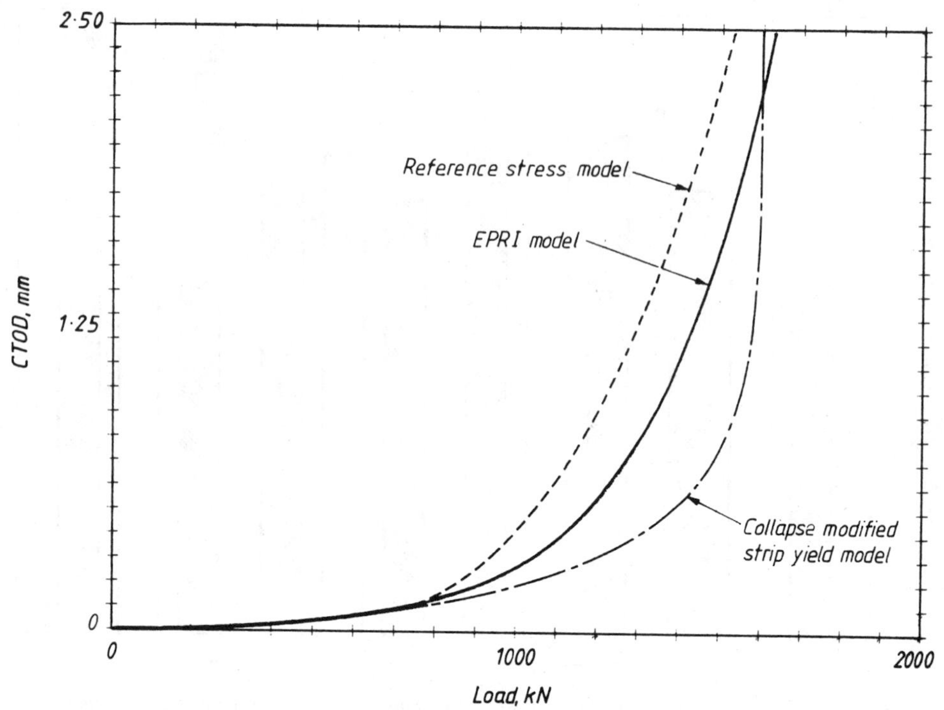

FIG. 8—*Comparison of fracture assessment models for wide-plate W12 (5556 weld metal) assuming a constant crack size* (a) *of 30 mm.*

or not this is the case when ductile instability occurs before net section yield is open to debate. For this reason it is recommended that residual stresses should be incorporated in ductile instability analyses to ensure conservative assessments. The advantage of the fracture models described in this paper is that the effect of residual stresses diminishes in a relative sense with increasing plastic strain.

In addition to performing driving force/R-curve assessments using the CTOD R-curves obtained from the wide-plate specimens, ductile instability analyses were also performed using the CTOD R-curves obtained from the small-scale fracture toughness tests. To enable driving force/R-curve diagrams to be constructed, the small-scale unloading compliance R-curve data were fitted with power law expressions which were then used to extrapolate the R-curves to larger amounts of crack extension. The results of the analyses are summarized in Tables 7 and 8. It can be seen that the predicted values of maximum far-field stress for wide plates W12 and W14 are significantly lower than those obtained using the appropriate wide-plate specimen R-curves, indicating that the extrapolated compact specimen R-curves are shallower than the corresponding wide-plate R-curves. However, in the case of wide-plate W16, the extrapolated compact specimen R-curve was virtually identical to the wide-plate specimen R-curve, resulting in similar estimates of maximum far-field stress.

Conclusions

The results of a series of welded and parent material 10-mm-thick aluminum wide-plate tests have been compared with predictions obtained using a collapse modified strip yield model, a reference stress model, and the CTOD estimation scheme published by EPRI.

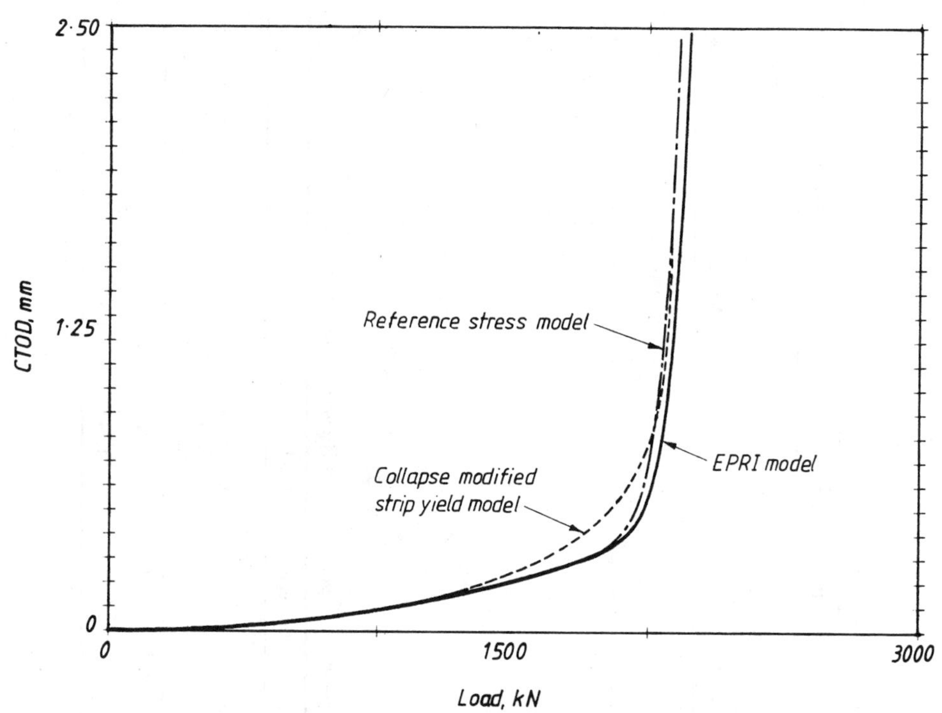

FIG. 9—*Comparison of fracture assessment models for low work-hardening 6061 parent material wide-plate test (MO2) assuming a constant crack size* (a) *of 30 mm.*

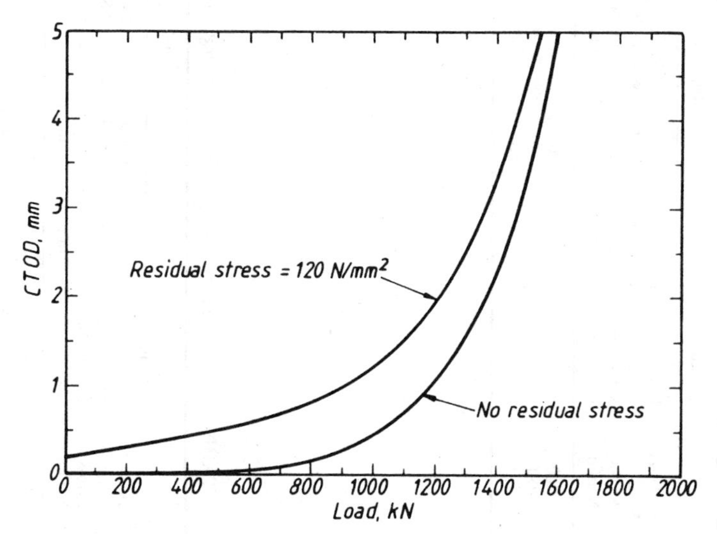

FIG. 10—*The influence of residual stresses on reference stress model prediction for wide plate W16* (a = *30 mm*).

TABLE 6—*Ductile instability predictions (including residual stresses).*

Wide Plate	Actual Maximum Stress, N/mm^2	Actual Crack Extension at Maximum Stress, mm	Assumed Residual Stress, N/mm^2	Strip Yield Model			Reference Stress Model			EPRI Model		
				Predicted Stress, N/mm^2	Predicted Crack Extension, mm	Actual Stress/ Predicted Stress	Predicted Stress, N/mm^2	Predicted Crack Extension, mm	Actual Stress/ Predicted Stress	Predicted Stress, N/mm^2	Predicted Crack Extension, mm	Actual Stress/ Predicted Stress
W12	200	≃20	38	200	6	1.0	177	10	1.13	180	11	1.11
W14	181	≃30	20	149	9	1.21	146	20	1.24	150	20	1.21
W16	154	≃40	25	143	13	1.08	128	23	1.20	129	23	1.19

TABLE 7—*Ductile instability predictions obtained using extrapolated compact specimen* R-*curves (ignoring residual stresses).*

Wide Plate	Actual Maximum Stress, N/mm^2	Actual Crack Extension at Maximum Stress, mm	Strip Yield Model			Reference Stress Model			EPRI Model		
			Predicted Stress, N/mm^2	Predicted Crack Extension, mm	Actual Stress/ Predicted Stress	Predicted Stress, N/mm^2	Predicted Crack Extension, mm	Actual Stress/ Predicted Stress	Predicted Stress, N/mm^2	Predicted Crack Extension, mm	Actual Stress/ Predicted Stress
W12	200	≃20	192	10	1.04	165	17	1.21	167	17	1.20
W14	181	≃30	146	9	1.24	134	16	1.38	140	17	1.29
W16	154	≃40	144	11	1.07	131	22	1.18	133	16	1.16

TABLE 8—*Ductile instability predictions obtained using extrapolated compact specimen* R-*curves (including residual stresses).*

				Strip Yield Model			Reference Stress Model			EPRI Model		
Wide Plate	Actual Maximum Stress, N/mm^2	Actual Crack Extension at Maximum Stress, mm	Assumed Residual Stress, N/mm^2	Predicted Stress, N/mm^2	Predicted Crack Extension, mm	Actual Stress/Predicted Stress	Predicted Stress, N/mm^2	Predicted Crack Extension, mm	Actual Stress/Predicted Stress	Predicted Stress, N/mm^2	Predicted Crack Extension, mm	Actual Stress/Predicted Stress
W12	200	≃20	38	187	16	1.07	157	20	1.27	159	21	1.26
W14	181	≃30	20	147	8	1.23	131	19	1.38	137	14	1.32
W16	154	≃40	25	141	11	1.09	126	21	1.22	128	21	1.20

The analyses on the welded wide-plate tests were performed both with and without residual stresses which were measured experimentally. It was found that:

1. All three fracture assessment models provided reasonably accurate predictions of maximum far-field stress (that is, within 20%) if the appropriate wide-plate CTOD R-curve was incorporated in the analysis.
2. The reference stress and EPRI models produced virtually identical predictions of maximum far-field stress and crack extension up to maximum load.
3. Conservative estimates of maximum far-field stress were obtained from all three models when extrapolated CTOD R-curves obtained from small-scale specimens were included in the analyses.

Finally, it is proposed that the influence of residual stresses in CTOD fracture assessment equations should be restricted to an elastic component of CTOD, including a first-order plastic zone correction factor. This method of including residual stresses ensures that their significance diminishes in a relative sense with increasing plastic strain.

Acknowledgments

This work was funded jointly by Research Members of The Welding Institute and the Minerals and Metals Division of the U.K. Department of Trade and Industry. The authors would also like to acknowledge Alcoa, who supplied the welded aluminum test panels.

APPENDIX

Fracture Assessment Models

Plastic Collapse Modified Strip Yield Model

Garwood [2] proposed a collapse modified strip yield model to assess ductile fracture

$$\delta = \frac{8\sigma_{pc}^{2}\bar{a}}{\pi E \sigma_{YS}} \ln \sec \left(\frac{\pi\sigma}{2\sigma_{pc}}\right) \tag{10}$$

where

$\bar{a}$ = equivalent crack size, equal to the half-length of a through-thickness crack in an infinite plate loaded in tension,
σ = applied far-field stress,
σ_{YS} = material's yield strength,
E = material's Young's modulus, and
σ_{pc} = far-field plastic collapse stress, calculated on the basis that collapse will occur when the stress on the remaining cross section reaches the material's flow strength σ_{flow}. It is generally assumed that σ_{flow} equals the average of the material's yield and tensile strengths.

Equation 10 can be reformulated in terms of effective primary and net section stress to give the following expression for the CTOD resulting from primary stresses (δ_{prim})

$$\delta_{prim} = \frac{\sigma_{YS}\pi a}{E}\left\{\left(\frac{\sigma_p}{\sigma_{YS}}\right)^2 + \left(\frac{\sigma_p}{\sigma_{YS}}\right)^2\left[\left(\frac{\sigma_{flow}}{\sigma_n}\right)^2 \frac{8}{\pi^2} \ln \sec\left(\frac{\pi}{2}\frac{\sigma_n}{\sigma_{flow}}\right) - 1\right]\right\} \tag{11}$$

where

$$\frac{\sigma_{YS}\pi a}{E}\left(\frac{\sigma_p}{\sigma_{YS}}\right)^2$$

represents the elastic component of CTOD due to primary stresses and

$$\frac{\sigma_{YS}\pi a}{E}\left(\frac{\sigma_p}{\sigma_{YS}}\right)^2\left[\left(\frac{\sigma_{flow}}{\sigma_n}\right)^2\frac{8}{\pi^2}\ell n \sec\left(\frac{\pi}{2}\frac{\sigma_n}{\sigma_{flow}}\right)-1\right]$$

represents the plastic component of CTOD due to primary stresses which implicitly includes a plastic zone correction for primary stresses.

If the influence of secondary stresses is restricted to an elastic component of CTOD including a plastic zone correction, the CTOD resulting from secondary stresses (δ_{sec}) is given by

$$\delta_{sec} = \frac{\sigma_{YS}\pi a}{E}\left[\left(\frac{\sigma_S}{\sigma_{YS}}\right)^2 + \frac{1}{2}\left(\frac{\sigma_S}{\sigma_{YS}}\right)^4\right] \tag{12}$$

where

$$\frac{\sigma_{YS}\pi a}{E}\left(\frac{\sigma_s}{\sigma_{YS}}\right)^2$$

represents the elastic component of CTOD produced by secondary stresses and

$$\frac{\sigma_{YS}\pi a}{E}\frac{1}{2}\left(\frac{\sigma_S}{\sigma_{YS}}\right)^4$$

represents the component of CTOD resulting from the plastic zone correction for secondary stresses.

Combining Eqs 11 and 12 results in the following expression for the total CTOD produced by primary and secondary stresses

$$\delta = \frac{\sigma_{YS}\pi a}{E}\left\{\left(\frac{\sigma_p + \sigma_s}{\sigma_{YS}}\right)^2 + \frac{1}{2}\left(\frac{\sigma_s}{\sigma_{YS}}\right)^4 + \left(\frac{\pi_p}{\sigma_{YS}}\right)^2\left[\left(\frac{\sigma_{flow}}{\sigma_n}\right)^2\frac{8}{\pi^2}\ell n \sec\left(\frac{\pi}{2}\frac{\sigma_n}{\sigma_{flow}}\right)-1\right]\right\} \tag{13}$$

Reference Stress Model

Anderson et al. [3] modified the Ainsworth [9] reference stress model to produce the CTOD reference stress equation

$$\delta = \frac{K_{I(a)}{}^2}{\sigma_{YS}E}\left[\frac{\bar{a}_e}{\bar{a}} + \frac{E\epsilon_{ref}}{\sigma_{ref}} - 1\right] \tag{14}$$

where $\bar{a}_e$ is the effective $\bar{a}$, including a plastic zone correction given by

$$\bar{a}_e = \bar{a} + \frac{1}{\beta\pi}\frac{n-1}{n+1}\frac{K_{I(a)}{}^2}{\sigma_{YS}{}^2}\frac{1}{1 + (\sigma_{ref}/\sigma_{YS})^2} \tag{15}$$

and $K_{I(a)}$ is the stress-intensity factor based on crack size $= a$. In Eq 15, β is 2 for plane stress and 6 for plane strain, and n is the Ramberg-Osgood strain hardening exponent.

The "reference stress" (σ_{ref}) in Eqs 14 and 15 is equal to the effective net section stress (σ_n). The reference strain is defined as the value of strain corresponding to σ_{ref} in a uniaxial tensile test. This method therefore requires the material's true stress-true strain curve. Since in ductile fracture assessments, the plastic component of CTOD is the dominant term, Eq 15 can be simplified if the Ramberg-Osgood strain hardening exponent is not known, by assuming that $(n - 1)/(n + 1)$ is equal to 1.0.

If Eqs 14 and 15 are rewritten in terms of effective primary and secondary stresses (restricting the influence of secondary stresses to an elastic component of CTOD including a plastic zone correction) and the μ term and the m factors which Anderson dropped to simplify his analysis are included, then the resulting expression is given by

$$\delta = \frac{\pi \sigma_{YS} a}{E} \left\{ \left[\frac{\sigma_p + \sigma_s}{\sigma_{YS}} \right]^2 \left(\frac{E}{E'} \frac{a_e}{m_{el} a} \right) + \left[\frac{\sigma_p}{\sigma_{YS}} \right]^2 \frac{\mu}{m_{FP}} \left(\frac{E \epsilon_{ref}}{\sigma_{ref}} - 1 \right) \right\} \quad (16)$$

where a_e is now given by

$$a_e = a + \frac{1}{\beta \pi} \frac{n - 1}{n + 1} \frac{(\sigma_p + \sigma_s)^2 \pi a}{\sigma_{YS}^2} \frac{1}{1 + (\sigma_{ref}/\sigma_{YS})^2} \quad (17)$$

and

m_{el} = 1 for plane stress and m_{el} = 2 for plane strain,
m_{FP} = 1.1 for tension and $m_{FP} \simeq$ 1.3 to 1.8 for bending,
μ = 0.75 for plane stress and μ = 1.0 for plane strain,
E' = E for plane stress and $E' = E/(1 - \nu^2)$ for plane strain, and
β = 2 for plane stress and β = 6 for plane strain.

EPRI CTOD Estimation Scheme

The J and CTOD estimation scheme published by EPRI [4] assumes that the true stress-true strain behavior of the material being studied obeys the Ramberg-Osgood power law hardening relationship given by

$$\frac{\epsilon}{\epsilon_0} = \frac{\sigma}{\sigma_0} + \alpha \left(\frac{\sigma}{\sigma_0} \right)^n \quad (18)$$

where

σ = applied stress,
ϵ = total strain produced by the applied stress,
σ_0 = reference value of stress,
ϵ_0 = given by σ_0/E, and
α and n = material constants.

In addition, it is assumed that the total applied J or CTOD is the sum of an elastic component (including a first-order plastic zone correction) and a fully plastic component.

Although not specified in the EPRI handbook, the elastic component of CTOD (δ_{el}) can be determined from

$$\delta_{el} = \frac{K_{I(ae)}^2}{m_{el} \sigma_{YS} E'} \quad (19)$$

where

$$a_e = a + \frac{1}{\beta\pi}\frac{n-1}{n+1}\left(\frac{K_{I(a)}}{\sigma_0}\right)^2 \frac{1}{1 + (\sigma_n/\sigma_0)^2} \tag{20}$$

and $K_{I(a_e)}$ is the stress-intensity factor based on a crack size ($= a_e$); σ_0 is the Ramberg-Osgood reference stress. Note that with the definitions given previously, the term (σ_n/σ_0) used in Eq 20 is equivalent to the term (P/P_0) used in the EPRI handbook.

In comparison, the fully plastic component of CTOD (δ_{FP}) is given by

$$\delta_{FP} = \alpha\epsilon_0 c d_n h_1 \left(\frac{\sigma_n}{\sigma_0}\right)^{n+1} \tag{21}$$

where

α and n = Ramberg-Osgood constants,
c = characteristic dimension (usually the remaining ligament),
d_n = material constant which relates the fully plastic J and CTOD's, and
h_1 = geometry-dependent function obtained using finite-element analysis.

Values of d_n and h_1 are presented in Ref *5* for a range of materials and geometries.

The total CTOD is obtained by combining Eqs 19 and 21 to give

$$\delta = \frac{K_{I(ae)}^{\ 2}}{m_{el}\sigma_{YS}E'} + \alpha\epsilon_0 c d_n h_1 \left(\frac{\sigma_n}{\sigma_0}\right)^{n+1} \tag{22}$$

If this expression is reformulated in terms of primary and secondary stresses assuming that the secondary stresses will only influence the elastic component of CTOD, the subsequent equation is given by

$$\delta = \frac{(\sigma_p + \sigma_s)^2\pi a_e}{m_{el}\sigma_{YS}E'} + \alpha\epsilon_0 c d_n h_1 \left(\frac{\sigma_n}{\sigma_0}\right)^{n+1} \tag{23}$$

where

$$a_e = a + \frac{1}{\beta\pi}\frac{n-1}{n+1}\frac{(\sigma_p + \sigma_s)^2\pi a}{\sigma_0}\frac{1}{1 + (\sigma_n/\sigma_0)^2} \tag{24}$$

It can be seen that the secondary stress CTOD component is limited to the elastic term, including a first-order plastic zone correction.

References

[*1*] Dawes, M. G., "The COD Design Curve," *Advances in Elasto-Plastic Fracture Mechanics,* L. H. Larsson, Ed., Applied Science Publishers, presented at the 2nd Advanced Seminar on Fracture Mechanics, Ispra, Italy, April 1979, pp. 279–300.

[*2*] Garwood, S. J., "A Crack Tip Opening Displacement (CTOD) Method for the Analysis of Ductile Materials," presented at the 18th National Symposium on Fracture Mechanics, Boulder, CO, June 1985.

[*3*] Anderson, T. L., Leggatt, R. H., and Garwood, S. J., "The Use of CTOD Methods in Fitness for Purpose Analyses," *The Crack Tip Opening Displacement in Elastic-Plastic Fracture Mechanics,* K. H. Schwalbe, Ed., Springer-Verlag, Berlin, Federal Republic of Germany, 1986, pp. 287–313.

[*4*] Kumar, V., German, M. D., and Shih, C. F., "An Engineering Approach for Elastic-Plastic Fracture Analysis," EPRI Report NP 1931, Electric Power Research Institute, July 1981.

[5] Beaney, E. M., "Accurate Measurement of Residual Stresses on Any Steel Using the Centre Hole Method," *Strain,* July 1976, pp. 99–106.
[6] Rosenthal, D. and Norton, J. T., "A Method of Measuring Triaxial Residual Stresses in Plates," *Welding Research Supplement,* May 1945, pp. 295–307.
[7] Gordon, J. R., "The Welding Institute Procedure for the Determination of the Fracture Resistance of Fully Ductile Metals," Welding Institute Research Report 275, June 1985.
[8] Dorn, W. S. and McCracken, D. D., *Numerical Methods with FORTRAN IV Case Studies,* John Wiley and Sons, New York, 1972.
[9] Ainsworth, R. A., "The Assessment of Defects in Structures of Strain Hardening Materials," *Engineering Fracture Mechanics,* Vol. 19, 1984, p. 633.

Microstructure and Micromechanical Modeling

Patrick T. Purtscher,[1] Richard P. Reed,[1] and David T. Read[1]

Effect of Void Nucleation on Fracture Toughness of High-Strength Austenitic Steels*

REFERENCE: Purtscher, P. T., Reed, R. P., and Read, D. T., **"Effect of Void Nucleation on Fracture Toughness of High-Strength Austenitic Steels,"** *Fracture Mechanics: Perspectives and Directions (Twentieth Symposium), ASTM STP 1020,* R. P. Wei and R. P. Gangloff, Eds., American Society for Testing and Materials, Philadelphia, 1989, pp. 433–446.

ABSTRACT: The fracture of seven austenitic stainless steels with varying nickel and nitrogen contents were studied at 4 K. Smooth, 6-mm-diameter tension specimens and 22-mm-thick compact specimens were used. Nitrogen content controlled the yield strength and influenced fracture toughness by its effect on yield strength. Increasing the nickel content increased the fracture toughness at a constant nitrogen content. Observations of the fracture surfaces and polished cross sections through the fracture surfaces of test specimens showed that nucleation controlled the dimpled rupture fracture process. A critical stress criterion for nucleation that depends on both the applied stress and strain was developed and applied to the fracture toughness test. This fracture criterion explained the increase in fracture toughness with increasing nickel and the decrease in fracture toughness with increasing yield strength for strengths over 600 MPa.

KEY WORDS: austenitic stainless steels, cryogenic temperatures, dimpled rupture, ductile fracture, fracture toughness, inclusions, void nucleation

The fracture toughness characterizes the initiation of the ductile fracture process ahead of a fatigue precrack. J_{Ic} and K_{Ic}, two measures of fracture toughness, have been used for applications where high strength and high reliability are required, such as aircraft parts and pressure vessels. However, the basic mechanisms that relate a material's microstructure to its fracture mechanics properties are not understood. It is extremely difficult to predict J_{Ic} from the microstructure and strength of an alloy [*1*].

Austenitic stainless steels are used in low-temperature structural applications where J_{Ic} is an important design consideration—liquefied gas storage tanks and supports for superconducting magnets [*2*]. At cryogenic temperatures, these alloys can have high yield strengths and retain a relatively high J_{Ic}.

The relationship between strength and fracture toughness of structural alloys has been studied extensively [*3–5*]. Wanhill [*6*] has summarized the results of numerous studies on the most common high-strength alloys. To Wanhill's data, we added the strength-versus-toughness data for a controlled series of austenitic stainless steels whose strength was raised systematically by increasing the concentration of interstitial alloying elements [*7*], see Fig.

* This work was partially supported by the U.S. Department of Energy, Office of Fusion Energy.

[1] Materials research engineering, metallurgist, and physicist, respectively, Fracture and Deformation Division, National Institute for Standards and Technology, MC 430, Boulder, CO 80303.

1. For these austenitic steels, the two properties are related empirically by

$$K_{Ic}(J) = 500 - 0.3\,\sigma_y \quad (1)$$

where $K_{Ic}(J)$ is in MPa$\sqrt{m}$, and σ_y, the yield stress, is in MPa. This equation predicts the fracture toughness to within ±20 MPa$\sqrt{m}$. All the curves in Fig. 1 have similar negative slopes, reflecting the significant influence of strength on the fracture toughness of high-strength, structural alloys.

Steels fracture in a ductile manner by a dimpled rupture process in which voids nucleate at discontinuities in the microstructure, typically nonmetallic inclusions, and grow under the influence of the local stress state. In the uniaxial tension test, the main direction of void growth should be along the tensile axis until a local neck is formed. The neck introduces

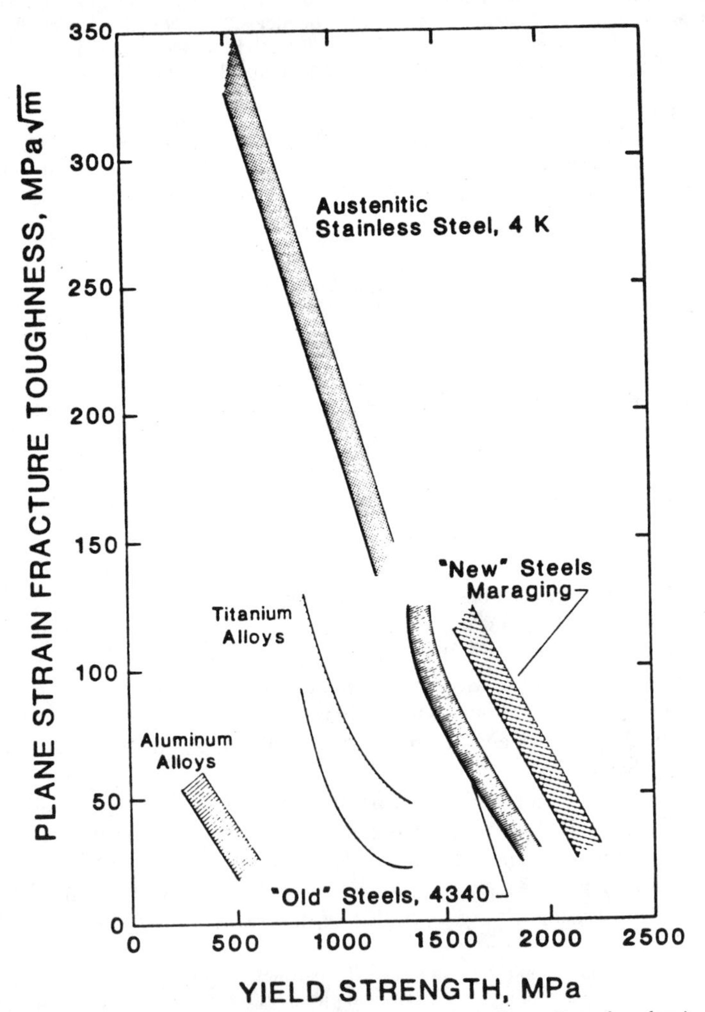

FIG. 1—*Fracture toughness versus yield strength for structural alloys. Data for aluminum, titanium, new steels, and old steels are from Ref 7. Austenitic stainless steel data are from Ref 8.*

triaxial stresses that cause any existing voids to expand uniformly in all directions. In the fracture toughness test, the region ahead of a sharp crack resembles the locally necked area of a tensile specimen.

Van Stone et al. [*1*] recently reviewed various models that have been proposed to describe fracture toughness in terms of the dimpled rupture fracture process. It is generally assumed that voids nucleate in the plastic zone ahead of the crack tip at relatively low plastic strains and that most of the energy measured in the J-integral test procedure is related to void growth [*8–10*]. Van Stone et al. argued that any successful modeling of ductile fracture ahead of a sharp crack must explicitly treat all three stages of the process: void initiation, growth, and coalescence.

In this paper, we examine the fracture process in a series of seven austenitic stainless steels broken at 4 K. A careful fractographic and metallographic study is conducted of the broken specimens to observe the dimpled rupture process. From these observations, a model for fracture toughness is developed that relates the void nucleation process to the mechanical properties which were previously presented in a separate report [*11*].

Materials

The seven laboratory heats of steel were supplied by the E. O. Paton Institute of Electrowelding, Kiev, U.S.S.R., in the form of 25-mm-thick plates in the hot-rolled condition. The plates were annealed at 1170°C for 2 h and then water quenched. The treatment produced coarse-grained, fully annealed austenitic structures. The composition, grain size, and inclusion spacing of the seven alloys are presented in Table 1. The base chemistry of these alloys is Fe-19Cr-4.5Mn, and the variables are the nickel and nitrogen contents.

The inclusion content of these alloys was nearly identical. The inclusion spacing was calculated by counting all of the inclusions >0.5 μm in diameter in a given area, normalizing that number to 1 mm^2, and taking the inverse square root to obtain an approximate value of spacing. Energy dispersive X-ray analysis determined that the inclusions found in the polished cross sections of these alloys were predominantly manganese-silicon and manganese-sulfur types.

Test Procedures

Smooth, 6-mm-diameter and 38-mm-gage tension specimens and 22-mm-thick compact specimens were machined from the annealed plates. The ASTM Methods of Tension Testing of Metallic Materials (E 8) and Test Method for J_{Ic}, a Measure of Fracture Toughness

TABLE 1—*Alloy compositions* (*in weight percent*).

Alloy No.	Cr	Ni	Mn	C	N	S	P	Si	Grain Size, μm	λ, μm
1	18.8	5.6	4.1	0.020	0.256	0.005	0.014	0.25	122	47
2	18.4	8.9	4.2	0.018	0.281	0.005	0.014	0.28	165	63
3	20.5	12.8	5.5	0.024	0.265	0.004	0.015	0.34	200	66
4	20.8	14.9	5.2	0.028	0.277	0.004	0.015	0.41	160	63
5	18.7	8.7	3.7	0.014	0.093	0.005	0.013	0.23	220	57
6	19.7	11.5	3.9	0.012	0.141	0.005	0.013	0.27	250	57
7	20.8	14.7	4.4	0.016	0.197	0.004	0.014	0.33	207	70

TABLE 2—*Summary of mechanical properties at 4 K.*

Alloy No.	Nickel Content, weight %	Nitrogen Content, weight %	σ_y,[a] MPa	σ_u,[a] MPa	σ_f,[a] MPa	ϵ_f,[a] $\ell n \cdot (A_0/A_f)$	J_{Ic},[b] kJ/m^2	$K_{Ic}(J)$, MPa$\sqrt{m}$
1	5.6	0.26	1050	1418	1710	0.21	27	75
2	8.9	0.28	950	1650	3160	0.55	99	143
3	12.8	0.27	1000	1580	3130	0.78	201	204
4	14.9	0.28	1165	1630	2990	0.66	276	239
5	8.7	0.09	460	1417	2550	0.55	360	273
6	11.5	0.14	720	1370	2770	0.70	288	244
7	14.7	0.20	870	1425	2790	0.67	312	254

[a] Average value from two or three tests; σ_y + σ_u are engineering stress values, σ_f is the true stress at fracture, and ϵ_f is the true strain at fracture.
[b] Measured value from one test.

(E 813) were followed for obtaining the tensile and fracture toughness properties of these alloys. Specific descriptions of the techniques are available elsewhere [7].

Results

The results of tension and fracture toughness tests at 4 K are listed in Table 2. The variation of the nickel and nitrogen contents produced a wide range of mechanical properties. Nitrogen, dissolved into the face-centered cubic (fcc) lattice at the interstitial sites, controls the yield strength of these austenitic alloys at cryogenic temperatures. Yield strength affects the fracture toughness, similar to the trend shown in Fig. 1, but at a constant yield strength, nickel content determines the fracture toughness of these alloys (Fig. 2). Increasing nickel and nitrogen concentrations suppress the strain-induced martensitic transformation, but the transformation does not directly affect the σ_y and J_{Ic} mechanical properties [11].

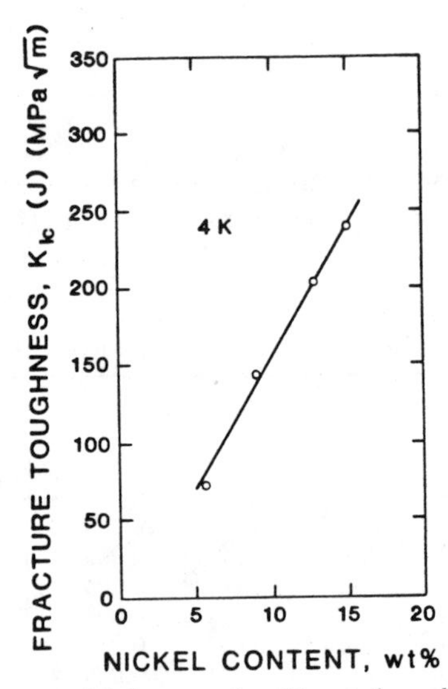

FIG. 2—*Fracture toughness versus nickel content for Alloys 1 through 4 with constant nitrogen content and similar yield strengths.*

Scanning electron microscopy (SEM) was performed on all the test specimens to determine the fracture morphology and the microstructural features that participate in the fracture process. The observations, summarized in Table 3 and illustrated in the next three figures, were always made in the region of the fracture surface with the highest triaxial stress. Most specimens fractured in a ductile manner, characterized on a macroscopic scale by large shear lips on the tension specimen and elastic-plastic behavior in the fracture toughness test. On the microscopic scale, the predominant features were large dimples that originated at the manganese-silicon and manganese-sulfur inclusions (Fig. 3). These inclusions were spherical in shape, between 1 and 2 μm in diameter, and identical to those found on the polished cross section (see Table 1).

Specimens from Alloys 1 and 5 exhibited a much smaller dimple size (1 to 2 μm in diameter) on the fracture surface (Fig. 4). No inclusions are visible in the smaller dimples; however, the existence of inclusions smaller than 0.5 μm in diameter, the smallest counted on the polished cross sections, cannot be discounted [*12*].

Fracture toughness specimens of Alloys 1 and 2 had a distinctly different fracture morphology than the other specimens in this test series. Their fracture surfaces were faceted; the facet size varied from 2 μm steps on the surface to a relatively smooth surface extending across a complete austenite grain. This type of faceted fracture, shown in Fig. 5, has been described by Tobler and Meyn [*13*] as a process of slipband cracking along {111} planes.

Selected tension specimens were sectioned along the tensile axis after testing to observe the microstructural processes that led to the formation of the large dimples (Fig. 3) on the fracture surface. Figure 6 shows a typical example. The rounded features of the dimples on the fracture surface are visible at the edge of the cross section. The area directly below the fracture surface appears undamaged despite the high local stresses and strains present in the area. There is no gradient of voids below the fracture surface. Void area fraction versus true strain for one representative tension specimen is shown in Fig. 7. An arbitrary function has been drawn through the data points to emphasize the limited amount of void growth present below the tensile fracture surface. A value of nucleation strain can be defined as the strain at zero area fraction of voids. In this example, the nucleation strain is about 0.75 of the true strain to fracture of the specimen, but approximately equal to the average true strain to fracture reported in Table 2 for the alloy.

TABLE 3—*Summary of fracture behavior.*

Alloy	Tensile	Fracture Toughness
No. 1, 5.6Ni-0.26N	2-μm dimples; small shear lips	facets; linear elastic
No. 2, 8.9Ni-0.28N	20- to 30-μm dimples; large shear lips	facets; elastic-plastic
No. 3, 12.8Ni-0.27N	20- to 30-μm dimples; large shear lips	30-μm dimples; elastic-plastic
No. 4, 14.9Ni-0.28N	20- to 30-μm dimples; large shear lips	20- to 40-μm dimples; elastic-plastic
No. 5, 8.7Ni-0.09N	duplex dimples: 2- and 10-μm size; large shear lips	duplex dimples: 2- and 10-μm size; elastic-plastic
No. 6, 11.5Ni-0.14N	20-μm dimples; large shear lips	20- to 30-μm dimples; elastic-plastic
No. 7, 14.7Ni-0.20N	20-μm dimples; large shear lips	20- to 30-μm dimples; elastic-plastic

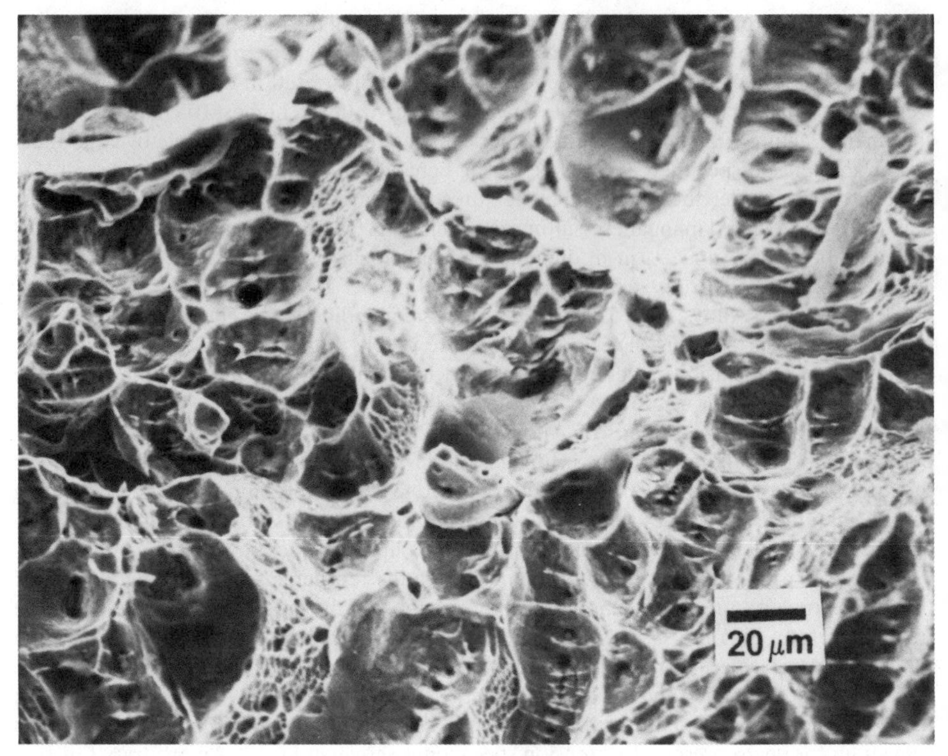

FIG. 3—*Fractograph (SEM) of tension specimen for Alloy 2 showing 20- to 30-μm dimples in center region of fracture surface.*

A cross section through the crack tip of a tested but unbroken compact tension specimen from the highest yield strength alloy is shown in Fig. 8. The applied J at the crack tip was approximately equal to J_{Ic} (zero physical crack growth). The crack tip was not blunted significantly, and damage was accumulating in the plastic zone before actual crack growth was detected. The damage was focused straight ahead, along the path of subsequent macroscopic crack growth. This suggests that a stress-controlled process is important [*14*].

Discussion

Development of Model

The fracture process in these steels can be interpreted in terms of two separate stages: a nucleation stage and a growth and coalescence stage. In the uniaxial tension test, the void nucleation stage occurs at strains very close to the fracture strain (Figs. 6 and 7). Growth and coalescence occur very rapidly and require nearly zero additional strain. The micrograph of the crack tip (Fig. 8) shows a similar trend for limited void initiation at applied energy levels less than J_{Ic}.

These observations indicate that the conditions for nucleation are difficult to reach and require a large plastic strain, but once nucleation occurs, the conditions for growth and coalescence have already been met so fracture can occur with little or no additional applied

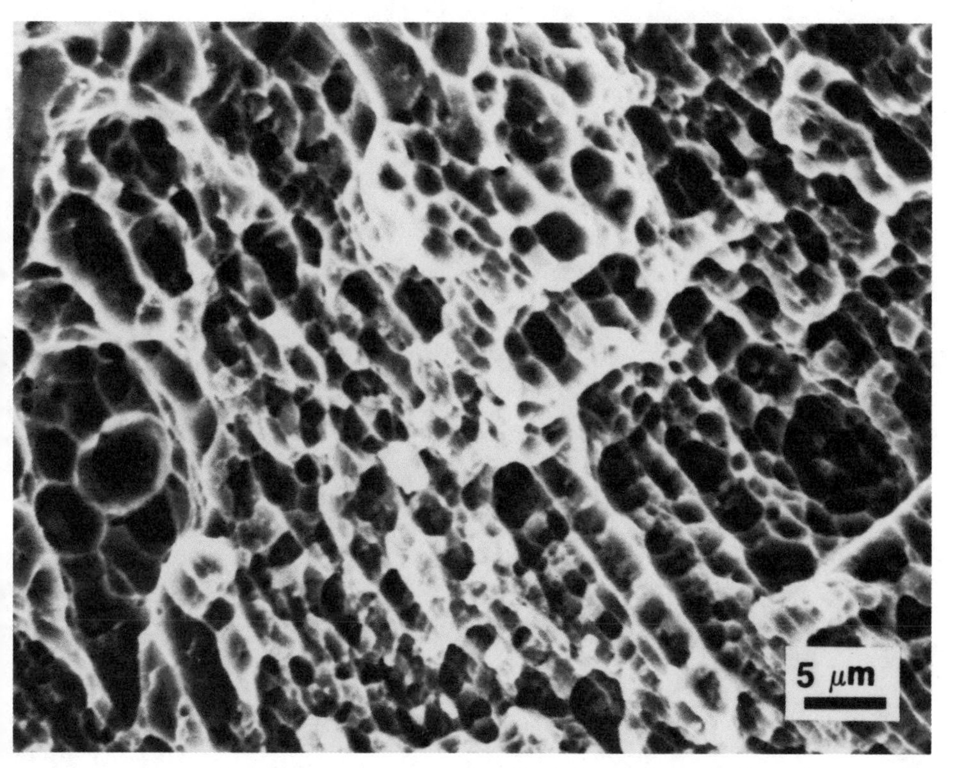

FIG. 4—*Fractograph* (*SEM*) *of tension specimen for Alloy 1 showing 2-μm dimples on fracture surface.*

strain. Broek [*15*] describes a very similar type of behavior in high-strength aluminum alloys where the nucleation stage dominates the fracture process.

From the observations, we assume that the nucleation of microvoids at the nonmetallic inclusions is the critical event that controls the fracture of austenitic stainless steels at 4 K. LeRoy et al. [*16*] modeled the nucleation condition in a tension test specimen by assuming that nucleation of a void at an inclusion occurs when the stress across the interface exceeds a critical value (σ_c) that is related to both the local stress (σ_L, a function of the strain), and the macroscopic stress (σ_m)

$$\sigma_c = \sigma_L + \sigma_m \tag{2}$$

and

$$\sigma_L = \sigma_H \sqrt{\epsilon_N} \tag{3}$$

where ϵ_N is the nucleation strain and σ_H is a parameter that relates the stress produced by dislocations piled up at an inclusion to the average strain. We modified the equation of LeRoy et al. to use the full value of the applied stress rather than the hydrostatic component of the applied stress for the macroscopic stress. Work of Cox and Low [*17*] on high-strength steels showed that void nucleation was dependent on the applied stress rather than the triaxial stress.

In the fracture toughness test, the stresses and strains ahead of a stationary crack sub-

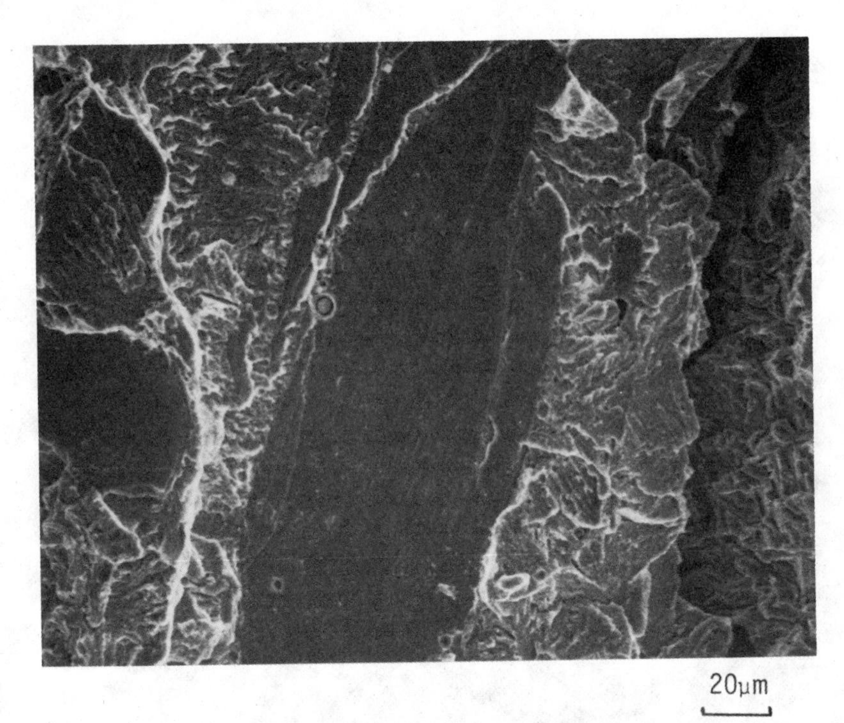

FIG. 5—*Fractograph (SEM) of fracture toughness specimen for Alloy 1 showing facets on fracture surface.*

jected to tensile (Mode I) opening can be predicted by continuum mechanics. For elastic-plastic conditions, a solution for the local strain (ϵ) at a distance ahead of a blunt crack has been calculated by Hutchinson [*18*] and Rice and Rosengren [*19*] (HRR) in terms of the J-integral (J)

$$\epsilon = \alpha\epsilon_0 \left(\frac{J}{\alpha\epsilon_0\sigma_0 I_n r}\right) n/n + 1\tilde{\epsilon} \tag{4}$$

where

α and n = constants of the Ramberg-Osgood fit to the material's true-stress versus true-strain curve,
ϵ_0 and σ_0 = yield strain and stress in the Ramberg-Osgood fit, and
r = distance from the crack tip.

The term I_n, a function of n, is about 4.5, and $\tilde{\epsilon}$ contains the angular dependence of the strain.

The constant n is of the order of 10, so for simplicity we assume $n/(n + 1) = 1$. For crack advance all along the crack front, we assume that the relevant distance, r, is the spacing of large inclusions, L. With $J = K^2/E$, then the HRR solution for strain ahead of a crack at the important inclusions is

$$\epsilon = \frac{K^2}{4.5\ \sigma_0\ EL} \tag{5}$$

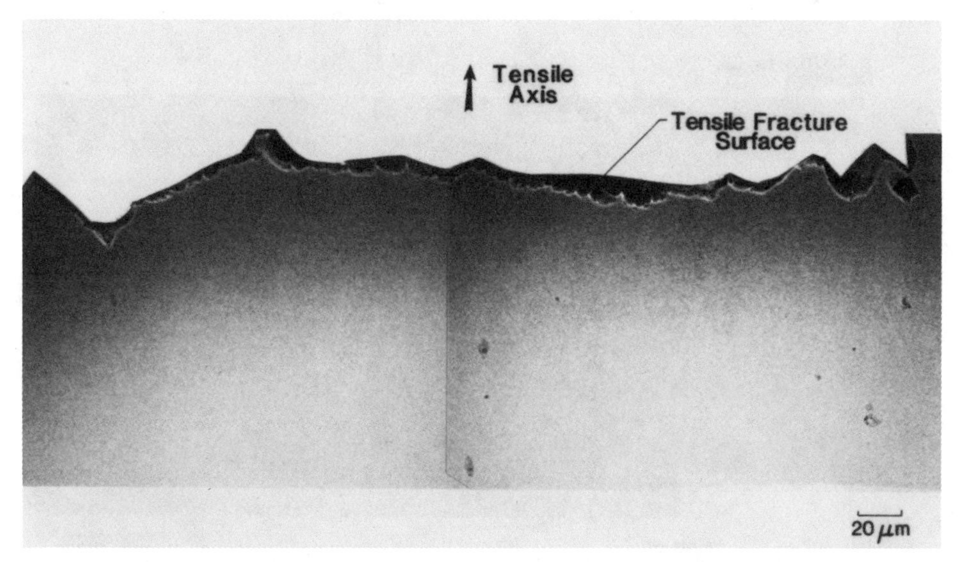

FIG. 6—*Micrograph (SEM) of polished cross section of tension specimen of Alloy 3 showing void initiation at inclusions with limited void growth.*

Nucleation was observed only very near the crack tip of the fracture toughness test specimens. Thus, nucleation strain is a major part of the plastic strain need to initiate tearing. Therefore, we equate the nucleation strain to the strain that is one inclusion spacing ahead of a blunt notch in the fracture toughness test and obtain an equation for the stress intensity factor associated with the nucleation process (K_{nuc})

$$K_{nuc} = \frac{(\sigma_c - 3.5\sigma_y)}{\sigma_H}(4.5\ \sigma_y\ EL)^{1/2} \tag{6}$$

We have also assumed that the $\sigma_m = \sigma_y$, increased by a factor of 3.5 by the crack-tip constraint.

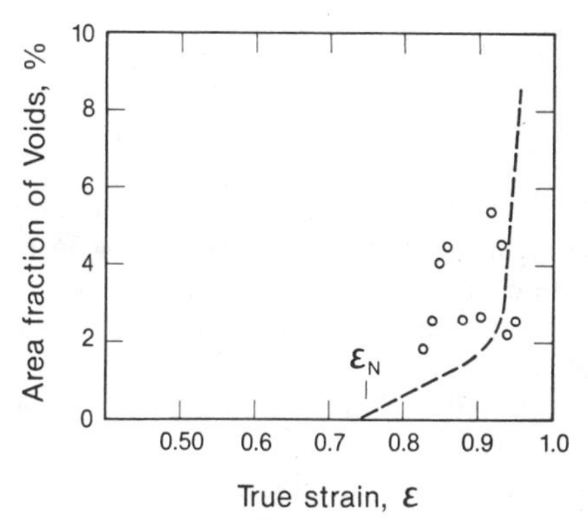

FIG. 7—*Area fraction of voids in polished cross section of tension specimen from Alloy 3 versus applied true strain.*

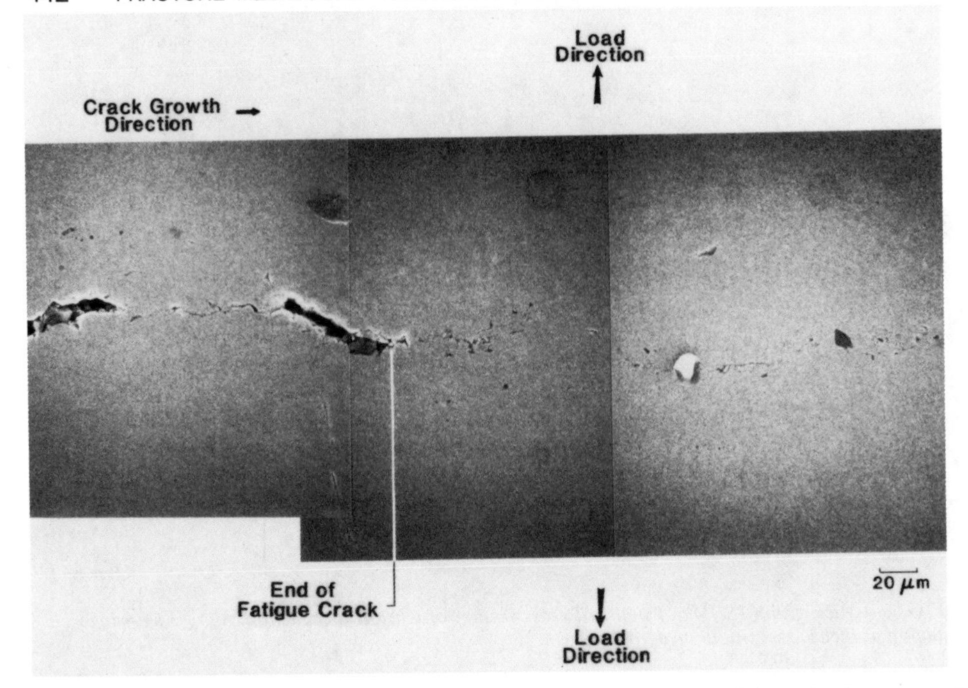

FIG. 8—*Micrograph (SEM) shows a cross section through crack tip of tested, but unbroken, compact tension specimen of Alloy 4, Applied* J *is approximately equal to* J_{Ic}.

The value of σ_c in this model is the cohesive energy of the interface between the defect and the matrix. We expect the value of σ_c to be relatively independent of the chemical composition of the matrix, but it can depend on the size, shape, and type of inclusions. Exact values of σ_c for these alloys are not known. LeRoy et al. [*16*] and Argon and Im [*20*] obtained σ_c values from 1200 to 1800 MPa for cementite particles in a ferritic steel matrix. In general, it appears that the σ_c value for a tightly bond particle in a ductile matrix is about $E/120$ (E = Young's modulus) [*20*]. In the terms of our model for fracture toughness, this value must be multiplied by 3 because we used the applied stress rather than the hydrostatic component of the stress in Eq 2. The elastic properties of austenitic stainless steels at cryogenic temperatures vary slightly, depending on the exact chemical composition, but are essentially the same as ferritic steels at room temperatures [*21*]. Therefore, we will assume that σ_c is $E/40$ or 5000 MPa for all the alloys.

The parameter σ_H in Eq 6 is also an unknown parameter that is a measure of the local stress exerted by the dislocation structures. An idealized model would be of a slip band in the austenitic matrix where dislocations have piled up against a barrier (a nonmetallic inclusion). In this case, the σ_H constant represents the stress concentration ahead of the pileup and is a function of the stacking-fault energy. The dislocation structures in the vicinity of the inclusions in these steels are difficult to characterize completely, so σ_H cannot be calculated from first principles. We assume that σ_H is a relatively strong function of the nickel content and not the nitrogen content. We do expect dislocation mobility and the ability to cross-slip to vary with stress state [*22*], so the value of σ_H determined from the tension test results cannot be used to predict the fracture toughness.

Evaluation of Model

To test this model for nucleation-controlled fracture toughness, we need to determine how σ_H varies with the nickel content. Therefore, considering only the highest strength alloys, Nos. 1 through 4 (nickel content varies and nitrogen content is constant), we set $K_{nuc} = K_{Ic}(J)$ and calculate the change in σ_H from Eq 6 for the range of nickel values at a constant yield strength using L (average value of 0.060 mm), σ_y (average value of 1040 MPa), σ_c (estimated value of 5000 MPa), and E (estimated value of 205 GPa). The calculated values are plotted versus nickel content in Fig. 9.

Now we can evaluate the model and our assumptions. First, we consider Alloys 5, 6, and 7 that have a range of both nitrogen and nickel contents. We can estimate σ_H from Fig. 9 from the nickel content and calculate a K_{nuc} using the measured yield strengths and the constant values of L, σ_c, and E. We can take the same approach to examine the prediction of K_{nuc} for Alloys 1 through 4, using the measured yield strengths rather than the average. The variables used in the calculations and the results for all seven alloys are found in Table 4.

The ratio of $K_{nuc}/K_{Ic}(J)$ for Alloys 1, 3, and 6 is very close to 1, as expected, if nucleation does control the measured fracture toughness. For Alloy 5 with the lowest yield strength, the model underpredicts the measured value by 30%. We interpret this to be a yield-strength effect. For lower-strength alloys, we expect nucleation strains to be less than the fracture strains so that additional plastic strain is required to grow the voids to the point of coalescence. For Alloys 2, 4, and 7, the ratio is not very close to 1, indicating that the model is sensitive to the values of the unknown parameters, σ_H and σ_c. If there is a variation in the σ_c value with nickel or nitrogen content or both, or if the true variation of σ_H is different from that shown in Fig. 9, then the model cannot accurately predict the fracture toughness.

A second, independent test of the model is found in Eq 1 for the Fe-18Cr-10Ni alloys with various nitrogen contents. From Fig. 9, we estimate σ_H to be 2400 MPa and assume the same values for σ_c and E. The average inclusion spacing for this alloy series was 0.055 mm. The predicted values of K_{nuc} and the calculated $K_{Ic}(J)$ from Eq 1 are plotted as a function of the yield strength in Fig. 10.

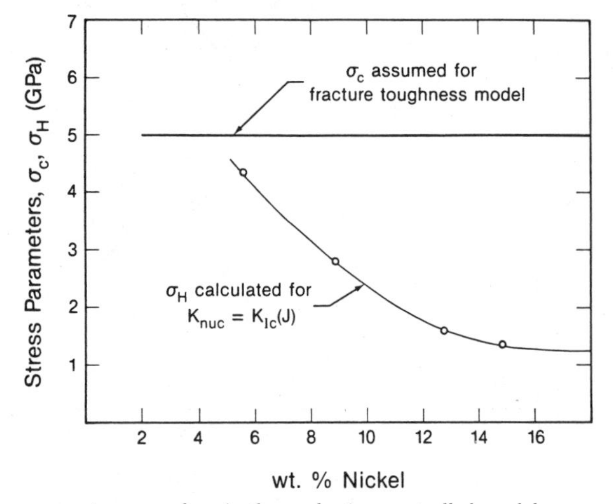

FIG. 9—*Stress parameters σ_c and σ_H in the nucleation-controlled model versus nickel content.*

TABLE 4—*Summary of data used in nucleation model for fracture toughness.*

Alloy No.	σ_H, MPa	σ_c, MPa	σ_y, MPa	K_{nuc}, MPa · $m^{1/2}$	$K_{Ic}(J)$, MPa · $m^{1/2}$	K_{nuc}, $K_{Ic}(J)$
1	4350	5000	1050	73	75	0.98
2	2280	5000	950	169	143	1.18
3	1600	5000	1000	221	204	1.08
4	1365	5000	1165	173	239	0.72
5	2900	5000	460	186	273	0.68
6	1900	5000	722	260	244	1.07
7	1375	5000	865	318	254	1.25

The model predicts a maximum toughness at a yield strength of 400 MPa and a decrease in toughness for higher yield strengths. The slope of the predicted curve for yield strengths greater than 600 MPa matches that for the experimental data. The different intercept for the predicted values from the model and measured values from Eq 1 could be due to a systematic error. In general, the model works reasonably well for the higher-strength alloys in which we expected nucleation to dominate the fracture process. For yield strengths below 600 MPa, the model underpredicts the fracture toughness, reflecting the lesser role of nucleation and a significant contribution of void growth and coalescence to the measured fracture toughness.

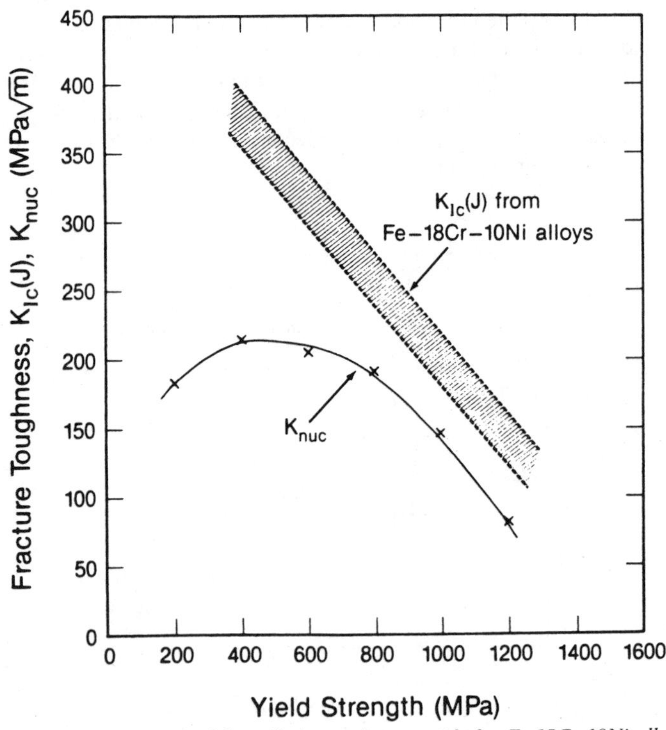

FIG. 10—*Fracture toughness, model prediction and measured, for Fe-18Cr-10Ni alloys versus yield strength* [8].

Summary

In high-strength austenitic stainless steels fractured at 4 K, large plastic strains nearly equal to the true fracture strain are needed to nucleate voids in the tension test. In the fracture toughness test, the void nucleation is limited to the region directly ahead of the advancing crack tip. On the basis on these observations, we assumed that tearing in the fracture toughness test is a stress-controlled process and begins when the stress at an inclusion-matrix interface ahead of the crack reaches the critical value required for nucleation. From this criterion for ductile fracture, we have developed a model that relates the stress intensity factor for nucleation to the inclusion spacing, the critical stress for void nucleation, the material's strength, and the local stress at inclusions produced by dislocation pileups. The model can predict reasonably well the relative effect of changes in the dislocation structure and yield strength on the fracture toughness.

The model represents an attempt to understand fracture toughness on the basis of the micromechanisms that control void nucleation. We have assumed that the fracture toughness represents an average nucleation strain. The model can be used to explain the results for high-strength austenitic steels at low temperatures where the inclusions are widely spaced in, and tightly bound to, the matrix. The model has shortcomings: It considers only the nucleation contribution to the fracture process; it is sensitive to the stress terms, so they must be known; and it ignores the effect of inclusion size, shape, and local spacing. However, the model does have two important advantages: It relates the effect of dislocations and interfacial stresses directly to the fracture toughness, and it does not require a detailed analysis of the fracture surface features, which we think are more representative of the tearing modulus than the fracture toughness of high-strength materials. The development of a more comprehensive model is certainly possible through more extensive testing and thorough material evaluation.

References

[1] Van Stone, R. H., Cox, T. B., Low, J. R., Jr., and Psioda, J. A., *International Metals Review*, Vol. 30, 1985, p. 157.
[2] McHenry, H. I. and Reed, R. P., *Nuclear Engineering Design*, Vol. 58, 1980, p. 219.
[3] Krafft, J. M., *Applied Materials Research*, Vol. 3, 1964, p. 88.
[4] Hahn, G. T. and Rosenfield, A. R., *Metallurgical Transactions A*, Vol. 6A, 1975, p. 653.
[5] Hirth, J. P. and Froes, F. H., *Metallurgical Transactions A*, Vol. 8A, 1977, p. 1165.
[6] Wanhill, R. J. H., *Engineering Fracture Mechanics*, Vol. 10, 1978, p. 337.
[7] Tobler, R. L., Read, D. T., and Reed, R. P. in *Fracture Mechanics, (13th Conference), ASTM STP 743,* R. Roberts, Ed., American Society for Testing and Materials, Philadelphia, 1981, p. 350.
[8] Rice, J. R. and Johnson, M. A. in *Inelastic Behavior of Solids,* M. F. Kanninen, W. G. Adler, A. R. Rosenfield, and R. I. Jaffee, Eds., McGraw-Hill, New York, 1970, p. 641.
[9] Ritchie, R. O. and Thompson, A. W., *Metallurgical Transactions A,* Vol. 16A, 1985, p. 223.
[10] Garrison, W. M., Jr., *Metallurgical Transactions A,* Vol. 17A, 1986, p. 669.
[11] Reed, R. P., Purtscher, P. T., and Yushchenko, K. A. in *Advances in Cryogenic Engineering Materials,* Vol. 30, R. P. Reed and A. F. Clark, Eds., Plenum Press, New York, 1986, p. 43.
[12] Thompson, A. W. and Weihrauch, P. F., *Scripta Metallurgica,* Vol. 10, 1976, p. 205.
[13] Tobler, R. L. and Meyn, D. in *Materials Studies for Magnetic Fusion Energy Applications at Low Temperatures—VIII,* R. P. Reed, Ed., National Bureau of Standards, Boulder, CO, 1985, p. 167; *Metallurgical Transactions,* Vol. 19A, 1988, p. 1626.
[14] McMeeking, R. M., *Journal of the Mechanics and Physics of Solids,* Vol. 25, 1977, p. 357.
[15] Broek, D., *Engineering Fracture Mechanics,* Vol. 5, 1973, p. 55.
[16] LeRoy, G., Embry, J. D., Edwards, G., and Ashby, M. F., *Acta Metallurgica,* Vol. 29, 1981, p. 1509.
[17] Cox, T. B. and Low, J. R., Jr., *Metallurgical Transactions A,* Vol. 5A, 1974, p. 145.

[18] Hutchinson, J. W., *Journal of the Mechanics and Physics of Solids,* Vol. 16, 1968, p. 13.
[19] Rice, J. R. and Rosengren, G. K., *Journal of the Mechanics and Physics of Solids,* Vol. 16, 1968, p. 1.
[20] Argon, A. S. and Im, J., *Metallurgical Transactions A,* Vol. 6A, 1975, p. 839.
[21] Ledbetter, H. M. in *Austenitic Steels at Low Temperature,* R. P. Reed and T. Horiuchi, Eds., Plenum Press, New York, 1983, p. 83.
[22] Floreen, S., Hayden, H. W., and Devine, T. M., *Metallurgical Transactions A,* Vol. 2A, 1971, p. 1403.

Er-Ping Chen[1]

Dynamic Brittle Fracture Analysis Based on Continuum Damage Mechanics

REFERENCE: Chen, E.-P., "**Dynamic Brittle Fracture Analysis Based on Continuum Damage Mechanics,**" *Fracture Mechanics: Perspectives and Directions* (*Twentieth Symposium*), *ASTM STP 1020,* R. P. Wei and R. P. Gangloff, Eds., American Society for Testing and Materials, Philadelphia, 1989, pp. 447–458.

ABSTRACT: The dynamic response of materials with existing flaw structures is the subject of the present investigation. In this paper, the cracked solid is described by a constitutive model based on continuum damage mechanics. The purpose of the analysis is to show that the constitutive model can predict many of the experimentally observed phenomena in these materials. This is demonstrated by treating a sample problem. The geometry and loading condition selected for this sample problem is that of a 45-deg inclined crack, centrally situated in a concrete panel and subjected to the action of a step tensile pulse. Damage localization and tension-softening have been observed. These results and their implications are discussed.

KEY WORDS: dynamic brittle fracture, continuum damage mechanics, concrete, inclined crack panel, localization of damage, tension softening, strain-rate effect

An important aspect of many mining and natural resources recovery techniques involves the fracture and fragmentation of rocks under rapidly applied loads. Typically, many rocks have an existing flaw structure. Accurate predictions of the dynamic responses of these materials require an understanding of their dynamic fracture behavior. Many observed phenomena in these materials, such as strain-rate effects, crack-tip damage, and tension softening, cannot be explained by classical elastodynamic fracture mechanics. In this paper, it is proposed to study the dynamic fracture of brittle rock by applying a continuum damage model developed by the author and his co-workers [*1–3*].

In this model, the rock is approximated as a continuum with a random distribution of subscale cracks. The moduli of the material follows those of a cracked solid given by the model of Budiansky and O'Connell [*4*]. The activation of these cracks by the applied load produces progressive damage to the material. In the continuum scale, accumulation of damage is reflected by the softening of the material moduli. Strain-rate effects have been explicitly included in the model. To demonstrate the utility of the current model, the problem of a 45-deg inclined crack, situated centrally in a concrete panel and subjected to the action of a step tensile pulse applied at the edges of the panel, has been treated. The analysis is based on the finite element method since the damage model has been implemented into the finite-element code PRONTO 2D [*5*]. Strain softening is predicted as a consequence of material damage accumulation due to subscale crack interaction and growth. Damage zones are seen to localize around the continuum crack. These results and their implications are discussed.

[1] Member of the Technical Staff, Applied Mechanics Division III, Sandia National Laboratories, Albuquerque, NM 87185.

A Brief Analytical Model Description

The basic assumption of the damage model is that the material is permeated by an array of randomly distributed cracks which grow and interact with one another under tensile loading. The model does not try to treat each individual crack, but rather treats the growth and interaction of cracks as an internal state variable which represents the accumulation of damage in the material. This damage, D, is assumed to degrade the material stiffness following the equations derived by Budiansky and O'Connell [*4*] for a random array of penny-shaped cracks in an isotropic elastic medium

$$\overline{K} = K(1 - D) \tag{1}$$

where K and $\overline{K}$ denote, respectively, the bulk modulus for undamaged and damaged material. The damage is related to the damaged Poisson's ratio $\overline{\nu}$ and crack density parameter C_d through

$$D = \frac{16}{9}\frac{(1 - \overline{\nu}^2)}{(1 - 2\nu)} C_d \tag{2}$$

The crack density parameter is related to the undamaged Poisson's ration, ν and $\overline{\nu}$, by the expression

$$C_d = \frac{45}{16}\frac{(\nu - \overline{\nu})(2 - \overline{\nu})}{(1 - \overline{\nu}^2)[10\nu - \nu(1 + 3\nu)]} \tag{3}$$

Thus, if the crack density parameter is known, the damaged Poisson's ratio can be calculated from Eq 3, and consequently, the damage parameter is found from Eq 2. The crack density parameter provides information about cracking in a given volume in that it is assumed to be proportional to the product of N, the number of cracks per unit volume, and a^3, the cube of the average crack dimension in the volume element under consideration, that is

$$C_d \sim N\, a^3 \tag{4}$$

Following the work of Grady and Kipp [*6*], N is expressed as a Weibull statistical distribution function activated by the current bulk strain measure $P/3\overline{K}$, where P is the pressure or mean stress $P = (\sigma_{xx} + \sigma_{yy} + \sigma_{zz})/3$, according to

$$N = k\left(\frac{P}{3\overline{K}}\right)^m \tag{5}$$

In Eq 5, k and m are material constants to be determined from strain-rate dependent fracture stress data. Since the size of the fragments is determined by the intersecting crack network within the rock volume, the crack dimension is thought to be proportional to the fragment size. Thus, the average crack dimension, a, is estimated from the nominal fragment diameter expression for dynamic fragmentation in a brittle material [*7*] as

$$2a \sim \left(\frac{\sqrt{20}\, K_{Ic}}{\rho\, C_s\, \dot{\epsilon}_{max}}\right)^{2/3} \tag{6}$$

where

ρ = mass density,
C_s = uniaxial wave speed ($\sqrt{E/\rho}$),
E = Young's modulus,
K_{Ic} = fracture toughness of the material, and
$\dot{\epsilon}_{max}$ = maximum volumetric strain rate experienced by the material throughout the fracture process.

Note that Eq 6 provides the average crack size that has been activated by the applied load in the volume element under consideration. Thus, when the strain rate is low, only large cracks have been activated and the material can only separate into a few large pieces. On the other hand, under high strain rates, smaller cracks would also have been activated and the specimen can break into many small fragments. The proportionality constants from Eqs 4 and 5 can be absorbed into the constant k. Hence, the additional material parameters for this constitutive model, aside from the commonly defined ones, are k and m as given in Eq 4. When bulk tension occurs in the material, it is possible to calculate at each time step the damage parameter, D, from the above expressions. The stiffness is then degraded by the factor $(1 - D)$. In this fashion, the post-damage responses of the material are represented. Note that the damage parameter, D, is an internal state variable which is evolutionary and irreversible in nature. In compression, the material is assumed to behave in an elastic/perfectly plastic manner.

Dynamic Material Behavior

The constants k and m in the analytical model are specific to the dynamic responses of the material under consideration. The material selected for illustration purposes in the current study is concrete and the representative values of its material properties are listed in Table 1. In order for us to determine the constants k and m in the damage model, strain-rate-dependent tensile strength data for concrete in the range of strain rates of 1 to 1000/s are required. In lieu of these test data, they can be estimated from the fracture toughness value by an equation given in Ref *8*

$$\sigma_c = \left(\frac{9\, E\, K_{Ic}^2}{16\, Y^2\, C}\right)^{1/3} \dot{\epsilon}^{1/3} \quad (7)$$

relating the tensile strength and strain rate. In Eq 7, E is the Young's modulus, Y is a crack geometric shape factor, and C is the shear wave speed. Assuming that the subscale cracks are penny-shaped, Y can be taken to be 1.12. Figure 1 shows the result from this estimation procedure. The constants k and m can be determined from strain-rate dependent fracture stress via a procedure given in Ref *1*. The values of k and m corresponding to the results in Fig. 1 are found to be $5.75 \times 10^{21}/m^3$ and 6.0, respectively.

TABLE 1—*Typical concrete properties.*

Mass density	2.4 Mg/m^3
Young's modulus	20.7 GPa
Poisson's ratio	0.18
Fracture toughness	2.75 MPa$\sqrt{m}$
Compressive strength	27.6 MPa

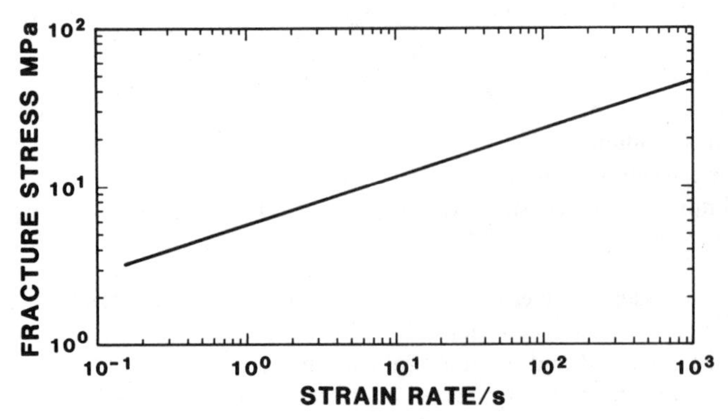

FIG. 1—*Strain rate versus fracture stress for concrete.*

For illustration purposes, the mean stress and damage versus time behavior of concrete under a constant bulk tensile strain rate of 100/s are shown in Fig. 2. This is obtained from a one-element idealization of a concrete block subjected to uniaxial tensile straining. Since the strain rate is constant, time represents a measure of strain, and the curves in Fig. 2 are essentially stress-strain and damage-strain plots. Figure 2 demonstrates that the mean stress and damage behavior are interactive. At low stress levels, damage accumulates at a slow rate and the bulk stress-strain relationship remains linear. As stress is increased, damage begins to accumulate at a faster rate resulting in the nonlinear stress-time response. The maximum rate of damage accumulation is attained when the stress reaches its maximum value. Beyond this point, the rate of damage accumulation decreases with decreasing stress level. The total damage moves gradually toward its asymptote as the stress approaches some constant level. Note that strain-softening behavior occurs naturally as consequences of the damage evolution. More details of the cyclic stress and damage responses have been covered previously [*1*] and will not be reiterated here.

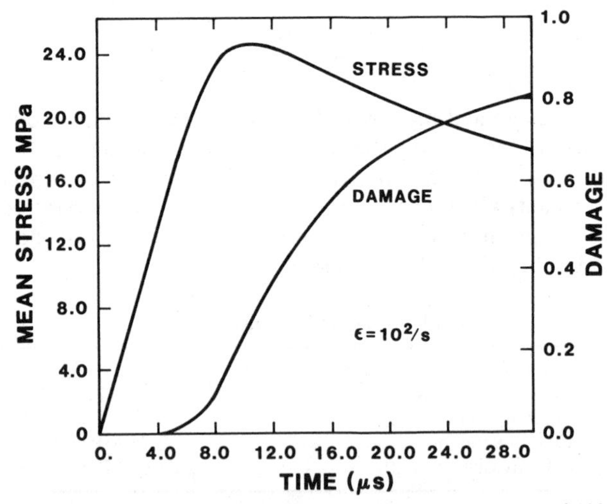

FIG. 2—*Time versus mean stress and damage for strain rate of 100/s.*

An Inclined Crack in a Concrete Panel

Consider the cracked panel in Fig. 3. The panel is subjected to uniform tension applied at the top and bottom edges in a direction normal to these surfaces. Plane strain conditions are assumed to prevail everywhere in the panel. The panel in Fig. 3 is 0.1 by 0.2 m in size, with a central crack 0.04 m in length. The crack is inclined at 45 deg to the horizontal axis. The load is applied as a step function in time and the magnitude is 10.0 MPa.

The finite element grid used in the calculation is shown in Fig. 4. The finite-element grid contains 720 four-node quadrilateral elements and 786 nodes. All runs were terminated at 0.4 ms since, at this time, at least two loading and unloading cycles have occurred near the elements around the crack tips. This is sufficient time to illustrate some of the salient features in the damage model. For comparison purposes, the solution in which the concrete panel is assumed to behave elastically (damage not included) has also been obtained. All results were obtained from the PRONTO 2D [5] code. The crack surfaces are taken to be contact surfaces to prevent them from flipping over under compressive stresses from reflected waves from the boundaries. No special consideration was given to the crack-tip regions.

Figure 5 depicts the pressure (mean stress) versus time plot for the element near the lower crack tip for both the elastic and damage models. The location of this element is given in Fig. 4 as a darkened square. This element, which is immediately ahead of the crack tip, is designated as Element 4. The element whose location corresponds to Element 4 for the upper crack tip has similar behaviors and thus will not be shown. Because of the interactions between the stress waves and the bounding surfaces, this element sees multiple loading and unloading cycles. As can be seen from Fig. 5, the loading characteristics are similar for both the elastic and the damage model. However, because of the subscale crack activation in the damage model, damage is being accumulated in the element. Therefore, the pressure was relaxed in the damage model and its magnitude began to decrease shortly after 0.05 ms, even though loading continues in the elastic case. The damage time-history for this element

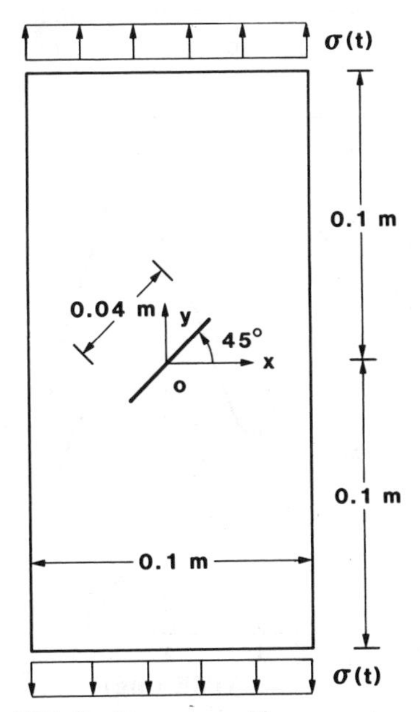

FIG. 3—*Example problem geometry.*

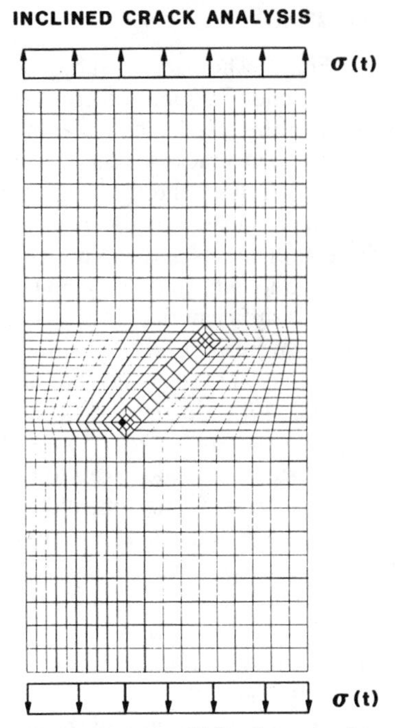

FIG. 4—*Finite element grid for the example problem.*

is given in Fig. 6. This figure shows that damage initially increases with time, following basically the stress curve in Fig. 5, and eventually reaches a level of about 0.66. Note that damage does not decrease after t = 0.05 ms even though unloading occurs due to wave interactions. This is because the damage accumulation process is an irreversible one. This may be explained from Figs. 7 and 8, respectively, in terms of normalized pressure and

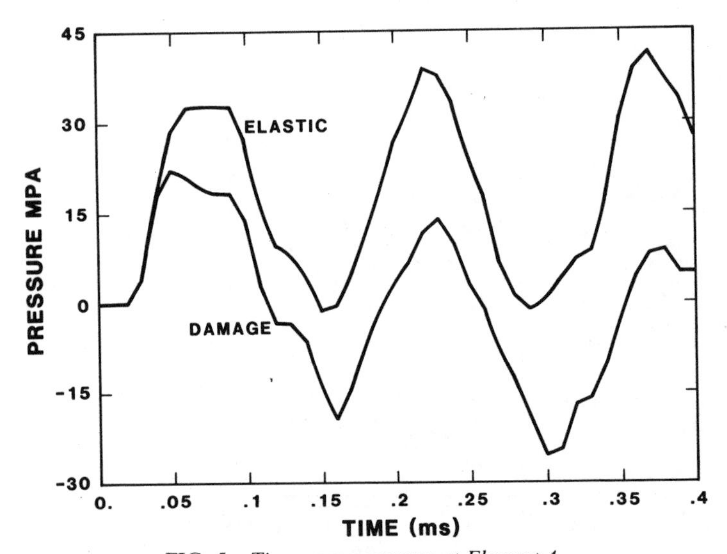

FIG. 5—*Time versus pressure at Element 4.*

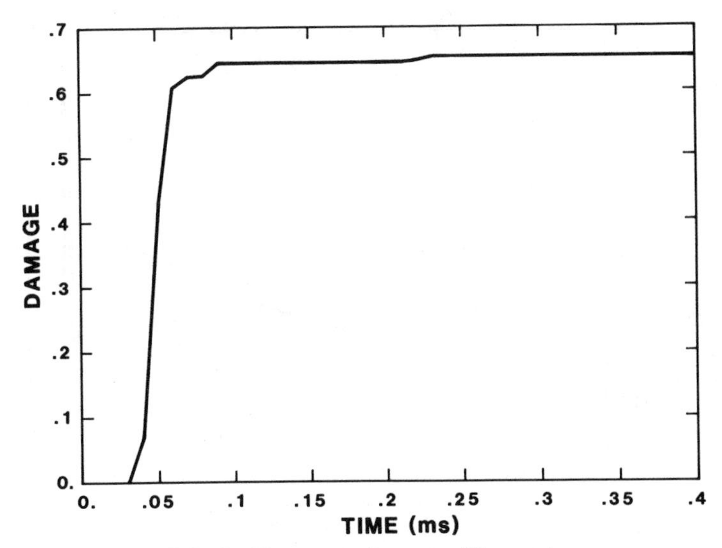

FIG. 6—*Time versus damage at Element 4.*

damage versus bulk strain plots, where bulk strain is defined to be $(\epsilon_{xx} + \epsilon_{yy} + \epsilon_{zz})/3$, for the same element as in Figs. 5 and 6. The pressure has been normalized against the compressive strength of the material, 27.6 MPa.

The elastic solution has also been included in Fig. 7. From Fig. 7, it is observed that the pressure in the damage model first increases linearly along the elastic bulk modulus, coinciding with the elastic solution. As damage is being accumulated, the damaged pressure-bulk strain curve begins to deviate from linearity. The pressure reaches its maximum level at a bulk strain value of 0.001, after which the damage accumulated is severe enough to relax the pressure, resulting in tension-softening behavior. This continues until the wave reflected from the opposite end arrives at the crack, the pressure becomes compressive in

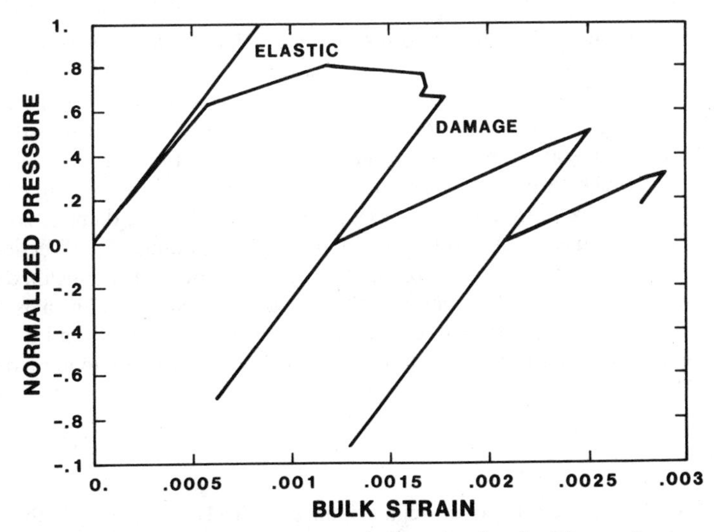

FIG. 7—*Normalized pressure—bulk strain plots for Element 4.*

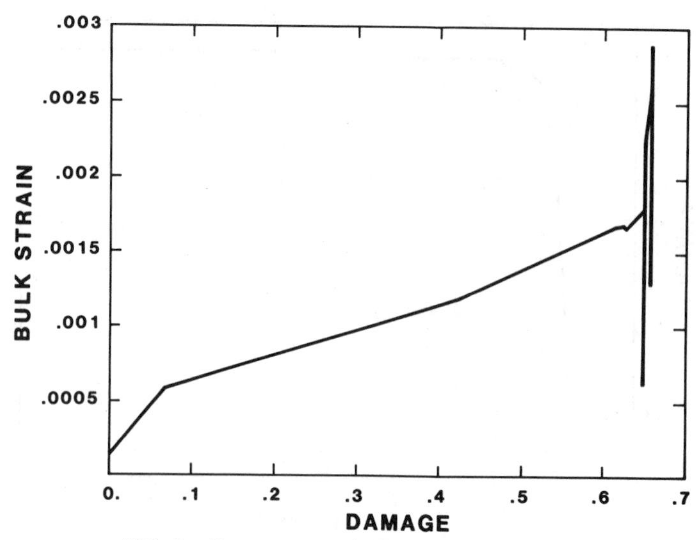

FIG. 8—*Damage versus bulk strain at Element 4.*

nature, and unloading occurs around a bulk strain level of 0.0016. In the current damage model, unloading is assumed to follow the original elastic bulk modulus. Upon reloading by subsequent wave reflection, the material remembers its level of damage and the pressure-bulk strain curve follows a corresponding damaged modulus. This pattern repeats itself as the material goes through loading and unloading cycles caused by wave reflections. These observations are also evident from the damage versus strain plot in Fig. 8. The damage level corresponding to the maximum pressure in Fig. 7 is around 0.3; this is consistent with that from Fig. 2. In Fig. 8, the first unloading occurs at approximately a bulk strain level of 0.0016. However, the damage stays constant as the strain is decreased. Upon reloading, the damage is increased gradually until unloading again occurs. The same behavior then repeats itself.

Finally, Figs. 9 through 12 exhibit deformed contour plots of the damage distribution in the concrete panel at times 0.1, 0.2, 0.3, and 0.4 ms, respectively. The deformations have been magnified 90 times in these plots. The damage contours represent areas where stiffnesses have been degraded by a factor $(1 - D)$ with respect to the original values. In general, the damage zone grows with time. The crack opens or closes depending on whether or not the crack surfaces are experiencing tension of compression at the time. It is clear that the damage is localized around the crack tip. Also, note that the damage zone extends in a direction perpendicular to the loading axis. This implicitly defines the direction of crack growth. In mixed-mode crack propagation problems, it is found that the inclined crack will initiate its growth at some angle different from the direction normal to loading axis. However, it will eventually align itself along the direction normal to the applied load direction [*9*]. Thus, reasonable predictions of the direction of fracture growth are obtained by the damage model.

Discussion

The inability of linear elastic fracture mechanics (LEFM) to deal with the failure of materials exhibiting nonlinear behavior is well understood. For rocks, concrete, ceramics,

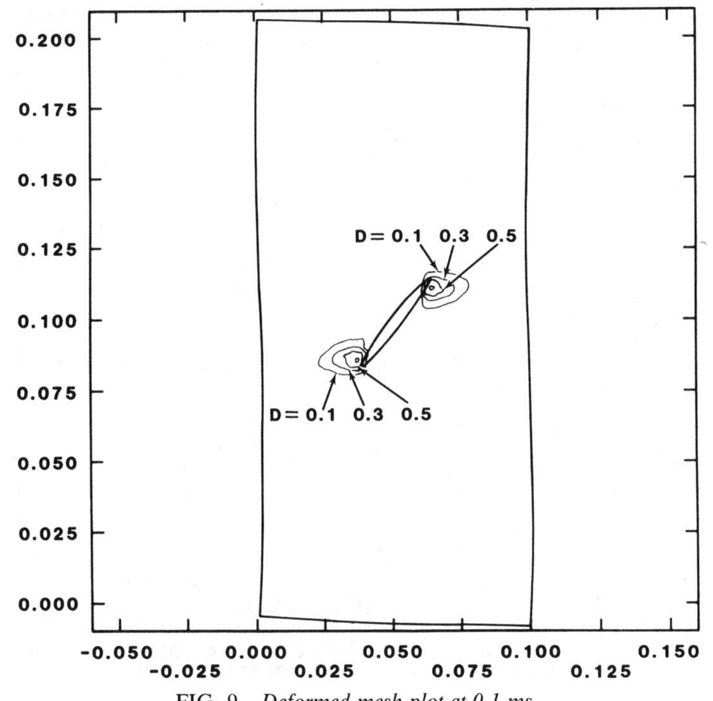

FIG. 9—*Deformed mesh plot at 0.1 ms.*

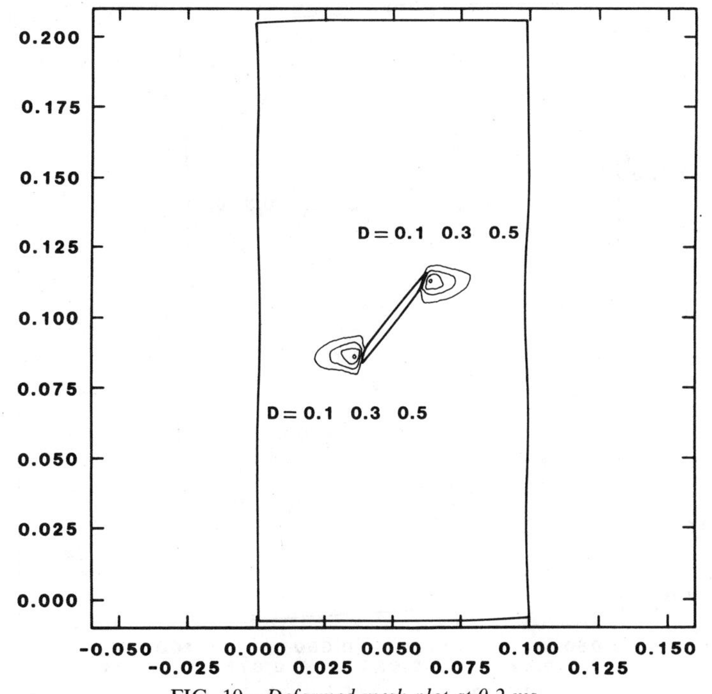

FIG. 10—*Deformed mesh plot at 0.2 ms.*

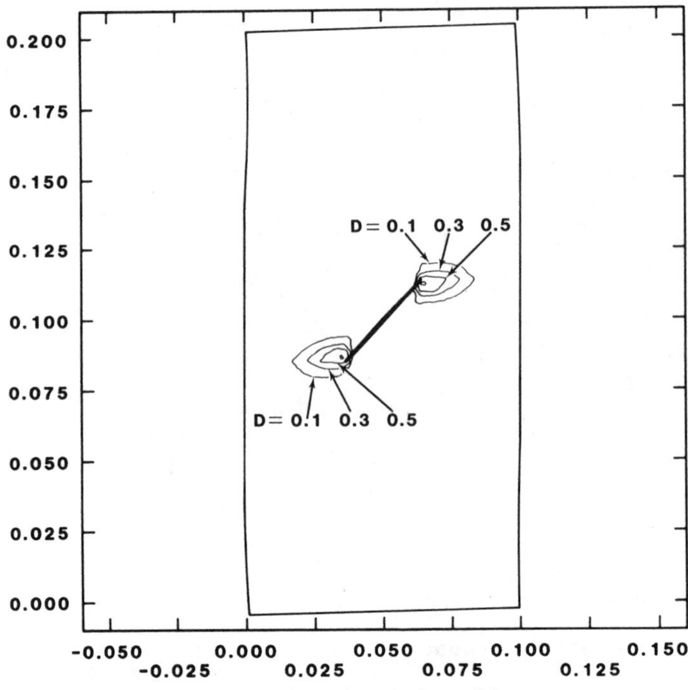

FIG. 11—*Deformed mesh plot at 0.3 ms.*

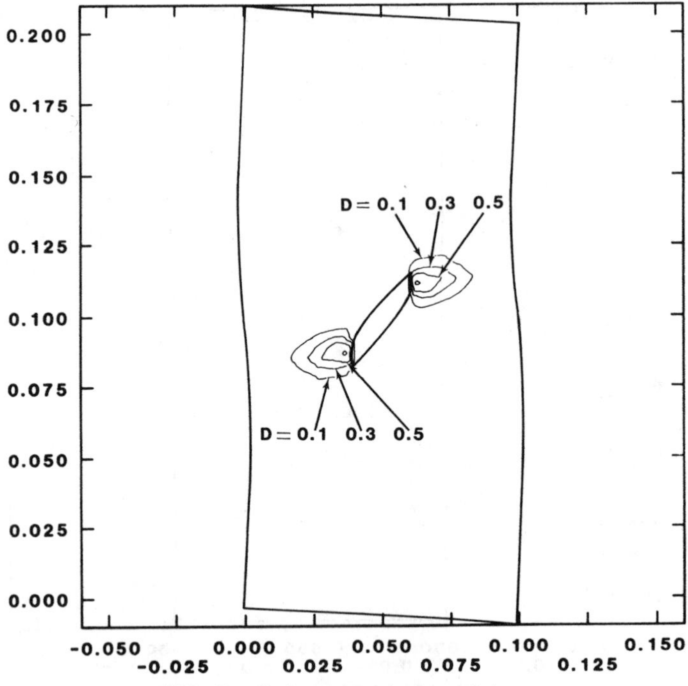

FIG. 12—*Deformed mesh plot at 0.4 ms.*

etc., the nonlinearity may be induced by the activation of subscale cracks which occur either naturally or are introduced during the manufacturing process. The difficulty of applying LEFM to these materials, in part at least, is due to the fact that local and global failure do not coincide as in the realm of LEFM. Thus, prior to the global failure, some consideration must be given to the occurrence of local fractures which dissipate part of the available energy. Disregarding the effect of these local fractures can result in erroneous predictions. The damage model presented here provides a convenient method to account for the preglobal failure responses through the accumulation of damage in the material. Moreover, the damage model may also be used to predict the global failure of the specimen. For example, say a damage level of 0.4 has been established experimentally for concrete to be the level at which material separation occurs. (Ideally, failure should occur when damage reaches 1. However, this implies total disintegration of the material element. At lower damage levels, the material may have been severely cracked and yet still hold together. At the slightest external disturbance, separation will occur. In many cases, this situation coincides with experimentally observed failure.)

Observing from Figs. 9 to 12, the load level at which the 0.4 damage contour comes into contact with the edge of the specimen would define the failure load for the panel. At present, the potential of the damage model to be used in fracture predictions for materials with subscale crack populations appears to be very good. Experimental verification is required before this potential can be realized.

Summary

A continuum damage model has been applied to study the dynamic response of brittle materials with subscale crack populations. The model treats the dynamic fracture process as a continuous accrual of damage, where the damage is defined to be the volume fraction of material that has been tension relieved by multiple crack growth and interaction. Numerical simulation of a center-cracked concrete specimen subjected to a step tensile pulse has been carried out. Tension-softening behavior in concrete has been predicted by the model as the consequence of the subscale cracking activities occurring in the material instead of an assumed constitutive behavior. The localization of the damage zone near the crack is also predicted by the model. The potential of using the damage model to predict both the local failure in terms of the process zone and global failure in terms of the degree of loss in load carrying capacity is also discussed.

Acknowledgment

This work is sponsored by the U.S. Department of Energy under Contract DE-AC04-76-DP00789.

References

[1] Taylor, L. M., Chen, E.-P., and Kuszmaul, J. S., "Microcrack Induced Damage Accumulation in Brittle Rock Under Dynamic Loading," *Journal of Computer Methods in Applied Mechanics and Engineering,* Vol. 55, 1986, pp. 301–320.

[2] Chen, E.-P., "Continuum Damage Mechanics Studies on the Dynamic Fracture of Concrete," *Proceedings,* Cement-Based Composites: Strain Rate Effects on Fracture, Materials Research Society Symposia, Vol. 64, S. Mindess and S. P. Shah, Eds., Materials Research Society, Pittsburgh, PA, 1986, pp. 63–67.

[3] Chen, E.-P. and Taylor, L. M., "Fracture of Brittle Rock Under Dynamic Loading Conditions,"

Fracture Mechanics of Ceramics, Vol. 7, R. C. Bradt, A. G. Evans, D. P. H. Hasselman, and F. F. Lange, Eds., Plenum Press, New York, 1986, pp. 175–186.
[4] Budiansky, B. and O'Connell, R. J., "Elastic Moduli of a Cracked Solid," *International Journal of Solids and Structures,* Vol. 12, 1976, pp. 81–97.
[5] Taylor, L. M. and Flanagan, D. P., "PRONTO 2D—A Two-Dimensional Transient Solid Dynamics Program," SAND86-0594, Sandia National Laboratories Report, Albuquerque, NM, March, 1987.
[6] Grady, D. E. and Kipp, M. E., "Continuum Modeling of Explosive Fracture in Oil Shale," *International Journal of Rock Mechanics and Mining Sciences,* Vol. 17, 1980, pp. 147–157.
[7] Grady, D. E.,"The Mechanics of Fracture Under High-Rate Stress Loading," *Proceedings,* William Prager Symposium on Mechanics of Geomaterials: Rocks, Concrete and Soils, Z. P. Bazant, Ed., Northwestern University, Evanston, IL, 1983.
[8] Kipp, M. E., Grady, D. E., and Chen, E.-P., "Strain-Rate Dependent Fracture Initiation," *International Journal of Fracture,* Vol. 16, 1980, pp. 471–478.
[9] *Mixed Mode Crack Propagation,* G. C. Sih and P. S. Theocaris, Eds., Sijthoff and Noordhoff Publishing, The Netherlands, 1981.

William J. Mills[1]

Effect of Loading Rate and Thermal Aging on the Fracture Toughness of Stainless-Steel Alloys

REFERENCE: Mills, W. J., **"Effect of Loading Rate and Thermal Aging on the Fracture Toughness of Stainless-Steel Alloys,"** *Fracture Mechanics: Perspectives and Directions* (*Twentieth Symposium*), *ASTM STP 1020*, R. P. Wei and R. P. Gangloff, Eds., American Society for Testing and Materials, Philadelphia, 1989, pp. 459–475.

ABSTRACT: The effect of loading rate on the fracture toughness of Types 304 and 316 stainless-steel plate and Type 308 weld before and after thermal aging was characterized using both fracture mechanics and Charpy specimens. Aging at 566°C for 10 000 h reduced static J_c initiation toughness values for both wrought and weld metals by 10 to 20%, and tearing moduli were reduced by 20 to 30%. Under semi-dynamic and dynamic loading conditions, the fracture resistance was not decreased below the static response for either unaged or aged materials. The present results also demonstrated that the large degradation in Charpy V-notch impact energy after aging was not representative of the impact resistance for stainless-steel components containing cracks or crack-life defects. Fractographic examinations revealed that the large loss in Charpy energy was associated with aging-induced microstructural changes that substantially decreased the energy required to initiate a crack from a blunt notch. At very slow strain rates, creep crack growth was found to cause a significant degradation in fracture toughness.

KEY WORDS: dynamic, elastic-plastic, fracture toughness, load displacement, semi-dynamic, stainless steel, static, temperature, thermal aging, welds

Austenitic stainless steel base metal and welds are used extensively in structural applications where high-temperature corrosion resistance, good ductility, and creep resistance are important considerations. The fracture resistance of these materials is exceptionally high, and cracks generally open by plastic bulging processes at very long critical crack lengths. Fracture control via a fracture-mechanics approach is not an important design requirement, since conventional stress and strain limits, such as those provided by the American Society of Mechanical Engineers (ASME) Boiler and Pressure Vessel Code [*1*], are typically sufficient to guard against ductile fracture.

However, a wealth of data [*2–14*] shows that thermal aging at elevated temperature significantly degrades the Charpy V-notch (CVN) impact energy of austenitic stainless steel (SS). Such information raises some concern that the fracture resistance of these materials might be seriously impaired by long-term exposure to high temperatures. By contrast, recent fracture-mechanics studies [*15,16*] revealed that thermal aging produced only a modest reduction in J_c fracture toughness.[2]

[1] Formerly, Westinghouse Hanford Company, Richland, WA 99352; presently, advisory engineer, Westinghouse Electric Corporation, P.O. Box 79, West Mifflin, PA 15122.

[2] Initiation toughness values are termed J_c rather than J_{Ic} because they do not strictly meet the ASTM E 813-81 requirements for plane-strain crack tip constraint.

The current study was undertaken in part to determine why the two tests yield different responses. Elastic-plastic J_c fracture toughness results obtained under static, semi-dynamic (test duration of 1 s) and dynamic (test duration of 20 to 40 ms) loading conditions on both as-received and thermally aged materials were compared with blunt-notch and precracked Charpy impact energy trends. In addition, limited slow loading rate testing (test durations of 1 and 20 h) was also performed to evaluate the potential for creep crack growth at 538°C.

Materials and Experimental Procedures

The materials used in this study are listed below:

(*a*) 60-mm-thick 304 SS plate,
(*b*) 51-mm-thick 316 SS plate, and
(*c*) 60-mm-thick 308 SS shielded-metal-arc weld.

The 316 and 304 SS plates were solution annealed, and the weld was tested in the as-welded condition. The thermal aged specimens were exposed to 566°C for 10 000 h. Chemical compositions and heat numbers are given in Table 1. Tensile properties for the unaged and 10 000-h aged conditions are presented in Table 2.

Microstructures for the wrought products displayed a duplex inclusion structure consisting of a few coarse carbide inclusions randomly distributed throughout the matrix and many smaller second-phase particles located inside each grain. Both wrought materials exhibited an American Society for Testing and Materials (ASTM) grain size number of 4. Thermal aging at 566°C produced a chromium-rich $M_{23}C_6$ carbide network along grain boundaries (Fig. 1). The weld consisted of an austenitic matrix with delta ferrite islands at substructural boundaries, as shown in Fig. 2*a*. Magnetic permeability testing revealed that the ferrite number for the weld was 6.4 (corresponding to a 6.3 volume percent ferrite level). Aging induced sigma phase and $M_{23}C_6$ carbide precipitation along the ferrite-matrix interface (Fig. 2*b*). The presence of intermetallic sigma phase was confirmed by selected area electron diffraction and energy dispersive X-ray analysis of thin-foil specimens.

Fracture tests were performed on deeply precracked compact specimens with a width of 50.8 mm. The thickness of 304 SS and 308 SS weld specimens was 25.4 mm; 316 SS specimens had a thickness of 22.9 mm. Notches in weld specimens were centered in the weld deposit and oriented parallel to the welding direction. Tests were performed at 538°C on an electro-hydraulic closed-loop machine under static (stroke rates of 1.3 mm/min for base metal and 0.5 mm/min for weld), semi-dynamic (600 mm/min for base metal and 250 mm/min for weld), and dynamic (15 m/min for both base metal and weld) loading conditions. Stroke rates for the base metal and weld differed by a factor of 2.5 so test duration times for the two materials were approximately the same (5 to 10 min for static tests and 1 s for semi-dynamic tests). Displacements were measured on the load-line by a high-temperature linear variable differential transformer (LVDT) displacement monitoring technique [*17*].

TABLE 1—*Chemical composition (percent by weight).*

Material	Heat Number	C	Mn	P	S	Si	Cr	Ni	Co	Ti	Cb + Ta	Cu	Mo
304 SS	300380	0.060	1.60	0.018	0.012	0.57	18.61	8.64	0.061	0.016	0.010	...	...
316 SS	8092297	0.057	1.86	0.024	0.019	0.58	17.25	13.48	0.02	0.02	...	0.10	2.34
308 SS Weld	...	0.07	1.95	0.010	0.015	0.62	20.30	9.82	0.08	...	0.04	0.21	0.21

TABLE 2—*Summary of tensile properties.*

Material	Aging Temperature, °C	Aging Time, h	Test Temperature, °C	Yield Strength, MPa	Ultimate Strength, MPa	Uniform Elongation, %	Total Elongation, %	Reduction in Area, %	α[a]	n[a]	Number of Specimens
304 SS	...	...	24	252	642	69	73	67	13.4	2.4	2
304 SS	...	...	538	140	421	33	39	62	12.5	2.3	2
304 SS	566	10 000	538	175	401	26	30	48	4.6	3.6	1
308 SS weld	...	...	24	455	634	38	44	48	15.2	3.5	4
308 SS weld	...	...	538	303	412	19	24	52	4.8	6.5	4
308 SS weld	566	10 000	538	293	401	11	15	48	...	...	2
316 SS	...	...	24	281	569	54	62	76	9.6	3.1	2
316 SS	...	...	538	182	466	35	40	60	11.1	2.6	2

[a] Ramberg-Osgood Strain-Hardening Law:

$$\frac{\epsilon}{\epsilon_{ys}} = \frac{\sigma}{\sigma_{ys}} + \alpha\left(\frac{\sigma}{\sigma_{ys}}\right)^n; \quad \sigma_{ys} = E\,\epsilon_{ys}$$

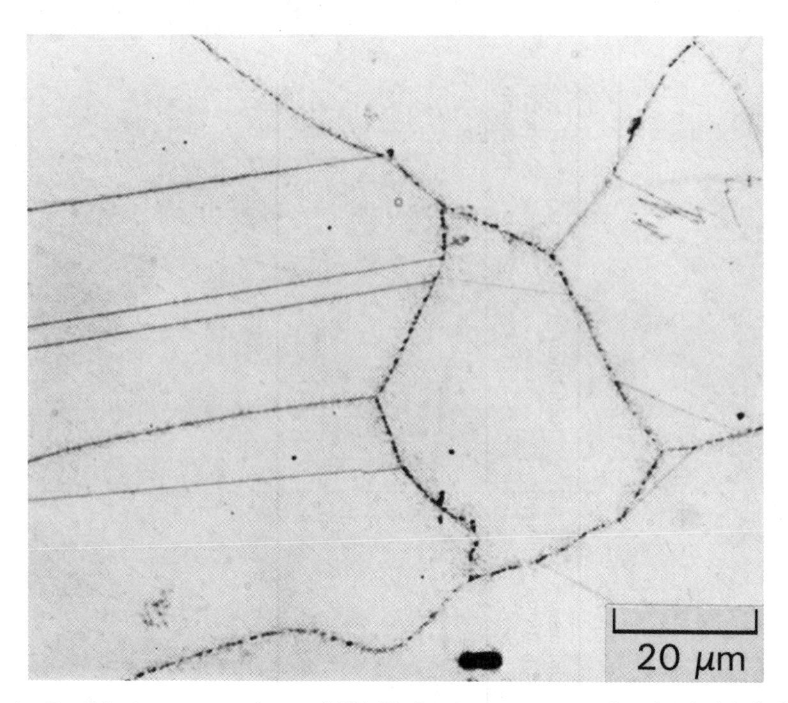

FIG. 1—*Typical microstructure for aged 304 SS. Specimen was immersion-sketched in boiling Murakami's reagent [150 mL H_2O; 30 g KOH; 30g K_3Fe $(CN)_6$], which preferentially attacked carbides without etching out grain boundaries. Note the intergranular carbide network in this figure. The microstructure for aged 316 SS was very similar to that for 304 SS.*

The fracture toughness behavior under static and semi-dynamic loading conditions was determined by the multiple-specimen J_R-curve technique. The analysis procedures differed slightly from those described in the ASTM Test Method for J_{Ic}, a Measure of Fracture Toughness (E 813-81), because the ASTM procedures and size requirements were generally not applicable to stainless-steel alloys. Specifically, compact specimens were loaded to various displacements producing different amounts of crack extension, Δa, and then unloaded. After unloading, each specimen was heat-tinted to discolor the crack growth region and subsequently broken open so that the amount of crack extension could be measured. The value of J for each specimen was determined from the load versus load-line displacement curve by the equation (E 813-81)

$$J = \frac{2A}{Bb}\frac{(1 + \alpha)}{(1 + \alpha^2)} \tag{1}$$

where

A = area under load versus load-line displacement curve,
b = unbroken ligament size,
α = $[(2a/b)^2 + 2(2a/b) + 2]^{1/2} - (2a/b + 1)$,
a = crack length, and
B = specimen thickness.

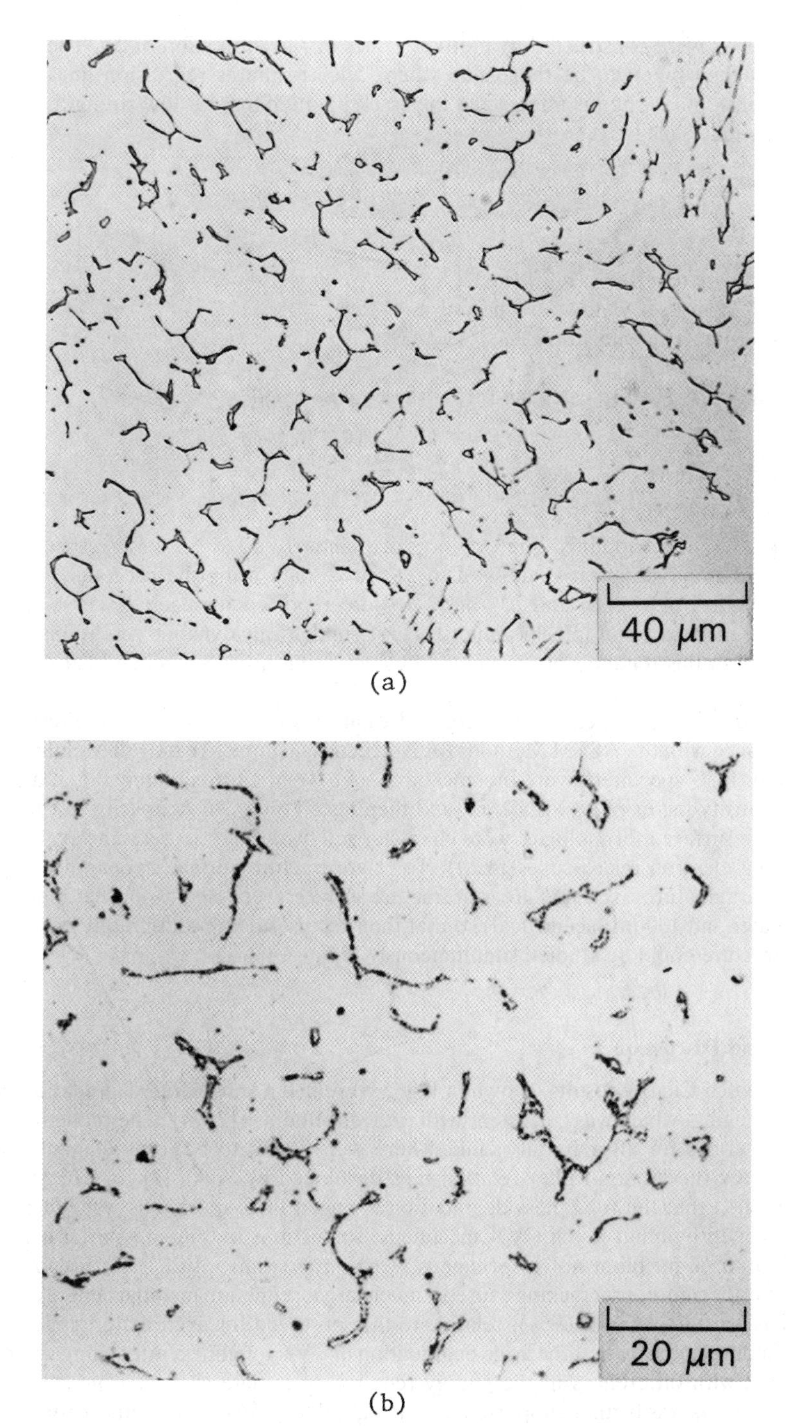

FIG. 2—*Typical microstructures for 308 SS weld* (*etchant: Murakami's reagent*): (a) *Unaged material had a duplex austenite-ferrite microstructure with residual delta ferrite located at the cores of primary and secondary dendritic branches;* (b) *Aged material exhibited partial transformation of delta ferrite into* $M_{23}C_6$ *carbides and sigma phase.*

The J_R curves were constructed by plotting values of J as a function of Δa. The initiation J_c value was then taken to be that value where a least-squares regression line through the crack extension data points intersected the crack blunting line for low strength, high strain-hardening materials [18]

$$J = 4\sigma_f(\Delta a) \quad (2)$$

where

σ_f = flow strength = $(\sigma_{ys} + \sigma_{uts})$,
σ_{ys} = 0.2% offset yield strength, and
σ_{uts} = ultimate tensile strength.

Tearing moduli, T, were calculated from the equation [19]

$$T = \frac{dJ_R}{da}\frac{E}{\sigma_f^2} \quad (3)$$

where E is Young's modulus. The variances of J_c and dJ_R/da (S_1^2 and S_2^2, respectively) were determined from statistical analysis of the J_R curve data using the procedures outlined in Ref *20*. Values of $J_c \pm S_1$ and $dJ_R/da \pm S_2$ were reported for each J_R curve. It was not possible to construct multiple-specimen J_R curves under fully dynamic conditions, so loading rate effects in this regime were evaluated by comparing single specimen load-displacement records.

Standard CVN specimens and precracked Charpy (PCC) specimens were tested at 24°C in accordance with the ASTM Methods for Notched Bar Impact Testing of Metallic Materials (E 23-86). PCC specimens were precracked to an a/W of approximately 0.5 at a maximum stress intensity factor of 28 MPa $\sqrt{m}$, and then tested on a 340 J capacity test machine.

Fracture surface morphologies were characterized by direct fractographic examination on a scanning electron microscope (SEM). To relate fracture surface appearance to key microstructural features, selected areas of fracture surfaces were electropolished (in 25-g CrO_3, 7-mL water and 130-mL acetic acid) so that the fracture surface topography and underlying microstructure could be studied simultaneously [21].

Results and Discussion

Blunt-notch Charpy results, shown in Fig. 3, revealed a substantial degradation in toughness after aging, which was consistent with previous findings [2–14]. The reductions in CVN impact energies for all materials studied here were found to be in agreement with those predicted by the Larson-Miller relationships developed by Sikka [2] for 316 SS. Figure 3 also revealed that the toughness degradation for aged PCC specimens was much less than that exhibited by blunt-notch CVN specimens. Reductions in toughness after aging ranged from 35 to 65% for blunt-notch specimens, versus approximately 25% for precracked specimens. Furthermore, precracking caused a much larger reduction in impact energy for unaged materials (on the order of 60%), relative to that observed for aged materials (20 to 45%). These findings indicate that the large degradation in CVN toughness after aging was primarily associated with differences in the energy required to initiate a crack from a blunt notch. This hypothesis was further supported by fractographic observations and fracture mechanics studies, as detailed below.

Fractographic examination of unaged base metal and weld CVN specimens revealed

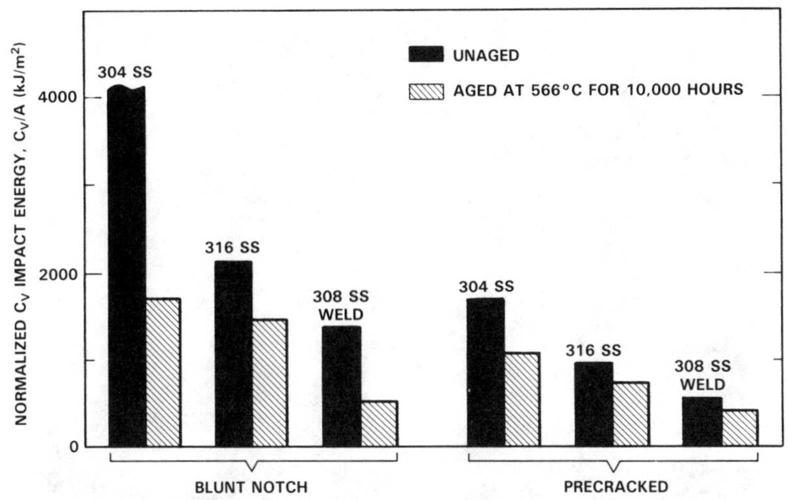

FIG. 3—*Effect of thermal aging on the normalized CVN and PCC impact energy levels (impact energy/remaining ligament area). The unaged 304 SS blunt-notch specimen did not fracture at an impact energy level of 325* J.

numerous tear-arrest markings parallel to the notch (Figs. 4*a* and 5), where cracks initiated but were quickly arrested by the fracture-resistant matrix. The extensive amounts of energy required for repeated reinitiation of cracking accounted for the high-impact energies exhibited by the unaged materials. After aging, however, no arrest markings were observed. In the aged base metal, grain boundaries were susceptible to cracking due to the presence of intergranular $M_{23}C_6$ carbides. During the early stages of plastic straining, grain boundaries in the vicinity of the notch separated and formed an almost continuous intergranular crack parallel to the notch (Fig. 4*b*). This served as a sharp starter crack and drastically reduced the energy to fracture. In weld specimens, delta ferrite transformation products, $M_{23}C_6$ carbides and sigma phase, provided effective microvoid nucleation sites, which also enhanced the initiation of a sharp crack. Hence, the large reduction in blunt-notch CVN impact energy resulted from aging-induced microstructural changes that markedly decreased the energy required to initiate a sharp crack.

Effects of aging on fracture-mechanics properties for SS base and weld metals are shown in Figs. 6 through 8. Under static loading conditions (open symbols), aging at 566°C for 10 000 h reduced J_c and tearing modulus values by approximately 10 to 20% and 20 to 30%, respectively.[3] The superior fracture resistance of these materials, even after 10 000-h exposures, precludes any possibility of non-ductile failure. Hence, conventional stress and strain design limits, such as those provided by the ASME Code, are adequate for guarding against ductile fracture, and they need not be supported by fracture mechanics analyses.

Figures 6 to 8 also revealed that the semi-dynamic fracture toughness responses for all materials except unaged 304 SS were superior to their static counterparts. Moreover, semi-dynamic J_R curves for the aged materials were almost identical to the corresponding static J_R curves for unaged materials. These results demonstrated that fracture resistance at intermediate loading rates was actually enhanced, rather than degraded. This behavior was consistent with the previous results for ferritic and austenitic steels [*22,23*], where dynamic

[3] Work in progress indicates that extending aging times to 20 000 to 50 000 h produces no further change in toughness properties.

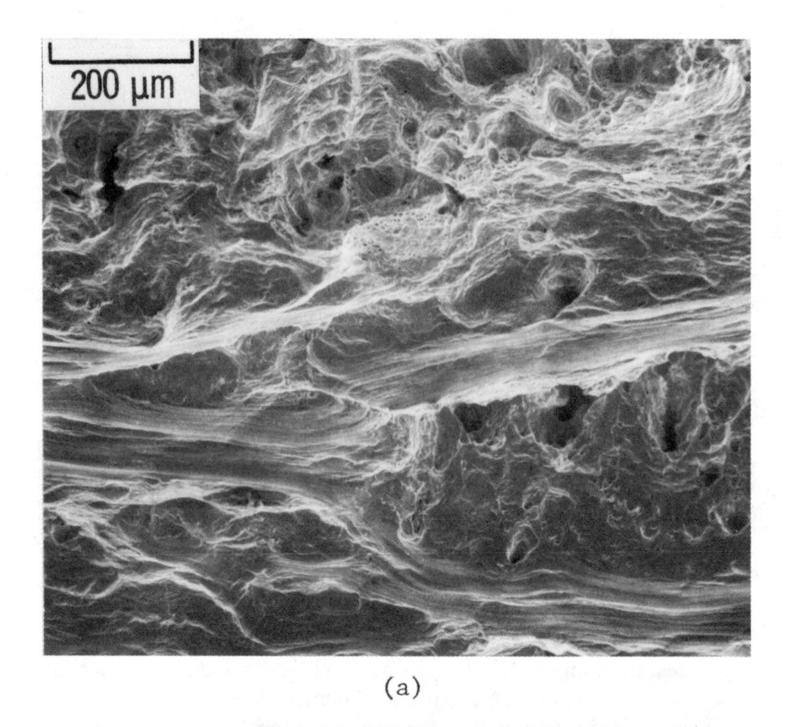

(a)

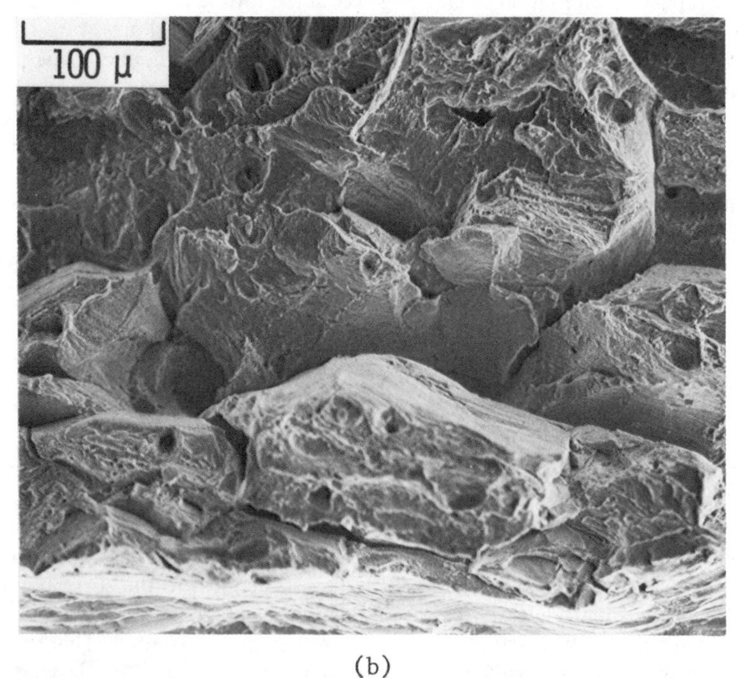

(b)

FIG. 4—*Fracture surface morphology in the blunt-notch region of 316 SS CVN specimens:* (a) *Unaged specimens exhibited numerous tear-arrest markings where cracks initiated but were quickly arrested by the fracture-resistant matrix. Note that sections of the notch (denoted by the parallel machining marks) were separated by the tear-arrest features.* (b) *Aged specimens displayed a continuous intergranular crack parallel to the notch (bottom of photograph). Away from the notch region, the fracture surface was primarily dimple rupture.*

FIG. 5—*Fracture surface appearance in the blunt-notch region of an unaged 308 SS weld CVN specimen. Note the tear-arrest markings separated by a portion of the notch.*

fracture toughness was found to increase with increasing crack velocity when the operative fracture mechanism was dimple rupture. An explanation for the absence of a rate effect in unaged 304 SS is lacking at this time.

The load-displacement records generated under static, semi-dynamic, and dynamic loading conditions were found to be very similar, indicating that the fracture behavior was relatively independent of loading rate. Typical load-displacement records for aged 316 SS are shown in Fig. 9; comparable behavior was also observed in the other unaged and aged materials. The minor differences in load-displacement curves reflected a combination of specimen-to-specimen variability and the slightly enhanced fracture resistance at intermediate loading rates, relative to the static behavior. The semi-dynamic and dynamic responses were almost identical for all test materials. These high-rate results confirmed that there were no detrimental loading rate effects on the fracture resistance of aged SS alloys. Furthermore, they demonstrated that conventional CVN tests were overly pessimistic in predicting aging effects on fracture initiating from sharp cracks or flaws, even under semi-dynamic and dynamic loading conditions.

Fracture surface morphologies for compact specimens were found to be insensitive to loading rate, which was consistent with the macroscopic behavior. Unaged base metal specimens exhibited microvoid coalescence exclusively (see Fig. 12 in Ref *24*), while aged specimens displayed dimple rupture coupled with limited intergranular fracture (Fig. 10). Microvoid coalescence dominated the fracture surfaces for both unaged and aged 308 SS welds (Fig. 11); however, aged welds exhibited a slightly smaller dimple size because aging-induced $M_{23}C_6$ carbides and sigma phase provided additional microvoid nucleation sites. The limited intergranular cracking in the aged base metal and more numerous microvoid nucleation sites in the aged weld accounted for the modest degradation in fracture resistance

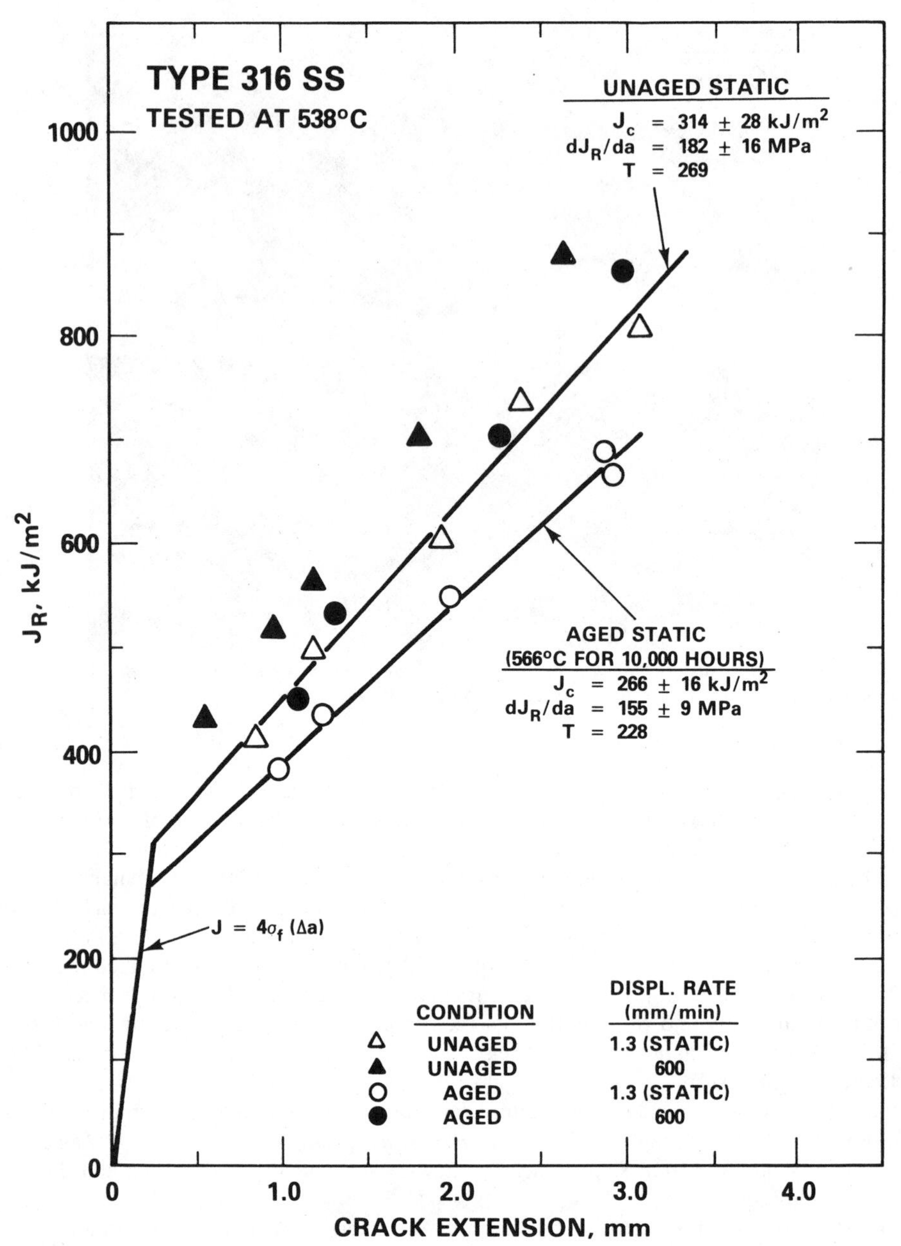

FIG. 6—*J_R curves for unaged and aged 316 SS.*

after thermal aging. Comparable fracture surface features were found in PCC specimens and in CVN specimens away from the blunt-notch region.

The foregoing J_c findings suggest that further toughness degradation may occur at slower strain rates. To examine this possibility, additional fracture-toughness testing of aged and unaged 304 SS was performed at 0.13 and 0.006 mm/min. Figure 12 indicates that the fracture resistance at 0.13 mm/min was essentially equivalent to the static behavior for the

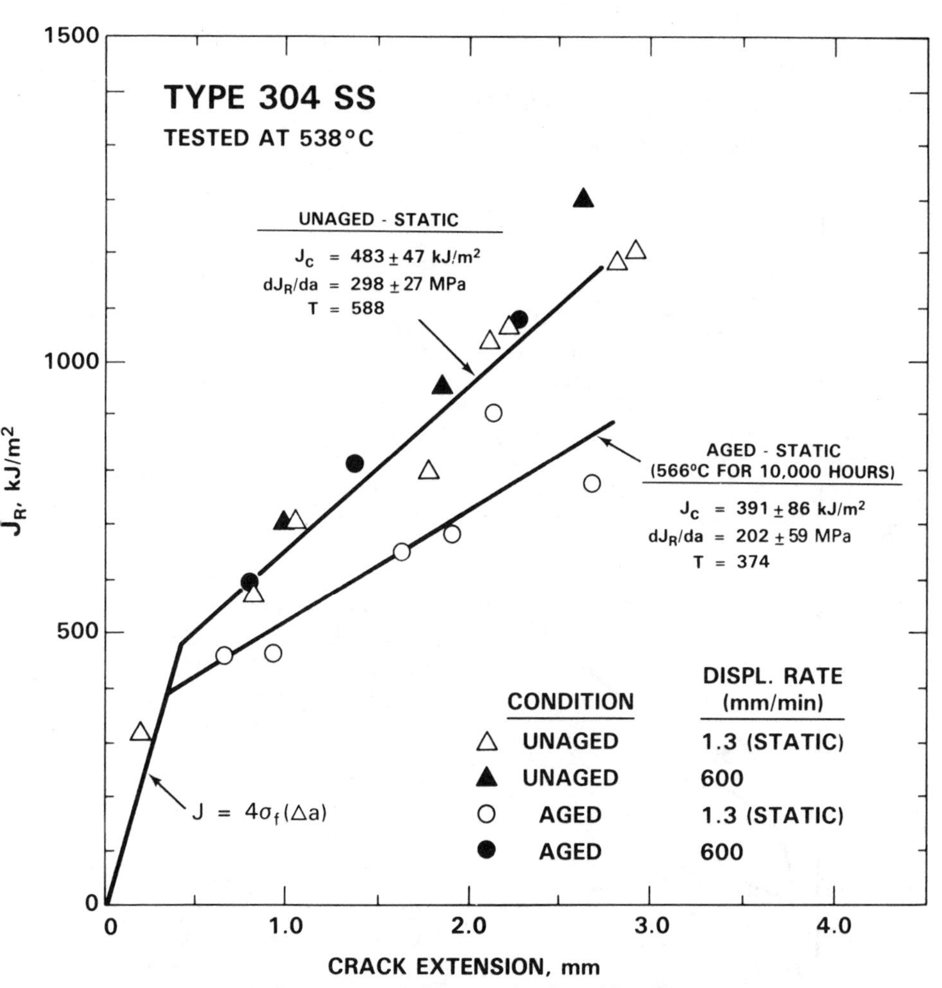

FIG. 7—*J_R curves for unaged and aged 304 SS.*

aged microstructure. At the slowest loading rate, however, the toughness for both the unaged and aged materials was significantly reduced. Extensive intergranular cracking was observed on the 0.006 mm/min fracture surfaces (Fig. 13), demonstrating that the reduction in fracture resistance was associated with creep crack growth. Hence, stainless-steel components containing crack-like defects may be susceptible to creep crack propagation, which could negate the very high tearing resistance displayed by this class of materials.

Conclusions

The effect of loading rate on the fracture toughness of unaged and aged 304 SS, 316 SS, and 308 SS weld was evaluated. Elastic-plastic J_c fracture toughness results obtained under static, semi-dynamic, and dynamic loading conditions were compared with blunt-notch and precracked Charpy impact energy trends. The results are summarized below.

1. Thermal aging at 566°C for 10 000 h caused a modest 10 to 20% reduction in J_c and a 20 to 30% reduction in tearing modulus for both wrought and weld metals. The superior

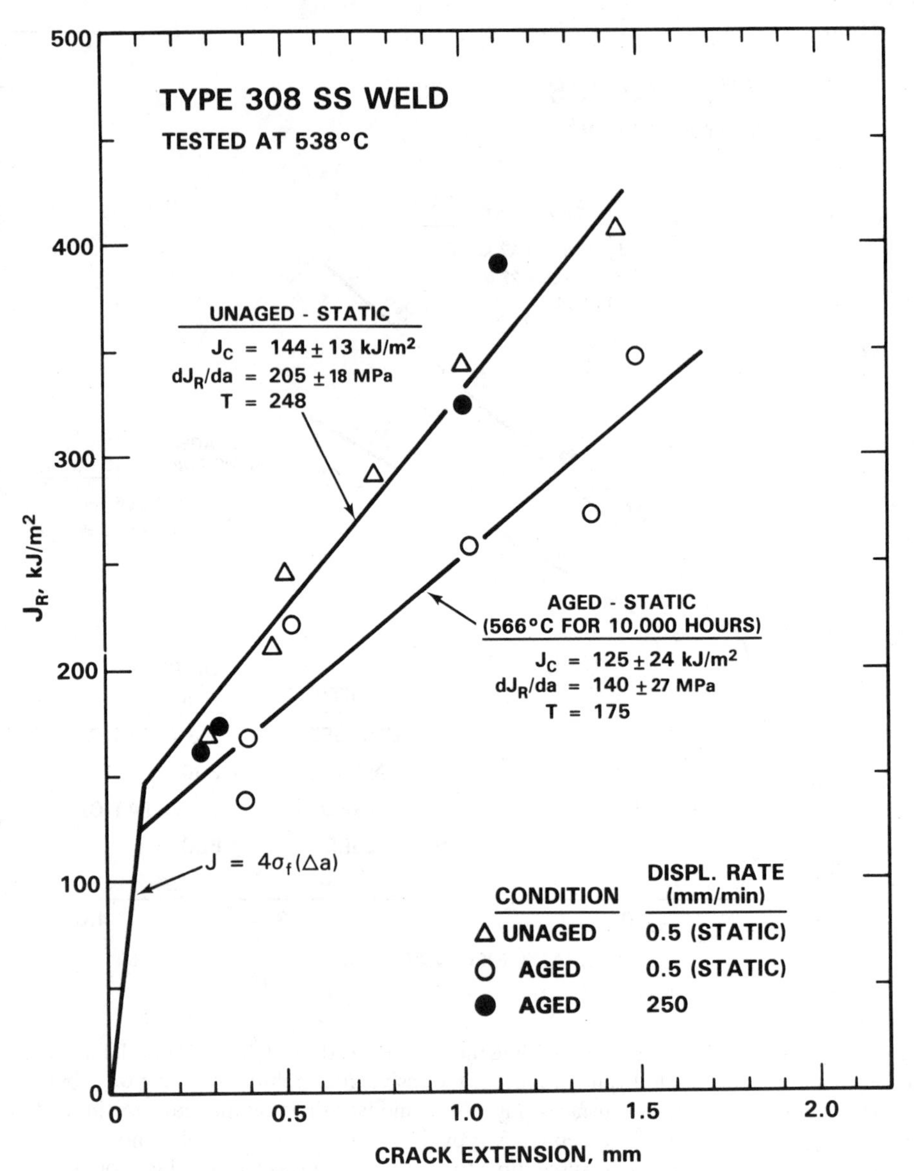

FIG. 8—*J_R curves for unaged and aged 308 SS weld.*

fracture resistance of SS alloys, even after long-term thermal aging, precludes any possibility of a non-ductile fracture.

2. The aging-induced toughness degradation in wrought materials was associated with localized tearing along grain boundaries decorated with $M_{23}C_6$ carbides. In the aged weld, delta ferrite transformation products, $M_{23}C_6$ carbides and sigma phase, reduced fracture resistance by providing effective microvoid nucleation sites.

3. Under semi-dynamic and dynamic loading conditions, fracture toughness properties

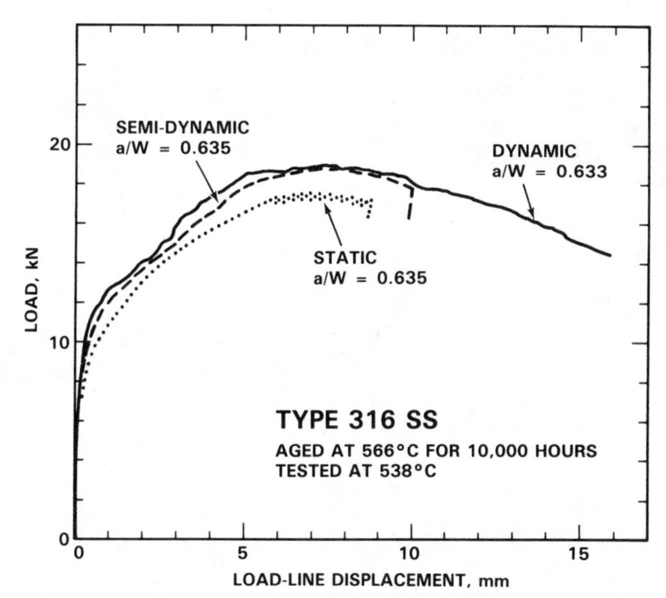

FIG. 9—*Typical static, semi-dynamic, and dynamic load-displacement records for aged 316 SS. These curves were selected on the basis of similar* a/W *ratios.*

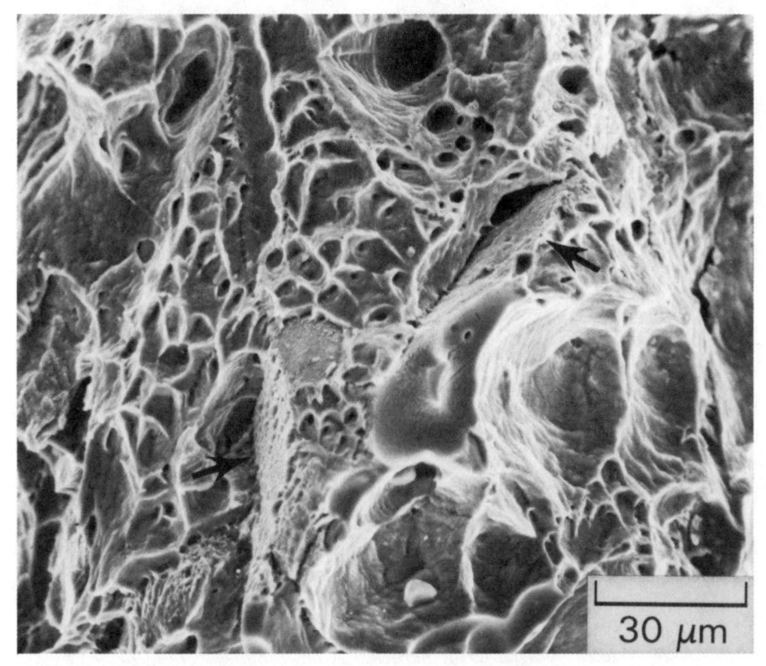

FIG. 10—*Typical fracture surface appearance for aged 304 SS. Dimple rupture was the primary fracture mechanism; however, limited evidence of intergranular cracking (denoted by arrows) was also observed. The fracture surface appearance for aged 316 SS was essentially the same as that for 304 SS.*

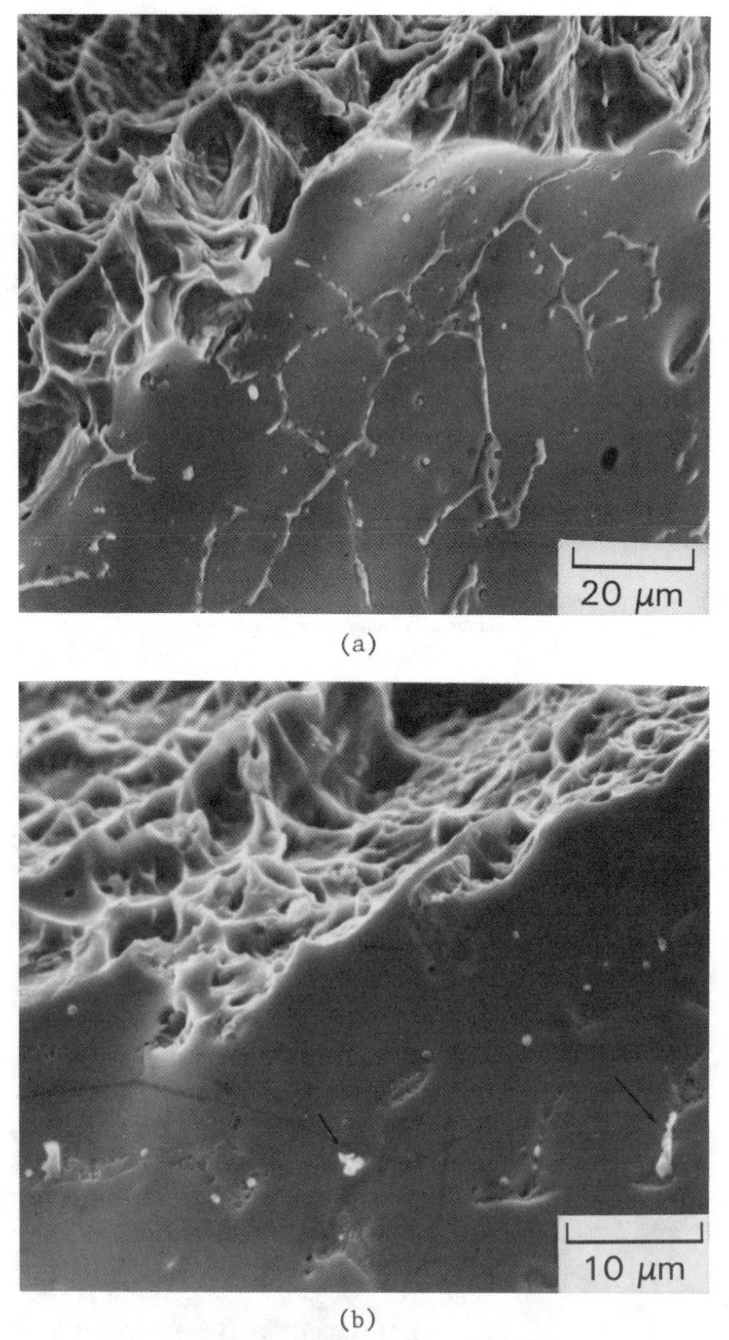

FIG. 11—*Metallographic/fractographic interfaces for 308 SS weld:* (a) *In the unaged condition, most of the microvoids were initiated by delta ferrite particles. A few dimples were nucleated by the spherical manganese silicide particles.* (b) *After aging, the delta phase transformation products, sigma phase (arrows) and $M_{23}C_6$ carbides, provided additional microvoid nucleation sites which reduced the average dimple size.*

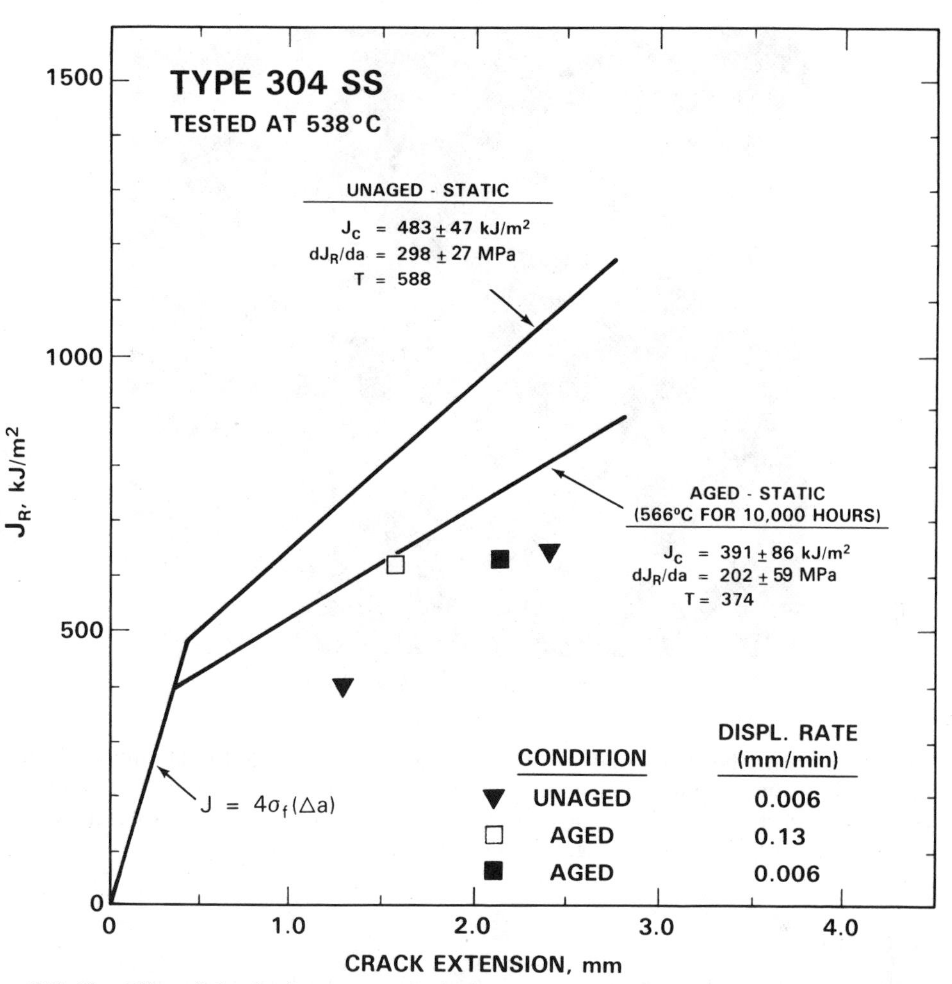

FIG. 12—*Effect of slow loading rates on the fracture resistance for unaged and aged 304 SS. The J_R curves for static loading conditions were taken from Fig. 7.*

for both unaged and aged materials were superior to properties under static loading for all materials except unaged 304 SS, where properties were independent of loading rate.

4. Blunt-notch Charpy impact energy was an overly pessimistic indicator of thermal aging effects for SS components containing cracks or crack-like defects. Fractographic evidence demonstrated that the large loss in CVN energy was associated with aging-induced microstructural changes that substantially decreased the energy required to initiate a crack from a blunt notch.

5. At very slow loading rates, creep crack growth along grain boundaries caused a large degradation in fracture toughness that could negate the exceptionally high fracture resistance exhibited by SS alloys.

Acknowledgments

This paper is based on work performed under U.S. Department of Energy Contract DE-AC06-76FF02170 with Westinghouse Hanford Company, a subsidiary of Westinghouse Elec-

FIG. 13—*Intergranular cracking at the very slow strain rate (0.006 mm/min).*

tric Corporation. The author wishes to gratefully acknowledge L. D. Blackburn for enlightening discussions. Appreciation is also extended to L. E. Thomas and B. Mastel for performing electron metallographic and fractographic examinations. The careful experimental work of W. D. Themar is greatly appreciated.

References

[1] *ASME Boiler and Pressure Vessel Code,* American Society of Mechanical Engineers, 1980.

[2] Horak, J. A., Sikka, V. K., and Raske, D. T., "Review of Mechanical Properties and Microstructure of Type 304 and 316 Stainless Steel After Long-Term Thermal Aging," *Proceedings of the IAEA Specialists Meeting on the Mechanical Properties of Structural Materials Including Environmental Effects,* IWGFR-49, Vol. 1, Chester, U.K., 1983, pp. 179–213.

[3] Spaeder, C. E., Jr. and Brickner, K. G., "Modified Type 316 Stainless Steel with Low Tendency to Form Sigma," *Advances in the Technology of Stainless Steels and Related Alloys, ASTM STP 369,* American Society for Testing and Materials, Philadelphia, 1964.

[4] Smith, G. V., Seens, W. B., Link, H. S., and Malenock, P. R., "Microstructural Instability of Steels for Elevated Temperature Service," *Proceedings of ASTM,* Vol. 51, 1951, pp. 895–917.

[5] Muchel, N. L., Ahlman, C. W., Wiedersum, G. C., and Zong, R. H., "Performance of Type 316 Stainless Steel Piping at 5000 psi and 1200°F," *Proceedings of the American Power Conference,* Vol. 28, 1966, pp. 556–568.

[6] Clark, C. L., Rutherford, J. J. B., Wilder, A. B., and Cordovi, M. A., "Metallurgical Evaluation of Superheater Tube Alloys After 12 and 18 Months' Exposure to Steam at 1200, 1350, and 1500°F," *Journal of Engineering for Power, Transactions,* American Society of Mechanical Engineers, 1962, pp. 258–288.

[7] Hoke, J., Eberle, F., and Wylie, R. D., "Embrittling Tendencies of Austenitic Superheater Materials at Elevated Temperatures," *Proceedings of ASTM,* Vol. 57, 1957, pp. 821–831.

[8] Hawthorne, J. R. and Watson, H. E., "Notch Toughness of Austenitic Stainless Steel Weldments with Nuclear Irradiation," *Welding Research Supplement,* 1973, pp. 255s–260s.

[9] Poole, L. K., "Sigma—An Unwanted Constituent in Stainless Weld Metal," *Metal Progress,* Vol. 65, June 1954, pp. 108–112.
[10] Smith, G. V., "Sigma Phase in Stainless—Part II," *The Iron Age,* Dec. 1950, pp. 127–132.
[11] Malcolm, V. T. and Low, S., "Sigma Phase in Several Cast Austenitic Steels," *Symposium on the Nature, Occurrence and Effects of Sigma Phase, ASTM STP 110,* American Society for Testing and Materials, Philadelphia, 1950, pp. 38–47.
[12] Binder, W. O., "Some Notes on the Structure and Impact Resistance of Columbium-Bearing 18-8 Steels after Exposure to Elevated Temperatures," *Symposium on the Nature, Occurrence and Effects of Sigma Phase, ASTM STP 110,* American Society for Testing and Materials, Philadelphia, 1950, pp. 146–164.
[13] Parker, T. D., "Strength of Stainless Steels at Elevated Temperature," *Source Book on Stainless Steel,* American Society for Metals, Metals Park, OH, 1976, pp. 80–99.
[14] Thomas, R. G. and Yapp, D., "The Effect of Heat Treatment on Type 316 Stainless Steel Weld Metal," *Welding Research Supplement,* 1978, pp. 361s–366s.
[15] Landermann, E. I. and Bamford, W. H., "Fracture Toughness and Fatigue Characteristics of Centrifugally Cast Type 316 Stainless Steel Pipe After Simulated Thermal Service Conditions," *Ductility and Toughness Considerations in Elevated Temperature Service,* MPC-8, G. V. Smith, Ed., American Society of Mechanical Engineers, New York, Dec. 1978, pp. 99–127.
[16] Mills, W. J., "Fracture Toughness of Aged Stainless Steel Primary Piping and Reactor Vessel," *Journal of Pressure Vessel Technology,* Vol. 109, 1987, pp. 440–448.
[17] Mills, W. J., James, L. A., and Williams, J. A., "A Technique for Measuring Load-Line Displacements of Compact Ductile Fracture Toughness Specimens at Elevated Temperatures," *Journal of Testing and Evaluation,* Vol. 5, 1977, pp. 446–451.
[18] Mills, W. J., "On the Relationship Between Stretch Zone Formation and the J-Integral for High Strain-Hardening Materials," *Journal of Testing and Evaluation,* Vol. 9, 1981, pp. 56–62.
[19] Paris, P. C., Tada, H., Zahoor, Z., and Ernst, H., "The Theory of Instability of the Tearing Mode for Elastic-Plastic Crack Growth," *Elastic-Plastic Fracture, ASTM STP 668,* American Society for Testing and Materials, Philadelphia, 1979, pp. 5–36.
[20] Mills, W. J., "Fracture Toughness of Stainless Steel Welds," *Fracture Mechanics: Nineteenth Symposium, ASTM STP 969,* American Society for Testing and Materials, Philadelphia, 1988, pp. 330–355.
[21] Chesnutt, J. C. and Spurling, R. A., "Fracture Topography Microstructure Correlations in the SEM," *Metallurgical Transactions,* Vol. 8A, 1977, pp. 216–218.
[22] Hahn, G. T., Hoagland, R. G., and Rosenfield, A. R., "Influence of Metallurgical Factors on the Fast Fracture Energy Absorption Rates," *Metallurgical Transactions,* Vol. 7A, 1976, pp. 49–54.
[23] Carlsson, J., Dahlberg, L., and Nilsson, F., "Experimental Studies of the Unstable Phase of Crack Propagation in Metals and Polymers," *Proceedings of an International Conference on Dynamic Crack Propagation,* 1972, Noordhoff International Publishing, Leyden, the Netherlands, pp. 165–181.
[24] Mills, W. J., "Heat-to-Heat Variations in the Fracture Toughness of Austenitic Stainless Steels," *Engineering Fracture Mechanics,* Vol. 30, 1988, pp. 469–492.

Fatigue Crack Propagation

Stefanie E. Stanzl,[1] *Maximilian Czegley,*[2] *Herwig R. Mayer,*[1] *and Elmar K. Tschegg*[2]

Fatigue Crack Growth Under Combined Mode I and Mode II Loading

REFERENCE: Stanzl, S. E., Czegley, M., Mayer, H. R., and Tschegg, E. K., **"Fatigue Crack Growth Under Combined Mode I and Mode II Loading,"** *Fracture Mechanics: Perspectives and Directions (Twentieth Symposium), ASTM STP 1020,* R. P. Wei and R. P. Gangloff, Eds., American Society for Testing and Materials, Philadelphia, 1989, pp. 479–496.

ABSTRACT: The influence of static Mode I and static Mode II loading on Mode I fatigue crack growth has been studied with 12 weight % chromium steel.

Static Mode I was superimposed as a tensile load on fatigue-loading with mean stress equal to zero, thus resulting in R-values of -1, -0.33, 0, $+0.25$, and $+0.5$. For these values, fatigue-crack growth curves were determined in the threshold regime with crack growth rates typically between 3×10^{-13} and 5×10^{-9} m/cycle.

In addition, a constant static Mode II load was superimposed on the same above-mentioned Mode I load. The resulting crack growth rates were found to vary systematically with crack length and load ratio. For R-ratios ≤ 0 enhanced fatigue crack growth rates were observed at crack lengths up to a defined value. After a certain crack length was attained, slightly reduced crack growth rates were detected. For high R-ratios, long as well as short fatigue cracks grew slower than under pure Mode I fatigue loading. The results are discussed in light of the crack closure phenomenon and the fracture morphology, which was studied in a scanning electron microscope (SEM).

In the Appendix, a basic procedure for determining stress-intensity values for high-frequency (21 kHz) fatigue loading is given in comparison with conventional loading.

KEY WORDS: mixed mode, combined mode, multiaxial loading, Mode II loading, near-threshold fatigue crack growth, crack closure, ultrasonic fatigue

The majority of crack growth studies has been performed in the crack opening mode (Mode I) during the last 20 years. In service, however, many components are not stressed in pure Mode I but in mixed-mode (multiaxial) polymodal loading condition, as in Mode I plus Mode II, Mode I plus Mode III, or a combination of all three modes. Examples for these cases are drive shafts, which are stressed in bending and torsion (a solid shaft is stressed in Mode I plus Mode III, whereas a tubular shaft is stressed in Mode I plus Mode II) or pre-cracked turbine blades, which are stressed in Mode III owing to centrifugal force and in Mode I because of gas and steam pressure.

These mixed-mode loading situations must be taken into account by the design engineer, and it must be remembered that material data that were obtained by laboratory tests under pure Mode I conditions cannot be applied easily to mixed-mode conditions. For these cases, criteria, rules, and laws have to be worked out and verified by experiments.

[1] Associate professor and research associate, respectively, University of Vienna, Austria, Institute for Solid State Physics, Baltzmanng 5, A-1090, Vienna, Austria. Professor Stanzl is presently with the University for Agriculture, Vienna, Austria, Institute for Meteorology and Physics, Türkenschanzstr. 18, A-1180, Vienna, Austria.

[2] Research associate and associate professor, respectively, Technical University of Vienna, Austria, Institute for Technical and Applied Physics, Karlspatz 13, A-1040, Vienna, Austria.

In recent times, more work has been published on the field of mixed-mode loading, which is interesting for scientists as well as engineers. A good literature review about these topics is given in the *Proceedings* of the Multiaxial Fatigue Conference, edited by Miller and Brown [*1*], where the influence of all combinations of the three loading modes on long and short cracks behavior is discussed.

Relatively few works exist on crack growth in metals under Mode II loading conditions. Superposition of static Mode II to static Mode I loading has been studied by Yokobori et al. [*2*] and Kordisch et al. [*3*], among others, and superposition of alternating Modes I and II (mixed-mode fatigue) by Hua et al. [*4,5*], Tanaka [*6*], Yokobori et al. [*7*], Truchon et al. [*8*], and Smith and Pascoe [*9*]. Otsuka et al. [*10*] investigated Mode II fatigue loading with superimposed static Mode I load and Hourlier and Pineau [*11–13*] examined the superposition of static Mode II loading to Mode I fatigue loading in tubular specimens. Apparently, only Pook studied fatigue crack propagation under pure Mode II loading [*14*].

Two aspects in particular have been studied theoretically and experimentally in the literature: crack path direction and crack growth rates. Several criteria have been proposed for the description of combined Mode I and II loading. The most important of these are

1. maximum-stress criterion [*15*],
2. the strain-energy density criterion [*16*], and
3. the maximum-energy-release-rate criterion [*17*].

In applying these criteria to static load conditions, Yokobori et al. [*2*] discovered that the experimentally found crack growth directions were far from those predicted by the criteria. Another experimental result is reported by Tanaka [*6*], who noticed that the direction of crack growth of an inclined crack is roughly perpendicular to the applied tensile axis for fatigue loading at stress intensities slightly above the threshold value for non propagation, whereas the crack grows in the same direction of the initial crack under stress ranges 1.6 times larger than the threshold values. An important aspect that could explain the high scatter of crack growth angles in the literature is discussed by Kordisch et al. [*3*]. They show that after the onset of crack propagation the complete stress-strain field at the crack tip will change, thus causing a change in the fracture path direction.

For cyclic Mode I plus static Mode II loading, Hourlier et al. [*11–13*] found that agreement of experimental results and theoretical calculations for the crack growth direction varied. The crack path direction is governed not only by the amplitude of the stress intensity factor, but also by its maximum value. Therefore, in their analysis especially the R-ratio sensitivity of fatigue cracking under pure Mode I loading is considered as an important parameter.

Concerning crack growth rates, Hua et al. [*4,5*] established that in the threshold regime, not only the Mode I displacement but also the Mode II component plays an important role for mixed Mode II and Mode I fatigue loading. The stress intensity factor required for crack growth and likewise the threshold cyclic stress intensity decreases with increasing $\Delta K_{II}/\Delta K_{I}$ ratios. Similar results are reported by Yokobori et al. [*7*] and by Otsuka et al. [*10*], who detected seven to eight higher crack growth rates for Mode II fatigue loading than for Mode I fatigue crack growth. As an explanation, one may consider the increase of the size of the crack tip reversed plastic zone owing to the Mode II component. In contrast to this crack growth enhancing effect, rubbing of the fracture surfaces is assumed to explain observations of crack growth retardation.

In the present paper fatigue crack propagation in the near-threshold regime was studied for combined Mode I cyclic and Mode II static loading.

Material and Experimental Procedures

Fatigue crack growth studies were performed for time-saving purposes with a newly developed universal-ultrasound testing machine, which is schematically drawn in Fig. 1. The equipment is described in detail in Ref *18*; therefore, in this paper, only the most important facts are reported. In principal the machine is a combination of a conventional hydraulic machine, tensile machine, and an ultrasound machine [*19,20*]. An ultrasonic transducer is mounted to the crosshead of the machine frame. The 21-kHz vibrations of the transducer are transmitted via coupling pieces to the specimen. Specimens vibrate in resonance and therefore must have a length equal to the half wave length of the longitudinal ultrasound wave. Maximum displacement is attained at the ends of the specimens and the maximum strain and stress in their center. At the second end of the specimen another coupling piece is attached which goes to the load cell and to the actuator of the hydraulic machine. Ultrasonic vibrations are transmitted by the ultrasound transducer to the two coupling pieces and the specimen; however, they are neither transmitted to other parts of the machine frame nor damped by these.

Defined static and dynamic tension or torsion loads that are measured with the load cell may be applied to the specimen by the actuator. The actuator has a floating piston and therefore may be rotated during tension or compression loading with no mentionable friction. Thus it is possible to apply a constant torque to the specimen with deadweights and with

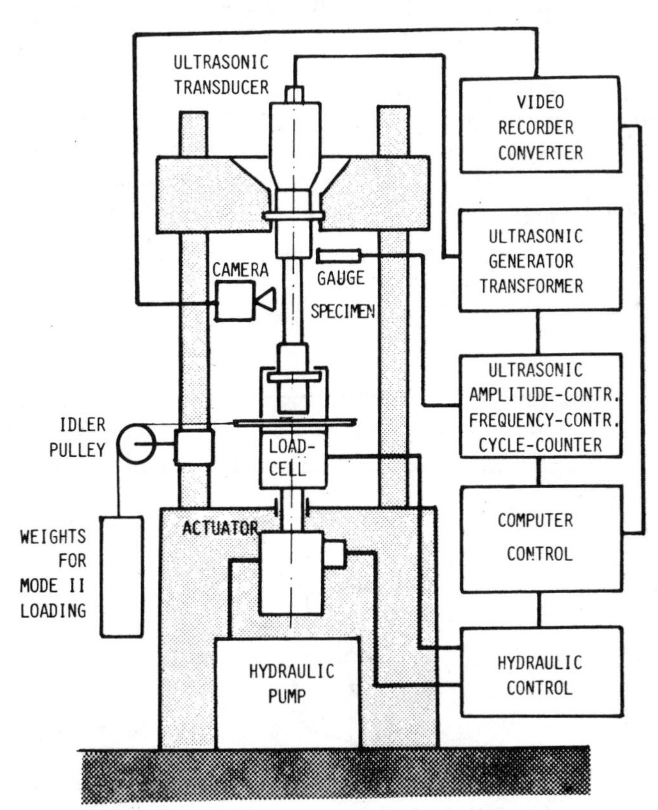

FIG. 1—*Universal ultrasound testing equipment consisting of ultrasound fatigue machine, conventional hydraulic, and tensile machine.*

the help of an idler pulley and a rotating disc that is fixed at the actuator and the load cell; static and dynamic tension or compression loading of the specimen are not influenced by this equipment in any way.

The whole testing system is fully computer-controlled. A computer supplies selected values for the ultrasonic control and for the hydraulic system, and it registers the actually performed ultrasonic displacement amplitudes as well as tension and compression values delivered by the hydraulic system via the load cell.

Crack propagation is observed with video equipment, as described earlier [*20*] with number of cycles, amplitudes, strain measurement, frequency, etc., introduced into the recording, so that observation and evaluation after the test is possible. In addition, crack lengths are measured by the potential drop technique. Thus fully automated crack growth studies, like ΔK-constant tests can be performed with this equipment.

Ultrasonic amplitudes or strains are controlled with a new electronic system, which includes a frequency control, so that these values are obtained with an accuracy equal to or better than 99%. The maximum deviation from the pre-given values produced by the hydraulics and measured by the load cell is likewise 1% at most. Two or more strain gages were attached to the specimen diameter in order to control the applied strains during crack propagation. By this procedure, the high stability of all load values was confirmed, and it was also ensured that static and cyclic loads were uniformly distributed around the whole diameter of the tubular specimens.

With the described universal machine, Mode I displacements can be applied statically or dynamically to specimens and a static or cyclic Mode II or Mode III load may be superimposed.

For the measurements of this work, tubular specimens with a length of 125 mm, a diameter of 18 mm, and thickness of 2 mm were used (Fig. 2.) The specimens were joined to the ultrasonic coupling pieces with threads screwed into both ends of the specimens. For tests measuring crack growth under (a) cyclic Mode I with a superimposed static tension load and (b) static and dynamic Mode I loading with a superimposed static Mode II loading, a

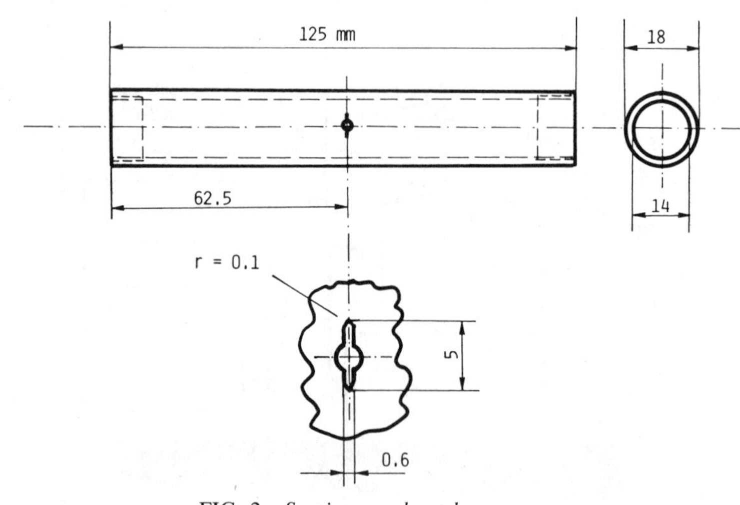

FIG. 2—*Specimen and notch geometry.*

notch was introduced. This was a hole through the specimen thickness plus a slit with a radius of 0.1 mm (Fig. 2).

Testing material was 13Cr steel (AISI 420, X20Cr13), which is usually used for steam turbines. Composition and mechanical properties are summarized in Table 1. Specimens were machined from round bars 25 mm in diameter.

Specimens were polished in the vicinity of the notch and crack initiation, and propagation were observed at ×130 magnification with a video camera (crack length measurement precision approximately 10 μm). For crack growth measurement, crack increments of approximately 150 μm were used. Crack propagation was considered to have stopped, when crack growth of at least 20 μm was not observed after at least 1×10^8 cycles.

Stress intensity factors K_I and K_{II} were calculated according to Erdogan and Ratwani [*21,22*] with

$$K_I = \sigma\sqrt{\pi \cdot a} \cdot Y \tag{1}$$

$$K_{II} = \tau\sqrt{\pi \cdot a} \cdot Y \tag{2}$$

where

σ = tensile stress,
τ = shear stress,
a = crack length including notch depth,
Y = geometry factor.

The geometry factor was determined according to Tada et al. [*23*] with a fitted polynomial function. In the following, c represents crack length without notch depth, d, that is, $a = c + d$.

For ultrasonic control purposes, the displacement amplitude at the specimen ends is used and also serves for calculating strain and stress in the center of the specimen [*20*]. In some cases, strain-gage measurements instead of displacement measurements serve as control signals. As shown in detail in the Appendix, crack increments modify the linear relationship between displacement at the end and stress in the center of a specimen. This has to be considered for calculating stresses. In general, the following equation is valid:

$$\sigma = \frac{\pi \cdot E}{l_V} \cdot A \cdot Z \tag{3}$$

TABLE 1—*Composition, heat treatments, and room temperature mechanical properties of X20Cr13 steel.*

Weight Percent					
C	Si	Mn	P	S	Cr
0.19	0.42	0.39	0.028	0.005	13.1
Austenitizing at 940°C 1 Std, followed by oil quenching and tempering for 1 h at 680°C					
Tensile yield strength monotonic	550 MPa				
Ultimate tensile strength	870 MPa				

where

E = Young's modulus,
A = displacement amplitude,
l_V = resonance length of the specimen, and
Z = factor that takes into account the changing specimen compliance due to crack extension.

Experimental Results

Influence of Mean Stress on Mode I Fatigue Crack Propagation

The influence of static tensile loads on fatigue crack propagation under Mode I loading with 21 kHz is shown in Figs. 3 and 4. In Fig. 3, the amplitude of the stress intensity value ΔK_I is plotted on the abscissa. With increasing R-values (mean stresses), the $(\Delta c/\Delta N)$ versus ΔK_I curves are shifted towards smaller ΔK_I values, as expected. For the measurements, crack increments of approximately 150 μm have been evaluated. The stress intensity values were reduced in steps of 7% to obtain reproducible near-threshold values. In the very low threshold regime, however, these steps seemed to be too high; therefore, the stress intensity values were reduced by only 5% until crack arrest occurred (no crack growth of at least 20 μm during at least 1×10^8 cycles). These values were considered as thresholds and are characterized by dots with arrows at about 5×10^{-13} m/cycle in Figs. 3 and 4. Afterwards, the load was again increased stepwise by 10% (increase of $\Delta K \geq 10\%$). Reproducible

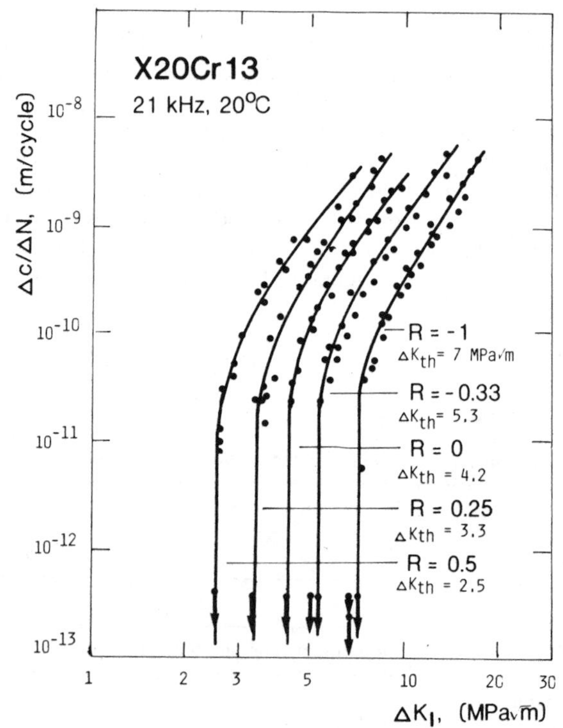

FIG. 3—*R-ratio influence on Mode I fatigue crack growth with* ΔK_I *on the abscissa* ($\Delta K = K_{max} - K_{min}$ *for all* R-*values*).

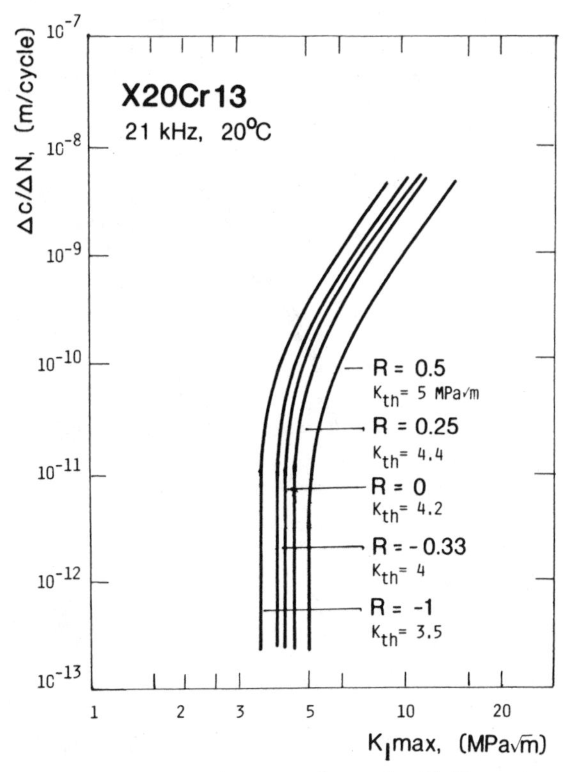

FIG. 4—R-*ratio influence on Mode I fatigue crack growth with* K_{Imax} *values on the abscissa.*

curves were obtained for different crack lengths (starting from crack length $c = 0.5$ mm up to 7 mm), though a slight sequence-effect was observed for the very low threshold regime. The small degree of scatter of the data points in Fig. 3 demonstrates the high accuracy of the control-system of our new equipment.

In Fig. 4, K_{max} values are plotted on the abscissa. The resulting $(\Delta c/\Delta N)$ curves for different R-values are rather close together with thresholds of K_{max} between 3.5 MPa$\sqrt{m}$ (for $R = -1$) and 5 MPa$\sqrt{m}$ (for $R = +0.5$). Plotting of individual data points has been abandoned in this figure for the sake of clarity.

Cyclic Mode I and Static Mode II Loading

A static Mode II load was superimposed on Mode I fatigue loading at R-values of -1 and $+0.5$ so that the K_{II} level was kept constant. For $R = +0.5$, superposition of Mode II loading resulted in slightly reduced crack growth rates when compared to pure Mode I crack growth; no influence of the crack length was observed (Fig. 5). For $R = -1$ on the other hand, a pronounced influence of the crack length was detected, as shown in Fig. 5. After accelerated crack propagation for crack lengths up to about 3 mm, the crack growth rate dropped beneath the Mode I curve with increasing crack length.

Therefore, additional tests were performed with constant values for cyclic ΔK_I, constant K_{Imax} and constant K_{II} values during the whole test for all crack lengths, ranging from approximately 0.5 to 7 mm. The experiments were performed for R-values of -1, 0, and

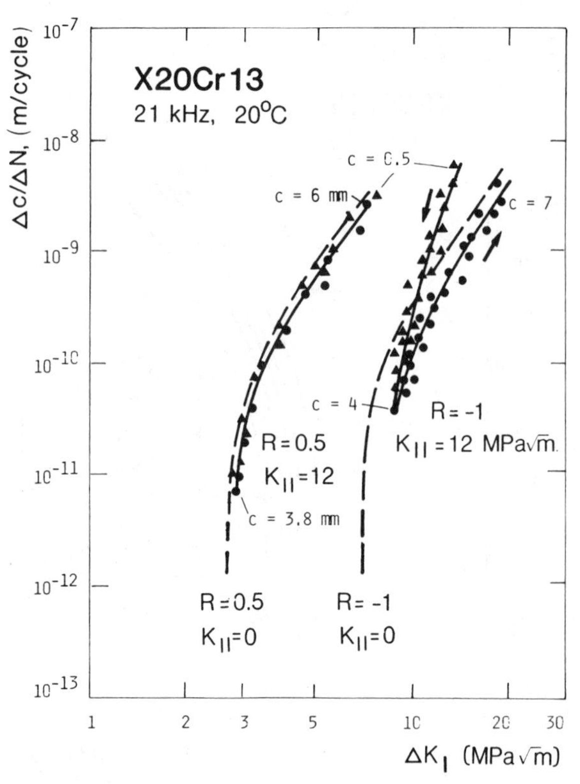

FIG. 5—*Influence of superimposed static Mode II load on crack propagation under Mode I fatigue loading at* R *values of −1 and +0.5.*

+0.5. K_{II} was 12 MPa$\sqrt{m}$ in all cases. For the choice of ΔK_I, a ΔK_I-value resulting in the same crack growth rate of approximately 7×10^{-10} m/cycle for all three R-ratios without superimposed Mode II was assumed as the guiding value. Thus, ΔK_I was 12 MPa$\sqrt{m}$ for $R = -1$, 7 MPa$\sqrt{m}$ for $R = 0$, and 4.7 MPa$\sqrt{m}$ for $R = +0.5$. The resulting crack growth rates for these constant applied stress intensities (sums and ratios also constant) are plotted in Fig. 6. For $R = +0.5$, a constant crack growth rate without any influence of the crack length is found. This crack growth rate is slightly lower than for cyclic Mode I loading without superimposed Mode II (its value is indicated by arrow and dash-dotted line for $K_{II} = 0$).

In contrast, higher initial crack growth rates for shorter crack lengths are observed for $R = 0$ and $R = -1$. The rates drop with increasing crack length towards a constant crack growth rate below the value for $K_{II} = 0$. The crack length where $(\Delta c/\Delta N)$ becomes constant is approximately 3 mm for $R = 0$ and approximately 5 mm for $R = -1$. However, no influence of crack length (in the range of approximately 0.5 to 7 mm) was detected for pure Mode I cyclic loading without superimposed static Mode II loading at all three mean values ($R = +0.5, 0, -1$). The resulting constant crack growth rates are characterized by horizontal dash-dotted lines in Fig. 6. These results demonstrate that superposition of static Mode II loads to cyclic Mode I fatigue cracking causes crack growth acceleration for relatively short cracks with their length depending on the loading condition, and reduced crack growth rates for long cracks in all loading conditions.

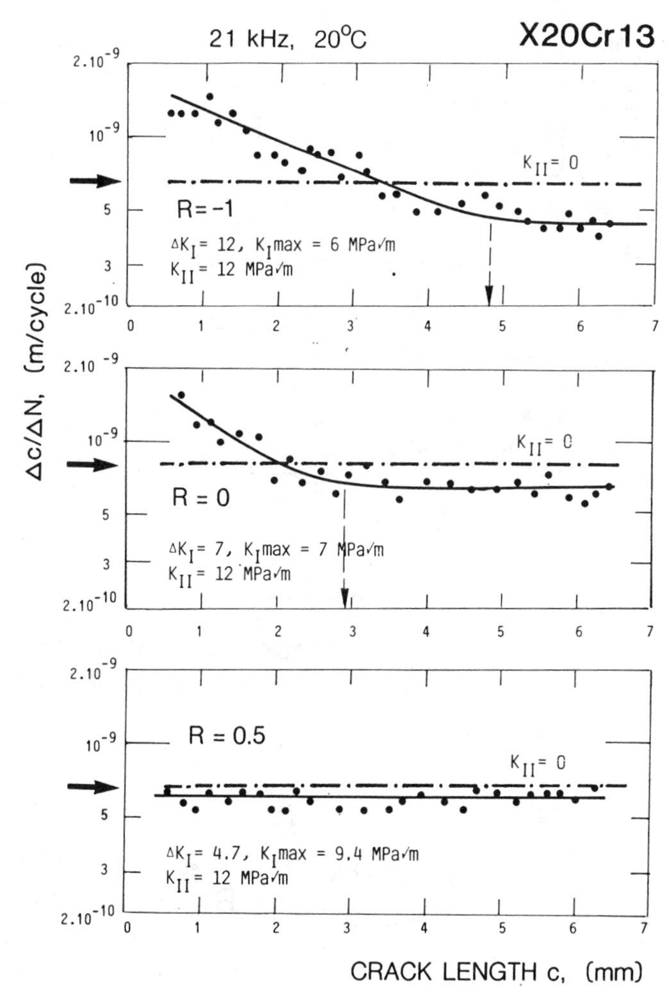

FIG. 6—*Fatigue crack growth rates at constant values for cyclic* ΔK_I, *static* K_{Imax} *and static* K_{II}. *Influence of crack length for different* R-*ratios.*

Similar retarding effects were obtained for a smaller superimposed static K_I value of 3 MPa$\sqrt{\text{m}}$.

Additional tests were conducted in order to investigate whether or not the same crack growth rate was obtained after Mode II testing and subsequent removal of the static Mode II load as for pure Mode I cyclic loading without superimposed Mode II. The results are shown in Figs. 7 and 8 for $R = +0.5$ and $R = -1$. After removal of static Mode II loading, crack arrest occurs first and afterwards the same high crack growth rate as for pure Mode I fatigue loading is indeed obtained. The increase is somewhat higher for $R = -1$ than for $R = +0.5$.

Having noticed that crack growth acceleration or retardation under a superimposed static shear load was governed by the crack length, and the above mentioned results led to the assumption that the newly created fracture surface asperities could be responsible for crack

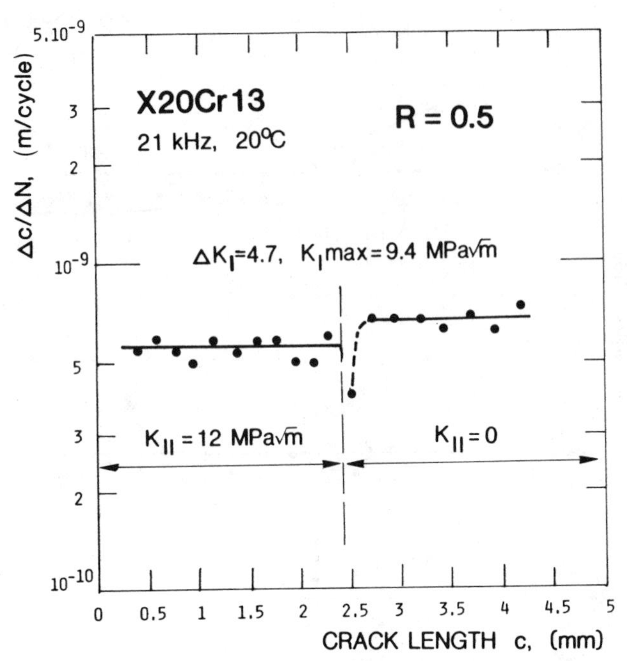

FIG. 7—*Fatigue crack propagation rates after removal of superimposed static Mode II loads;* R = *+0.5.*

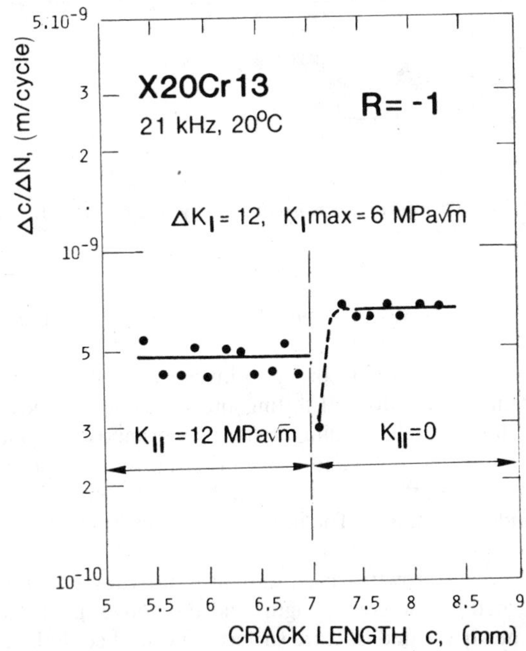

FIG. 8—*Fatigue crack propagation rates after removal of superimposed static Mode II loads;* R = *−1.*

growth retardation with increasing fracture surface area. Therefore, in another test with $R = -1$, $\Delta K_I = 12$ MPa$\sqrt{m}$ and $K_{II} = 12$ MPa$\sqrt{m}$, the experiment was interrupted when a reduced constant crack growth rate of approximately 5×10^{-10} m/cycle was attained at a crack length of 7 mm (Fig. 9). The fracture surfaces were removed with a fine saw, then the experiment continued with the same nominal stress intensities as before. After this procedure, a crack growth rate of 10^{-9} m/cycle was found, which is approximately the same value as for a crack length of 0.5 mm. The crack growth rate again decreased gradually to about 5×10^{-10} m/cycle with increasing crack length. The slope of the decreasing crack growth curve was found to be roughly the same as for the short cracks at the beginning of the test (see Fig. 6).

Crack path orientations were determined for all cases of superimposed Mode II loads. In general, the average deviation from the direction perpendicular to the tensile axis was not more than 5 deg.

Fractography

The fracture surfaces were examined in a scanning electron microscope (SEM) in order for the eventual crack closure effects due to static Mode II loading to be localized or quantified. Note that the fracture morphology is similar in principle for all loading conditions examined in this study (Figs. 10 and 11). Estimating the fracture surface roughness reveals values up to about 10 μm.

In addition, no substantial difference in flattening of the fracture surface asperities for different R-values could be detected; neither the amount nor the degree of flattening is essentially different, though one might expect more and stronger flattening for $R = -1$ than for $R = +0.5$.

Where static Mode II loads were superimposed, clear shear features can be recognized, as shown in Fig. 10, that comes from a test in which a Mode II load with $K_{II} = 12$ MPa$\sqrt{m}$ was superimposed on fatigue loading with $R = +0.25$. In addition, an enhanced amount of abrasion debris can be seen. Figure 10 gives evidence of extensive surface roughness flattening. During crack growth measurement, pronounced crack retardation was observed in this specimen area. Obviously, part of the applied energy was consumed for deformation of the fracture surface asperities, and it should be assumed that extensive crack closure became effective.

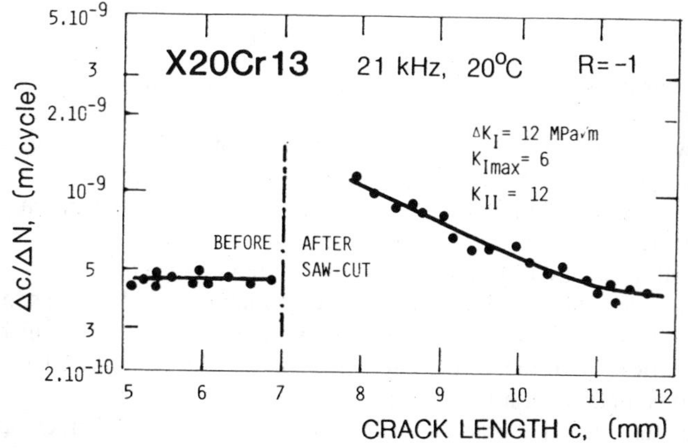

FIG. 9—*Fatigue crack growth rates under combined Mode I fatigue plus static Mode II loading, after removal of fracture surfaces (generated in a previous fatigue test) by a fine saw-cut.*

FIG. 10—*Fracture surface after Mode I fatigue cracking with* $\Delta K_I = 6.2$ $MPa\sqrt{m}$ $K_{II} = 12$ $MPa\sqrt{m}$, *and* $R = +0.25$.

Discussion

The curves in Fig. 5 demonstrate two important results for Mode I fatigue cracking with superimosed Mode II loads:

1. The crack growth rate is influenced by the mean load (static Mode I load).
2. Crack growth propagation is usually slightly retarded by superposition of a Mode II load; it is accelerated only if the *R* ratio is less than or equal to zero and only as long as the crack length does not exceed a defined value.

With these results in mind, it was assumed that crack closure effects play an important role. For more detailed discussion of this question, the mean-stress dependence of the crack growth rates without superimposed shear load seemed of interest. Plotting $(\Delta c/\Delta N)$ versus ΔK or K_{max} in Figs. 3 and 4 shows that fatigue crack growth of this material under pure Mode I loading is governed less by the range, ΔK, than by the maximum values, K_{max}, of cyclic stress intensities.

Another result that may yield some information about an eventual crack closure effect is the crack path angle. The experimentally observed crack growth directions forming an angle θ of 3 to 5 deg with the normal to the tensile axis are far from those which are predicted by several theories. In Table 2, the predicted values are summarized for three loading conditions and compared with the measurements.

A theory that assumes that crack growth depends only on Mode I crack opening and that crack growth rates $(\Delta c/\Delta N)$ become maximum for specified angles θ, reveals angles between 43 and 48 deg for the discussed loading conditions. The calculations were performed using

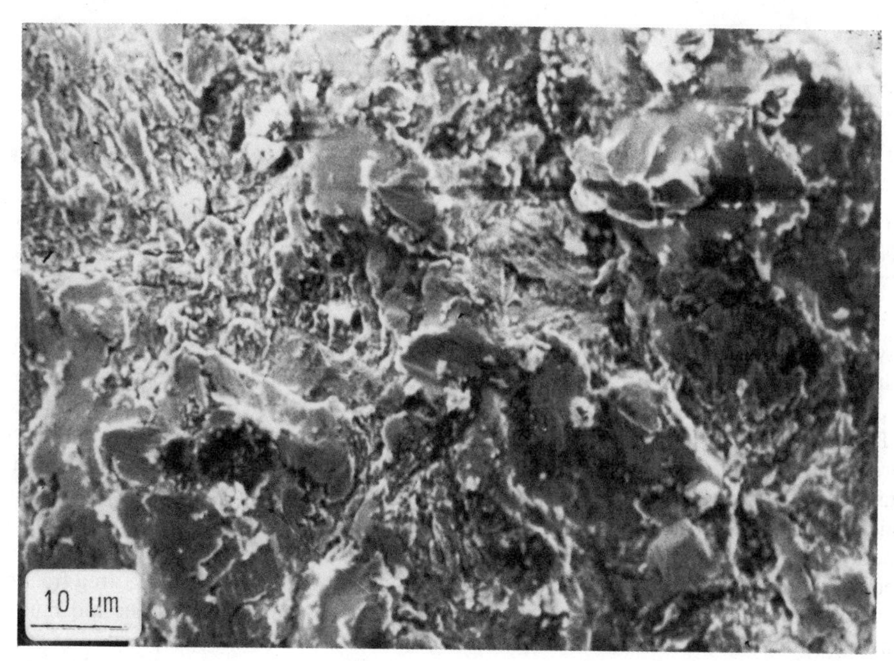

FIG. 11—*Fracture surface after Mode I fatigue cracking with* $\Delta K_I = 3$ *MPa*$\sqrt{m}$, $K_{II} = 12$ *MPa*$\sqrt{m}$, *and* R = *0.5.*

the equations [*11*]

$$\Delta K_\text{I}\,(\theta) = \Delta K_\text{I}\,(\theta = 0)\cos^3\theta/2 \tag{4}$$

$$K_\text{Imax}\,(\theta) = \cos^2\theta/2\,(K_\text{Imax}\cos\theta/2 + 3\,K_\text{II}\sin\theta/2) \tag{5}$$

If the angle is calculated according to the maximum normal stress theory, θ-values of 56 to 61 deg are found. The angles θ for crack growth propagation are obtained, setting $\tau = 0$ in equation [*4*]

$$\tau = \frac{K_\text{I}}{\sqrt{2\tau r}}\left(\frac{1}{4}\sin\theta/2 + \frac{1}{4}\sin 3\theta/2\right) + \frac{K_\text{II}}{\sqrt{2\tau r}}\left(\frac{1}{4}\cos\theta/2 + \frac{3}{4}\cos 3\theta/2\right) \tag{6}$$

Table 2 demonstrates that the experimental results for the crack path directions of this work cannot be explained by the above-mentioned theoretical considerations.

Additional results that cannot be explained with the above-mentioned models are those of Fig. 6. The main results of this figure are (partly shown in Fig. 5 already)

1. crack growth acceleration for rather short cracks at R-ratios ≤ 0,
2. crack growth retardation for longer cracks at $R = -1$ and 0, and for all crack lengths at $R = +0.5$, and
3. reduced constant crack growth rates for long enough cracks.

TABLE 2—*Crack path angles θ (deviation from direction normal to tensile axis); theoretical predictions and measurements.*

R	K_{II}/K_{Imax}	θ From $\Delta c/\Delta N$ = max in Eqs 4, 5	θ With σ_N = max from Eq 6	θ Observed
−1	2	48 deg	61 deg	5 deg
0	1.7	46 deg	60 deg	5 deg
+0.5	1.3	43 deg	56 deg	3 deg

In view of these partly surprising results, it was supposed that fracture-surface-roughness-induced crack closure [24,25] played an important role. The following explanation seems to describe the observed phenomena:

Two opposing mechanisms are assumed to govern fatigue crack propagation under superimposed shear loads. First, Mode II loading increases the resulting effective stress intensity by increasing the plastic zone size. This should result in higher crack tip displacements, which facilitate fatigue crack growth (a similar argumentation is given in Ref *4* for mixed-mode loading). With increasing crack length, however, this effect is opposed by another mechanism that is assumed to be related with the rough fracture surfaces. The area fraction of rough fracture surfaces increases causing an increasing amount of friction and mutual support of the fracture surfaces. This results in a reduction of the crack promoting cyclic stress intensity to a lower effective value at the crack tip, in a similar manner as discussed in Ref *26*.

It is worth noting that initial acceleration does not take place for $R = +0.5$ and is the more pronounced the lower R is. For us to understand this effect, it is assumed that the compressive part of cycling causes sufficient flattening of the rough fracture surfaces at low and negative R-ratios, removing the barriers for the Mode II displacements. As soon as the area of rough fracture surfaces becomes larger with increasing crack length or, similarly, if the compressive part is small enough in tests with higher R-ratios, the applied compressive load components obviously are not great enough to flatten the impeding asperities effectively.

The results of Figs. 7 and 8 give further evidence for the effect of a retarding crack closure mechanism. The initial decrease of $(\Delta c/\Delta N)$ after removal of the applied Mode II loads, in tests with constant ΔK_I and K_{Imax} values, may be attributed to the existence of the previously generated relatively large Mode II plastic zone. After the crack has overcome this area, a higher rate for pure Mode I fatigue crack growth is obtained. Obviously, the relative mismatch between the fracture surfaces is now removed and the asperities do not mutually support each other any longer, thus reducing the cyclic stress intensity at the crack tip. Removing fracture surface asperities by a saw cut (Fig. 9) acts in a similar manner.

The SEM results in Figs. 10 and 11 demonstrate that the fracture surface roughness of the tested specimens is in the range of approximately 10 μm. The crack-tip opening displacement (CTOD) values, on the other hand, are approximately one or two orders of magnitude lower than the fracture surface roughness, whereas the crack opening displacement (COD) values are in the same range as the surface roughness. Therefore, reduction of fatigue crack growth rates under superimposed static Mode II loads is not surprising, since the relative mismatch of such relatively high surface asperities certainly reduces the cyclic stress intensity effective at the crack tip. A more detailed quantitative treatment of the influence of Mode II displacements on ΔK_I reduction and crack growth retardation will be given when more data on Mode II displacements become available.

The attainment of constant (reduced) crack growth rates (Fig. 6) at a certain crack length that depends on the R-ratio may be explained in the following way; in any case, only a limited fracture surface area will be effective for reducing ΔK_I and K_{II}. This area is a priori smaller at high R-ratios than at low ones. Therefore, the reduction of $(\Delta c/\Delta N)$ to a constant value is obtained at a smaller crack length with a high R-ratio (for example, $+0.5$) than with lower and negative R-ratios.

The observed effect of initial fatigue crack acceleration under superimposed static Mode II loading and subsequent retardation at greater crack length has not been observed to the authors knowledge until now. The reasons for this are

1. Most works in the literature have been performed at R-ratios equal to or greater than zero.
2. The effect is especially effective in the near-threshold regime. This regime has been studied in this work, whereas most other results were obtained at higher crack growth rates.

Conclusions

Superposition of static Mode II loads on Mode I fatigue loads may accelerate or retard fatigue crack propagation. What will actually occur depends on the mean Mode I load level. At higher mean loads, lower crack growth rates than under pure Mode I fatigue loading result that are caused by the relative mismatch between the rough fracture surfaces. This mismatch causes friction and mutual support of the fracture surface asperities and thus leads to some reduction of the effective stress intensity value acting at the crack tip. If, however, these retarding asperities are reduced or removed, enhanced fatigue crack growth values occur. Such enhancements are observed if

1. a compressive component during fatigue loading is effective, resulting in flattening the fracture topography (at R-ratios typically less than or equal to zero);
2. cracks are short enough so that friction and mutual support of the fracture surface asperities cannot become effective; or if
3. fracture surface roughness is reduced or removed by some procedure such as mechanical or chemical degradation.

Crack growth retardations for Mode I fatigue loading plus superimposed static Mode II loads are most pronounced in the near-threshold regime where the CTOD values are smallest and thus the ratio of these values and the surface roughness minimum.

Crack growth acceleration and retardation may be similarly expected for other cases of complex or mixed-mode loading.

APPENDIX

One special feature of the ultrasonic method is that a specimen of constant cross section is not stressed uniformly along the whole specimen length. Strain and stress maximum values are located in the center of the specimen length, where the displacement amplitude is zero (vibration node). The ends of the specimen are antinodes of vibration (maximum of displacement amplitude), and strain and stress are zero there.

Stress values in the center of a specimen cannot be measured directly. Therefore, they are usually determined by measuring the displacement amplitudes at the ends and are then calculated with the equation for a standing wave according to

$$\sigma = \frac{\pi \cdot E}{l_r} \cdot s \tag{7}$$

where

σ = stress in the specimen center,
E = Young's modulus,
s = displacement amplitude of the specimen without crack, and
l_r = resonance length of the specimen, which is given by

$$l_r = \frac{1}{2 \cdot f} \sqrt{(E/\rho)} \tag{8}$$

where ρ is equal to density of specimen material and f to frequency. Usually the displacement amplitude is used for control purposes, therefore these tests, as in this work, are displacement controlled.

If a crack is present in the specimen center, the displacement amplitude is given by $s + s^*$. The quantity s^* comes from the crack-induced increase of the specimen compliance. (After a load is attached to a specimen containing a crack, it becomes longer by an additional amount of s^*). Thus a displacement amplitude $(s + s^*)$ is measured with a specimen that contains a crack, with s^* increasing with increasing crack length.

If σ is determined for fracture mechanical calculations with the measured displacement amplitude from Eq 7, this calculation has to take account of this additional displacement amplitude s^*. This has to be introduced into Eq 7 by multiplying it by a factor

$$Z = \frac{s}{s + s^*} \tag{9}$$

Equation 7 then becomes

$$\sigma = \frac{\pi \cdot E}{l_r} \cdot Z \cdot (s + s^*) \tag{10}$$

For $s^* = 0$ (specimen without crack), Z is equal to 1 and for $s^* > 0$ (specimen with a crack), Z is less than 1, which shows that the factor Z becomes important for long cracks.

The factor Z can be calculated in the following way: s of a specimen with constant cross section and without crack is given by

$$s = \frac{1}{2} \int_0^{l_r} E \cdot \frac{F}{A} \cdot \sin kx \cdot dx = \frac{l_r}{\pi} \frac{F}{E \cdot A} \tag{11}$$

where $F = \sigma A$ is load, and A = cross section of specimen. One can calculate s^* from the compliance C at the crack according to

$$s^* = \frac{1}{2} F \cdot C \tag{12}$$

The compliance C is given by integrating the following differential equation over the whole crack length a

$$\frac{1}{2 \cdot B} \cdot F^2 \cdot \frac{dC}{da} = \frac{K^2}{E^*} \tag{13}$$

where

B = specimen thickness,
a = crack length, and
$E^* = E/(1 - \nu^2)$ for plane-strain condition,
and ν = Poisson ratio:

$$C = 2 \cdot B \int_a \frac{K^2}{E^* \cdot F^2} da \tag{14}$$

Z then is obtained from Eqs 9, 11, 12, and 14 as

$$Z = \frac{1}{1 + \frac{B \cdot (1 - \nu^2) \cdot \pi}{l_r \cdot A} \int_a \frac{K^2}{\sigma^2} da} \tag{15}$$

The integral in Eq 15 has to be determined numerically for each crack length. Equations for determination of s^* are sometimes given in compendia of K-factor calculations. For tubular specimens with notches, for example, as used in this work, the factor is given by Tada et al. [23]. For flat specimens with constant cross section, Z is calculated for several crack geometries in Ref 27.

Acknowledgment

Financial support by the Fonds zur Förderung der wissenschaftlichen Forschung, Wien, is gratefully acknowledged.

The authors thank Professor Sawaki, Tohoku University, Sendai, Japan, for interesting discussions.

References

[1] *Multiaxial Fatigue, ASTM STP 853,* K. J. Miller and M. W. Brown, Eds., American Society for Testing and Materials, Philadelphia, 1985.

[2] Yokobori, T., Maekawa, I., Yokobori, T. A., Jr., Sato, K., and Ishazaki, Y. in *Proceedings,* Symposium on Absorbed Spectrometry Energy/Strain Energy Density, Budapest, G. C. Sih, E. Czoboly, and F. Gillemont, Eds., 1982, pp. 45–55.

[3] Kordisch, H., Riedmüller, J., and Sommer, E. in *Proceedings,* Symposium on Absorbed Spectrometry Energy/Strain Energy Density, Budapest, G. C. Sih, E. Czoboly, and F. Gillemont, Eds., 1982, pp. 33–43.

[4] Hua, G., Brown, M. W., and Miller, K. J., *Fatigue of Engineering Materials and Structures,* Vol. 5, No. 1, 1982, pp. 1–17.

[5] Hua, G., Alagok, N., Brown, M. W., and Miller, K. J. in *Multiaxial Fatigue, ASTM STP 853,* K. J. Miller and M. W. Brown, Eds., American Society for Testing and Materials, Philadelphia, 1985, pp. 184–202.

[6] Tanaka, K., *Engineering Fracture Mechanics,* Vol. 6, 1974, pp. 493–507.
[7] Yokobori, T., Jr., Yokobori, T., Sato, K., and Syon, K., *Fatigue of Engineering Materials and Structures,* Vol. 8, No. 4, 1985, pp. 315–325.
[8] Truchon, M., Amestoy, M., and Dang-Van, K. in *Proceedings,* 5th International Congress on Fracture, Vol. 4, 1981, pp. 1841–1849.
[9] Smith, E. W. and Pascoe, K. J., *Fatigue of Engineering Materials and Structures,* Vol. 6, No. 3, 1983, pp. 201–224.
[10] Otsuka, A., Mori, K., Ohshima, T., and Tsuyama, S. in *Proceedings,* 5th International Congress on Fracture, Vol. 4, 1981, pp. 1851–1858.
[11] Hourlier, F. and Pineau, A. in *Proceedings,* 5th International Congress on Fracture, Vol. 4, 1981, pp. 1833–1840.
[12] Hourlier, F. and Pineau, A., *Fatigue of Engineering Materials and Structures,* Vol. 5, No. 4, 1982, pp. 287–302.
[13] Hourlier, F., d'Hondt, H., Truchon, M., and Pineau, A. in *Multiaxial Fatigue, ASTM STP 853,* K. J. Miller and M. W. Brown, Eds., American Society for Testing and Materials, Philadelphia, 1985, pp. 228–247.
[14] Pook, L. P., *International Journal of Fracture,* Vol. 13, 1977, pp. 867–869.
[15] Erdogan, F. and Sih, G. C., *Journal of Basic Engineering,* Vol. 85D, 1963, pp. 519–527.
[16] Sih, G. C., *International Journal of Fracture,* Vol. 10, 1974, pp. 305–313.
[17] Anderson, G. P., Ruggles, V. L., and Stibor, G., *International Journal of Fracture Mechanics,* Vol. 7, 1971, pp. 63–76.
[18] Tschegg, E. K., Stanzl, S. E., and Czegley, M., *Vorrichtung zur Överlagerung einer konstanten oder veranderlichen Kraft und einer Ultraschallschwingung, Österreichisches* Patent AT 384754B, Int. Cl. B06B3/00 vom 11.1.1988, Ammeldetag 17.12.1985.
[19] Mitsche, R., Stanzl, S. E., and Burkert, D., *Wissenschaftlichen Film,* Vol. 14, 1973, pp. 3–11.
[20] Stanzl, S. E. and Tschegg, E. K., *Metal Science,* Vol. 14, April 1980, pp. 137–142.
[21] Erdogan, F. and Ratwani, M., *International Journal of Fracture Mechanics,* Vol. 6, No. 4, 1970, pp. 379–392.
[22] Erdogan, F. and Ratwani, M., *International Journal of Fracture Mechanics,* Vol. 8, 1972, pp. 87–95.
[23] Tada, H., Paris, P. C., and Irwin, G. R., *The Stress Analysis of Cracks Handbook,* Del Research Corporation, Hellertown, PA, 1973.
[24] Suresh, S. and Ritchie, R. O. in *Proceedings,* International Symposium on Fatigue Crack Growth Threshold Concepts, L. D. Davidson and S. Suresh, Eds., American Institute for Mining, Metallurgical, and Petroleum Engineers, Warrendale, PA, 1984, pp. 227–261.
[25] Suresh, S., *Metallurgical Transactions A,* Vol. 16A, 1985, pp. 249–260.
[26] Tschegg, E. K., *Materials Science and Engineering,* Vol. 54, 1982, pp. 127–136.
[27] Schoeck, G., *Zeitschrift für Metallkunde,* Vol. 73, No. 9, 1982, pp. 576–578.

Leslie Banks-Sills[1] *and Daniel Schur*[1]

On the Influence of Crack Plane Orientation in Fatigue Crack Propagation and Catastrophic Failure

REFERENCE: Banks-Sills, L. and Schur, D., **"On the Influence of Crack Plane Orientation in Fatigue Crack Propagation and Catastrophic Failure,"** *Fracture Mechanics: Perspectives and Directions (Twentieth Symposium), ASTM STP 1020,* R. P. Wei and R. P. Gangloff, Eds., American Society for Testing and Materials, Philadelphia, 1989, pp. 497–513.

ABSTRACT: During fatigue crack propagation in sufficiently thin metallic sheets, shear lips develop which spread throughout the material thickness. This leads to a situation where, in addition to a Mode I component of deformation, there is also a Mode III component. Finite-element calculations are carried out on several geometries in an attempt to assess the effect of this additional deformation upon catastrophic failure and fatigue crack propagation. For these particular geometries, it is shown that the slant crack is less dangerous than the flat crack with regard to catastrophic failure and, when propagating in fatigue, may grow faster or slower than the flat crack.

KEY WORDS: fracture mechanics, fatigue, Mode III, mixed-mode, slant crack, shear lips, finite elements

The expected direction of crack growth in a uniaxially loaded sheet is normal to the applied load, with the fracture face 90 deg with respect to the plane of the body, that is, Mode I growth. Indeed, in relatively thick sheets subjected to comparatively low loads, this is precisely what occurs. In other circumstances, however, with materials such as aluminum alloys, titanium, austenite and mild steel, and copper, shear lips have been observed to develop. If these sheets are sufficiently thin, the lips spread throughout the entire surface as shown schematically in Fig. 1 [*1–8*], so that both Modes I and III deformation are present. The crack continues to propagate with this new geometry, which has been reported to be either slower than [*9,10*] or at the same rate as [*11–13*] that of a flat crack. In this investigation, an attempt is made to characterize the behavior shown in Fig. 1*a*, considering its effect on both fast fracture and fatigue crack propagation. Currently in engineering calculations, this geometry is considered two-dimensional; that is, the out-of-plane Mode III component is neglected. In order for the effect of neglecting this deformation component to be assessed, the three-dimensional version of this problem is investigated.

Two geometries are considered: a central crack and an edge crack passing through a finite-thickness plate at an angle of 45 deg with respect to the thickness direction. Although shear lip formation is a plastically induced phenomena, the material is modeled as being linearly elastic and isotropic in order to simplify the analysis. A more realistic treatment of the

[1] Associate professor and graduate student, respectively, Department of Solid Mechanics, Materials and Structures, Faculty of Engineering, Tel Aviv University, 69978, Ramat Aviv, Israel.

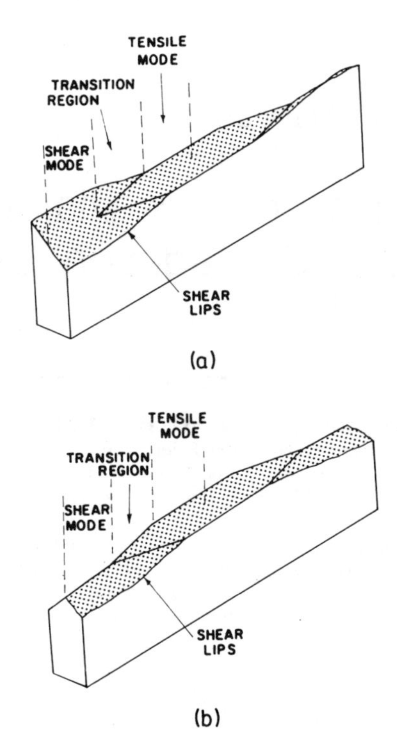

FIG. 1—*Schematic view of shear lip development (steady state is achieved at end of transition region).*

problem should include plastic deformation. Perhaps, once the shear lips have reached a steady state form as in Fig. 1, plastic behavior again may be neglected to examine continuing fatigue crack growth. This is the approach taken in this study.

For this problem to be addressed, analysis of edge and central slant cracks (Fig. 2) is carried out by means of the finite-element method. At several locations on the crack front, the stress-intensity factors K_I and K_{III} are determined by means of displacement extrapolation along the crack face; details of the numerical analysis are presented. In order for us to check the accuracy of the solutions, several two and three-dimensional geometries in Modes I and III with known solutions are examined with results agreeing to within 3%. Then, the analysis of finite-thickness plates containing a slant crack is considered and results are presented. Finally, the effect of the crack plane orientation upon fast fracture and fatigue crack propagation is examined.

Finite-Element Analysis

In this study, the finite-element method is used to determine the Mode I and Mode III stress-intensity factors. The program ADINA [*14*] is employed with solid, isoparametric elements containing between eight and twenty nodal points. Those surrounding the crack tip are solid, 20-noded, quarter-point elements which enable modeling of the linear elastic, square-root singularity. It has been shown that square-root singular stresses may be obtained in part of the element by moving the mid-side nodes closest to the crack tip to the quarter-point [*15–18*]. Illustrated in Fig. 3 are the approximately known regions such that on all

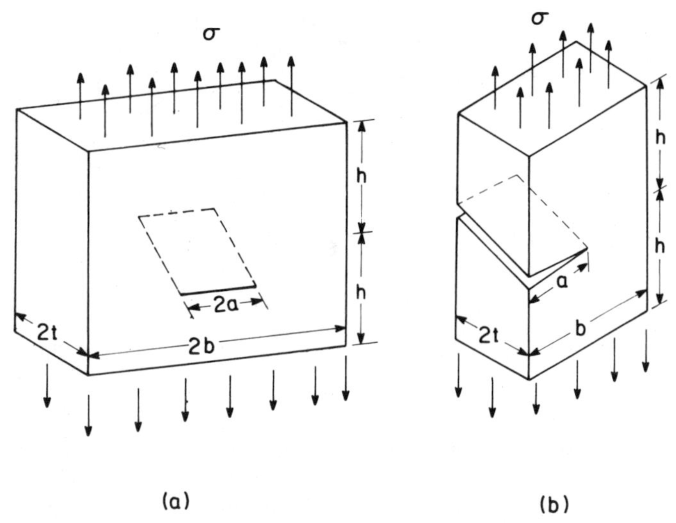

FIG. 2—*Geometry of* (a) *central slant crack and* (b) *edge slant crack in a finite-thickness plate.*

rays emanating from the crack-tip singular behavior is obtained. These elements are compatible with adjoining elements, ensuring convergence of the finite-element scheme.

To calculate the stress-intensity factors, the method of displacement extrapolation is employed. Expressions are written from the first term of the asymptotic displacement expansion as

$$K_{\mathrm{I}}^* = \frac{\sqrt{2\pi}\,\mu}{\kappa + 1} \frac{v_2(r) - v_1(r)}{\sqrt{r}} \tag{1a}$$

and

$$K_{\mathrm{III}}^* = \frac{\sqrt{2\pi}\,\mu}{8} \frac{w_2(r) - w_1(r)}{\sqrt{r}} \tag{1b}$$

where

μ = shear modulus,
κ = $3 - 4\nu$ for plane strain and $(3 - \nu)/(1 + \nu)$ for generalized plane stress,
ν = Poisson's ratio,
r = distance from the crack tip,
$v(r)$ and $w(r)$ = y and z direction displacements, respectively, and
1 and 2 = as subscripts, the lower and upper crack faces, respectively (see Fig. 4).

The values of v and w are taken from the finite-element results. From theoretical considerations, one can show that there is a range of r for which K_{I}^* and K_{III}^* in Eq 1 are straight lines and

$$K_{\mathrm{I}} = \lim_{r \to 0} K_{\mathrm{I}}^* \tag{2a}$$

$$K_{\mathrm{III}} = \lim_{r \to 0} K_{\mathrm{III}}^* \tag{2b}$$

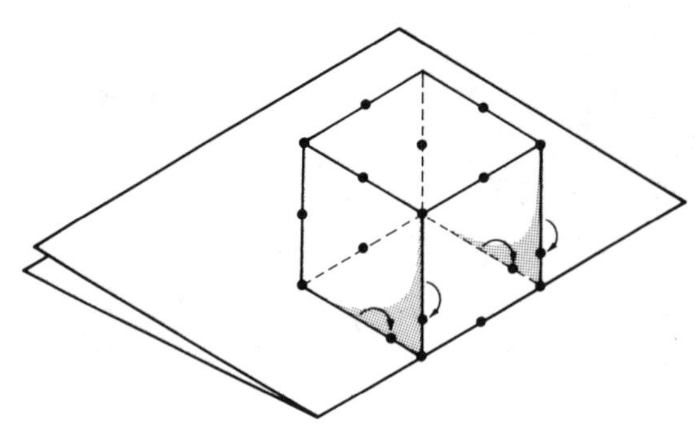

FIG. 3—*Singular, three-dimensional, quarter-point, isoparametric finite element. Approximate singular region is cross-hatched.*

This method was first suggested in Ref *19*. With the values of K_I^* and K_{III}^* along the crack face, linear regression is employed to determine a "best" straight line [*20*] in the crack-tip region. It should be noted that all calculated stress-intensity factors are given in nondimensional form with respect to $\sigma\sqrt{\pi a}$, where σ is the remote applied stress and a is crack length.

In order to demonstrate the accuracy of this method in both Modes I and III, several geometries with known solutions are considered. The first is a two-dimensional central crack in a rectangular plate. The solution of Isida [*21*] which is accurate to four significant figures is chosen for comparison. From symmetric considerations, one-fourth of the body is modeled. The finite-element mesh contains 100 square elements of equal size. For crack length $a/b = 0.5$ and height $h/b = 1$, the best results are obtained with a mesh in which the first two rows of elements adjacent to the crack contain eight-noded elements, the next row contains seven-noded elements, the rest of the elements are four-noded (in the sequel, this type of mesh is called a graduated mesh), and reduced integration is employed, that is, two-point Gaussian integration. In that case, there is an error of 1.1% as compared to [*21*]. A careful study was carried out for this geometry: both two- and three-point Gaussian integration were considered in combination with the mesh described and one in which all elements were eight-noded. It was found that a graduated mesh with two-point Gaussian quadrature to calculate the stiffness matrix led to the most accurate results. However, since this is a specific geometry, general conclusions cannot be made.

Next, Mode III deformation is examined. Since with the slant crack geometry, in addition to the Mode I deformation component there is also a Mode III component, a finite width strip with an edge crack loaded in Mode III is also considered. For Mode III deformation

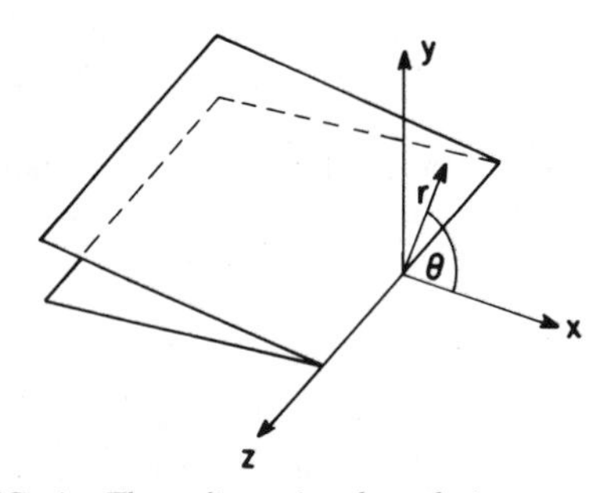

FIG. 4—*Three-dimensional crack tip geometry.*

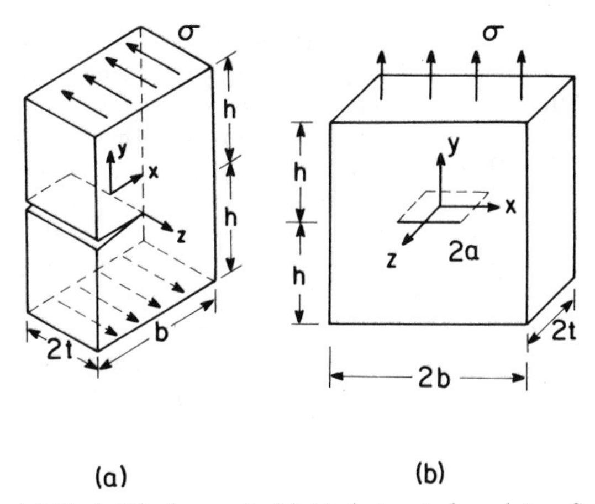

FIG. 5—(a) *Mode III edge crack;* (b) *Mode I central crack in a finite plate.*

to be treated with the finite-element method, either a special program is required or the problem must be regarded as three-dimensional. For crack length $a/b = 0.5$ and plate height $h/b = 3$ (Fig. 5*a*), half of the body is modeled with a graduated mesh of 600 cubic elements: 10 along the width of the body, 30 along the height, and 2 along the thickness, yielding 1278 nodal points. Since all nodal points are constrained to remain in the xy-plane, the thickness is irrelevant in this analysis. In three-dimensions, a graduated mesh is one in which the two rows of elements adjacent to the crack and throughout the thickness contain 20 nodal points, the next row contains 16 nodal points, and the rest of the elements contain 8 nodal points. As in the remainder of this investigation, two-point Gaussian quadrature is employed to determine the stiffness matrix. The stress-intensity factor K_{III} is computed by means of Eqs 1*b* and 2*b* at three locations along the crack front, namely, $z/t = 0, 0.5$, and 1, and found to be 1.135 at each one. A solution to this problem for $h/b \rightarrow \infty$ is 1.128 [*22*], resulting in a difference of 0.6%.

Next, a central crack in a finite thickness, rectangular plate loaded in tension is considered (Fig. 5*b*). For crack length $a/b = 0.5$, plate height $h/b = 1$, and thickness $2t/b = 1$, one-eighth of the body is modeled by a graduated mesh containing 500 cubic elements (ten elements along the plate width and height, and five elements along the thickness). The stress-intensity factor K_{I} is calculated by means of Eqs 1*a* and 2*a* at the six element boundaries along the crack front. For the internal boundaries, a plane-strain assumption is employed and at the surface, plane stress.

Of course, neither of these assumptions is completely correct. It was shown in Ref *23* that when $a/b = 0.1$, $h/b = 2$, and $2t/b = 0.4$, plane-strain conditions exist at the midplane and for about 30% of the plate thickness; for $2t/b = 0.06$, plane-strain conditions did not exist anywhere. In either case, the accuracy of the plane-strain assumption along the crack front within the body depends upon specimen and crack dimensions and does not necessarily hold along the entire crack front. On the other hand, the state of stress at the intersection point of the crack and the free surface of the plate is more problematic. It seems to be fairly well accepted now that the stress singularity is less than or equal to one-half at the plate surface [*23–26*]. Hence, from the definition

$$K_{\text{I}} = \lim_{r \to 0} \sqrt{2\pi r}\, \sigma_{yy}(r,\theta = 0) \tag{3}$$

K_I goes to zero where the crack intersects the free surface. An assumption of plane stress at this surface in Eq 1*a* permits some freedom in the existing *z*-direction deformation, allowing for a drop in the value of K_I. But this approximation is clearly insufficient to provide the correct solution of zero at the surface. Another approach has been to employ the method of forces to compute K_I [*27*]. Since this technique does not require a plane-stress or plane-strain assumption, there is a certain credibility to the values of K along the crack front within the body. But although the stress-intensity factor does decrease at the plate surface, it does not go to zero. A further possibility as suggested in Ref *23* is to cease employing stress-intensity factors for three-dimensional applications and to use instead energy parameters, such as Griffith's energy. Since the energy is not expected to approach zero on the plate surface and is not related to the stress-intensity factor as it is within the body, this approach is indeed more logical. Nonetheless, since presently both catastrophic and fatigue crack propagation predictions are based upon stress-intensity factors, in this study, continued use of these concepts is made.

In Fig. 6, the stress-intensity factor for the central crack in a finite-thickness plate is plotted along the crack front; numerical results are also presented in Table 1. Since the value of Poisson's ratio ν influences somewhat the K_I values [*28*], it should be noted that $\nu = 0.3$ is used in all calculations. Consider the results for $2t/b = 1$ and denoted as fine mesh; that is the mesh which has already been described as containing 500 elements. It may be observed that the three-dimensional values of K_I are larger than the two-dimensional value. Moreover, the stress-intensity factor varies as a function of position along the crack front increasing to a maximum value near the free surface. These trends have been observed previously in various other studies [*27,29–31*]. Also plotted in Fig. 6 are graphical results taken from Ref

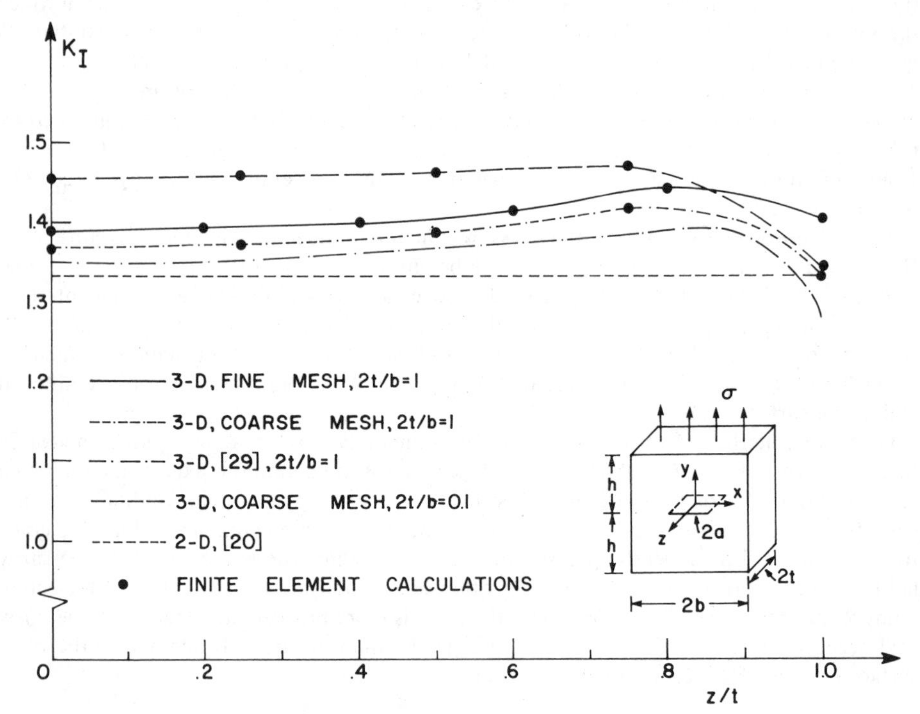

FIG. 6—*Nondimensional stress-intensity factor for a central flat crack in a finite plate as a function of position* z/t *along the crack front* (a/b = *0.5,* h/b = *1*).

TABLE 1—*Values of nondimensional stress-intensity factor* K_I *at various locations along crack front for a central crack in a finite-thickness plate* ($a/b = 0.5$, $h/b = 1$, *and* $2t/b = 1$), *as calculated with the finer mesh* (*10 × 10 × 5 elements*).

z/t	K_I
0	1.388
0.2	1.391
0.4	1.399
0.6	1.414
0.8	1.439
1.0	1.401

29 in which energy methods were employed in combination with Richardson's extrapolation. The difference between the results determined here and those presented in Ref *29* is not more than 3%. Moreoever, two-dimensional K-values in Ref *29* were generally smaller than the exact solution taken for comparison. Thus, the methods and mesh employed here appear to yield reasonable results.

It may be concluded from the several examples considered that use of quarter-point singular elements together with a relatively coarse, graduated-mesh, two-point Gaussian quadrature, and the displacement extrapolation method, produce accurate Mode I and Mode III stress-intensity factors in both two- and three-dimensional geometries. In the next section, slant crack geometries are considered with these methods.

The Slant Crack

Remote tensile stress is applied to a plate containing either a central or an edge crack at an angle of 45 deg with respect to the thickness direction (Fig. 2). Because of the geometry, both Mode I and III deformation components are induced. The Mode I and Mode III stress-intensity factors are calculated along the crack front for both geometries. Since there is only one plane of symmetry for the central crack geometry and none for the edge crack, a mesh equivalent to the one employed in the previous section for the central crack in a finite-thickness plate contains 2000 elements. Because this requirement necessitates more computer space than that available on the CDC 855 which is employed, a coarser mesh is constructed.

For us to evaluate the loss in accuracy resulting from a coarser mesh, the central crack in a finite-thickness plate which was considered in the previous section is reanalyzed. In Fig. 7, the cross section of the three-dimensional mesh employed is exhibited. The mesh contains four elements in the thickness direction; so that, there are 256 elements and 768 nodal points. Both the length and width of the elements adjacent to the crack tip are the same as in the finer mesh, that is $L = 0.05b$. Values of K_I plotted in Fig. 6 are seen to be approximately 1.5% lower than the values calculated with the finer mesh and closer to the result presented in Ref *29*.

In addition, since the slant cracks are examined in plates thinner than the one analyzed here, the flat crack geometry is again considered for several thinner plates with the mesh in Fig. 7 and four element layers through half the thickness. Values of K_I at the midplane are plotted for different plate thicknesses in Fig. 8. Since the mesh is the same for each plate thickness, the element aspect ratio varies. This variation is generally expected to influence the results. For the singular element, this ratio changes from 1:1:0.0625, when $2t/b = 0.025$, to 1:1:2.5, when $2t/b = 1$. To examine this point, with $2t/b = 0.1$, the mesh in Fig. 7 is employed with between one to five element layers along the thickness direction;

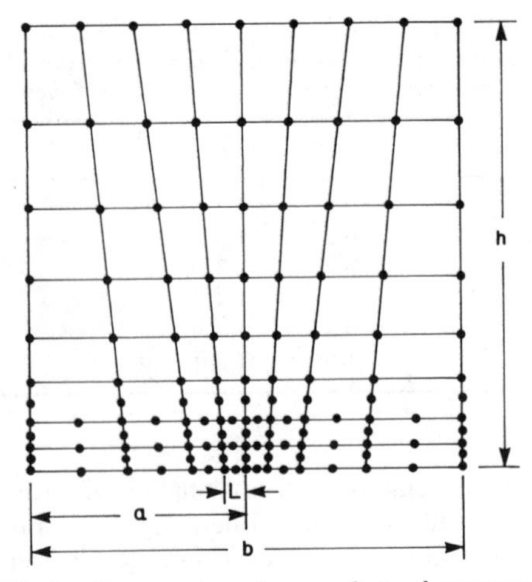

FIG. 7—*Cross section of coarse finite-element mesh.*

so that, in addition to the change in the number of nodal points (from 291 to 927), the aspect ratio of the singular element varies from 1:1:0.2 to 1:1:1. It is seen that the stress-intensity factor at the midplane does not change. Apparently the aspect ratio in the thickness direction is not important in determining the stress-intensity factor at the midplane, and hence the results determined in Fig. 8 are not influenced by this parameter. Moreover, in other studies it was seen that when the apsect ratio changes from 1:1:1 to 1:1:10, again the value of K at the midplane is unaffected. This is not the case for other values of the stress-intensity factor along the crack front. Returning to Fig. 8, as plate thickness decreases, the value of the stress-intensity factor increases. It may be noted in passing that this would seem to imply that the values of K_c determined for thin plates from a two-dimensional analysis are somewhat underestimated. Of course in this analysis we have considered a

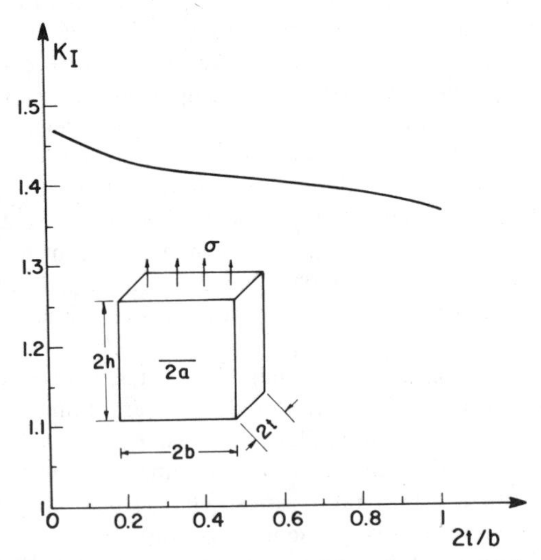

FIG. 8—*Nondimensional stress-intensity factor at midplane for a central flat crack in a finite plate as a function of thickness* (a/b = *0.5,* h/b = *1). Calculations carried out with coarse mesh.*

central crack rather than an edge crack. One should consider edge cracked bodies to see if the same conclusion is possible.

Moreover, in Fig. 6, results for $2t/b = 0.1$ are also illustrated. The mesh employed is that shown in Fig. 7 with four element layers along half the thickness, containing 256 elements and 768 nodal points. It may be observed from Fig. 6 that as plate thickness decreases, the stress intensity factor along the crack front increases in the center of the body and decreases at the plate surface. This phenomenon was also observed in Refs *23* and *27*.

Next, both an edge crack and a central crack at an angle of 45 deg with respect to the thickness direction are considered with crack length $a/b = 0.5$, height $h/b = 1$, and three thicknesses, $2t/b = 0.025, 0.05$, and 0.1 (see Fig. 2). These thicknesses were chosen in order to study the effect of thickness on the stress-intensity factors of a slant crack in a thin body. All bodies are modeled with 512 elements; for the central crack, this is a model of half the body, for the edge crack, the entire body (Fig. 9). The condition of antisymmetry about the plane $z = 0$ is not employed. Again, a cross section of the mesh in the xy-plane is shown in Fig. 7. With four element layers throughout the entire thickness $2t$, and a graduated mesh, the model consists of 1536 nodal points. The stress-intensity factors K_I and K_{III} are calculated at three locations along the crack front, $z/t = 0, 0.5$, and 1. For K_I, at the midplane ($z/t = 0$) and half way to the surface ($z/t = 0.5$), a plane-strain assumption is employed; at the surface, plane stress is assumed. Calculations were also carried out at $z/t = -0.5$ and -1. These values were in agreement to at least five significant figures with those at the antisymmetrical position along the crack front, demonstrating the accuracy of the results.

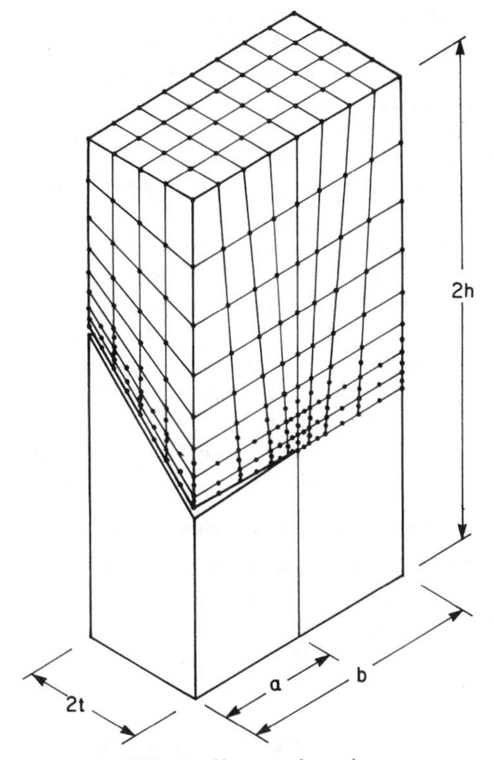

FIG. 9—*Slant crack mesh.*

TABLE 2—*Nondimensional values of Mode I and Mode III stress-intensity factors for central slant crack in a finite-thickness plate* 2t, *turned 45 deg with respect to the thickness direction;* a/b = *0.5,* h/b = *1.*

		z/t		
$2t/b$	Mode	0	0.5	1.0
0.025	K_I	1.44	1.44	1.30
	K_{III}	0.29	0.29	0.29
	K_{III}/K_I	0.20	0.20	0.22
0.05	K_I	1.43	1.41	1.30
	K_{III}	0.25	0.26	0.27
	K_{III}/K_I	0.17	0.18	0.21
0.1	K_I	1.35	1.35	1.23
	K_{III}	0.15	0.15	0.15
	K_{III}/K_I	0.11	0.11	0.12

In Table 2, values of the nondimensional stress-intensity factors for a central slant crack are presented. It may be observed that the Mode III deformation component increases as thickness decreases and that for each thickness this component is virtually constant throughout the thickness. It may be also seen, that for a central flat crack with $2t/b = 0.025$ and 0.05, that at the midplane, K_I is 1.47 and 1.46, respectively, 2.0% larger than its value for the slant crack; for $2t/b = 0.1$, K_I is 1.45, 6.9% larger.[2] For the slant crack, the behavior of K_I along the crack front for each thickness is virtually constant within the body, decreasing at the surface. From our previous experience, an exact description of the change of K along the crack front requires a finer mesh. The value of the stress-intensity factor at the midplane, however, should be well described by a mesh as coarse as this. The trends for Mode I in the slant cracked body, nevertheless, are somewhat similar to that for a plate with a straight-through crack with only Mode I deformation. That is, K_I increases as plate thickness decreases and drops more radically at the surface for the thin plate (see Figs. 6 herewith and Fig. 10 from Ref *27*). Since the mesh is rather coarse in the thickness direction, the expected increase near the plate surface is not captured.

Next, an edge slant crack of the same geometry is examined. Values of the stress-intensity factor are presented in Table 3. It is observed that as with the central slant crack the ratio of K_{III} to K_I increases as thickness decreases and that the greatest value of this ratio occurs at the plate surface. The value of K_{III} remains constant within the body, increasing at the surface; a finer mesh of course, is required to validate this behavior.

The results presented in Tables 2 and 3 demonstrate that for the geometries studied, the Mode III stress-intensity factor may be as much as 24% of the Mode I component. Its effect upon catastrophic failure and fatigue crack propagation is examined in the next section.

Catastrophic Failure and Fatigue Crack Propagation

In the previous section, the ratio of the Modes I and III stress-intensity factors were presented for a central and an edge crack at an angle of 45 deg with respect to the thickness direction in a finite-thickness plate. It should be noted that only a specific geometry was analyzed, that of $a/b = 0.5$ and $h/b = 1$ for three thicknesses. In this section, several Modes

[2] For completeness, the values of the stress-intensity factor K_I for central and edge flat cracks at the midplane, calculated with the mesh shown in Figs. 7 and 9 for crack length $a/b = 0.5$ and plate height $h/b = 1$ are presented in Table 4.

TABLE 3—*Nondimensional values of Mode I and Mode III stress-intensity factors for edge slant crack in a finite-thickness plate* 2t, *turned 45 deg with respect to the thickness direction;* a/b = *0.5,* h/b = *1.0.*

		z/t		
$2t/b$	Mode	0	0.5	1.0
	K_I	3.01	3.00	2.74
0.025	K_{III}	0.64	0.64	0.66
	K_{III}/K_I	0.21	0.21	0.24
0.05	K_I	2.95	2.91	2.70
	K_{III}	0.58	0.58	0.59
	K_{III}/K_I	0.20	0.20	0.22
0.1	K_I	2.76	2.77	2.52
	K_{III}	0.44	0.44	0.46
	K_{III}/K_I	0.16	0.16	0.18

I and III mixed-mode fracture criteria are considered in an attempt to assess the influence of the Mode III deformation component upon catastrophic failure and fatigue crack propagation.

First the effect of the Mode III deformation component upon catastrophic failure is examined. This situation occurs in thin plates in which cracks have turned out of their original straight plane during fatigue crack propagation, propagate self-similarly at an angle of 45 deg with respect to the thickness direction, and then fail catastrophically when some critical load is reached. To this end, several mixed-mode criteria are considered.

If a crack in mixed-modes I and III propagates in a self-similar manner, which occurs for the slant crack, it is possible to show that

$$\left(\frac{K_I}{K_{Ic}}\right)^2 + \left(\frac{K_{III}}{K_{IIIc}}\right)^2 = 1 \tag{4}$$

Again under self-similar conditions and from a consideration of Griffith's energy, it is possible to show for plane strain conditions that

$$\left(\frac{K_I}{K_{Ic}}\right)^2 + \frac{1}{1-\nu}\left(\frac{K_{III}}{K_{Ic}}\right)^2 = 1 \tag{5}$$

Equations 4 and 5 are identical if

$$K_{IIIc} = \sqrt{1-\nu}\, K_{Ic} \tag{6}$$

It was seen in Ref *32* for 4340 steel, heat treated to an ultimate strength of 270 to 280 ksi, that $K_{IIIc}/K_{Ic} = 1.2$; for $\nu = 0.3$, Eq 6 yields 0.84 for this ratio, which does not agree with the experimental result. It should be noted, however, that there was quite a bit of nonlinearity in the Mode III experiments, which would tend to produce K_{IIIc} values larger than the actual ones. The next criterion considered is derived from Sih's strain energy density theory [*33*], again assuming self-similar crack propagation

$$\left(\frac{K_I}{K_{Ic}}\right)^2 + \frac{1}{1-2\nu}\left(\frac{K_{III}}{K_{Ic}}\right)^2 = 1 \tag{7}$$

In this case, Eqs 4 and 7 are identical if

$$K_{\mathrm{IIIc}} = \sqrt{1 - 2\nu}\, K_{\mathrm{Ic}} \tag{8}$$

For the 4340 steel of Ref *32*, Eq 8 produces a smaller fracture toughness ratio than Eq 6. The last criterion considered was presented in Ref *32* from the test results on 4340 steel and is given as

$$\left(\frac{K_{\mathrm{I}}}{K_{\mathrm{Ic}}}\right)^2 + \left(\frac{K_{\mathrm{III}}}{K_{\mathrm{IIIc}}}\right)^{4.75} = 1 \tag{9}$$

This is an empirically determined upper bound of the test results, whereas Eq 4 served as a lower bound.

For both the central and edge crack at an angle of 45 deg with respect to the thickness direction, these criteria are employed to compute the ratio of the critical far-field stress σ_{sc}, as determined by a three-dimensional analysis of the slant crack, to its value σ_{fc}, established from a three-dimensional analysis of a flat crack under Mode I loading. In all cases, the same mesh is employed shown in Fig. 9. Since the values of the stress-intensity factors are most accurate at the midplane, this ratio is calculated on the basis of the stress-intensity factors computed there. Given $K_{\mathrm{III}}/K_{\mathrm{I}}$ from Tables 2 and 3, the ratio $K_{\mathrm{I}}/K_{\mathrm{Ic}}$ is calculated from Eqs 5, 7, and 9, and the stress ratio is computed with the aid of Table 4. In Eq 9, the value for $K_{\mathrm{IIIc}}/K_{\mathrm{Ic}}$ is taken from Ref *32* as 1.2; determination of $K_{\mathrm{I}}/K_{\mathrm{Ic}}$ requires iteration for this criterion. Results are presented in Table 5 for the central crack geometry and Table 6 for the edge crack. It may be observed that trends for both geometries are similar; that is, for each criterion, as thickness increases the ratio σ_{sc}/σ_{fc} increases. In calculating this ratio, the values of the nondimensional stress-intensity factors K_{I} for the slant and flat cracks are the dominant constituent. For both geometries, as thickness increases, the critical stress at failure for the slant crack becomes larger than that of the flat crack. It appears in general that the flat crack is more dangerous than the slant crack. It must be emphasized that these results are for a specific geometry and may change for different conditions.

In Ref *34*, the energy release rate as a function of position along the crack face in a compact tension specimen was calculated for a flat and slant crack. It was seen that the ratio G_s/G_f, where G_s is the energy release rate for the slant crack and G_f is that for the flat crack, was ~0.65 at the specimen midplane. A similar calculation may be carried out from the analysis performed here. In this case, for $2t/b = 0.1$, the energy ratio for the edge crack is 0.84 when Eq 5 is employed. In Ref *34*, the thickness $2t/b$ was probably between 0.04 and 0.08 and the crack length was not given, so that it is difficult to make this comparison.

Finally, the effect of crack plane orientation upon fatigue crack growth is examined. Paris' fatigue crack propagation law [*35*] is employed with ΔK replaced by ΔK_{eff}, an effective stress-intensity factor, namely

$$\frac{da}{dN} = C\,(\Delta K_{\mathrm{eff}})^n \tag{10}$$

TABLE 4—*Values of* K_I *for central and edge flat cracks at midplane of body for several thicknesses* $2t/b$ ($a/b = 0.5$, $h/b = 1.0$).

$2t/b$	0.025	0.05	0.1
Central	1.47	1.46	1.45
Edge	3.09	3.08	3.06

TABLE 5—*Ratio of critical stress* σ_{sc}/σ_{fc} *evaluated at midplane for a central crack as determined from three-dimensional analyses of a slant and flat crack, respectively, for various failure criteria; plate dimensions,* a/b = *0.5,* h/b = *1.0.*

Failure Criterion	2t/b 0.025	0.05	0.1
Eq 5	0.99	1.00	1.07
Eq 7	0.97	0.99	1.06
Eq 9	1.02	1.03	1.07

where C and n are material parameters. A ΔK_{eff} may be determined from the mixed-mode criteria mentioned above as follows: from Eq 5

$$\Delta K_{\text{eff}} = \Delta K_{\text{I}} \left[1 + \frac{1}{1 - \nu}\left(\frac{K_{\text{III}}}{K_{\text{I}}}\right)^2\right]^{1/2} \tag{11}$$

and from Eq 7,

$$\Delta K_{\text{eff}} = \Delta K_{\text{I}} \left[1 + \frac{1}{1 - 2\nu}\left(\frac{K_{\text{III}}}{K_{\text{I}}}\right)^2\right]^{1/2} \tag{12}$$

This type of approach has been employed by many authors to treat mixed-mode fatigue crack propagation, mainly, for Modes I and II [*36–40*]. A further possibility is a model based upon the theory of continuously distributed dislocations and is given in Ref *41* as

$$\Delta K_{\text{eff}} = \Delta K_{\text{I}} \left[1 + \frac{8}{1 - \nu}\left(\frac{K_{\text{III}}}{K_{\text{I}}}\right)^4\right]^{1/4} \tag{13}$$

Employing Eq 10 together with the expressions given in Eqs 11 through 13, ratios of the crack increment for a slant crack da_s to that of a flat crack da_f, for the same number of cycles, at the crack length $a/b = 0.5$, are determined and presented in Tables 7 and 8 for a central and edge crack, respectively. In Eq 10, the material parameter n is taken as both 2 and 4, which may be looked upon as bounds for most engineering materials. For both the edge and central cracks, it may be seen that as plate thickness increases, the crack propagation rate for the slant crack is slower than that for the flat crack. This observation is magnified with increasing n.

These numerical results may be qualitatively compared to experimental findings presented

TABLE 6—*Ratio of critical stress* σ_{sc}/σ_{fc} *evaluated at midplane for an edge crack as determined from three-dimensional analyses of a slant and flat crack, respectively, for various failure criteria; plate dimensions,* a/b = *0.5,* h/b = *1.0.*

Failure Criterion	2t/b 0.025	0.05	0.1
Eq 5	0.99	1.02	1.09
Eq 7	0.97	1.00	1.07
Eq 9	1.02	1.04	1.11

TABLE 7—*Ratio of crack increment* da_s/da_f (da_s *is determined from a three-dimensional analysis of a central slant crack and* da_f *from a three-dimensional analysis of a central flat crack*) *evaluated at the midplane for several crack propagation rules;* n *is a material parameter, plate dimensions,* $a/b = 0.5$, $h/b = 1.0$.

Crack Propagation Law	n	$2t/b$		
		0.025	0.05	0.1
Eq 11	2	1.02	0.99	0.88
	4	1.04	0.99	0.77
Eq 12	2	1.06	1.03	0.89
	4	1.13	1.05	0.79
Eq 13	2	0.97	0.96	0.87
	4	0.95	0.91	0.75

in the literature. In Ref *9* for aluminum 7075-T6 and Ti-8Al-1Mo-1V, and in Ref *10* for aluminum 2024-T3, flat cracks were seen to grow more rapidly than slant cracks. Considering Tables 7 and 8, for most cases considered, da_s/da_f is less than unity so that flat cracks propagate faster than slant cracks. It was reported in Refs *11* and *12* for aluminum, mild steel, and copper that there was no difference in propagation rate between slant and flat cracks. This is essentially what is calculated for the central crack with Eq 11 when $2t/b$ = 0.05, and for the edge crack with Eq 12 at this thickness. It should be noted, however, that perhaps in all of these calculations, average stress intensity factors rather than K values at the midplane would yield more realistic results.

Another interesting facet of the influence of plate thickness upon fatigue crack propagation rate may be examined with the results calculated here. In Refs *13* and *42*, fatigue tests on mild steel revealed that cracks in thinner sheets propagate more quickly than those in thicker sheets; for this material, it was reported in Ref *12* that thickness does not influence fatigue crack rate. For Ti-8Al-1Mo-1V [*9*], aluminum 7075-T6, and aluminum 2024-T3 [*43*], cracks in thinner sheets grew more slowly than those in thicker ones. In these tests, differentiation was not made with respect to slant and flat cracks. If it is assumed that crack propagation rate is governed by ΔK at the specimen midplane, for a central flat crack, Fig. 8 reveals that thinner plates have higher stress-intensity factors, so that thinner plates should promote more rapid crack growth as with the mild steel experiments in Refs *13* and *42*. Moreover,

TABLE 8—*Ratio of crack increment* da_s/da_f (da_s *is determined from a three-dimensional analysis of an edge slant crack and* da_f *from a three-dimensional analysis of an edge flat crack*) *evaluated at the midplane for several crack propagation rules;* n *is a material parameter, plate dimensions,* $a/b = 0.5$, $h/b = 1.0$.

Crack Propagation Law	n	$2t/b$		
		0.025	0.05	0.1
Eq 11	2	1.01	0.97	0.84
	4	1.03	0.94	0.71
Eq 12	2	1.06	1.01	0.87
	4	1.12	1.02	0.75
Eq 13	2	0.96	0.93	0.82
	4	0.93	0.86	0.67

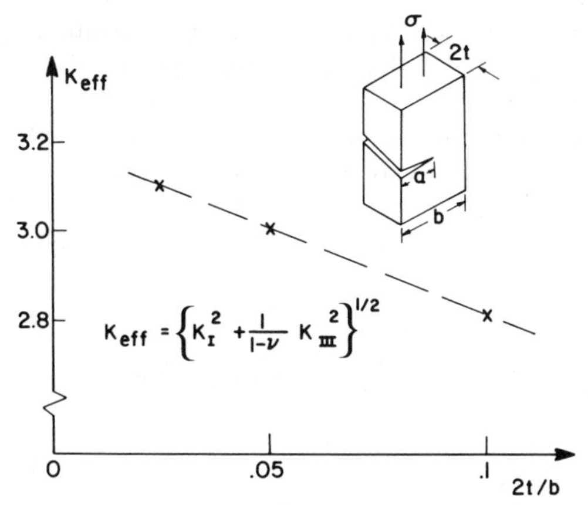

FIG. 10—*Nondimensional effective stress-intensity factor for edge slant crack in a finite plate as function of thickness* (a/b = *0.5,* h/b = *1). Crosses represent values calculated from finite-element results.*

for an edge slant crack, the nondimensional effective stress-intensity factor obtained from Eq 5 is plotted in Fig. 10. Again on the basis of this graph and the assumptions of Eq 10, slant cracks in thinner plates will propagate more rapidly than those in thicker ones. The same conclusion can be made for central slant cracks. These calculations explain the behavior of mild steel only. From the tests presented in the literature, aluminum and titanium appear to promote slower crack growth in thinner sheets. Apparently, Eq 10, which is an empirical formula, requires some adjustment to account for the thickness effect, such as

$$\frac{da}{dN} = C(\Delta K_{\text{eff}})^n \, (t)^m \tag{14}$$

where t is some nondimensional thickness and m is a material parameter.

Conclusions

Since it has been observed that cracks in tensile fields under fatigue loading in sufficiently thin sheets propagate out of their plane, the importance of the Mode III deformation component upon fatigue crack propagation and catastrophic failure was assessed. To this end, edge and central cracks in finite-thickness plates at an angle of 45 deg with respect to the thickness direction were analyzed by the finite-element method. It was seen that for the geometry considered of nondimensional crack length $a/b = 0.5$ and plate height $h/b = 1$, the ratio of K_{III} to K_{I} was as much as 24%.

The values of the stress-intensity factors for these two geometries and three thicknesses were employed in several mixed-mode criteria to evaluate the effect of a slant crack upon catastrophic failure and fatigue crack propagation rate. It was seen that the critical stress at failure for the slant crack σ_{sc} was sometimes greater than, less than, and equal to that for the flat crack σ_{fc}. It appears, however, that the ratio of σ_{sc}/σ_{fc} approaches a limit near unity, with most values greater than one, so that the slant crack geometry seems less dangerous than the flat crack. With respect to fatigue crack propagation rate, the rate of crack growth of the slant crack is sometimes slower, faster, and the same depending upon specimen thickness, fatigue crack growth law, and type of crack. Of course, all of these conclusions are based upon a particular geometry and are not general, but rather indicative.

It may be observed from all of the calculations, that it is the difference in K_I for the slant and flat cracks which is most influential in affecting changes in catastrophic and fatigue crack propagation (see Tables 2 to 4).

In experimental studies on different materials, qualitatively similar findings have been obtained. Inclusion of these effects may help to alleviate some of the differences between fatigue crack growth analyses and testing. More extensive numerical and experimental studies are required to incorporate these effects into routine engineering calculations.

Acknowledgment

We would like to thank Mr. Zvi Granot of the Israel Aircraft Industries for many helpful discussions and interest in this work.

References

[1] Frost, N. E., *Journal of Mechanical Engineering Science,* Vol. 1, 1959, pp. 151–170.
[2] Frost, N. E., *Journal of Mechanical Engineering Science,* Vol. 4, 1962, pp. 22–35.
[3] Yusuff, S., *Aircraft Engineering,* Vol. 34,1962, pp. 140–148.
[4] Frost, N. E., Marsh, K. J., and Pook, L. P., *Metal Fatigue,* Clarendon Press, Oxford, U.K., 1974, Chapter 5.
[5] Pook, L. P., *Metal Science,* Vol. 10, 1976, pp. 334–335.
[6] Rhodes, D., Culver, L. E., and Radon, J. C. in *Fracture and Fatigue—Elasto-Plasticity, Thin Sheet and Micromechanisms Problems,* J. C. Radon, Ed., Pergamon Press, Oxford, U.K., 1980, pp. 287–296.
[7] Schijve, J., *Engineering Fracture Mechanics,* Vol. 14, 1981, pp. 789–800.
[8] Edwards, R. A. H. and Zuidema, J., *Engineering Fracture Mechanics,* Vol. 22, 1986, pp. 751–758.
[9] Wilhem, D. P. in *Fatigue Crack Propagation, ASTM STP 415,* American Society for Testing and Materials, Philadelphia, 1967, pp. 363–383.
[10] Swain, M. H. and Newman, J. C., Jr., "On the Use of Marker Loads and Replicas for Measuring Growth Rates for Small Cracks," *Fatigue Crack Topography,* AGARD-CP-376, Advisory Group for Aerospace Research and Development, 1984, pp. 12-1–12-7.
[11] Frost, N. E. and Dugdale, D. S., *Journal of the Mechanics and Physics of Solids,* Vol. 6, 1958, pp. 92–110.
[12] Frost, N. E. and Denton, K., *Journal of Mechanical Engineering Science,* Vol. 3, 1961, pp. 295–298.
[13] Ritchie, R. O., Smith, R. F., and Knott, J. F., *Metal Science,* Vol. 9, 1975, pp. 485–492.
[14] Bathe, K. J., *ADINA-Automatic Dynamic Incremental Nonlinear Analysis System Theory and Modeling Guide,* Adina Engineering, 1983.
[15] Barsoum, R. S. *International Journal of Fracture,* Vol. 10, 1974, pp. 603–605.
[16] Henshell, R. D. and Shaw, K. G., *International Journal for Numerical Methods in Engineering,* Vol. 9, 1975, pp. 495–507.
[17] Barsoum, R. S., *International Journal for Numerical Methods in Engineering,* Vol. 10, 1976, pp. 25–37.
[18] Banks-Sills, L. and Bortman, Y., *International Journal of Fracture,* Vol. 25, 1984, pp.169–180.
[19] Chan, S. K., Tuba, I. S., and Wilson, W. K., *Engineering Fracture Mechanics,* Vol. 2, 1970, pp. 1–17.
[20] Banks-Sills, L. and Einav, O., *Journal of Computers and Structures,* Vol. 25, 1987, pp. 445–449.
[21] Isida, M., *International Journal of Fracture,* Vol. 7, 1971, pp. 301–316.
[22] Tada, H., Paris, P. C., and Irwin, G. R., *The Stress Analysis of Cracks Handbook,* Del Research Corporation, Hellertown, PA, 1973, pp. 2.1–2.3.
[23] Burton, W. S., Sinclair, G. B., Solecki, J. S., and Swedlow, J. L., *International Journal of Fracture,* Vol. 25, 1984, pp. 3–32.
[24] Benthem, J. P., *International Journal of Solids and Structures,* Vol. 13, 1977, pp. 479–492.
[25] Kawai, T., Fujitani, Y., and Kumagai, K. in *Proceedings of the International Conference on Fracture Mechanics and Technology,* G. C. Sih and C. L. Chow, Eds., Sijhoff and Noordhoff Publishers, Amsterdam, The Netherlands, 1977, pp. 1157–1163.

[26] Bazant, Z. P. and Estenssoro, L. F., *International Journal of Solids and Structures,* Vol. 15, 1979, pp. 405–426.
[27] Raju, I. S. and Newman, J. C., Jr., "Three-Dimensional Finite-Element Analysis of Finite-Thickness Fracture Specimens," NASA TN D-8414, National Aeronautics and Space Administration, Washington, DC, 1977.
[28] Ingraffea, A. R. and Manu, C., *International Journal for Numerical Methods in Engineering*, Vol. 15, 1980, pp. 1427–1445.
[29] Yagawa, G., Ichimiya, M., and Ando, Y. in *Numerical Methods in Fracture Mechanics*, A. R. Luxmoore and D. R. J. Owen, Eds., University College Swansea, Swansea, West Glamorgan, U.K., 1978, pp. 249–267.
[30] Hilton, P. D. and Kiefer, B. V., "The Enriched Element for Finite Element Analysis of Three Dimensional Elastic Crack Problems" presented at the Pressure Vessels and Piping Division Conference of the American Society of Mechanical Engineers, San Francisco, 1979.
[31] Moyer, E. T., Jr., and Liebowitz, H. in *Application of Fracture Mechanics to Materials and Structures*, G. C. Sih, E. Sommer, and W. Dahl, Eds., Martinus Nijhoff Publishers, Amsterdam, The Netherlands, 1984, pp. 595–606.
[32] Shah, R. C. in *Fracture Analysis (Eighth Conference), ASTM STP 560*, American Society for Testing and Materials, Philadelphia, 1974, pp. 29–52.
[33] Sih, G. C., *International Journal of Fracture*, Vol. 10, 1974, pp. 305–321.
[34] Neale, B. K. and Adams, N. J. in *Numerical Methods in Fracture Mechanics*, A. R. Luxmoore and D. R. J. Owen, Eds., University College Swansea, Swansea, West Glamorgan, U.K., 1978, pp. 207–217.
[35] Paris, P. C. and Erdogen, F., *Journal of Basic Engineering*, Vol. 85, 1963, pp. 528–534.
[36] Roberts, R. and Kibler, J. J., *Journal of Basic Engineering*, Vol. 93, 1971, pp. 671–680.
[37] Patel, A. B. and Pandey, R. K., *Fatigue of Engineering Materials and Structures*, Vol. 4, 1981, pp. 65–77.
[38] Hourlier, F. and Pineau, A. in *Avances in Fracture Research*, D. Francois, Ed., Pergamon Press, New York, 1981, pp. 1833–1840.
[39] Badaliance, R., *Engineering Fracture Mechanics*, Vol. 13, 1980, pp. 657–666.
[40] Hua, G., Brown, M. W., and Miller, K. J., *Fatigue of Engineering Materials and Structures*, Vol. 5, 1982, pp. 1–17.
[41] Tanaka, K., *Engineering Fracture Mechanics*, Vol. 6, 1974, pp. 493–507.
[42] Jack, A. R. and Price, A. T., *Acta Metallurgica*, Vol. 20, 1972, pp. 857–866.
[43] Broek, D. and Schijve, J., *Aircraft Engineering*, Vol. 38, 1966, pp. 31–33.

Yoichi Tanaka[1] *and Isao Soya*[1]

Fracture Mechanics Model of Fatigue Crack Closure in Steel

REFERENCE: Tanaka, Y. and Soya, I., "**Fracture Mechanics Model of Fatigue Crack Closure in Steel,**" *Fracture Mechanics: Perspectives and Directions* (*Twentieth Symposium*), *ASTM STP 1020,* R. P. Wei and R. P. Gangloff, Eds., American Society for Testing and Materials, Philadelphia, 1989, pp. 514–529.

ABSTRACT: Fatigue crack propagation tests were performed on a high-strength steel with yield strength of 598 MPa under various stress ratios. A fatigue crack opening model, which contains both the effects of R and ΔK, was investigated based upon the cyclic elasto-plastic finite-element analysis (FEA) and the fracture mechanics approach with the small-scale yielding concept. Assuming the existence of the residual deformation in the wake of a crack, the crack opening stress-intensity factor can be calculated as the value at which the fatigue crack opening displacement becomes zero. Using this model, how R and ΔK affect U was investigated. The resultant formula for U contains two parameters, one for the plastic deformation at a crack tip, and the other for the oxide and the roughness in a fatigue crack. The effective stress-intensity factor range, based on the measurements of crack opening loads, gives the same crack propagation behavior for various stress ratios. It is observed in experiments, however, that U depends on the stress ratio R and the stress-intensity factor range ΔK. The model successfully explains the experimental results. The fatigue crack propagation behavior under various stress ratios is described by the formula with five parameters, which can be obtained with only one specimen. The predicted crack propagation rates and threshold stress-intensity factor range agreed with empirical results. Therefore, the formula derived from this model may be applied well to the evaluation of fatigue crack extension.

KEY WORDS: fatigue crack propagation, stress-intensity factor, crack closure, stress ratio, fracture mechanics, finite-element analysis (FEA), high tensile strength steel, residual deformation

As the environmental and service conditions of structures have become stringent, establishment of a strict safety evaluation method has been urgently needed. Generally, in welded structures where fatigue crack propagation distances are relatively large, their fatigue lives have been said to depend mostly on crack propagation [*1*]. It is particularly important to understand precisely the propagation behavior of fatigue cracks by taking into account many related factors, such as residual stress and stress ratios. Many works [*2,3*] are found on fatigue crack propagation behavior; Elber's crack closure concept [*4*] is successfully adaptable to them [*5,6*].

For the effects of mean stresses on closure in particular, several researchers have reported, because of practical importance, empirical equations for predicting crack opening loads, showing the dependency of the crack opening ratio U ($= \Delta K_{eff}/\Delta K$, where ΔK_{eff} is effective stress-intensity factor range and ΔK is stress-intensity factor range; see also Eq 2) only on the stress ratio R (= minimum load/maximum load) [*4,7*]. The crack opening ratio has also

[1] Research engineer and senior researcher, respectively, Research and Development Laboratories-II, Nippon Steel Corporation, 5-10-1 Fuchinobe, Sagamihara, Kanagawa-ken, Japan, 229.

been known to depend on ΔK as well as on R, decreasing remarkably near the fatigue crack growth threshold ΔK_{th} [8,9]. This means that the dependency on both R and ΔK must be taken into consideration when evaluating crack propagation behavior near the threshold.

In this study, fatigue crack propagation experiments were carried out on a steel plate under various stress ratios, and finite-element analysis (FEA) was conducted to get a better understanding of crack closure phenomena. Furthermore, a small-scale yielding model was proposed to analyze the effects of R and ΔK for crack closure quantitatively. The model was applied to the experimental results, and the general empirical equation of U was investigated.

Test Materials and Experimental Methods

High-strength 16-mm-thick HT60 steel which was made by the quenched and tempered (QT) process was used as the test material for fatigue crack propagation tests. Table 1 shows its chemical composition and mechanical properties. The microstructure of HT60 steel was tempered martensite with an average grain size of 7.39 μm. In addition, SM41A (16-mm-thick low-carbon steel, $\sigma_Y = 269$, $\sigma_B = 453$ MPa) was also used to verify the results of FEA.

Compact type (CT) and center-cracked tension (CCT) specimens were used for the experiment. Figure 1 shows the shapes of the specimens. Electroservo hydraulic fatigue testing machines with 50- and 100-kN capacities were used. Crack lengths were evaluated by the compliance method, whereas crack opening displacements were measured by the clip-on gage attached to the specimens, as shown in Fig. 1.

Testing frequency ranged from 5 to 40 Hz, but was lowered to 1 Hz during compliance measurement, taking into consideration the frequency characteristics of the clip-on gage.

To convert compliances into crack lengths, the equation which was proposed by Saxena et al. [10] and corrected by the benchmark measurement, was used, whereas the five-point averaging method was employed to evaluate crack lengths by the beach mark method. The values of K for CT and CCT were calculated in accordance with the ASTM Standard Test Method for Constant-Load-Amplitude Fatigue Crack Growth Rates Above 10^{-8} m/cycle (E 647).

Crack propagation rates were evlauated by the secant method. To obtain data for a small da/dN region, tests were carried out by gradually decreasing ΔK, with R kept constant.

TABLE 1—*Chemical composition (in weight %) and mechanical properties of the material.*

CHEMICAL COMPOSITION	
Carbon	0.13
Silicon	0.29
Manganese	1.31
Phosphorus	0.020
Sulfur	0.007
Copper	0.015
Nickel	0.007
Chromium	0.009
Molybdenum	<0.005
Vanadium	0.004
MECHANICAL PROPERTIES	
σ_Y, MPa	598
σ_B, MPa	647
Elongation, %	37

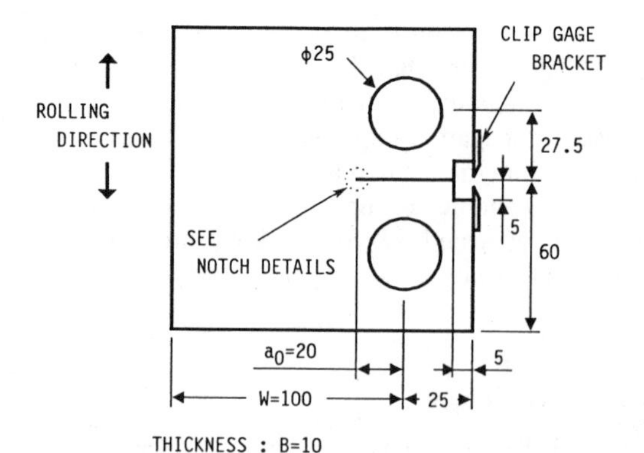

(a) COMPACT TYPE SPECIMEN

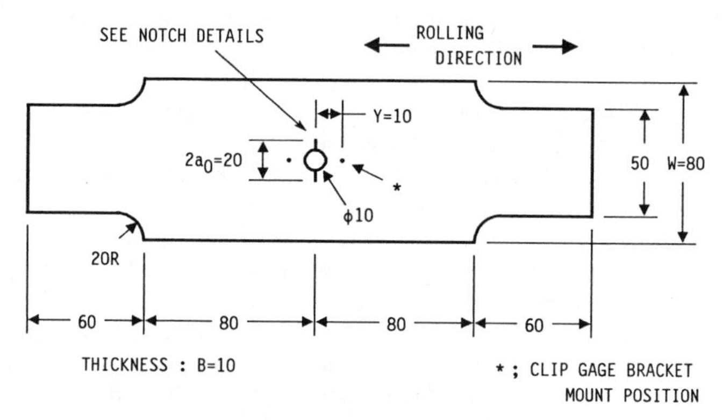

(b) CENTER CRACKED TENSION SPECIMEN

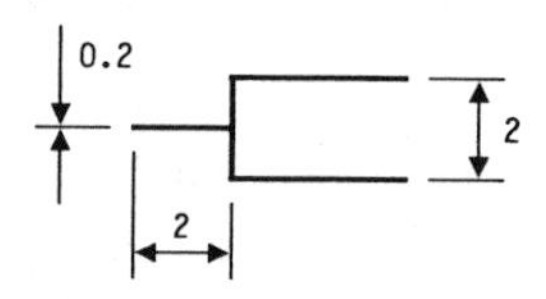

(c) NOTCH DETAILS

FIG. 1—*Crack propagation test specimens (dimensions in millimeters).*

Decreasing ΔK was controlled according to [11]

$$\Delta K = \Delta K_0 \exp[C(a - a_0)] \tag{1}$$

where

a_0 = initial crack length,
ΔK_0 = initial stress-intensity factor range,
C = decreasing rate of ΔK [$= (d\Delta K/da)/\Delta K$] ($C$ was taken as -0.05 mm^{-1} in this study).

The load was reduced gradually until da/dN became less than 1×10^{-7} mm/cycle, and, after that, increased 1.2- to about 1.5-fold and then kept constant during testing. The value of ΔK during the decreasing ΔK test was partly overlapped by that during the constant ΔP (load range) test. Then it was confirmed that there was no effect of decreasing load on da/dN.

Further, closure load (P_{op}) was determined as the point in the compliance (load versus specimen deflection curve) below which the load-displacement relationship deviated from the linear elastic compliance [12]. In this case, in order to improve accuracies, closure load was determined using the curve which was obtained by deducting the displacement in the linear relationship from the load-displacement curve.

Results of Experiment

Stress Ratio Dependency of Fatigue Crack Propagation Characteristics

Figure 2 shows the results of the fatigue crack propagation tests when R is changed from -1 to 0.8. Here, the ΔK is defined as $K_{max} - K_{min}$ even for $R < 0$. The CT specimens were used when $R \geq 0.1$, while the CCT specimens were used when $R \leq 0.1$. In other words, both the CT and CCT specimens were used when $R = 0.1$, and there was no significant difference between them. The data for the CCT specimens with 2 mm or more difference between right and left crack lengths were omitted.

In Fig. 2, the R dependency of crack propagation rate is not observed in the larger ΔK region when $R > 0$, but only near ΔK_{th}. When $R < 0$, on the other hand, R dependency is observed over the whole range of ΔK, and shows a parallel relationship, as given in the log-log scale.

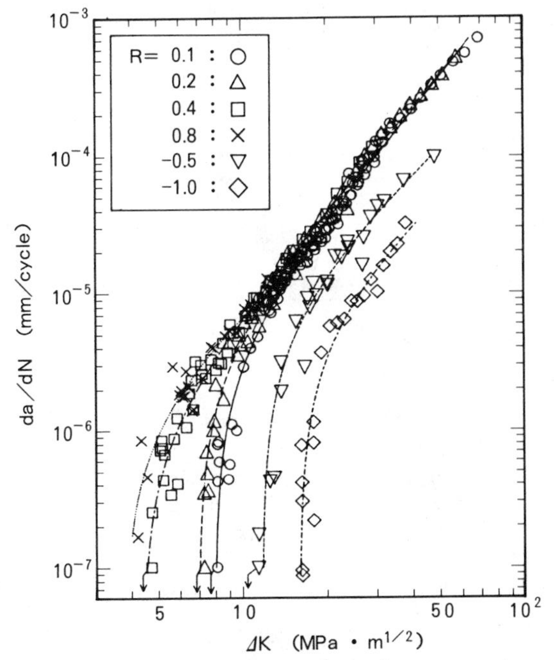

FIG. 2—*Crack propagation rate for various stress ratios.*

Arrangement Based on Effective Stress-Intensity Factor Range

As reported in many papers, the stress ratio dependency of fatigue crack propagation rates can be shown by a single curve, in accordance with the relationship of $\Delta K_{eff} = U\Delta K$ [4,5,7,8], where U is the opening ratio, which can be defined by the crack opening stress-intensity factor range K_{op}, as shown in

$$U = (K_{max} - K_{op})/(K_{max} - K_{min}) \tag{2}$$

Figure 3 shows the results rearranged on the basis of ΔK_{eff}. All the results almost agree, regardless of R. Then, in accordance with the modified Paris rule, they can be expressed by a single curve as [3,13]

$$da/dN = C(\Delta K_{eff}{}^{m} - \Delta K_{eff,th}{}^{m}) \tag{3}$$

where $\Delta K_{eff,th}$ is the threshold stress-intensity factor range, expressed in terms of ΔK_{eff}. The constants obtained by regression of experimental data, such as C, m, $\Delta K_{eff,th}$, are shown in the figure.

ΔK *Dependency of Crack Opening Ratio* U

Many researchers, including Elber [4], have proposed equations estimating the crack opening ratio U. Although most of them consider only R as a factor which contributes to U, it is known that U actually depends on ΔK. Near ΔK_{th}, U is also known to approach zero [5,8,9]. Figure 4, which shows the ΔK dependency of U measured in the present study,

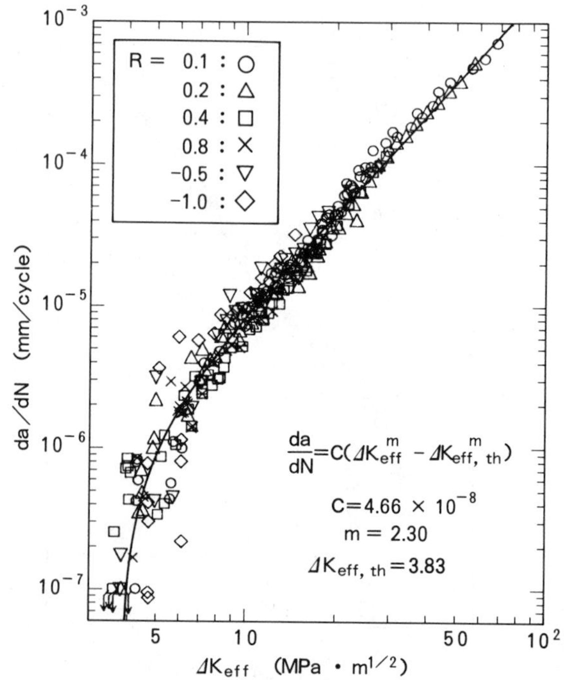

FIG. 3—*Relationship between crack propagation rate and effective stress-intensity factor range.*

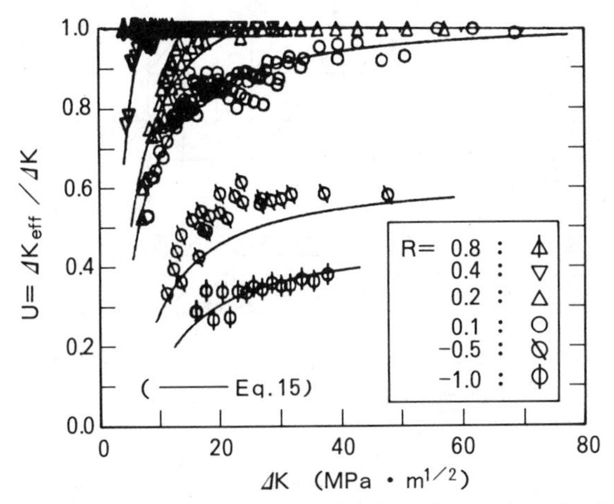

FIG. 4—*Change of crack opening ratio* U *as a function of stress-intensity factor range* ΔK.

also gives the same trend. The solid line in Fig. 4 indicates the results of analysis by using the model explained in the following sections.

Finite-Element Analysis of Plasticity-Induced Closure

It is well known that plasticity-induced closure can be reproduced by cyclic elastic-plastic FEA [*14*], but most of the analyses were performed under plane stress conditions. In the case of low fatigue crack growth rate, such as near threshold, however, the cracks are considered to propagate under plane strain condition [*15*], as described later. Therefore, plasticity-induced closures under plane strain condition were analyzed.

Method of Analysis

Analysis was carried out for 50-mm-wide CT specimens. Figure 5 shows finite-element subdivisions used for analysis. The minimum mesh size was 0.125 mm. A crack tip was first positioned at Node A (Fig. 5*b*), and the crack was made propagate by one mesh for every loading-unloading cycle. This procedure was repeated until the crack reached Node *K*, followed by the repetition of the loading-unloading cycle without making the crack propagate. In this analysis, the mixed hardening theory was employed to reproduce the cyclic elastic-plastic behavior of the materials [*16*], together with a mixed hardening parameter of 0.5. The stress-strain curve was input through multilinear approximation. Analysis was carried out for the mild steel (σ_Y = 284 MPa, hereinafter referred to as M steel) and the high-strength steel (σ_Y = 565 MPa, hereinafter referred to as H steel), and the effect of yield stresses on plasticity-induced closures were also investigated.

In general, the analysis of closure behavior leads to solving contact problems, which is extremely complicated. In this analysis, utilizing the symmetry of the specimen, the crack was positioned on the *x*-axis, and the closure was treated as a boundary problem; namely, in the course of loading, the boundary condition of nodes on the crack laid on the *x*-axis are changed from fixed to free condition for the crack to open when the reaction force acting on the fixed nodes become positive, while in the course of unloading, the displacement of the node in the *y* direction is constrained for the crack to carry compressive reaction force when it becomes zero. The crack is made propagate at the minimum load. The applied

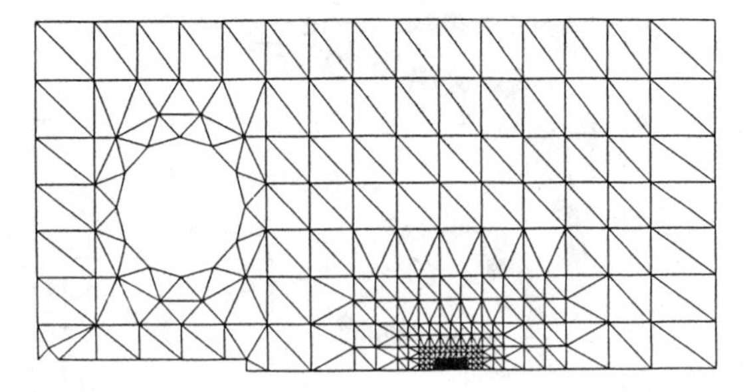

(a) Finite element subdivision (630 elements, 360 nodes)
CT specimen (B=10mm, W=50mm)

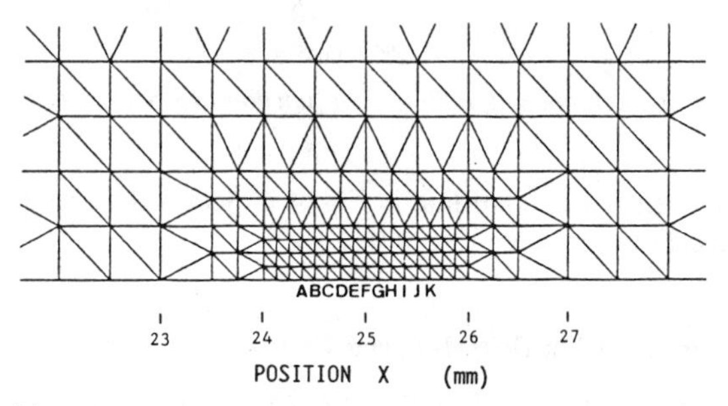

(b) Details around the crack tip

FIG. 5—*Finite-element subdivision for CT specimen.*

loads were from 0 to 9.8 kN. This load range corresponds to 40 to 45 MPa $m^{1/2}$ in terms of ΔK. Thus ΔK was selected at a high level in the experimental condition to realize cyclic yielding around the crack tip and to clarify the role of plastic deformation on crack closure. Moreover, Young's modulus and Poisson's ratio were 206 GPa and 0.3, respectively.

Forms of Crack Surfaces

Figure 6 shows the calculated displacements of crack surfaces of the M steel: the solid line shows the displacements of fatigue cracks, whereas the dotted line shows those of ideal cracks without cyclic crack extension. In addition, the dashed line gives the results of calculation by using the model explained later. The most outstanding difference between the fatigue crack and ideal crack is the residual stretch formed on the wake of fatigue crack. In the fatigue crack, therefore, its crack surfaces come into contact with each other (plasticity-induced closure) [*4,5,14,17*] before the load becomes zero. The residual stretch thicknesses, obtained as the difference between the fatigue crack and ideal crack, were 3.85 μm at the crack tip for the M steel and 1.45 μm for the H steel, but the thicknesses became thinner with the increasing distance from the crack tip.

Behavior of Closures

Figure 7 depicts the change in crack opening loads as the crack propagates. One triangular wave form corresponds to one loading-unloading cycle. In Figs. 7*a* and 7*b*, opening loads and closure loads are greatly different in their values up to Step 11, and this may be because

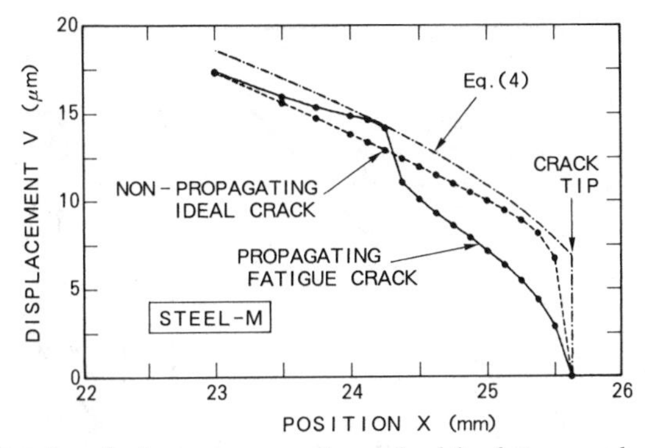

FIG. 6—*Crack surface displacements at maximum load for fatigue crack and ideal crack.*

the amount of crack propagation per cycle (0.125 mm) is greater than actual crack propagation [*14*]. The first cycle loading is responsible for the gradual increase in the crack opening load from the initial stage of crack propagation. When crack propagation is arrested after Step 12, the opening and closure loads approach each other to a negligible extent, with the former being a little higher. These normalized opening and closure loads of about 0.1 for steel M are a little lower than the measured opening load to maximum load ratios of about 0.2 for SM41A steel at high ΔK region comparable to FEA, and tend to decrease thereafter with the increasing number of loading after crack arrest.

Yielding Zones at Crack Tips and Specimen Compliances

The yielding zone widths lying ahead of the crack on the x-axis were 0.875 and 0.250 mm for the M and H steels, respectively. Since the yielding zone widths of FEA results are sufficiently smaller than the sizes of the specimens, their compliances give almost straight lines, as shown in Fig. 8*a*. However, when the linear component due to elastic deformation is deducted (Fig. 8*a*, right), the entire curve shows small hysteresis and bend at the load corresponding to the closure load. Their shapes are convex to the right because of possible existence of a re-yielding zone. The changes in compliance due to closure is smaller than that observed by ordinary experiments, which may be ascribed to a smaller closing part of

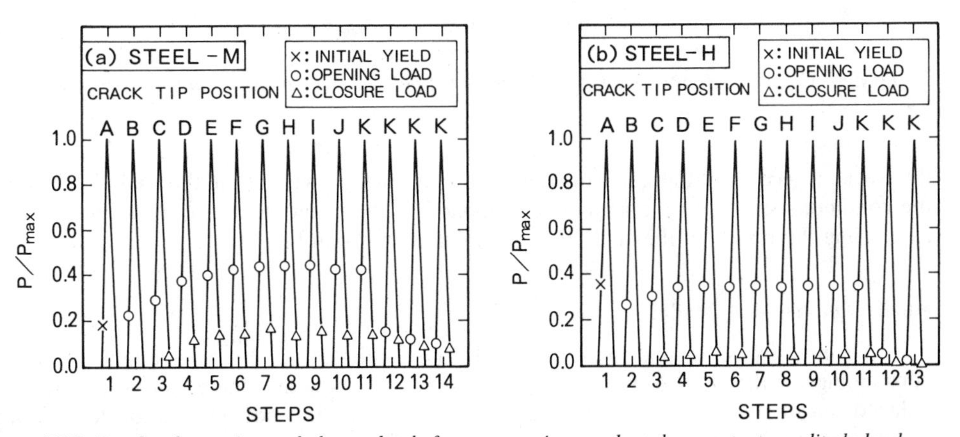

FIG. 7—*Crack opening and closure loads for propagating crack under constant amplitude load.*

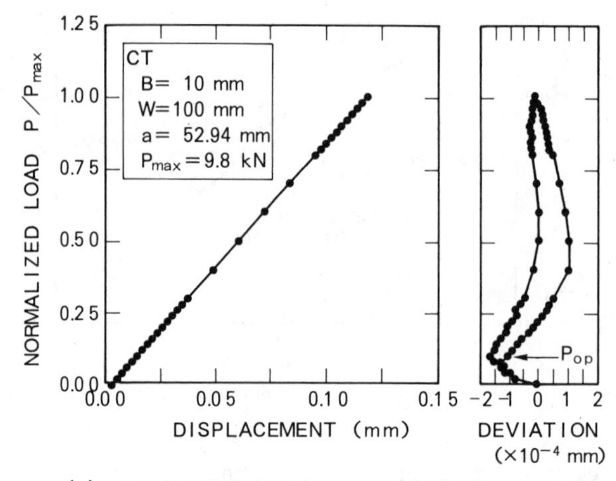

(a) Result of finite element analysis for steel-M

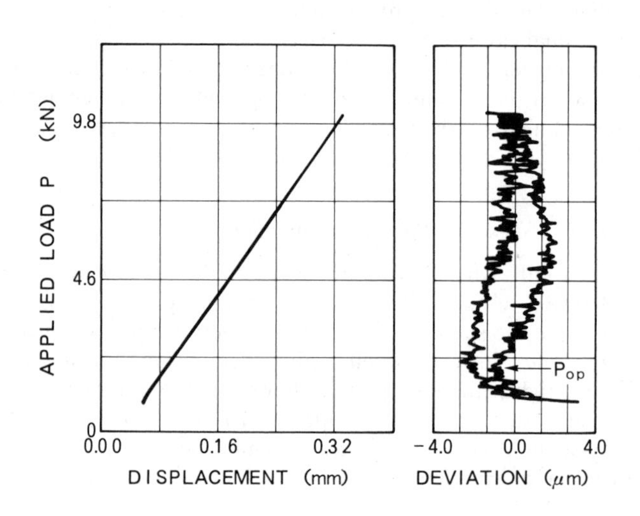

(b) Measured result for SM41A steel in high ΔK region

FIG. 8—*Compliance curves for CT specimen by* (a) *finite-element analysis and* (b) *experiment.*

the crack on account of the small amount of fatigue crack propagation in the analysis. When the conditions similar to the analysis are applied to the experiment, the hysteresis curve resembling the results of the analysis is obtained as shown in Fig. 8*b*.

Small-Scale Yielding Model

FEA can easily be applied to an idealized closure problem, and it gives qualitative understandings of the closure. For example, FEA results show that there can be plasticity-induced closure even under the plane-strain condition. But it is not suitable for deriving a quantitative result including many factors affecting the closure. For instance, the phenomenon is said to be affected to crack surface roughness, and the presence of oxides on the wake of fatigue crack [*5,18,19*], as well as plasticity, but FEA of their effects is rather difficult. To perform analysis including them and to investigate the effect of R and ΔK, an attempt was made to construct a model regarding the closure phenomenon.

Correction of Elastic Cracks by Small-Scale Yielding

The description of stress fields and displacements near crack tips on the basis of stress-intensity factor is correct only when a substance is completely elastic. Practically, a crack-tip yielding zone extends. When the yielding zone at the crack tip is small in scale, however, it is known that an elastic analysis can be applied reasonably by shifting the actual crack tip by a half (r_p) of the yielding zone width ω, as shown in Fig. 9 [*20*]. In this case, the displacement of the ideal crack surface at the position of x is given by

$$V = 4K^* \{(r_p - x)/2\pi\}^{1/2}/E^* \tag{4}$$

where $r_p = (K_{max}/\alpha\sigma_Y)^2/2\pi$ and $E^* = E/(1 - \nu^2)$ (plane strain) or E (plane stress). In Eq 4, E, ν, and K_{max} are Young's modulus, Poisson's ratio, and maximum K value, respectively. K^* is a corrected K for the shifted crack tip, but little difference is found between K^* and the original K. So, in the analysis performed hereinafter, K is looked upon as K^*. The constraint factor of plastic deformation is represented by α and equals 1 under plane stress condition. According to the slipband theory, α is equal to 3 under plane strain condition, but the value of 1.7 is also used empirically.

Figure 6 shows the crack surface displacements calculated from Eq 4, with ΔK and α being 43.4 MPa $m^{1/2}$ and 3, respectively. In this case, the plastic zone width, $2r_p$, for the M and H steels are 0.824 and 0.210 mm, respectively, coinciding well with the results of FEA. Equation 4 gives displacements for ideal cracks, which are a little larger than those obtained by FEA, but these two kinds of displacements are, as a whole, in fairly good agreement.

Since the results of Eq 4 are based on elastic calculation, they do not include plastic deformation remaining on fatigue crack surfaces. Assuming the residual plastic deformation thickness δ_{res}, displacement V_f along the fatigue crack surface can be written as

$$V_f = V - \delta_{res} = 4K\{(r_p - x)/2\pi\}^{1/2}/E^* - \delta_{res} \tag{5}$$

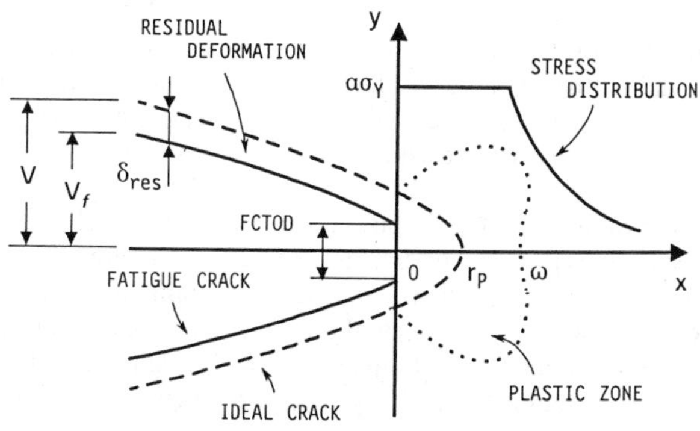

FIG. 9—*Small-scale yielding model for crack closure analysis.*

When $x = 0$ in Eq 5, the fatigue crack-tip opening displacement (FCTOD), ϕ_f, can be calculated as

$$\phi_f = 2V_f(x = 0) = 4KK_{max}/\pi\alpha\sigma_Y E^* - 2\delta_{res} \quad (6)$$

If $\phi_f = 0$, K_{op} can be obtained as

$$K_{op}^* = \pi\alpha\sigma_Y E^*\delta_{res}/2K_{max} \quad (7)$$

In Eq 7, since δ_{res} seems to depend on the materials, the effects of materials on K_{op} are still unknown. When K_{op} values are calculated from Eq 7 by using the same material constants as in the FEA, and on the assumption that δ_{res} for the M and H steels are 3.85 and 1.45 μm, respectively, the resultant K_{op} of the M and H steels are 24.7 MPa $m^{1/2}$ and 18.5 MPa $m^{1/2}$, respectively, for $\Delta K = 43$ MPa $m^{1/2}$.

In FEA, a crack closure was detected at the node which is 0.125-mm mesh size (r_0) before the crack tip, so that the effect of r_0 must be taken into consideration when the analytical result and the result of FEA are compared. Simple calculation leads to the following opening K value K_{op}^* at a distance of r_0 from the crack tip:

$$K_{op}^* = K_{op}/(1 + r_0/r_p)^{1/2} \quad (8)$$

Calculation of K_{op}^* by using the above-mentioned values results in 21.6 MPa $m^{1/2}$ for the M steel and 12.5 MPa $m^{1/2}$ for the H steel, and these values are still higher than those obtained by FEA. Equation 8 indicates that K_{op} tends to be underestimated by FEA with the increasing mesh size, and this trend is in good agreement with the results obtained by Newman [*14*].

Estimation of Residual Stretch Thickness

The small-scale yielding model mentioned above does not involve the effect of re-yielding, and it is taken into account in δ_{res}. Further, δ_{res} is affected by crack surface roughness and fretting oxide debris, so that δ_{res} can be expressed as

$$\delta_{res} = \delta_{plas} + \delta_{ox} + \delta_{rough} \quad (9)$$

where δ_{plas}, δ_{ox}, and δ_{rough} indicate the contributions by plastic deformation, taking into consideration the effect of re-yielding, by oxides, and by crack surface roughness, respectively.

Budiansky et al. [*17*], by the use of the Dugdale-Barenblatt model, found that δ_{plas} at $R = 0$ is proportional to the opening displacement at the maximum load. Their analyses did not make clear how δ_{ox} and δ_{rough} depend on R and ΔK, but generally the crack surface roughness becomes greater with increasing da/dN [*21*], and the contribution of oxides is remarkable near ΔK_{th} [*22*].

The authors then assume that δ_{plas} is plastic deformation (a half of crack-tip opening displacement of ideal crack) corresponding to $f(R)\Delta K$, and summation of δ_{ox} and δ_{rough} corresponds to plastic deformation when K reaches a constant value K_0. The function $f(R)$ varies only with R and shows R dependency of δ_{plas}, including the effect of re-yielding. Since the concrete form of $f(R)$ is not easily obtainable by analytical means, it will be empirically determined. From these assumptions and Eq 4, δ_{res} can be represented as

$$\delta_{res} = 4(f(R)\Delta K + K_0)(r_p/2\pi)^{1/2}/E^* \quad (10)$$

Substitution of Eq 10 for Eq 7 leads to

$$K_{op} = f(R)\Delta K + K_0 \tag{11}$$

where the properties of materials are included in $f(R)$ and K_0. From Eqs 11 and 2, the value of U when $K_{op} \geq K_{min}$ can be written as

$$U = 1/(1 - R) - f(R) - K_0/\Delta K = g(R) - K_0/\Delta K \tag{12}$$

where the sum of R-dependent terms is shown as $g(R)$.

The second term on the right side of Eq 12 is negligible in high ΔK region. The empirical equations for U reported so far are considered to be applicable when ΔK is relatively great, so that $g(R)$ can be determined by referring to them. Kato et al. proposed the following equation for the wide R range [7]

$$U = 1/(1.5 - R) \tag{13}$$

Then, a generalized Eq 13 is assumed as $g(R)$, as shown in

$$g(R) = 1/(R_0 - R) \tag{14}$$

where R_0 is a constant and may reflect the material properties such as the tensile strength. Consequently, U can be expressed, by using Eqs 12 and 14, as

$$U = 1/(R_0 - R) - K_0/\Delta K \text{ (for } K_{op} \geq K_{min}\text{) or 1 (for } K_{op} < K_{min}) \tag{15}$$

In other words, U can be represented as the function of R and ΔK, which includes two independent parameters, R_0 and K_0.

Comparison of Analytical Results with Experimental Data

Figure 10 shows the relationship between U and $1/\Delta K$. The dotted lines in the figure were drawn as follows: two parameters, R_0 and K_0, were determined by regressing data of

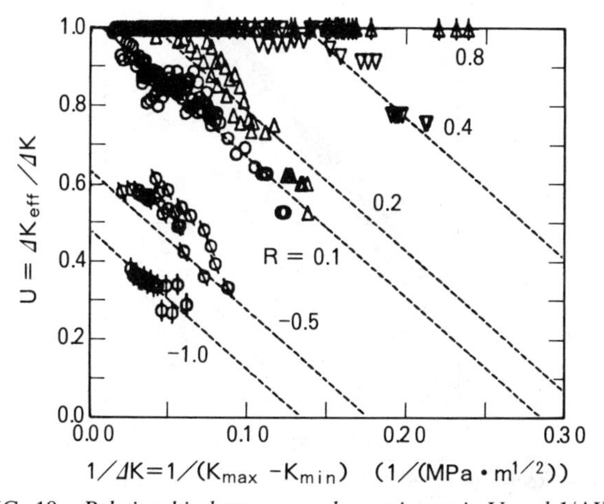

FIG. 10—*Relationship between crack opening ratio* U *and* $1/\Delta$K.

$R = 0.1$ by means of the least square. In this case, $K_0 = 3.60$ MPa $m^{1/2}$ and $R_0 = 1.07$ were obtained. These values were then applied to other stress ratios. In Fig. 10, the outstanding linearity of the experimental data for $1/\Delta K$ proves the reasonableness of the second term on the right side of Eq 12; and further, good agreement between Eq 15 and the experiment results, irrespective of R, verifies that the assumption represented by Eq 14 is also reasonable. Equation 15 can also be expressed as the function of ΔK, and shown by solid lines in Fig. 4, too.

When the parameters R_0 and K_0 which depend on material and can be determined by closure measurement in the test of $R \approx 0$ are known, ΔK in the test of any stress ratio can be converted to ΔK_{eff} by using these equations even without closure measurement for each stress ratio. Figure 11 shows the experimental results rearranged this way using the parameters obtained from the test of $R = 0.1$, while the solid line is the best-fit curve for the data in Fig. 3.

Discussion

Equations 3 and 15 enable the behavior of fatigue crack propagation, when R varies, to be calculated by five parameters, such as C, m, R_0, K_0, and $\Delta K_{eff.th}$, all of which can be obtained by only one specimen. Figure 12 shows calculated da/dN for various R by using Eqs 3 and 15 with the parameters obtained by the regression in Fig. 3 for C, m, and $\Delta K_{eff.th}$ and by the line of $R = 0.1$ in Fig. 10 for R_0 and K_0. The dependency of da/dN on R is observed only near ΔK_{th} when $R \geq 0$, but no R dependency when ΔK is greater, with da/dN being on one curve. On the other hand, when $R < 0$, the effect of R is observed over the whole range of ΔK and shows the parallel relationship in the log-log scale. These results coincide well with those in Fig. 2.

If one assumes that $\Delta K_{eff}(= U\Delta K)$ is constant, Eqs 3 and 15 enable the calculation of ΔK, which gives the same da/dN for different R, resulting in

$$\Delta K = (\Delta K_{eff} + K_0)(R_0 - R) \text{ (for } K_{op} \geq K_{min}\text{) or } \Delta K_{eff} \text{ (for } K_{op} < K_{min}\text{)} \tag{16}$$

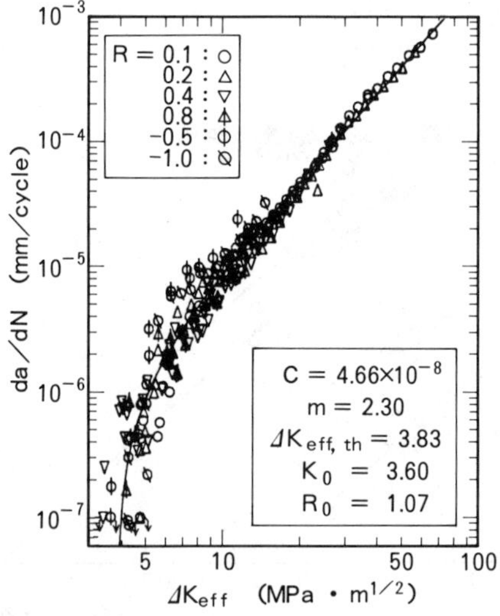

FIG. 11—*Rearrangement of crack propagation data using estimated* U.

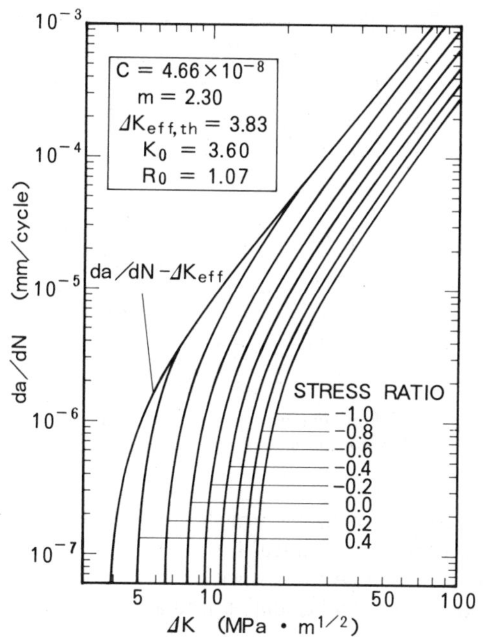

FIG. 12—*Calculated result of crack propagation rates for various stress ratios.*

where $\Delta K_{eff} = [(da/dN)/C + \Delta K_{eff,th}{}^{m}]^{1/m}$. The results of calculation are given in Fig. 13, in which the dotted line shows the limit for the R dependency, to be represented as

$$\Delta K = K_0(R_0 - R)/(1 - R_0 + R) \tag{17}$$

In addition, the R dependency of ΔK_{th} is obtainable from Eq 16 when $da/dN = 0$, or $\Delta K_{eff} = \Delta K_{eff,th}$, and can be written as

$$\Delta K_{th} = \max[(\Delta K_{eff,th} + K_0)(R_0 - R), \Delta K_{eff,th}] \tag{18}$$

The results are also shown in Fig. 13. As shown in this figure, the difference between the ΔK_{th} line and the line for $da/dN = 1 \times 10^{-7}$ mm/cycle is extremely small. In this case, therefore, the value of ΔK at which da/dN becomes less than 1×10^{-7} mm/cycle in the experiment may well be looked upon as ΔK_{th}. The dependency of crack growth rate on stress ratio such as shown in Fig. 13 has already been shown experimentally [*2,23,24*]. The formula for U expressed by Eq 15 is very similar to the one which was derived experimentally by Bignonnet et al. [*25*].

Summary and Conclusions

For precise evaluation of the fatigue crack propagation lives of structures, the effects of external forces or of stress ratios due to welding residual stresses must be taken into account. Since the effects of the stress ratio can be expressed by the crack opening ratio U, it is useful to obtain general equations considering factors affecting U. In the present study, crack propagation tests of HT60 steel were carried out under various stress ratios. And finite-

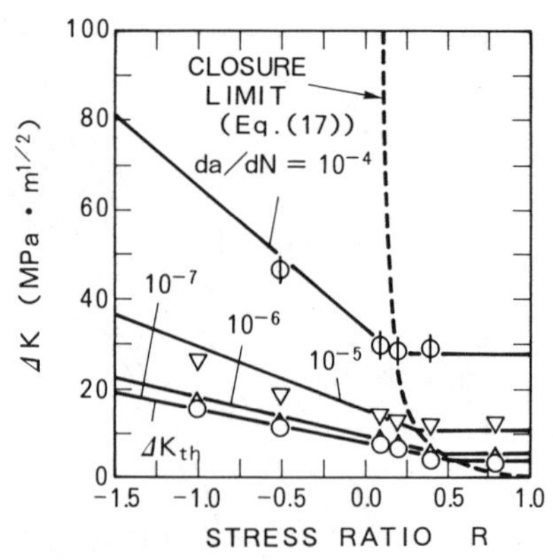

FIG. 13—*Relationship between stress-intensity factor range and stress ratio for various crack propagation rates (dimensions of crack propagation rates in mm/cycle).*

element analyses were conducted to confirm the role of plastic deformation in the crack closure phenomenon. Furthermore, a linear fracture mechanics model was derived to quantitatively study the phenomenon. An investigation was then carried out on the effects of R and ΔK on U. The conclusions can be summarized as follows:

1. The empirical model assuming small-scale yielding and residual stretch can reproduce the closure phenomenon. This model includes plasticity-induced closure, which is shown to be important even under plane strain condition by FEA.
2. The fatigue crack propagation rates can be rearranged by the modified Paris rule, shown in Eq 3, despite the stress ratios. However, C, m, and $\Delta K_{eff.th}$ are the constants determined by the experiment and do not depend on the stress ratios.
3. The crack opening ratio U depends on ΔK as well as on R and can be represented by Eq 15, where R_0 and K_0 are the material constants, and are 1.07 and 3.60 MPa m$^{1/2}$ for HT60 in this work, respectively.
4. Equations 3 and 15 show the dependency of da/dN on both R and ΔK. The effects of R appear only near ΔK_{th} when $R \geq 0$, but the R dependency is found over the whole ΔK range when $R < 0$.
5. Five parameters in Eqs 3 and 15 can be obtained by one specimen.
6. The R dependency of ΔK_{th} can be represented by Eq 18.
7. In the above-mentioned case, the value of ΔK at which da/dN becomes less than 1×10^{-7} mm/cycle in the experiment may be considered as ΔK_{th}.

The theoretical equation of U proposed in this paper includes the material constants R_0 and K_0, which must be determined empirically. Those constants were obtained for HT60 steel. But it is still unknown whether or not they can be applied to steels other than HT60. How these constants depend on material remains a problem to be investigated.

References

[1] Vosikovsky, O., Bell, R., Burns, D. J., and Mohaupt, U. H. in *Behaviour of Offshore Structures (BOSS'85), Proceedings,* 4th International Conference on Behavior of Offshore Structures, J. A. Battjes, Ed., Elsevier, Amsterdam, 1985, pp. 453–464.

[2] Ohta, A., Sasaki, E., and Kosuge, M., *Transactions,* Japan Society of Mechanical Engineers, Vol. 43, No. 373, 1977, pp. 3179–3189 (in Japanese).

[3] Klesnil, M. and Lukas, P., *Materials Science and Engineering,* Vol. 9, 1972, pp. 231–240.
[4] Elber, W. in *Damage Tolerance in Aircraft Structures, ASTM STP 486,* American Society for Testing and Materials, Philadelphia, 1971, pp. 230–242.
[5] Kobayashi, H. and Mura, T., *Proceedings,* 27th Japan National Symposium on Strength, Fracture and Fatigue, 1982, pp. 97–126 (in Japanese).
[6] Dill, H. D. and Saff, C. R. in *Fatigue Crack Growth under Spectrum Loads, ASTM STP 595,* American Society for Testing and Materials, Philadelphia, 1976, pp. 306–319.
[7] Katoh, A., Kurihara, M., and Kawahara, M., *Journal of the Society of Naval Architects of Japan,* Vol. 153, 1983, pp. 336–343 (in Japanese).
[8] Jono, M., Song, J., Mikami, S., and Ohgaki, M. *Journal of the Society of Materials Science, Japan,* Vol. 33, No. 367, 1984, pp. 468–474 (in Japanese).
[9] Okamoto, T., Toyosada, M., Fujiwara, H., and Hamada, H., *Journal of the Society of Naval Architects of Japan,* Vol. 153, 1983, pp. 344–351 (in Japanese).
[10] Saxena, A. and Hudak, S. J., Jr., *International Journal of Fracture,* Vol. 14, No. 5, 1978, pp. 453–468.
[11] Saxena, A., Hudak, S. J., Jr., Donald, J. K., and Schmidt, D. W., *Journal of Testing and Evaluation,* Vol. 6, No. 3, 1973, pp. 167–174.
[12] Kikukawa, M., Jono, M., Tanaka, K., and Takatani, M., *Journal of the Society of Materials Science, Japan,* Vol. 25, 1976, pp. 899–903 (in Japanese).
[13] Ohta, A., Soya, I., Nishijima, S., and Kosuge, M., *Engineering Fracture Mechanics,* Vol. 24, No. 6, 1986, pp. 789–802.
[14] Newman, J. C., Jr., in *Mechanics of Crack Growth, ASTM STP 590,* American Society for Testing and Materials, Philadelphia, 1976, pp. 281–301.
[15] Suzuki, Y., Masumoto, T., and Ogura, T., *Journal of the Japan Institute of Metals,* Vol. 44, No. 9, 1980, pp. 1076–1083 (in Japanese).
[16] Soya, I., Takashima, H., and Tanaka, Y. in *Fracture Mechanics: Seventeenth Volume, ASTM STP 905,* J. H. Underwood, R. Chait, C. W. Smith, D. P. Wilhem, W. A. Andrews, and J. C. Newman, Eds., American Society for Testing and Materials, Philadelphia, 1986, pp. 202–225.
[17] Budiansky, B. and Hutchinson, J. W., *Journal of Applied Mechanics,* Vol. 45, 1978, pp. 267–276.
[18] Ritchie, R. O., Suresh, S., and Moss, C. M., *Journal of Engineering Materials and Technology,* Vol. 102, 1980, pp. 293–299.
[19] Minakawa, K. and McEvily, A. J., *Scripta Metallurgica,* Vol. 15, 1981, pp. 633–636.
[20] Okamura, H. in *Introduction of Linear Fracture Mechanics,* Baifukan, Japan, 1976 (in Japanese).
[21] Horng, J. L. and Fine, M. E. in *Fatigue Crack Threshold Concepts,* D. L. Davidson and S. Suresh, Eds., The Metallurgical Society of American Institute for Mining, Metallurgical, and Petroleum Engineers, 1984, pp. 115–129.
[22] Suresh, S. and Ritchie, R. O. in *Fatigue Crack Threshold Concepts,* D. L. Davidson and S. Suresh, Eds., The Metallurgical Society of American Institute for Mining, Metallurgical, and Petroleum Engineers, 1984, pp. 227–261.
[23] Schmidt, R. A. and Paris, P. C. in *Progress in Flow Growth and Fracture Toughness Testing, ASTM STP 536,* American Society for Testing and Materials, Philadelphia, 1973, pp. 79–94.
[24] Nakai, Y., Tanaka, K., and Kawashima, R., *Journal of the Society of Materials Science, Japan,* Vol. 33, No. 371, 1984, pp. 85–91 (in Japanese).
[25] Bignonnet, A., Dias, A., and Lieurade, H. P. in *Advances in Fracture Research, Proceedings,* 6th International Conference on Fracture, S. R. Valluri, D. M. R. Taplin, P. Ramarao, J. F. Knott, and R. Dubey, Eds., Pergamon Press, Oxford, 1984, pp. 1861–1868.

Anthony Palazotto[1] *and E. Bednarz*[2]

A Finite-Element Investigation of Viscoplastic-Induced Closure of Short Cracks at High Temperatures

REFERENCE: Palazotto, A. and Bednarz, E., "**A Finite-Element Investigation of Viscoplastic-Induced Closure of Short Cracks at High Temperatures,**" *Fracture Mechanics: Perspectives and Directions (Twentieth Symposium), ASTM STP 1020,* R. P. Wei and R. P. Gangloff, Eds., American Society for Testing and Materials, Philadelphia, 1989, pp. 530–547.

ABSTRACT: The authors have developed a finite-element program with constant-strain capabilities that can handle cyclic loading. The constitutive relation is considered to be elastic-viscoplastic incorporating the Bodner-Partom flow rule. Incalloy 718 is the alloy considered at a temperature of 670°C. The computer algorithms are discussed for consideration of crack closure and extension.

The program is applied to a single-edge-cracked specimen with an initial crack length of 25 μm acting under a load distribution of 90% of the yield stress for the material considered. A plane-stress finite element is used to evaluate the effect of an *R*-value equal to 0.1, considering frequencies of 0.01 and 1.0 Hz. The crack is allowed to grow to a final length of 46 μm. The plastic zone in front of the crack tip is approximately equal to the initial crack length, yielding conditions associated with short cracks.

The results indicate the effect viscoplasticity plays when considering and comparing two separate frequencies. A plastic wake, due to crack closure, develops behind the propagating crack and increases for the lower frequency. In order to realistically predict plasticity-induced closure, one must mathematically allow the crack to grow. Stress and strain fields in front of the crack show a noticeable change with time due to the viscoplastic effect.

KEY WORDS: fatigue, viscoplasticity, finite elements, short cracks

The use of finite elements in fracture mechanics is quite extensive. A certain amount of research has been published in recent years in the use of this numerical technique in characterizing viscoplasticity effects within a crack-tip region [*1–3*]. The authors have recently participated in work depicting the fatigue effect on short cracks considering elastic-plastic material [*4*]. This last reference is another publication in short-crack considerations which include crack closure and extension. Newman [*5,6*] has carried out finite-element studies of a propagating crack under cyclic loading. Reference *4* lists several more references on the short-crack problem.

To the authors' knowledge, Ref *4* is the first paper that attempted to approach the fatigue loading of a propagating crack at high temperature by elastic-viscoplastic analysis. Focus was directed toward the viscoplastic-induced closure on cracks that may be considered short.

[1] Professor of aeronautics and astronautics, Air Force Institute of Technology, Wright Patterson Air Force Base, Dayton, OH 45433.

[2] Ph.D. candidate, Air Force Institute of Technology, Wright Patterson Air Force Base, Dayton, OH 45433.

The primary interest of the present paper is in the numerical features of the development of the viscoplastic programs, including the concepts of crack closure and extension. Major details of this work can be found in Ref *7*.

Approach

This section provides some numerical characteristics present within a finite-element program referred to as Visco II, a program that has the capability of solving two-dimensional plane-stress or plane-strain problems by constant triangles. (A plane-stress analysis is carried out in subsequent applications.) The main feature of Visco II is the consideration of elastic-viscoplasticity incorporating the Bodner-Partom flow law within a fracture mechanic environment. A revision of an initial program [*8*] had to be carried out to include the phenomenon of fatigue loading referred to as "closure." This fatigue characteristic has been discussed in several references noted previously, and thus will not be elaborated upon herein. The results subsequently are presented for a short-edge crack specimen that is acted upon by a varying load which depicts properties of not only crack growth but also crack closure. The closure feature is a by-product of the plasticity-induced response of a superalloy, IN718, at a temperature of 670°C; thus the need for a viscoplasticity approach.

For completeness, the Bodner-Partom flow law is discussed [*9*]. This viscoplastic flow law satisfies the requirements for time-dependent and time-independent behavior, and has been used successfully to study viscoplasticity in a variety of analyses, including cyclically loaded specimens. This law, based on dislocation dynamics, suggests a continuous flow relation between stress and viscoplastic strain commencing at the onset of loading. As used herein, the law assumes that materials involved are isotropic and exhibit no kinematic hardening.

Formulation of the constitutive equations begins by recognizing that total strain can be separated into elastic, as well as viscoplastic, strains such that

$$\epsilon_{ij} = \epsilon_{ij}^{e} + \epsilon_{ij}^{p} \tag{1}$$

Taking the time derivative of this equation, the total strain rate is then expressed as

$$\dot{\epsilon}_{ij} = \dot{\epsilon}_{ij}^{e} + \dot{\epsilon}_{ij}^{p} \tag{2}$$

where the elastic strain rate, $\dot{\epsilon}_{ij}^{e}$, is related to the stress rate through the time derivation of Hooke's law.

The determination of the viscoplastic strain rate, however, begins by squaring the rate form of the Prandtl-Ruess plastic flow equation

$$\dot{\epsilon}_{ij}^{p} = \lambda \dot{S}_{ij} \tag{3}$$

to obtain

$$\dot{\epsilon}_{ij}^{p}\, \dot{\epsilon}_{ij}^{p} = \lambda^{2} S_{ij}\, S_{ij} \tag{4}$$

where

$\dot{\epsilon}_{ij}^{p}$ = components of the deviatoric viscoplastic strain rate tensor,
S_{ij} = components of the deviatoric stress tensor, and
λ = a proportionality constant.

Then, by substituting the relations

$$J_2 = \frac{1}{2} S_{ij} S_{ij} \tag{5}$$

and

$$D_2^p = \frac{1}{2} \epsilon_{ij}^p \epsilon_{ij}^p \tag{6}$$

one obtains

$$D_2^p = \lambda^2 J_2 \tag{7}$$

D_2^p is defined as the second invariant of the plastic strain rate and J_2 the second invariant of the deviatoric stress tensor.

Based on experimental work, Bodner and Partom expressed D_2^p as

$$D_2^p = D_0^2 \exp\left\{-\left(\frac{Z^2}{3J_2}\right)^n \left(\frac{n+1}{n}\right)\right\} \tag{8}$$

where

D_0 = limiting value of the plastic strain rate in shear,
Z = a measure of material hardness, and
n = a rate sensitivity parameter.

Furthermore, Z depends on the deformation history of the material and is assumed to be a function of plastic work, $\overline{W}_p$, as follows

$$Z = Z_1 - (Z_1 - Z_0) \exp[-m\overline{W}p] \tag{9}$$

where

Z_0 = initial value of hardness prior to any plastic deformation,
Z_1 = maximum value attainable,
m = a parameter controlling the rate of material work hardening, and
$\overline{W}_p$ = relative amount of plastic work done relative to some initial state expressed in the form

$$\overline{W}p = \int S_{ij}\epsilon_{ij}^p dt = 2(D_2^p J_2)^{1/2} \tag{10}$$

Most materials exhibit some thermal recovery of hardening or relaxation of accumulated plastic work at high temperatures. To model this particular behavior, plastic work may be redefined as

$$\overline{W}p = \int S_{ij}\epsilon_{ij}^p dt + \int [\dot{Z}_{rec}/[m(Z_1 - Z)]dt \tag{11}$$

where the thermal hardness recovery term is

$$\dot{Z}_{rec} = -A((Z - Z_2)/Z_1)^r Z_1 \tag{12}$$

Here, Z_2 is the value of Z in a completely non-work-hardened condition, indicating the minimum value of material hardness, whereas A and r are the hardening recovery coefficient and exponent, respectively. The appropriate constants for IN718 are presented in Table 1.

The specific procedure used to solve the Bodner relations is a Euler extrapolation scheme which integrates the equations with respect to time. This is accomplished for each element in the finite-element code as follows:

$$Z^i = Z_1 + (Z_0 - Z_1) \exp [-m\overline{W}_p^{i-1} - 1/Z_0] \tag{13}$$

$$(D_2^p)^i = D_0^2 \exp [-((Z^i)^2/(3\, J_2^{i-1}))^n((n + 1)/n)] \tag{14}$$

$$(\epsilon_{ij}^p)^i = [(D_2^p)^i/J_2^{i-1}]^{1/2} (S_{ij})^{i-1} \tag{15}$$

$$\{d\epsilon_{ij}^p\} = \{\epsilon_{ij}^p\}^i\, dt^i \tag{16}$$

$$\dot{Z}_{\rm rec}^i = A[(Z^i - Z_2)/Z_1]^r Z_1 \tag{17}$$

$$\overline{W}_p^i = \overline{W}_p^{i-1} + \{S_{ij}\}^{i-1}\{d\epsilon_{ij}^p\}^i + \dot{Z}_{\rm rec}^i\, dt^i/[m(Z_1 - Z^i)] \tag{18}$$

where the superscript i refers to the current time step.

Visco II is the computer program used in this research. The overall solution technique incorporated is the residual force method. Key features within this program are a Gauss-Siedel iterative procedure coupled with a crack closure algorithm to solve for equilibrium, the Euler extrapolation scheme previously discussed to model elastic-viscoplastic material behavior, a data storage/restart capability to handle extensive cyclic loading, and a procedure to simulate crack growth. It should be pointed out that Visco II is an extension of the Visco program developed by Hinnerichs [8], modified to meet the needs of this research.

The basic concept in the residual force method is that the elastic stiffness matrix remains constant throughout an analysis, and plasticity effects are incorporated into the solution through the inclusion of a plastic load vector. The resultant equilibrium equations then take on the form

$$[K]\{u\}^i = \{P\}^i + \{Q\}^{i-1} \tag{19}$$

TABLE 1—*Bodner coefficients for IN718 at 670°C.*

Parameter	Description	Value
E	elastic modulus	16.250×10^4 MPa
n	strain rate exponent	3.0
D_0	limiting value of strain rate	10^6 s
Z_0	limiting value of hardness	1622 MPa
Z_1	maximum value of hardness	1795 MPa
Z_2	minimum value of hardness	718 MPa
A	hardening recovery coefficient	1.5×10^{-3} s^{-1}
r	hardening recovery exponent	7.00
σ_{ys}	material yield stress	896 MPa
m	hardening rate exponent	0.427 MPa

where

$[K]$ = constant elastic stiffness matrix,
$\{u\}^i$ = nodal displacement vector,
$\{P\}^i$ = a vector of externally applied loads, and
$\{Q\}^{i-1}$ = vector of internal loads developed due to the accumulation of plastic deformations.

The superscript i in this relation refers to the current time step indicated previously. The algorithm employed takes on the following form:

1. Compute the current time

$$t^i = t^{i-1} + dt^i \tag{20}$$

2. Calculate the plastic strain rate using the Bodner model/Euler explicit extrapolation scheme

$$\{\dot{\epsilon}_{ij}^p\}^i = [(D_2^p)^i / J_2^{i-1}]\{S_{ij}\}^{i-1} \tag{21}$$

3. Compute the plastic strain increments

$$\{d\epsilon_{ij}^p\}^i = \{\dot{\epsilon}_{ij}^p\}^i \, dt^i \tag{22}$$

4. Calculate the plastic load vector

$$\{Q\}^{i-1} = \int [B]^T [D] \{\epsilon_{ij}^p\}^i \, \text{dvol} \tag{23}$$

where $[B]$ relates strain to displacement; $[D]$ = plane stress-strain relation.

5. Compute the current external loads

$$\{P\}^i = \{P\}^{i-1} + \{P\}^i \, dt^i \tag{24}$$

6. Calculate nodal displacements using the Gauss-Siedel technique

$$\{u\}^i = [K]\,\{\{P\}^i + \{Q\}^{i-1}\} \tag{25}$$

7. If closure has occurred, determine the resultant nodal forces and modify the external load vector accordingly

$$\{P\}^i = \{P\}^i + \{P_{\text{clo}}\}^i \tag{26}$$

8. Calculate total strains

$$\{\epsilon_{ij}^p\}^i = [B]\,\{u\}^i \tag{27}$$

9. Compute total stresses

$$\{\sigma_{ij}\}^i = [D]\,\{\{\epsilon_{ij}\}^i - \{\epsilon_{ij}^p\}^i\} \tag{28}$$

10. Modify current time step size and continue procedure until analysis is completed.

With this procedure, time is incremented directly and the resultant displacements, stresses, and strains are subsequently evaluated. The particular time step, dt, is determined subject to stress and strain tolerances and to an algorithm relating these constraints to changes in the current time step size. This approach thereby maximizes the allowable time step and thus minimizes computational requirements (see subsequent paragraph).

The stress and strain parameters are functions of averaging expressions that take on values related to effective stress and strain for each element, as well as stress and strain tolerance values. By controlling the size of the time step, the level of accuracy inherent in the equilibrium and compatability relations is determined. A very thorough investigation of Δt was carried out [7,8], and the results show that if the amount of change in stress or strain components in any element is not allowed to exceed 10% during a single time increment, then the overall results are acceptable. It is therefore possible to vary Δt throughout an analysis to produce an accurate as well as a computer-efficient solution.

Crack Closure

A major feature of the Visco II program is its ability to handle fatigue loading that leads to possible closure behind the crack tip. In order to do this, an algorithm consistent with the overall program philosophy of not changing the stiffness array is carried out. Figure 1 shows the method graphically.

If negative displacements occur along the crack face (indicating crack closure), Visco II performs an iterative closure routine. This procedure first resets applicable nodal displacements to zero (as negative values are physically impossible) and then calculates a resultant nodal force required to insure a zero displacement using the following relations

$$\{P_{\text{clo}}\}^i = \sum_{e=1}^{3} \int [B]\{\sigma_{ij}\}\text{dvol} \tag{29}$$

$\{P_{\text{clo}}\}^i$ is the vector of external forces at the closed nodes and is a function of stresses in the three adjacent elements. These forces are then added to the external load vector and the solution for equilibrium is repeated. This procedure thus incorporates the changing boundary conditions involved in crack closure without modifying the stiffness matrix. Since nodal displacements will change with the addition of these closure loads, this crack closure algorithm is repeated after the equilibrium solution until the difference in closure loads on subsequent iterations is less than a specified tolerance value (1%). Because this procedure is performed only when negative displacements occur at nodes along the crack surface, subsequent crack opening is easily handled. Crack opening is indicated by a lack of negative displacements and the crack closure algorithm is, thus, not performed. Additionally, by compensating for crack closure through the inclusion of an internal load vector, the global stiffness matrix remains constant and does not require any modifications during the above routine.

Crack Growth

A second characteristic present within this Visco II program is its ability to follow the growth of a crack by "popping" or releasing nodes at designated time intervals. Figure 2 visually displays the phenomenon. As shown, when a node is to be released, the forces at that node required to maintain a zero displacement are evaluated using equation (29). This is equivalent to an oppositely directed force, and it is then applied to the node and reduced

I) No Closure:

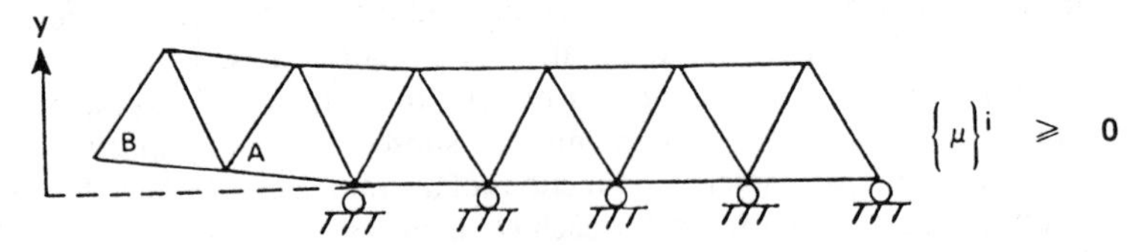

II) Negative Displacements:

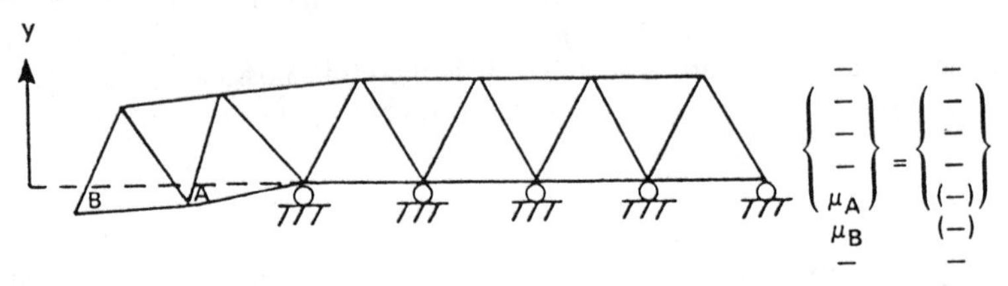

III) Reset to Zero Displacements:

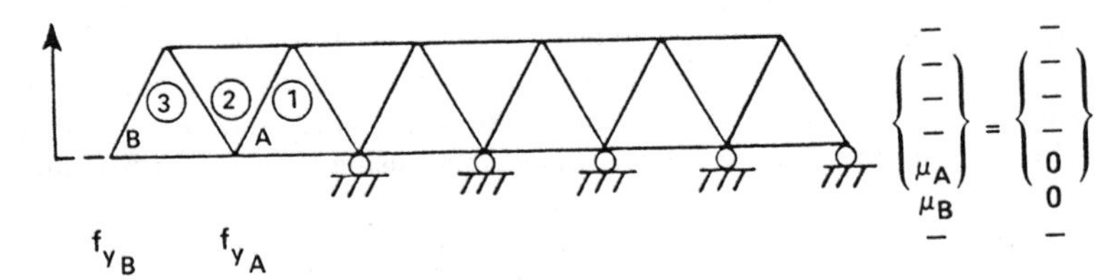

IV) Determine Closure Loads:

(e = element)

$$\{\epsilon_y\}^i = [B]\{\mu\}^i$$
$$\{\sigma_{ij}\}^i = [D]\left(\{\epsilon_y\}^i - \{\epsilon_y{}^P\}^i\right)$$
$$\{P_{CLO}\}^i = \sum_{e=1}^{3} \int [B]\{\sigma_{ij}\}^i dvol$$

V) Resolve Equilibrium:

$$\{\mu\}^i = [K]^{-1}\left(\{P\}^i + \{P_{CLO}\}^i + \{Q\}^{i-1}\right)$$

FIG. 1—*Closure model.*

linearly over time. The boundary condition that had previously been in effect is conveniently handled in the Gauss-Siedel equation solver, which makes use of the fact that the stiffness matrix is initially developed on an individual nodal basis in which only adjacent nodes contribute to displacement values. When a node is fixed, its equilibrium equation is skipped over in the iterative solution procedure. As that node is released, however, its equilibrium equation is then included. By handling crack growth and changing boundary conditions in this fashion, one need not refactor the global stiffness matrix.

In order to further detail the nature of the node release mechanism used in the program, one must consider the load-time curve for a given situation. Figure 3 indicates such a function for the study subsequently discussed. It can be observed that the R ratio (P_{min}/P_{max}) is 0.1. Two frequencies are being considered, 1 Hz and 0.01 Hz. The nature of modal release can be physically interpreted keeping in mind the function shown in Fig. 3.

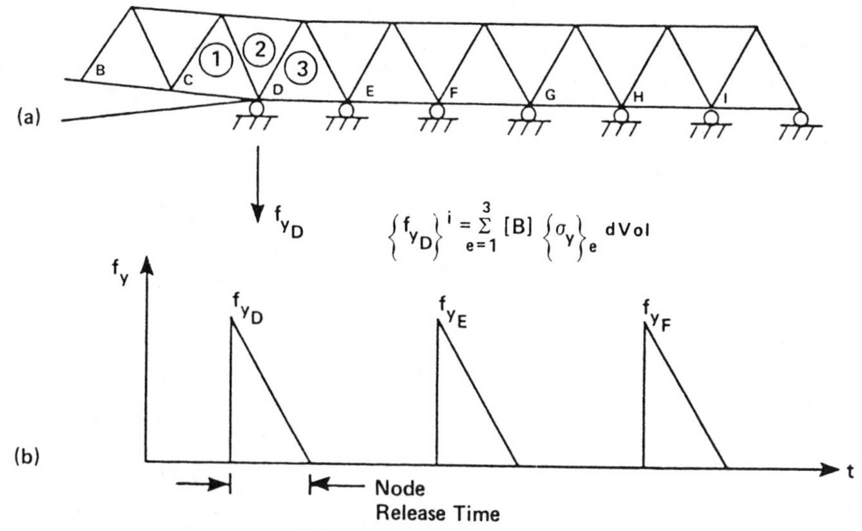

FIG. 2—*Crack growth model.*

Node release (crack growth) is assumed to occur at the point of maximum load application, at the one-half cycle point for $R = 0.1$ as shown in Fig. 4. This allows the forces to be reduced at the released node as the total external load is also being reduced. The result of this is that stresses near the crack tip do not increase significantly during the node release process. This condition may have resulted had node release occurred during an increase in external loads. As explained previously, when a node is released, a force is initially calculated and internally applied to that node to maintain a zero displacement. This force is then reduced over a specific time interval—the node release time—until that node is no longer restrained. For the finite-element simulation herein, the node release time was arbitrarily set equal to one-half the time from maximum load application to the point of minimum positive load. Hence, for $R = 0.1$ and 1 Hz, release time was set at 0.25 s. In this manner, the node release algorithm was completed prior to any externally applied compressive force or the beginning of a subsequent load cycle.

The investigation, to be discussed subsequently, allowed the release of five nodes during the computer analysis. This is called node popping and is the numerical technique representing crack growth. The release of a node was assumed to be a function of a steady-state condition in front of the crack tip. Since it was not feasible to attain these conditions prior to the releases of each node, a preliminary study was carried out to determine an optimal interval between subsequent node releases. A single-edge-cracked specimen was loaded at approximately 60% yield strength and load ratios of both -1.0 and 0.1 were applied. Crack-tip stress distributions ahead of the crack tip and displacements behind the crack tip remained virtually unchanged after one complete load cycle. Plastic strains ahead of the crack tip, however, required several cycles to reach a near steady-state condition. In particular, after two complete load cycles, the change in the average amount of plastic strains in the three elements immediately in front of the crack tip was less than 4%. Additional cycles were therefore deemed unnecessary when considering the amount of crack growth anticipated. For this reason, during the computer analysis of the single-edge-cracked specimens, a node is released during one load cycle and another complete cycle is run to allow for the stabi-

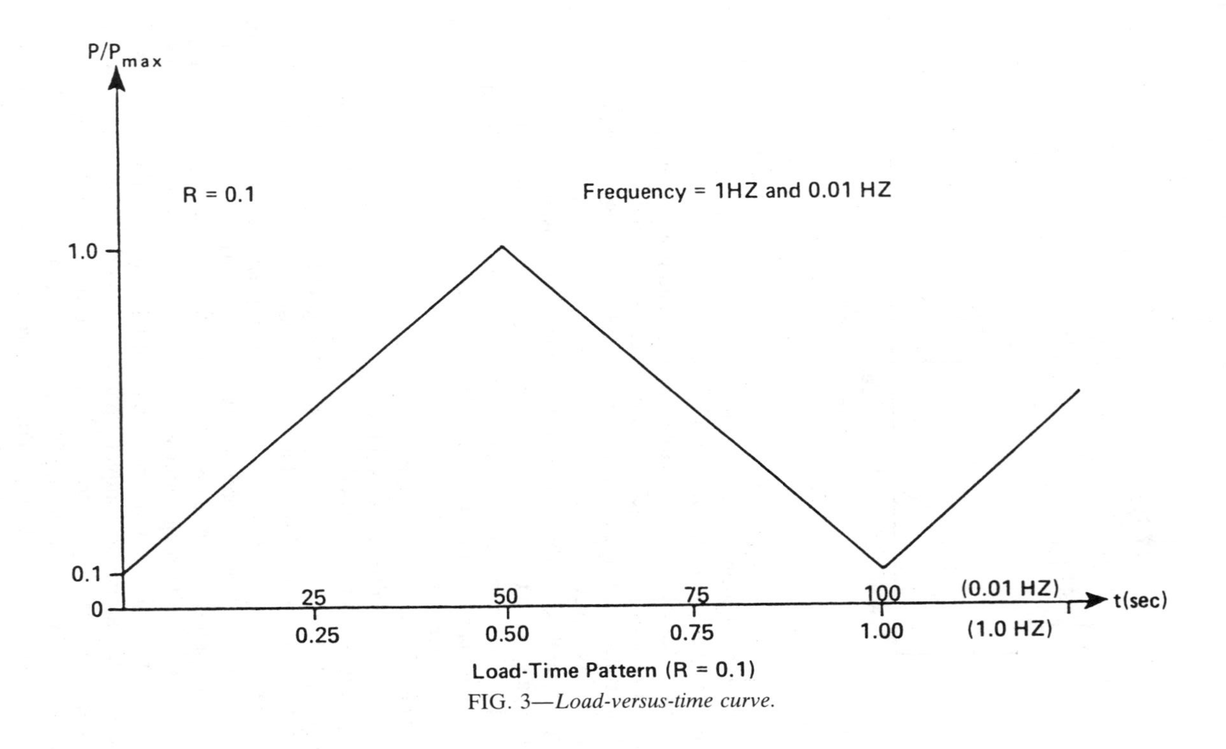

FIG. 3—*Load-versus-time curve.*

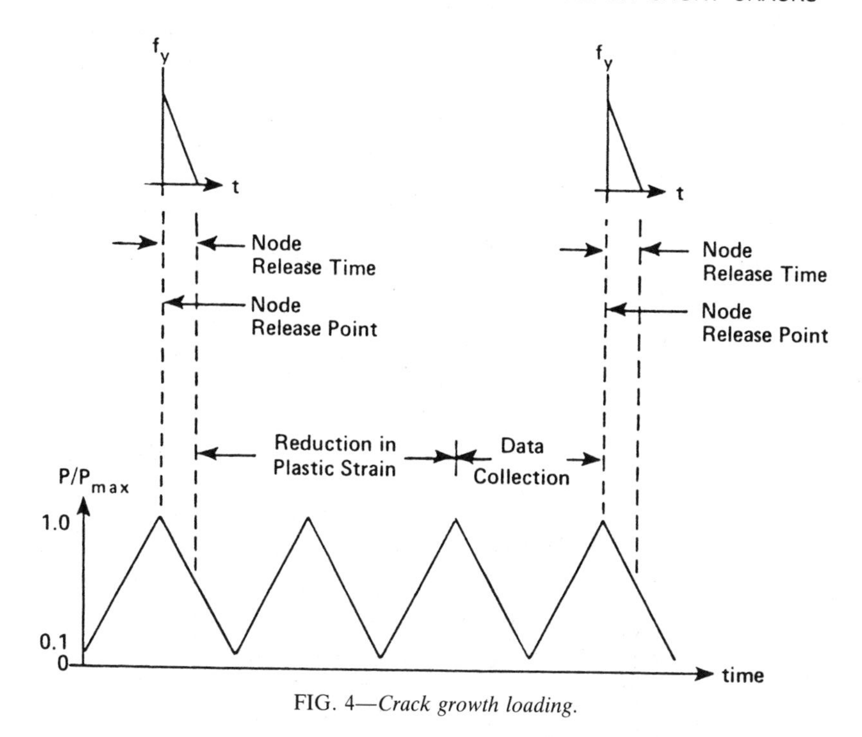

FIG. 4—*Crack growth loading.*

lization of stress and strain conditions. On the following cycle, data are collected until the desired crack growth is attained.

Application

The previous section discussed some of the major segments within Visco II. The authors applied this program to the investigation of viscoplastic effects of a cracked specimen. They have chosen to evaluate a short edge-notch specimen (Fig. 5). Short cracks have shown a variation in behavior from long cracks as fatigue loading is considered. Several investigations, alluded to previously, have examined this feature. The authors have also presented and published a paper specializing in plasticity-induced closure [*4*]. It is not their purpose to go back over the literature review, but only to present the application of the newly developed program in considering viscoplasticity-induced closure in a specimen at a high temperature.

As indicated previously, a short edge crack is considered. Figure 5 represents the finite-element model of half the specimen with symmetry taken into account. A boundary load equivalent to $0.9\sigma_{ys}$ (σ_{ys} = yield stress for IN718 equal to 896 MPa) is applied with an initial crack length assumed to be 25 μm. Thus, an initial K_{I} = 8.08 MPa $\text{m}^{1/2}$ is assumed. Figure 6 depicts the element model in the vicinity of the crack tip. Enough refinement is constructed to allow for a final crack length of 45 μm. Thus, one can see from Fig. 6 that the crack can propagate through five nodes without any refinement difficulties. The global model consists of 389 elements and 231 nodes. Using the features of loading and unloading for R = 0.01 and the sequence indicated in Fig. 4, the crack length versus cycle curve shown in Fig. 7 can be produced. A total of 17 cycles are considered in this study. Two frequencies, 1 Hz and 0.01 Hz, are also considered. Thus, 1700 s represents the length of time for the second frequency compared to only 17 s for the first. IN718 will start to display viscoplastic effects

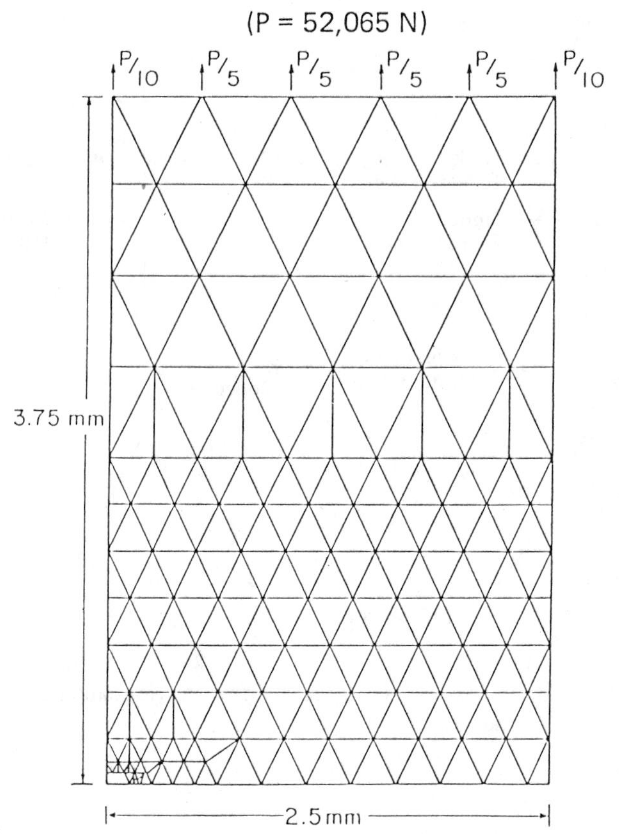

FIG. 5—*Finite-element model.*

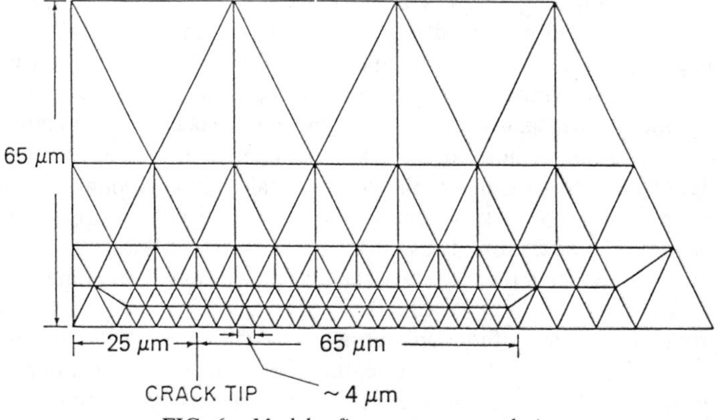

FIG. 6—*Model refinement near crack tip.*

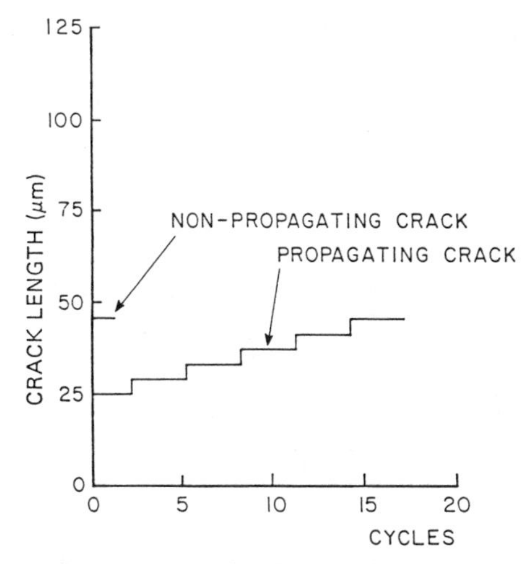

FIG. 7—*Crack length versus load cycle.*

at a high temperature. As will be illustrated, this investigation is a comparison between a propagating and a nonpropagating crack characterizing short crack effects displayed by viscoplasticity-induced closure.

Figure 8 is a plot of the displacement profile behind the crack tip at minimum load. For the nonpropagating crack, one may initially observe that the crack is fully opened and that the maximum displacement occurs at the node directly behind the crack tip. The propagating crack displacement profile shows that the crack is closed for a small length behind the tip and then starts to open. One may further observe a discontinuity at the node with an approximate location of the initial crack length (25 μm). This seems to suggest that under cyclic loading the displacements take on high values directly behind the crack tip when the history effect produced by prior cycling and crack extension is not included. Thus, this study reveals that a finite-element generated displacement profile behind a propagating crack tip is unrealistic unless the cumulative history effect of propagating is included.

Figure 8 further indicates the effect of time. The smaller frequency (0.01 Hz versus 1.0 Hz) produces a displacement profile, relative to the crack tip, with larger ordinate values. One can see a slight but recognizable difference both for the propagating as well as for the nonpropagating crack.

Figure 9 shows the displacement profile behind the crack tip at maximum load. The large displacements correspond to the nonpropagating crack as well as to the lower frequency. The shaded area corresponds to the plastic wake remaining behind the propagating crack. The viscoplastic time effect does not seem to be significant in developing the amount of wake, at least for this study.

The next feature examined is the plastic strain profile in the *y*-direction along the crack length. Figure 10 shows the results for the maximum load during each cycle. There are obviously no residual strains behind the crack tip for the nonpropagating crack. The propagating crack generates residual strain behind the tip that can be closely associated with the plastic wake previously discussed. One can also observe that the measurable plastic strain region ahead of the crack tip is approximately equal to the initial crack length. This is typical of short cracks. The characteristics of viscoplastic are starting to be observed in this figure as in the previous plots. The 0.01-Hz frequency has a plastic strain plot slightly higher than the 1-Hz frequency.

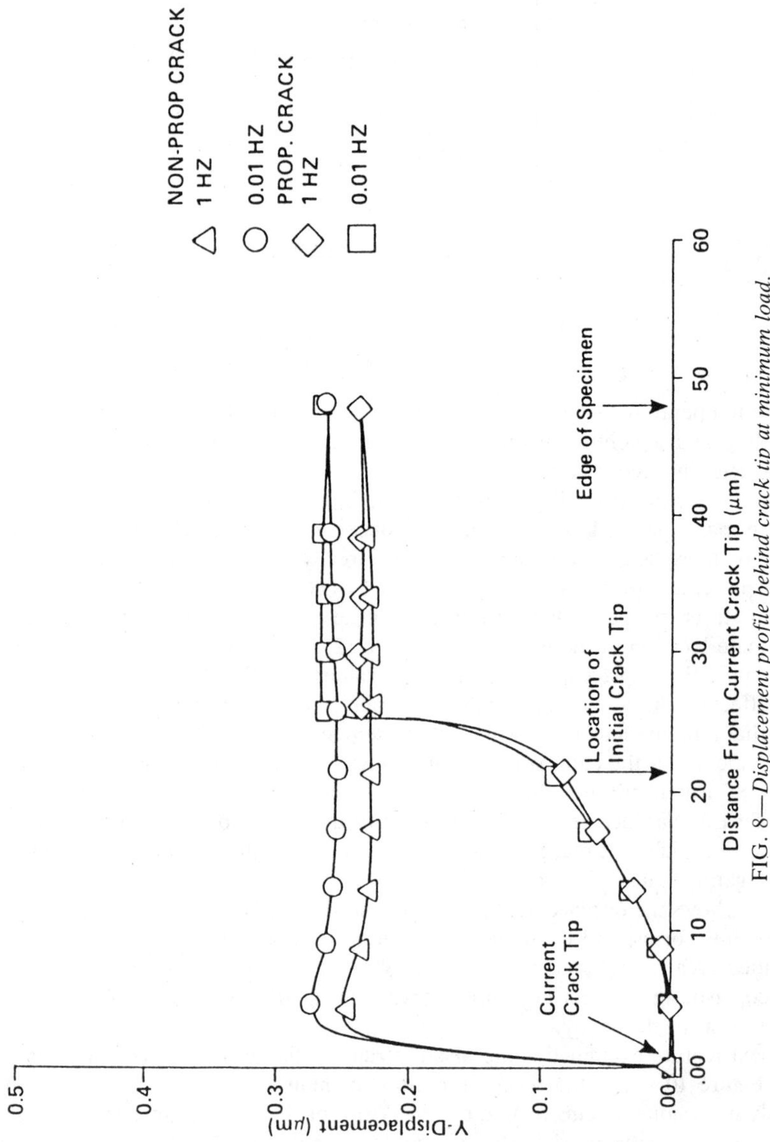

FIG. 8—*Displacement profile behind crack tip at minimum load.*

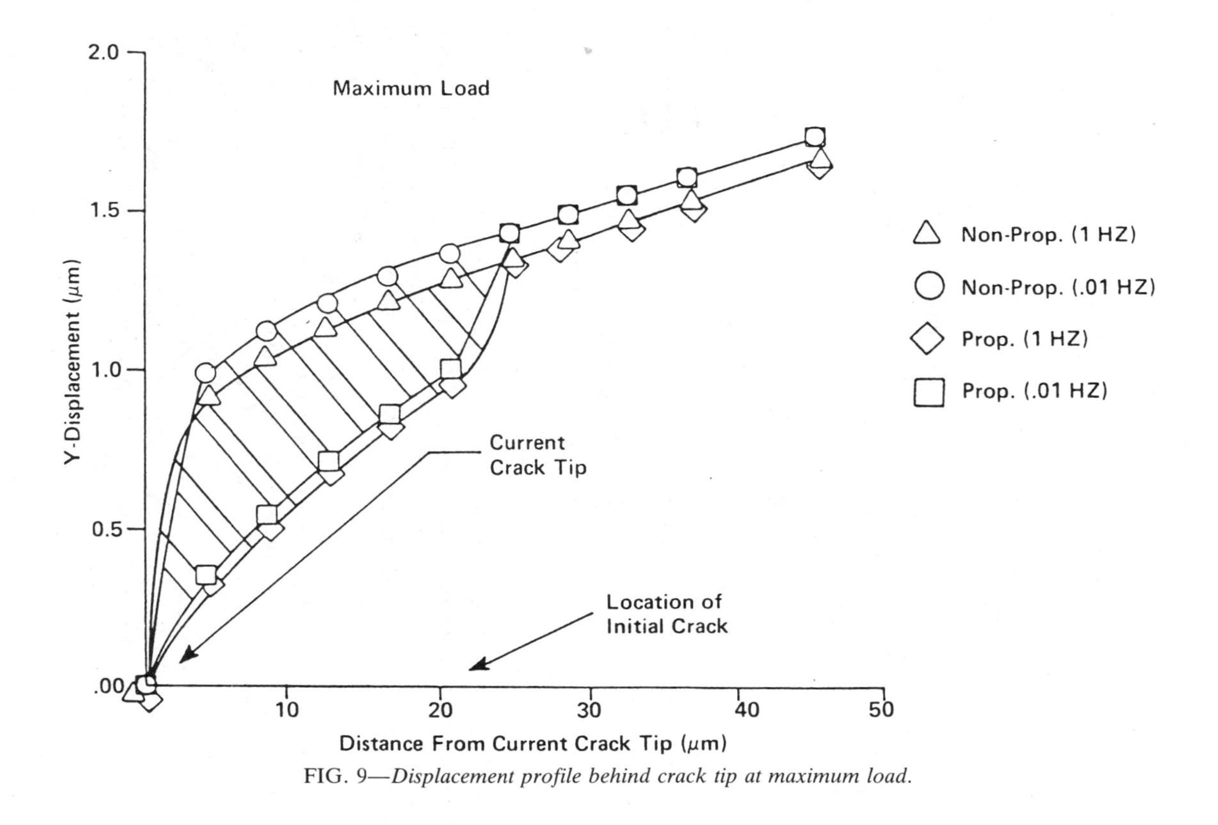

FIG. 9—*Displacement profile behind crack tip at maximum load.*

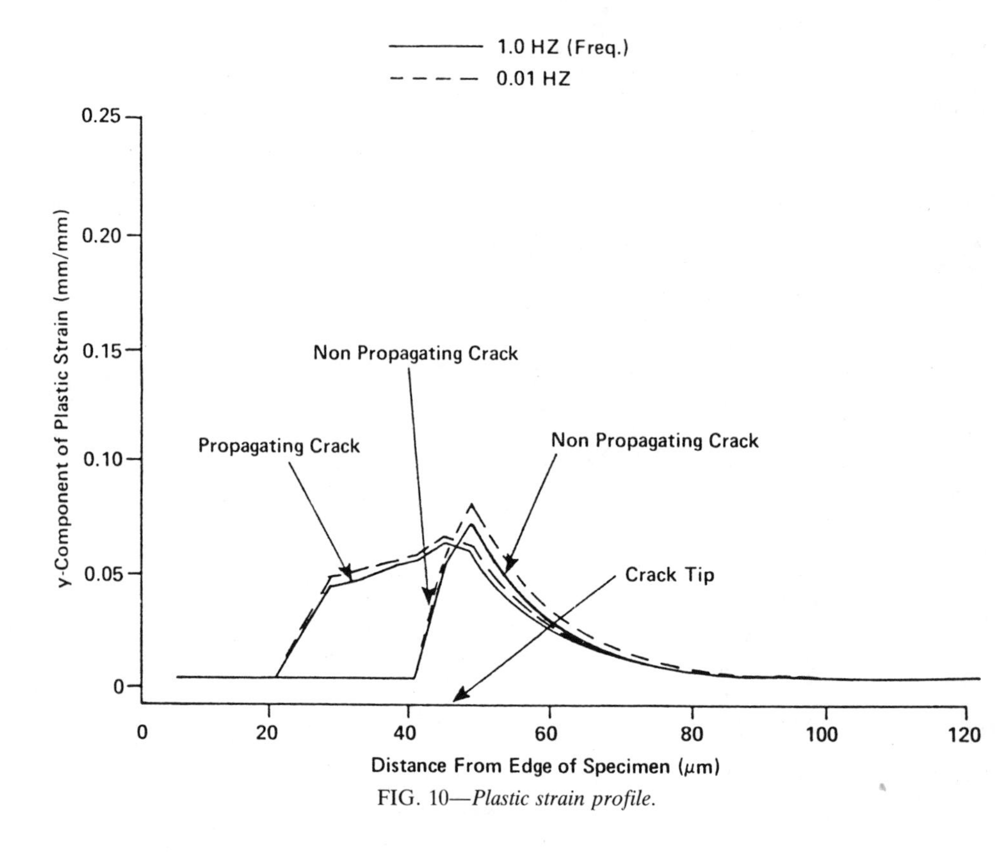

FIG. 10—*Plastic strain profile.*

The σ_y stress profile ahead of the crack tip (Fig. 11) at a full crack length is nearly identical for propagating and nonpropagating cracks, indicating that the stress levels ahead of the crack tip are not affected by fatigue (growth of the plastic wake). Once the stress reaches a saturated value it does not change, thus explaining the steady-state conditions mentioned earlier. One can also observe that the stress component relaxes with the time effect indicated by the slight differences perpetrated by the different frequencies.

As a last consideration, let us direct attention to K_{op} and K_{CL} shown in Fig. 12. The plot is one of comparing K_I values for propagating cracks at the two frequencies. K_{op} is defined as K_I evaluated at a load in which the last node opens, while K_{CL} is determined at the load in which the first node closes. The higher frequency produces a somewhat higher set of curves (as would be expected). If ΔK_{eff} ($\Delta K_{eff} = K_{max} - K_{op}$) used by Elber [*10*] and Newman [*11*] is evaluated from Fig. 12, one can make the statement, as suggested by the stated references, that the small crack effect is more pronounced for lower frequencies; that is, the crack grows faster with the larger ΔK_{eff}.

Conclusion

This paper presents the features of a viscoplastic finite-element program which makes use of constant strain elements in evaluating cracks under cyclic loading. The algorithms asso-

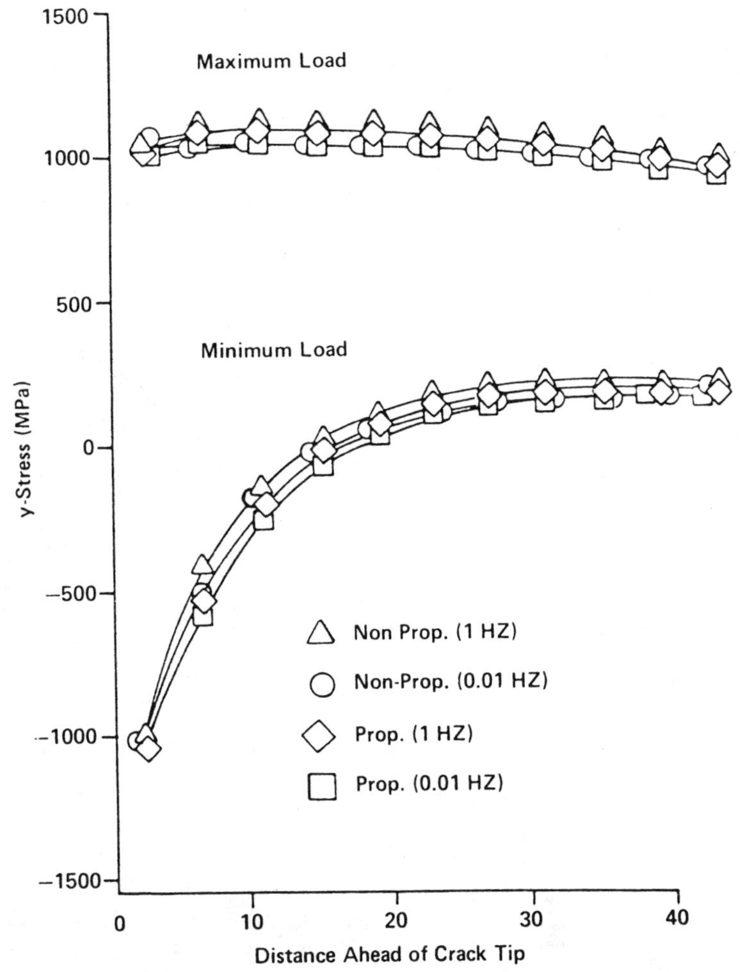

FIG. 11—*Stress ahead of crack tip for total crack length.*

ciated with crack closure and crack growth are discussed. The code is applied to the study of a short edge crack considering the alloy IN718 at 670°C, comparing propagating and nonpropagating cracks at 1 and 0.1 Hz frequencies. Several conclusions can be made from this study.

1. In order to model plasticity-induce closure under cyclic loading, one needs to carry out crack extension. A nonpropagating crack cannot properly simulate the actual phenomenon.
2. The characteristics of viscoplasticity at high temperature were reflected in this study in (*a*) closure effects, (*b*) plastic strain functions, (*c*) stress functions, and (*d*) plastic wake, even though the growth was for a relatively short time.

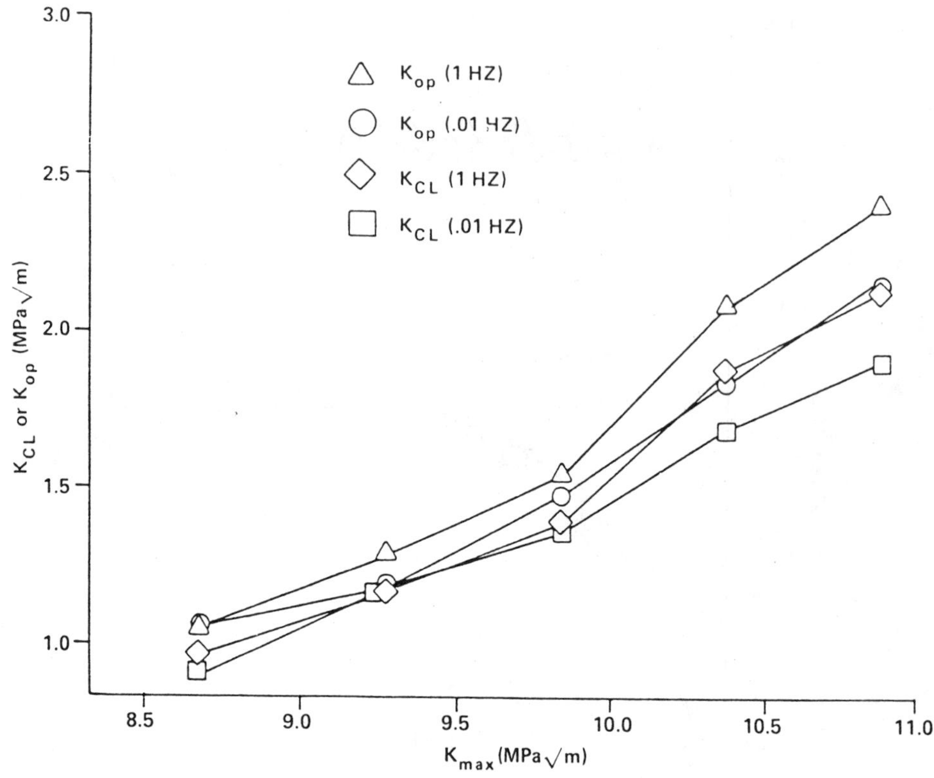

FIG. 12—*Crack stress intensity* K_{CL} *or* K_{op}.

References

[1] Smail, J. and Palazotto, A. N., "The Viscoplastic Crack Growth Behavior of a Compact Tension Specimen Using the Bodner-Parton Flow Law," *Engineering Fracture Mechanics,* Vol. 19, No. 1, 1984, pp. 137–158.

[2] Wilson, R. E. and Palazotto A. N., "Viscoplasticity in a Superalloy at Elevated Temperatures Considering Tension and Compressive Loading," *Engineering Fracture Mechanics,* Vol. 22, No. 6, 1985, pp. 927–937.

[3] Henkel, C. and Palazotto, A. N., "The Effects of Low Cycle Fatigue Comparing Compact Tension and Center Cracked Specimens at Elevated Temperatures," *Engineering Fracture Mechanics,* Vol. 24, No. 4, 1986, pp. 483–494.

[4] Nicholas, T., Palazotto, A., and Bednarz, E., "An Analytical Investigation of Plasticity Induced Closure Involving Short Cracks" in *Mechanics of Fatigue Crack Closure, ASTM STP 982,* American Society for Testing and Materials, Philadelphia, 1988, pp. 361–379.

[5] Newman, J. C., Jr., "A Finite Element Analysis of Fatigue Crack Closure," *Mechanics of Crack Growth, ASTM STP 590,* American Society for Testing and Materials, Philadelphia, 1976, pp. 280–301.

[6] Newman, J. C., Jr., and Armen, H., Jr., "Elastic-Plastic Analysis of a Propagating Crack Under Cyclic Loading," *AIAA Journal,* American Institute of Aeronautics and Astronautics, Vol. 13, No. 8, 1975, pp. 1017–1023.

[7] Bednarz, E., "A Numerical Study of Plasticity Induced Closure in Short Cracks by the Finite Element Method," presented to the Aeronautics and Astronautics Department, Air Force Institute of Technology, in partial fulfillment of the requirements of Ph.D. degree, Wright Patterson Air Force Base, Dayton, OH, 1987.

[8] Hinnerichs, T., "Viscoplastic and Creep Crack Growth Analysis by the Finite Element Method,"

presented to the Aeronautics and Astronautics Department, Air Force Institute of Technology in partial fulfillment of the requirements of Ph.D. degree, Wright Patterson Air Force Base, Dayton, OH, 1980.

[9] Bodner, S. R. and Partom, Y. "Constitutive Equation for Elastic Viscoplastic Strain Hardening Materials," *Journal of Applied Mechanics,* Vol. 42, 1975, pp. 385–389.

[10] Elber W., "Fatigue Crack Closure Under Cycle Tension," *Engineering Fracture Mechanics,* Vol. 2, No. 1, 1970, pp. 37–45.

[11] Newman, J. C., Jr., Swain, M. H., and Phillip, E. P., "An Assessment of the Small Crack Effect for 2024-T3 Aluminum Alloy" in *Small Fatigue Cracks,* R. O. Ritchie and J. Lankford, The Metallurgical Society of the American Institute of Mining, Metallurgical and Petroleum Engineers, 1986, pp. 427–452.

Farrel J. Zwerneman[1] *and Karl H. Frank*[2]

Crack Opening Under Variable Amplitude Loads

REFERENCE: Zwerneman, F. J. and Frank, K. H., **"Crack Opening Under Variable Amplitude Loads,"** *Fracture Mechanics: Perspectives and Directions (Twentieth Symposium), ASTM STP 1020,* R. P. Wei and R. P. Gangloff, Eds., American Society for Testing and Materials, Philadelphia, 1989, pp. 548–565.

ABSTRACT: Crack growth rate tests were conducted on eleven compact tension specimens of A588A steel plate to assess the ability of linear cumulative damage models to estimate fatigue life under load-time histories composed of cycles of variable amplitude and mean. Crack opening was monitored using a clip gage and through transmission ultrasonics. Crack growth rate results show that closely spaced overloads can cause acceleration; retardation occurs only if overloads are separated by many small cycles. Comparison of crack opening data to crack opening estimates based on Elber's closure model shows that the influence of crack opening on growth rate in constant-amplitude tests is less than predicted by the model, while the influence of crack opening in variable-amplitude tests is greater than predicted by the model.

KEY WORDS: crack closure, crack propagation, fatigue (materials), loads (forces), overload, predictions, retardation, stress ratio, stress cycle, underload

It is generally assumed that the fatigue life of a component operating under variable-amplitude loads can be conservatively estimated using a linear cumulative damage model, such as Miner's rule [*1,2*]. High load peaks within the variable-amplitude load-time history are expected to retard growth from subsequent small cycles [*3–5*]. The present work shows that Miner's rule is conservative only if overloads are separated by many small cycles; if overloads are separated by few small cycles, Miner's rule unconservatively estimates growth rate. The growth rate estimate becomes increasingly unconservative as the mean load level of the small cycles in the load-time history increases relative to the mean of the overall history.

Crack opening was monitored during application of variable-amplitude loads in order to determine the cause of the observed high growth rates. The load-time histories used in these tests include constant-amplitude histories at three different load ratios, histories with closely spaced overloads and underloads, histories with isolated overloads and underloads, and histories representing service loads on a highway bridge. During application of these load-time histories, through-transmission ultrasonics and a clip gage were used to establish separate records of crack opening versus time. These records indicate a fundamental difference between crack opening behavior in the vicinity of isolated overloads and behavior in the vicinity of closely spaced overloads. When overloads are separated by many small cycles,

[1] Assistant professor, Civil Engineering Department, Oklahoma State University, Stillwater, OK 74078-0327.

[2] Associate professor, Department of Civil Engineering, University of Texas, Austin, TX 78758-1076.

the overload causes the crack to be propped open behind the crack tip and retardation occurs as expected. When overloads are separated by few small cycles, propping open of the crack is not observed and no retardation occurs.

Experimental Program

Load-Time Histories

The variable-amplitude load-time histories used in this research are shown in Fig. 1. Isolated overloads (OL) and underloads (UL) have been used by many researchers, and crack opening behavior associated with these load types has been described [*6–9*]. Isolated overloads and underloads were used in the present work to provide verification for the experimental procedure and data acquisition techniques employed. In addition, crack opening behavior observed following isolated overloads and underloads provides a contrast to behavior observed following closely spaced overloads and underloads.

The load-time histories shown in Figs. 1*b*, 1*c*, and 1*d* are treated in this work as single complex cycles for purposes of crack growth rate data reduction. The closely spaced overload and underload histories (MS) were used to isolate the effect of the mean level of minor cycles on fatigue damage produced by these histories. The superimposed sine (SS) history is constructed with constant minor cycle amplitudes and varying minor cycle mean levels. This history represents a service loading composed of one major load cycle with minor vibrational cycles superimposed. The test truck (TT) history was recorded on a bridge as a

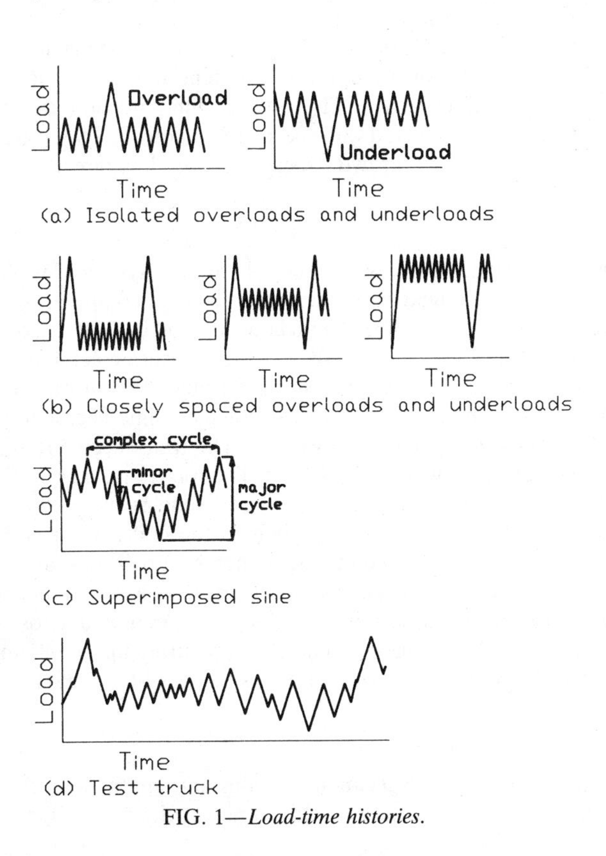

FIG. 1—*Load-time histories.*

truck passed over the bridge. In addition to the variable-amplitude tests, constant-amplitude tests were conducted to provide data for comparison with work from other researchers and to establish a base for estimating damage due to variable amplitude loads.

Load ratios of 0.1, 0.2, 0.3, 0.4, and 0.7 were used in crack growth rate tests. In constant amplitude tests, load ratio is defined as the minimum load in a cycle divided by the maximum load in that cycle. In variable amplitude tests, load ratio is calculated on the complex cycle; cycle load ratio is defined as the minimum load in a complex cycle divided by the maximum load in that same complex cycle.

Specimens

The standard compact-type (CT) specimen for fatigue crack growth rate testing was used in this work. The specimen was designed according to specifications listed in the ASTM Test Method for Constant-Load Amplitude Fatigue Crack Growth Rates Above 10^{-8} m/Cycle (E 647-81). Stress intensity at the crack tip was calculated using the equation provided in Article 9.3.1 of ASTM E 647-81. A drawing of the specimen is provided in Fig. 2*a*, and the specimen is shown mounted in the test frame in Fig. 2*b*.

Specimen width was chosen to comply with ASTM E 647-81, Article 8.6.3, requiring that specimen thickness-to-width ratio be less than 0.15 in order for crack length measurements to be made on only one side. The large width also helps reduce crack length measurement error relative to specimen width and minimizes the increase in stress intensity range over a measurable amount of crack growth. This leads to an improved experimental representation of the continuous crack growth rate curve.

Test specimens were machined from ASTM A588A weathering steel. Chemical composition is given in Table 1. Mechanical properties from tensile tests conducted with specimens oriented longitudinal (L) and transverse (T) to the rolling direction are shown in Table 2. Crack growth rate tests were conducted with specimens in both TL (crack growth longitudinal to rolling direction) and LT (crack growth transverse to rolling direction) orientations.

Apparatus

Loads were applied using a closed-loop servo-hydraulic system. The MS, SS, and TT load-time histories were produced with the aid of a microcomputer substituted for the standard function generator. A $\times 32$ magnification microscope attached to a micrometer slide accurate to ± 1.27 μm (0.00005 in.) was used to measure crack length.

Crack opening measurements were made ultrasonically [*10*] and with a clip gage. Ultrasonic transducers can be seen mounted in a clear polymethylmethacrylate (Plexiglas) frame on the top and bottom of the specimen in Fig. 2*b*. The transducers are 6.35-mm-diameter (0.25 in.), 5-MHz compression wave transducers. The top transducer acts as a transmitter, and the bottom transducer acts as a receiver. The Plexiglas frame provides a means of aligning the transducers 0.38 mm (0.015 in.) behind the crack tip at midthickness of the specimen. The Plexiglas frame is held on the specimen by a pair of weak springs. This form of attachment holds the transducers on the specimen with a repeatable pressure without significantly restraining specimen motion. The clip gage is mounted across the crack on the front face of the specimen. This gage is capable of measuring up to 3.81 mm (0.150 in.) of relative displacement between two points on opposite sides of the crack.

Procedure

Fatigue cracks were initiated in specimens by constant amplitude cycling at a high load range. When a crack was observed in a machined notch, the load range was decreased by

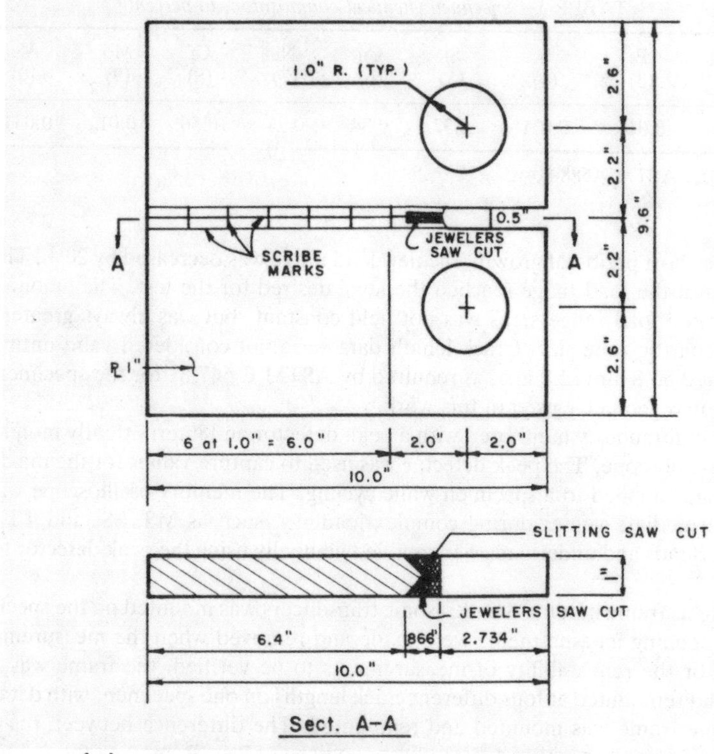

(a) Compact tension specimen.

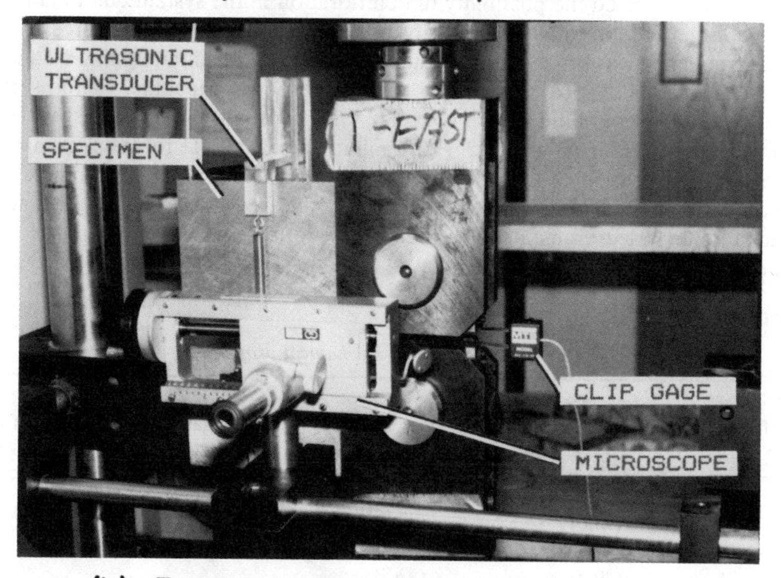

(b) Experimental apparatus.

FIG. 2—*Compact tension specimen installed in test frame (1 in. = 2.54 cm).*

TABLE 1—*Specimen chemical composition, in percent.*[a]

C (1)	Mn (2)	P (3)	S (4)	Si (5)	Cu (6)	Ni (7)	Cr (8)	Mo (9)	V (10)	Al (11)
0.14	0.99	0.014	0.031	0.37	0.34	0.21	0.56	0.01	0.034	0.024

[a] Steel Grade: ASTM A588-80A.

20%. After a short period of growth, the new load range was decreased by 20%. This pattern continued until the load range reached the level desired for the test. The amount of crack growth between load range steps was not held constant, but was always greater than the plane stress plastic zone size. Crack length data were not considered valid until the crack length reached 50.8 mm (2.0 in.), as required by ASTM E 647-81 for the specimen size and machined notch geometry used in this work.

Load was continuously monitored with a peak detector and intermittently monitored with a memory oscilloscope. The peak detector was used to capture values for the maximum and minimum loads applied to a specimen while cycling. The memory oscilloscope was used to capture intermediate values during complex loadings, such as MS, SS, and TT loadings. Isolated overloads and underloads were applied manually using the peak detector to monitor load.

The Plexiglas frame holding the ultrasonic transducers was mounted on the specimen each time crack opening measurements were made and removed when the measurements were complete. For the repeatability of measurements to be verified, the frame was mounted, removed, and remounted at four different crack lengths on one specimen, with data recorded each time the frame was mounted and remounted. The difference between the mounting and remounting measurements differed by an average of 7%. Removing the frame between measurements minimized the possibility of deterioration in the system due to abrasion and fatigue and largely removed electronic drift in the instrumentation as a source of error.

Cycling frequency was set at the maximum obtainable. Maximum cycling frequency was limited by the flexibility of the specimen and the response capability of the servovalve. Frequencies ranged from a maximum of 35 Hz for a constant amplitude test with a crack-length-to-specimen-width ratio of 0.25, to a minimum of 0.2 Hz for a superimposed sine specimen with a crack-length-to-specimen-width ratio of 0.54. For tests conducted at high frequencies, it was necessary to reduce periodically the frequency to 1 Hz for crack length and crack opening measurements. The total test time for a specimen ranged from eleven days to six months.

TABLE 2—*Specimen mechanical properties.*

Orientation (1)	Static Yield Stress, MPa (2)	Ultimate Stress, MPa (3)	Elongation,[a] % (4)	Reduction of Area, % (5)
L	348 (50.0)	572 (82.9)	33.0	67.5
T	341 (49.4)	569 (82.5)	27.0	47.0

[a] 2 in. (50.8 mm) gage length.

Data Reduction

Crack growth rate is calculated by the secant method. The secant method was chosen over the polynomial method because the primary concern in this work is crack growth under variable amplitude load-time histories. The arrangement of the load-time histories in some tests is such that crack growth rate will vary drastically between consecutive sets of crack length measurements. Use of the polynomial method to calculate growth rate in these tests would obscure the very phenomena being investigated.

Crack growth rate and stress-intensity range values were plotted on a log-log scale and a line was fit to the data by the least-squares method. The equation for this fitted line has the general form

$$da/dN = C\Delta K^m \tag{1}$$

where

da/dN = crack growth rate,
C = da/dN axis intercept,
ΔK = stress intensity range, and
m = slope of crack growth rate curve.

The accuracy of the reduction technique described above was evaluated by rearranging Eq 1 and integrating to calculate total fatigue life. This integration was performed for three constant-amplitude specimens. Calculated fatigue lives were within 3% of experimental fatigue lives for all three specimens, thus verifying the data reduction technique.

In reducing the MS, SS, and TT data, each of the load-time histories shown in Figs. 1*b*, 1*c*, and 1*d* is treated as one complex cycle. For plotting the data points for these tests, stress-intensity range and crack growth rate are calculated as if the minor cycles are not present. The stress-intensity range is calculated on the basis of the major cycle load range. Growth rate is calculated on the basis of inches of growth per complex cycle. The extent to which variable-amplitude crack growth rate data plots above constant amplitude data is an indication of damage caused by the presence of minor cycles in the complex load-time history.

Measured growth rates are compared to estimates based on rainflow counting [*11*] and Miner's rule. Rainflow counting is used to break the complex cycle into component constant amplitude cycles. The growth caused by these individual cycles is added according to Miner's rule. Miner's rule calls for a simple linear summation of growth rates. The growth rate for the MS history is estimated as shown below:

$$(da/dN)_{\text{complex}} = (1)\,(C_1\,\Delta K_{\text{major}}{}^m) + (9)\,(C_2\,\Delta K_{\text{minor}}{}^m)$$

$$\Delta K_{\text{minor}} = 0.30\,\Delta K_{\text{major}}$$

$$(da/dN)_{\text{complex}} = [1 + (9)\,(C_2/C_1)\,(0.30)^m](C_1\,\Delta K_{\text{major}}{}^m)$$

The ratio C_2/C_1 accounts for differences in constant-amplitude growth rate due to differences in major and minor cycle load ratios. It is estimated that an MS complex cycle will cause growth at a rate $[1 + (9)\,(C_2/C_1)\,(0.30^m)]$ times faster than a constant-amplitude cycle with a stress-intensity range of ΔK_{major}.

Analysis of ultrasonic transmission data required simultaneous plotting of load versus time and ultrasonic transmission versus time. The method of data reduction is illustrated in Fig. 3. When a transmitted signal hits the minimum level, the crack is fully open. The time

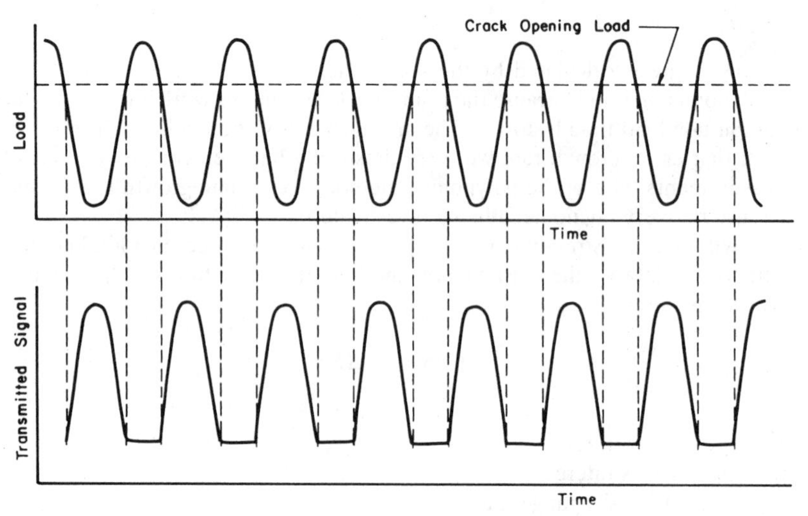

FIG. 3—*Reduction of ultrasonic transmission data.*

at which the crack comes fully open can be determined by projecting a line down to the time axis on the transmitted signal versus time plot. Once the time at which the crack opens is known, crack opening load can be determined from the load-versus-time plot.

Crack opening loads reported as a part of this work are not the result of direct observations of working cracks. Variation in ultrasonic transmission with load was used as an indirect indication of crack opening load. This does not detract from the present data, since this same statement can be made in regard to crack opening loads inferred from load-displacement data, from potential drop measurements, or from examination of striations.

The clip gage employed in this work was not sufficiently sensitive to record the nonlinear variation of crack opening with load. Crack opening, measured at the crack mouth, varied linearly with load from minimum load to maximum load in a cycle for all cycles recorded. The area involved in closure was not sufficient to produce a change in specimen flexibility measurable with the available clip gage. Therefore, use of the clip gage was limited to measurement of crack opening on the face of the specimen when load was at the maximum peak or minimum valley in a cycle. Clip gage data were used to determine specimen compliance and to quantify the extent to which the crack was propped open following overloads.

Experimental Results

Constant Amplitude

Constant-amplitude tests were conducted with specimens in the TL and LT orientations. Results from specimens tested in the TL orientation are shown in Fig. 4. Data are compared with a best-fit line derived by Wilson from constant-amplitude tests on A588A steel [*12*]. Wilson's tests were conducted at a load ratio of 0.1 with specimens in the TL orientation. Data from specimens tested in the LT orientation were compared with a best-fit line derived by Wilson from tests with specimens in the LT orientation and to a line defining the upper bound of crack growth rate data obtained by Barsom in tests with four different ferrite-pearlite steels [*13*]. The present data are in good agreement with data from both Wilson and Barsom.

Crack opening data from two constant-amplitude tests are shown in Fig. 5. The top section

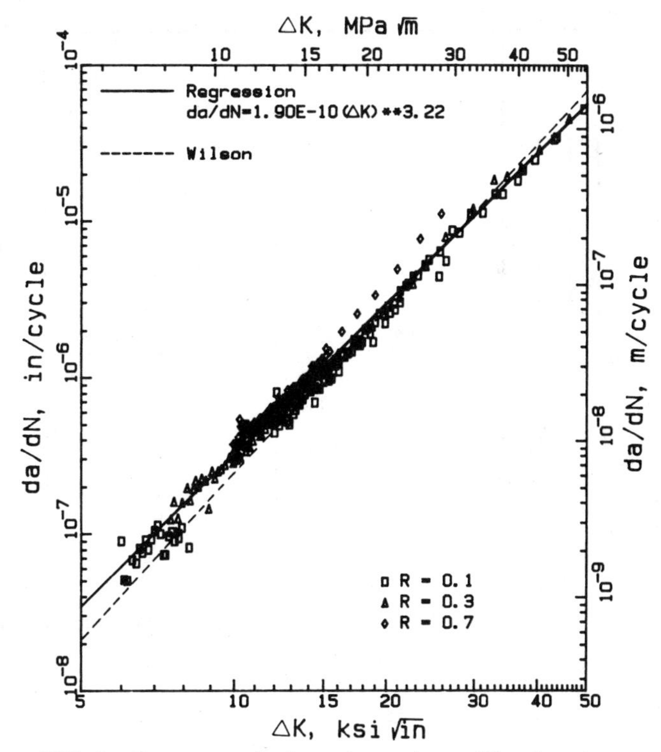

FIG. 4—*Constant-amplitude crack growth rate, TL orientation.*

of each figure depicts the load history, the middle sections is a plot of the percentage of the load range over which the crack is open, and the bottom section shows the variation of crack growth rate with crack length. The results shown in Fig. 5*a* are from a test conducted at a load ratio of 0.3. The vertical line reaching from the bottom of the load blocks down to zero load at each change in load range represents an underload. It was necessary to apply an underload following each load range reduction to reestablish crack growth. Three different load ranges were used in this test so that crack opening data could be obtained for one growth rate at three different crack lengths.

In the middle and bottom sections of Fig. 5*a*, three distinct groups of data are evident. Each group of data corresponds to a different load range used in this test. Both percent open and growth rate increase with increasing crack length for a constant load range. When load range decreases, both percent open and growth rate decrease. Both percent open and growth vary with stress-intensity range for the load ratio used in this test.

Comparison of crack opening data in Figs. 5*a* and 5*b* provides evidence that crack opening varies with load ratio as well as stress intensity range. Data shown in Fig. 5*a* were obtained in a test conducted at a load ratio of 0.3, while data shown in Fig. 5*b* were obtained in a test conducted at a load ratio of 0.7. The increase in load ratio from 0.3 to 0.7 causes a definite increase in the percentage of the load range over which the crack is open. At $R = 0.7$, the crack is open over the full load range, except for several measurements taken immediately after a load range reduction. At $R = 0.3$, the crack is never open over the full load range. There is a definite upper limit to the variation of percent open with stress intensity range.

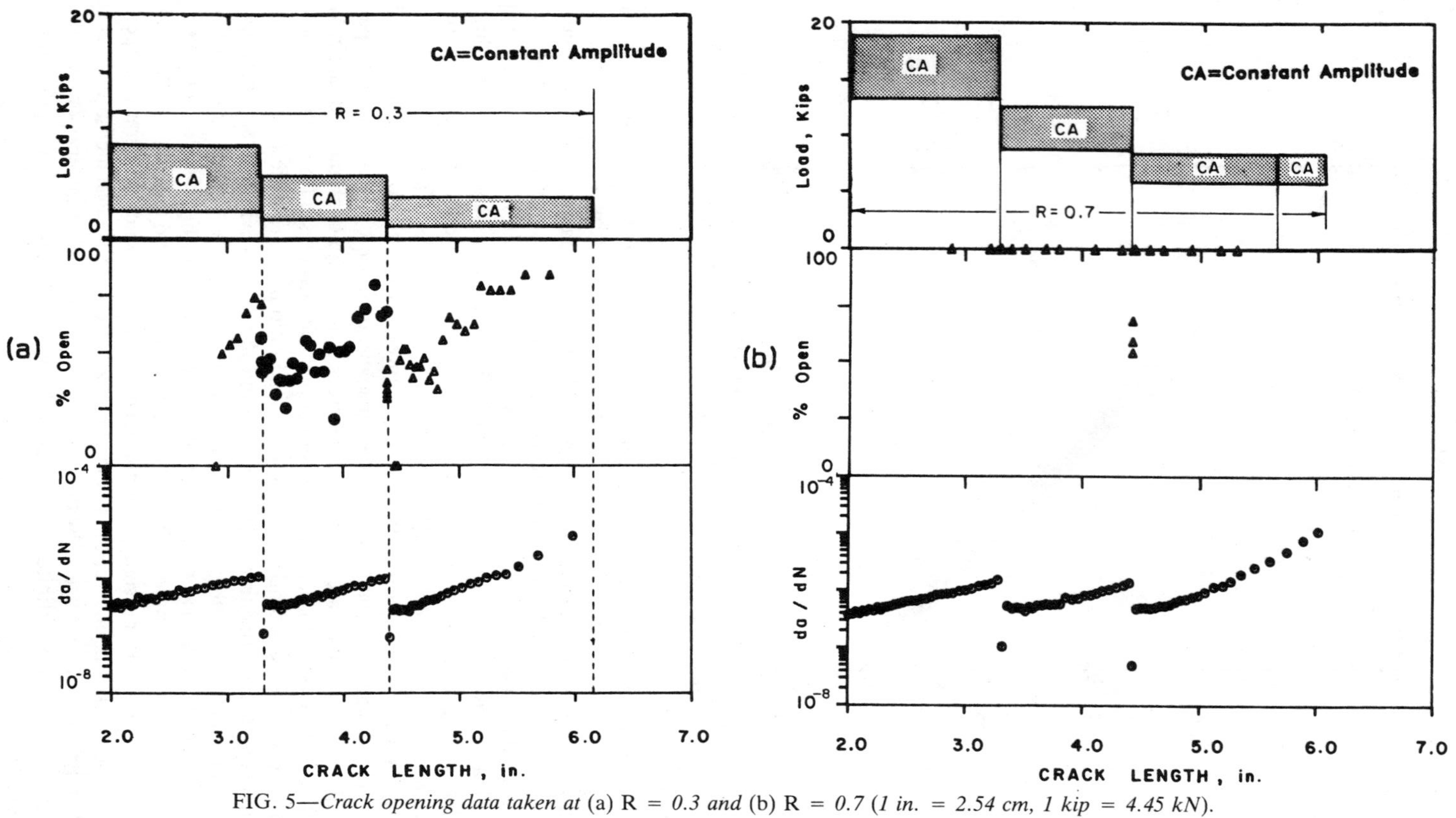

FIG. 5—*Crack opening data taken at* (a) R = 0.3 *and* (b) R = 0.7 *(1 in. = 2.54 cm, 1 kip = 4.45 kN).*

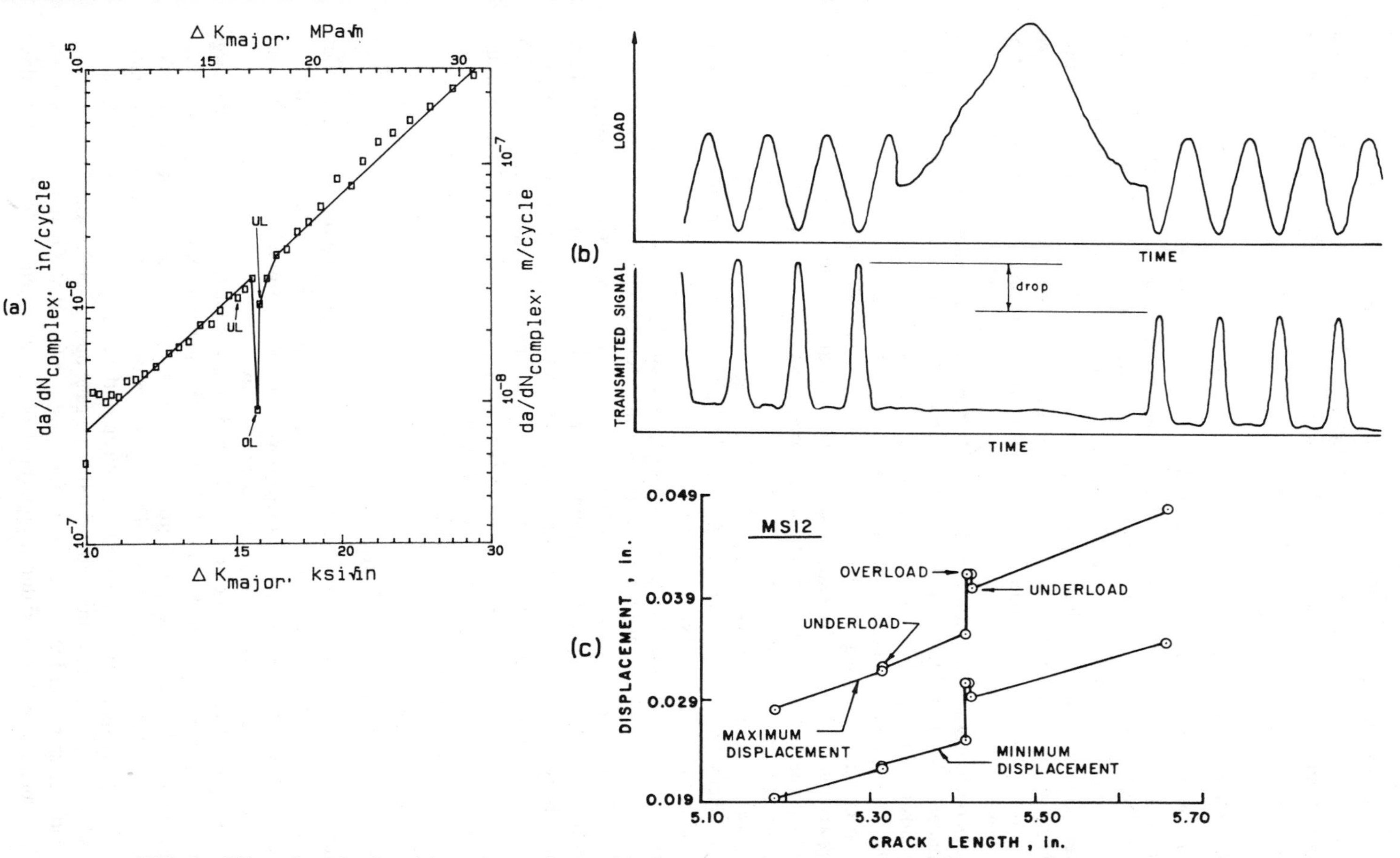

FIG. 6—*Effect of overload on* (a) *crack growth rate,* (b) *ultrasonic transmission, and* (c) *clip gage displacement (1 in. = 2.54 cm).*

Separate regression analyses of the three different sets of data plotted in Fig. 4 show that growth rate increases 9% as load ratio increases from 0.1 to 0.3 and 23% as load ratio increases from 0.3 to 0.7. These increases in growth rate can likely be attributed to increases in the percentage of the load cycle over which the crack is open.

Isolated Overloads and Underloads

Crack growth rate data from one overload/underload test are plotted in Fig. 6*a*. Data taken immediately following the application of an overload or underload are indicated by arrows. Overloads are abbreviated as OL and underloads as UL. No isolated overload was applied closer than 2.4 times the calculated plastic zone size from a preceding overload. The crack growth rate variations seen in Fig. 6*a* conform to behavior reported by other investigators [*6–9*]. Isolated underloads have no effect on growth rate; overloads cause crack growth retardation; an underload applied after an overload reduces the retarding effect of the overload.

Crack opening was monitored with ultrasonic transmission during application of overloads and underloads. Data obtained in the vicinity of one overload is presented in Fig. 6*b*. Neither overloads nor underloads cause a change in the load level at which the crack opens. The overload does, however, cause a decrease in the amount of transmitted signal during subsequent cycling, indicating the crack is not closed as tightly after the overload as it was before the overload. This observation is consistent with the idea that the crack is propped open by plastic deformation at the crack tip. Application of the isolated underload had no effect on transmitted signal.

Crack opening measurements made with the clip gage are plotted in Fig. 6*c*. In this figure, the line labeled "MAXIMUM DISPLACEMENT" identifies clip gage displacements recorded with the load at the maximum level in a cycle, and the line labeled "MINIMUM DISPLACEMENT" identifies clip gage displacements recorded with the load at the minimum level in a cycle. The underload produces no significant change in crack opening when it precedes the overload. The overload produces an abrupt increase in absolute displacement. The underload following the overload produces an abrupt, but smaller, decrease in absolute displacement. The gap between maximum and minimum displacements is not changed by application of the overload and subsequent underload, indicating that specimen compliance is unaffected by discontinuities in the load-time history. The measured ranges between minimum and maximum displacements just before and after the overload were within 1.5% of values calculated using an elastic solution from Tada, Paris, and Irwin [*14*].

Closely Spaced Overloads and Underloads

Crack growth rate data taken during applications of closely spaced overloads and underloads are shown in Fig. 7*a*. Experimental data are compared with constant-amplitude data and with the Miner's rule estimate. Lines drawn through the variable-amplitude data are drawn parallel to the constant-amplitude line and fit by eye.

If crack growth caused by the small cycles had been completely retarded by the large cycles, the variable-amplitude data would have plotted along the same line as the constant-amplitude data. Since the measured growth rate plots above Miner's estimate, growth is being accelerated rather than retarded. The degree of acceleration is dependent on the mean load level of the small cycles relative to the mean of the large cycle.

Crack opening measurements made with the clip gage during application of closely spaced overloads are shown in Fig. 7*b*. Data are for a test conducted at a major cycle load ratio of 0.7. Displacements recorded during application of closely spaced overloads do not exhibit

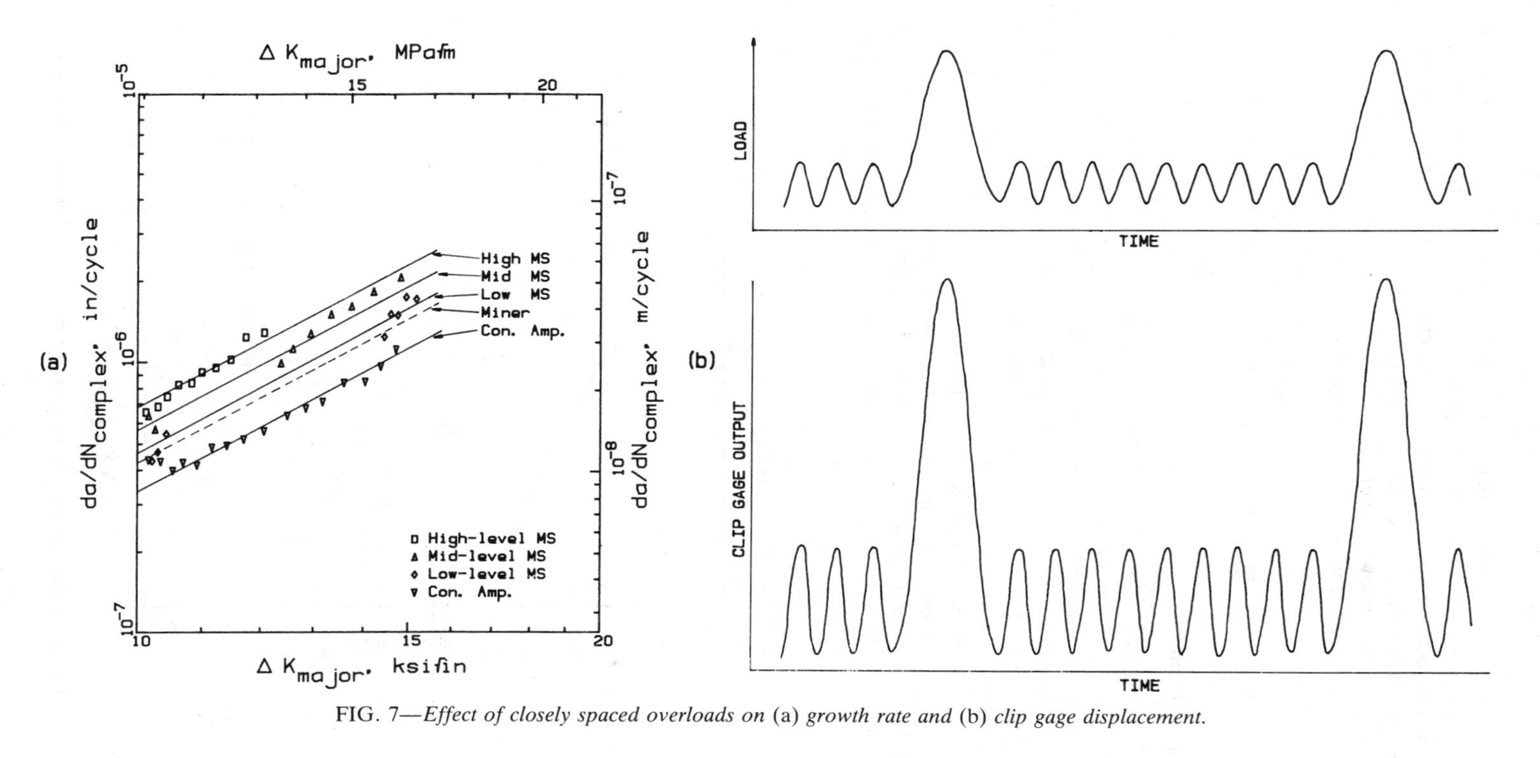

FIG. 7—*Effect of closely spaced overloads on* (a) *growth rate and* (b) *clip gage displacement.*

the same variation as those recorded during application of isolated overloads. In tests with closely spaced overloads, the crack is not propped open following the large cycle. Ultrasonic transmission measurements also indicate that the crack is not propped open by closely spaced overloads; records from a test conducted at a major cycle load ratio of 0.2 showed no drop in transmitted signal following overloads.

Ultrasonic measurements were also recorded during a closely spaced overload test conducted at a major cycle load ratio of 0.7. At this major cycle load ratio, measurements indicate the crack is open over the entire load cycle. Even so, crack growth rate increases as the mean level of the minor cycles increases relative to the major cycle.

Superimposed Sine

Two different SS histories were used in crack growth rate tests: for one history, the ratio of minor cycle size to major cycle size, p, is 0.66 and for the other p is 0.30. In Fig. 8, SS data are compared with constant-amplitude data and with the Miner's rule estimated growth rate. Lines drawn through SS data points are drawn parallel to the constant-amplitude line and fit by eye. The small cycles in the SS load-time history have basically the same effect as the small cycles in the MS load-time history. Growth rate is not retarded by the large cycle in the history: growth is actually accelerated above the Miner's rule estimate.

Ultrasonic transmission and clip gage data show that crack opening is unaffected by the presence of small cycles in the SS history. Ultrasonic transmission measurements demonstrate the percentage of the load range over which the crack is open depends on major cycle load

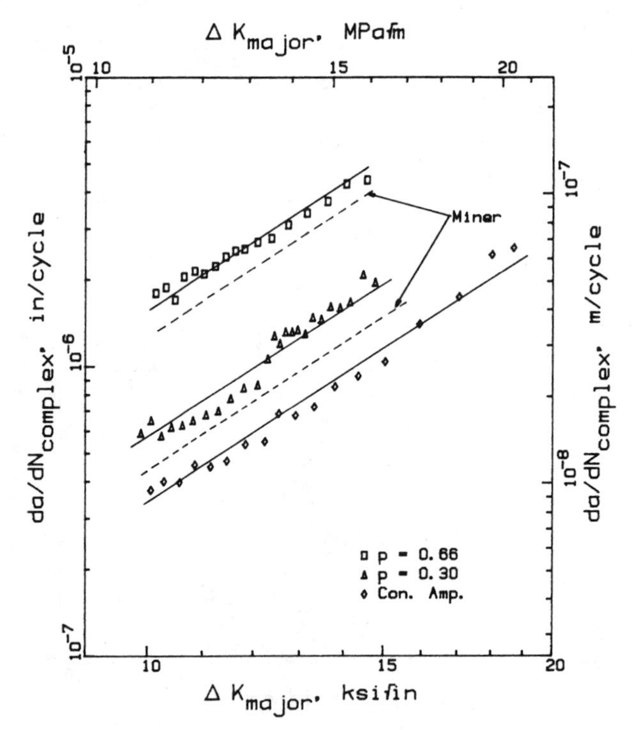

FIG. 8—*Crack growth rate under superimposed sine load-time histories.*

ratio and stress-intensity range, just as if the small cycles were not present. Clip gage measurements show the crack to be opening and closing elastically for all cycles.

Test Truck

Test truck crack growth rate data are plotted in Fig. 9. The initial data from this specimen were taken with TT loading. At ΔK_{major} = 8.9 MPa$\sqrt{m}$ (8.1 ksi $\sqrt{in.}$), loading was switched to constant amplitude. The CA ΔK was progressively decreased to 6.59 MPa$\sqrt{m}$ (6.0 ksi $\sqrt{in.}$) in an attempt to return to the threshold indicated by the initial TT data. Crack growth became very erratic, and consistent growth could not be reestablished until the stress intensity range was increased to 11.2 MPa$\sqrt{m}$ (10.2 ksi $\sqrt{in.}$). Once consistent growth was established, the test was continued with CA loading until ΔK reached 17.7 MPa$\sqrt{m}$ (16.1 ksi $\sqrt{in.}$). At this point, CA loading was replaced with TT loading. TT loading continued until ΔK_{major} reached 22.0 MPa$\sqrt{m}$ (20.0 ksi $\sqrt{in.}$), where CA loading was returned. Later in the test, the load ratio was increased from 0.10 to 0.42.

At the completion of this test, a corrosion product was found on the crack surface at the point where growth became erratic. The test truck history was applied at a relatively low frequency to minimize attenuation of the small cycles. This, coupled with the low stress-intensity range and low load ratio, produced real time growth at a rate slow enough to allow corrosion deposits to build up inside the crack. The corrosion deposits closed the crack, resulting in the low growth rates described above. Similar behavior was noted by Wilson in near threshold crack growth rate tests with A588A steel [*12*]. Examination of crack surfaces from other specimens used in the present test program showed that the heavy buildup of corrosion products occurred only on Specimen TT7.

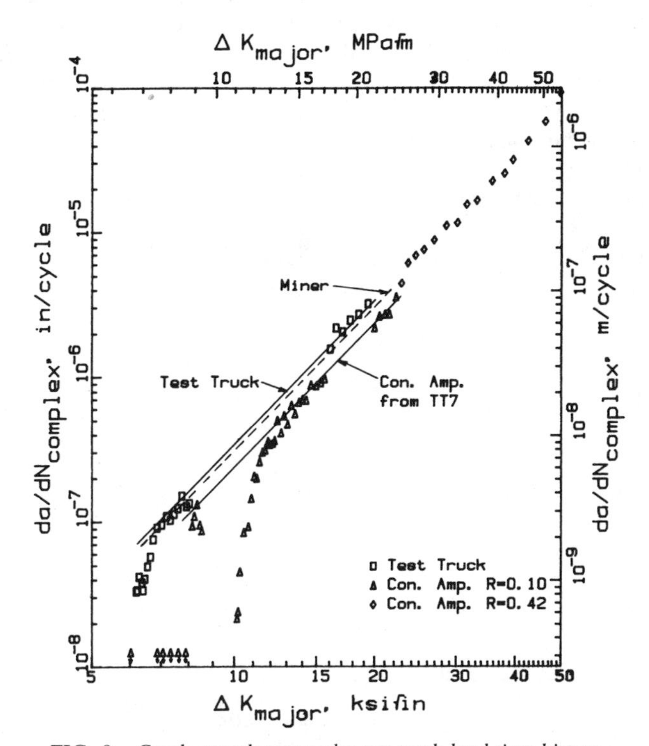

FIG. 9—*Crack growth rate under test truck load-time history.*

Other than behavior associated with the corrosion deposit, behavior observed during TT tests is basically the same as behavior observed during MS and SS tests. Growth rate is accelerated above the Miner's rule estimate and crack opening is unaffected by the presence of small cycles in the load-time history.

Discussion

Crack growth rate data just presented demonstrate that the application of high loads within variable amplitude load-time histories does not insure conservatism in fatigue life estimates based on linear cumulative damage models. This is especially apparent in the results of the MS tests with the small cycles at the bottom of the load-time history. Conventional damage models assume that overloads retard growth from following small cycles, with the result that the overall rate of growth is estimated below a linear sum of the rates of growth caused by large and small cycles [*3,4,15,16*]. Observed behavior contradicts conventional models in that the measured growth rate is slightly greater than the Miner's rule estimate. Research reported by others [*17–25*] indicates that the degree of acceleration declines as more small cycles are inserted between high load peaks. Retardation occurs only when overloads are separated by many small cycles.

In Fig. 10, the present crack opening data are compared to Elber's crack closure model [*5*]. The data presented in Fig. 10 represent three specimens or portions of specimens tested at $R = 0.1$, one at $R = 0.2$, two at $R = 0.3$, one at $R = 0.4$, one at $R = 0.426$, three at $R = 0.7$, and one at $R = 0.73$. There is a great deal of scatter in the data because of the previously noted variation of percentage open with stress-intensity range. Elber's closure model does not include stress-intensity range as a variable. The present data and Elber's model agree qualitatively in that percentage open increases with increasing load ratio. Quantitatively, however, Elber's model predicts a much slower increase in percentage open with load ratio than indicated by the means of the present data. Also, Elber does not show the crack to be open over the full load range for any load ratio, whereas the present data indicate the crack to be open over the entire load range for load ratios greater than approximately 0.5.

There is also quantitative disagreement between the crack closure concept of effective stress-intensity range and the observed rate of increase in growth rate with increasing load ratio. In constant-amplitude tests, the increase in growth rate with increasing load ratio is

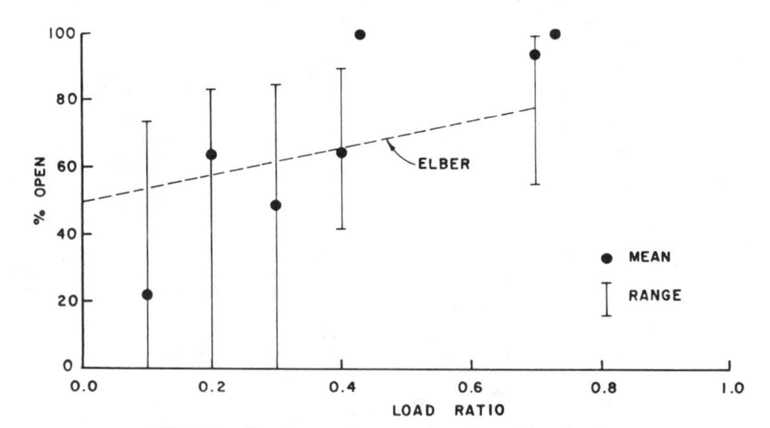

FIG. 10—*Crack opening as a function of load ratio.*

much less than predicted by Elber's model. In variable-amplitude tests, the increase in growth rate with increasing load ratio is much greater than predicted by Elber's model. In constant amplitude tests, growth rate increased only 35% as R increased from 0.1 to 0.7. The present data indicate that percentage open increases by a factor of 5 in going from $R = 0.1$ to 0.5. Given that, for steel, growth rate varies with stress-intensity range cubed, a fivefold increase in percentage open should lead to a growth rate increase by a factor of 125. For MS tests conducted at a major cycle load ratio of 0.7, minor cycle load ratios are 0.89, 0.90, and 0.91 for small cycles at the bottom, middle, and top, respectively, of the load history. This is not sufficient variation in load ratio to cause the observed growth rate fluctuations. Also, crack opening measurements indicate the crack to be open over the entire load cycle when load ratio exceeds 0.5. There should be no increase in percent open, and according to the model, no increase in growth rate as the small cycles move up in the waveform if the crack is open over the entire load cycle.

Available crack growth models do not provide a satisfactory explanation for observed variations in growth rate. Since available empirical models are based primarily on studies using aluminum specimens, an explanation for some of the differences in behavior observed in the present work and that predicted by the models may be due to differences in the stress-strain behavior of steel and aluminum.

Summary and Conclusions

Crack growth rate and crack opening data have been presented for A588A steel specimens tested under constant-amplitude, overload, underload, and simulated service load-time histories. Crack opening was monitored ultrasonically and with a clip gage. Growth rate data show that closely spaced high load cycles can cause crack growth rate acceleration. Crack opening measurements show the crack is not propped open by plastic deformations near the crack tip if overloads are closely spaced.

The observed high growth rates are not explained by available crack growth models. Models employing crack tip blunting or residual compressive stresses ahead of the crack tip as the means of retarding growth are not capable of predicting acceleration. Elber's crack closure model, while conceptually capable of predicting acceleration, does not accurately predict changes in growth rate with changing load ratio for steel specimens. Also, the crack closure model provides no insight as to the cause of accelerated growth rates observed in the present variable-amplitude tests.

Additional work is needed to define more precisely the cause of the observed crack growth rate acceleration. Variables such as the number of minor cycles between overloads and the minor cycle to major cycle load range ratio require investigation. Also, confidence in crack opening data could be improved by using another means of measurement, such as load-displacement or potential-drop measurements, in conjunction with the ultrasonic transmission method employed here.

Acknowledgments

Preliminary phases of this research were sponsored by the Texas Department of Highways and Public Transportation and the Federal Highway Administration. Financial support was also provided to the first author by the American Institute of Steel Construction and Armco in the form of a fellowship. Experimental work was performed in The University of Texas at Austin Civil Engineering laboratories.

References

[1] Miner, M. A., "Cumulative Damage in Fatigue," *Journal of Applied Mechanics Transactions,* American Society of Mechanical Engineers, Vol. 67, Sept. 1945, pp. A149–A164.
[2] The Committee on Fatigue and Fracture Reliability of the Committee on Structural Safety and Reliability of the Structural Division, "Fatigue Reliability: Variable Amplitude Loading," *Proceedings, Journal of the Structural Division,* American Society of Civil Engineers, Vol. 108, No. ST1, 1982, pp. 3–23.
[3] Hudson, C. M. and Hardrath, H. F., "Investigation of the Effects of Variable-Amplitude Loadings on Fatigue Crack Propagation Pattern," NASA Technical Note D-1803, National Aeronautics and Space Administration, Aug. 1963.
[4] Rice, J. R., "Mechanics of Crack Tip Deformation and Extension by Fatigue," *Fatigue Crack Propagation, ASTM STP 415,* American Society for Testing and Materials, Philadelphia, 1967, pp. 247–311.
[5] Elber, W., "The Significance of Fatigue Crack Closure," *Damage Tolerance of Aircraft Structures, ASTM STP 486,* American Society for Testing and Materials, Philadelphia, 1971, pp. 230–242.
[6] Mills, W. J. and Hertzberg, R. W., "Load Interaction Effects on Fatigue Crack Propagation in 2024-T3 Aluminum Alloy," *Engineering Fracture Mechanics,* Vol. 8, 1976, pp. 657–667.
[7] Von Euw, E. F. J., Hertzberg, R. W., and Roberts, R., "Delay Effects in Fatigue Crack Propagation," *Stress Analysis and Growth of Cracks, Proceedings,* 1971 Symposium on Fracture Mechanics, Part 1, ASTM *STP 513,* American Society for Testing and Materials, Philadelphia, 1972, pp. 230–259.
[8] Trebules, V. W., Jr., Roberts, R., and Hertzberg, R. W., "Effect of Multiple Overloads on Fatigue Crack Propagation in 2024-T3 Aluminum Alloy," *Progress in Flaw Growth and Fracture Toughness Testing, ASTM STP 536,* American Society for Testing and Materials, Philadelphia, 1973, pp. 115–146.
[9] Broek, D., *Elementary Engineering Fracture Mechanics,* Sijthoff and Noordhoff, The Netherlands, 1978, pp. 252–258.
[10] Mahulikar, D. S. and Marcus, H. L., "Fatigue Crack Closure and Residual Displacement Measurements on Aluminum Alloys," *Fatigue of Engineering Materials and Structures,* Vol. 3, 1981, pp. 257–264.
[11] Endo, T., Mitsunaga, K., Takahashi, K., Kobayashi, M., and Matsuishi, M., "Damage Evaluation of Metals for Random or Varying Load—Three Aspects of Rainflow Method," *Mechanical Behavior of Metals, Proceedings,* 1974 Symposium on the Mechanical Behavior of Materials, The Society of Materials Science, Japan.
[12] Wilson, A. D., "Influence of Inclusions on the Fracture Properties of A588A Steel," *Fracture Mechanics (Fifteenth Symposium), ASTM STP 833,* R. J. Sanford, Ed., American Society for Testing and Materials, Philadelphia, 1984, pp. 412–435.
[13] Barsom, J. M., "Fatigue Crack Propagation in Steels of Various Yield Strengths," *Transactions, Journal of Engineering for Industry,* American Society of Mechanical Engineers, Series B, 93, No. 4, Nov. 1971, pp. 1190–1196.
[14] Tada, H., Paris, P. C., and Irwin, G. R., *The Stress Analysis of Cracks Handbook,* Del Research Corp., St. Louis, MO, 1973.
[15] Wheeler, O. E., "Spectrum Loading and Crack Growth," *Transactions, Journal of Basic Engineering,* American Society of Mechanical Engineers, March 1972, pp. 181–186.
[16] Engle, R. M. and Rudd, J. L., "Analysis of Crack Propagation Under Variable Amplitude Loading Using the Willenborg Retardation Model," AIAA Paper No. 74-369, AIAA/ASME/SAE 15th Structures, Structural Dynamics and Materials Conference, Las Vegas, NV, 17–19 April 1974.
[17] Albrecht, P. and Friedland, I. M., "Fatigue Limit Effect on Variable-Amplitude Fatigue of Stiffeners," *Proceedings, Journal of the Structural Division,* American Society of Civil Engineers, Vol. 105, No. ST12, Dec. 1979, pp. 2657–2675.
[18] Fisher, J. W., Mertz, D. R., and Zhong, A., "Steel Bridge Members Under Variable Amplitude Long Life Fatigue Loading," NCHRP Report 267, Transportation Research Board, Lehigh University, National Cooperative Highway Research Program, Dec. 1983.
[19] Gurney, T. R., "Some Fatigue Tests on Fillet Welded Joints Under Simple Variable Amplitude Loading," The Welding Institute, May 1981.
[20] Schilling, C. B., Klippstein, K. H., Barsom, J. M., and Blake, G. T., "Fatigue of Welded Steel Bridge Members Under Variable Amplitude Loadings," NCHRP Report 188, Transportation Research Board, U.S. Steel Corporation, National Cooperative Highway Research Program, 1978.
[21] Tilly, G. P. and Nunn, D. E., "Variable Amplitude Fatigue in Relation to Highway Bridges,"

Proceedings, The Institution of Mechanical Engineers, Applied Mechanics Group, Vol. 194, No. 27, 1980, pp. 259–267.

[22] Barsom, J. M., "Fatigue Crack Growth Under Variable Amplitude Loading in A514-B Steel," *Progress in Flaw Growth and Fracture Toughness Testing, ASTM STP 536,* American Society for Testing and Materials, Philadelphia, 1973, pp. 147–167.

[23] Joehnk, J. M., "Fatigue Behavior of Welded Joints Subjected to Variable Amplitude Stresses," M.S. thesis, The University of Texas at Austin, May 1982.

[24] Zwerneman, F. J., "Influence of the Stress Level of Minor Cycles on Fatigue Life of Steel Weldments," M.S. thesis, The University of Texas at Austin, May 1983.

[25] Swensson, K. D., "The Application of Cumulative Damage Theory to Highway Bridge Fatigue Design," M.S. thesis, The University of Texas at Austin, May 1984.

Environmentally Assisted Cracking

Shu-Jun Gao,[1] Han-Zhong Xiao,[1] and Xiao-Jing Wan[1]

Strain-Induced Hydrides and Hydrogen-Assisted Crack Growth in a Ti-6Al-4V Alloy

REFERENCE: Gao, S.-J., Xiao, H.-Z., and Wan, X.-J., "**Strain-Induced Hydrides and Hydrogen-Assisted Crack Growth in a Ti-6Al-4V Alloy,**" *Fracture Mechanics: Perspectives and Directions (Twentieth Symposium), ASTM STP 1020,* R. P. Wei and R. P. Gangloff, Eds., American Society for Testing and Materials, Philadelphia, 1989, pp. 569–580.

ABSTRACT: The formation of strain-induced hydrides and the interaction of hydrides with crack growth in a Ti-6Al-4V alloy containing 790-ppm hydrogen have been studied by transmission electron microscopy (TEM) of thin-foil specimens stretched *in situ*. Three types of strain-induced hydrides were found: a crack tip hydride and two plate-like hydrides. The hydrides have either a face-centered tetragonal (fct) or a face-centered cubic (fcc) structure and are metastable. The formation and subsequent rupture of the hydrides at and in front of the crack tip were observed during crack growth. Hydrogen-assisted crack growth in Ti-6Al-4V alloys is discussed in terms of a hydride mechanism for cracking.

KEY WORDS: strain-induced hydrides of titanium, hydrogen assisted crack growth, *in situ* TEM study, the hydride mechanism for cracking, Ti-6Al-4V alloy

Hydrogen in metals has attracted much attention during the past two decades because of the embrittling effect of hydrogen and its role in delayed failure (that is, in hydrogen-induced subcritical crack growth) in high-strength alloys. During this period, fracture mechanics has evolved as an important tool for studying hydrogen-induced cracking. Crack growth behavior is characterized in terms of the crack-tip stress-intensity factor (K), which has been identified as the mechanical driving force for crack growth [*1*]. The behavior is affected by many other parameters, such as temperature and hydrogen content. The temperature dependence of crack growth rate is of particular importance because it can provide insight for the mechanism of hydrogen-induced cracking.

In recent studies of subcritical crack growth in Ti-6Al-4V alloys, an unusual temperature dependence was observed [*2,3*]: the rate of crack growth was found to decrease, rather than increase, with increases in temperature. More specifically, the experimental results showed that, in hydrogen-charged Ti-6Al-4V alloys, the crack growth rate under sustained loading at a given K decreased with increasing temperature over the range 223 to 353 K [*2*]. A similar temperature dependence for fatigue crack growth was observed for uncharged Ti-6Al-4V alloy tested in distilled water from 272 to 353 K [*3*]. For some other α and $\alpha + \beta$ titanium alloys containing hydrogen [*4–6*], the crack growth rate was also found to decrease with increasing temperature over a certain temperature region. It was proposed that hydrides might be responsible for the observed temperature dependence. Hence, it is of interest to examine the interactions of hydrides with crack growth. This interaction can be studied advantageously by transmission electron microscopy (TEM) because the interaction between

[1] Associate professor, graduate student, and professor, respectively, Institute of Metal Research, Academia Sinica, Shenyang, People's Republic of China.

hydride and cracking can be followed step-by-step by *in situ* observations of a specimen that is being stretched on a tensile stage inside the electron microscope.

The occurrence of hydrides in hydrogen-charged Ti-6Al-4V alloy during cracking has been demonstrated in a preliminary investigation [7]. In this paper, a detailed study on strain-induced hydrides is reported. The relationship between hydrides and hydrogen-assisted crack growth, especially the unusual temperature dependence for crack growth rate, is discussed.

Material and Experimental Methods

An industrial grade Ti-6Al-4V alloy was used for this study. The alloy was received as a plate of 12 mm thick in the rolled condition. The base composition of this alloy in weight percent are as follows: aluminum 6.05, vanadium 4.07, iron 0.12, oxygen 0.11, carbon 0.02, nitrogen 0.01, hydrogen 0.0045, and titanium balance. Specimen blocks were cut from the plate and thermally charged with hydrogen by heating them at 973 K in a hydrogen atmosphere at pressure of several hundreds torr. The absorbed hydrogen content was determined by using two methods, the pressure method and the weight method. For the first method, the hydrogen pressure difference Δp before and after charging the specimen was measured, then the mol fraction Δn of absorbed hydrogen was calculated by the relationship $\Delta p \cdot V = \Delta n \cdot RT$, where V is volume, R is gas constant, and T is absolute temperature (V and T were kept constant during hydrogen charging). For the second method, the specimen weight before and after charging was measured by means of an analytical balance with an accuracy of 10^{-4} g, then the absorbed hydrogen content was determined directly. The experimental results have shown that the mean deviation of hydrogen content determined by these two methods was 3%. The hydrogen content of 790 ppm (by weight) was used in this study.

After the hydrogen charging treatment, the specimen block was heated to 1373 K in air, held at the temperature for 2 h to achieve a uniform distribution of hydrogen, and then cooled in air. To prepare the specimen for TEM, we cut thin slices 0.2 to 0.3 mm thick from the block by electrodischarge machining, then mechanically ground them down to a thickness of about 80 μm, followed by electrolytic thinning until perforation. The electrolyte used for thinning consisted of 64 parts methanol, 30 parts butanol, and 6 parts perchloric acid by volume. To minimize the absorption of hydrogen, the electrolyte was kept at 238 K during electrolytic thinning. The electrolytic thinning process introduced microcracks into the specimens. Those specimens, with cracks suitably oriented for easy growth, were selected for *in situ* observations.

In situ observations were carried out in a JEM 200 CX electron microscope equipped with a tensile stage. During tensile loading (stretching), dislocation movement at the crack tip was used to monitor the loading process. Whenever dislocation movement occurred, loading was interrupted, and the microstructural changes in the crack tip region were examined and recorded. To keep the dissolution of hydrides caused by electron beam heating to a minimum, examinations of microstructural changes were made only periodically and the duration of observation was kept to no more than 3 min each time. Because of crack growth and other processes, the stresses in the specimens tended to relax with time. After some stress relaxation, the load on the specimen was increased slightly to start a new cycle.

Results

After hydrogenation and subsequent heat treatment, the Ti-6Al-4V alloy has an $\alpha + \beta$ Widmanstatten (basketweave) structure, consisting of plates of α phase and interplate layers of β phase. It is noted that there is an interface phase along the α/β boundaries (see F in

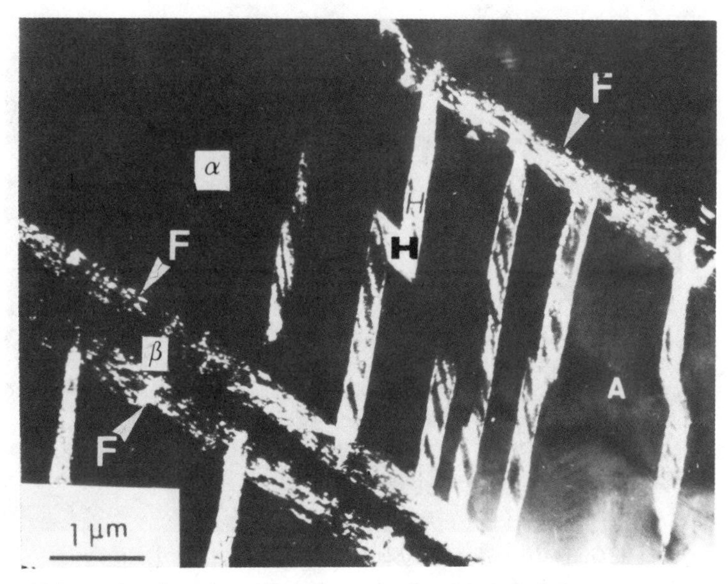

FIG. 1—*Interface phase* (F) *and strain-induced hydride* (H). *DF image.*

Fig. 1). There are no spontaneous hydrides in the alloy before stretching the specimens in the tensile stage. Strain-induced hydrides were observed after loading the specimen for about several hours depending on load level, crack orientation, etc.

The strain-induced hydrides may be classified into three types. The first type is hydrides formed at or in front of the crack tip (Figs. 2 and 3). These hydrides do not have a definitive shape and are more readily seen under dark field imaging using a strong reflection of the

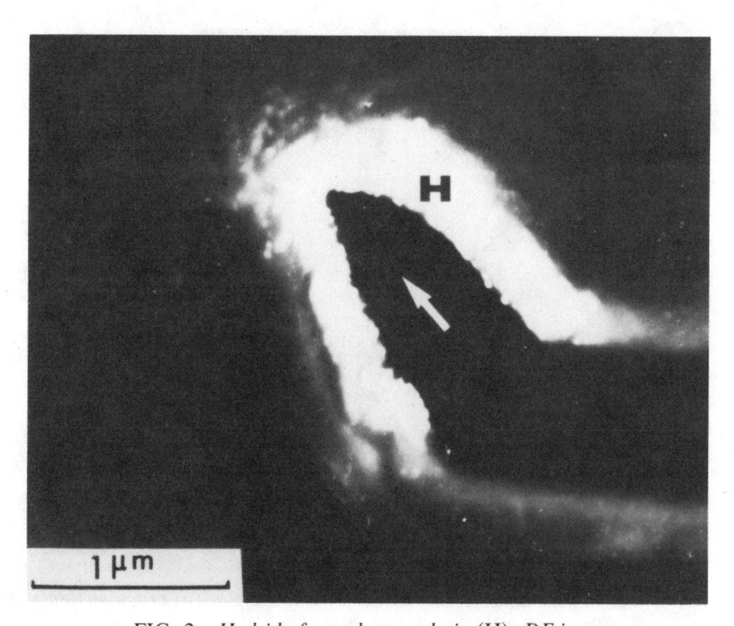

FIG. 2—*Hydride formed at crack tip* (H). *DF image.*

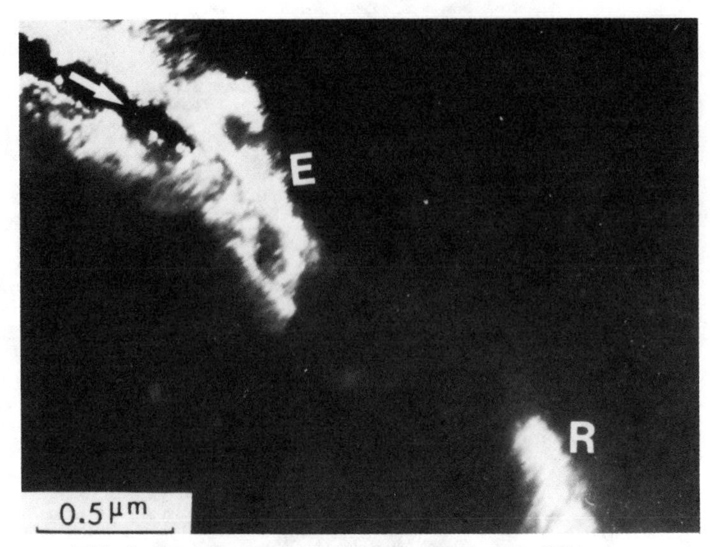

FIG. 3—*Hydride formed in front of crack tip* (R). *DF image.*

hydride. Electron diffraction analysis showed that this type of hydride has a face-centered tetragonal (fct) lattice, with an axial ratio less than unity ($c/a < 1$). The other two types appear in the α phase away from the crack tip. Both are plate-like, nucleate at the interface phase, and grow into the α matrix. The second type of hydrides consists of either thin slices (Fig. 4*a*) or bright and dark strips (Fig. 4*b*) and has an fct structure also with $c/a < 1$. Sometimes dislocations can be seen within these hydrides. The third type of hydrides is

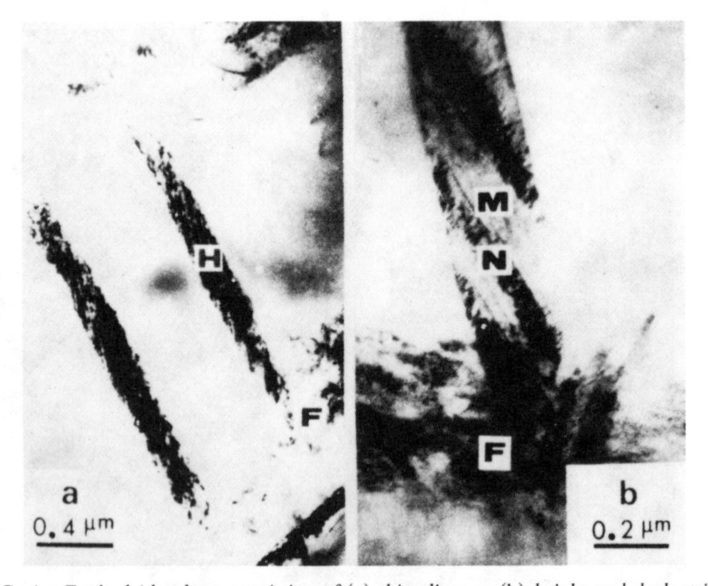

FIG. 4—*Fct hydride plates consisting of* (a) *thin slices or* (b) *bright and dark stripes.*

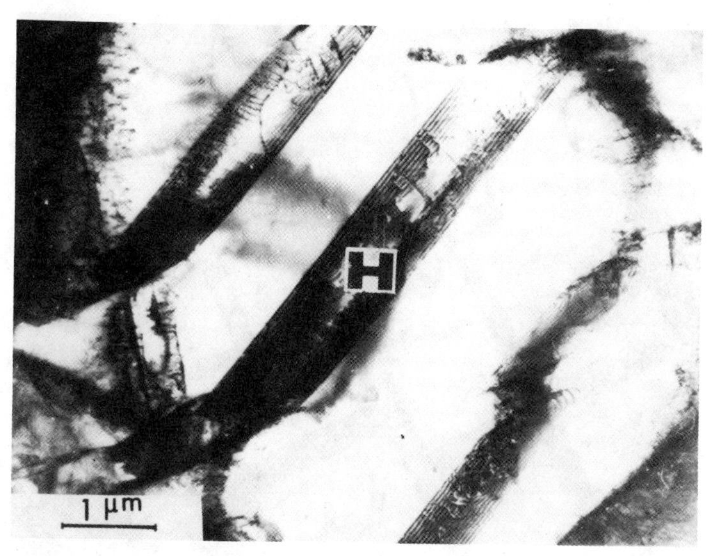

FIG. 5—*Fcc hydride plates.*

face-centered cubic (fcc) and shows fringe contrast, in which dislocation arrays are readily visible (Fig. 5).

All three types of strain-induced hydrides are metastable. They dissolve when examined for extended periods of time, from electron beam heating, or when the stress field is relieved. These hydrides can dissolve in the matrix even at room temperature. An example of the dissolution process for hyride plates can be seen by comparing Figs. 6*a* and 6*b*. After the specimens were kept unloaded for seven days outside the electron microscope, hydrides at C, D, E, M, and N in Fig. 6 completely vanished. Those at the other sites partially disappeared or decreased in size.

The interaction of strain-induced hydrides with crack growth was studied. Three modes of hydrogen or hydride-induced cracking were found by *in situ* observations and are illustrated by Figs. 7 and 8.

1. After the crack has been extended by loading, stress relaxation will inhibit further crack growth. However, the stresses in the crack-tip region remain high and will cause enrichment of hydrogen and nucleation and growth of new hydrides at the crack tip. After growing to a critical size, these hydrides will break and produce an additional increment of crack advance. New hydride particles will again form at the crack tip, and the process repeats itself to produce stepwise crack growth. The whole cracking process by the formation, growth, and fracture of strain-induced hydrides is illustrated in Fig. 7. In Fig. 7*a*, the crack (indicated by the arrow) has passed through a broken hydride in the prior cracking event. There is as yet no new hydride at the new crack tip. Two hours later, a hydride particle has formed and grown at the crack tip (Fig. 7*b*). After an additional 2.5 h, this hydride particle has broken and a new hydride has formed at the new crack tip (Fig. 7*c*).
2. In the second cracking mode, a hydride particle is nucleated in front of the crack tip (particle *R* in Fig. 8*a*). When this hydride grows beyond a critical size, a microcrack is observed to form in it (Fig. 8*b*). This microcrack eventually links up with the main crack to effect crack advance, as illustrated in Fig. 8*c*.

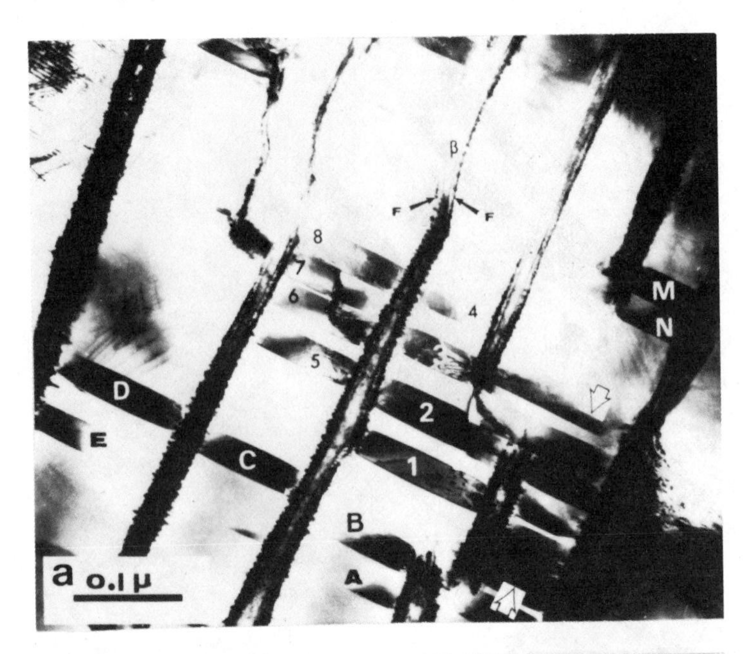

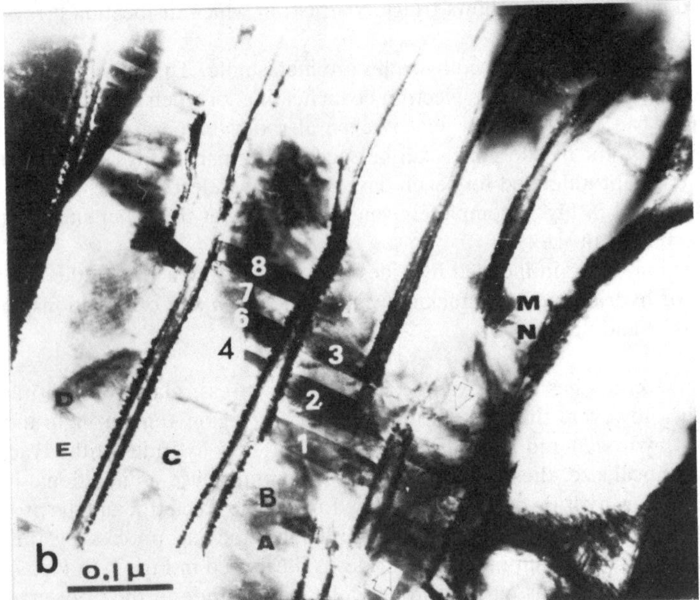

FIG. 6—*Dissolution of fcc hydride plates. Seven days have elapsed from* (a) *to* (b).

3. The third mode of cracking involves the second and third types of strain-induced hydride plates. Since these hydride plates provide easy paths for crack growth, the advancing crack simply grows along them.

The results show clearly that hydrogen-induced cracking of Ti-6Al-4V alloys is closely identified with the formation of strain-induced hydrides.

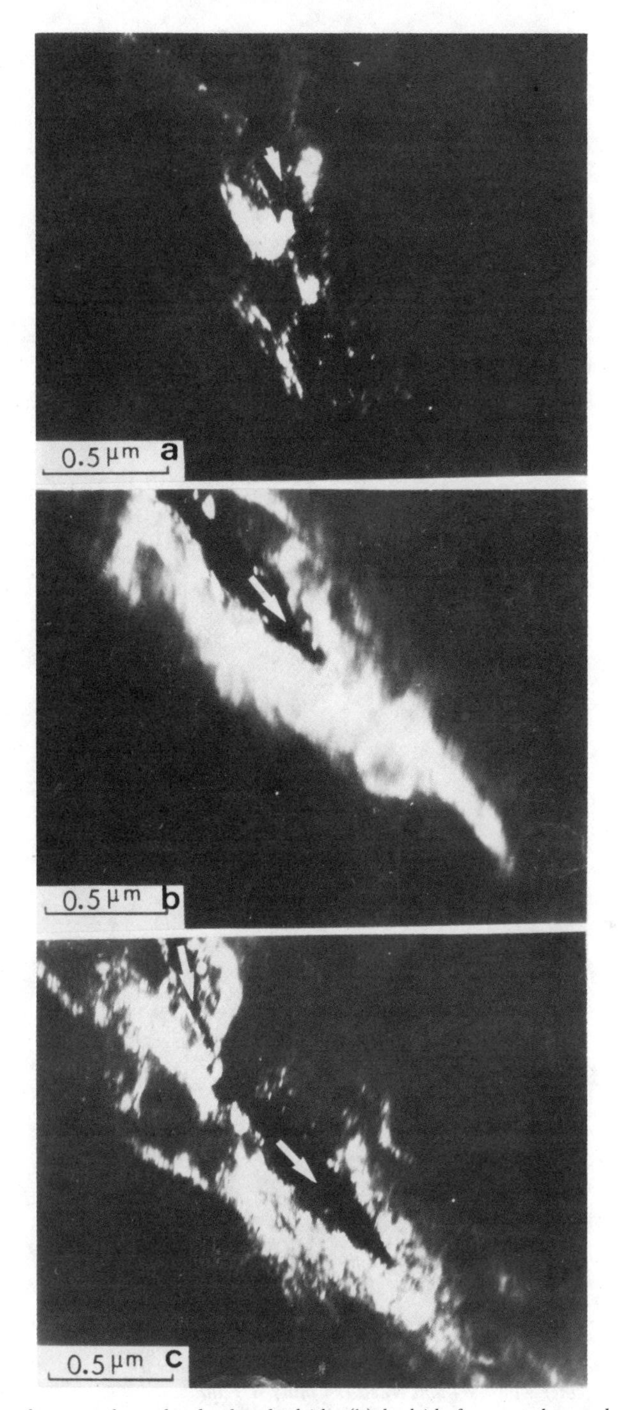

FIG. 7—(a) *Crack passes through a broken hydride,* (b) *hydride forms at the crack tip (2 h), and* (c) *old hydride breaks and new hydride forms (4.5 h). DF images.*

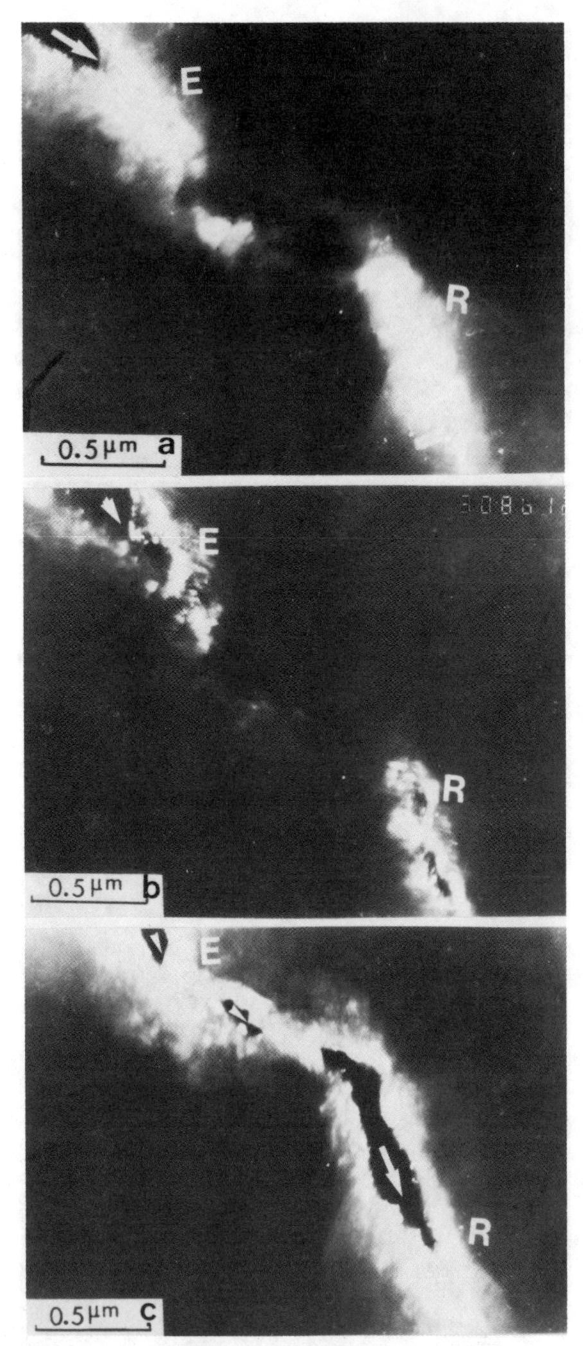

FIG. 8—(a) *Hydride forms in front of the crack tip,* (b) *microcrack occurs in it* (*35 min*), *and* (c) *microcrack joints together with the main crack* (*55 min*). *DF images.*

Discussion

Strain-Induced Hydrides

Recent studies [*8,9*] have shown that titanium hydrides (TiH_x) can form with one of three crystalline structures, depending on the hydrogen concentration. At low hydrogen concentrations, the hydride has a fct structure, with an axial ratio $c/a > 1$ and an ordered arrangement of hydrogen [*8,9*]. The medium hydrogen-bearing hydride has a fcc unit cell with respect to the titanium atoms. The hydrogen atoms are randomly distributed over the tetrahedral interstitial sites. At high hydrogen concentrations, near the limiting composition TiH_2, the hydrides transform to a fct structure (with $c/a < 1$) at temperatures below about 290 to 310 K [*10,11*].

In this study, both medium and high hydrogen-bearing hydrides were observed, but not the low concentration hydrides. This is not surprising because of the relatively high bulk hydrogen concentration used here and the buildup of hydrogen in the crack tip region by the local stresses. The question of whether or not the first and second types of hydrides are identical has not been resolved, because they have the same crystal structure but form at different places and exhibit different morphologies.

Some hydrides of the second type show many bright and dark strips, as seen in Fig. 4*b*. Hydride precipitates with a similar morphology have been observed previously in titanium [*9*] and its alloy [*12*]. The third type of hydride shows fringe contrast (Fig. 5), which is consistent with the observations made in Ti-5Al-2.5Sn alloy [*13*]. These fringes are the displacement α fringes that arise from the interface between the hydride precipitate and the titanium matrix [*14*]. The presence of interfacial dislocation array (Fig. 5) implies that the interface between the hydride and the α-phase matrix is either coherent or partially coherent [*14*].

It was noted that the strain-induced hydrides are metastable, and they can dissolve even at room temperature. *In situ* heating experiments on hydrogen-charged Ti-5Al-2.5Sn alloy, carried out on a heating stage inside the electron microscope, indicated that there are two types of relatively unstable hydrides [*13*]. These hydrides are similar to the second and third types of hydrides in the Ti-6Al-4V alloy and can be decomposed by heating to approximately 370 K. The stability of strain-induced hydrides is of particular interest because it relates to the temperature dependence for hydrogen-assisted crack growth rate. This aspect is discussed in the following subsection.

Hydrogen-Assisted Crack Growth

Recently, the kinetics of subcritical crack growth in Ti-6Al-4V alloy have been studied by using fracture-mechanics-based methods. The results have been reported in detail elsewhere [*2,3*], but may be summarized as follows. For hydrogen-charged Ti-6Al-4V alloy [*2*], the dissolved hydrogen can enhance subcritical crack growth under sustained load. The crack growth rate (da/dt) increased with hydrogen concentration and decreased with increasing temperature from 223 to 353 K. This decrease, at a given stress-intensity (K) level, can be expressed by the relationship $da/dt \propto \exp(Q/RT)$, where Q = 23.9 kJ/mol (Fig. 9). For uncharged Ti-6Al-4V alloy, tested in distilled water under cyclic loading, crack growth rate was increased over that in air [*3*]. The growth rate (da/dN) decreased with increasing temperature and loading frequency. The temperature dependence at a given frequency can be described by the expression $da/dN \propto \exp(Q/RT)\ (\Delta K)^{2.7}$, where Q = 6.2 kJ/mol and ΔK is stress-intensity range (Fig. 10).

These behaviors of hydrogen-assisted crack growth can be understood in terms of the interaction between cracking and strain-induced hydrides observed in this study. The *in situ*

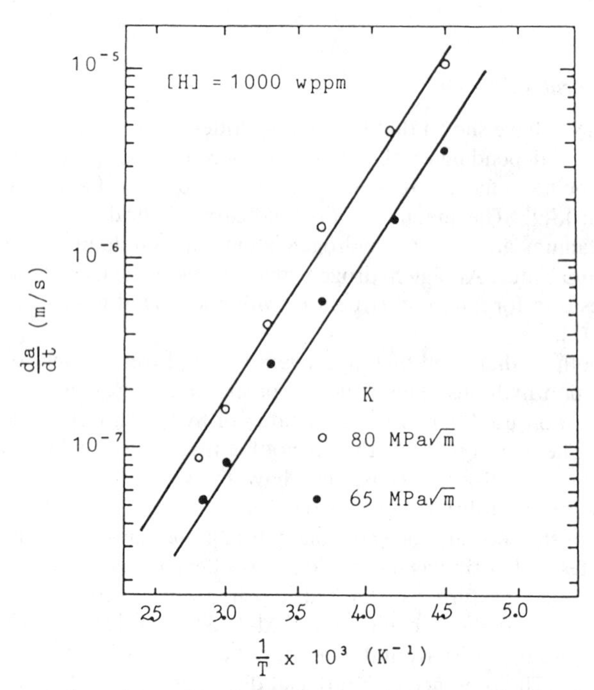

FIG. 9—*Temperature dependence of crack growth rate under sustained load for hydrogen-charged Ti-6Al-4V alloy.*

observations clearly indicate that the formation and subsequent rupture of the strain-induced hydrides at and in front of the crack tip are responsible for enhancing crack growth in the hydrogen-charged Ti-6Al-4V alloy. Since hydrides form more readily in the alloys that contain more hydrogen, crack growth rates are expected to increase with hydrogen content. Because of the instability of hydrides, especially at the higher temperature, it is reasonable to suppose that the amount of hydrides at or in front of the crack tip would decrease with increasing temperature and, hence, an accompanying decrease in crack growth rates. This could explain the observed temperature dependence for crack growth rate in the hydrogen-charged Ti-6Al-4V alloy.

Previous study of the reactions of water vapor with fresh surfaces of titanium alloy indicated that the reactions are very rapid and result in the formation of a monolayer of TiO_2 on the surface and the release of hydrogen [*15*]. It is believed that this hydrogen can enter the material and result in the formation of strain-induced hydrides at the crack tip in conjunction with applied loading. Because the reactions producing hydrogen are very rapid, the overall process may be controlled by the rate of formation of the strain-induced hydrides. Thus, the temperature dependence for fatigue crack growth in Ti-6Al-4V alloy in distilled water may be understood in the same way as that of the hydrogen-charged alloy. As for the frequency effect, it may be explained also on the basis of the strain-induced hydride mechanism. At the lower frequencies, because there is more time, more hydrides can be formed at the crack tip. Hence, the crack could be expected to grow faster.

The formation and subsequent fracture of hydrides at the crack tip, accompanied by crack advance, also have been found in vanadium [*16*], zirconium [*17*], and niobium [*18,19*]. The strain-induced hydride mechanism for cracking, therefore, appears to be generally applicable to all the hydride-forming metals.

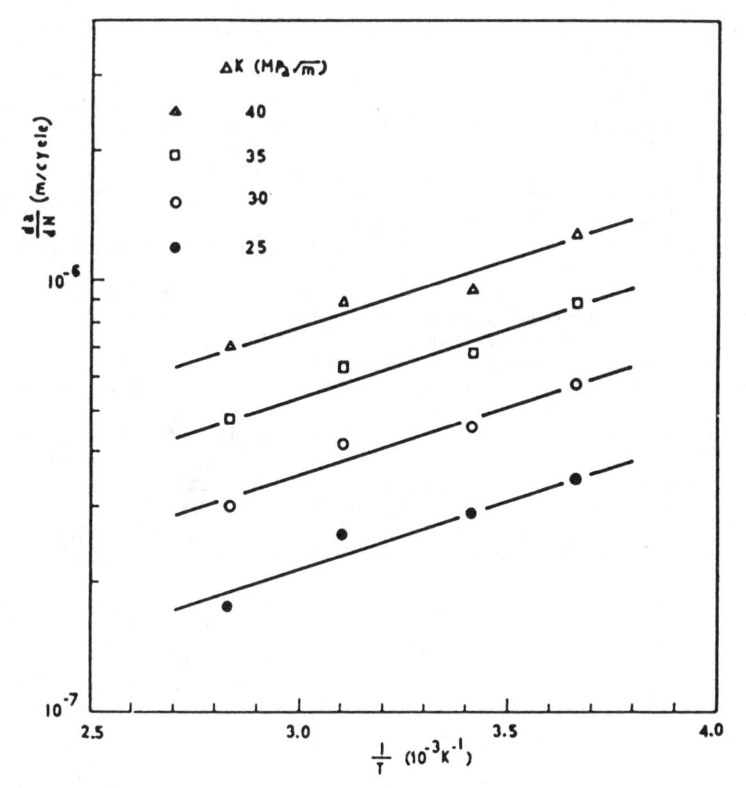

FIG. 10—*Temperature dependence of crack growth rate under cyclic load for uncharged Ti-6Al-4V alloy in water.*

Conclusions

Three types of strain-induced hydrides have been found in Ti-6Al-4V alloy, namely, the crack-tip hydride and two plate-like hydrides with fct and fcc crystalline structure. All of the hydrides are metastable and can be dissolved at room temperature. Formation and subsequent rupture of strain-induced hydrides at and in front of the crack tip during crack growth were observed; they provide support for the strain-induced hydride mechanism for hydrogen-assisted crack growth in the hydride-forming metals and alloys. This mechanism can explain the observed crack growth behavior in hydrogen-charged Ti-6Al-4V alloy, and in uncharged Ti-6Al-4V alloy tested in distilled water.

Acknowledgments

The authors wish to thank Professor R. P. Wei for reading the manuscript, and Professors K. H. Kuo and B. X. Zhou for advice and support.

References

[1] Smith, H. R., Piper, D. E., and Downey, F. K., Jr., *Engineering Fracture Mechanics,* Vol. 1, 1968, p. 123.
[2] Wan, X.-J., Qin, R.-S., and Xiao, H.-Z., *Proceedings of the 4th International Conference on Hydrogen and Materials*, Beijing, China, 9–13 May 1988.

[3] Gao, S. J., Qin, R. S., Zhang, S. S., and Wan, X. J., *Acta Metallurgica Sinica,* Vol. 22, 1986, p. A195.
[4] Moody, N. R. and Gerberich, W. W., *Metallurgical Transactions A,* Vol. 11A, 1980, p. 973.
[5] Pardee, W. J. and Paton, N. E., *Metallurgical Transactions A,* Vol. 11A, 1980, p. 1391.
[6] Pao, P. S. and Wei, R. P., "Titanium, Science, and Technology," *Proceedings,* 5th International Conference on Titanium, Munich, Federal Republic of Germany, Sept. 1984, Vol. 4, G. Lütjering, U. Zwicker, and W. Bunk, Eds., DGM, Oberursel, 1985, p. 2503.
[7] Xiao, H.-Z., Gao, S.-J., and Wan, X.-J., *Scripta Metallurgica,* Vol. 21, 1987, p. 265.
[8] Numakura, H. and Koiwa, M., *Acta Metallurgica,* Vol. 32, 1984, p. 1799.
[9] Woo, O. T., Weatherly, G. C., Coleman, C. E., and Gilbert, R. W., *Acta Metallurgica,* Vol. 33, 1985, p. 1897.
[10] Sidhu, S. S., Heaton, L., and Zauberis, D. D., *Acta Crystallographica,* Vol. 9, 1956, p. 607.
[11] Yakel, H. L., Jr., *Acta Crystallographica,* Vol. 11, 1958, p. 46.
[12] Hall, I. W., *Scandinavian Journal of Metallurgy,* Vol. 7, 1978, p. 277.
[13] Hall, I. W., *Scandinavian Journal of Metallurgy,* Vol. 7, 1978, p. 187.
[14] Edington, J. W., *Practical Electron Microscopy in Materials Science,* Van Nostrand Reinhold, New York, 1976.
[15] Gao, S. J., Simmons, G. W., and Wei, R. P., *Materials Science and Engineering,* Vol. 62, 1984, p. 65.
[16] Takano, S. and Suzuki, T., *Acta Metallurgica,* Vol. 22, 1974, p. 265.
[17] Cann, C. D. and Sexton, E. E., *Acta Metallurgica,* Vol. 28, 1980, p. 1215.
[18] Gahr, S., Grossbeck, M. L., and Birnbaum, H. K., *Acta Metallurgica,* Vol. 25, 1977, p. 125.
[19] Grossbeck, M. L. and Birnbaum, H. K., *Acta Metallurgica,* Vol. 25, 1977, p. 135.

P. K. Liaw,[1] *T. R. Leax*,[1] and *J. K. Donald*[2]

Gaseous-Environment Fatigue Crack Propagation Behavior of a Low-Alloy Steel

REFERENCE: Liaw, P. K., Leax, T. R., and Donald, J. K., **"Gaseous-Environment Fatigue Crack Propagation Behavior of a Low-Alloy Steel,"** *Fracture Mechanics: Perspectives and Directions (Twentieth Symposium) ASTM STP 1020,* R. P. Wei and R. P. Gangloff, Eds., American Society for Testing Materials, Philadelphia, 1989, pp. 581–604.

ABSTRACT: Fatigue crack growth behavior of 4340 steel was investigated in gaseous environments: laboratory air, hydrogen, and helium. No significant effect of specimen orientation on the rates of fatigue crack propagation was observed. In each environment, increasing load ratio decreased the threshold stress-intensity range (ΔK_{th}). The effect of load ratio on ΔK_{th} in the air environment was, however, much more significant than that in wet hydrogen and dry environments (dry hydrogen and dry helium). At low load ratios ($\lesssim 0.5$), the values of ΔK_{th} in the air environment were larger than those in dry environments. The ΔK_{th} values in wet hydrogen were in between those in air and dry environments. At a high load ratio of 0.8, ΔK_{th} was, however, insensitive to test environment.

The effects of load ratio and environment on fatigue crack growth rate properties are discussed in light of crack closure and hydrogen embrittlement mechanisms. While the effects of roughness-induced crack closure and hydrogen embrittlement on crack growth rates were found to be negligible, oxide-induced crack closure governed the kinetics of gaseous-environment, near-threshold crack propagation behavior. However, thick oxide deposits in wet hydrogen did not cause high levels of crack closure.

KEY WORDS: fatigue crack growth, stress intensity, low alloy steel, crack closure, laboratory air, hydrogen, helium, hydrogen embrittlement, environments

It is well recognized that the successful application of fracture mechanics for design against fatigue damage requires crack propagation rate properties of intended service conditions. Up to the present time, a significant amount of fatigue crack growth rate results have been generated for a wide range of structural materials. However, most of these crack growth rate data have been developed at relatively high stress-intensity ranges and address what is called Paris-region crack propagation. Nevertheless, many machine components are frequently subjected to low-stress and high-cycle loading. Thus, knowledge of near-threshold fatigue crack growth behavior is of prime importance for life prediction and integrity assessment of these components.

Near-threshold crack growth rate properties are strongly affected by factors, such as, load ratio ($R = P_{min}/P_{max}$, where P_{min} and P_{max} are the applied minimum and maximum loads, respectively), microstructure, and environment [*1–11*]. Detailed information regarding the effects of these variables on near-threshold crack propagation rates is necessary to provide sufficient data for material selection and design of structural components. Moreover, mech-

[1] Materials Analysis Department, Westinghouse R&D Center, Pittsburgh, PA 15235.
[2] Professional Service Group, Inc., 728 Main St., Hellertown, PA 18055. Presently, director, Fracture Technology Associates, STAR Route, Pleasant Valley, PA 18951.

anistic understanding of near-threshold crack growth performance is important since it allows us to interpret and to extrapolate the experimentally determined data.

In this paper, near-threshold fatigue crack growth behavior of a 4340 steel was investigated in gaseous environments. The crack growth experiments were conducted at various load ratios. The level of surface roughness and the thickness of oxide deposits present on the fracture surface were quantitatively characterized. The gaseous-environment fatigue crack propagation behavior is discussed in light of crack closure measurements and hydrogen embrittlement mechanisms.

Experimental Procedure

Material and Specimen

Fatigue crack propagation rate experiments were performed on a quenched and tempered 4340 steel. The chemical composition of this steel in weight percent was 0.43 carbon, 0.67 manganese, 0.006 phosphorus, 0.012 sulfur, 0.29 silicon, 0.72 chromium, 1.72 nickel, 0.23 molybdenum, and 0.02 vanadium. Mechanical properties are presented in Table 1. The microstructure was a tempered martensite with a prior austenite grain size of approximately 25 μm.

Crack growth rate testing was conducted by using compact-type (CT) specimens, which were machined from a cylindrical 4340 steel forging. The CT specimens had one of two orientations, CR and LR-RL, as defined in the ASTM Method for Plane-Strain Fracture Toughness of Metallic Materials (E 399-83). The dimensions of a CT specimen for CR orientation were 50.8 mm wide, 30.5 mm high, 11.5 mm thick; and 38.1 mm wide, 22.9 mm high, 11.4 mm thick for LR-RL orientation. Note that material strengths in the circumferential orientation were slightly lower than those in the LR-RL orientation (Table 1).

Test Environment

Test environments included (*a*) laboratory air with relative humidity of 30% to 40%, (*b*) 655-kPa wet hydrogen gas (99.95% purity), (*c*) 655-kPa dry hydrogen gas (99.95% purity), and (*d*) 655-kPa dry helium gas (99.95% purity). For the wet hydrogen environment, hydrogen gas was bubbled through a water reservoir to maintain 100% relative humidity. The wet hydrogen gas (continuous flow) was contained in stainless-steel, O-ring sealed chambers clamped to each side of the CT specimen. For dry hydrogen and dry helium gas environments, gas was passed through an on-line calcium chloride desiccator, and then flushed through the chambers to remove residual impurities. Dry hydrogen or dry helium (continuous flow) was also contained in the O-ring sealed chambers. Based on the purity of the bottle gas and the experimental procedure, moisture levels in the two dry environments were estimated to be less than 10 ppm.

TABLE 1—*Mechanical properties.*[a]

Orientation	0.2% Yield Strength, MPa	Ultimate Tensile Strength, MPa	Elongation, %	Reduction in Area, %
Circumferential	696	852	22.8	63.0
LR-RL	716	868	20.0	49.5

[a] Specimens were fabricated from the cylindrical forging.

Fatigue Crack Growth Testing

Fatigue crack growth experiments were conducted using a computerized electrohydraulic fatigue machine [*12,13*]. Fatigue testing was performed under sinusoidal loading at a frequency of 100 Hz. The investigated load ratios were 0.1, 0.5, and 0.8. Crack length was determined by the compliance technique [*12–14*]. An MTS clip-on extensometer, specifically designed for high-frequency applications, was attached on the front face of the CT specimens for measuring crack opening displacement.

To develop near-threshold fatigue crack growth rate properties, the stress-intensity range, ΔK ($= K_{max} - K_{min}$, where K_{max} and K_{min} are the applied maximum and minimum stress intensities, respectively) was continuously decreased according to the equation [*12,13*]

$$\Delta K = \Delta K_0 \exp[c(a - a_0)]$$

where

ΔK_0 = initial stress-intensity range,
a = instantaneous crack length,
a_0 = initial crack length, and
c = negative constant (-0.059 mm^{-1}) with a dimension of reciprocal length, which specifies the ΔK gradient for the test.

The stress-intensity range expression for the CT specimen can be found in Ref *14*. Following decreasing ΔK crack propagation rate testing, the value of c was changed to a positive value of 0.059 mm^{-1} for obtaining higher ΔK crack growth rates.

A modified secant method [*15*] was employed to convert a versus N (cycle) data to da/dN (crack growth rate) results. Each crack length increment used to calculate da/dN spans the adjacent first and third a-versus-N data points with ΔK corresponding to the crack length midway through the increment.

The ΔK threshold, ΔK_{th}, was operationally defined as the stress-intensity range corresponding to a fatigue crack growth rate of 1.0×10^{-10} m/cycle. The threshold was established by fitting a least-squares line through the crack growth rate data between 5×10^{-10} and 5×10^{-11} m/cycle and determining ΔK_{th} at a propagation rate of 1.0×10^{-10} m/cycle.

During crack growth testing, crack closure levels were determined by using an unloading compliance method [*16*]. In this technique, the unloading linear-elastic displacement was subtracted from the load-versus-displacement curve to increase the sensitivity to detect crack closure levels. The crack closure level was defined as the point where the load-versus-displacement curve begins to deviate from the unloading linear-elastic compliance line. The value of (K_{cl}) is, therefore, measured and defined as the stress-intensity corresponding to the crack closure point.

Fracture Surface Characterizations

Fracture surfaces were first examined using optical microscopy. Moreover, surface roughnesses and oxide deposits present on fracture surfaces were characterized. Following that, fracture modes were investigated by scanning electron microscopy (SEM).

The roughness levels of fracture surfaces were measured using a surface analyzer (Surfanalyzer 360 system, Clevite Corporation). Four roughness profiles were traced and recorded along the crack growth direction of the fracture surface. The level of roughness at a given ΔK is defined as the average of the four measured roughness profiles. The roughness recorded in each trace is the arithmetic average (AA) deviation from the mean surface, this is, the average of numerous measurements at the heights of the surface peaks and valleys (measured from the mean surface) (American Standard "Surface Roughness, Waviness and Lay," ASA B461-1955, UDC621.9.016, published by the American Society of Mechanical Engineers, 1955). The mean surface is defined as a perfect surface that would be formed if all of the roughness peaks were cut off and used up in filling the valleys below this surface.

Auger spectroscopy was used to measure the thickness of oxide deposits present on fracture surfaces. In the Auger spectrometer, Ar^+ ions bombarded oxide debris with a sputtering rate of 80 Å/min. As Ar^+ ions continued sputtering fracture surfaces, the concentration of oxygen decreased and that of iron increased. Thus, the depth profiles of oxygen and iron concentrations can be determined. Oxide thickness is defined as the depth where the concentration of oxygen equals that of iron [*5,6,8,9*].

Experimental Results

Effect of Orientation and Load Ratio

Fatigue crack growth rate data in CR and LR-RL orientations are presented in Fig. 1. In both laboratory air and dry helium environments, no significant influence of specimen orientation on the rates of near-threshold crack propagation was observed.

The crack growth rate properties at load ratios of 0.1, 0.5, and 0.8 are presented in Figs. 2*a* and 2*b* for laboratory air and dry helium environments, respectively. In laboratory air, Fig. 2*a*, increasing the load ratio from 0.1 to 0.8 significantly increases near-threshold crack growth rates. Decreasing ΔK levels increases the influence of load ratio on crack propagation rates. Interestingly, in wet hydrogen and dry (helium or hydrogen) environments, crack propagation rates at $R = 0.8$ are only slightly faster than those at $R = 0.1$. In particular, dry helium environment crack growth rates at load ratios of 0.1 and 0.8 are essentially identical (Fig. 2*b*). Thus, the influence of load ratio on near-threshold crack growth rates in the air environment is much more significant than that in wet hydrogen and dry environments.

Effect of Environment

Near-threshold crack propagation rate results in the four environments are compared in Fig. 3*a* at a load ratio of 0.1. It was observed that the crack growth rates in laboratory air were the slowest in the four environments investigated while those in the two dry environments were the fastest. The rates of crack growth in wet hydrogen are located between those in laboratory air and dry environments. Thus, near-threshold crack growth rates in moisture-containing environments (air and wet hydrogen) are slower than those in dry environments. Surprisingly, these results are in contrast with traditional corrosion fatigue data where it is usually expected that fatigue cracks propagate faster in wet environments than in dry environments [*17–19*]. In Fig. 3*a*, decreasing ΔK levels increases the effect of environment on crack growth rates.

Figure 3*b* presents near-threshold crack growth rate properties in the various environments at a high load ratio of 0.8. Note that the difference in crack propagation rates for the four environments is much reduced, as compared with the data at $R = 0.1$. Therefore, increasing load ratio decreases the influence of environment on fatigue crack growth behavior. It should be mentioned that in Fig. 3*b*, dry helium environment crack propagation rates are, however, slightly slower than those in the other three environments.

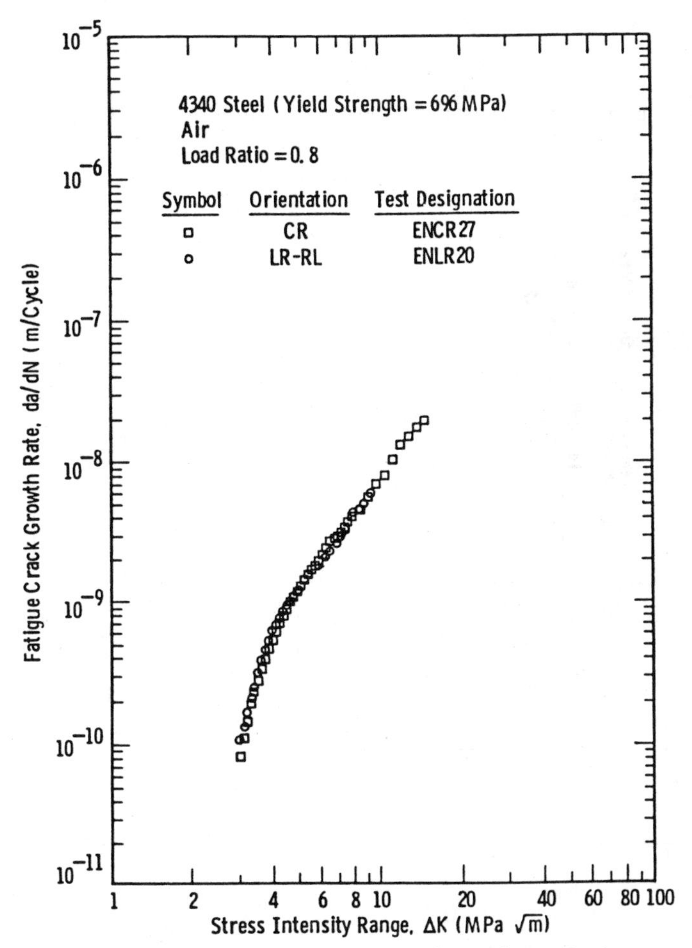

FIG. 1—*Effect of orientation on crack growth rates in air.*

Threshold Stress-Intensity Range, ΔK_{th}

The values of ΔK_{th} are plotted versus load ratio in Fig. 4 for the four gaseous environments, and are listed in Table 2. In the laboratory air environment, increasing load ratio significantly decreases ΔK_{th}. For instance, in laboratory air, ΔK_{th} at $R = 0.1$ is 2.8 times larger than that at $R = 0.8$. In the wet hydrogen and two dry environments, a similar trend was observed. However, the effect of load ratio on ΔK_{th} in these three environments is much less significant than that in the laboratory air environment. For example, in dry helium, ΔK_{th} at $R = 0.1$ is only 1.1 times larger than that at $R = 0.8$.

At a low load ratio of 0.1, ΔK_{th} in laboratory air is the largest of the four environments, while in dry environments it is the smallest. As an example, at $R = 0.1$, ΔK_{th} in laboratory air is 2.0 times larger than that in dry helium. The value of ΔK_{th} in wet hydrogen is in between those for laboratory air and dry environments. Note that increasing load ratio decreases the difference in ΔK_{th} in the four environments. Moreover, at a high load ratio of 0.8, the values of ΔK_{th} in the various environments converge to a narrow range of 3 to 4 $MPa\sqrt{m}$ although the values of ΔK_{th} in dry environments appear to be slightly larger than those in wet environments.

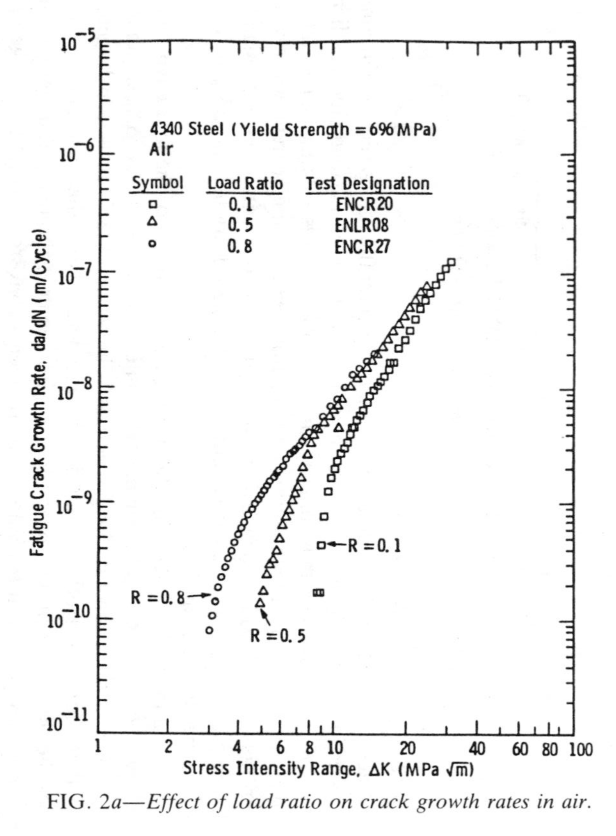

FIG. 2a—*Effect of load ratio on crack growth rates in air.*

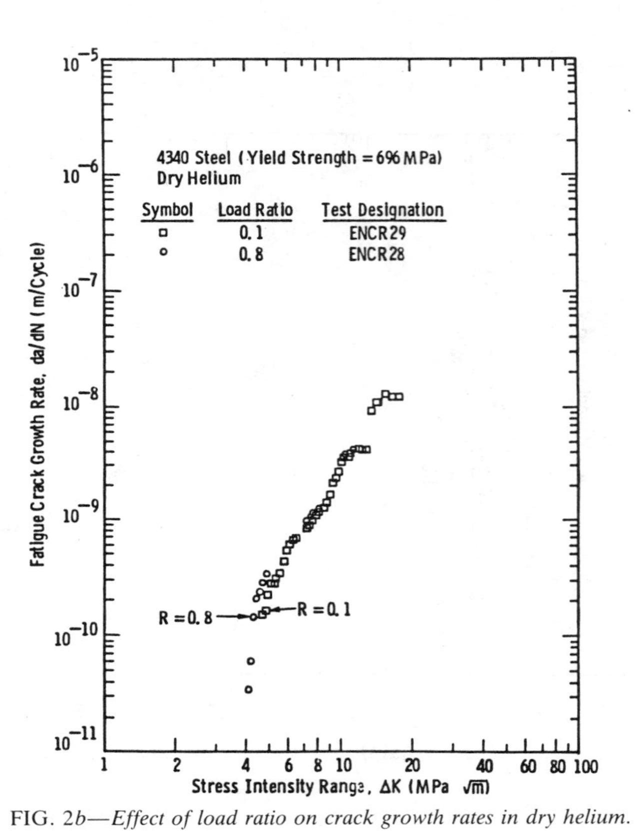

FIG. 2b—*Effect of load ratio on crack growth rates in dry helium.*

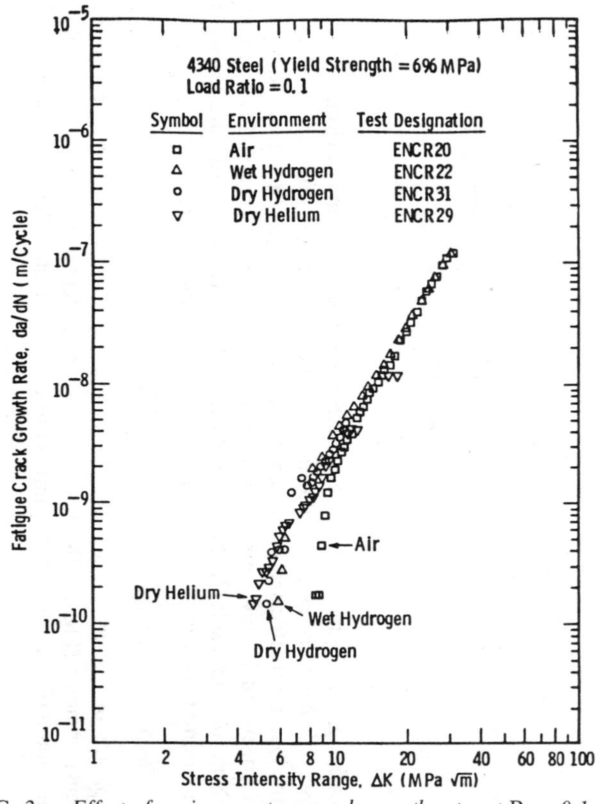

FIG. 3a—*Effect of environment on crack growth rates at* R = 0.1.

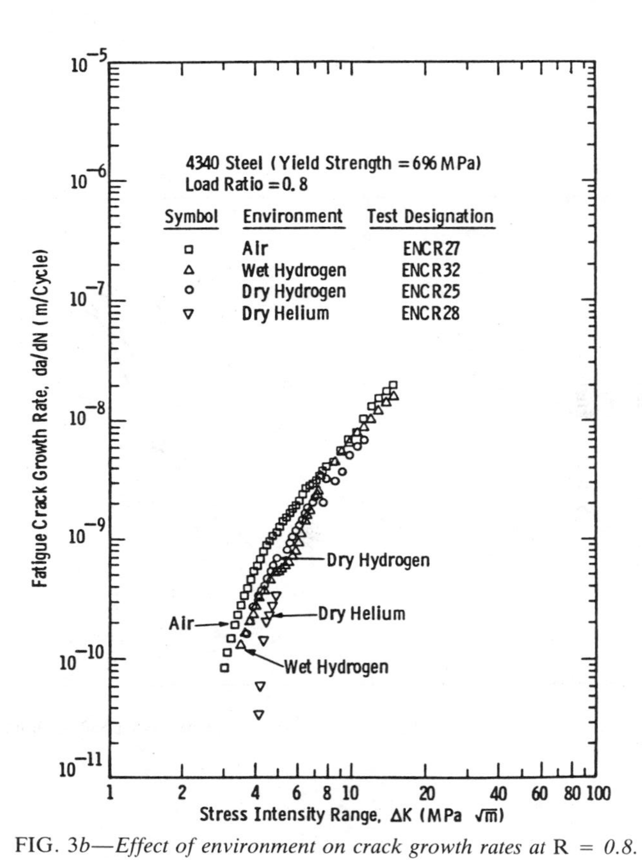

FIG. 3b—*Effect of environment on crack growth rates at* R = 0.8.

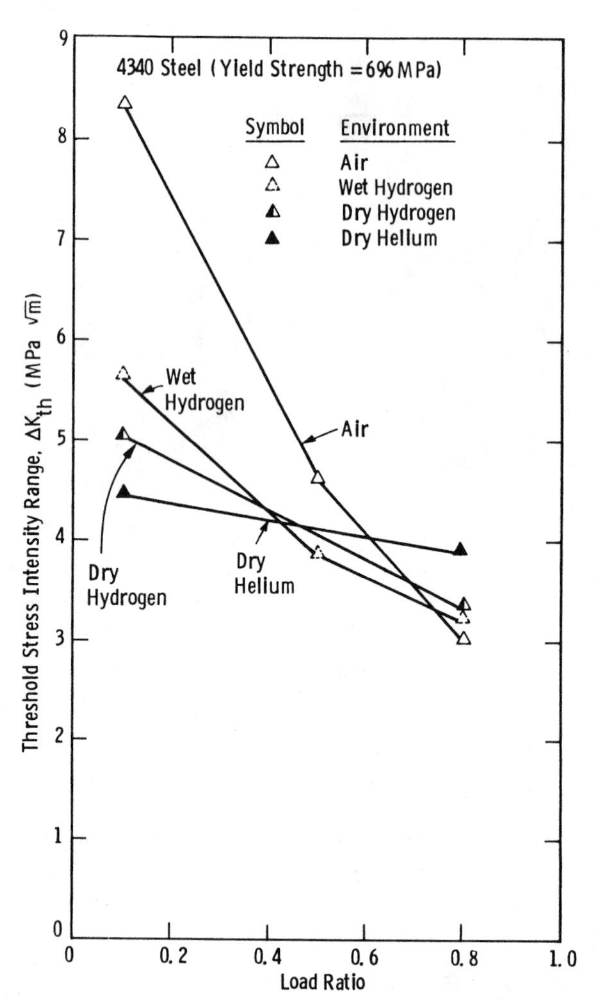

FIG. 4—*Effects of load ratio and environment on* ΔK_{th}.

Summarizing the results in Fig. 4, great effects of load ratio and environment on ΔK_{th} were observed. Nevertheless, the influence of load ratio and environment is interrelated; that is, the load ratio effects in the air environment are much more pronounced than those in wet hydrogen and dry environments, and increasing load ratio decreases the environmental effects. Note that in Table 2, the values of ΔK_{th} at a given load ratio are insensitive to specimen orientation in both air and dry helium environments.

Scanning Electron Microscopy

SEM photos of fracture surfaces are shown in Figs. 5 and 6. In laboratory air and dry helium environments (Fig. 5), the fracture mode is predominantly transgranular although a minute amount of intergranular fracture is present. At threshold levels, it is a transgranular fracture mode.

In wet and dry hydrogen environments (Fig. 6), a mixture of intergranular and transgranular fracture was observed. Nevertheless, the amount of intergranular fracture seems

TABLE 2—*Threshold stress-intensity range* (ΔK_{th}), *effective threshold stress-intensity range* ($\Delta K_{th,eff}$), *oxide thickness, and crack-tip opening displacement (CTOD) at threshold.*

Test Designation	Orientation	Load Ratio	Environment	ΔK_{th}, $MPa\sqrt{m}$	$\Delta K_{th,eff}$, $MPa\sqrt{m}$	Oxide Thickness, Å	$CTOD/_{max}$[a], Å	$CTOD/_{cyclic}$[b], Å
ENCR20	CR	0.1	air	8.36	4.43	3634	3021	1223
ENLR08	LR-RL	0.5	air	4.63	...	1140	3022	375
ENCR27	CR	0.8	air	3.02	3.02	810	7982	160
ENLR20	LR-RL	0.8	air	2.95	2.95	523	7667	153
ENCR22	CR	0.1	wet H_2	5.70	3.82	8550	1404	569
ENLR06	LR-RL	0.5	wet H_2	3.90	...	...	2144	266
ENCR32	CR	0.8	wet H_2	3.30	3.30	3563	9531	191
ENCR31	CR	0.1	dry H_2	5.10	4.25	499	1124	455
ENCR25	CR	0.8	dry H_2	3.39	3.39	437	10 058	201
ENCR29	CR	0.1	dry He	4.50	3.28	404	875	354
ENLR21	LR-RL	0.1	dry He	3.82	2.70	855	635	255
ENCR28	CR	0.8	dry He	4.02	4.02	380	14 143	283

[a] $CTOD/_{max} = 0.49 \frac{K^2_{max,th}}{\sigma_y E}$

where σ_y is yield strength, E is Young's modulus and $K_{max,th}$ is the maximum stress intensity at threshold [Ref *27*].

[b] $CTOD/_{cyclic} = 0.49 \frac{\Delta K^2_{th}}{2\sigma_y E}$ [Ref *27*].

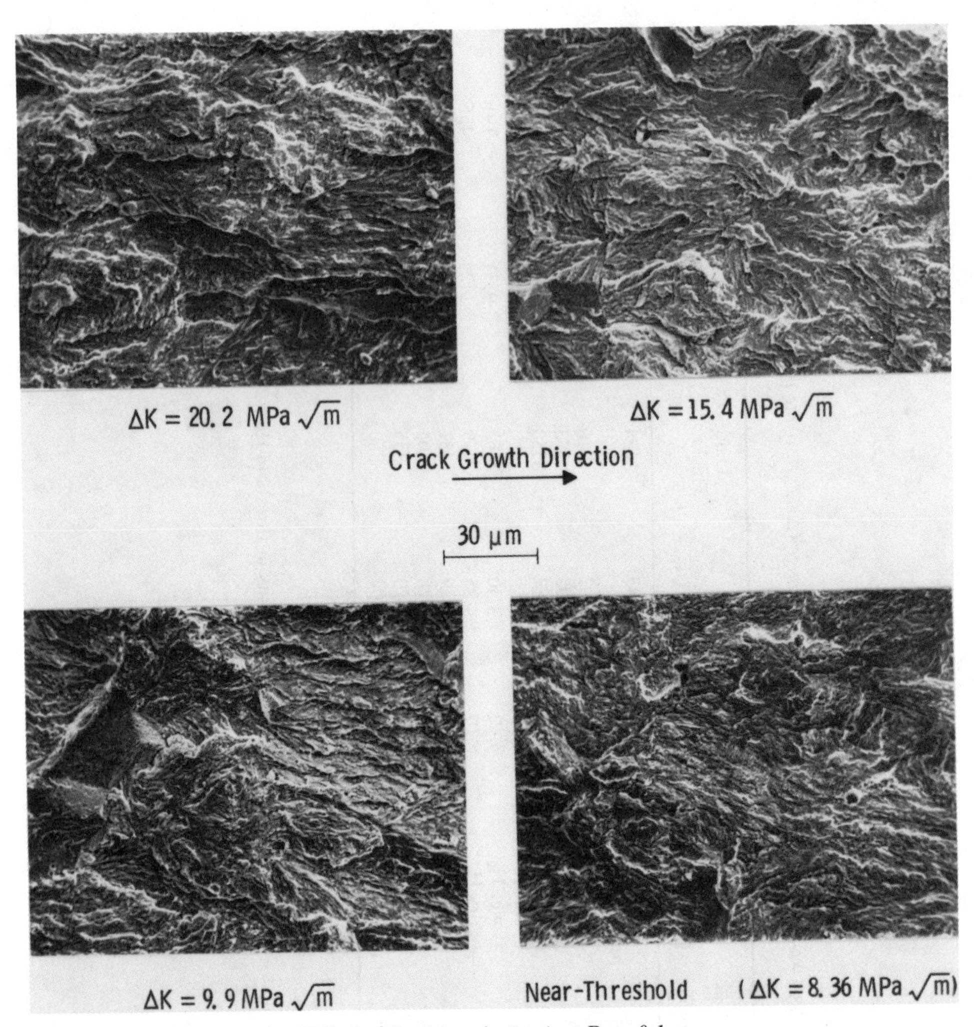

FIG. 5—*Fractography in air at* R = *0.1.*

to be dependent upon ΔK level. At high ΔK values (> 16 $\text{MPa}\sqrt{\text{m}}$), intergranular fracture is absent. As ΔK decreases, intergranularity appears, reaches a maximum, and then decreases. The same behavior was also reported in other steels [*4,5,7,20,21*]. At threshold levels (Fig. 6), a mixed mode of intergranular and transgranular fracture was observed in wet hydrogen while a completely transgranular fracture mode was found in dry hydrogen. It should be noted that the extent of intergranular fracture in wet and dry hydrogen is much greater than that in laboratory air and in dry helium. Since crack growth rates in dry hydrogen and dry helium are comparable at $R = 0.1$ (Fig. 3*a*), no direct correlation between crack propagation rate and the extent of intergranular fracture is evident.

Fracture modes at $R = 0.8$ are transgranular regardless of test environment and ΔK level, which is in contrast with the result at $R = 0.1$. Thus, increasing load ratio tends to decrease the extent of intergranularity. At $R = 0.8$, secondary cracking was also observed on fracture surfaces. However, decreasing ΔK levels decreases the extent of secondary cracking.

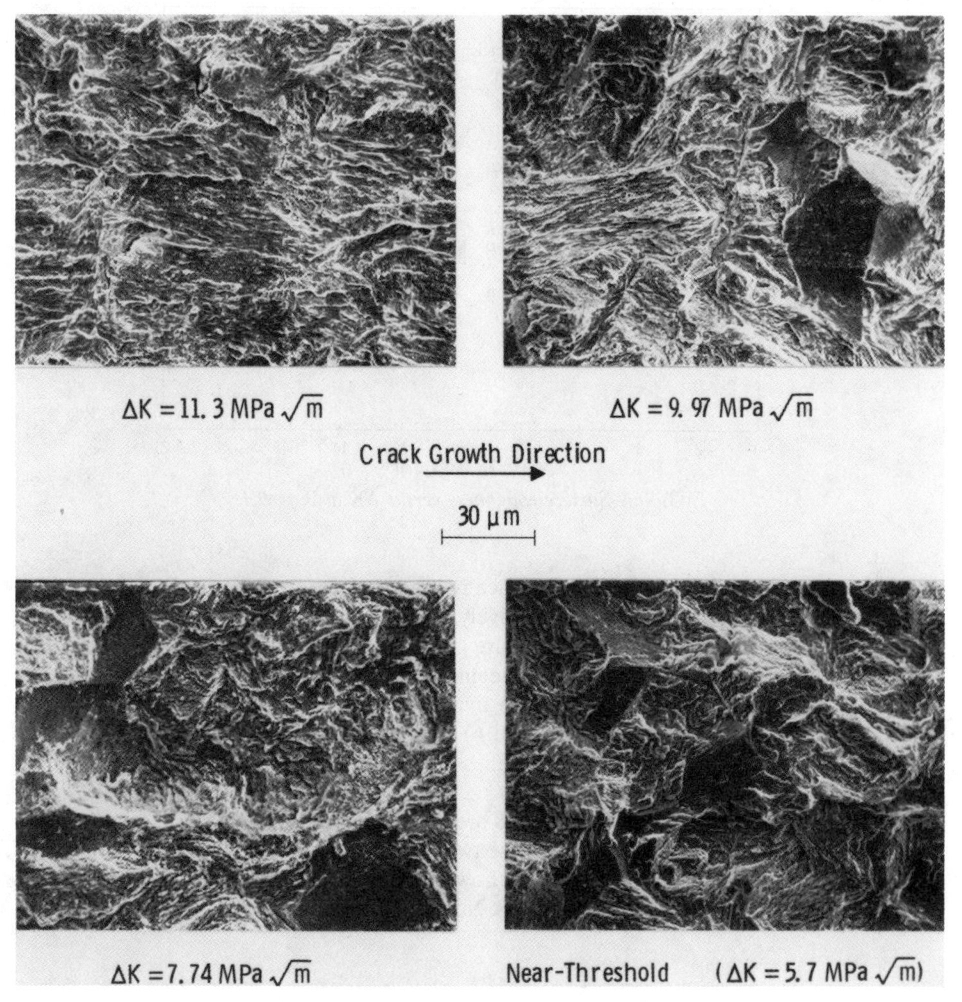

FIG. 6—*Fractography in wet hydrogen at* R = *0.1.*

Discussion

In this investigation, significant effects of load ratio and environment on fatigue crack propagation behavior have been found in a 4340 steel. The rates of near-threshold crack growth in the air environment are more sensitive to load ratio than in wet hydrogen and dry environments. At low load ratios, crack propagation rates in moisture-containing environments are slower than those in dry environments. Nevertheless, increasing load ratio decreases the environmental effects. These crack growth rate properties will be explained in light of the characteristics of surface roughness and oxide deposits observed on fracture surfaces.

Surface Roughness

Surface roughness has been shown to affect near-threshold fatigue crack growth rate properties because of the rough fracture surfaces associated with low-ΔK crack growth

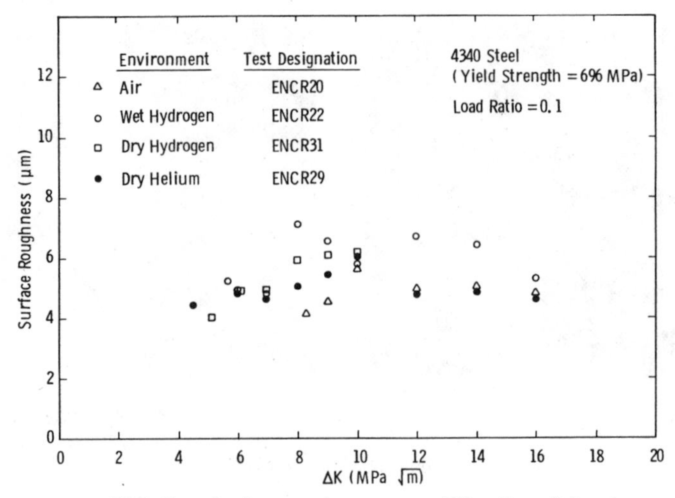

FIG. 7a—*Surface roughness versus* ΔK *at* R = *0.1.*

characteristics [*3,5,22*]. The results of surface roughness measurements at R = 0.1 and 0.8 are presented in Figs. 7*a* and 7*b*, respectively. At R = 0.1, the roughness levels in wet hydrogen and dry hydrogen environments are generally greater than those in laboratory air and dry helium environments. This trend seems to be consistent with the greater extent of intergranular fracture observed in the two hydrogen environments than in laboratory air and dry helium environments (Figs. 5 and 6) since intergranularity can promote surface roughness [*5,10,21*]. At R = 0.8, the roughness levels in the four environments were found to be similar, as presented in Fig. 7*b*.

The roughness levels at R = 0.1 and 0.8 are compared in Fig. 7*b*. It appears that there are comparable surface roughnesses at these two load ratios. The upper-bound dash-line of the roughness data scatterband at R = 0.1, however, generally exceeds the roughness levels in the wet hydrogen environment at R = 0.8. Note that the upper-bound dash-line represents

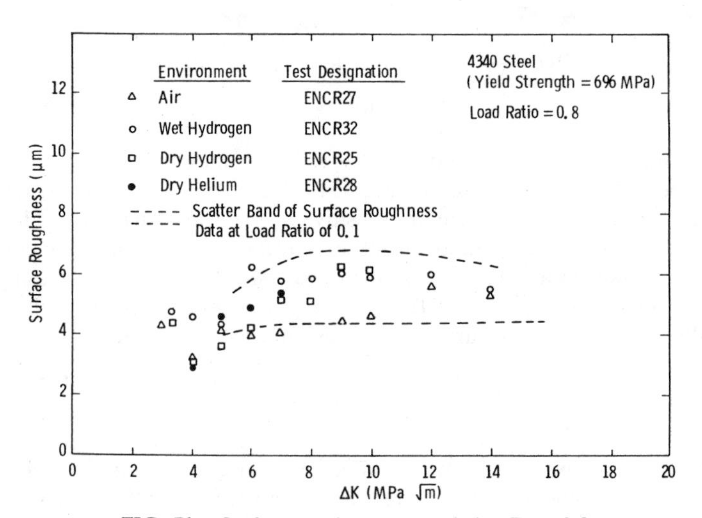

FIG. 7b—*Surface roughness versus* ΔK *at* R = *0.8.*

the surface roughness in the wet hydrogen environment at $R = 0.1$. Therefore, the roughness in wet hydrogen at $R = 0.1$ is generally greater than that at $R = 0.8$. This behavior is in agreement with the finding that increasing load ratio decreases the extent of intergranular fracture.

Oxide Deposits

Recently, it has been reported that oxide deposits are present during near-threshold crack propagation [*2,4,5–9*]. The formation of oxide deposits has been attributed to fretting oxidation which results from plasticity-induced crack closure and Mode II displacement (fretting) [*4,8,9,23–25*]. Optical photos of fracture surfaces at load ratios of 0.1 and 0.8 are shown in Figs. 8*a* and 8*b* for moisture-containing and dry environments, respectively. In laboratory air and wet hydrogen environments at $R = 0.1$, a significant amount of black deposits observed at threshold levels indicate the presence of oxide, Fig. 8*a*. At $R = 0.8$, the extent of oxide debris is much less pronounced than that at $R = 0.1$. Thus, increasing load ratio decreases the amount of oxide deposits. In dry hydrogen and dry helium environments at $R = 0.1$, some oxide deposits are visible, Fig. 8*b*. However, they are not so pronounced as observed in moisture-containing environments. At $R = 0.8$, oxide debris is minimal. In both moisture-containing and dry environments, the amount of oxide debris increases as the crack growth rate approaches the threshold level. Note that the oxide deposits have been identified as Fe_2O_3 (see Ref *4*).

Increasing load ratio decreases plasticity-induced crack closure and Mode II displacement [*4,8,9,23–25*], thereby resulting in less fretting oxidation and a decreased amount of oxide deposits at higher load ratios, as found in Figs. 8*a* and 8*b*. The lack of oxide debris in dry environments results apparently from the shortage of oxygen and moisture levels in dry environments. In Figs. 8*a* and 8*b*, decreasing ΔK increases the amount of oxide debris. This behavior results from larger Mode II displacements, and thus more intensive fretting oxidation with decreasing ΔK levels. Moreover, the slower crack growth rate with decreasing ΔK prolongs the time for fretting oxidation at each decreased ΔK level, which also contributes to thicker oxide layers at lower ΔK levels.

The results of oxide thickness measurements are shown in Table 2. A plot of oxide thickness at threshold versus load ratio is presented in Fig. 9. In laboratory air and wet hydrogen environments, increasing load ratio significantly decreases oxide thickness. In laboratory air and wet hydrogen environments, oxide thicknesses at $R = 0.1$ are 4.5 and 2.4 times larger than that at $R = 0.8$, respectively. It should be noted that in the laboratory air, the oxide thicknesses naturally formed were measured to be approximately 300 Å, which was about 12 to 3 times smaller than that determined at the threshold level depending on the load ratio. This trend clearly indicates that the formation of thick oxide deposits at threshold is due to fretting oxidation rather than to thermal oxidation.

In dry hydrogen and dry helium environments, oxide thickness is much less sensitive to load ratio than observed in moisture-containing environments. In dry hydrogen and dry helium environments, oxide thicknesses at $R = 0.1$ are only 1.3 and 1.1 times larger than that at $R = 0.8$, respectively.

At a low load ratio of 0.1, oxide thicknesses in moisture-containing environments are much larger than those in dry environments. As an example, at $R = 0.1$, oxide thickness in laboratory air is 9.0 times larger than that in dry helium. It should be noted that oxide thicknesses at two load ratios in wet hydrogen are the largest among the four environments. In Fig. 9, increasing load ratio decreases the difference in oxide thicknesses in the moisture-containing and dry environments. Furthermore, at a high load ratio of 0.8, oxide thicknesses in the laboratory air and dry environments appear to converge to a single value of approximately 500 Å.

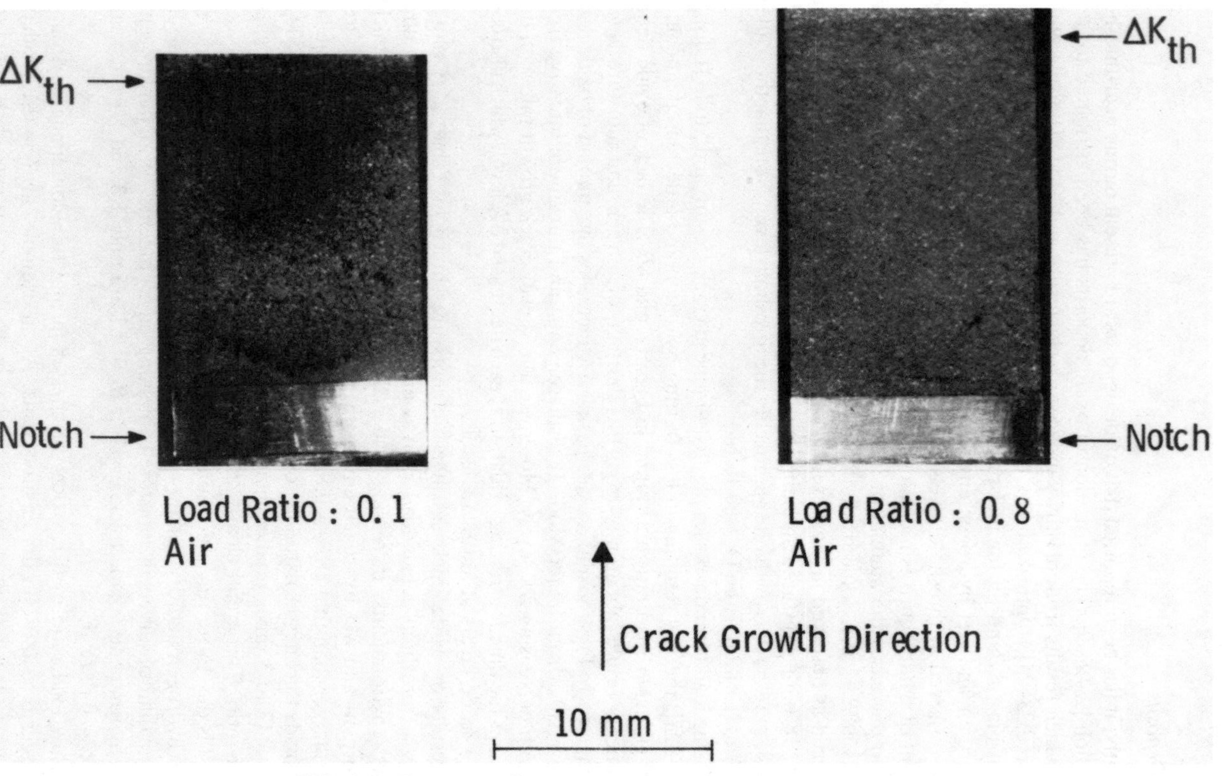

FIG. 8a—*Fracture surfaces in moisture-containing environments.*

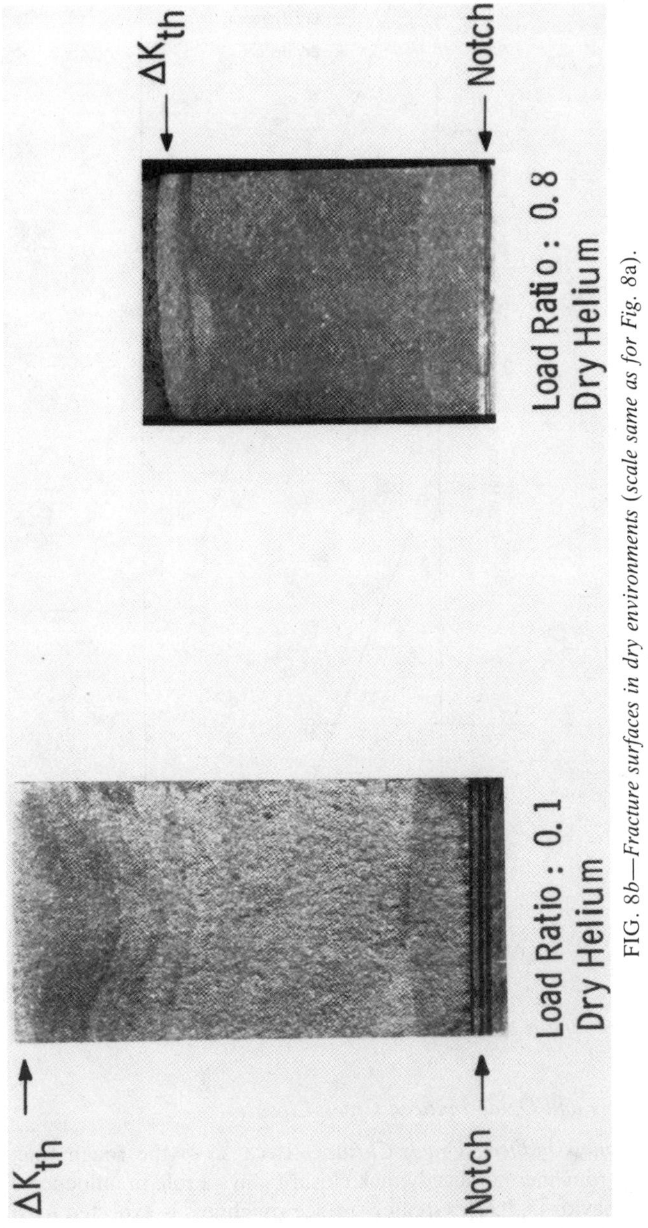

FIG. 8*b*—*Fracture surfaces in dry environments (scale same as for Fig.* 8a).

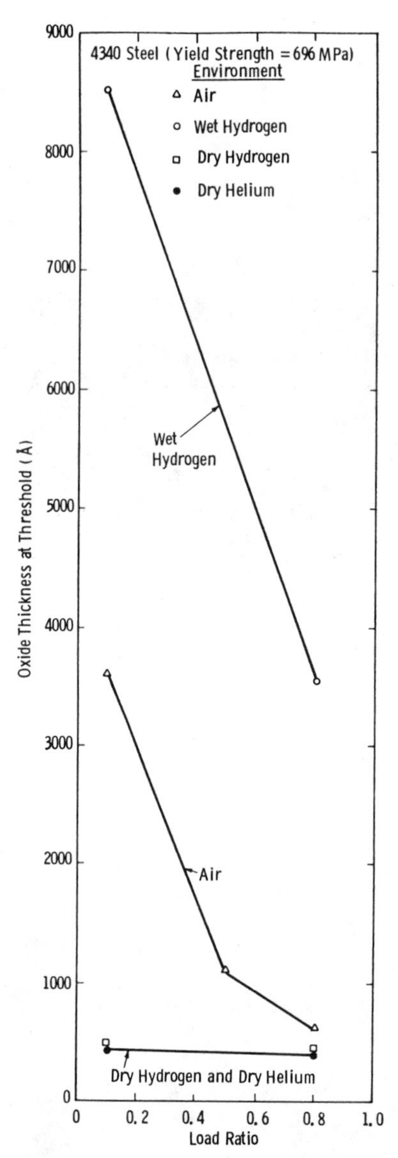

FIG. 9—*Effects of load ratio and environment on oxide thickness at threshold.*

Surface Roughness and Oxide-Induced Crack Closure

Surface Roughness-Induced Crack Closure—Because of the rough fracture surfaces at threshold levels, roughness-induced crack closure plays a role in influencing near-threshold crack growth behavior [*3,5,22*]. Greater surface roughness is expected to give higher crack closure levels [*3,5,22*]. The results of crack closure measurements are presented in Fig. 10 where the values of K_{cl}/K_{max} at $R = 0.1$ are plotted against ΔK for the four environments, respectively. There is no direct correlation between surface roughnesses and crack closure levels [Figs. 7*a* and 10]. For example, the dry hydrogen environment gives the greater level of surface roughness than the laboratory air or dry helium environment, and yet the crack closure levels in the dry hydrogen environment seem to be the lowest in the four environments. Moreover, in wet and dry hydrogen environments there are comparable roughness levels, and yet the crack closure levels in the wet hydrogen environment are much greater than those in the dry hydrogen environment. While surface roughness in the air environment appears to be the lowest in the four environments, the crack closure levels in air are the highest.

If roughness-induced crack closures were important in governing near-threshold crack growth kinetics, a reasonably good correlation between surface roughness and crack closure level should exist. Based on the above comparison of the results presented in Figs. 7*a* and 10, no correlation between surface roughness and crack closure level was observed. Therefore, it is suggested that roughness-induced crack closure is not a dominant factor in affecting the rates of near-threshold crack propagation.

Oxide-Induced Crack Closure—It is quite interesting to find that the effects of load ratio and environment on oxide thickness (Fig. 9) are similar to those on ΔK_{th} (Fig. 4) except in the wet hydrogen environment. These results suggest that oxide deposits affect and control near-threshold crack growth behavior by the mechanism of oxide-induced crack closure. The essence of the oxide-induced crack closure concept is that oxide deposits wedge-open the crack tip and promote early contact of fracture surfaces, thereby increasing the crack closure level and decreasing crack driving force.

In this investigation, the significance of oxide-induced crack closure can be further appreciated by comparing oxide thickness with crack-tip opening displacement (CTOD). The values of oxide thickness and CTOD at threshold are compared in Table 2. At $R = 0.1$, oxide thicknesses in the air environment are comparable with CTODs. On the other hand, oxide thicknesses in dry environments are generally smaller than CTODs. Moreover, as mentioned before, oxide thicknesses in the air environment are approximately 7 to 9 times larger than those in dry environments. Thus, oxide-induced crack closure is expected to be more significant in the air environment than in dry environments (Fig. 10). Note that in wet hydrogen, oxide thickness at $R = 0.1$ is approximately 6 to 14 times larger than CTODs. Oxide-induced crack closure in wet hydrogen was, however, found to be less significant than in laboratory air (Fig. 10).

As reported before, the value of K_{cl}/K_{max} at $R = 0.1$ is plotted versus ΔK for the four environments [Fig. 10]. In Fig. 10, the values of K_{cl}/K_{max} in laboratory air and wet hydrogen

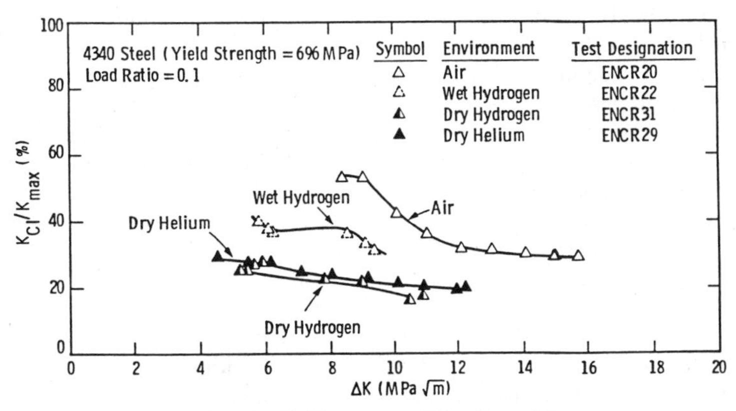

FIG. 10—K_{cl}/K_{max} *versus* ΔK *at* $R = 0.1$.

environments tend to increase with decreasing ΔK levels, while those in dry hydrogen and dry helium environments are relatively insensitive to ΔK levels. In moisture-containing environments, the larger value of K_{cl}/K_{max} with decreasing ΔK correlates with thicker oxide deposit, and thus a larger extent of oxide-induced crack closure at lower ΔK levels. In dry environments, oxide deposits are not thick enough to cause significant oxide-induced crack closure, which results in relatively constant K_{cl}/K_{max} values with decreasing ΔK.

The above good correlations between oxide thicknesses and crack closure levels strongly indicate that in the present study, oxide-induced crack closure is the dominant factor in influencing near-threshold crack growth behavior.

At $R = 0.8$, oxide thicknesses in both moisture-containing and dry environments are approximately 3 to 40 times smaller than CTODs. Furthermore, increasing load ratio decreases oxide thickness. Therefore, oxide-induced crack closure will be less pronounced with increasing load ratio.

Effects of Load Ratio and Environment on Fatigue Crack Propagation Rate Properties

Based on the above discussion, oxide-induced crack closure has been shown to be much more important in affecting near-threshold crack growth behavior than roughness-induced crack closure. In the following subsections, the effects of load ratio and environment on crack growth behavior are explained in light of the oxide-induced crack closure and hydrogen embrittlement mechanisms.

Influence of Load Ratio—As reported previously, decreasing load ratio increases the extent of oxide-induced crack closure. The much more significant amount of oxide-induced crack closure at lower load ratios ($\lesssim 0.5$) promotes earlier contact of fracture surfaces, and elevates the crack closure level. This behavior decreases crack driving force, ΔK_{eff} ($= K_{max} - K_{cl}$, where ΔK_{eff} is the effective stress-intensity range, and K_{max} and K_{cl} are the maximum and crack closure stress intensities, respectively) at lower load ratios, thereby resulting in slower near-threshold crack propagation rates and thus a higher ΔK_{th} with decreasing load ratio.

The fact that oxide-induced crack closure accounts for the effect of load ratio on near-threshold crack growth rates can be substantiated by plotting da/dN versus ΔK_{eff} at various load ratios. In each environment, the crack propagation rates at $R = 0.1$ and 0.8 are essentially identical, as exemplified in Fig. 11. Therefore, oxide-induced crack closure rationalizes the influence of load ratio on near-threshold crack growth behavior.

The near-threshold crack propagation rate properties in dry environments are less sensitive to load ratio than in the air environment (Figs. 2*a*, 2*b*, and 4). This trend can also be explained by the oxide-induced crack closure model. The extent of oxide-induced crack closure in dry environments was found to be less pronounced than that in the air environment. Thus, the decreased oxide-induced crack closure in dry environments results in a decreased effect of load ratio on near-threshold crack growth rate properties, as compared to the results in the air environment. Even though oxide thicknesses in wet hydrogen are much greater than those in laboratory air, the extent of oxide-induced crack closure in wet hydrogen was found to be less significant, as presented in Fig. 10. This behavior yields a reduced effect of load ratio on crack growth rate results in wet hydrogen than in laboratory air (Fig. 4).

In Fig. 2*a*, decreasing ΔK levels increase the influence of load ratio on near-threshold crack propagation rates. This behavior is also consistent with the concept of oxide-induced crack closure. As noted in Figs. 8*a* and 8*b*, the amount of oxide debris increases with decreasing ΔK. The extent of oxide-induced crack closure is expected to increase with

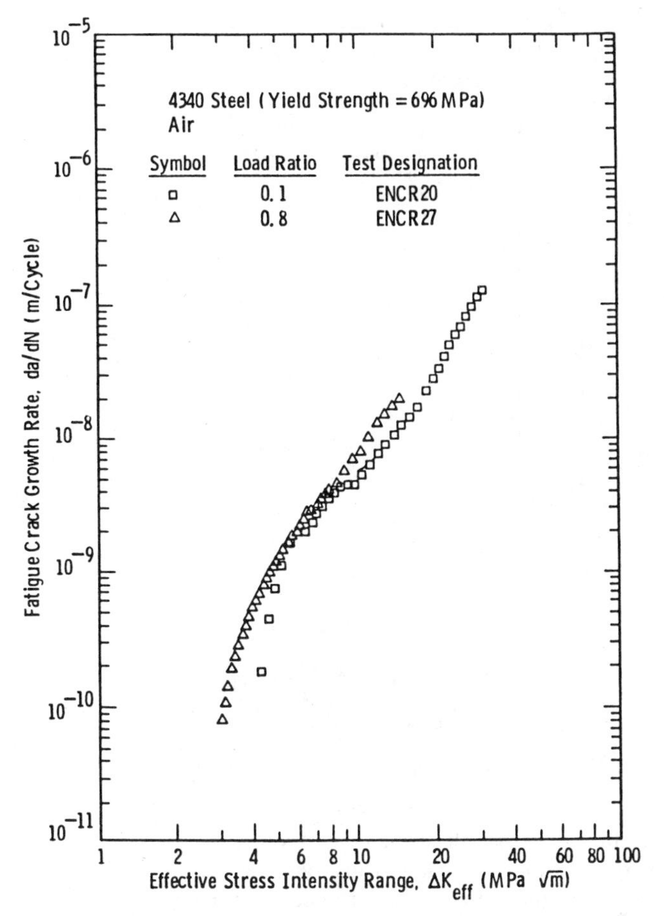

FIG. 11—*Crack growth rates at* R = *0.1 and 0.8 versus* ΔK_{eff}.

decreasing ΔK levels (as shown in Fig. 10), thereby resulting in a more pronounced influence of load ratio on crack growth rates at lower ΔK values.

Influence of Environment—In Fig. 10, at a given ΔK level, the value of K_{cl}/K_{max} in the laboratory air or wet hydrogen environment is larger than that in the dry hydrogen or dry helium environment; that is, crack closure levels in moisture-containing environments are higher than those in dry environments. This trend correlates with larger oxide thicknesses, and thus a greater extent of oxide-induced crack closure in moisture-containing environments. In Fig. 10, the higher crack closure levels in moisture-containing environments decrease ΔK_{eff}, which yields slower crack propagation rates and larger ΔK_{th} values, relative to dry environments. Moreover, the value of K_{cl}/K_{max} in wet hydrogen is smaller than that in laboratory air, and larger than that in the two dry environments. This observation leads to the experimental result that the crack growth rate properties in wet hydrogen are between those for laboratory air and dry environments (Fig. 3*a*).

Direct verification of oxide-induced crack closure to rationalize the influence of environment on near-threshold crack propagation behavior is provided by plotting da/dN versus ΔK_{eff}. Figure 12 presents the results of da/dN versus ΔK_{eff} for the four environments at

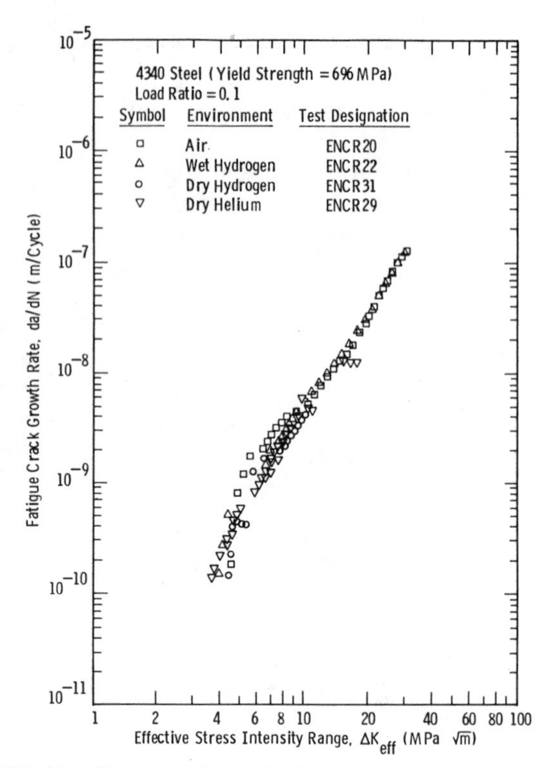

FIG. 12—*Crack growth rates in four environments versus* ΔK_{eff}.

R = 0.1. It was observed that all of the crack growth rate properties in the various environments converged to a narrow band. Thus, oxide-induced crack closure offers an explanation to account for the effect of environment on near-threshold crack propagation behavior.

The present investigation shows that increasing load ratio decreases the influence of environment on near-threshold crack growth rate results (Figs. 3 and 4). This trend can also be rationalized by the oxide-induced crack closure mechanism. As mentioned previously, increasing load ratio decreases the extent of oxide-induced crack closure because of thinner oxide layers at higher load ratios. Consequently, the decreased oxide-induced crack closure at higher load ratios gives a decreased effect of environment on near-threshold crack propagation behavior.

In Fig. 3*a*, decreasing ΔK increases the effect of environment on crack growth rates. This behavior is also in agreement with the oxide-induced crack closure model. The more extensive oxide-induced crack closure at lower ΔK levels (Figs. 8 and 10) promotes the influence of environment on near-threshold crack propagation rate properties.

It should be noted that, in Fig. 9, oxide thicknesses in wet hydrogen are greater than those in air. In contrast, the levels of crack closure, K_{cl}/K_{max}, in wet hydrogen are lower than those in air. These results suggest that in wet hydrogen the oxide thicknesses which effectively wedge the crack tip and introduce crack closure be smaller than those measured on fracture surfaces. Similar behavior was observed in a higher-strength 4340 steel [*17*]. During wet-hydrogen environment fatigue crack propagation, moisture will condense at the crack tip because of the capillary effects [*6,7,26*]. At lower ΔK levels, smaller CTODs will further promote the extent of capillary effects and enhance moisture condensation at the crack tip. After the tests were finished and the specimens were broken apart, the condensed moisture at the crack tip stayed on the fracture surfaces and resulted in thick oxide deposits which were not totally effective in elevating crack closure levels. Moreover, it is likely that during the wet-hydrogen environment test, oxide deposits grew thicker far away from the crack tip because of incomplete penetration of moisture condensation at the crack tip. This trend may also reduce the effectiveness of oxide-induced crack closure. Thus, thick oxide deposits measured on the fracture surfaces of moisture-containing environments may not give high levels of crack closure.

At $R = 0.8$, oxide thicknesses in wet hydrogen are expected to be small because of decreased fretting oxidations, as mentioned before. On the contrary, thick oxide deposits at $R = 0.8$ were observed in wet hydrogen (Table 2 and Fig. 9). This trend further proves the concept of moisture condensation in wet hydrogen.

One may argue that the faster crack growth rates in wet hydrogen than in air (Fig. 3*a*) be related to a greater extent of hydrogen embrittlement in wet hydrogen. If this argument were substantial at $R = 0.1$, the wet-hydrogen environment crack growth rates of da/dN versus ΔK_{eff} should be faster than those in laboratory air since ΔK_{eff} is the actual mechanical crack driving force. This behavior was not found in the present investigation (Fig. 12). Therefore, the influence of hydrogen embrittlement on wet-hydrogen environment crack propagation rates is suggested to be minimal for the present steel and test conditions.

Effective Threshold Stress-Intensity Range

To summarize the crack closure results, the value of effective threshold stress-intensity range, $\Delta K_{th,eff}$ (= $K_{max,th} - K_{cl,th}$, where $K_{max,th}$ and $K_{cl,th}$ are the maximum and crack closure stress intensities at threshold, respectively), is plotted against load ratio in the four environments in Fig. 13. The values of $\Delta K_{th,eff}$ are insensitive to load ratio in each environment. Moreover, at a fixed load ratio, the values of $\Delta K_{th,eff}$ converge to a narrow range of approximately 3 to 4 MPa$\sqrt{m}$. Therefore, crack closure provides a rationale to explain the effects of load ratio and environment on ΔK_{th}.

Conclusions

1. Specimen orientation does not affect near-threshold fatigue crack propagation rates of 4340 steel. In the four environments (laboratory air, wet hydrogen, dry hydrogen, and dry helium), increasing load ratio increases near-threshold crack growth rates. The influence of load ratio on the rates of near-threshold crack propagation, however, is much more significant in moisture-containing environments than in dry environments. Moreover, decreasing ΔK levels increases the effect of load ratio on crack growth rates.

2. In moisture-containing environments, near-threshold crack propagation rates are slower than those in dry environments. Increasing load ratio decreases the effect of environment on crack growth rates. Furthermore, decreasing ΔK values increases the environmental effect.

3. In laboratory air and dry helium environments, a transgranular mode of fracture was dominant although a minute amount of intergranular fracture was present. In wet and dry hydrogen environments, a mixture of intergranular and transgranular fracture was observed. Increasing load ratio decreases the extent of intergranular fracture.

4. Surface roughnesses in wet and dry hydrogen environments were found to be greater than those in laboratory air and dry helium environments. No direct correlation between surface roughness and crack closure level was observed. Roughness-induced crack closure

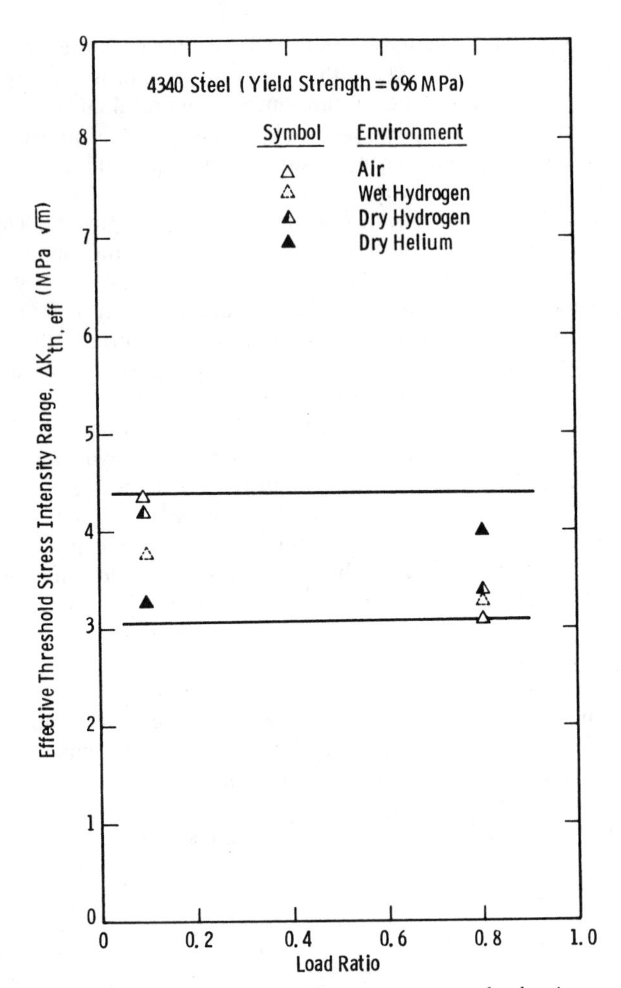

FIG. 13—$\Delta K_{th,eff}$ *in four environments versus load ratio.*

plays a much less significant role in governing near-threshold crack growth behavior than oxide-induced crack closure.

5. At a low load ratio of 0.1, oxide deposits in moisture-containing environments are much thicker than those in dry environments. Oxide thicknesses in moisture-containing environments are comparable to crack-tip opening displacements at threshold levels, thereby resulting in a significant extent of oxide-induced crack closure. Increasing load ratio decreases oxide thickness. The effects of load ratio and environment on oxide thickness are essentially identical to those on threshold stress-intensity range, ΔK_{th}.

6. Oxide-induced crack closure levels in moisture-containing environments are significantly higher than those in dry environments. The higher crack closure levels in moisture-containing environments decrease effective stress-intensity range (ΔK_{eff}), which yields slower crack propagation rates relative to dry environments. The greater decrease of oxide-induced crack closure in dry environments than in moisture-containing environments correlates with the decreased effect of load ratio on the rates of near-threshold crack growth in dry envi-

ronments. At lower ΔK levels, the greater extent of oxide-induced crack closure increases the influence of load ratio and environment on near-threshold crack propagation behavior.

7. Oxide-induced crack closure accounts for the effects of load ratio and environment on near-threshold fatigue crack growth rate properties of 4340 steel. Hydrogen embrittlement is suggested to be minimal in affecting the crack growth behavior of the present 4340 steel. Moreover, thick oxide deposits present on the fracture surfaces of the wet-hydrogen environment do not warrant high crack closure levels.

Acknowledgments

The authors wish to thank Dr. J. M. Wells for reviewing this manuscript. We are grateful to D. Detar and A. Karanovich for Auger analysis and SEM work, respectively. We appreciate the reviewers' comments. Optical microscopy was conducted by B. Sauka and P. J. Yuzawich. The financial support was provided by the Westinghouse Steam Turbine-Generator Division. The program monitor was M. Schneider of the Westinghouse Steam Turbine-Generator Division.

References

[*1*] Paris, P. C., Bucci, R. J., Wessel, E. T., Clark, W. G., Jr., and Mager, T. R. in *Stress Analysis and Growth of Cracks, ASTM STP 513,* American Society for Testing and Materials, Philadelphia, 1972, p. 141.
[*2*] Stewart, A. T., *Engineering Fracture Mechanics,* Vol. 13, 1980, p. 463.
[*3*] Minakawa, K. and McEvily, A. J., *Scripta Metallurgica,* Vol. 15, 1981, p. 633.
[*4*] Suresh, S., Zamiski, G. F., and Ritchie, R. O., *Metallurgical Transactions,* Vol. 12A, 1981, p. 1435.
[*5*] Liaw, P. K., Saxena, A., Swaminathan, V. P., and Shih, T. T., *Metallurgical Transactions,* Vol. 14A, 1983, p. 1631.
[*6*] Liaw, P. K., Hudak, S. J., Jr., and Donald, J. K., *Metallurgical Transactions,* Vol. 13A, 1982, p. 1633.
[*7*] Liaw, P. K., Hudak, S. J., Jr., and Donald, J. K. in *Fracture Mechanics: Fourteenth Symposium. Volume II: Testing and Application, ASTM STP 791,* American Society for Testing and Materials, Philadelphia, 1983, p. 370.
[*8*] Liaw, P. K., Leax, T. R., Williams, R. S., and Peck, M. G., *Acta Metallurgica,* Vol. 30, 1982, p. 2071.
[*9*] Liaw, P. K., Leax, T. R., Williams, R. S., and Peck, M. G., *Metallurgical Transactions,* Vol. 13A, 1982, p. 1607.
[*10*] Esaklul, K. A., Wright, A. G., and Gerberich, W. W., *Scripta Metallurgica,* Vol. 17, 1983, p. 1073.
[*11*] Horng, J. L. and Fine, M. E., *Materials Science and Engineering,* Vol. 67, 1984, p. 185.
[*12*] Williams, R. S., Liaw, P. K., Peck, M. G., and Leax, T. R., *Engineering Fracture Mechanics,* Vol. 18, 1983, p. 953.
[*13*] Liaw, P. K., Logsdon, W. A., and Attaar, M. H. in *Fatigue at Low Temperatures, ASTM STP 857,* American Society for Testing and Materials, Philadelphia, 1985, p. 173.
[*14*] Saxena, A. and Hudak, S. J., Jr., *International Journal of Fracture,* Vol. 14, 1978, p. 453.
[*15*] Clark, W. G., Jr., and Hudak, S. J., Jr., *Journal of Testing and Evaluation,* Vol. 3, 1975, p. 454.
[*16*] Kikukawa, M., Jono, M., and Tanaka, K. in *Proceedings,* Second International Conference on Mechanical Behavior of Materials, Boston, Mass., N. Promisel and V. Weiss, Eds., American Society for Metals, Metals Park, Ohio, 1976, p. 716.
[*17*] Liaw, P. K., Leax, T. R., and Donald, J. K., *Acta Metallurgica,* Vol. 35, 1987, p. 1415.
[*18*] McEvily, A. J. and Wei, R. P. in *Corrosion Fatigue: Chemistry, Mechanics and Microstructure,* NACE-2, O. F. Devereux, A. J. McEvily, and R. W. Staehle, Eds., National Association of Corrosion Engineers, 1972, p. 381.
[*19*] Endo, K., Komai, K., and Matsuda, Y., *Memo on Fracture Engineering,* Vol. 31, Kyoto University, Japan, 1969, p. 25.

[20] Liaw, P. K., Anello, J., and Donald, J. K., *Metallurgical Transactions,* Vol. 13A, 1982, p. 2177.
[21] Liaw, P. K. and Logsdon, W. A., *Transactions,* American Society of Mechanical Engineers, *Journal of Engineering Materials and Technology,* Vol. 107, 1985, p. 26.
[22] Suresh, S. and Ritchie, R. O., *Metallurgical Transactions,* Vol. 13A, 1982, p. 1627.
[23] Benoit, D., Ramdar-Tixier, R., and Tixier, R., *Materials Science and Engineering,* Vol. 45, 1980, p. 1.
[24] Elber, W. in *Damage Tolerance in Aircraft Structure, ASTM STP 486,* American Society for Testing and Materials, Philadelphia, 1971, p. 230.
[25] Davidson, D. L., *Fatigue of Engineering Materials and Structures,* Vol. 3, 1980, p. 229.
[26] Johnson, H. H. and Willner, A. W., *Applied Materials Research,* 1965, p. 34.
[27] Tracey, D. M., *Transactions,* American Society of Mechanical Engineers, *Journal of Engineering Materials and Technology,* Vol. 98, 1976, p. 146.

Ronald A. Mayville,[1] *Thomas J. Warren,*[1] *and Peter D. Hilton*[1]

The Crack Velocity-K_I Relationship for AISI 4340 in Seawater Under Fixed and Rising Displacement

REFERENCE: Mayville, R. A., Warren, T. J., and Hilton, P. D., **"The Crack Velocity-K_I Relationship for AISI 4340 in Seawater Under Fixed and Rising Displacement,"** *Fracture Mechanics: Perspectives and Directions (Twentieth Symposium), ASTM STP 1020,* R. P. Wei and R. P. Gangloff, Eds., American Society for Testing and Materials, Philadelphia, 1989, pp. 605–614.

ABSTRACT: An experimental study was conducted to determine if the type of loading, excluding cyclic loading, has an effect on the crack velocity/stress-intensity relationship for environmentally assisted cracking. Compact tension specimens made of AISI 4340, σ_0 = 1095 MPa (159 ksi), were tested in synthetic seawater under constant load, constant crack mouth opening displacement (CMOD), constant stroke rate (rising displacement) ranging over four orders of magnitude, and dual stroke rate. The results show that when crack growth is assisted by the environment the same $\dot{a}$-K_I curve is obtained for each type of loading within the bounds of data scatter. These observations are supported by calculations which show that the crack-tip opening displacement rate, a measure of crack-tip strain rate, is insensitive to the different types of loading for this system.

KEY WORDS: environmentally assisted cracking, fracture, AISI 4340, seawater, resistance curve, monotonic loading

Fracture mechanics has been applied to several forms of subcritical crack growth, including fatigue, creep, and environmentally assisted cracking (EAC). Much success has been achieved in quantifying fatigue crack growth under nominally inert conditions; account can be made for load history effects and, to some extent, plasticity. This form of crack growth is treated in the design of several types of engineering structures.

Subcritical crack growth in the presence of an aggressive environment has not yet lent itself to such a consistent description. Experiments performed under the conventional fixed load or displacement usually provide one of the two crack velocity ($\dot{a}$)/stress intensity (K_I) relationships shown in Fig. 1. However, the crack velocity at a given stress intensity can depend on initial stress-intensity factor for constant load tests [*1*], and there is a currently unpredictable incubation period that depends on such factors as prior load history [*2*]. Likewise, the fatigue crack growth rate in environment is observed to depend on the frequency as well as the alternating stress intensity.

The applicability of the $\dot{a}$-K_I curves obtained from conventional tests to the tests performed under monotonically increasing load or displacement that are of current interest has not been established. It is possible, for example, that these curves depend on monotonic loading rate. Abramson, Evans, and Parkins [*3*] have observed differences in the $\dot{a}$-K_I relationship

[1] Staff consultants and manager, Mechanics and Materials, respectively, Arthur D. Little, Inc., Cambridge, MA 02140.

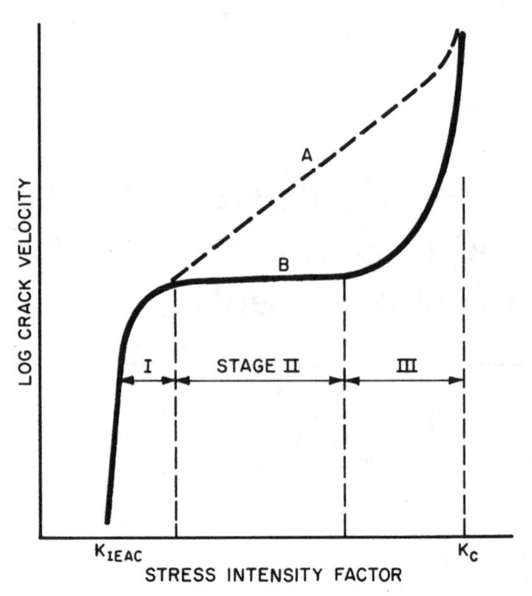

FIG. 1—*Two forms of conventional crack velocity/stress-intensity curves for environmentally assisted cracking under sustained load.*

for experiments performed at various displacement rates for a magnesium alloy in a potassium chloride solution. This behavior is also exhibited in smooth specimens for this material/environment system.

Current theories of environmentally assisted cracking [*4*] place major significance on the crack-tip strain as a controlling parameter. Higher crack-tip strain rates are thought to accelerate crack growth, either by hydrogen embrittlement or dissolution, by maximizing fresh surface exposure and hence reaction rates.

The difficulty in determining the effect of type of loading on crack velocity by using crack-tip strain rate arguments is quantifying this parameter and resolving the dilemma that it depends on crack velocity itself. Rice and colleagues [*5*] have provided relationships between various crack-tip parameters, applied loading rate, and material properties for a growing crack in plane, elastic-plastic small-scale yielding. The expressions for strain rate components near the crack tip contain as yet undetermined coefficients, but the expression for crack-tip opening displacement rate is

$$\dot{\delta} = \alpha(\dot{J}/\sigma_0) + \beta(\sigma_0/E)\dot{a}\ell n(R/r) \tag{1}$$

where

J = J-integral,
$\alpha = \Delta\delta\sigma_0/\Delta J$,
$\beta \simeq 5$,
E = Young's modulus,
$R \simeq 0.2E(J/\sigma_0^2)$, and
r = distance behind the crack tip.

Equation 1 can be expressed in a form that is more useful for comparing $\dot{\delta}$ for different experiments

$$\dot{\delta} = \beta(\sigma_0/E)\dot{a}\ell n(\rho/r) \tag{2}$$

where ρ is equal to $R \exp(\alpha T/\beta)$ and T equals $(E/\sigma_0^2)dJ/da$. This form clearly shows the strong dependence of $\dot{\delta}$ on crack velocity.

The objective of the investigation reported here was to establish the effect of loading type, primarily for monotonically increasing, or rising, displacement, on the $\dot{a}$-K_I curve for a mildly susceptible, medium-strength steel in seawater. Experiments were also performed under constant displacement and constant load. In addition, Eq 2 is evaluated for each case to assess the difference in $\dot{\delta}$ as a measure of crack-tip strain rate.

Experiments

The material, environment, and test system used in this investigation is the same as that reported earlier [6]. AISI 4340 steel was heat-treated to a hardness of R_c36, which gave transverse yield and tensile strengths, σ_0 = 1095 MPa (159 ksi) and σ_u = 1145 MPa (166 ksi). The maximum load toughness as determined with side-grooved, 25-mm-thick three-point bend specimens is 160 MPa$\sqrt{\text{m}}$. K_{IJ}, as derived according to the formula, $K_{IJ} = \sqrt{J_{Ic}E'}$, is 120 MPa$\sqrt{\text{m}}$ and K_{IEAC}, also known as K_{ISCC}, for this material in ASTM D1141 synthetic seawater at 35°C is 98 MPa$\sqrt{\text{m}}$; $E' = E/(1 - \nu^2)$; and ν = Poisson's ratio. The fracture mode in air and seawater is void coalescence.

All experiments in this study were performed with 25-mm-thick compact tension specimens containing 10% side grooves and precracked to a/W = 0.68, as shown in Fig. 2. Specimen

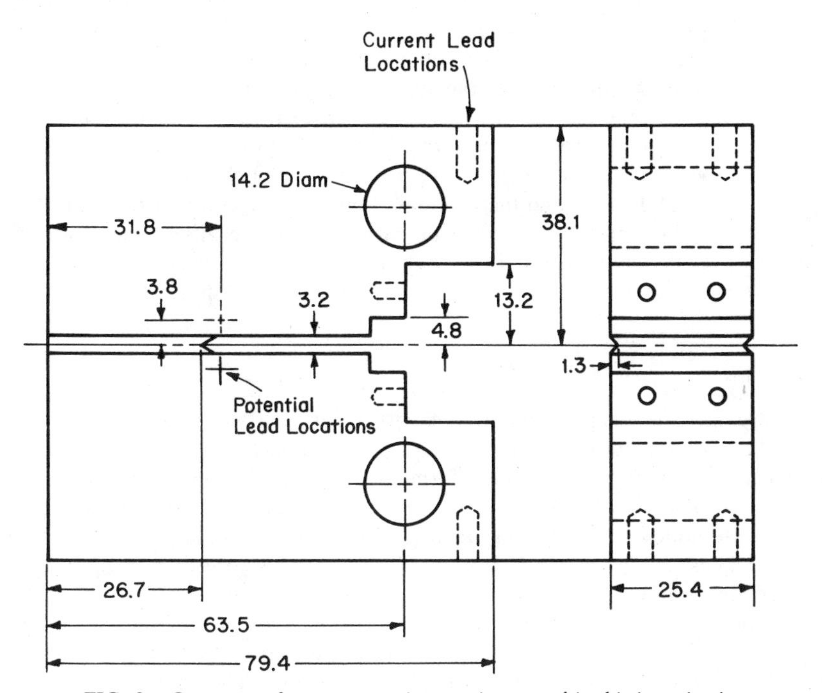

FIG. 2—*Geometry of compact tension specimen used in this investigation.*

geometry differs from the conventional plan view to provide more material for insertion in the environmental cell, $W = 2.5B$, and to permit use of ceramic sleeves in the pinholes for electrical insulation through enlarged holes.

Experiments were conducted in a horizontally oriented, electromechanical, closed-loop testing machine that can be used at stroke rates as low as 1 μm/h. The d-c potential drop method was used to monitor crack length, the specimen lead connections for which are shown in Fig. 2. The crack length-potential drop calibration determined by tests in air was adjusted by a multiplication factor to account for the difference in predicted and actual final crack lengths in the environment. This difference is due to crack face shorting in the presence of the environment and can be accounted for by a numerical procedure (see Ref 7). A simpler, single multiplication factor, applied only to the amount of crack extension Δa, was considered accurate enough for the experiments in this investigation. A factor of 1.15 was found to work quite well.

Tests were monitored and controlled by computer and the system was protected by an uninterruptable power supply because of the ten-day duration of some of the experiments and the frequent occurrence of brown-outs.

Experiments were performed under constant load, constant crack mouth opening displacement (CMOD), constant stroke rate (rising displacement), and dual stroke rate conditions. The constant load and constant CMOD tests represent conventional environmentally assisted cracking tests, usually performed with the cantilever beam and wedge-opening-loaded (WOL) specimens. In the latter case, the load and the stress intensity decrease with crack length. Constant CMOD tests were controlled by using the clip gage as the transducer in the closed-loop. The occurrence of crack growth under these loadings can be clearly attributed to environmental effects, with the possible exception of constant loads in excess of 90% of the plastic collapse load, for which time-dependent plastic flow can lead to slow crack extension [8].

Constant stroke rate experiments, in which the machine actuator displacement rate is held fixed, were run at four different rates ranging from 1.25×10^{-1} to 1.25×10^{-4} mm/min; 2.5×10^{-1} mm/min is the rate usually used to generate crack growth resistance curves.

Dual stroke rate experiments were performed in an effort to change the loading rate term of Eq 2 for one value of crack velocity. The stroke rate was held fixed at 1.25×10^{-4} mm/min until $K_I = 120$ MPa$\sqrt{m}$, and then increased quickly to a constant rate of 6.25×10^{-4} mm/min, or five times the initial rate. Comparison of the results of these tests to tests performed at a single, constant rate of 1.25×10^{-4} mm/min provided a means of changing the loading term without changing crack velocity, at least for a short time.

Results

Results are presented as plots of crack velocity against the Mode I stress-intensity factor, which has been corrected for plasticity using a single application of the Irwin plane-strain plastic zone correction factor. Use of this formulation is considered adequate for comparison of the effects of loading on EAC response, even though the plane strain requirements of ASTM Test Method for Plane-Strain Fracture Toughness of Metallic Materials (E 399-83) require a thickness of 47 mm for $K_I \leq 150$ MPa$\sqrt{m}$. The use of side grooves and a plastic zone correction should greatly reduce this thickness requirement.

Crack velocity versus K_I data for the constant load and constant CMOD tests are shown in Fig. 3. Constant load data for $P/P_L > 0.9$ are not included, and a line has been drawn through all of the data to represent it in subsequent plots. The crack velocity is seen to increase sharply with K_I for stress intensities near the sustained load threshold value of 98

MPa$\sqrt{m}$. There is reasonably good agreement between results for these two types of conventional loading in which, in one case, K_I increases with crack length, and in the other, K_I decreases with crack length. Agreement is particularly good near the threshold, but there is some deviation at higher crack velocities. This may be due to scatter, as will be reflected by subsequent results.

Susceptibility to EAC is not as easily revealed in the constant stroke rate experiments as in the constant load and CMOD tests. In the former, a nonzero crack velocity results even in inert environments. EAC under these conditions can be shown in two ways. The first method is to compare the crack velocity-K_I relationships obtained under inert and environmental conditions. An elevation in crack velocity at a given K_I demonstrates susceptibility to the environment. Another convenient method is to compare resistance curves [9]; a reduction in the resistance curve is evidence of susceptibility.

Comparisons are shown in Fig. 4 of the crack growth resistance curves for the range of

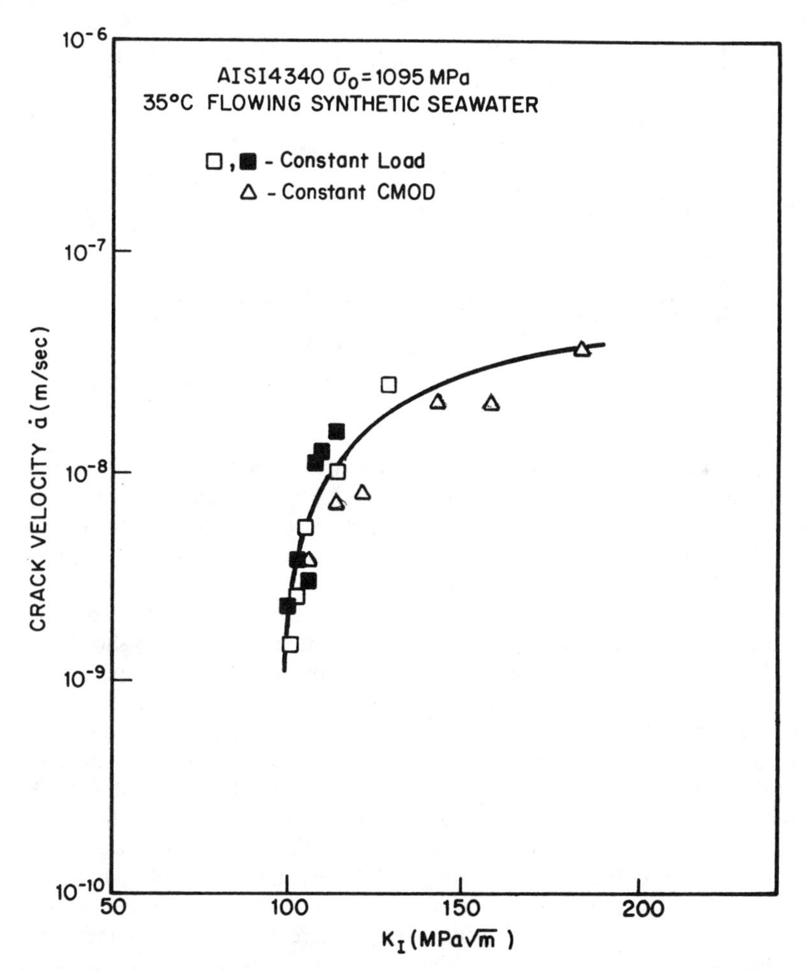

FIG. 3—*Crack velocity-*K_I *data for constant load and constant CMOD experiments; curve represents trend of data for use in subsequent plots.*

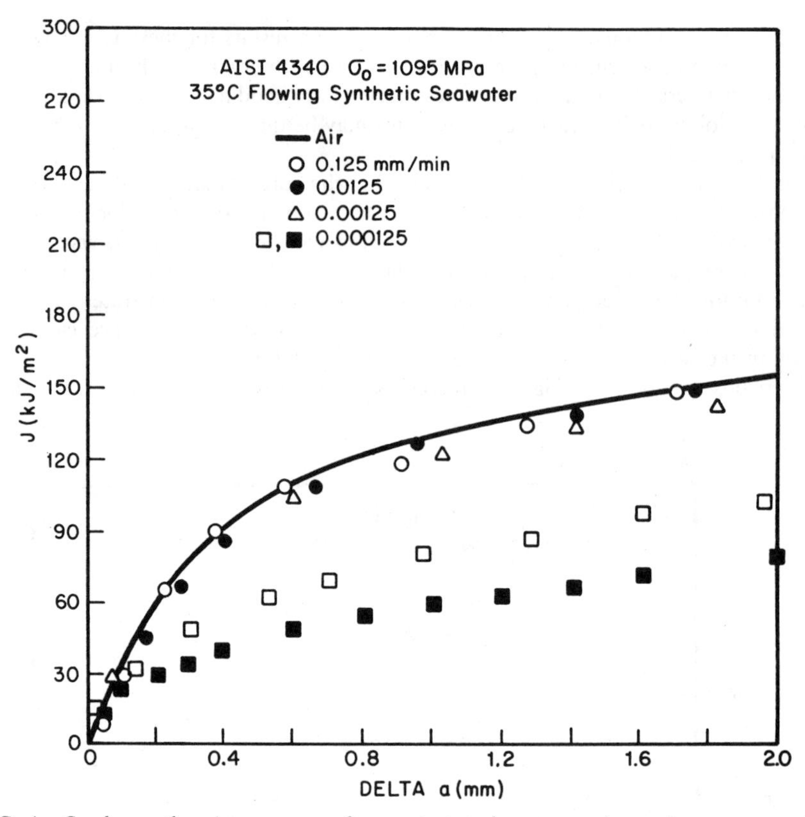

FIG. 4—*Crack growth resistance curves for constant stroke rate experiments in seawater and air.*

constant stroke rates in environment to the R-curve determined in air. The resistance curve in air was found to be independent of stroke rate for this material [*6*]. The results show that only the tests performed at the slowest rate, 1.25×10^{-4} mm/min, exhibited susceptibility to the environment. There is also some difference in the R-curves obtained at this rate under nominally identical conditions, which will result in scatter in the $\dot{a}$-K_I curves.

The crack velocity-K_I data are shown in Fig. 5 for the constant stroke rate tests in comparison to the constant load and CMOD curve presented earlier. Also shown in Fig. 5 is the $\dot{a}$-K_I curve for the test performed in air at the slowest rate and the expected air curve for a test performed one order of magnitude higher in rate. The crack velocity at a given K_I under inert conditions scales approximately with the stroke rate (see Ref *6*). The crack velocity data for the slowest environmental experiments are considerably higher than the corresponding air data, supporting the conclusion of susceptibility revealed by the R-curves. The data from the two constant stroke rate tests at 1.25×10^{-4} mm/min show some scatter and generally fall below the trend curve representing the constant load and CMOD data. The reason that susceptibility is not exhibited by the experiment performed in environment at a rate of 1.25×10^{-3} mm/min, as demonstrated by the R-curves in Fig. 4, is explained by the expectation that inert crack velocities would be greater than environmentally induced crack velocities.

A rate of 6.25×10^{-4} mm/min was chosen as the second stroke rate in the dual rate experiments. This is approximately the highest rate at which EAC is expected to occur in

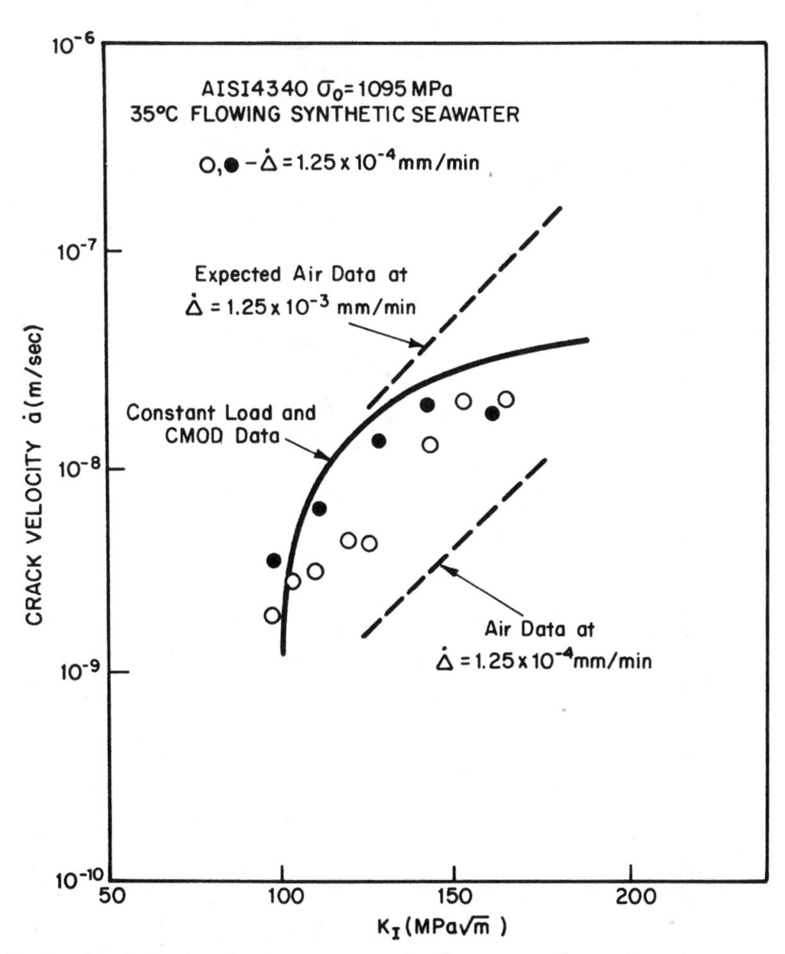

FIG. 5—*Crack velocity data for slowest constant stroke rate experiments in environment and air in comparison to sustained load data.*

the rising displacement test, since at higher rates the $\dot{a}$-K_I curve obtained under inert conditions would exceed the EAC data derived from conventional tests. If the loading rate has no effect on the EAC response, one would expect to see no abrupt change in the $\dot{a}$-K_I curve in changing from 1.25 to 6.25 $\times$ 10^{-4} mm/min in the dual stroke rate experiments.

The crack velocity-K_I data for the dual stroke rate experiments are shown in Fig. 6 in comparison with the constant stroke rate data and the curve representing constant load and CMOD data. The data do not show a clear discontinuity at K_I = 120 MPa$\sqrt{m}$, the stroke rate transition point. The experiment represented by the open diamonds appears to suggest a jump at 120 MPa$\sqrt{m}$, but these data lie above all of the other data even for the initial rate of 1.25 $\times$ 10^{-4} mm/min. Considering the scatter in the data, there appears to be no difference in the $\dot{a}$-K_I curves for the different types of loading.

Analysis

Application of the crack-tip opening displacement Eq 2 supports the observation that little or no discontinuity should occur in the crack velocity-stress intensity curve for the dual

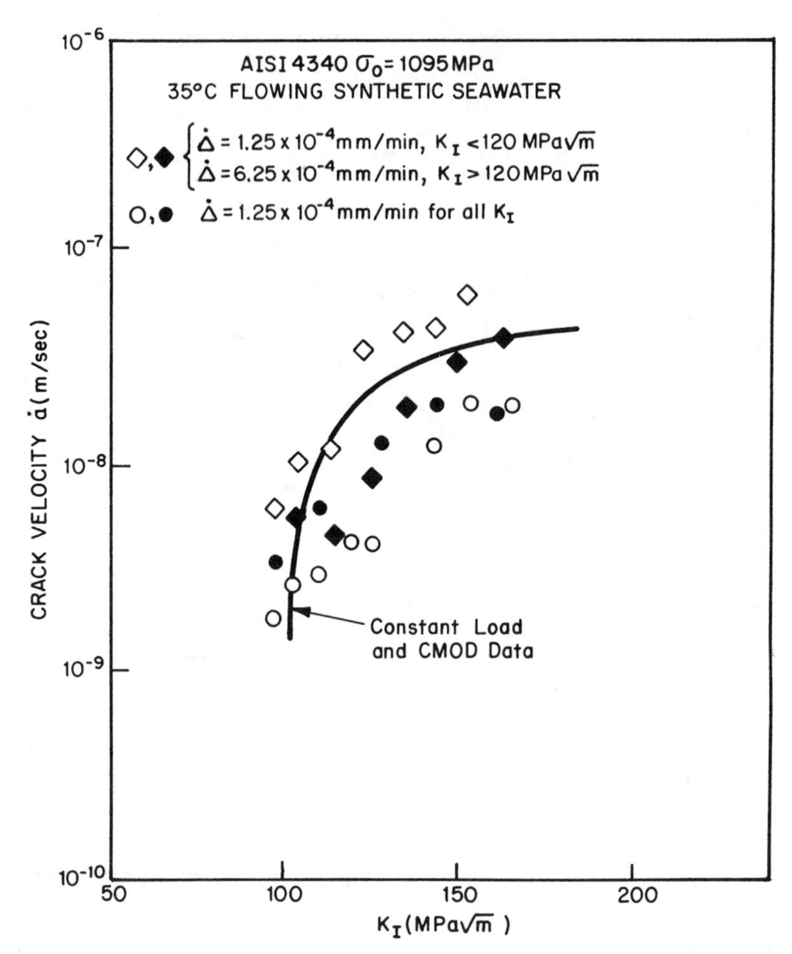

FIG. 6—*Crack velocity data for dual stroke rate experiments in comparison to constant stroke rate and sustained load data.*

rate experiments even if strain rate is important. This can be shown by calculating ρ for the different types of loading using an analysis similar to that presented in the appendix to Ref *10* to calculate dJ/da. The loading term was calculated in all cases at $K_I = 120$ MPa$\sqrt{m}$ using the same crack velocity. Maximum load had not been attained at this point in any of the experiments, with the exception of the constant CMOD test. Strictly speaking, Eq 2 does not apply to the constant CMOD test, since there is unloading ahead of the crack tip, but the result is included for comparison. The results are presented in Table 1.

The load parameter of Eq 2, represented by ρ, does not vary substantially for the different load cases. An even smaller variation can be expected when ρ is substituted in Eq 2 to evaluate crack tip opening displacement rate. It is not possible to compute the ratio of $\dot{\delta}$ for any two load cases without choosing a value for r. There is no physical basis on which to make this choice, but if one uses a relatively large value, say $r = 1$ mm, then the ratio of $\dot{\delta}$ for the two stroke rates in Table 1 is

$$\dot{\delta}(6.25 \times 10^{-4})/\dot{\delta}(1.25 \times 10^{-4}) = \ln(5.6)/\ln(3.6) = 1.3$$

TABLE 1—*Evaluation of ρ for various load cases.*

Load Case	dJ/da, kJ/m²/mm	T	ρ, mm
Constant load	2.2	0.4	2.5
Constant CMOD	−4.5	−0.8	2.2
Stroke rate = 1.25×10^{-4} mm/min	17.	2.9	3.6
Stroke rate = 6.25×10^{-4} mm/min	36.	6.1	5.6

If smaller values of r are used, say on the order of the crack-tip opening displacement, then the ratio is even smaller. The conclusion is that the crack-tip opening displacement rate and, most likely, the crack tip strain rate do not differ substantially for the different load cases. Thus, even if crack-tip strain rate is important in determining EAC response, different types of loading should result in the same $\dot{a}$-K_I curves. This is supported by the experimental results, even though there is some scatter in the data.

Conclusions

Experiments performed under different types of constant and monotonically increasing loading provide essentially the same crack velocity-K_I relationship for environmentally assisted cracking of a medium-strength steel in seawater. This independence of EAC response on loading type is supported by the relatively small differences in calculated crack-tip opening displacement rate, which is considered a measure of the crack-tip strain rate. The latter parameter is thought to be a controlling in the EAC process. The results suggest that the rising displacement test provides the same EAC data as conventional tests, provided EAC is induced.

Acknowledgment

The authors gratefully acknowledge the support of this investigation by the National Science Foundation under Grant MSM 8412643.

References

[1] Wei, R. P., Novak, S. R., and Williams, D. P., "Some Important Considerations in the Development of Stress Corrosion Cracking Test Methods," *Materials Research and Standards,* Vol. 12, No. 9, 1972, pp. 25–30.

[2] Benjamin, W. D. and Steigerwald, E. A., "An Incubation Time for the Initiation of Stress-Corrosion Cracking in Precracked 4340 Steel," *Transactions,* American Society for Metals, Vol. 60, 1967, pp. 547–548.

[3] Abramson, G., Evans, J. T., and Parkins, R. N., "Investigation of Stress Corrosion Crack Growth in Mg Alloys Using J-Integral Estimations," *Metallurgical Transactions A,* Vol. 16A, Jan. 1985, pp. 101–108.

[4] Ford, F. P., "Mechanisms of Environmental Cracking in Systems Peculiar to the Power Generation Industry," EPRI Report NP-2589, Electric Power Research Institute, Sept. 1982.

[5] Rice, J. R., Drugan, W. J., and Sham, T.-L., "Elastic-Plastic Analysis of Growing Cracks," *Fracture Mechanics (12th Conference), ASTM STP 700,* American Society for Testing and Materials, Philadelphia, 1980, pp. 189–221.

[6] Mayville, R. A., Warren, T. J., and Hilton, P. D., "The Influence of Displacement Rate on Environmentally Assisted Cracking of Precracked Ductile Steel Specimens," *Journal of Engineering Materials and Technology,* Vol. 109, July 1987, pp. 188–193.

[7] Gangloff, R. P., "Electrical Potential Monitoring of the Formation and Growth of Small Fatigue

Cracks in Embrittling Environments," *Advances in Crack Length Measurement,* C. J. Beevers, Ed., Engineering Materials Advisory Services, LTD., West Midlands, U.K., 1982, pp. 175–229.
[*8*] Ingham, T. and Morland, E., "Influence of Time-Dependent Plasticity on Elastic-Plastic Fracture Toughness," *Elastic-Plastic Fracture: Second Symposium, Volume I—Inelastic Crack Analysis, ASTM STP 803,* C. F. Shih and J. P. Gudas, Eds., American Society for Testing and Materials, Philadelphia, 1983, pp. I-721–I-746.
[*9*] Anderson, D. R. and Gudas, J. P., "Stress Corrosion Evaluation of Titanium Alloys Using Ductile Fracture Mechanics Technology," *Environment-Sensitive Fracture: Evaluation and Comparison of Test Methods, ASTM STP 821,* American Society for Testing and Materials, Philadelphia, 1984, pp. 98–113.
[*10*] Paris, P. C., Tada, H., Zahoor, A., and Ernst, H., "The Theory of Instability of the Tearing Mode of Elastic-Plastic Crack Growth," *Elastic-Plastic Fracture, ASTM STP 668,* J. D. Landes, J. A. Begley, and G. A. Clarke, Eds., American Society for Testing and Materials, Philadelphia, 1979, pp. 5–36.

Véronique Tremblay,[1,2] *Phuc Nguyen-Duy,*[1] *and J. Ivan Dickson*[2]

Influence of Cathodic Charging on the Tensile and Fracture Properties of Three High-Strength Steels

REFERENCE: Tremblay, V., Nguyen-Duy, P., and Dickson, J. I., **"Influence of Cathodic Charging on the Tensile and Fracture Properties of Three High-Strength Steels,"** *Fracture Mechanics: Perspectives and Directions (Twentieth Symposium) ASTM STP 1020*, R. P. Wei and R. P. Gangloff, Eds., American Society for Testing and Materials, Philadelphia, 1989, pp. 615–627.

ABSTRACT: The influence of cathodic charging in sulfuric acid solution on mechanical properties of AISI 4340, AISI 300M, and AMS 6340 steels was studied for different tempering temperatures. Cathodic charging decreased the tensile elongation, the reduction in area, and the ultimate tensile stress as a result of hydrogen-induced surface cracks having brittle features, which then transformed into cracks propagating by microvoid nucleation and coalescence along planes of maximum shear stress.

The curves of the J-integral as a function of the crack advance Δa presented a change in slope at a small Δa value. For the lower tempering temperatures, this Δa value was relatively large and similar to that required to obtain a single crack propagating across the thickness. Cathodic charging decreased J_{Ic} as a result of bands of brittle cracking near each lateral surface and of facilitated initiation of cracking, in an alternating fashion along planes of maximum shear stress.

KEY WORDS: J-integral, hydrogen embrittlement, steel

Much attention has been paid in recent years to the hydrogen embrittlement of steels, especially of high-strength steels. While much progress has been realized in identifying the manner in which hydrogen influences the tensile and fracture behavior of steels, this complex phenomenon remains only partially understood. It has been shown that hydrogen can favor the occurrence of crack propagation modes such as intergranular cracking [*1–3*], cleavage [*4*] or quasi-cleavage [*1*], and microvoid nucleation and growth [*1,5*]. Hydrogen can also produce microvoid nucleation with little growth, giving rise to a microfractographic feature, which has been referred to [*6*] as "tearing topography surface" (TTS). A number of studies [*7–10*] have also shown that hydrogen charging can favor the occurrence of plastic instability by facilitating cracking on planes of maximum shear stress. A study [*11*] has also been reported in which the decrease in resistance to fracture resulting from hydrogen charging of an AISI 4340 (American Iron and Steel Institute) steel having a yield strength of 1202 MPa was evaluated by J-integral measurements.

The objective of the present study was to evaluate and study the behavior during tensile

[1] Research assistant and senior scientist, respectively, Institut de Recherche d'Hydro-Québec (IREQ), Varennes, PQ, JOL 2PO, Canada.

[2] Graduate student and professor, respectively, Département de génie métallurgique, École Polytechnique, C.P. 6079, Succ. "A," Montreal, PQ H3C 3A7, Canada.

and J-integral tests of three different high-strength steels tempered at different temperatures and submitted to hydrogen charging.

Experimental Procedure

The study was carried out on three molybdenum-bearing steels of similar carbon contents, AISI 4340, AISI 300M, and AMS 6304 (Aerospace Material Specification), the chemical analyses of which are given in Table 1. These steels were received in the form of normalized bars of circular cross section, with diameters of 127 mm, 152 mm, 51 mm, respectively. The AISI 4340 and AISI 300M steels were cut into sections of 40 by 80 to 100 by 200 mm and 50 by 50 by 200 mm, respectively, prior to heat treatment. Lengths of 200 mm were austenitized for 1 h either at 830°C (4340 steel) or at 865°C (300M and 6304 steels), quenched in oil, and immediately tempered for 1 h at a selected temperature. Specimens, with long axis parallel to the rolling direction, were machined from the mid-thickness portions of the heat-treated sections. Tension specimens were prepared with a 10-mm-diameter reduced gage length section. Single-edge notch bend specimens for J_{Ic} testing were machined with length of 170 mm, width W of 30 mm, thickness B of 15 mm, and with a 6-mm-deep central notch. The span lengths for the four-point bend tests employed were S_1 = 120 mm and S_2 = 60 mm.

Tests were performed in both air and in 0.1 N H_2SO_4 (sulfuric acid) solution to which thiourea, a poison for the cathodic recombination reaction, had been added. In this solution, cathodic charging was performed for 1 h prior to testing and continued during the test at a constant current density of 180 A/m^{-2}. Two strips of platinum, 10 by 40 mm, were employed as anodes and positioned 5 mm from opposite sides of the specimens. For the J_{Ic} tests, the four-point bend specimens were fatigue precracked in air to an a_0/W value between 0.45 and 0.65. The tests in H_2SO_4 solution were performed in an environmental cell attached to the specimen employing a silicone adhesive. A region 30 (parallel to the crack) by 12 mm on each lateral surface was exposed to the H_2SO_4 solution, with the mid-width of the platinum anodes positioned opposite the precrack tip.

The J_{Ic} tests were performed following the ASTM Test Method for J_{Ic}, a Measure of Fracture Toughness (E 813-81) recommended procedure, with periodic partial unloadings, employing a computer-controlled Instron servohydraulic machine. The procedure was to first perform two elastic loadings-unloadings of 2.5 and 3.5% of the full 3.66-mm crack opening displacement (COD) range allowed for the extensometer, in order to measure the initial crack length from the compliance. This COD range employed in testing corresponded to the central 90% of the displacement range allowed for the extensometer. The tests were performed at a stroke speed of 0.01 mm/min with an elastic unloading of 15% of the load after every 2.5% of the maximum total displacement (3.66 mm) allowed for the extensometer. With this procedure a maximum of 36 periodic unloadings could be performed during the tests. Specimens which had not broken during the tests were heat-tinted (5 min light heat-treating with a bunsen burner) to produce a light gray tint, immersed in liquid nitrogen, and broken in two. The tension tests were carried out at a nominal strain rate of 5.5 × 10^{-4} s^{-1}.

TABLE 1—*Chemical analysis in weight percent of the three steels studied.*

Steel	C	Mn	P	Si	Ni	Cr	Mo	V	Ti	W	Nb	Co
AISI 4340	0.456	0.72	0.02	0.24	1.5	0.75	0.21	0.007	0.003	0.003	0.007	0.017
AISI 300M	0.424	0.81	0.007	1.12	1.6	0.66	0.35	0.067	0.006	0.006	0.008	0.01
AMS 6304	0.408	0.58	0.007	0.25	0.18	0.88	0.50	0.23	0.002	0.005	0.008	0.009

Results and Discussion

Tension Tests

For the tension tests, the yield strength σ_y and the initial portion of the stress-strain curve for a given steel and heat treatment were, within experimental accuracy, identical for the tests in air and in H_2SO_4. The tensile elongation e_f, the ultimate tensile strength σ_u, and the reduction in area *RA* were decreased by testing in H_2SO_4 solution. The effect of this solution on the values of e_f and *RA* was especially important for the lower tempering temperatures. All specimens with $\sigma_y > 1250$ MPa tested in the H_2SO_4 solution broke at $e_f < 3.5\%$. A few broke at zero permanent elongation. For the lower tempering temperatures for which the effect of hydrogen charging on e_f was particularly important, since the difference between σ_y and σ_u for tests in air was relatively small, the decrease in σ_u caused by testing in H_2SO_4 under hydrogen-charging conditions was not very marked. The influence of the test environment on σ_y, σ_u, e_f, and *RA* is shown as a function of the tempering temperature in Figs. 1 to 4. The curves drawn in these figures refer to the results obtained in air. Each point representing a result in air corresponds to the average result for two tests: each point representing a result in the H_2SO_4 solution corresponds to the result of a single test for the AMS 6304 steel and for the 300M steel tempered at 500°C and the average result from three tests in all other cases. Any significant variation obtained between the repetitive tests is reported by a scatter band indicating the maximum and minimum values obtained in individual tests.

For the tension tests in air, fracture occurred generally by necking and crack initiation by microvoid nucleation and coalescence in the center of the necked region. For one of two specimens of AISI 300M steel tested with tempering temperatures of 400 and 500°C, cracking

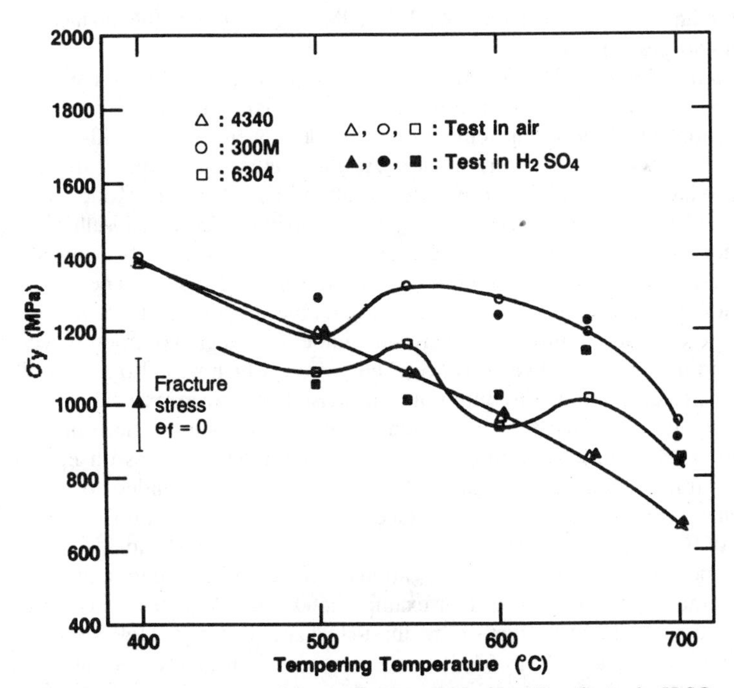

FIG. 1—*Yield stress σ_y as a function of the tempering temperature. Results in the* H_2SO_4 *solution are indicated by full symbols when these differed by more than 4% from those obtained in air.*

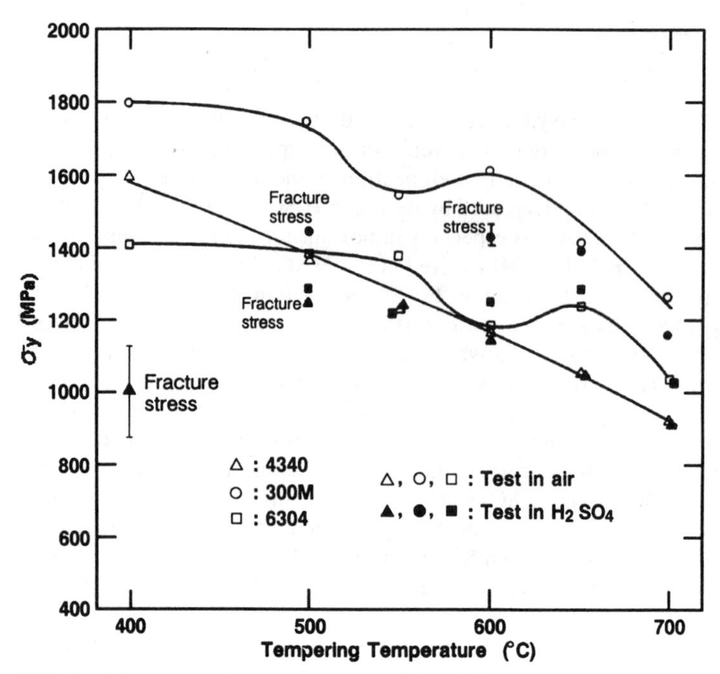

FIG. 2—*Ultimate tensile stress* σ_u *as a function of the tempering temperature.*

initiated for one of two specimens of AISI 300M steel tested at the surface by microvoid nucleation and growth apparently at a fine machining mark.

For the tension tests in H_2SO_4, cracks always initiated at the external surface and propagated initially in Mode I, perpendicular to the tensile axis. The only tests performed in the H_2SO_4 solution for a tempering temperature of 400°C were on AISI 4340, and the entire fracture surface was brittle intergranular. As the tempering temperature increased, the cracking produced from the initiation sites resulted in semi-elliptic Mode I cracks (Fig. 5) with a macrofractographic aspect which appeared brittle. These semi-elliptic cracks were concentrated in the necked region and were most numerous near its thinnest portion, although a few were present completely outside the necked region. The number of semi-elliptic cracks present increased with increasing tempering temperature. In general, one or several of these crack initiations sites were surrounded by a region of fracture which occurred by microvoid nucleation and coalescence. This region, which was also perpendicular to the tensile axis, presented some discontinuous lines visible at low magnification (Fig. 5). Their aspect clearly indicated that they corresponded to intermediate positions of portions of the crack front. High- and low-magnification stereographic observations on matching regions of opposite fracture surfaces indicated that these lines corresponded to a hill-and-valley surface relief, with the peaks on one surface matching the valleys on the opposite surface. The slope of the hills and valleys were usually approximately at 60 and 30 deg to the average fracture plane, in agreement with observations reported by Beachem and Yoder [*11*]. For the lower tempering temperatures (for example, 500°C for AMS 6304 steel), this zone was followed by a zone of brittle fracture by quasi-cleavage. For $\sigma_y < 1160$ MPa and for two of the steels (AISI 4340 and AMS 6304), 50% or more of the fracture surface corresponded to a set (because of the cylindrical specimen shape and the multi-initiation sites) of planes of maximum shear stress, all to the same side (Fig. 6) of the Mode I cracks responsible for the initial crack propagation. For $\sigma_y < 950$ MPa, this transition to a set of planes of maximum shear stress occurred immediately at the boundary with the Mode I semi-elliptic crack. This extensive amount of crack propagation on a single set of planes of maximum shear was obtained neither in the AISI 300M tension specimens tested in H_2SO_4 solution nor in any sample tested in air.

The microfractographic aspect of the semi-elliptic initial cracks presented, for intermediate tempering temperatures (500 to 600°C), a mixture of intergranular facets and of TTS features. For higher tempering temperatures, the intergranular facets disappeared and the amount of TTS fracture surface increased. For all tempering temperatures, the region closest to the external surface appeared more brittle, with less TTS surface present and either more intergranular fracture or, for higher tempering temperatures, more fracture surface having an aspect similar to quasi-cleavage.

The fractographic observations showed that the decrease in e_f, RA, and σ_u associated with testing in the H_2SO_4 solution was the result of the initiation of cracks at or very near the surface and their subsequent growth. The tests on the 4340 steel tempered at 400°C indicated fracture prior to yielding. For specimens tempered to lower strengths, the presence of some surface cracks outside of the necked region showed that some crack initiation occurred prior to necking. The occurrence of necking, however, resulted in a concentration of the semi-

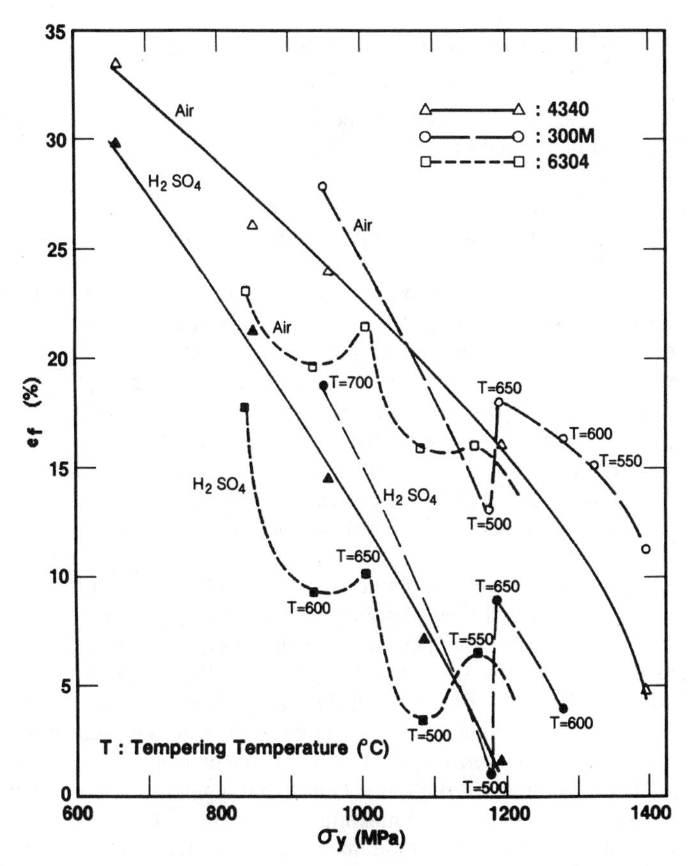

FIG. 3—*Tensile elongation* e_f *as a function of the tempering temperature.*

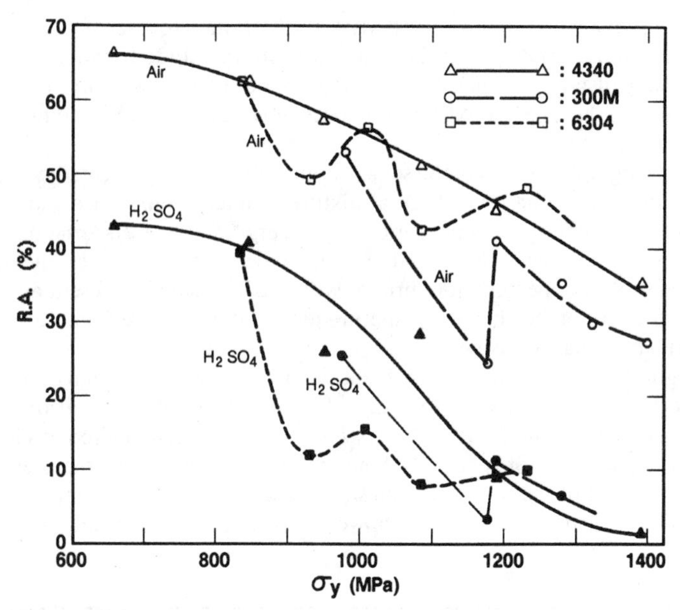

FIG. 4—*Reduction in area* RA *as a function of the tempering temperature.*

elliptic surface cracks near the center of the necked region. This observation suggested that the strain concentration associated with necking strongly favored the initiation of additional surface cracks.

The triaxial stresses associated with necking result in the maximum stress being in the interior of the necked region. The stress concentration should be especially important ahead of the tip of the surface cracks which have previously formed in this region. The degree of embrittlement, however, should be most pronounced in the immediate vicinity of the external surface. In the case of hydrogen associated with an external source, crack initiation is generally at the surface [*12*]. In one test on AISI 300M tempered at 700°C, the fracture did not correspond to the thinnest zone of the necked section. While the site of necking appeared determined by the site where one or more cracks caused the greatest decrease in specimen section, the site of fracture should be determined by the surface crack which first grows to a critical stress intensity factor. The two sites can differ if the center of the necked region is determined by the presence of several cracks and if the occurrence of the necking does not strongly accelerate the growth of the existing cracks in the thinnest region of the necked section.

In a number of previous studies, evidence has been presented in different steels [*7–10*] including AISI 4340 [*8*] that hydrogen charging can favor the occurrence of plastic instability by favoring crack propagation along planes of maximum shear stress. The clearest evidence of this plastic instability effect was observed for AISI 4340 and AMS 6304 steels tempered to $\sigma_y > 1160$ MPa, in that more than 50% of the fracture surface in the tension tests in the H_2SO_4 solution corresponded to cracking along one set of planes of maximum shear stress. The discontinuous lines observed on the fracture surfaces (Fig. 5), which permitted the identification of intermediate positions of the crack front segments, also reflect this tendency for cracking to occur locally on planes of maximum shear stress when the fracture mechanism is microvoid nucleation and growth. These lines correspond to the peaks and valleys of the hill-and-valley fracture surface, as a result of decohesion occurring in an alternate fashion along the planes of maximum shear stress intersecting the crack front. Although testing under hydrogen charging conditions appeared to favor its occurrence, this fractographic feature was also present on the fracture surfaces produced by microvoid initiation and coalescence during the J_{Ic} tests in air (see next section). This result confirms the tendency noted previously [7–10] for such crack growth to occur on planes of maximum shear stress in some steels even in the absence of hydrogen charging. This fractographic aspect was not observed for the tension samples tested in air.

J_{Ic} *Determination*

The J_{Ic} value was determined from the intersection point of the crack blunting line, $J = 2\sigma_f\Delta a$, with the regression line for the J-Δa curve determined from the load-crack mouth opening displacement (P-COD) measurements and calculated employing the equation proposed by Ernst et al. [13] to account for crack propagation. The value of σ_f was taken as $(\sigma_y + \sigma_u)/2$ for the tension tests in air. For tests in air and particularly in the H_2SO_4 solution on the steels tempered at the lower temperatures, the experimentally determined J-Δa curves, which lay to the right of the crack blunting line, indicated a change in slope at a small value of Δa, which will be referred to as Δa^*. The procedure suggested by ASTM E 813-81 was accordingly modified, as in the study of Hackett et al. [14], to separate the J versus Δa curves into two portions of different slopes (Fig. 7). The value of J_{Ic} was taken at the intersection of the blunting line with the higher slope portion of the J-Δa curve, corresponding to the lower values of Δa.

Figure 8 illustrates the manner in which J_{Ic} in the two test environments varied with the yield stress for the three steels. Tests were performed only for tempering temperatures equal or greater than 550 to 500°C, the latter value applicable to the 6304 steel tested in air and

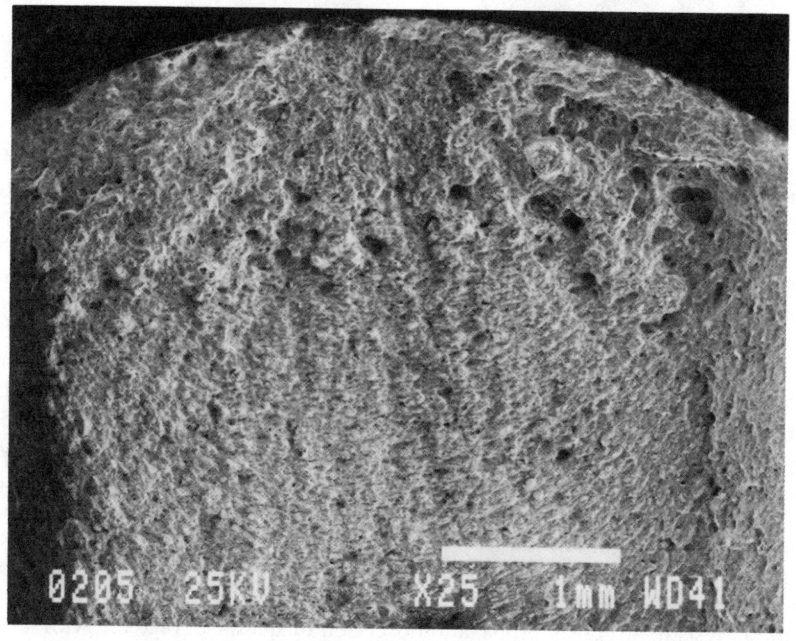

FIG. 5—*Portion of fracture surface of cathodically charged specimen of AISI 4340 (500°C temper) showing region having brittle aspect surrounded by lines showing intermediate positions of crack front.*

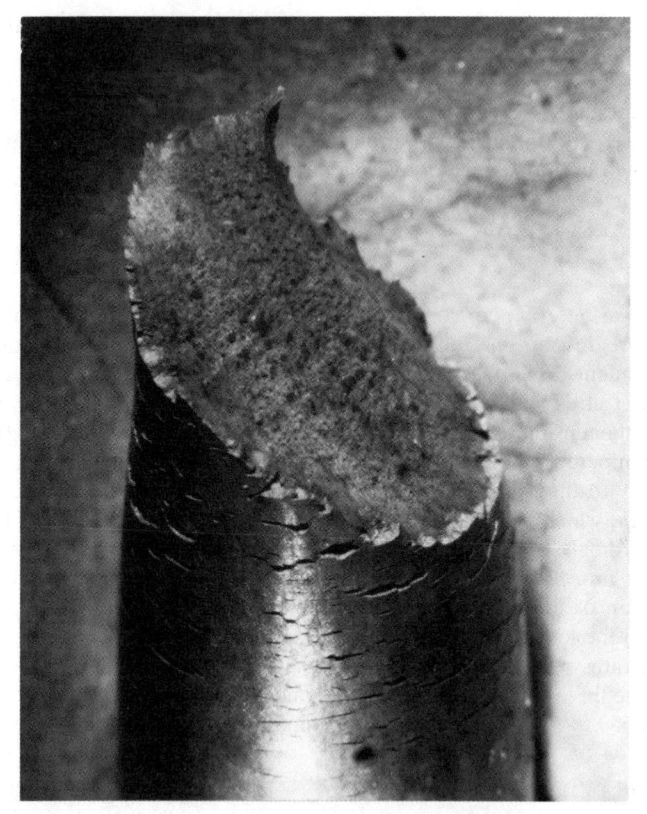

FIG. 6—*Fracture surface of cathodically charged AISI 4340 steel (700°C temper).*

the AISI 4340 steel tested in both environments. The results presented refer to the average for two tests in air and of three tests in the H_2SO_4 solution, except for the AMS 6304 steel, for which only a single test was performed in the H_2SO_4 solution for each tempering temperature.

Especially for the lower tempering temperatures, the tests in H_2SO_4 solution are seen to result in lower values of J_{Ic} than in air. The increase in σ_y associated with the secondary hardening resulting from tempering the AMS 6304 at 650°C appears particularly interesting in that the hydrogen charging procedure employed did not result in a decrease in the J_{Ic} value at this temperature compared with that obtained in air. The highest J_{Ic} value obtained in the tests in the H_2SO_4 solution for a yeild stress of 1000 MPa was that for the AMS 6304 steel tempered at 650°C. The beneficial effect of tempering the AMS 6304 steel at 650°C in order to obtain a decreased sensitivity to hydrogen embrittlement while maintaining a relatively high yield stress is also indicated by the e_f and *RA* values obtained in H_2SO_4 being higher for a tempering temperature of 650°C than of 600°C, even though the former resulted in higher σ_y and σ_u values. For the tests in H_2SO_4, e_f, and *RA* values obtained for the AMS 6304 steel tempered at 650°C, however, are below those for the other two steels at a similar σ_y of approximately 1000 MPa.

For the J_{Ic} tests in air, crack propagation occurred by quasi-cleavage for AISI 4340 steel tempered at 400°C and for AISI 300M tempered at or below 600°C and by microvoid formation and coalescence in the other tests. On the fracture surfaces produced by the latter mechanism, hill-and-valley fracture surface relief and discontinuous lines indicating prior positions of the crack front were observed (Fig. 9), as for the tension tests in H_2SO_4 solution. In the H_2SO_4 solution, the crack propagation also occurred except near the lateral edges by microvoid nucleation and coalescence with the presence of similar crack front lines. Near these edges, the material had been particularly embrittled by the hydrogen charging procedure employed and, except for the AISI 4340 steel tempered at 700°C, a band of brittle cracking 0.5 to 1.5 mm in width extended 1 to 6 mm ahead of the precrack tip (Fig. 9). The portion of this band nearest the surface had a brittle transgranular aspect somewhat resembling quasi-cleavage, with some intergranular or TTS fractographic features present in this zone further from the lateral surface.

A few tests were interrupted after a small number of unloadings. The crack advance at the surface was verified by microscopic observation prior to the specimens being immersed in liquid nitrogen and broken open for subsequent fractographic observation. These tests showed that the band of brittle cracking near the lateral surface occurred very early (within the first three or four unloadings) and prior to the point in the test which corresponded to the measured J_{Ic} value. The indication was that the occurrence of this cracking at the edges was not detected on the P-COD curve and that the J_{Ic} value measured corresponded to the start of crack propagation in the region between these bands of brittle fracture.

The reason for the change in slope of the J versus Δa curve was investigated. A similar change in slope has been studied previously by Hackett et al. [*14*] for hydrogen-charged 4340 steel (σ_y = 1202 MPa), who attributed the portion of higher dJ/da to the occurrence of highly branched intergranular cracking, with a resulting tendency for reduced constraint, in the first 0.5 to 0.7 mm of crack propagation. In the present study, a brittle layer next to

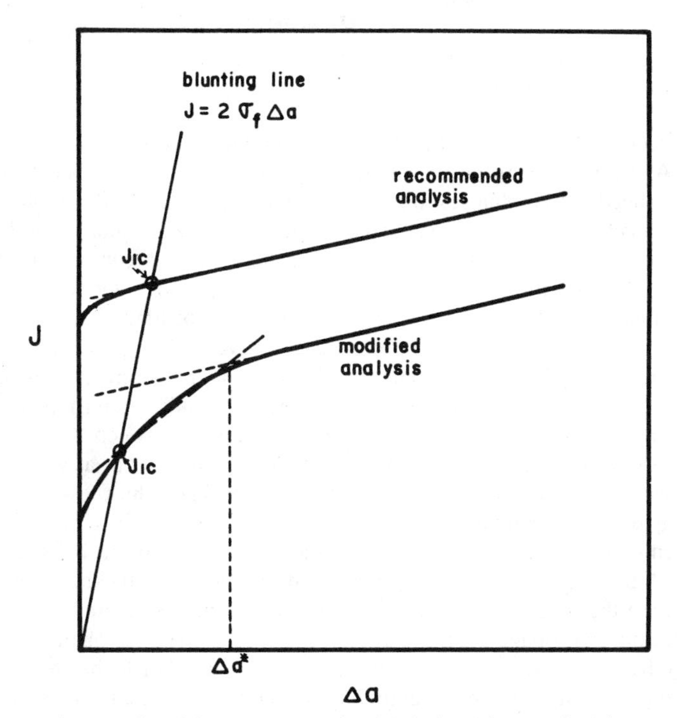

FIG. 7—*Procedure employed for measuring* J_{Ic} *and* Δa^* *depending on the* J-Δa *curve obtained.*

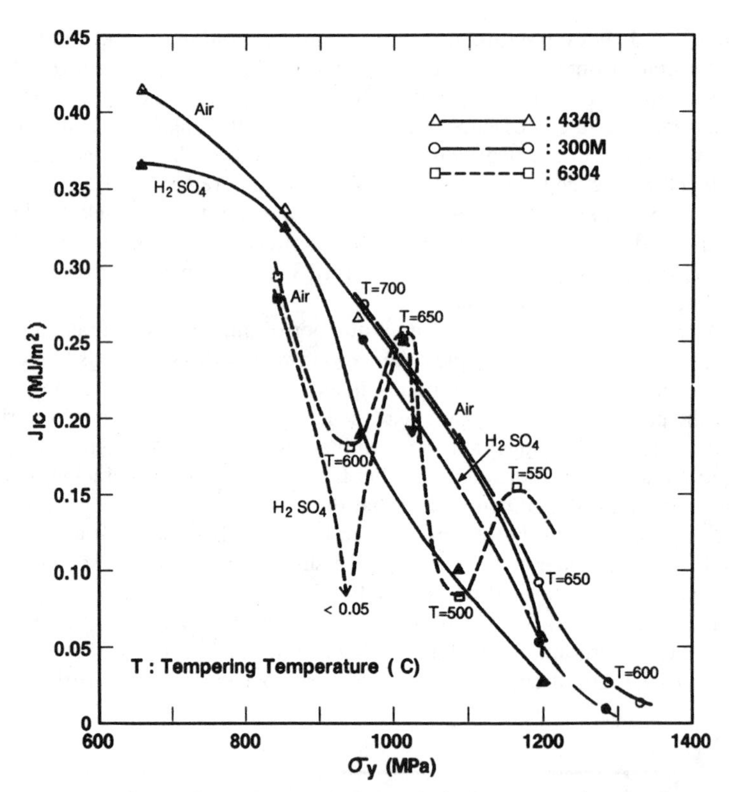

FIG. 8—*Variation of* J_{Ic} *with* σ_y. *For AMS 6340, which shows secondary hardening, some of the tempering temperatures are indicated.*

the fatigue precrack did not form other than the layer very near the lateral surfaces described previously. As well, the J-Δa curves obtained for the tests in air for steels with $\sigma_y > 950$ MPa also produced similar changes in slope. Metallographic and fractographic observations indicated the occurrence of some secondary cracking perpendicular to both the fracture surface and to the primary crack propagation direction. These secondary cracks were more numerous for the tests in H_2SO_4 and became shorter but more opened as the tempering temperature increased. Their occurrence, however, was not limited to the Δa values corresponding to the higher slope portion of the J-Δa curve.

From the discontinuous fractographic lines corresponding to crack front positions, the manner in which early crack propagation had proceeded could be determined. For the lower tempering temperatures, fair-to-good agreement was found between Δa^*, the value of Δa at which the slope of the J-Δa curve changed, and the crack depth at which a continuous crack front was first obtained across the portion of specimen thickness in which cracking occurred. For these lower tempering temperatures and tests in air, crack propagation was generally seen to initiate from two or three sites along the crack front. For the tests in H_2SO_4 solution, crack propagation in the main portion of the specimen thickness initiated near the corners between the fatigue precrack and the bands of brittle fracture at the sides (Fig. 9). For some specimens, propagation also initiated at one or two sites along the fatigue precrack, further away from the lateral edges of the fracture surface. For higher tempering temperatures, the crack front lines on a number of specimens indicated that a band-like crack through the central (plane-strain) portion of the thickness was obtained for very small values of Δa, which were in fair agreement with the small values of Δa^* obtained from the J-Δa curves. In contrast, the fractographic observations on other specimens indicated that such a band-like crack was only obtained for a Δa value two to three times larger than the Δa^* value from the J-Δa curve.

From these observations, a different explanation for the change in slope of the propagation portion of the J versus Δa curve can be suggested. The lower tempering temperatures result in significant variation of the resistance to the start of crack propagation along the crack front. Initiation of crack propagation begins in regions of relatively low local toughness. The lateral spread of cracking into the neighboring regions of higher toughness or the relaxation of constraint associated with mutual interference between the individual propagating cracks or both results in a relatively high dJ/da value. Once the different cracks initiated have coalesced to form one crack, the dJ/da value obtained is related to the resistance to the propagation of this crack. The value of dJ/da falls to a lower value, because the factors causing the value of J to increase rapidly with Δa no longer exist.

The interrupted tests performed in H_2SO_4 indicated that the technique employed did not permit the detection of the occurrence of the thin layer of brittle cracking near the lateral surfaces. The presence of this band of brittle cracking and the presence of embrittled material in the vicinity strongly favored further crack propagation to initiate near the corners between the brittle lateral cracks and the fatigue precrack, although some initiation also started nearer the mid-thickness region. That cracking initiated in the embrittled corners and spread into tougher material also favors an initially high dJ/da value. For a sufficiently high J value, the full crack width could propagate in the relatively tough material, resulting in a lower value of dJ/da. Thus, by producing embrittlement especially near the sides, the hydrogen charging procedure employed favored a relatively high value of Δa^*, in agreement with the result that for the lower tempering temperatures Δa^* was generally larger for tests in H_2SO_4 than in air (Fig. 10).

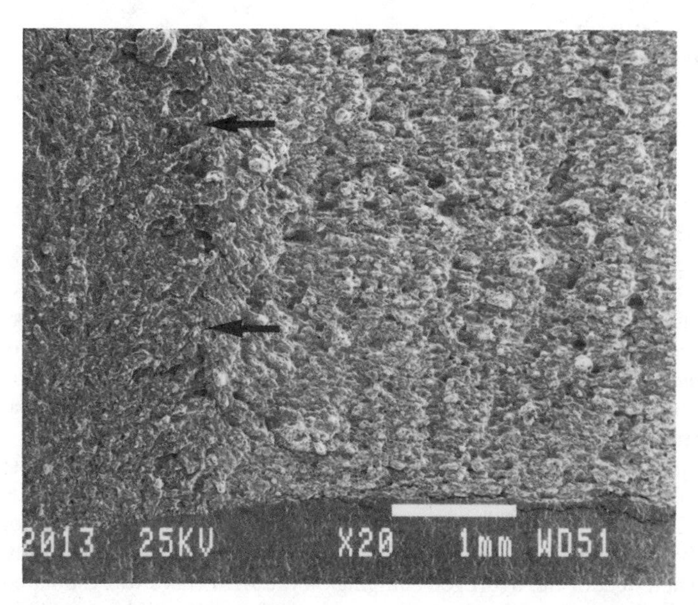

FIG. 9—*Brittle band at lateral edges (to the left of arrows) and lines showing intermediate positions of crack front in remaining portion of cathodically charged* J_{Ic} *specimen of AISI 4340 (500°C temper).*

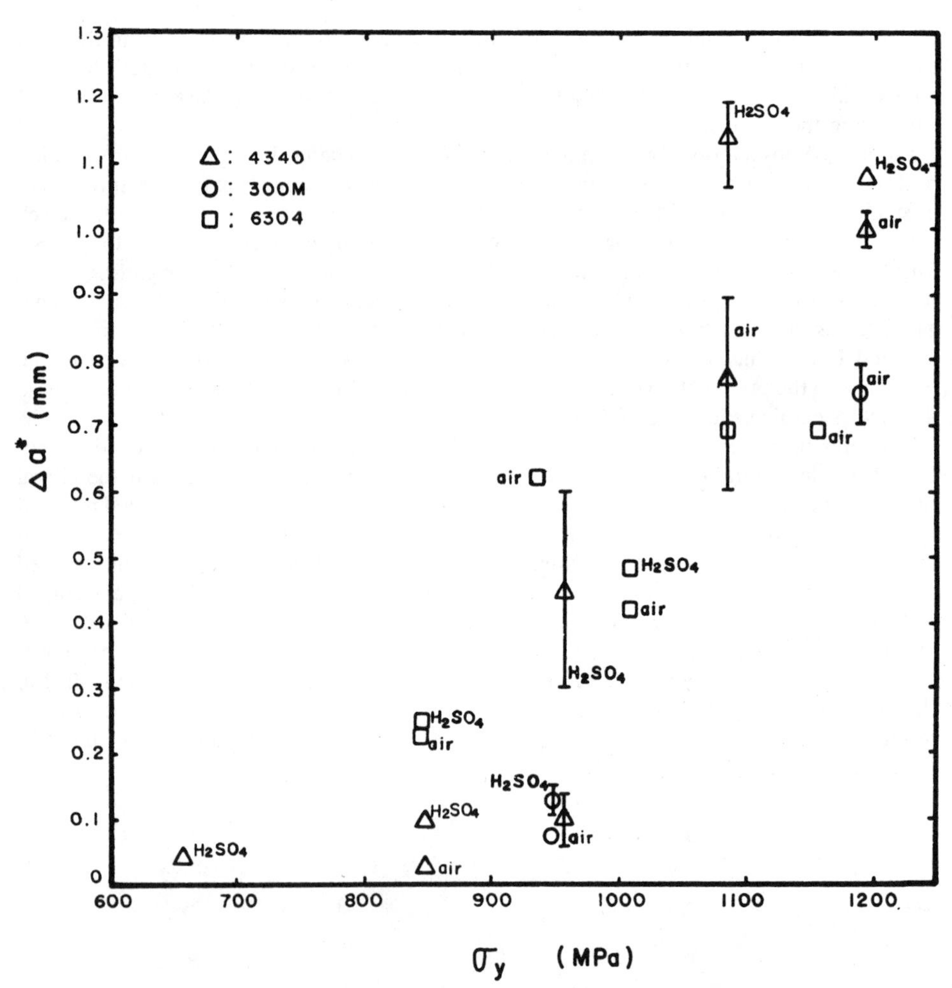

FIG. 10—*Variation of* a_o *with yield stress, including values of* a_o *which lie to left side of the blunting line.*

For the higher tempering temperatures employed, the lower values of Δa^* obtained and the short crack growth stage of higher dJ/da slope on the J-Δa curve appears related to a smaller variation of resistance to the start of crack propagation along the crack front. Even when cracking initiated at several individual sites, there was little tendency to obtain a high dJ/da portion of the J-Δa curve. For these higher tempering temperatures, hydrogen charging tended to cause relatively little embrittlement and at most produced a small band of brittle cracking near the lateral surfaces.

For the J-integral measurements in the H_2SO_4 solution, with the relatively short time of precharging employed, the fracture mechanism did not differ from the tests in air, other than within the band-like brittle region at the sides. Also, the J_{Ic} tests performed in the present study are generally for higher tempering temperatures than in the study by Hackett et al. [*14*]. The effect of the hydrogen charging on the J-Δa curve was estimated by treating these brittle cracks, which form prior to cracking in the interior, as side grooves and correcting for their presence. Comparison of the corrected results with those obtained in air indicated that hydrogen-charging still resulted in an important reduction of the J_{Ic} value even though the crack propagation mechanism remained microvoid nucleation and coalescence. The second or lower slope portion of the J-Δa curve after correction was, however, within experimental accuracy similar for the tests in air and in H_2SO_4 solution. Hirth and co-workers [7–10] have demonstrated that the strain to initiate cracking along planes of maximum shear stress by a mechanism of microvoid nucleation and coalescence is decreased considerably by hydrogen charging. The present results suggest that it is this effect which is responsible for the portion of the decrease in J_{Ic} which cannot be attributed to the occurrence of the bands of brittle cracking near the lateral surfaces.

Conclusions

Tension testing in 0.1 N H_2SO_4 solution, for the hydrogen-charging conditions employed, resulted in a decrease with respect to results in air in the values of e_f, RA, and σ_u for the three steels studied, with the influence being particularly marked for the lower tempering temperatures employed. This decrease in tensile properties was shown to be the result of the initiation of cracks at the external surface and their subsequent growth. A decrease in J_{Ic} was also obtained for testing in H_2SO_4 solution, a portion of which was associated with the occurrence of a band of brittle crack growth near each lateral surface. The remainder was attributed to hydrogen facilitating the initiation of microvoid nucleation and coalescence along planes of maximum shear stress. For the lower tempering temperatures, a two-stage J-Δa curve was obtained for tests in air and especially for tests in the H_2SO_4 solution. The Δa value associated with the change in slope of this curve was found to be consistent with that which produced a single crack propagating across the thickness.

References

[1] Beachem, C. D., *Metallurgical Transactions*, Vol. 3, No. 2, 1972, pp. 437–451.

[2] Banerji, S. K., McMahon, C. J., Jr., and Feng, H. C., *Metallurgical Transactions A*, Vol. 9A, No. 2, 1978, pp. 237–247.

[3] Craig, B. D. and Krauss, G., *Metallurgical Transactions A*, Vol. 11A, No. 11, 1980, pp. 1799–1808.

[4] Kaczorowksi, M., Lee, C. S., and Gerberich, W. W., *Materials Science and Engineering*. Vol. 81, 1986, pp. 305–315.

[5] Garber, R. I., Bernstein, I. M., and Thompson, A. W., *Metallurgical Transactions A*, Vol. 12A, No. 2, 1981, pp. 225–234.

[6] Thompson, A. W. and Chesnutt, J. C., *Metallurgical Transactions A*, Vol. 10A, No. 8, 1979, pp. 1193–1196.

[7] Lee, T. D., Goldenberg, T., and Hirth, J. P., *Metallurgical Transactions A*, Vol. 10A, No. 2, 1979, pp. 199–208.

[8] Lee, T. D., Goldenberg, T., and Hirth, J. P., *Metallurgical Transactions A*, Vol. 10A, No. 4, 1979, pp. 439–448.

[9] Chang, S. C. and Hirth, J. P., *Metallurgical Transactions A*, Vol. 16A, No. 8, 1985, pp. 1417–1425.

[10] Rajan, V. B. and Hirth, J. P., *Metallurgical Transactions A*, Vol. 18A, No. 2, 1987, pp. 335–340.

[11] Beachem, C. D. and Yoder, G. R., *Metallurgical Transactions A*, Vol. 4, No. 4, 1973, pp. 1145–1153.

[12] Page, R. A. and Gerberich, W. W., *Metallurgical Transactions A*, Vol. 13A, No. 2, 1982, pp. 305–311.

[13] Ernst, H. A., Paris, P. C., and Landes, J. D. in *Fracture Mechanics (13th Conference), ASTM STP 743*, American Society for Testing and Materials, Philadelphia, 1981, pp. 476–502.

[14] Hackett, E. M., Moran, P. J., and Gudas, J. P. in *Fracture Mechanics: Seventeenth Volume, ASTM STP 905*, American Society for Testing and Materials, Philadelphia, 1986, pp. 512–541.

Noel E. Ashbaugh[1] *and Theodore Nicholas*[2]

Threshold Crack Growth Behavior of Nickel-Base Superalloy at Elevated Temperature

REFERENCE: Ashbaugh, N. E. and Nicholas, T., **"Threshold Crack Growth Behavior of Nickel-Base Superalloy at Elevated Temperature,"** *Fracture Mechanics: Perspectives and Directions (Twentieth Symposium), ASTM STP 1020,* R. P. Wei and R. P. Gangloff, Eds., American Society for Testing and Materials, Philadelphia, 1989, pp. 628–638.

ABSTRACT: An experimental program was conducted to evaluate the effects of frequency and R on the near-threshold crack growth behavior of Inconel 718 at 649°C in laboratory air. Frequencies from 0.01 to 400 Hz and R from 0.1 to 0.9 were applied to compact tension [C(T)] and middle- or center-cracked tension [M(T)] specimens under decreasing-K conditions using computer-controlled test machines. Digital load-displacement data were obtained to determine crack length and closure load. The fatigue crack growth threshold in Inconel 718 at 649°C obtained using decreasing ΔK testing was generally associated with a crack arrest phenomenon which could be attributed to the buildup of oxides with time. Over the ranges of R and ν used in this investigation, the growth rate behavior at the onset of crack arrest appears to be a combination of time-dependent and cyclic-dependent behavior. Even at 400 Hz, purely cyclic behavior was apparently never reached. For crack growth rate modeling, both frequency and stress ratio have to be incorporated in the characterization of ΔK_{th}. Over the ranges of parameters tested, a cyclic threshold was approached at high frequencies and low R and a sustained load time-dependent threshold was obtained at high R, indicating that the cyclic contribution to the growth rate was negligible.

KEY WORDS: threshold, fatigue crack growth, nickel-base superalloy, elevated temperature, crack arrest, stress ratio, frequency

Damage-tolerant design requires the ability to accurately predict crack growth rates under conditions representative of those encountered in service. In turbine engine components, this involves wide variations in temperature, frequency (ν), and stress ratio (R = ratio of minimum to maximum stress). Further, initial flaw sizes in engine components are generally small so that crack growth occurs starting at very low stress-intensity factor ranges, ΔK. Because much of the life of a component can be spent at very low growth rates in the near-threshold regime, it is important to accurately model growth rates in this regime. Crack growth modeling procedures, using such interpolative models as hyperbolic sine (SINH) or modified sigmoidal equation (MSE) [*1–4*], require numerical values for the threshold stress-intensity factor. Further, the threshold value is not a constant, but rather a function of temperature, frequency, and stress ratio. For accurate modeling, therefore, these functional relationships have to be established.

[1] Senior research engineer, University of Dayton Research Institute, 300 College Park, Dayton, OH 45469.

[2] Senior scientist, Air Force Wright Aeronautical Laboratories (AFWAL/MLLN), Wright-Patterson Air Force Base, OH 45433.

The application of a threshold stress intensity to design based on the existence of an initial flaw has not been attempted. This procedure would be of greatest value for high-frequency fatigue in the presence of very small initial flaws. There, stable crack growth from an initial flaw would provide insufficient life because of the large number of cycles accumulated during very short times. Thus, the stress levels would have to be below some threshold where the crack does not propagate. For low-cycle fatigue, on the other hand, the combination of stress level and flaw size would have to result in an initial growth rate such that failure does not occur during the life of the component. This initial growth rate could be above the one determined in a standard threshold test, conventionally 10^{-10} m/cycle. Two major obstacles need to be resolved when using threshold values in design. One is the existence of initial defects which are small, whereas threshold data are obtained on nominally large cracks. The anomalously high growth rates of small cracks in some materials makes the use of long crack data both questionable and nonconservative for small initial flaws. The second obstacle is the application of data obtained on a propagating crack to a condition of an initial flaw or defect which has not been propagating previously. Thus, the main use for threshold values as determined in this investigation is for the calibration of interpolative crack growth models for predicting growth rates in the near threshold regime over a range of frequencies and R values.

Prior studies in crack growth behavior in nickel-base superalloys have shown several significant features which have to be considered in modeling. First, at elevated temperatures, crack growth can range anywhere from purely cycle-dependent to purely time-dependent, depending on the values of ΔK, frequency (ν), stress ratio (R), or wave shape [*5–9*]. These studies have concentrated on behavior at ΔK values and growth rates well above threshold. Each type of behavior can be associated with an observed fracture surface appearance: cycle-dependent behavior appears as transgranular fracture, while time-dependent behavior is intergranular. There is also a broad region of mixed-mode behavior with a combination of intergranular and transgranular fracture features. In turn, each region has to be modeled using the appropriate relationships which are unique to that region [*5*]. Thus, an awareness of the governing fracture mechanisms is important in developing models for fatigue crack growth.

A second feature which has been noted in nickel-base superalloys is the phenomenon of crack arrest in conventional threshold testing at elevated temperature. Using a decreasing ΔK loading history, abrupt crack arrest due to oxide buildup has been observed at growth rates well above the conventional definition of threshold of 10^{-10} m/cycle [*10*]. No such arrest is observed in tests in vacuum. The abrupt arrest has also been observed in titanium alloys and has been attributed to the crossover between the competing effects of oxide buildup with time and decrease in applied stress intensity in the load-shedding procedure [*11*]. At some point, the crack is wedged open by the oxides to a sufficient extent such that the driving force is insufficient for further crack advance. We have observed that the level has to be increased significantly to restart the crack in a subsequent constant load range test. An alternate explanation for crack arrest in elevated temperature tests in air proposed by McEvily [*12*] attributes the phenomenon to the strength and thickness of the oxide layer and the rupture of this layer due to the stress-strain state at the crack tip.

The third feature observed in that both mean load and cyclic amplitude affect the crack growth rate in superalloys [*13*]. Cracks grow not only under fatigue loading, but also under sustained loading and combinations of the two [*14*]. Under sustained load, the growth is due primarily to environmental degradation of the material rather than to creep. It is akin to stress corrosion cracking in other materials; here oxygen is the aggressive environment.

The present investigation was undertaken to determine the influence of stress ratio and frequency on the threshold in a nickel-base superalloy. Inconel 718, which has been studied extensively in prior investigations, was chosen as the test material because of the existing database, knowledge of the micro-mechanisms, and its ideal behavior as a model material which can produce cycle- or time-dependent behavior over a range of frequencies and R which are readily produced in the laboratory. Threshold values are reported, covering a wide range of test conditions, and are interpreted and discussed in terms of both the observed phenomenology and the underlying mechanisms.

Experiments

Inconel 718 was used as the test material at a temperature of 649°C for all tests. Two types of specimens were employed in this investigation. One group of compact tension [C(T)] specimens was machined from 12.5-mm-thick plate to a thickness, $B = 10$ mm, with $H/W = 0.6$ and $W = 40$ mm. Middle- or center-cracked tension [M(T)] specimens of width, $W = 50$ mm, and a second group of C(T) specimens of width, $W = 40$ mm, were cut from 2.3-mm-thick sheet and machined to final dimensions without reducing the thickness. All specimens were given the standard heat-treatment as detailed in Table 1. The C(T) specimens were tested in a servo-hydraulic test machine under computer control using a combination of extensometer and back-face strain gage to monitor the deformation. The foil gages for back-face strain were protected with a ceramic coating to prevent oxidation at the test temperature. Compliance values obtained from the extensometer and strain gage were used to determine crack length and control the test. Tests were conducted over a range of frequencies from 0.01 to 30 Hz.

M(T) specimens were tested in a hybrid test system especially built for high-frequency testing [*15*]. It employs an electro-dynamic shaker to apply the vibratory loading and a pneumatic system for the mean load. Crack lengths were monitored using the a-c electric potential-drop technique. Tests on this system covered a range of frequencies from 1 to 400 Hz. The test matrix for the two specimen types covered a range of stress ratios from 0.1 to 0.9.

Threshold testing was conducted on fatigue precracked specimens using a standard computer-controlled load shedding procedure. The maximum value of K was controlled as a function of crack length according to the equation

$$K = K_0 \exp[C(a - a_0)] \tag{1}$$

where K_0 and a_0 are the reference K and crack length values, respectively, at the start of the test, and C is the constant determining the rate of load shedding. Values of C of -0.08 and -0.2 mm^{-1} (-2 and -5 in.$^{-1}$) were used. Prior work has shown that values as low as $C = -1.2$ mm^{-1} (-30 in.$^{-1}$) do not adversely influence the value of threshold obtained in this class of material [*10*].

Tests were conducted until crack arrest was observed. In the few tests where no arrest occurred, the test was continued until a growth rate of approximately 10^{-10} m/cycle was achieved.

TABLE 1—*Heat treatment for Inconel 718.*

Anneal at 967.4°C (1775°F) for 1 h
Air cool to below 717.6°C (1325°F)
Age at 717.6°C (1325°F) for 8 h
Furnace cool at 37.7°C/h (100°F/h) to 620.5°C (1150°F)
Age at 620.5°C (1150°F) for 8 h

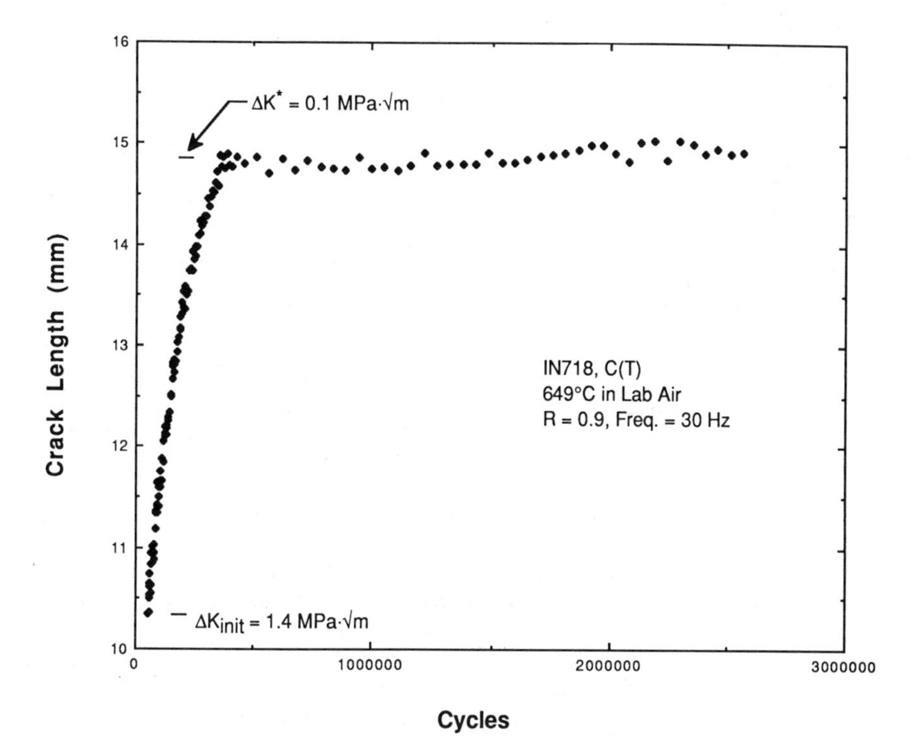

FIG. 1—*Typical* a-N *data from threshold test.*

A typical crack length versus number of cycles curve is presented in Fig. 1. The behavior can be seen to be divided into two distinct regions. Up to 340 000 cycles, the conventional reduction in growth rate (slope of the $a-N$ curve) as ΔK is reduced is observed. (Note that the initial cycle count for the data is 50 000.) For applied cycles beyond 340 000, the crack length ceases to increase. The data in the growth rate regime are reduced in the conventional manner using a seven-point sliding polynomial. The value of crack length at arrest is used to calculate ΔK^*, the value of ΔK at the point of crack arrest. In the following paragraphs, the term *threshold* is used to describe a crack arrest value or a threshold value at 10^{-10} m/cycle.

Results and Discussion

Experimental data are presented for threshold values obtained from each of the decreasing K tests for both C(T) and M(T) geometries. Since no significant differences were observed in data obtained from the two different specimen geometries (and thicknesses), no indication of specimen type is presented in the figures. With the exception of only a few data points as noted, the threshold values represent quantities immediately prior to crack arrest as discussed in the previous section.

The maximum value of the stress intensity at threshold, K_{th}, is plotted in Fig. 2 as a function of frequency. The data at the highest R, 0.9, appear to be relatively independent of frequency, even though no data were obtained for this value of R at the lower frequencies. Since the stress intensity range is very small, ($\Delta K = 2.4$ MPa $\sqrt{\text{m}}$), and probably below a purely cyclic threshold, the material behavior is assumed to be essentially time-dependent.

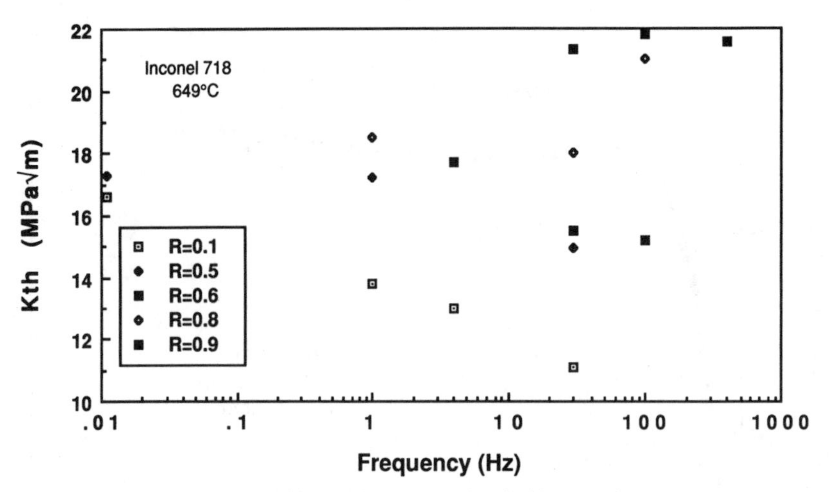

FIG. 2—*Maximum stress intensity at threshold versus frequency.*

The threshold value at high R is thus associated with a mean load or sustained load growth mechanism and does not appear to be influenced by the superimposed cycling. At low R of 0.1, on the other hand, the value of K_{th} decreases with increasing frequency. This could be attributed to the greater buildup of oxide debris at the lower frequencies than at the higher frequencies because of the greater time per cycle spent exposed to the environment. If the mechanism of oxide-induced closure is responsible for the observed trend, then for the same effective driving force, $K_{max} - K_{cl}$, this requires a higher K_{max} as frequency decreases as observed in the data.

Another aspect of the low R threshold behavior which must be considered is that as frequency increases, the relative contribution of the cyclic mechanism to the sustained load mechanism increases. It is to be expected, then, that at the two extremes of frequency at $R = 0.1$, purely cyclic- and purely time-dependent behavior would be observed. The data of Fig. 2 show that for purely cyclic behavior, K_{max} appears to approach a value of approximately 10 MPa $\sqrt{m}$ at very high frequencies. On the other hand, at a very low frequency, K_{max} should ultimately approach a value similar to that observed at high R under purely time-dependent crack growth (approximately 22 MPa $\sqrt{m}$). Figure 2 shows that most of the data obtained in this investigation represent thresholds obtained at conditions which represent some mix of time-dependent and cycle-dependent behavior.

The data of Fig. 2 are replotted as K_{th} against R in Fig. 3. The data show a continuous increase of K_{th} with R for a given frequency for all of the data, converging at the highest R of 0.9. The values for frequency for $R = 0.1$ data are indicated on the curve, and it can be seen that the highest value of K_{th} corresponds to the lowest value of frequency. For the remainder of the data, the lower bound of the data represents the highest frequency and vice versa. At high R, there is a constant K_{th} of approximately 21 to 22 MPa $\sqrt{m}$ which represents time-dependent behavior. At low R, the data spread out with frequency showing that cycle-dependent behavior becomes less significant as frequency decreases. As deduced from Fig. 2, the data of Fig. 3 indicate that most of the conditions covered in this investigation represent conditions somewhere between purely time-dependent and purely cycle-dependent behaviors.

The stress-intensity range at threshold, ΔK_{th}, is plotted in Fig. 4 as a function of R. No symbols are used to distinguish data at the different frequencies. It can be seen that except for $R = 0.1$, the data at all frequencies consolidate into a narrow band. Further, the trend

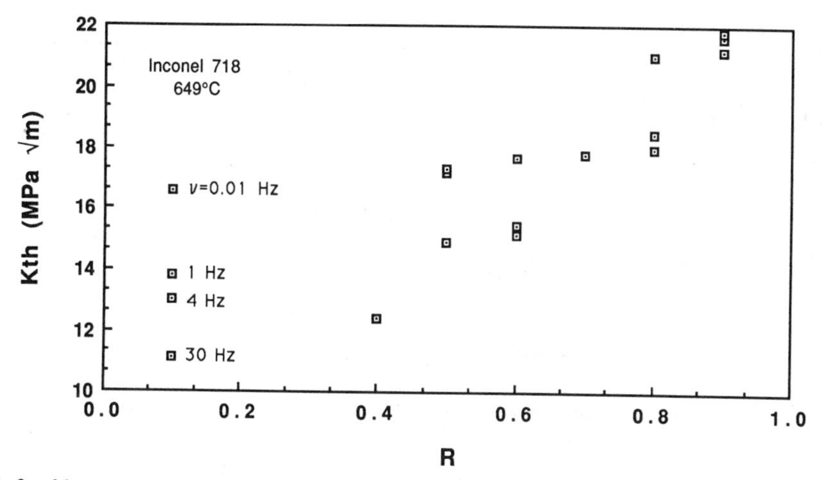

FIG. 3—*Maximum stress intensity at threshold versus* R. *Frequency is indicated for data at* R = *0.1.*

of the band is towards ΔK_{th} approaching zero for $R = 1$, which represents purely sustained loading with no cyclic contribution. This provides additional evidence of the essentially time-dependent behavior of this material near threshold at high R.

The growth rate at crack arrest, da/dt^*, is plotted against frequency in Fig. 5. With the exception of three data points shown with arrows, all of the data lie along a single band. For analytical modeling purposes, the growth rate at crack arrest can be represented by the straight line drawn in Fig. 5 having the equation

$$\frac{da^*}{dt} = \nu \frac{da}{dN} = A\nu^{0.4} \tag{2}$$

with da/dt^* having units of mm/s, ν in Hz, and $A = 2 \times 10^{-5}$. All of the data points represented by this expression represent some combination of sustained load cracking and

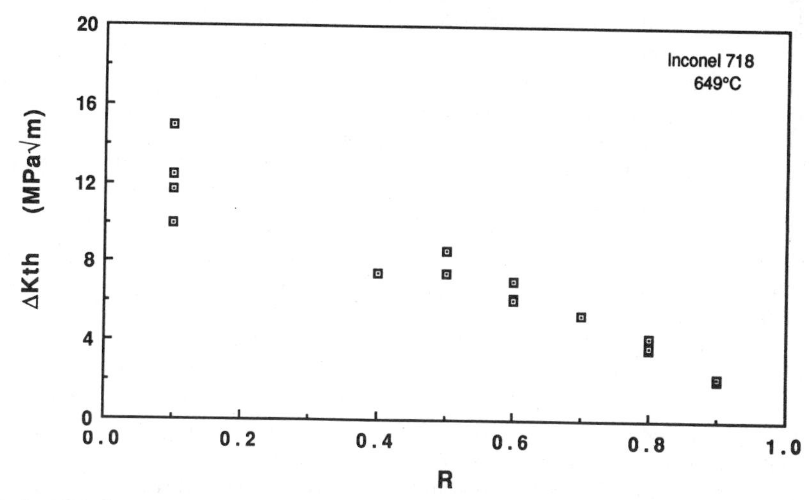

FIG. 4—*Threshold stress-intensity range as a function of* R. *Data obtained at all frequencies are shown.*

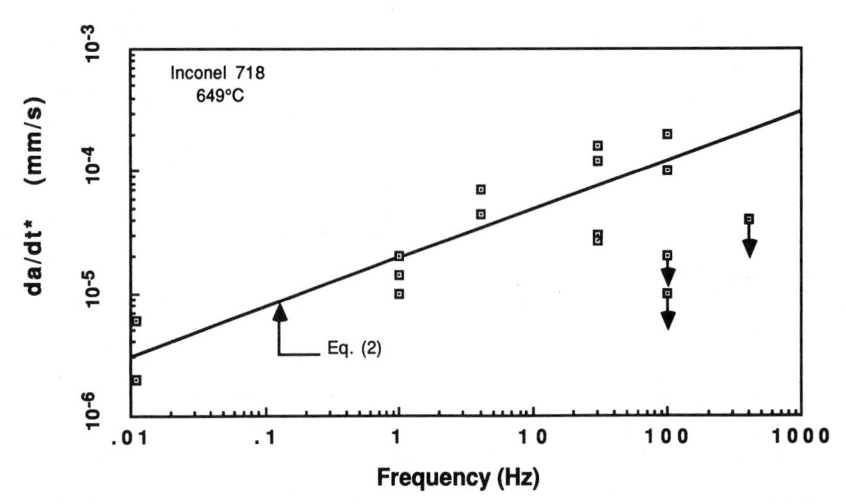

FIG. 5—*Crack growth rate at threshold versus frequency. Arrows indicate test was halted before crack arrest occurred.*

fatigue. For purely sustained load cracking, da/dt^* for a given R would be independent of cyclic frequency. Conversely, for purely cycle-dependent behavior, da/dt^* should increase linearly with frequency to provide a constant value of da/dN at threshold. This would result in a curve with a slope of one in Fig. 5, or an exponent on frequency, ν, of 1.0 in Eq 2. Clearly, the data of Fig. 5 demonstrate that, at threshold, there is a combination of cycle-dependent and time-dependent behavior governing crack growth behavior.

The three data points in Fig. 5 which have arrows indicating an upper limit of da/dt^* correspond to tests in which there was no crack arrest. In these three tests, the test was terminated at the growth rates indicated because they achieved a threshold growth rate of 10^{-10} m/cycle. One of the data points at 100 Hz represents a test at $R = 0.8$; the other two indicated points represent tests at $R = 0.9$. These three tests, as previously discussed, appear to represent a condition of purely time-dependent growth due solely to the mean load with no superimposed influence of the very low amplitude cycling. Using the argument that purely sustained load crack growth would result in a horizontal line in Fig. 5, and noting that as the frequency is lowered, the time-dependent component becomes more dominant, it can be speculated that the purely sustained load value of da/dt^* would be at or below 10^{-6} mm/s based on the data at 0.01 Hz. It is expected, then, to see the three data points in Fig. 5 outside the trend line and to expect them to be much lower if the material behavior was purely time-dependent.

Another observation from the high R threshold tests is that crack arrest occurred at low frequencies but not at higher frequencies above approximately 100 Hz. Although the growth behavior appears to be primarily time-dependent, the superimposed high-frequency cycling on the high mean load appears to keep the crack propagating even though it does not influence the growth rate or the value of the maximum value of K at threshold. When lower frequencies are superimposed on high mean loads, the smaller contribution of the cycling is inadequate to keep the crack growing until threshold, that is, crack arrest occurs. In this case, oxide buildup does not affect closure because of the high mean load. Although crack arrest in this case might be attributed to phenomena such as crack branching or crack blunting due to creep, this explanation is unlikely at the very low growth rates and K values near threshold. The mechanism of fracture of oxide layers proposed by McEvily [*12*] could also be responsible. Fractography and examinations of the crack tip profiles at threshold or crack arrest are expected to shed further light on these matters in future work.

Values for the closure stress intensity, K_{cl}, are plotted as a function of frequency in Fig. 6. It can be seen that for any given value of R, the value of K_{cl} is approximately constant. For the higher values of R between 0.6 and 0.9, however, the data do not represent closure but, rather, the value of K_{min} ($K_{min} = R\ K_{max}$). This simply indicates that in most cases, closure was at or below the minimum load in the cycle. For values of R of 0.1 and 0.5, however, the closure values are above K_{min}. This can be seen by comparing the K_{cl} values of Fig. 6 with the K_{th} values of Fig. 2. The results for $R = 0.1$ and 0.5 indicate that the closure K increases with increasing R, indicating that mean load has an effect on the closure level. If closure is attributed to the buildup of oxides behind the crack tip, then it appears that oxide buildup is inhibited more at low values of R where the cyclic range is greater than at higher R where the crack is open a larger portion of the time. Cyclic frequency does not appear to play a role in the amount of oxide buildup and the subsequent closure level.

The effective stress intensity range at threshold, ΔK_{eff}^*, is plotted against R in Fig. 7. For purely cycle-dependent behavior, this quantity would be expected to be constant, representing an inherent material property. The data show a definite trend towards a decrease in ΔK_{eff}^* with increasing R, indicating that mean load has a marked effect on the threshold behavior. This implies that environmental degradation ahead of the crack tip is responsible for the observed behavior.

The data for ΔK_{eff}^* at $R = 0.1$ show a definite trend with frequency which the other data do not. This result contradicts the conclusions drawn from Fig. 2, where it was assumed that variations in threshold K were due to differences in closure level due to oxide buildup. That ΔK_{eff}^* decreases with increasing frequency implies that the crack arrest phenomenon is due not to closure but rather to changes in the materials inherent resistance to crack growth with frequency or environmental exposure time. In this case, this would say that the material is more resistant to crack growth for *lower* frequencies or longer exposure times. This does not appear to make physical sense. An alternate explanation, which appears more reasonable, is that the closure loads which were determined using a mechanical displacement gage at the crack mouth may not be representative of the closure values that consolidate crack growth rates.

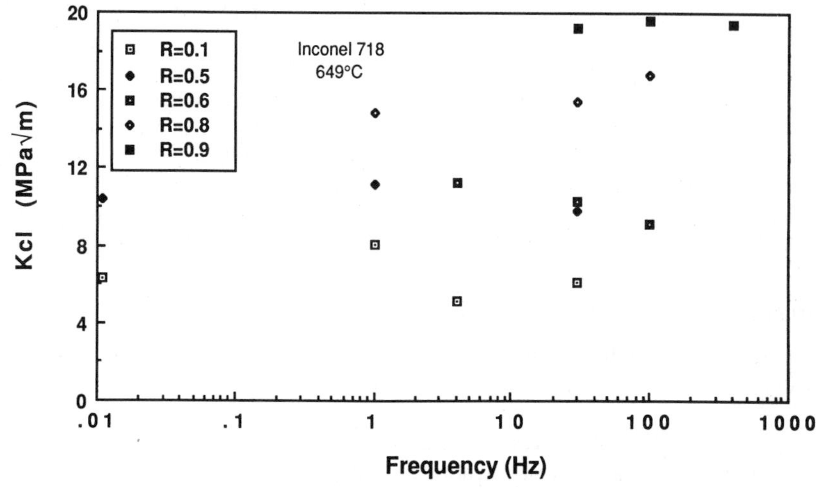

FIG. 6—*Closure stress intensity as a function of frequency.*

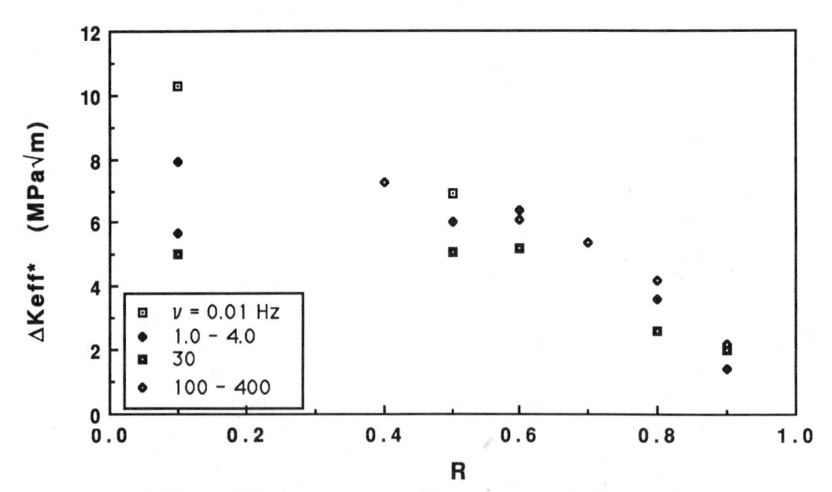

FIG. 7—*Effective stress intensity range at threshold versus* R.

From the reported observations on threshold, it becomes apparent that a formal definition of threshold as the ΔK corresponding to a growth rate of 10^{-10} m/cycle is inadequate for characterizing threshold in materials which exhibit time-dependent crack growth at elevated temperature. If the case of purely sustained load crack growth is eliminated for the present, cyclic behavior at high mean loads is more appropriately described in terms of da/dt than da/dN. Further, the superposition of low-amplitude cycling at high mean loads over a broad range of frequencies has little effect on da/dt, but a significant effect on da/dN because of the large differences in frequency which, in turn, have little influence on the growth rate. Finally, since crack arrest is observed to occur in many instances, there is no condition which corresponds to a stable crack growth rate of 10^{-10} m/cycle. We thus propose a new, additional definition of threshold to describe the condition where crack arrest occurs, $\Delta K_{th}{}^{*}$. Further, we would suggest that the point at crack arrest be identified in da/dN versus ΔK plots with a unique symbol and an arrow pointing down to signify that crack arrest occurred. Finally, for situations where no crack arrest occurs, we suggest that both da/dt as well as da/dN be considered in defining a threshold for a given material or application.

The potential application of the threshold values determined in this investigation to damage-tolerant design is limited to cases where the initial flaw sizes are above the small crack regime, say 1 mm or more. Further, these flaws must have developed the same magnitude of crack closure as is developed in the decreasing ΔK tests. For a crack which has not been propagating previously, closure would not have been expected to develop. Finally, since crack growth is time-dependent, the numerical value of the threshold crack growth rate, either da/dN or da/dt, should be chosen such that failure does not occur within the design life if a reasonable size initial flaw starts growing at the selected threshold rate.

Conclusions

Several conclusions can be drawn based on the results and observations in this experimental investigation:

1. The fatigue crack growth threshold in Inconel 718 at 649°C obtained using decreasing ΔK testing is really a crack arrest phenomenon in most cases which is attributed to the buildup of oxides with time. Crack arrest can frequently occur at growth rates well above the conventional definition of threshold of 10^{-10} m/cycle.

2. Over the ranges of R and ν used in this investigation, the growth rate behavior at the onset of crack arrest appears to be a combination of sustained load and cyclic phenomena. Only in a few cases at high R was the behavior purely time-dependent. Purely cyclic behavior was apparently never reached even at frequencies as high as 400 Hz.

3. Both frequency and stress ratio influence the value of ΔK_{th} and, thus, have to be considered in crack growth rate modeling.

4. It appears that there are two distinct thresholds: a cyclic threshold and a sustained load threshold. The cyclic threshold is approached at high frequencies and low R (low mean stress). The sustained load threshold is obtained at high R when the cyclic contribution becomes negligible.

5. Far-field closure measurements produce values of ΔK_{eff}^* which are inconsistent with the proposed explanation of crack arrest due to oxide buildup.

Acknowledgment

The authors wish to express the appreciation of Messrs. David Maxwell and Richard Goodman for their help in preparing the specimens, conducting the tests, and reducing the data.

References

[*1*] Annis, C. G., Jr., Wallace, R. M., and Sims, D. L., "An Interpolative Model for Elevated Temperature Fatigue Crack Propagation," Technical Report AFML-TR-76-176, Part I, Air Force Materials Laboratory, Wright-Patterson Air Force Base, OH, 1976.

[*2*] Utah, D. A., "Crack Growth Modeling in an Advanced Powder Metallurgy Alloy," Technical Report AFWAL-TR-80-4098, Air Force Wright Aeronautical Laboratories, Wright-Patterson Air Force Base, OH, 1980.

[*3*] Larsen, J. M. and Nicholas, T., "Cumulative-Damage Modeling of Fatigue Crack Growth in Turbine Engine Materials," *Engineering Fracture Mechanics,* Vol. 22, No. 4, 1985, pp. 713–730.

[*4*] Nicholas, T., Haritos, G. K., and Christoff, J. R., "Evaluation of Cumulative Damage Models for Fatigue Crack Growth in an Aircraft Engine Alloy," *Journal of Propulsion and Power,* Vol. 1, No. 2, 1985, pp. 131–136.

[*5*] Weerasooriya, T., "Effect of Frequency on Fatigue Crack Growth Rate of Inconel 718 at High Temperatures," AFWAL-TR-87-4038, Air Force Wright Aeronautical Laboratories, Wright-Patterson Air Force Base, OH, June 1987.

[*6*] Scarlin, R. B., "Effects of Loading Frequency and Environment on High Temperature Fatigue Crack Growth in Nickel-Base Alloys," *Fracture 1977, Advances in Research on the Strength and Fracture of Materials,* Vol. 2B, Fourth International Conference on Fracture, University of Waterloo, Canada, June 1977, pp. 849–857.

[*7*] Clavel, M. and Pineau, A., "Frequency and Waveform Effects on the Fatigue Crack Growth Behavior of Alloy 718 at 298 K and 823 K," *Metallurgical Transactions,* Vol. 9A, 1978, pp. 471–480.

[*8*] Venkataraman, S. and Nicholas, T., "Mechanisms of Elevated Temperature Fatigue Crack Growth in Inconel 718 as a Function of Stress Ratio," *Effects of Load and Thermal Histories on Mechanical Behavior of Materials,* P. K. Liaw and T. Nicholas, Eds., The Metallurgical Society of the American Institute of Mining, Metallurgical, and Petroleum Engineers, 1987, pp. 81–99.

[*9*] Nicholas, T. and Ashbaugh, N. E., "Fatigue Crack Growth at High Load Ratios in the Time-Dependent Regime," *Fracture Mechanics: Nineteenth Symposium, ASTM STP 969,* American Society for Testing and Materials, Philadelphia, 1989, pp. 800–817.

[*10*] Zawada, L. P. and Nicholas, T., "The Effect of Closure on the Near-Threshold Fatigue Crack Propagation Rates of a Nickel-Base Superalloy," *Mechanics of Fatigue Crack Closure, ASTM STP 982,* J. C. Newman, Jr. and W. Elber, Eds., American Society for Testing and Materials, Philadelphia, 1988, pp. 548–567.

[*11*] Allison, J. E. and Williams, J. C., "Near-Threshold Fatigue Crack Growth Phenomena at Elevated Temperature in Titanium Alloys," *Scripta Metallurgica,* Vol. 19, 1985, pp. 773–778.

[*12*] McEvily, A. J., "An Overview of Fatigue," *Fatigue 87,* Vol. 3, R. O. Ritchie and E. A. Starke,

Jr., Eds., Engineering Materials Advisory Services, Ltd., West Midlands, U.K., 1987, pp. 1503–1516.

[13] Shahinian, P. and Sadananda, K., "Effects of Stress Ratio and Hold Time on Fatigue Crack Growth in Alloy 718," *Journal of Engineering Materials and Technology,* Vol. 101, July 1979, p. 224.

[14] Nicholas, T. and Weerasooriya, T., "Hold-Time Effects in Elevated Temperature Fatigue Crack Propagation," *Fracture Mechanics: Seventeenth Volume, ASTM STP 905,* J. H. Underwood, R. Chait, C. W. Smith, D. P. Wilhem, W. A. Andrews, and J. C. Newman, Eds., American Society for Testing and Materials, Philadelphia, 1986, pp. 155–168.

[15] Goodman, R. C. and Brown, A. M., "High Frequency Fatigue of Turbine Blade Material," AFWAL-TR-82-4151, Air Force Wright Aeronautical Laboratories, Wright-Patterson Air Force Base, OH, Oct. 1982.

Fracture Mechanics of Nonmetals and New Frontiers

Jia-Min Bai[1] *and Tsu-Tao Loo*[1]

Strength of Stress Singularity and Stress-Intensity Factors for a Transverse Crack in Finite Symmetric Cross-Ply Laminates Under Tension*

REFERENCE: Bai, J.-M. and Loo, T.-T., **"Strength of Stress Singularity and Stress-Intensity Factors for a Transverse Crack in Finite Symmetric Cross-Ply Laminates Under Tension,"** *Fracture Mechanics: Perspectives and Directions (Twentieth Symposium), ASTM STP 1020,* R. P. Wei and R. P. Gangloff, Eds., American Society for Testing and Materials, Philadelphia, 1989, pp. 641–658.

ABSTRACT: A theoretical investigation into the mechanics of the failure phenomenon of a composite laminate is presented. The study is specialized to the problem of a finite $[0/90]_s$ symmetric cross-ply composite laminate with a transverse through-crack under uniaxial tension. The stress field of such laminates is analyzed with particular emphasis on the stress singularity of the crack and the corresponding stress-intensity factors. The Lekhnitskii's method of complex stress functions for anisotropic bodies [*1*] is employed in the analysis of the strength of crack singularity. The crack problem is solved by analytic method, and the solution is expressed in the form of the sum of two infinite series of Fourier-hyperbolic functions supplemented by a singular term with predetermined order. Through numerical computation, the effects of the material constants and geometric parameters on the stress-intensity factors are obtained. They seem to furnish a reasonable explanation to the observed failure phenomenon.

KEY WORDS: transverse cracks, delamination, finite composite laminates, stress singularity, stress-intensity factors, material constants, geometric effects

As is well known from experiments on fiber-reinforced composite laminates, the phenomenon of an aggregation of transverse cracks often occurs before delamination leading to ultimate failure [*2*]. Any attempt from the theoretical aspect to explain the mechanism of such a phenomenon will help improve the design of composite structures. Hence, an accurate analysis of the crack problem of the composite laminates has its practical significance.

The transverse crack problem was first studied by Tsai and Hahn in 1975 [*3*]. Since then, much research on this subject has been done [*4–7*]. Nevertheless, most of these works did not consider rigorously the local effect and the nature of stress singularity of the crack, which are essential for initiating further cracks and delamination in a composite laminate. In recent years, Delale and Erdogan and others [*8,9*] treated the problems of infinite isotropic or orthotropic strips bonded together with the crack normal to and on the interfacial boundary. The strength of the singularity and stress-intensity factors or stress coefficients for such cracks has been computed. Since the technique of the infinite integral transform was used

* This work is part of a research sponsored by the National Science Foundation of China.

[1] Graduate student, professor and director, respectively, Shanghai Jiao Tong University, 1954 Hua Shan Road, Shanghai 200030, People's Republic of China.

for the solution in these works, the class of problems that can be treated is restricted to those with infinite medium.

In the present paper, however, the finite laminate problem is treated. The study is specialized to the problem of a finite $[0/90]_s$ symmetric cross-ply composite laminate with a transverse crack under uniaxial tension in order to access the effects of both physical and geometrical parameters of the laminate on the strength of singularity of the crack and related stress-intensity factors. The power of stress singularity at the crack tip is found, as expected, depending notably on the material constants. This result seems to reveal the cause for the crack in the 90-deg ply turning into local delamination, rather than further extension along the crack at the interfacial boundary.

An analytic solution for the problem is given in the paper. It is expressed in the series of orthogonal functions supplemented by an additional singular term with predetermined order so that the stress-intensity factors can be computed. Results from the numerical computation indicate a trend of the stress-intensity factor decreasing with a decrease of the length-thickness ratio of the laminate and gradually approaching a stable value. It is just this trend of variation that is probably responsible for the appearance of an aggregation of transverse cracks prior to the interlaminar delamination observed in experiments.

Formulation of Problem

For simplicity, a finite $[0/90]_s$ symmetric cross-ply composite laminate with a transverse through-crack in either layer is studied in the paper. The laminate, with length $2a$, width $2b$, and thickness $2h$, is subjected to a uniaxial tension such that both ends of it undergo the same uniform displacement across the whole section. Let h_1,h_2 be the thickness of the 0-deg ply and 90-deg ply in the laminate, respectively. Assume the displacement at the end section ($x = a$) be $\epsilon_0 a$, where ϵ_0 is the average axial strain of the laminate in the longitudinal direction. Since the thickness-width ratio (h/b) is much smaller than unity, we may treat the problem as a pair of plane-strain problems of elasticity for orthotropic bodies bonded together. Due to symmetry, one quarter of the laminate is considered, as shown in Fig. 1. A Cartesean coordinate system $oxyz$ is chosen so that the oxz-plane coincides with the interface plane between 0-deg ply and 90-deg ply of the laminate, and the y-axis with the line of transverse crack front. Thus, the plane-strain problem to be solved for each lamina has the related boundary conditions as follows: For 0-deg ply or the first layer

$$x = 0: \; u^{(1)}(0,y) = 0, \quad \tau_{xy}^{(1)}(0,y) = 0 \tag{1a,b}$$

$$x = a: \; u^{(1)}(a,y) = \epsilon_0 a, \; \tau_{xy}^{(1)}(a,y) = 0 \tag{1c,d}$$

$$y = h_1: \; \sigma_y^{(1)}(x,h_1) = 0, \; \tau_{xy}^{(1)}(x,h_1) = 0 \tag{1e,f}$$

For 90-deg ply or the second layer,

$$x = 0: \quad \sigma_x^{(2)}(0,y) = 0, \quad \tau_{xy}^{(2)}(0,y) = 0 \tag{2a,b}$$

$$x = a: \quad u^{(2)}(a,y) = \epsilon_0 a, \; \tau_{xy}^{(2)}(a,y) = 0 \tag{2c,d}$$

$$y = -h_2: \; v^{(2)}(x,-h_2) = 0, \; \tau_{xy}^{(2)}(x,-h_2) = 0 \tag{2e,f}$$

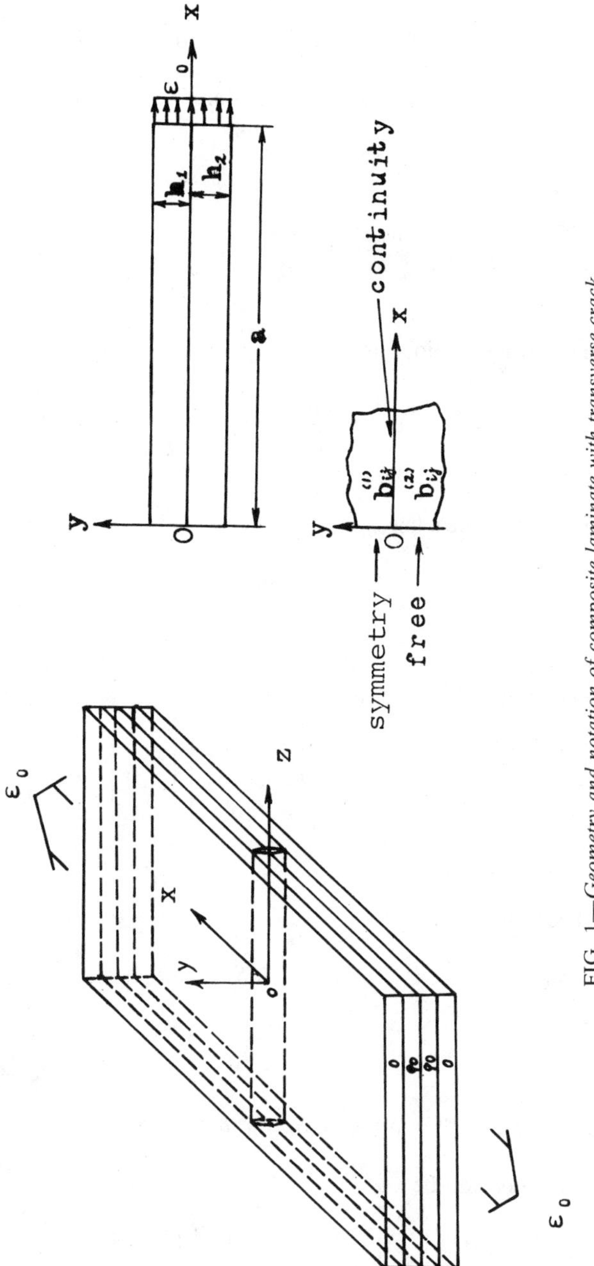

FIG. 1—*Geometry and notation of composite laminate with transverse crack.*

In addition, the continuity conditions at the interface are

$$y = 0: u^{(1)}(x,0) = u^{(2)}(x,0),\ v^{(1)}(x,0) = v^{(2)}(x,0) \tag{3a,b}$$

$$\sigma_y^{(1)}(x,0) = \sigma_y^{(2)}(x,0),\ \tau_{xy}^{(1)}(x,0) = \tau_{xy}^{(2)}(x,0) \tag{3c,d}$$

The equations of equilibrium for the plane problems are

$$\frac{\partial \sigma_x^{(i)}}{\partial x} + \frac{\partial \tau_{xy}^{(i)}}{\partial y} = 0$$
$$\frac{\partial \tau_{xy}^{(i)}}{\partial x} + \frac{\partial \sigma_y^{(i)}}{\partial y} = 0 \tag{4}$$
$$(i = 1,2)$$

and the displacement-strain relations are

$$\epsilon_x^{(i)} = \frac{\partial u^{(i)}}{\partial x},\ \epsilon_y^{(i)} = \frac{\partial v^{(i)}}{\partial y},\ \gamma_{xy}^{(i)} = \frac{\partial u^{(i)}}{\partial y} + \frac{\partial v^{(i)}}{\partial x} \tag{5}$$
$$(i = 1,2)$$

Assume the generalized Hooke's law

$$\epsilon_i = S_{ij}\sigma_j \tag{6}$$
$$(i,j = 1,2, \ldots, 6)$$

where S_{ij} represents the compliance coefficients of the composite. In case of plane strain, the stress-strain relation is reduced to the form as

$$\begin{bmatrix} \epsilon_x^{(i)} \\ \epsilon_y^{(i)} \\ \gamma_{xy}^{(i)} \end{bmatrix} = \begin{bmatrix} b_{11}^{(i)} & b_{12}^{(i)} & 0 \\ b_{12}^{(i)} & b_{22}^{(i)} & 0 \\ 0 & 0 & b_{66}^{(i)} \end{bmatrix} \begin{bmatrix} \sigma_x^{(i)} \\ \sigma_y^{(i)} \\ \tau_{xy}^{(i)} \end{bmatrix} \tag{7}$$
$$(i = 1,2)$$

where the reduced compliance tensor

$$b_{11}^{(1)} = S_{11} - S_{12}^2/S_{22},\ b_{12}^{(1)} = S_{13} - S_{12}S_{23}/S_{22},\ b_{22}^{(1)} = S_{33} - S_{23}^2/S_{22}$$

$$b_{66}^{(1)} = S_{66}$$

and

$$b_{11}^{(2)} = S_{33} - S_{13}^2/S_{11},\ b_{12}^{(2)} = S_{23} - S_{12}S_{13}/S_{11},\ b_{22}^{(2)} = S_{22} - S_{12}^2/S_{11},\ b_{66}^{(2)} = S_{44}$$

The inverse of this relation will be

$$\begin{bmatrix} \sigma_x^{(i)} \\ \sigma_y^{(i)} \\ \tau_{xy}^{(i)} \end{bmatrix} = \begin{bmatrix} c_{11}^{(i)} & c_{12}^{(i)} & 0 \\ c_{12}^{(i)} & c_{22}^{(i)} & 0 \\ 0 & 0 & c_{66}^{(i)} \end{bmatrix} \begin{bmatrix} \epsilon_x^{(i)} \\ \epsilon_y^{(i)} \\ \gamma_{xy}^{(i)} \end{bmatrix} \tag{9}$$
$$(i = 1,2)$$

Determination of Stress Singularity

In order for the strength of stress singularity of the transverse crack to be determined, the Williams method of eigenfunctions expansion is employed, as Wang and Choi did earlier [10]. Because of the anisotropic properties of the fiber-reinforced composite materials, the Lekhnitskii's complex stress function $F^{(i)}(x,y)$ is introduced such that

$$\sigma_x^{(i)} = \frac{\partial^2 F^{(i)}}{\partial y^2},\ \sigma_y^{(i)} = \frac{\partial^2 F^{(i)}}{\partial x^2},\ \tau_{xy}^{(i)} = \frac{\partial^2 F^{(i)}}{\partial x \partial y} \qquad (i = 1,2) \tag{10}$$

Then the equations of equilibrium are satisfied identically, and the governing equation for the stress function from the condition of compatibility will be

$$b_{22}^{(i)} \frac{\partial^4 F^{(i)}}{\partial x^4} + (2b_{12}^{(i)} + b_{66}^{(i)}) \frac{\partial^4 F^{(i)}}{\partial x^2 \partial y^2} + b_{11}^{(i)} \frac{\partial^4 F^{(i)}}{\partial y^4} = 0 \qquad (i = 1,2) \tag{11}$$

The solution of Eq 11 can be expressed in the functions of complex variable $z_k = x + \mu_k y$, where μ_k are the complex roots of the characteristic equation

$$b_{11}^{(i)}\mu_k^4 + (2b_{12}^{(i)} + b_{66}^{(i)})\mu_k^2 + b_{22}^{(i)} = 0 \qquad (i = 1,2) \tag{12}$$

In the present case, μ_k are the complex conjugates of pure imaginary roots. If for first layer $i = 1$ we take $k = 1 - 4$, for $i = 2$, $k = 5 - 8$, then we have $\mu_1 = \bar{\mu}_3 = is_1$, $\mu_2 = \bar{\mu}_4 = is_2$ and $\mu_5 = \bar{\mu}_7 = is_3$, $\mu_6 = \bar{\mu}_8 = is_4$ while s_i $(i = 1, \ldots, 4)$ are real numbers. Hence

$$F^{(i)}(x,y) = \sum_k F_k(z_k) \qquad (i = 1,2) \tag{13}$$

Accordingly, the stresses and displacements will be

$$\sigma_x^{(i)} = \sum_k \mu_k^2 F_k''(z_k),\ \sigma_y^{(i)} = \sum_k F_k''(z_k),\ \tau_{xy}^{(i)} = -\sum_k \mu_k F_k''(z_k) \tag{14a,b,c}$$

and

$$u^{(i)} = \sum_k p_k F_k'(z_k),\ v^{(i)} = \sum_k q_k F_k'(z_k) \tag{15a,b}$$

where

$$p_k = b_{11}^{(i)}\mu_k^2 + b_{12}^{(i)},\ q_k = b_{12}^{(i)}\mu_k + b_{22}^{(i)}/\mu_k \qquad (i = 1,2) \tag{16}$$

here the prime denotes the differentiation with respect to the argument inside the parenthesis.

Take the function

$$F_k(z_k) = C_k z_k^{\delta+2}/(\delta + 1)(\delta + 2) \tag{17}$$

where C_k and δ are the unknown constants. Analogous to Williams's method, the boundary conditions on $x = 0$ for both layers, Eq 1*a*,*b* and Eq 2*a*,*b*, together with the conditions of continuity at their interface ($y = 0$), Eqs 3*a* to 3*d* are imposed. Eight homogeneous complex algebraic equations are expected. However, in view of the fact that the stress functions $F^{(i)}(x,y)$ are real functions, the functions $F_k(z_k)$ should be in pairs of conjugated functions, and δ must be real. Thus

$$F^{(1)}(x,y) = 2 \text{ Re } [F_1(z_1) + F_2(z_2)]$$
$$F^{(2)}(x,y) = 2 \text{ Re } [F_5(z_5) + F_6(z_6)] \tag{18}$$

with

$$F_1(z_1) = \overline{F}_3(\overline{z}_3) = C_1 z_1^{\delta+2}/(\delta + 1)(\delta + 2) = (a_1 + ia_2)(x + is_1 y)^{\delta+2}/(\delta + 1)(\delta + 2)$$
$$F_2(z_2) = \overline{F}_4(\overline{z}_4) = C_2 z_2^{\delta+2}/(\delta + 1)(\delta + 2) = (a_3 + ia_4)(x + is_2 y)^{\delta+2}/(\delta + 1)(\delta + 2) \tag{19}$$

The expressions for $F_5(z_5)$ and $F_6(z_6)$ are similar except for $a_5. . .a_8$ replacing $a_1. . .a_4$; and s_3,s_4 replacing s_1,s_2. Consequently, these homogeneous algebraic equations should be

$$p_1 s_1^{\delta+1}[a_1 \cos(\delta + 1)\pi/2 - a_2 \sin(\delta + 1)\pi/2] + p_2 s_2^{\delta+1}[a_3 \cos(\delta + 1)\pi/2 - a_4 \sin(\delta + 1)\pi/2] = 0$$

$$s_1^{\delta+1}[a_1 \cos(\delta + 1)\pi/2 - a_2 \sin(\delta + 1)\pi/2] + s_2^{\delta+1}[a_3 \cos(\delta + 1)\pi/2 - a_4 \sin(\delta + 1)\pi/2] = 0$$

$$s_3^{\delta+2}(a_5 \cos \delta\pi/2 + a_6 \sin \delta\pi/2) + s_4^{\delta+2}(a_7 \cos \delta\pi/2 + a_8 \sin \delta\pi/2) = 0$$

$$s_3^{\delta+1}[a_5 \cos(\delta + 1)\pi/2 + a_6 \sin(\delta + 1)\pi/2] + s_4^{\delta+1}[a_7 \cos(\delta + 1)\pi/2 + a_8 \sin(\delta + 1)\pi/2] = 0 \tag{20}$$

$$p_1 a_1 + p_2 a_3 - p_5 a_5 - p_6 a_7 = 0$$

$$iq_1 a_2 + iq_2 a_4 - iq_5 a_6 - iq_6 a_8 = 0$$

$$a_1 + a_3 - a_5 - a_7 = 0$$

$$s_1 a_2 + s_2 a_4 - s_3 a_6 - s_4 a_8 = 0$$

where p_k,q_k are given in Eq 16. In order for the non-trival solution of a_k to be assured, the determinate of their coefficients in Eq 20 is set equal to zero. The eigenvalues of δ are thus obtained from its transcendental equation. And the condition of the finiteness of the strain energy in a laminate requires that the parameter δ, that is, the power of stress singularity at the crack tip, must be in a range

$$-1 < \delta < 0 \tag{21}$$

As illustration, the numerical examples for the power of crack singularity, that is, $-1 < \delta < 0$, are computed for the following composite materials:

(1) $E_{22} = E_{33} = 14.5$ GPa, $G_{12} = G_{23} = 5.9$ GPa, $\nu_{12} = \nu_{23} = \nu_{13} = 0.21$

(2) $E_{22} = E_{33} = 14.5$ GPa, $G_{12} = 6.4$ GPa, $G_{23} = 3.2$ GPa, $\nu_{12} = \nu_{23} = \nu_{13} = 0.21$

(3) $E_{22} = E_{33} = 14.5$ GPa, $G_{12} = 6.4$ GPa, $G_{23} = 3.2$ GPa, $\nu_{12} = \nu_{13} = 0.21$, $\nu_{23} = 0.42$

The results are plotted against the ratio of E_{11}/E_{22} in graphs, as shown in Fig. 2. It is interesting to note that when the crack appears in the 90-deg ply, the order of singularity is always less than 0.5 and is decreasing with an increase of the ratio E_{11}/E_{22} and vice versa. The result seems to provide a possible explanation that the crack may be arrested at the interface of the layer with higher rigidity and the local delamination may be initiated instead due to the singular normal and shearing stresses along the interface. On the contrary, an unstable growth of the crack along its line would appear if there is a crack in the 0-deg ply. Note also that the power of singularity will become ½ again as $b_{ij}^{(1)} = b_{ij}^{(2)}$, and that the results agree nicely with those from Williams and Zak when both layers are degenerated respectively to isotropic materials [*11*].

Method of Solution

To solve the problem analytically, the method of orthogonal series expansion [*12*] is used. First let the dimensionless parameters

$$R = C_{11}^{(1)}/C_{22}^{(1)},\ G = C_{66}^{(1)}/C_{22}^{(1)},\ K = C_{12}^{(1)}/C_{22}^{(1)},\ S = C_{12}^{(2)}/C_{22}^{(1)},\ Q = C_{66}^{(2)}/C_{22}^{(1)} \tag{22}$$

Then by substituting Eqs 9 and 5 into Eq 4 and using Eq 22, we obtain the equations of equilibrium in terms of displacements for each layer as:

For 0-deg ply layer

$$\begin{aligned} &\left(\frac{\partial^2}{\partial x^2} + \frac{G}{R}\frac{\partial^2}{\partial y^2}\right) u^{(1)} + \frac{K+G}{R}\frac{\partial^2 v^{(1)}}{\partial x \partial y} = 0 \\ &\frac{G+K}{G}\frac{\partial^2 u^{(1)}}{\partial x \partial y} + \left(\frac{\partial^2}{\partial x^2} + \frac{1}{G}\frac{\partial^2}{\partial y^2}\right) v^{(1)} = 0 \end{aligned} \tag{23}$$

For 90-deg ply layer

$$\begin{aligned} &\left(\frac{\partial^2}{\partial x^2} + Q\frac{\partial^2}{\partial y^2}\right) u^{(2)} + (Q+S)\frac{\partial^2 v^{(2)}}{\partial x \partial y} = 0 \\ &\frac{Q+S}{Q}\frac{\partial^2 u^{(2)}}{\partial x \partial y} + \left(\frac{\partial^2}{\partial x^2} + \frac{1}{Q}\frac{\partial^2}{\partial y^2}\right) v^{(2)} = 0 \end{aligned} \tag{24}$$

If we introduce the strain functions $\Phi^{(i)}(x,y)$ $(i = 1,2)$ such that

$$u^{(1)} = -\frac{G+K}{R}\frac{\partial^2}{\partial x \partial y}\Phi^{(1)}(x,y),\quad v^{(1)} = \left(\frac{\partial^2}{\partial x^2} + \frac{G\partial^2}{R\partial y^2}\right)\Phi^{(1)}(x,y) \tag{25}$$

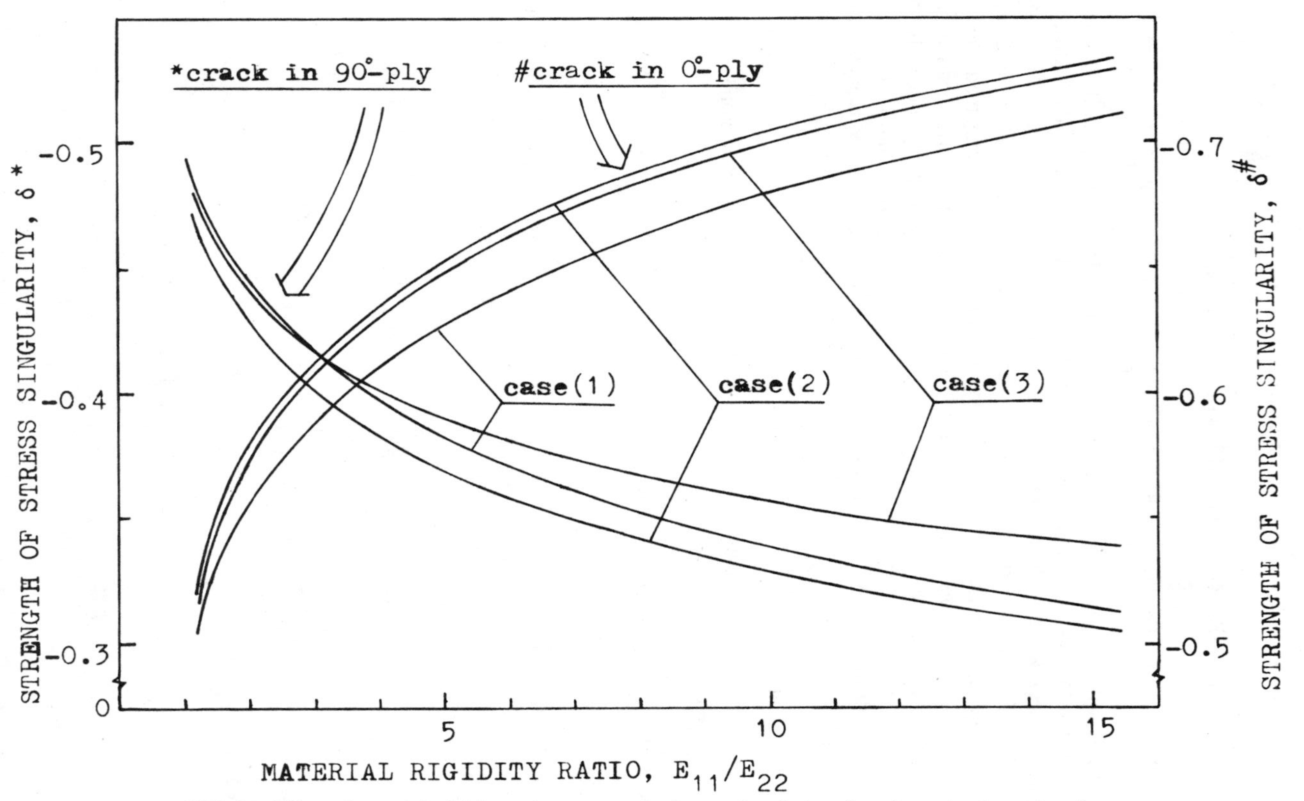

FIG. 2—*Effect of materials rigidity ratio on strength of stress singularity of crack terminating at interface.*

and

$$u^{(2)} = -(Q + S)\frac{\partial^2}{\partial x \partial y}\Phi^{(2)}(x,y),\ v^{(2)} = \left(\frac{\partial^2}{\partial x^2} + Q\frac{\partial^2}{\partial y^2}\right)\Phi^{(2)}(x,y) \quad (26)$$

the equations of equilibrium will be satisfied completely provided that $\Phi^{(i)}(x,y)$ fulfill the following differential equations, respectively

$$\left(\frac{\partial^2}{\partial x^2} + \lambda_1\frac{\partial^2}{\partial y^2}\right)\left(\frac{\partial^2}{\partial x^2} + \lambda_2\frac{\partial^2}{\partial y^2}\right)\Phi^{(1)}(x,y) = 0 \quad (27)$$

and

$$\left(\frac{\partial^2}{\partial x^2} + \lambda_3\frac{\partial^2}{\partial y^2}\right)\left(\frac{\partial^2}{\partial x^2} + \lambda_4\frac{\partial^2}{\partial y^2}\right)\Phi^{(2)}(x,y) = 0 \quad (28)$$

with

$$\lambda_1 + \lambda_2 = \frac{1}{G} - \frac{K}{R}\frac{2G + K}{G};\ \lambda_1\lambda_2 = \frac{1}{R}$$
$$\lambda_3 + \lambda_4 = \frac{1}{Q} - \frac{S(2Q + S)}{Q};\ \lambda_3\lambda_4 = 1 \quad (29)$$

By the method of separation of variables, the solution of above differential equations may be expressed in the forms as

$$\Phi^{(1)}(x,y) = \sum_{n=0}^{\infty}\cos\alpha_n x[A_{1n}Sh\alpha_n y/\sqrt{\lambda_1} + A_{2n}Ch\alpha_n y/\sqrt{\lambda_1} + A_{3n}Sh\alpha_n y/\sqrt{\lambda_2} + A_{4n}$$
$$\times Ch\alpha_n y/\sqrt{\lambda_2}] + \sum_{m=0}^{\infty}[Ch\sqrt{\lambda_1}\omega_m x(B_{1m}\sin\omega_m y + B_{2m}\cos\omega_m y) + Ch\sqrt{\lambda_2}\omega_m x$$
$$\times (B_{3m}\sin\omega_m y + B_{4m}\cos\omega_m y)] - \frac{\epsilon_0}{2(G + K)/R}x^2 y \quad (30)$$

and

$$\Phi^{(2)}(x,y) = \sum_{n=0}^{\infty}\cos\alpha_n x[D_{1n}Sh\alpha_n y/\sqrt{\lambda_3} + D_{2n}Ch\alpha_n y/\sqrt{\lambda_3} + D_{3n}Sh\alpha_n y/\sqrt{\lambda_4} + D_{4n}$$
$$\times Ch\alpha_n y/\sqrt{\lambda_4}] + \sum_{m=0}^{\infty}\cos\beta_m y[L_{1m}Sh\sqrt{\lambda_3}\beta_m x + L_{2m}Ch\sqrt{\lambda_3}\beta_m x + L_{3m}Sh\sqrt{\lambda_4}\beta_m x + L_{4m}$$
$$\times Ch\sqrt{\lambda_4}\beta_m x] - \frac{\epsilon_0}{2(Q + S)}x^2 y \quad (31)$$

where α_n, β_m, ω_m, and A_{in}, B_{im}, D_{in}, L_{im} are arbitrary constants.

With the consideration of the prescribed boundary conditions, we choose

$$\alpha_n = n\pi/a;\ \omega_m = m\pi/h_1;\ \beta_m = m\pi/h_2 \quad (32)$$

The corresponding stress and displacement fields can thus be obtained from these strain functions by using Eqs 25, 26 and Eqs 5, 6, respectively. They are

$$u^{(1)}(x,y) = \frac{G+K}{R}\left\{\sum_{n=1}^{\infty} \alpha_n^2 \sin \alpha_n x[A_{1n}/\sqrt{\lambda_1}Ch\alpha_n y/\sqrt{\lambda_1} + A_{2n}/\sqrt{\lambda_1}Sh\alpha_n y/\sqrt{\lambda_1}\right.$$

$$+ A_{3n}/\sqrt{\lambda_2}Ch\alpha_n y/\sqrt{\lambda_2} + A_{4n}/\sqrt{\lambda_2}Sh\alpha_n y/\sqrt{\lambda_2}] - \sum_{m=1}^{\infty} \omega_m^2[\sqrt{\lambda_1}Sh\sqrt{\lambda_1}\omega_m x(B_{1m} \cos \omega_m y$$

$$\left. - B_{2m} \sin \omega_m y) + \sqrt{\lambda_2}Sh\sqrt{\lambda_2}\omega_m x(B_{3m} \cos \omega_m y - B_{4m} \sin \omega_m y)]\right\} + \epsilon_0 x \quad (33a)$$

$$v^{(1)}(x,y) = \sum_{n=1}^{\infty} \alpha_n^2 \cos \alpha_n x[(\lambda_2 G - 1)(A_{1n}Sh\alpha_n y/\sqrt{\lambda_1} + A_{2n}Ch\alpha_n y/\sqrt{\lambda_1})$$

$$+ (\lambda_1 G - 1)(A_{3n}Sh\alpha_n y/\sqrt{\lambda_2} + A_{4n}Ch\alpha_n y/\sqrt{\lambda_2})] + \sum_{m=1}^{\infty} \omega_m^2[\lambda_1(1 - \lambda_2 G)Ch\sqrt{\lambda_1}$$

$$\times \omega_m x(B_{1m} \sin \omega_m y + B_{2m} \cos \omega_m y) + \lambda_2(1 - \lambda_1 G)Ch\sqrt{\lambda_2}\omega_m x(B_{3m} \sin \omega_m y$$

$$+ B_{4m} \cos \omega_m y)] - \epsilon_0 R/(G + K)y \quad (33b)$$

$$\sigma_x^{(1)}(x,y) = c_{66}^{(1)}\left\{\sum_{n=1}^{\infty} \alpha_n^3 \cos \alpha_n x[(1 + \lambda_2 K)/\sqrt{\lambda_1}(A_{1n}Ch\alpha_n y/\sqrt{\lambda_1} + A_{2n}Sh\alpha_n y/\sqrt{\lambda_1})\right.$$

$$+ (1 + \lambda_1 K)/\sqrt{\lambda_2}(A_{3n}Ch\alpha_n y/\sqrt{\lambda_2} + A_{4n}Sh\alpha_n y/\sqrt{\lambda_2})] - \sum_{m=1}^{\infty} \omega_m^3[(1 + \lambda_2 K)\lambda_1$$

$$\times Ch\sqrt{\lambda_1}\omega_m x(B_{1m} \cos \omega_m y - B_{2m} \sin \omega_m y) + (1 + \lambda_1 K)\lambda_2 Ch\sqrt{\lambda_2}\omega_m x(B_{3m} \cos \omega_m y$$

$$\left. - B_{4m} \sin \omega_m y)] + \epsilon_0 R/(G + K)\right\} \quad (33c)$$

$$\sigma_y^{(1)}(x,y) = c_{22}^{(1)}\left\{\sum_{n=1}^{\infty} a_n^3 \cos \alpha_n x[((G + K)K/R + \lambda_2 G - 1)/\sqrt{\lambda_1}(A_{1n}Ch\alpha_n y/\sqrt{\lambda_1}\right.$$

$$+ A_{2n}Sh\alpha_n y/\sqrt{\lambda_1}) + ((G + K)K/R + \lambda_1 G - 1)/\sqrt{\lambda_2}(A_{3n}Ch\alpha_n y/\sqrt{\lambda_2}$$

$$+ A_{4n}Sh\alpha_n y/\sqrt{\lambda_2})] + \sum_{m=1}^{\infty} \omega_m^3[\lambda_1(1 - \lambda_2 G - (G + K)K/R)Ch\sqrt{\lambda_1}\omega_m x$$

$$\times (B_{1m} \cos \omega_m y - B_{2m} \sin \omega_m y) + \lambda_2(1 - \lambda_1 G - (G + K)K/R)Ch\sqrt{\lambda_2}\omega_m x$$

$$\left. \times (B_{3m} \cos \omega_m y - B_{4m} \sin \omega_m y)] \; \epsilon_0(K - R/(G + K))\right\} \quad (33d)$$

$$\tau_{xy}^{(1)}(x,y) = c_{66}^{(1)} \Bigg\{ \sum_{n=1}^{\infty} \alpha_n^3 \sin \alpha_n x[(1 + \lambda_2 K)(A_{1n} Sh\alpha_n y/\sqrt{\lambda_1} + A_{2n} Ch\alpha_n y/\sqrt{\lambda_1})$$

$$+ (1 + \lambda_1 K)(A_{3n} Sh\alpha_n y/\sqrt{\lambda_2} + A_{4n} Ch\alpha_n y/\sqrt{\lambda_2})] + \sum_{m=1}^{\infty} \omega_m^3 [\sqrt{\lambda_1}(\lambda_1 + K/R) Sh\sqrt{\lambda_1}\omega_m x$$

$$\times (B_{1m} \sin \omega_m y + B_{2m} \cos \omega_m y) + \sqrt{\lambda_2}(\lambda_2 + K/R) Sh\sqrt{\lambda_2}\omega_m x(B_{3m} \sin \omega_m y$$

$$+ B_{4m} \cos \omega_m y)] \Bigg\} \quad (33e)$$

and

$$u^{(2)}(x,y) = (Q + S) \Bigg\{ \sum_{n=1}^{\infty} \alpha_n^2 \sin \alpha_n x[D_{1n}/\sqrt{\lambda_3} Ch\alpha_n y/\sqrt{\lambda_3} + D_{2n}/\sqrt{\lambda_3} Sh\alpha_n y/\sqrt{\lambda_3}$$

$$+ D_{3n}/\sqrt{\lambda_4} Ch\alpha_n y/\sqrt{\lambda_4} + D_{4n}/\sqrt{\lambda_4} Sh\alpha_n y/\sqrt{\lambda_4}] + \sum_{m=1}^{\infty} \beta_m^2 \sin \beta_m y[L_{1m}\sqrt{\lambda_3} Ch\beta_m\sqrt{\lambda_3}x$$

$$+ L_{2m}\sqrt{\lambda_3} Sh\beta_m\sqrt{\lambda_3}x + L_{3m}\sqrt{\lambda_4} Ch\beta_m\sqrt{\lambda_4}x + L_{4m}\sqrt{\lambda_4} Sh\beta_m\sqrt{\lambda_4}x] \Bigg\} + \epsilon_0 x \quad (34a)$$

$$v^{(2)}(x,y) = \sum_{n=1}^{\infty} \alpha_n^2 \cos \alpha_n x[(\lambda_4 Q - 1)(D_{1n} Sh\alpha_n y/\sqrt{\lambda_3} + D_{2n} Ch\alpha_n y/\sqrt{\lambda_3})$$

$$+ (\lambda_3 Q - 1)(D_{3n} Sh\alpha_n y/\sqrt{\lambda_4} + D_{4n} Ch\alpha_n y/\sqrt{\lambda_4})] + \sum_{m=1}^{\infty} \beta_m^2 \cos \beta_m y[(\lambda_3 - Q)$$

$$\times (L_{1m} Sh\beta_m\sqrt{\lambda_3}x + L_{2m} Ch\beta_m\sqrt{\lambda_3}x) + (\lambda_4 - Q)(L_{3m} Sh\beta_m\sqrt{\lambda_4}x$$

$$+ L_{4m} Ch\beta_m\sqrt{\lambda_4}x)] - \epsilon_0/(Q + S)y \quad (34b)$$

$$\sigma_x^{(2)}(x,y) = c_{66}^{(2)} \Bigg\{ \sum_{n=1}^{\infty} \alpha_n^3 \cos \alpha_n x[(1 + \lambda_4 S)/\sqrt{\lambda_3}(D_{1n} Ch\alpha_n y/\sqrt{\lambda_3}$$

$$+ D_{2n} Sh\alpha_n y/\sqrt{\lambda_3}) + (1 + \lambda_3 S)/\sqrt{\lambda_4}(D_{3n} Ch\alpha_n y/\sqrt{\lambda_4} + D_{4n} Sh\alpha_n y/\sqrt{\lambda_4})]$$

$$+ \sum_{m=1}^{\infty} \beta_m^3 \sin \beta_m y[(\lambda_3 + S)(L_{1m} Sh\beta_m\sqrt{\lambda_3}x + L_{2m} Ch\beta_m\sqrt{\lambda_3}x) + (\lambda_4 + S)$$

$$\times (L_{3m} Sh\beta_m\sqrt{\lambda_4}x + L_{4m} Ch\beta_m\sqrt{\lambda_4}x)] + \epsilon_0/(Q + S) \Bigg\} \quad (34c)$$

$$\sigma_y^{(2)}(x,y) = c_{22}^{(2)} \left\{ \sum_{n=1}^{\infty} \alpha_n^3 \cos \alpha_n x[((Q + S)S + Q\lambda_4 - 1)/\sqrt{\lambda_3}(D_{1n} Ch\alpha_n y/\sqrt{\lambda_3} \right.$$

$$+ D_{2n} Sh\alpha_n y/\sqrt{\lambda_3}) + ((Q + S)S + Q\lambda_3 - 1)/\sqrt{\lambda_4}(D_{3n} Ch\alpha_n y/\sqrt{\lambda_4} + D_{4n} Sh\alpha_n y/\sqrt{\lambda_4})$$

$$+ \sum_{m=1}^{\infty} \beta_m^3 \sin \beta_m y[((Q + S)S\lambda_3 + Q - \lambda_3)(L_{1m} Sh\beta_m \sqrt{\lambda_3} x + L_{2m} Ch\beta_m \sqrt{\lambda_3} x)$$

$$+ ((Q + S)S\lambda_4 + Q - \lambda_4)(L_{3m} Sh\beta_m \sqrt{\lambda_4} x + L_{4m} Ch\beta_m \sqrt{\lambda_4} x)]$$

$$\left. + \epsilon_0(S - 1/(Q + S)) \right\} \quad (34d)$$

$$\tau_{xy}^{(2)}(x,y) = c_{66}^{(2)} \left\{ \sum_{n=1}^{\infty} \alpha_n^3 \sin \alpha_n x[(1 + \lambda_4 S)(D_{1n} Sh\alpha_n y/\sqrt{\lambda_3} + D_{2n} Ch\alpha_n y/\sqrt{\lambda_3}) \right.$$

$$+ (1 + \lambda_3 S)(D_{3n} Sh\alpha_n y/\sqrt{\lambda_4} + D_{4n} Ch\alpha_n y/\sqrt{\lambda_4})] + \sum_{m=1}^{\infty} \beta_m^3 \cos \beta_m y[\sqrt{\lambda_3}(\lambda_3 + S)$$

$$\times (L_{1m} Ch\beta_m \sqrt{\lambda_3} x + L_{2m} Sh\beta_m \sqrt{\lambda_3} x) + \sqrt{\lambda_4}(\lambda_4 + S)(L_{3m} Ch\beta_m \sqrt{\lambda_4} x$$

$$\left. + L_{4m} Sh\beta_m \sqrt{\lambda_4} x)] \right\} \quad (34e)$$

The arbitrary constants involved in the expressions may be determined from the related boundary conditions Eqs 1*a* to 1*f* and Eqs 2*a* to 2*f* and continuity conditions Eqs 3*a* to 3*d*. Consequently, the solution obtained is exact even though there is a singular point at the crack tip. Nevertheless, in order to ensure the appropriate power of stress singularity of the crack and to make the calculation of the stress-intensity factors possible, an additional singular term of predetermined order is therefore supplemented to the series solution. This singular term is derived from the stress functions $F^{(1)}(x,y)$ and $F^{(2)}(x,y)$ in the preceding section with a value of δ within $[-1,0]$. It has been shown that only one of the eight constants involved in $F^{(i)}(x,y)$, say a_1, remains to be arbitrary when a part of prescribed conditions is imposed, as formulated in Eq 20. With stress functions Eqs 18 and 19, the corresponding expressions for stresses and displacements can be easily found through Eqs 14, 15, and 16. Combined with this singular term, this solution is taken to satisfying all prescribed conditions from Eq 1 to Eq 3.

Due to the orthogonality property of the functions in the series, the equations derived from the various conditions are to be normalized respectively so that the undermined coefficient in each term of a series can be expressed in terms of the coefficients of the other series. The involved arbitrary constants may thus be determined accordingly.

With the choice of the parameters α_n, β_m, ω_m in Eq 32, the boundary conditions of Eqs 1*a* and 1*b* are satisfied identically. Moreover, from the condition Eq 2*b* we get the relation

$$L_{3m} = -\frac{\sqrt{\lambda_3}(\lambda_3 + S)}{\sqrt{\lambda_4}(\lambda_4 + S)} L_{1m} \quad (35)$$

$$(m = 1,2, \ldots M, \ldots)$$

According to conditions Eq 1c and 1d, the equations thus obtained are to be normalized by multiplying on both sides of the equations with Cos $\omega_m y$ and Sin $\omega_m y$, respectively, and integrating with respect to y in the internal $[0,h_1]$, yielding

$$a_1\, ISI(u^{(1)})_c - (G + K)/R(\sqrt{\lambda_1}B_{1m}^* + \sqrt{\lambda_2}B_{3m}^*)h_1/2 = 0 \quad (m = 1,2,\ldots,M,\ldots) \tag{36}$$

$$a_1\, ISI(u^{(1)})_s + (G + K)/R(\sqrt{\lambda_1}B_{2m}^* + \sqrt{\lambda_2}B_{4m}^*)h_1/2 = 0 \quad (m = 1,2,\ldots M,\ldots) \tag{37}$$

$$a_1\, ISI(\tau_{xy}^{(1)})_c + c_{66}^{(1)}\omega_m[\sqrt{\lambda_1}(\lambda_1 + K/R)B_{2m}^* + \sqrt{\lambda_2}(\lambda_2 + K/R)B_{4m}^*]h_1/2 = 0 \quad (m = 1,2,\ldots M,\ldots) \tag{38}$$

$$a_1\, ISI(\tau_{xy}^{(1)})_s + c_{66}^{(1)}\omega_m[\sqrt{\lambda_1}(\lambda_1 + K/R)B_{1m}^* + \sqrt{\lambda_2}(\lambda_2 + K/R)B_{3m}^*]h_1/2 = 0 \quad (m = 1,2,\ldots M,\ldots) \tag{39}$$

Here we denote $SI(\ )$ as the expression inside the parenthesis derived from the singular stress functions and $ISI(\ )_s$ or $ISI(\ )_c$ be the normalized integration with respect to sine or cosine function. Similarly, the equations obtained from the conditions Eqs 1f, 2f, 3a, and 3d are multiplied by Sin $\alpha_n x$, and then integrated with respect to x in the interval $[0,a]$, the following equations are obtained

$$\begin{aligned} &a_1\, ISI(\tau_{xy}^{(1)})_s + c_{66}^{(1)}a/2\alpha_n[(1 + \lambda_2 K)(A_{1n}^* + A_{2n}^*) + (1 + \lambda_1 K)(A_{3n}^* + A_{4n}^*)] \\ &+ c_{66}^{(1)}\sum_{m=1}^{\infty}\omega_m[\sqrt{\lambda_1}(\lambda_1 + K/R)(-1)^{1+n+m}\alpha_n B_{2m}^*/(\alpha_n^2 + \lambda_1\omega_m^2) + \sqrt{\lambda_2}(\lambda_2 + K/R) \\ &\times (-1)^{1+n+m}\alpha_n B_{4m}^*/(\alpha_n^2 + \lambda_2\omega_m^2)] = 0 \quad (n = 1,2,\ldots N,\ldots) \end{aligned} \tag{40}$$

$$\begin{aligned} &a_1\, ISI(\tau_{xy}^{(2)})_s + c_{66}^{(2)}a/2\alpha_n[(1 + \lambda_4 S)(-D_{1n}^* + D_{2n}^*) + (1 + \lambda_3 S)(-D_{3n}^* + D_{4n}^*)] \\ &+ c_{66}^{(2)}\sum_{m=1}^{\infty}\beta_m(-1)^m\{\sqrt{\lambda_3}(\lambda_3 + S)\alpha_n/(\alpha_n^2 + \beta_m^2\lambda_3)[L_{1m}^*(1/Ch\beta_m\sqrt{\lambda_3}a - (-1)^n) \\ &+ L_{2m}^*(-1)^{n+1}] + \sqrt{\lambda_4}(\lambda_4 + S)\alpha_n/(\alpha_n^2 + \beta_m^2\lambda_4)[L_{3m}^*(1/Ch\beta_m\sqrt{\lambda_4}a - (-1)^n) \\ &+ L_{4m}^*(-1)^{n+1}]\} = 0 \quad (n = 1,2,\ldots N,\ldots) \end{aligned} \tag{41}$$

$$\begin{aligned} &(G + K)/Ra/2[A_{1n}^*/(\sqrt{\lambda_1}Sh\alpha_n h_1/\sqrt{\lambda_1}) + A_{3n}^*/(\sqrt{\lambda_2}Sh\alpha_n h_1/\sqrt{\lambda_2})] - (Q + S)a/2 \\ &\times [D_{1n}^*/(\sqrt{\lambda_3}Sh\alpha_n h_2/\sqrt{\lambda_3}) + D_{2n}^*/(\sqrt{\lambda_4}Sh\alpha_n h_2/\sqrt{\lambda_4})] - (G + K)/R\sum_{m=1}^{\infty}(-1)^{n+1} \\ &\times [\sqrt{\lambda_1}B_{1m}^*\alpha_n/(\alpha_n^2 + \omega_m^2\lambda_1) + \sqrt{\lambda_2}B_{3m}^*\alpha_n/(\alpha_n^2 + \omega_m^2\lambda_2)] = 0 \quad (n = 1,2,\ldots N,\ldots) \end{aligned} \tag{42}$$

$$c_{66}^{(1)}\alpha_n a/2[(1 + \lambda_2 K)A_{2n}^*/Ch\alpha_n h_1/\sqrt{\lambda_1} + (1 + \lambda_1 K)A_{4n}^*/Ch\alpha_n h_1/\sqrt{\lambda_2}]$$

$$+ c_{66}^{(1)} \sum_{m=1}^{\infty} \omega_m[\sqrt{\lambda_1}(\lambda_1 + K/R)\alpha_n(-1)^{n+1}B_{2m}^*/(\alpha_n^2 + \lambda_1\omega_m^2) + \sqrt{\lambda_2}(\lambda_2 + K/R)\alpha_n(-1)^{n+1}$$

$$\times B_{4m}^*/(\alpha_n^2 + \lambda_2\omega_m^2)] - c_{66}^{(2)}\alpha_n a/2[(1 + \lambda_4 S)D_{2n}^*/Ch\alpha_n h_2/\sqrt{\lambda_3} + (1 + \lambda_3 S)D_{4n}^*$$

$$\div Ch\alpha_n h_2/\sqrt{\lambda_4}] - c_{66}^{(2)} \sum_{m=1}^{\infty} \beta_m[\sqrt{\lambda_3}(\lambda_3 + S)\alpha_n/(\alpha_n^2 + \lambda_3\beta_m^2)(L_{1m}^*(1/Ch(\sqrt{\lambda_3}\beta_m a)$$

$$- (-1)^n) + (-1)^{n+1}L_{2m}^*) + \sqrt{\lambda_4}(\lambda_4 + S)\alpha_n/(\alpha_n^2 + \lambda_4\beta_m^2)(L_{3m}^*(1/Ch(\sqrt{\lambda_4}\beta_m a)$$

$$- (-1)^n) + (-1)^{n+1}L_{4m}^*)] = 0 \quad (n = 1,2, \ldots N, \ldots) \quad (43)$$

And multiplying Cos $\alpha_n x$ to the equations from the conditions Eqs 1e, 2e, 3b, 3c and integrating with respect to x within $[0,a]$, we have

$$a_1\, ISI(\sigma_y^{(1)})_c + c_{22}^{(1)}\alpha_n a/2[((G + K)K/R + \lambda_2 G - 1)/\sqrt{\lambda_1}(A_{1n}^* cth\alpha_n h_1/\sqrt{\lambda_1}$$

$$+ A_{2n}^* th\alpha_n h_1/\sqrt{\lambda_1}) + ((G + K)K/R + \lambda_1 G - 1)/\sqrt{\lambda_2}(A_{3n}^* cth\alpha_n h_1/\sqrt{\lambda_2}$$

$$+ A_{4n}^* th\alpha_n h_1/\sqrt{\lambda_2})] + c_{22}^{(1)} \sum_{m=1}^{\infty} \omega_m[\lambda_1(1 - \lambda_2 G - K(G + K)/R)(-1)^{m+n}\sqrt{\lambda_1}\omega_m B_{1m}^*$$

$$\div (\alpha_n^2 + \lambda_1\omega_m^2) + \lambda_2(1 - \lambda_1 G - K(G + K)/R)(-1)^{m+n}\sqrt{\lambda_2}\omega_m B_{3m}^*/(\alpha_n^2$$

$$+ \lambda_2\omega_m^2)] = 0 \quad (n = 1,2, \ldots N, \ldots) \quad (44)$$

$$a_1\, ISI(v^{(2)})_c + a/2[(\lambda_4 Q - 1)(-D_{1n}^* + D_{2n}^*) + (\lambda_3 Q - 1)(-D_{3n}^* + D_{4n}^*)]$$

$$+ \sum_{m=1}^{\infty} (-1)^m\{(\lambda_3 - Q)[L_{1m}^*((-1)^n - 1/Ch(\beta_m\sqrt{\lambda_3}a)) + L_{2m}^*(-1)^n]\beta_m\sqrt{\lambda_3}$$

$$\div (\alpha_n^2 + \beta_m^2\lambda_3) + (\lambda_4 - Q)[L_{3m}^*((-1)^n - 1/Ch(\beta_m\sqrt{\lambda_4}a)) + L_{4m}^*(-1)^n]\beta_m\sqrt{\lambda_4}$$

$$\div (\alpha_n^2 + \beta_m^2\lambda_4)\} = 0 \quad (n = 1,2, \ldots N, \ldots) \quad (45)$$

$$a/2[(\lambda_2 G - 1)A_{2n}^*/Ch\alpha_n h_1/\sqrt{\lambda_1} + (\lambda_1 G - 1)A_{4n}^*/Ch\alpha_n h_1/\sqrt{\lambda_2}] - a/2[(\lambda_4 Q - 1)D_{2n}^*$$

$$\div Ch\alpha_n h_2/\sqrt{\lambda_3} + (\lambda_3 Q - 1)D_{4n}^*/Ch\alpha_n h_2/\sqrt{\lambda_4}] + \sum_{m=1}^{\infty} [\lambda_1(1 - \lambda_2 G)(-1)^n\sqrt{\lambda_1}\omega_m B_{2m}^*$$

$$\div (\alpha_n^2 + \lambda_1\omega_m^2) + \lambda_2(1 - \lambda_1 G)(-1)^n\sqrt{\lambda_2}\omega_m B_{4m}^*/(\alpha_n^2 + \lambda_2\omega_m^2)] - \sum_{m=1}^{\infty} \{(\lambda_3 - Q)[L_{1m}^*$$

$$\times ((-1)^n - 1/Ch(\beta_m\sqrt{\lambda_3}a)) + L_{2m}^*(-1)^n]\beta_m\sqrt{\lambda_3}/(\alpha_n^2 + \beta_m^2\lambda_3) + (\lambda_4 - Q)$$

$$\times [L_{3m}^*((-1)^n - 1/Ch(\beta_m\sqrt{\lambda_4}a)) + L_{4m}^*(-1)^n]\beta_m\sqrt{\lambda_4}/(\alpha_n^2 + \beta_m^2\lambda_4)\} = 0$$

$$(n = 1,2, \ldots N, \ldots) \quad (46)$$

$$\alpha_n a/2[((G + K)K/R + \lambda_2 G - 1)/\sqrt{\lambda_1}A_{1n}^*/Sh\alpha_n h_1/\sqrt{\lambda_1} + ((G + K)K/R + \lambda_1 G - 1)$$

$$\div \sqrt{\lambda_2}A_{3n}^*/Sh\alpha_n h_1/\sqrt{\lambda_2}] - \alpha_n a/2[((Q + S)S + Q\lambda_4 - 1)D_{1n}^*/\sqrt{\lambda_3}/Sh\alpha_n h_2/\sqrt{\lambda_3}$$

$$+ ((Q + S)S + Q\lambda_3 - 1)D_{3n}^*/\sqrt{\lambda_4}/Sh\alpha_n h_2/\sqrt{\lambda_4}] + \sum_{m=1}^{\infty} \omega_m[\lambda_1(1 - \lambda_2 G - K(G$$

$$+ K)/R)(-1)^n\sqrt{\lambda_1}\omega_m B_{1m}^*/(\alpha_n^2 + \lambda_1\omega_m^2) + \lambda_2(1 - \lambda_1 G - K(G + K)/R)(-1)^n\sqrt{\lambda_2}$$

$$\times \omega_m B_{3m}^*/(\alpha_n^2 + \lambda_2\omega_m^2)] = 0 \quad (n = 1,2, \ldots N, \ldots) \quad (47)$$

Finally, multiplying Cos $\beta_m y$ to the equation from Eq 2*d*, Sin $\beta_m y$ to equations from Eqs 2*a* and 2*c* and integrating with respect to y in $[-h_2,0]$, yields

$$a_1\, ISI(\tau_{xy}^{(2)})_c + c_{66}^{(2)}\beta_m[\sqrt{\lambda_3}(\lambda_3 + S)(L_{1m}^* + L_{2m}^*) + \sqrt{\lambda_4}(\lambda_4 + S)$$

$$(L_{3m}^* + L_{4m}^*)]h_2/2 = 0 \quad (m = 1,2, \ldots M, \ldots) \quad (48)$$

$$\sum_{n=1}^{\infty} \alpha_n\{(1 + \lambda_4 S)[D_{1n}^*((-1)^m cth\alpha_n h_2/\sqrt{\lambda_3} - 1/Sh\alpha_n h_2/\sqrt{\lambda_3}) - D_{2n}^*$$

$$\times (-1)^m th\alpha_n h_2/\sqrt{\lambda_3}]\beta_m\sqrt{\lambda_3}/(\alpha_n^2 + \lambda_3\beta_m^2) + (1 + \lambda_3 S)[D_{3n}^*((-1)^m cth\alpha_n h_2/\sqrt{\lambda_4} - 1$$

$$\div Sh\alpha_n h_2/\sqrt{\lambda_4}) - D_{4n}^*(-1)^m th\alpha_n h_2/\sqrt{\lambda_4}]\beta_m\sqrt{\lambda_4}/(\alpha_n^2 + \lambda_4\beta_m^2)\} + \beta_m[(\lambda_3 + S)L_{2m}^*$$

$$\div Sh(\sqrt{\lambda_3}\beta_m a) + (\lambda_4 + S)L_{4m}^*/Sh(\sqrt{\lambda_4}\beta_m a)]h_2/2 = \epsilon_0(1 - (-1)^m)/\beta_m/(Q + S)$$

$$(m = 1,2, \ldots M, \ldots) \quad (49)$$

$$a_1\, ISI(u^{(2)})_s + (Q + S)[L_{1m}^*\sqrt{\lambda_3} + L_{2m}^*\sqrt{\lambda_3} + L_{3m}^*\sqrt{\lambda_4}$$

$$+ L_{4m}^*\sqrt{\lambda_4}]h_2/2 = 0 \quad (m = 1,2, \ldots M, \ldots) \quad (50)$$

For numerical computation, we truncate the series at M and N terms, respectively, thus Eq 35 to Eq 50 form $(8M + 8N)$ simultaneous linear algebraic equations. However, there are A_{in}^*, D_{in}^*, B_{im}^*, L_{im}^* and $a_1 (i = 1,2,3,4; m = 1,2, \ldots M; n = 1,2,3, \ldots N)$ all together $(8N + 8M + 1)$ unknown constants, hence one more equation is needed. A self-equilibrium condition of the normal stresses along the interface ($y = 0$) is further imposed so that a good approximation may be expected with few terms of the series taken in the computation, that is

$$\int_o^a \sigma_y^{(1)}(x,0)\, dx = 0 \quad (51)$$

Furthermore, in the preceding equations the following relations between the symbols A_{in}

and A_{in}^* etc. are used

$$
\begin{aligned}
&A_{1n}^* = \alpha_n^2 Sh\alpha_n h_1/\sqrt{\lambda_1}A_{1n},\ A_{2n}^* = \alpha_n^2 Ch\alpha_n h_1/\sqrt{\lambda_1}A_{2n},\ A_{3n}^* = \alpha_n^2 Sh\alpha_n h_1/\sqrt{\lambda_2}A_{3n} \\
&A_{4n}^* = \alpha_n^2 Ch\alpha_n h_1/\sqrt{\lambda_2}A_{4n},\ B_{1m}^* = \omega_m^2 Sh\sqrt{\lambda_1}\omega_m aB_{1m},\ B_{2m}^* = \omega_m^2 Sh\sqrt{\lambda_1}\omega_m aB_{2m} \\
&B_{3m}^* = \omega_m^2 Sh\sqrt{\lambda_2}\omega_m aB_{3m},\ B_{4m}^* = \omega_m^2 Sh\sqrt{\lambda_2}\omega_m aB_{4m} \\
&D_{1n}^* = \alpha_n^2 Sh\alpha_n h_2/\sqrt{\lambda_3}D_{1n},\ D_{2n}^* = \alpha_n^2 Ch\alpha_n h_2/\sqrt{\lambda_3}D_{2n},\ D_{3n}^* = \alpha_n^2 Sh\alpha_n h_2/\sqrt{\lambda_4}D_{3n} \\
&D_{4n}^* = \alpha_n^2 Ch\alpha_n h_2/\sqrt{\lambda_4}D_{4n},\ L_{1m}^* = \beta_m^2 Ch\sqrt{\lambda_3}\beta_m aL_{1m},\ L_{2m}^* = \beta_m^2 Sh\sqrt{\lambda_3}\beta_m aL_{2m} \\
&L_{3m}^* = \beta_m^2 Ch\sqrt{\lambda_4}\beta_m aL_{3m},\ L_{4m}^* = \beta_m^2 Sh\sqrt{\lambda_4}\beta_m aL_{4m}
\end{aligned} \quad (52)
$$

Calculation of Stress-Intensity Factors

The stress field in the present problem is in the form

$$\sigma_{ij}^{(k)} = a_1 SI(\sigma_{ij}^{(k)}) + \text{non-singular terms} \quad (53)$$

To characterizing the behavior of the stresses in the neighborhood of the crack tip, we define the stress-intensity factors such that along the crack line (that is, the y-axis)

$$k_I^c = \lim_{y \to o} \sqrt{2}y^{-\delta}\sigma_x^{(1)}(0,y) \quad (54)$$

or in non-dimensional form

$$K_I^c = k_I^c / h_2^{-\delta}\sigma_{ave} \quad (54a)$$

and along the interface (that is, the x-axis)

$$k_I^d = \lim_{x \to o} \sqrt{2}x^{-\delta}\sigma_y^{(1)}(x,0);\ k_{II}^d = \lim_{x \to o} \sqrt{2}x^{-\delta}\tau_{xy}^{(1)}(x,0) \quad (55)$$

or in the non-dimensional form

$$K_I^d = k_I^d / h_2^{-\delta}\sigma_{ave};\ K_{II}^d = k_{II}^d / h_2^{-\delta}\sigma_{ave} \quad (55a)$$

where σ_{ave} represents the average applied stress on the laminate, that is, $\sigma_{ave} = \epsilon_0(E_{11}h_1 + E_{22}h_2)/h$.

For illustration, we take the graphite-epoxy composite for an example, the material constants of which are [*13*]

$$E_{11} = 138 \text{ GPa},\ E_{22} = E_{33} = 14.5 \text{ GPa},\ G_{12} = G_{23} = G_{31} = 5.9 \text{ GPa}$$

$$\nu_{12} = \nu_{23} = \nu_{31} = 0.21$$

With these constants, we find $\delta = -0.341138$. The stresses and then stress-intensity factors are computed accordingly, and the results are plotted in Figs. 3 and 4. In Fig. 3, the stress-

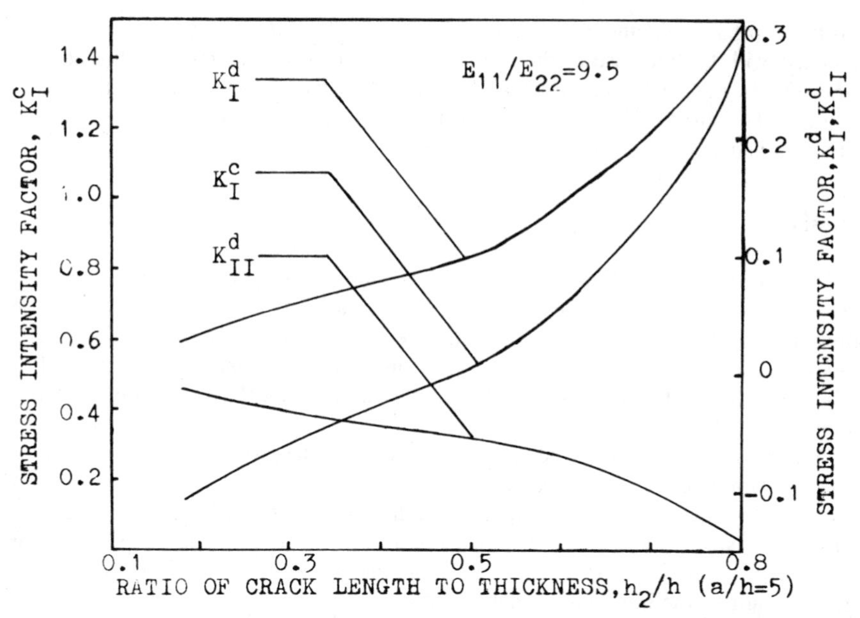

FIG. 3—*Dependence of stress-intensity factors on ratio of transverse crack length to thickness of the laminate.*

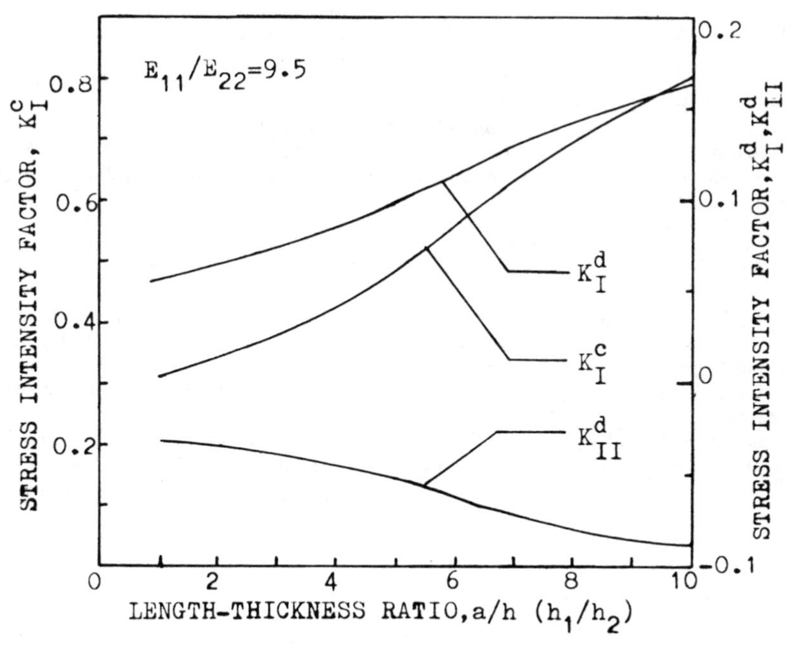

FIG. 4—*Dependence of stress-intensity factors on length-to-thickness ratio of laminate.*

intensity factors K_I^c, K_I^d and K_{II}^d are plotted versus (h_2/h) for a given ratio of E_{11}/E_{22}. They all increase with the increase of the ratio (h_2/h) which is in agreement with that observed in experiments. Figure 4 shows that the stress-intensity factors decrease with the decreasing (a/h) ratio. Therefore, it is natural to expect an increase of transverse crack density before the interlaminar delamination takes place.

Conclusion

1. The strength of singularity of a transverse crack terminating at the interface between the layers of a laminate depends notably on the material constants, particularly on the relative rigidity of the adjacent layers in the direction normal to the crack. The result may provide a reasonable explanation to the phenomenon that crack will be arrested at the interface confronting the layer with higher rigidity and the local delamination may then be initiated.
2. The stress-intensity factors generally increase with increasing crack length to thickness ratio of the laminate for a given value of E_{11}/E_{22}.
3. The stress-intensity factors decrease with the decrease of length-thickness ratio of the laminate for given E_{11}/E_{22} ratio. It is just this trend that is probably responsible for the appearance of an increase of crack density in a laminate and in the meantime a descent of its rigidity prior to the interlaminar delamination.
4. The solution presented in the paper is applicable to the problem of any finite composite laminate with special orthotropic layers.

References

[1] Lekhnitskii, S. G., *Theory of Elasticity of an Anisotropic Body,* Mir Publishers, Moscow, 1981.
[2] Liechti, K. M., Reifsnider, K. L. et al., "Cumulative Damage Modal for Advanced Composite Materials," AFWAL-TR-82-4094, Air Force Wright Aeronautical Laboratories, Wright-Patterson Air Force Base, OH, July 1982.
[3] Tsai, S. W. and Hahn, H. T., *Inelastic Behavior of Composite Materials,* American Society of Mechanical Engineers, New York, 1975, pp. 73–96.
[4] Garrett, K. W. and Baily, J. E. *Journal of Material Science,* Vol. 12, 1977, pp. 157–168.
[5] Parvizi, A., Garrett, K. W. et al., *Journal of Material Science,* Vol. 13, 1978, pp. 195–201.
[6] Reifsnider, K. L. and Talug, A. in *Proceedings,* ARO-NSF Research Workshop on Mechanics of Composite Materials, 1978, pp. 130–161.
[7] Wang, A. S. D. and Crossman, F. W., *Journal of Composite Materials,* Vol. 14, 1980, pp. 71–87.
[8] Delale, F. and Erdogan, F., *International Journal of Fracture,* Vol. 15, 1979, pp. 343–364.
[9] Lu, M. C. and Erdogan, F., *Engineering Fracture Mechanics,* Vol. 18, 1983, pp. 491–506, 507–528.
[10] Wang, S. S. and Choi, I., *Journal of Applied Mechanics,* Vol. 49, 1982, pp. 541–548.
[11] Sih, G. C. and Chen, E. P., *Mechanics of Fracture, Cracks in Composite Materials,* Vol. 6, Martinus Nijhoff Publishers, Amsterdam, the Netherlands, 1981.
[12] Wang, R. J. and Loo, T. T. in *Proceedings,* 9th U.S. National Congress on Applied Mechanics, 1982, p. 112.
[13] Crossman, F. W. and Wang, A. S. D. in *Damage in Composite Materials, ASTM STP 775,* K. L. Reifsnider, Ed., American Society for Testing and Materials, Philadelphia, 1980.

Hsai-Yang Fang,[1] *G. K. Mikroudis,*[1] *and Sibel Pamukcu*[1]

Fracture Behavior of Compacted Fine-Grained Soils

REFERENCE: Fang, H.-Y., Mikroudis, G. K., and Pamukcu, S., **Fracture Behavior of Compacted Fine-Grained Soils,"** *Fracture Mechanics: Perspectives and Directions* (*Twentieth Symposium*), *ASTM STP 1020,* R. P. Wei and R. P. Gangloff, Eds., American Society for Testing and Materials, Philadelphia, 1989, pp. 659–667.

ABSTRACT: Existing cracks, when triggered to advance, often contribute to the progressive failure mechanism and erosion of slopes, embankments, excavations, river banks, and other earthen structures of compacted fine-grained soils. To assess the effect of existing cracks on these earth structures it is necessary to quantitatively measure the magnitude of load necessary to advance the crack to failure. This load has been defined as the fracture load of soils in previous investigations.

This study reports the results of an experimental procedure used to measure the fracture load and its variation with precrack length and also evaluate the load deflection relations for compacted clays and marine clays. A simple procedure, based on concepts of linear elastic fracture mechanics (LEFM) theory, was utilized. Soil specimens were prepared and tests were conducted according to ASTM Method E 399. Fracture load and fracture toughness were evaluated. Results indicate that reasonable estimations of fracture load in clay soils could be made using the fundamentals of LEFM theory, which constitutes the basis of the fracture toughness analysis.

KEY WORDS: soil cracking, soil fracture, tensile strength of soils, cracking of compacted clays

Nomenclature

- a Fracture length, mm
- P Fracture load, N
- B Specimen thickness, mm
- w Specimen width, mm
- E Modulus of elasticity, MPa
- K_c Stress-intensity factor, MPa · m$^{1/2}$
- G_c Strain energy release rate, J/m^2
- ν Poisson's ratio

Soil cracks are frequently observed in many natural and man-made earthen structures. These cracks are a result of an internal energy imbalance in the soil mass caused by non-uniform moisture and temperature distribution or distribution of compaction energy during

[1] Professor and director, Geotechnical Engineering Division, research associate, Geotechnical Engineering Division, and assistant professor, Department of Civil Engineering, Fritz Engineering Laboratory 13, Lehigh University, Bethlehem, PA 18015.

construction. Cracking in flood plain clays results from deposition followed by cyclic expansion and contraction from seasonal wetting and drying. Many preconsolidated clays also exhibit cracking and fissures due to unloading or dessication. These closely spaced, small cracks contribute to progressive erosion or landslides in excavations, slopes, dams, highway embankments, river banks, and other earthen structures [*1–9*].

Due to seasonal ground-water table fluctuations, rainfall, or melting snow, water fills the cracks and fissures, thus softening the clay. Depending on the environmental conditions, however, this occurrence can also advance crack formation by the action of capillary tension or compression of air within the pores. Changing conditions of the environment also contribute to the magnitude of the pressures created and the effectiveness of these actions [*10*]. When rain falls on the dessicated surface of a clay deposit, it is absorbed by capillarity, and the air in the clay pores may become so compressed as to cause tension cracking. When saturated soil dries, menisci develop in the voids of the soil structure, which produce the tension in the soil-water system and the corresponding compressive stresses on the soil skeleton. For fine-grained soil, these compressive stresses can reach up to 300 kPa. The internal cyclic load caused by combination of shrinkage, thermal expansion or contraction, capillarity action, and fluctuation of stresses in the pore space is called the "fracture load" [*11*] in the soil.

A simple method, based on the concepts of linear fracture mechanics theory, was developed and utilized to quantitatively measure the fracture load and evaluate cracking behavior of the soil. Four types of clay samples were prepared and subjected to fracture tests. The fracture test specimens were prepared and tested according to a procedure similar to the ASTM Test Method for Plane-Strain Fracture Toughness of Metallic Materials (E 399-83). The results of the tests are discussed both from the geotechnical and fracture mechanics points of view. Interpretation of the results indicates that linear fracture mechanics can assist in evaluating fracture behavior of soils as it relates to slope stability and landslide problems. It must be noted, however, that the method is not applicable to swelling type of clays.

Theoretical Considerations

The basic theory of linear elastic fracture mechanics (LEFM) was developed about half a century ago by Griffith [*12*]. It has acquired a considerable amount of success in predicting failure caused by crack propagation in metals. More recently, it has been applied for predicting fracture behavior in concrete and rock [*13*], stabilized construction materials [*14–16*], and for compacted fine-grained soils [*17*].

Although LEFM theory is limited in principle to linear-elastic materials, it has been used successfully for studying fracture of other materials, provided that the zone of plastic or nonlinear strains in the vicinity of flaws or cracks are small. For short-term analysis (instantaneous loading), the soil is assumed to behave as a linear-elastic material. Therefore, the LEFM theory does provide a fast and simple way of estimating the amount of energy required to fracture, or create free surface in, the material. Soils can exist in a stable condition with some degree of flaws and cracks. With changes in the environmental conditions and fluctuations in the states of stress, these flaws and cracks can grow to cause failure along zones at which shear resistance is minimized. LEFM theory gives an adequate description of the gross features of the stresses and strains near a crack in soil and helps estimate their magnitude.

In particular, one can determine the critical strain energy release rate G_c, or fracture toughness, K_c, which is the conjugate to the force driving the fracture process [*18*]. In practice, the value of G_c is measured in a simple laboratory test using ASTM E 399-83. It uses a precracked specimen which is pulled apart by a load, P, as shown in Fig. 1. The

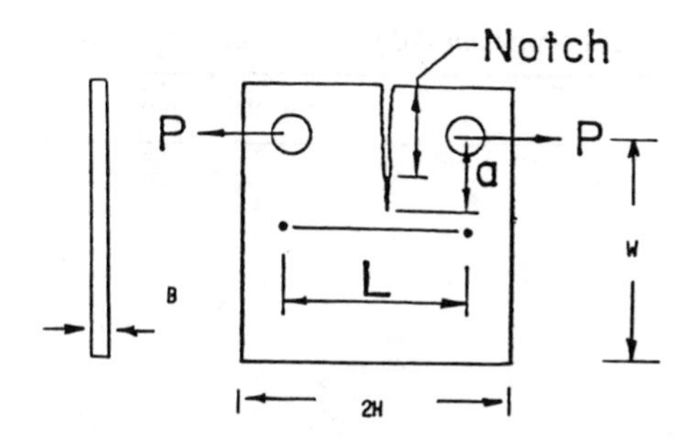

FIG. 1—*Sketch of testing specimen for the fracture load of the compacted soil. Specimen size:* w = *63.5 mm;* 2H = *76.2 mm;* B = *6.35 mm.* L = *gage length,* P = *fracture load, and* a = *crack length.*

fracture load P can also be calculated from Eq 1 once the crack dimensions and geometry and G_c or fracture toughness, K_c, are known

$$K_c = Y(a/w)\,\frac{P\,(a)^{1/2}}{Bw} \tag{1}$$

where, B, w, and a are given in the Fig. 1 caption. The term designated as $Y(a/w)$ is a function of a/w, and it is the finite width correction [*19*].

G_c and K_c are related by

$$G_c = \frac{K_c^2}{E}\,(1 - \nu^2) \tag{2}$$

where E and ν are elastic constants.

This paper deals only with estimating the fracture load, P, and fracture toughness, K_c. The experimental study reported includes the method of soil specimen preparation, the development of the appropriate equipment, and the interpretation of test results in terms of geotechnical applications.

Experimental Study

Soil Specimens

Four different types of fine-grained soil samples were used in the study. Two of them were soft marine clays obtained from Gulf of Mexico and Gulf of Maine. The index properties of these marine clays are given in Table 1. The other two were obtained from local deposits near Lehigh University grounds and were classified as silty clay. These local soils were

TABLE 1—*Summary of properties of soft marine clays.*

Soil Properties	Gulf of Mexico	Gulf of Maine
Saturated unit weight	...	1327 kg/m^3 [a]
Natural water content	...	163% [a]
Liquid limit, LL	72	121
Plastic limit, PL	34	51
Field vane shear		
0.005 rad/s	[b]	2.41 kN/m^2 [a]
0.002 rad/s		2.55 kN/m^2
Classification	...	silty-clay [a]
Field moisture equivalent	78.2%	65.2%
Centrifuge moisture equivalent	67.9%	55.6%
Shrinkage limit	7.4	10.1

[a] Average values reported by Perlow and Richards (1973).
[b] Laboratory vane sheer strength determined on gravity cores averaged about 4 kN/m^2.

further mixed with various percentages of bentonite or sand to prepare additional specimens. All specimens were passed through a No. 40 sieve. The molding water content for all specimens was constant at 25%. The molded dry unit weight varied for the two marine clays.

Preparation of Soil Specimens

Preparation of the soil specimens for the fracture tests conformed to that of the compact tension (CT) specimens in ASTM E 399. The sample sizes were w = 63.5 mm, $2H$ = 76.2 mm, B = 6.35 mm (see Fig. 1). The mold used in preparing the test specimens was developed at the Geotechnical Laboratory of Lehigh University. Figure 2 shows this equipment. The

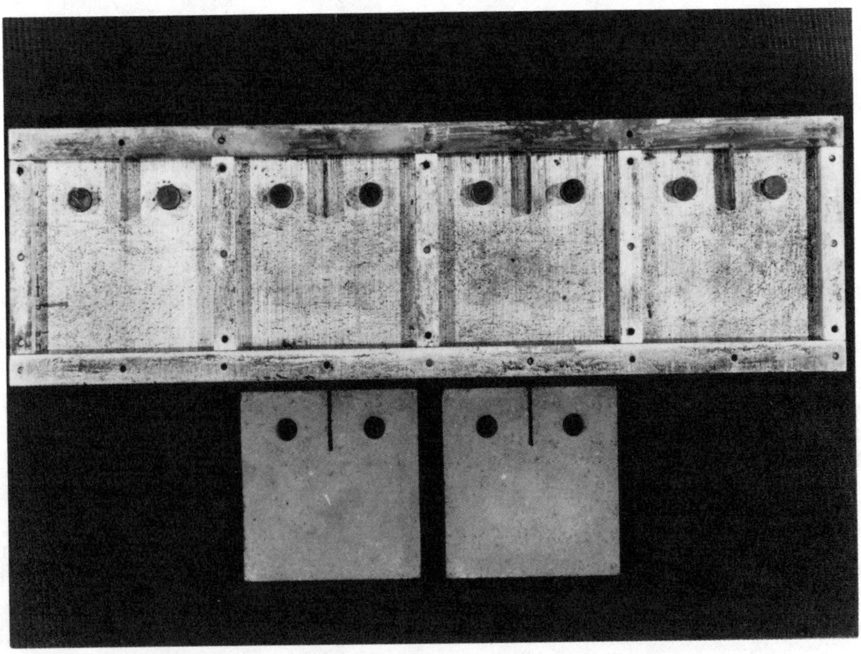

FIG. 2—*Metal molds for preparation of soil testing specimens for the fracture load test.*

mold consisted of a steel base with removable sides and interior walls. It could accommodate up to four specimens at the same time. The interior of the mold was coated with a lubricating agent to facilitate the removal of the specimens. After the specimens were extracted from the mold, they were air dried. A glass plate cover was placed over them to prevent warping. The notch formed on the specimens was 25.4 mm long. They were precracked at various crack lengths of a. The test procedure followed ASTM E 399. Figure 3 shows a typical soil specimen tested to failure.

Results

The results of fracture tests on three types of laboratory prepared silty clay specimens are given in Table 2. The measured values of fracture load, P, and the calculated values of K_c, are given for different crack lengths, a, for identically prepared specimens in each group. The data show an inverse relationship between crack length and fracture load. The trend of data is consistent and falls within the same range for the three types of silty clay specimens tested.

The load deflection trace for three different dry density specimens of Maine Clay are shown in Fig. 4. The fracture load results for all marine clays are summarized in Table 3. As observed from Fig. 4 and Table 3, the fracture load at failure increases with increasing molding dry density of the soil specimens. However, this increase is not linear, and the data indicate an upper limit for the fracture load. The results on the deflection at failure are not as consistent. Observing Fig. 3, as the dry density of the specimen increases from 14.7 kN/m^3 to 15.3 kN/m^3, the crack growth resistance of material seems to increase, as indicated by the increased deflection at maximum load. However, as the dry density is increased further to 16.1 kN/m^3, the deflection at maximum load is decreased, indicating a loss in toughness. These data suggest that there may be a range of densities for some soil specimens where there is increased plasticity in the fracture process.

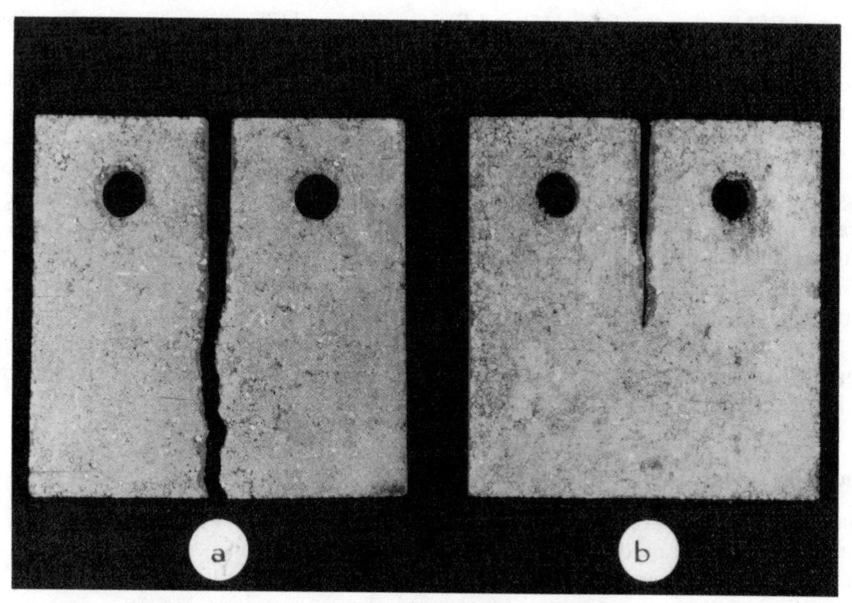

FIG. 3—*Typical soil samples:* (left) *before test;* (right) *after failure.*

TABLE 2—*Summary of fracture toughness data on three compacted fine-grained soils.*

Sample	a, mm	P, N	$Y(a/w)$	K_c,[a] MPa · $m^{1/2}$
1	24.13	22.46	11.31	0.0979
	27.94	20.91	12.24	0.1064
	31.75	16.46	13.58	0.0988
	35.56	15.57	15.58	0.1134
	39.37	10.68	18.67	0.0981
2	26.67	18.24	11.89	0.0878
	27.94	19.57	12.24	0.0993
	33.02	15.57	14.17	0.0994
	35.56	12.90	15.58	0.0940
	38.10	9.79	17.48	0.0828
3	20.32	22.24	10.76	0.0846
	24.13	22.24	11.31	0.0969
	24.13	21.35	11.31	0.0930
	30.48	15.12	13.08	0.0856
	31.75	13.34	13.58	0.0801

[a] $K_c = Y(a/w)\dfrac{P(a)^{1/2}}{BW}$

The consistency of data obtained from laboratory compacted soil specimens and the marine soil specimens indicate that the test procedure can be applied to make reasonable estimations of the magnitude and variation of the fracture load in soils in which there are existing cracks of various length. Clearly, further testing is needed to assess some of the trends of data and to exploit the technique in this new capacity.

Discussion

Using LEFM concepts, Fang [*15*] and Fang and Owen [*17*] presented an explanation of progressive cracking mechanism which can lead to slope failures in soil deposits. It was discussed that the existing cracks in soil will eventually fill up with water due to nonuniform soil particle size, various temperature and moisture gradients in soil matrix, nonuniform compaction during construction, and seasonal fluctuation of the ground-water table and surface water intake. These activities create within pore water, either tensile stresses compressing the soil skeleton or compressive stresses loosening the soil skeleton. Periodic occurrence of these stresses constitutes low-frequency cyclic loading on the existing cracks and fissures in the soil deposit which eventually lead to the growth of these cracks.

The existing crack in the soil is modeled by a precracked specimen, and the pressure fluctuation is analogous to the loading in the test devised for LEFM theory. Therefore, both the theory and the test procedure can be used in an analogous fashion to measure the magnitude of the load necessary to create failure in soil specimens with existing crack condition.

Conclusions

Based on the results reported here, the following conclusions were made:

1. LEFM may be used to study fracture behavior of certain soils.
2. Fracture toughness of soils can be readily measured in the laboratory.

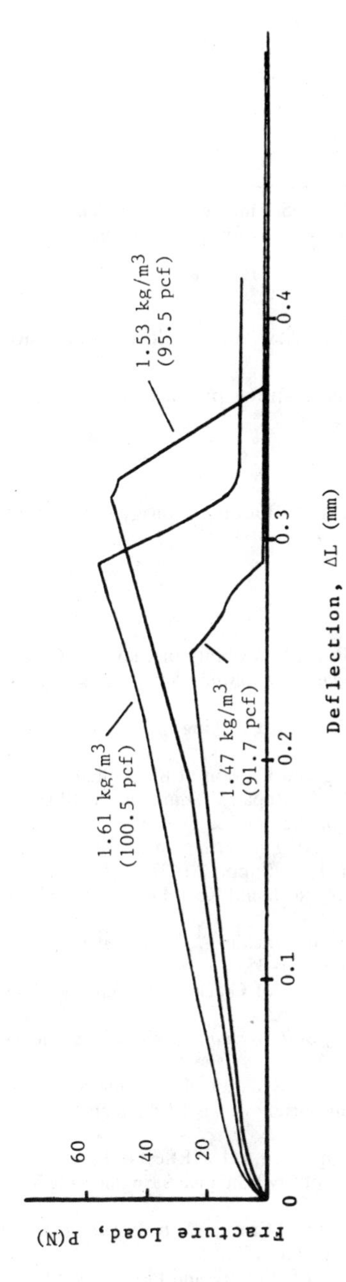

FIG. 4—*Fracture load versus change in gage length with various molded dry densities, Gulf of Maine marine clay.*

TABLE 3—*Summary of fracture test results on marine clays.*[a]

Location	Molded Dry Unit Weight, kN/m^3	Fracture Load, *N*
Gulf of Mexico	14.5	40.5
Gulf of Mexico	15.6	44.5
Gulf of Mexico	16.3	44.9
Gulf of Maine	14.7	26.7
Gulf of Maine	15.3	51.1
Gulf of Maine	16.1	55.6

[a] Specimen size = 76.2 by 63.5 by 6.35 mm. Length of notch = 2.54 cm. Test procedure follows ASTM E 399. Molding water content = 25% for all specimens.

3. The results may be used to assess slope failure and erosion potential of soil deposits in the field.
4. The approach presented here shows promise; however, further research is needed.

Acknowledgment

This research was funded by Enviro-Geotechnology Consultants, Inc., Bethlehem, Pennsylvania.

References

[1] Leonards, G. A. and Rawat, P. C., "Flexibility of Clay and Cracking of Earth Dams," *Journal of the Soil Mechanics and Foundations Division,* American Socity of Civil Engineers, Vol. 89, No. SM2, 1963, pp. 47–98.
[2] Vaughan, P. R., "Introductory Notes: Cracking of Clay Cores of Dams," *British Geotechnical Society,* Vol. 28, Jan. 1970.
[3] Vaughan, P. R. et al., "Cracking and Erosion of the Rolled Clay Core of Balderhead Dam and the Remedial Works Adopted for Its Repair," *Transactions*, 10th International Congress on Large Dams, Montreal, Vol. 1, 1970, pp. 73–93.
[4] Walker, F. C., "Prevention of Cracking in Earth Dams," *Transactions,* 10th International Congress on Large Dams, Montreal, Vol. 1, 1970, pp. 361–370.
[5] Covarrubias, S. W., "Cracking of Earth and Rock Fill Dams," *Havard Soil Mechanics Series,* No. 82, 1970.
[6] Sherard, J. L., "Embankment Dam Cracking," *Embankment Dam Engineering, Casagrande Volume,* Wiley, New York, 1973, pp. 271–353.
[7] Ajaz, A., "Detection and Prevention of Cracking of Clay Cores in Dams," *Geotechnical Engineering,* Vol. 9, No. 1, 1978, pp. 39–62.
[8] Sukje, L., *Rheological Aspects of Soil Mechanics,* Wiley-Interscience, New York, 1969, pp. 456–473.
[9] Spencer, E., "Effect of Tension of Stability of Embankment, *Journal of Soil Mechanics and Foundations Division,* American Society of Civil Engineers, Vol. 94, No. SM5, 1968, pp. 1159–1173.
[10] Fang, H. Y., Evans, J. C., and Kugelman, I. J., "Effect of Pore Fluid on Soil Cracking Mechanism" in *Proceedings,* American Society of Civil Engineers Engineering Mechanics Specialty Conference, Vol. 2, 1984, pp. 1292–1295.
[11] Fang, H. Y. et al., "Mechanism of Soil Cracking" in *Proceedings,* 20th Annual Meeting of Society of Engineering Sciences, 1983, pp. 156–157.
[12] Griffith, A. A., "The Phenomena of Rupture and Flow in Solids," *Philosophical Transactions of the Royal Society of London,* Series A221, 1921, pp. 163–198.
[13] George, K. P. and Cheng, P. C., "Crack Propagation Studies in Pavement Slab," *New Horizons in Construction Materials,* Vol. 1, Envo, 1976, pp. 567–581.

[14] George, K. P., "Theory of Brittle Fracture Applied to Soil Cement," *Journal of the Soil Mechanics and Foundations Division,* American Society of Civil Engineers, Vol. 96, No SM3, 1970, pp. 991–1010.
[15] Fang, H. Y., "Discussion of Mechanistic Approach to the Solution of Cracking Pavements," HRB Special Report 140, Highway Research Board, 1973, pp. 154–156.
[16] Sih, G. C. and Fang, H. Y., *Fracture Toughness Values of Highway Pavement Materials,* Institute of Fracture and Solid Mechanics, Lehigh University, Bethlehem, PA, 1972.
[17] Fang H. Y. and Owen, T. D., Jr., "Fracture-Swelling-Shrinkage Behavior of Soft Marine Clays" in *Proceedings,* International Symposium on Soft Clay, Asian Institute of Technology, Bangkok, Thailand, 1977, pp. 15–25.
[18] Irwin, G. R., "Fracture Mechanics," *Structural Mechanics,* Pergamon Press, London, 1960.
[19] Sih, G. C., *Handbook of Stress-Intensity Factors for Researchers and Engineers,* Institute of Fracture and Solid Mechanics, Lehigh University, Bethlehem, PA, 1973.

Yukitaka Murakami[1] *and Motohiro Kaneta*[2]

Fracture-Mechanics Approach to Tribology Problems

REFERENCE: Murakami, Y. and Kaneta, M., "**Fracture-Mechanics Approach to Tribology Problems,**" *Fracture Mechanics: Perspectives and Directions,* (*Twentieth Symposium*), *ASTM STP 1020,* R. P. Wei and R. P. Gangloff, Eds., American Society for Testing and Materials, Philadelphia, 1989, pp. 668–687.

ABSTRACT: Two tribology problems are studied analytically from the viewpoint of fracture mechanics. One is an inclined semicircular surface crack under Hertzian contact loading and the other is an elliptical crack embedded parallel to the surface under Hertzian contact loading.

The behavior of crack opening/closure for a surface crack formed on lubricated rolling/sliding line-contact surface is analyzed theoretically on the basis of fracture mechanics. The crack opening displacement is controlled mainly by surface traction, contact pressure, and oil hydraulic pressure. Both the direction and the magnitude of the surface traction govern the oil seepage into the crack. The oil hydraulic pressure on crack faces is induced by two types of mechanism depending on the movement of contact pressure: the contact pressure transmitted directly by the oil to the crack faces and the oil blocking phenomenon caused by closure of mouth of crack. It was concluded by the analytical results that the oil seepage into a surface crack is the crucial factor which causes the pitting phenomena in rolling/sliding contact fatigue. From this point of view, the reason why pitting phenomena are more frequently observed on the follower surface than on the driver surface can be clearly explained.

The variations of three-dimensional, mixed-mode stress-intensity factors for an embedded elliptical crack under a Hertzian contact pressure moving over the surface are analyzed. The crack opening occurs near the leading crack tip as well as near the trailing crack tip during one pass of contact load. This crack opening behavior was confirmed by the experimental measurements. The directions of crack extension are also discussed.

KEY WORDS: surface cracks, subsurface cracks, tribology, fracture mechanics, stress-intensity factors, rolling/sliding contact, Hertzian contact, crack propagation, lubrication, crack opening, crack closure

Applications of fracture mechanics to tribology problems are not as common as to other fields of strength of materials. Only a few institutions have recently started to study tribology problems from the viewpoint of fracture mechanics [*1–14*]. Although many unsolved problems of tribology are expected to be solved by the application of fracture mechanics, it seems that most investigators of tribology are not active yet in the introduction of fracture mechanics to their field.

In this study, we treat two important problems of tribology. One is the analysis of pit nucleation mechanism on the lubricated rolling/sliding contact surface. This problem has been very popular in tribology, and many hypotheses have been proposed to explain the mechanism. However, because of the complication of the problem, that is, because of many related factors, no conclusive theory has been established yet.

[1] Professor, Department of Mechanics and Strength of Solids, Faculty of Engineering, Kyushu University, Higashi-ku, Fukuoka, 812 Japan.

[2] Professor, Department of Mechanical Engineering, Kyushu Institute of Technology, Tobata-ku, Kitakyushu, 804 Japan.

The second problem is the numerical and experimental analysis of opening/closing behavior of a three-dimensional embedded crack parallel to the free surface under a moving line contact load. This problem is correlated with the nucleation of spalling of a roll in steelmaking process and shelling of railroad rails caused by the contact with wheels. These two phenomena have essentially the same characteristic, namely, that a crack grows below the free surface until it becomes a critical length, extends towards free surface, and finally causes a large surface fracture.

In this paper, the crack opening/closing behavior and stress-intensity factors in these two problems are studied by a three-dimensional numerical analysis and experimental measurements. The possible mechanism of fracture is discussed on the basis of the results of analysis and experiments.

Method of Analysis

The method of analysis is explained by taking the first problem (Fig. 1) as an example. It is assumed that the surface of an elastic half-space is loaded by normal and tangential stresses, defined by Eqs 1 and 2, which arise from a Hertzian contact that moves across the half-space from left to right as shown in Fig. 1.

$$p(x) = p_0(1 - x^2/c^2)^{1/2} \quad \text{(normal pressure distribution)} \tag{1}$$

$$q(x) = fp(x) \quad \text{(tangential stress distribution)} \tag{2}$$

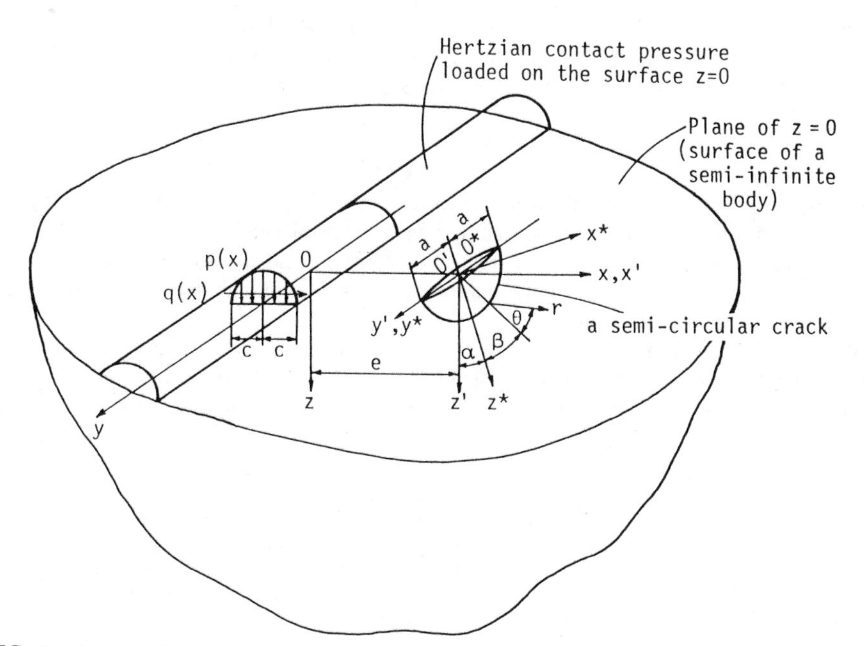

FIG. 1—*Analytical model and coordinate system:* a = *radius of semicircular crack;* c = *half-width of Hertzian contact;* α = *crack inclination angle;* (r,θ) = *polar coordinate with origin at crack tip; and* β = *angle from deepest crack tip of crack front.*

where

p_0 = maximum normal Hertzian pressure,
c = half-length of the Hertzian contact, and
f = mean coefficient of friction.

A surface crack inclined at an angle $90° - \alpha$ to the surface is assumed to be a semicircle with radius a, and to be in the $y^* - z^*$ plane.

Murakami and Nemat-Nasser [*15*] and Murakami [*16*] applied the body-force method [*17*] to the analysis of stress intensity factors for arbitrarily shaped three-dimensional cracks. The essence of the method is to make the boundary conditions of a crack by distributing body forces in a solid which contains no crack. In this paper, the method is extended to the problem of a crack under contact stress field.

In the analysis of the three-dimensional crack problem shown in Fig. 1, three kinds of pairs of body forces, shown in Fig. 2, must be distributed at the boundary which is to be a crack in order to satisfy the boundary conditions associated with the normal stress in the x^* direction and the shearing stresses in the y^* and z^* directions. If the shape of the crack is circular or elliptic, we can introduce pairs of body forces (doublets) of a definite form suitable for numerical analysis. However, in this paper, the pairs of body forces $\vec{p}$ are assumed in the form of Eq 3, because they can be extended to crack problems with various shapes.

$$\vec{p} = \vec{f}(\vec{q})\sqrt{2a_0\epsilon - \epsilon^2} \tag{3}$$

where $\vec{f}(\vec{q})$ is a function of location $\vec{q}$ on the crack boundary where pairs of body forces are distributed, that is, $\vec{f}(\vec{q})$ is the vector which indicates the intensities of the three kinds of pairs of body forces shown in Fig. 2. The pairs of body forces $\vec{p}$ are produced by the limiting value of the multiplication of the densities $\vec{\rho}$ of point forces and the distance ξ^{**} of the pair, that is, $\vec{P} = \lim_{\xi^{**}\to 0}\vec{\rho} \cdot \xi^{**}$. The details of the analysis is explained in Ref *16*.

$$\sqrt{2a_0\epsilon - \epsilon^2}$$

indicates the standard shape of crack surface displacement, and accordingly, $\vec{f}(\vec{q})$ is also the measure of relative (nondimensional) change in the displacement relative to the standard case. ϵ is the distance from $\vec{q}$ to the nearest crack contour, and a_0 is the representative crack dimension (Fig. 3). In case of a semicircular crack $a_0 = a$, and in case of an elliptical crack, we choose a_0 = the minor radius.

Although the problem of Fig. 1 is essentially expressed by integral equations, the solution is hardly obtainable in the closed form. Therefore, in order to solve the integral equations numerically, we divide the crack area into finite triangular subregions and replace the integral

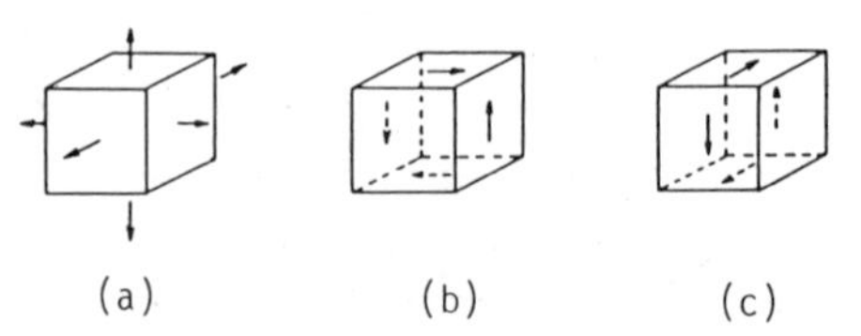

FIG. 2—*Three kinds of pair of body forces (doublets). These doublets are continuously embedded in a body which contains no crack. The intensities of doublets are determined so that the boundary conditions of crack are satisfied. Pairs of body forces to control* (a) *normal stresses and* (b, c) *shearing stress at crack boundary.*

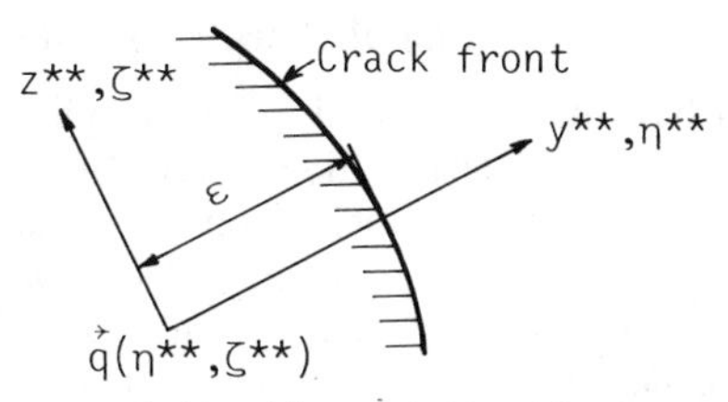

FIG. 3—*Local coordinate system:* (y**, z**) *and* (η**, ζ**) *planes are on crack plane;* (η**, ζ**) *indicates the coordinate of the point where a pair of body force is applied.*

equations by linear simultaneous equations by regarding $\vec{f}(\vec{q})$ as constant over each subregion [*15*]. Thus, solving the problem is reduced to determining the values of $\vec{f}(\vec{q})$ so as to satisfy the boundary conditions for a crack under the contact stress field. The values of $\vec{f}(\vec{q})$ at the triangular subregions at the crack contour correspond to dimensionless stress-intensity factors, because Eq 3 is expressed in the form relative to unit external loading [*16*].

The basic equations for a surface crack inclined to the surface of a semi-infinite body are derived from Green's function obtained by Mindlin [*18*]. The details of the equations and the practical procedure for numerical computations are explained in Refs *8* and *16*.

The mesh pattern used in this numerical analysis is shown in Fig. 4.

The stress-intensity factors K_I, K_{II}, and K_{III} are

$$K_I = F_I p_0 \sqrt{\pi a},\ K_{II} = F_{II} p_0 \sqrt{\pi a} \text{ and } K_{III} = F_{III} p_0 \sqrt{\pi a} \tag{4}$$

where F_I, F_{II}, and F_{III} are dimensionless stress-intensity factors.

The crack-opening displacement (COD) is given by

$$\text{COD} = 4(1 - \nu^2) p_0 F_{Ij} \sqrt{2a\epsilon_j - \epsilon_j^2}/E \tag{5}$$

where ν is Poisson's ratio ($\nu = 0.3$) and E is Young's modulus. As can be understood from the explanations related to Eq 3, F_{Ij} is the opening mode component of $\vec{f}(\vec{q})$. ϵ_j is the minimum distance from the center of gravity of the jth triangular element to the nearest crack front.

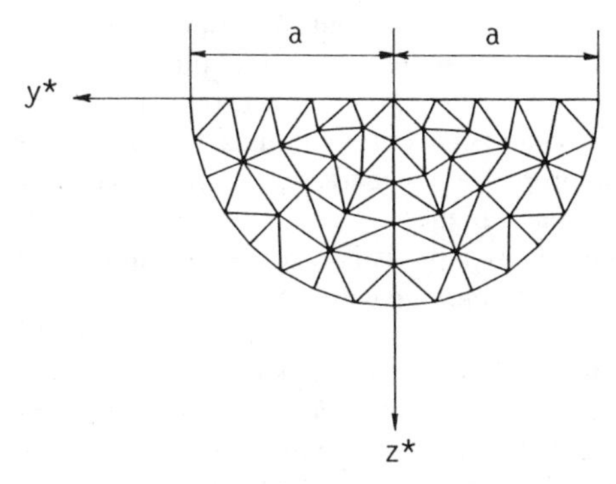

FIG. 4—*Mesh pattern* (N = 72) *for a semicircular crack. Crack is on* (y*, z*) *plane. The intensities of a pair of body forces distributed in a triangular area indicate the degree of crack-surface displacement.*

Direction of Crack Extension and Criterion of Crack Propagation

Although several criteria have been proposed for the condition of crack propagation in mixed-mode conditions and for the determination of growth direction, the criterion given by Erdogan and Sih [*19*] for tensile mode crack extension is adopted in the present investigation. This theory states that if the stress-intensity factor which determines the stress distribution on the fracture plane exceeds a critical value K_{Ic}, the crack extends macroscopically by choosing the plane perpendicular to the direction along which the tangential stress σ_θ defined in the polar coordinates (r, θ) (Fig. 1) becomes the maximum. The angle of crack extension, θ_0, is given by one of two roots of the equation

$$\tan\frac{\theta_0}{2} = \frac{1 \pm \sqrt{1 + 8\gamma^2}}{4\gamma}, \quad \gamma = \frac{K_{II}}{K_I} = \frac{F_{II}}{F_I} \tag{6}$$

We must choose the angle θ_0 which gives the maximum σ_θ. Equation 6 can be approximately applied to predict the direction of fatigue crack growth if K_{Ic} is replaced by the threshold stress-intensity factor range $\Delta K_{\sigma th}$ [*20*]. The stress-intensity factor $K_{\sigma max}$, which prescribes the distribution of $\sigma_{\theta max}$ at the plane $\theta = \theta_0$, is given by

$$K_{\sigma max} = F_{\sigma max}\, p_0 \sqrt{\pi a} \tag{7}$$

where $F_{\sigma max}$ is determined from F_I and F_{II} at the crack tip considering stress transformation from the crack plane to the plane of $\theta = \theta_0$.

Recent experiments have shown that $\Delta K_{\sigma th}$ (threshold stress-intensity factor range in opening mode) in fatigue fracture for various steels has the values $\Delta K_{\sigma th} \cong 3 - 7.5$ MPa $\cdot$ $m^{1/2}$ regardless of material [*21–25*].

In similar way, we also define similar quantities $K_{\tau max}$ and $\Delta K_{\tau th}$ (threshold stress-intensity factor range in shear mode) for the crack growth by shear mode. Therefore, by focusing our attention on the relationship between the magnitudes of $K_{\sigma max}, \Delta K_{\sigma th}$ and $K_{\tau max}, \Delta K_{\tau th}$, we will discuss possible mechanisms for the propagation of surface and subsurface crack.

Growth Mechanism of a Surface Crack in Lubricated Rolling/Sliding Contact (Fig. 1)

Stress-Intensity Factors and Crack Opening/Closing Behavior

Since Way's pioneering work [*26*], many investigations on the occurrence of pits due to a rolling-contact fatigue have been carried out by various researchers from different points of view. However, nowadays, it is well recognized that this phenomenon is very complicated.

According to the recent studies, it is commonly accepted that the process of fatigue consists of two stages, crack initiation and crack growth. Since the pitting phenomenon is associated with cracking, it may be appropriate to use fracture mechanics in order to study theoretically the problems of rolling-contact fatigue.

In the present study, Way's hypothesis is reconsidered in detail from the viewpoints of crack opening displacement and oil seepage into the crack. Crack growth behavior is analyzed; the initiation process of the crack will not be discussed, that is, we assume the existence of a small, semicircular surface crack from the beginning.

The relationships among the inclined angle of a crack, the direction of movement of contact pressure and the surface traction force are schematically illustrated in Fig. 5. In this analysis, the direction of inclination of the crack is fixed to the positive x-direction in Fig. 1 ($\alpha = 45$ deg) and the direction of surface traction is changed. In relation to the direction of tangential traction force, terms "positive sliding surface or driver surface" and "negative sliding surface or follower surface" are used in this paper. For example, when the contact pressure moves from the left of the mouth of a crack to the right, the conditions $f > 0$ and $f < 0$ correspond to the positive and the negative sliding surface, respectively (see Fig. 5).

For the sake of simplicity, the oil hydraulic pressure p_f applied on crack faces by oil in the crack is assumed to decrease linearly from crack mouth to crack front. Namely, the fluid pressure in the crack is expressed in the region $|e| \leqq c$ as

$$p_f(z^*) = p(e)(1 - z^*/a) \tag{8}$$

Figure 6 shows the relationship between the dimensionless stress-intensity factors $F_{\rm I}$ and $F_{\rm II}$ for the crack tip at the deepest point and the normalized distance e/c from the crack mouth to the center of Hertzian contact pressure. These figures can be considered to simulate the variations of stress-intensity factors during a loading cycle. If we follow the variations of curves from left to right, it means that the load comes from left and leaves to right, and vice versa. In the region of $|e/c| < 1$, the fluid pressure of Eq 8 is applied on the crack faces. The value of $F_{\rm III}$ at the deepest crack tip is equal to zero.

Figure 7 shows the sequential variations of the crack opening/closure due to the movement of contact pressure for the case $a/c = 0.5$ in Fig. 6. The figure shows the half of crack face and the values of $F_{\rm Ij}$ corresponding to the j-triangular subregion are indicated in it. Bold lines indicate closed parts of the crack. The absolute value of crack opening displacement can be calculated using Eq 5. When $F_{\rm Ij} > 0$, the crack surfaces at the j-subregion are separated from each other, that is, the crack is open. When $F_{\rm Ij} < 0$, they are closed (in our mathematical

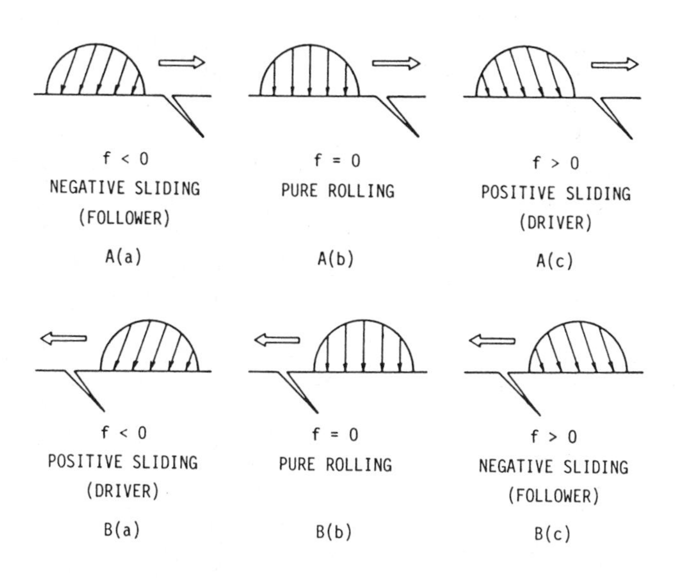

FIG. 5—*Schematic illustrations showing relationship among directions of crack inclination, movement of Hertzian contact, and surface traction. The calculations of one pass of contact load for* A(a), A(b), *and* A(c) *are also regarded as those for* B(a), B(b), *and* B(c) *if we look at the results for* A(a), A(b), *and* A(c) *in the reversed order. However, it should be noted that the history of the change of stress-intensity factors and crack opening displacements for* A *and* B *is completely different.*

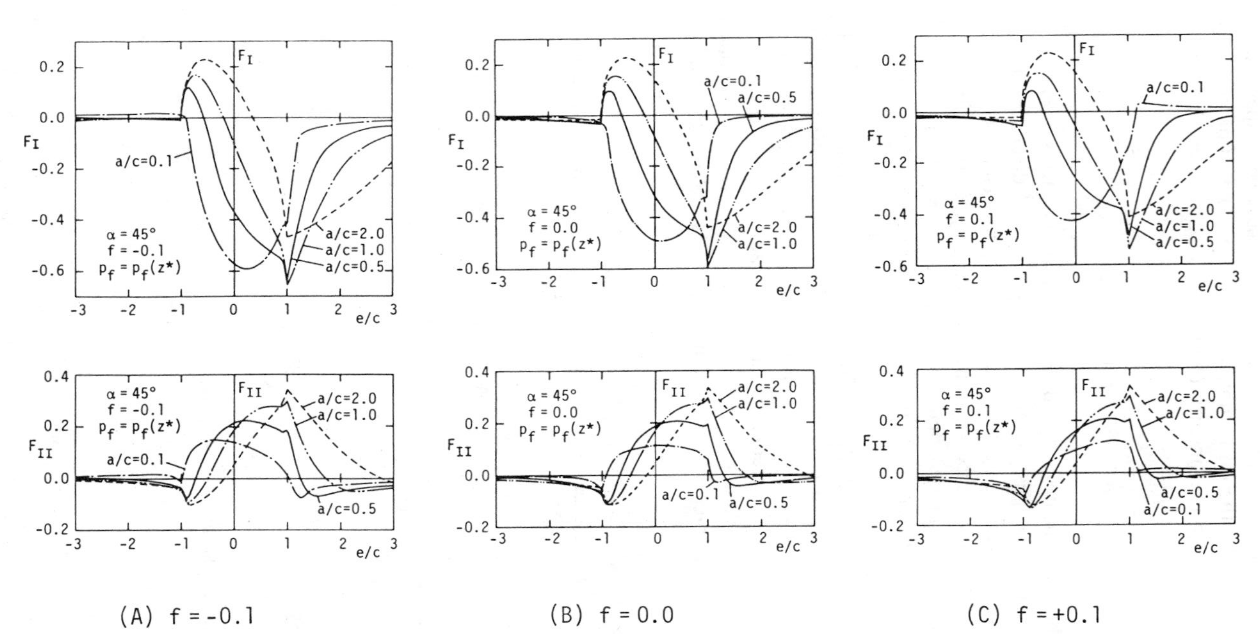

FIG. 6—*Variation of dimensionless stress-intensity factors* F_I *and* F_{II} *at the deepest crack tip during a loading cycle. If the calculation curve is traced from left to right, it means the variation of* F_I *and* F_{II} *when the load comes from left and leaves to right. If the calculation curve is traced from right to left, it means the reversed case.*

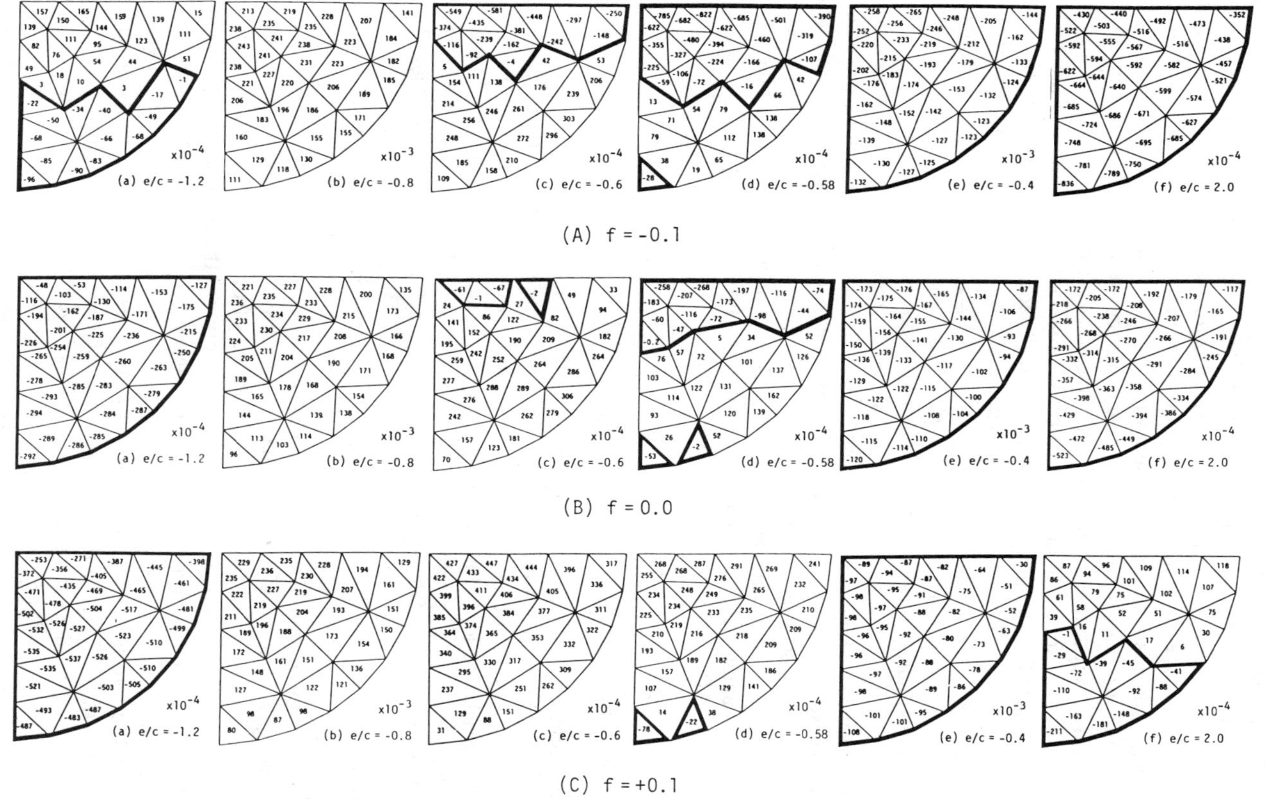

FIG. 7—*Variations of* F_{ij} *at triangular subregions;* a/c = 0.5. *Each figure shows the half of crack face. The regions surrounded by the bold lines indicate the closed part of crack. If the illustrations are traced from left to right, it means the variation of crack-surface displacement for the load coming from left and leaving to right, and vice versa. The sequence of crack opening and closure from left to right or right to left is very important, because it controls the possibility of oil seepage into the crack.*

approximation, they cross each other, that is, the crack-opening displacement becomes negative, although the condition $F_{Ij} < 0$ is never achieved in a real situation).

Figure 8 shows the variations of $K_{\sigma max}$ due to the movement of the contact pressure which are calculated from the results of Fig. 6. In order for $K_{\sigma max}$ to be compared with $K_{\sigma th}$ (threshold stress-intensity factor range in opening mode) of actual materials, $K_{\sigma max}$ is evaluated by using the experimental data of Ref *27*, that is, the maximum Hertzian pressure p_0 = 1.1 GPa and the half contact width c = 0.2 mm. The angle θ_0 denoted in these figures shows the direction of crack growth associated with the maximum value of $K_{\sigma max}$. Although Fig. 8 gives correct values $K_{\sigma max}$ when the crack is open in the whole crack region, that is, $F_{Ij} > 0$ in Fig. 6, the values of $K_{\sigma max}$ are overestimated when the part of the crack is closed, that is $F_{Ij} < 0$, because in case of $F_{Ij} < 0$ the effect of frictional force between crack faces are not considered.

When there is no oil hydraulic pressure on the crack faces, the crack growth by tensile mode is restrained as described in the previous paper [*8*]. Moreover, when the size of crack is small as a/c = 0.1, the condition for crack growth ($K_{\sigma max} > \Delta K_{\sigma th}$) is never satisfied as shown in Fig. 8, and consequently, the crack growth by tensile mode is unlikely to occur, even if the oil hydraulic pressure is applied on crack faces.

It has been clarified [*9*] that if the surface traction forces are large and the frictional coefficient between crack faces is small, a crack extends by shear mode along the original crack plane. These conditions can be brought by the collapse of elastohydrodynamic lubrication (EHL) film, oil seepage into the crack, etc. As the crack extends to a critical length by shear mode, the condition of crack growth in tensile mode will be satisfied, if the oil fills the crack. For example, the maximum value of $K_{\sigma max}$ for a/c = 0.5 in Fig. 8 reaches approximately 3.2 MPa $\cdot \sqrt{m}$; this value exceeds $\Delta K_{\sigma th}$.

Possibility of Oil Seepage into Crack

When the contact pressure moves as shown in Fig. 5*A* in the region of $e/c < -1$, the faces of the crack at the negative sliding surface ($f < 0$) are separated or open ($F_{Ij} > 0$) at least in the neighborhood of the mouth of crack, and oil seepage into the crack is possible; oil enters the crack interior before the contact region reaches the crack mouth. In the case of pure rolling ($f = 0$) and positive sliding ($f > 0$), the possibility of oil seepage into crack is extremely low, because F_{Ij} is negative at the whole crack face, as shown in Figs. 7*B*(*a*) and 7*C*(*a*).

On the other hand, when the contact pressure moves from right to left against the inclined crack as shown in Fig. 5*B*, only the crack formed on the negative sliding surface ($f > 0$) becomes open in the region of $e/c > 1$ [see Fig. 7(f)]. However, the oil is expected to be removed through the mouth of crack, because the crack is closed from its deepest point to crack mouth as the contact approaches [see the changes from Figs. 7*C*(f) to 7*C*(e)].

Now, if the oil is assumed to be contained in the interior of the crack with relatively large size ($a/c > 1$), F_{Ij} becomes positive as seen from Figs. 6 and 7 in certain region of $|e/c| < 1$ for all the cases of $f < 0$, $f = 0$, and $f > 0$. For all these cases, the maximum value of $K_{\sigma max}$ exceeds the threshold stress-intensity factor $\Delta K_{\sigma th}$ as shown in Fig. 8. Moreover, the maximum value of $K_{\sigma max}$ is larger in magnitude for the case $f > 0$ than for the case $f < 0$. However, we must consider the possibility of oil seepage into crack. If the mouth of the crack is closed, the oil will not enter. As already described, when the crack is inclined as in Fig. 5*A*(*a*), the oil is most likely to penetrate into the crack [compare Fig. 7*A*(*a*) with Fig. 7*C*(*a*)]. As a result, it is expected that the crack inclined as in Fig. 5*A*(*a*) on the follower surface ($f < 0$) grows at higher rate than that on the driver surface ($f > 0$).

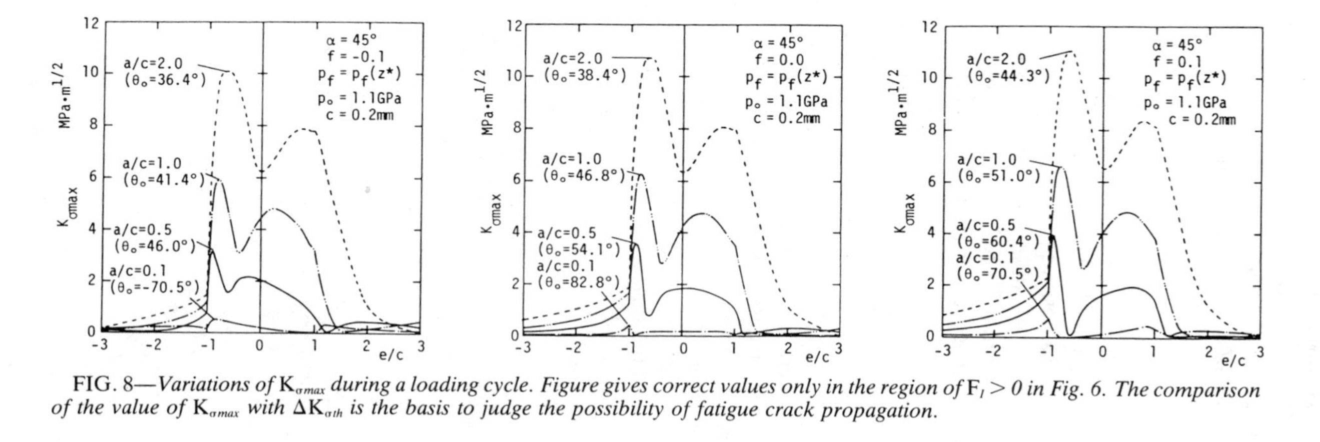

FIG. 8—*Variations of* $K_{\sigma max}$ *during a loading cycle. Figure gives correct values only in the region of* $F_I > 0$ *in Fig. 6. The comparison of the value of* $K_{\sigma max}$ *with* $\Delta K_{\sigma th}$ *is the basis to judge the possibility of fatigue crack propagation.*

Experimental observations [27,28] show that the occurrence of pitting or the growth rate of crack is accelerated especially at negative sliding surface, that is, fatigue life is shorter for negative sliding (follower) surface than for positive (driver) surface. The analytical results of this study are consistent with experimental observations.

Consequently, it can be concluded that both the direction and the magnitude of frictional force between two contact surfaces are the crucial factors of oil seepage into crack and accordingly of crack growth.

Oil Blocking Action

Considering Fig. 6 in connection with Fig. 5*A*, we can recognize that as Hertzian contact moves towards right, F_{I} attains a positive peak because of the oil hydraulic pressure effect and then decreases and takes a negative value. If we do not consider the oil hydraulic pressure for the boundary condition of crack face, F_{I} do not have positive values [8]. When F_{I} becomes negative before the center of Hertzian contact pressure reaches the mouth of crack (this is the case when a/c is relatively small as $a/c \leqq 1$), the crack formed on the positive sliding surface ($f > 0$) closes from the deepest point. This can be understood from the variation of the closed part of crack shown in Figs. 7*C*(*c*) to 7*C*(*e*). Although the process of excluding oil from the deepest point of the crack is not actually calculated, it can be expected easily from Fig. 7*C*(*c*). However, the crack on the negative sliding surface ($f < 0$) closes from the mouth of crack [see the changes from Figs. 7*A*(*b*) to *A*(*e*)] and the oil is sealed in the crack. This mechanism is schematically explained in Fig. 9. If the movement of contact pressure seals the mouth of the crack, the oil is compressed in the crack, the oil pressure becomes very high and consequently the crack may extend further by tensile mode. It should be noted that if the combinations of the directions of crack inclination, the movement of contact pressure, and the surface traction force are not like the left portion of Fig. 9, there will be no possibility of oil seepage into the crack.

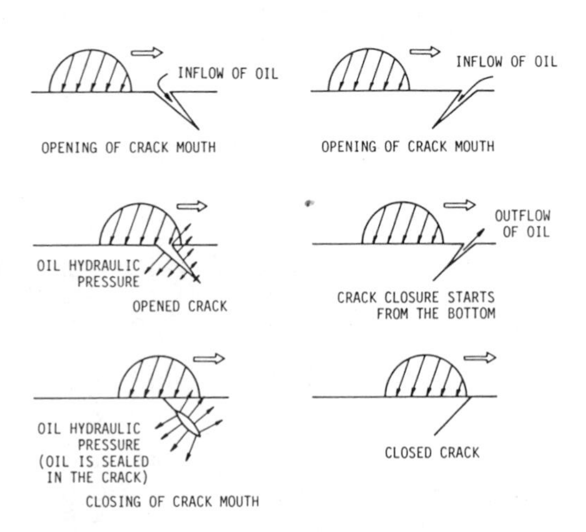

FIG. 9—*Oil seepage into crack and occurrence of oil hydraulic pressure action. Look at the figures in comparison with Figs. 5 and 7.*

Mechanism of Opening and Closing of Subsurface Cracks Due to Moving Hertzian Loading

Analytical Model

An elliptical crack with size defined in Fig. 10 is assumed to exist parallel to the surface of an elastic half-space at a depth d beneath the surface. The crack size is changed by changing the minor and major radius a and b, though only the analytical results for $a/b = 0.5$ are discussed in this paper. It is assumed that the surface is loaded by a Hertzian contact that moves perpendicularly to the major axis of the elliptical crack over the half-space surface. A tangential stress proportional to the Hertzian contact pressure is also assumed within the contact region. The tangential force is directed towards the right in Fig. 10. Accordingly, these normal and tangential stresses are defined by Eqs 1 and 2. The crack region is divided into 128 triangular subregions, as shown in Fig. 11. The frictional force between the crack faces is considered only when the crack faces receive a compressive stress.

Analytical Results

Figure 12 shows the variations of the crack opening/closure due to the movement of contact load. They show the results calculated for frictional coefficients between contacting surfaces $f = 0.1$, 0.4, and 1.0 under conditions $a/b = 0.5$, $a/c = 1.0$, $d/c = 0.25$, and $f_c = 0.5$, where f_c is the coefficient of friction between the crack faces. The black areas correspond to the closed parts of crack ($F_{Ij} \leqq 0$), and other regions indicate the open parts ($F_{Ij} > 0$). The open parts are classified into three by the level of the magnitude of F_{Ij}; no hatching portions correspond to $F_{Ij} \geqq 10^{-1}$, hatched portions $10^{-2} \leqq F_{Ij} < 10^{-1}$, and cross-hatched portions $0 < F_{Ij} < 10^{-2}$.

Figure 13 shows the variation of F_I and F_{II} during one loading cycle at the left (trailing) crack tip (A in Fig. 11) and the right (leading) crack tip (B in Fig. 11) under the conditions of $a/b = 0.5$, $a/c = 1.0$, $d/c = 0.25$, and $f_c = 0.5$. The values of $|F_I|$ and $|F_{II}|$ decrease, in general, with increasing the frictional coefficient (f_c) between crack faces. However, the tendencies of relative variations of F_I and F_{II} are less influenced by f_c within the conditions of the present calculations. Hence, the effect of f_c will not be considered in the following discussion.

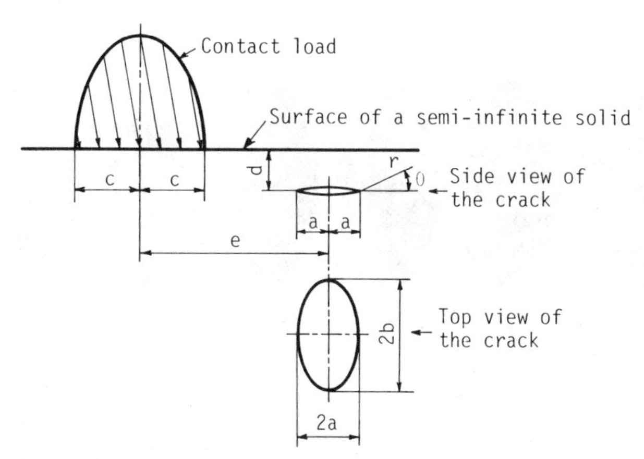

FIG. 10—*Analytical model of a subsurface elliptical crack parallel to the surface.*

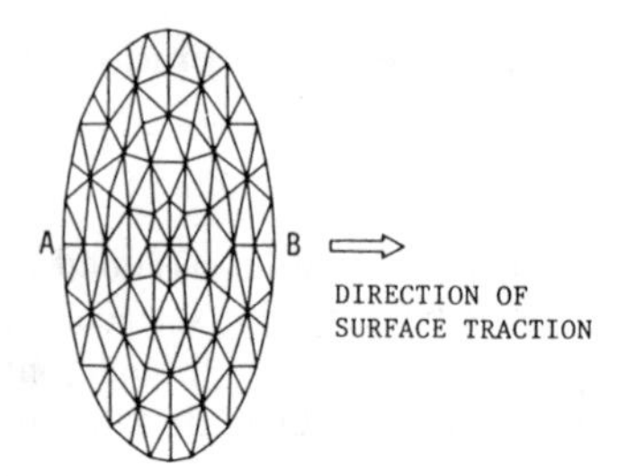

FIG. 11—*Mech pattern for elliptical crack.*

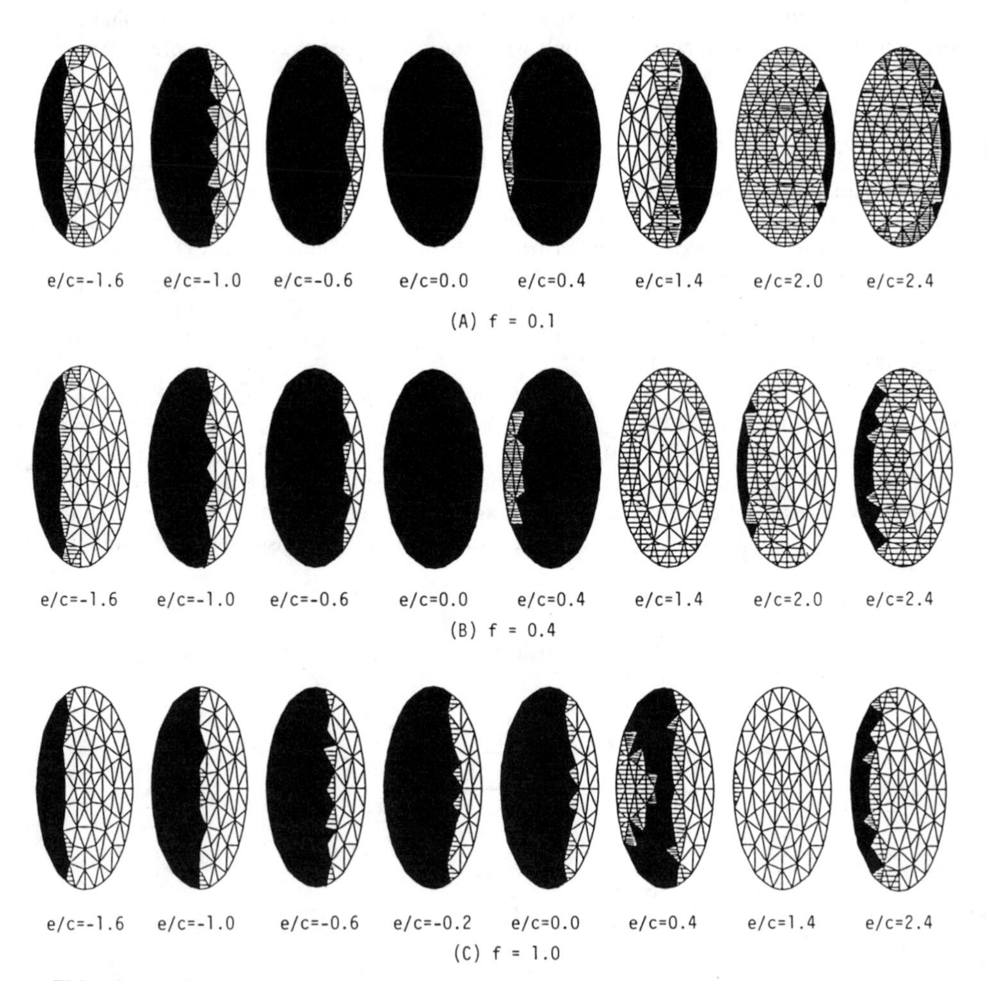

FIG. 12—*Analytical results of variations of crack opening and closure due to movement of contact load* (a/b = *0.5,* a/c = *1.0,* d/c = *0.25,* f_c = *0.5*). *Black area indicate closed part of crack. White areas indicate open part of crack, that is, crack-surface displacement is positive even under compressive loading. Cross-lined areas indicate the part where crack-surface displacement is very small.*

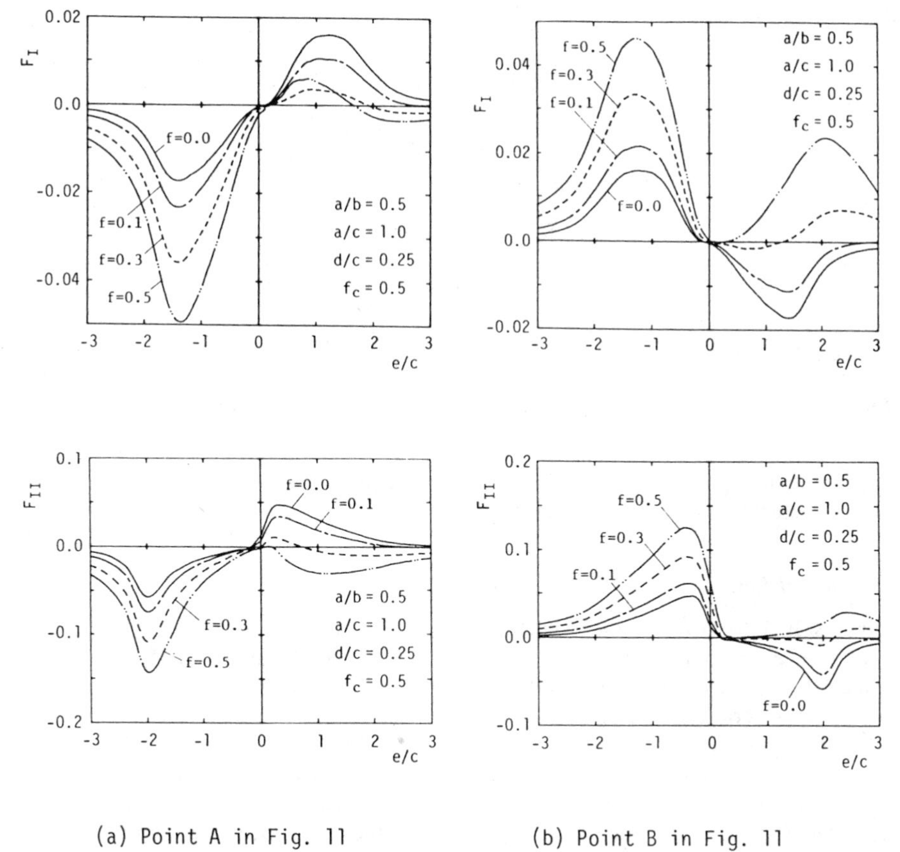

FIG. 13—*Variations of* F_I *and* F_{II} *during a loading cycle:* (a) *Point A in Fig. 11;* (b) *Point B in Fig. 11.*

Method of Experiment

Figure 14 shows a schematic view of the testing device. The lower specimen which contains a two-dimensional through crack is made of a plastic eraser available in the market. The shape and dimensions are illustrated in Fig. 15. The face having 7 mm width was used as the contact surface and the opposite face adhered to a foundation. The crack was produced by piercing with a blade with sharp edge parallel to the contact surface and at a distance d from the surface. The blade had width $2a$ and thickness of 0.2 mm. The upper cylindrical specimen, which is 20 mm in width, 62 mm in outer diameter, and 27 mm in inner diameter, is made of a transparent acrylic resin.

A contact load W was applied by suspending weights on both sides of the rod of 5 mm in diameter, which was inserted through a hole drilled in the upper specimen. The rod is at a distance 10 mm from the contact surface and parallel to the cylinder axis. A tangential force F was applied by pulling the thread hooked on both sides of the rod. After both specimens are set in position and the forces were applied, the back side of the testing device shown in Fig. 14 was illuminated with a lamp. The crack opening was recorded photograph-

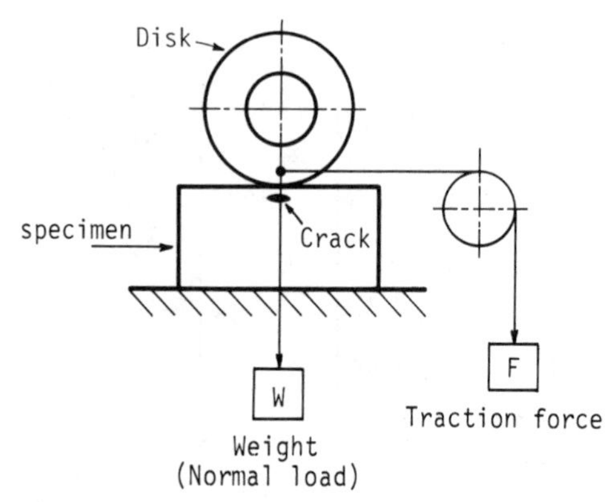

FIG. 14—*Schematic view of testing device. A two-dimensional through crack is made in an eraser.*

ically by the light passing through the crack. Before test it was checked that no transmitted light was detected under zero contact load.

Experimental Results

Figures 16*A* to 16*F* are the schematic illustrations of the crack opening and closure behavior at each contact position. The notation e in these figures indicates the distance between the center of crack and loading position; $e/c < 0$ means that the loading position is at the left of crack center and $e/c > 0$ at the right of it. The tangential force is directed toward the right.

As seen in these figures, when the contact load is located at the left side of the crack, the right-side region of the crack opens irrespective of magnitude of tangential force. When the contact load is above the center of the crack, the crack faces are completely closed. When the contact load moves towards the right of the crack, the opened region moves depending on the magnitude of tangential force and the loading position.

When the magnitude of the tangential force was small compared to those in Fig. 16, the crack opening was detected only near the right tip and only when the contact load was at the left of crack. The detection of crack opening was also difficult when the crack size was decreased.

When the calculated values of $F_{\mathrm{I}j}$ are positive but very small, the experimental detection of crack opening is very difficult. Taking this into consideration, it may be concluded that the tendencies of the experimental results in Fig. 16 agree very well with the theoretical ones.

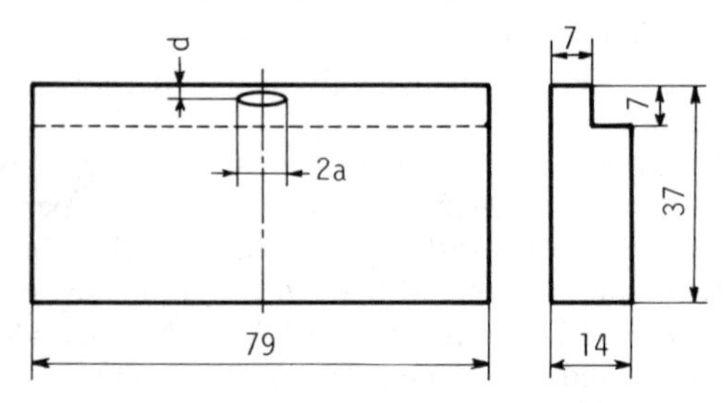

FIG. 15—*Shape and size of specimen in Fig. 14 (dimensions in mm).*

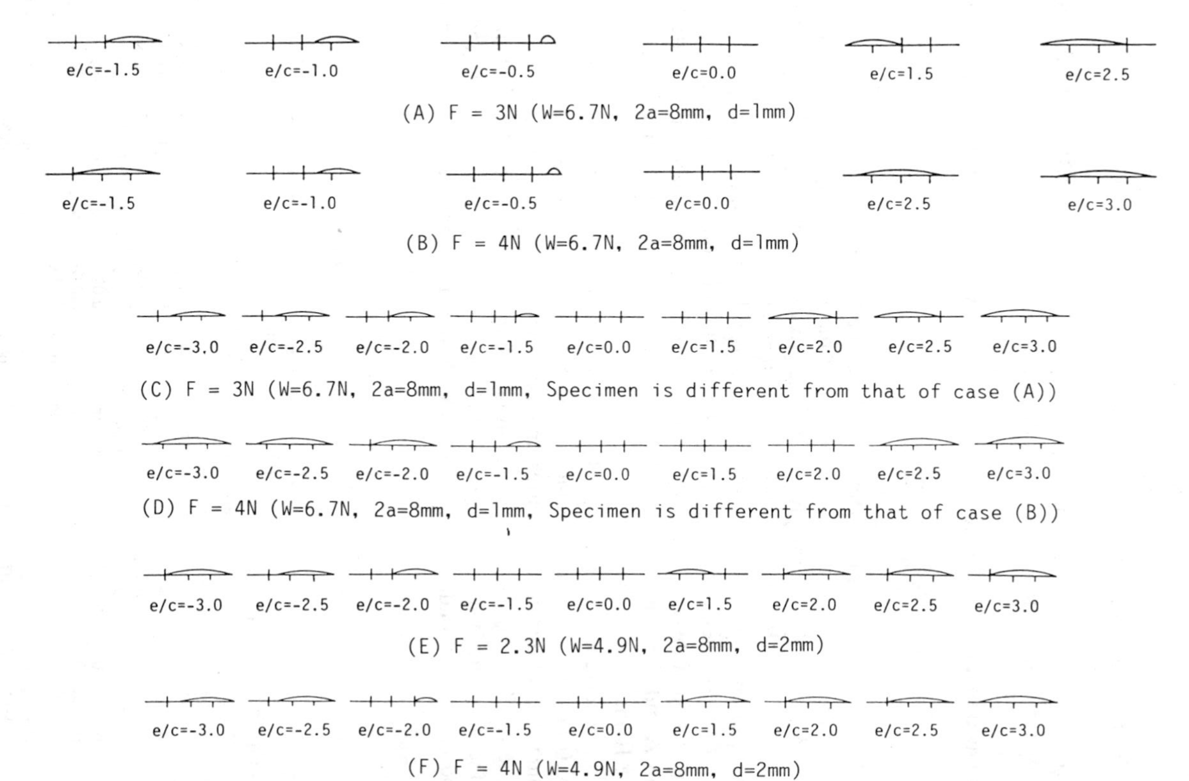

FIG. 16—*Schematic illustrations of crack opening and closure of the subsurface crack (the crack in Fig. 14) under contact loading (experimental results).*

Mode and Direction of Crack Growth

The critical condition and the direction of crack growth were determined by the same criteria as for a surface crack. We assume that if the relation

$$K_{\tau max} \geqq \Delta K_{\tau th} \tag{9}$$

is satisfied, shear mode fatigue crack growth occurs macroscopically in the direction $\theta = \theta_0$

[θ_0 is the angle at which $K_\tau(\theta)$ takes the maximum value ($K_{\tau max}$), and $\Delta K_{\tau th}$ is the threshold stress-intensity factor range for the shear-mode fatigue crack growth.]

It is also assumed that the tensile-mode fatigue crack growth occurs when the condition

$$K_{\sigma max} \geqq \Delta K_{\sigma th} \tag{10}$$

is satisfied. In this case, the crack extends to the plane $\theta = \theta_0$, on which $K_\sigma(\theta)$ has the maximum value ($K_{\sigma max}$). It is empirically known that the value of $\Delta K_{\tau th}$ is somewhat lower than that of $\Delta K_{\sigma th}$ [*21,29*]: accordingly, the fatigue growth for small cracks occurs at first in shear mode [*21,29*]. The stress-intensity factors for the present problems not only vary along the front of an elliptical crack, but also in general have three components, namely, K_I, K_{II}, and K_{III}. However, the criteria for the critical condition of crack growth and for the growth direction in this case are not yet established. Hence, we will focus our attention on the Points A and B in Fig. 11, where the stress-intensity factor (K_{III}) in antiplane shear mode (Mode III) is zero.

Table 1 represents crack growth angles θ_0 in the instant when the stress intensity factors $K_\tau(\theta)$ and $K_\sigma(\theta)$ have the maximum values during one pass of the contact load. We denote the maximum values of dimensionless stress intensity factors at the crack tip A or B by $F_{\tau max}$ and $F_{\sigma max}$. The values of $F_{\tau max}$ and $F_{\sigma max}$ vary during one pass of the contact load. Therefore, we denote the maximum values of $F_{\tau max}$ and $F_{\sigma max}$ *during one pass of the contact load* by $(F_{\tau max})_{max}$ and $(F_{\sigma max})_{max}$. In addition, for the shake of practical applications, the quantities

TABLE 1—*Values of* $F^*_{\tau max}$ *and* $F^*_{\sigma max}$ (f_c = *0.5*, b/c = *2.0*, A *and* B *correspond to left and right crack tips in Fig. 11*).

	d/c	0.25		0.5		0.5		0.25		0.5		0.5	
	a/c	1.0		1.0		0.5		1.0		1.0		0.5	
	f	θ_0	$F^*_{\tau max}$	θ_0	$F^*_{\tau max}$	θ_0	$F^*_{\tau max}$	θ_0	$F^*_{\sigma max}$	θ_0	$F^*_{\sigma max}$	θ_0	$F^*_{\sigma max}$
A	0	0	−0.0581	0	−0.0767	0	−0.0670	−68.9	0.0562	−69.2	0.0753	−69.3	0.0726
A	0.1	0	−0.0750	0	−0.0929	0	−0.0804	−69.0	0.0398	−69.3	0.0587	−69.3	0.0573
A	0.2	0	−0.0920	0	−0.1092	0	−0.0938	−69.0	0.0242	−69.3	0.0428	−69.8	0.0425
A	0.3	0	−0.1088	0	−0.1254	0	−0.1072	66.7	0.0117	−70.2	0.0277	−69.9	0.0288
A	0.4	0	−0.1257	0	−0.1417	0	−0.1206	67.8	0.0233	68.5	0.0165	−69.9	0.0156
A	0.5	0	−0.1426	0	−0.1579	0	−0.1340	67.1	0.0376	67.2	0.0288	66.7	0.0242
A	1.0	0	−0.2338	0	−0.2392	0	−0.2016	67.4	0.1508	64.0	0.1328	64.3	0.1124
B	0	0	−0.0581	0	−0.0767	0	−0.0670	0	0	−69.2	0.0753	−69.3	0.0726
B	0.1	1.51	0.0612	1.20	0.0771	1.11	0.0738	−68.8	0.0738	−69.1	0.0921	−69.2	0.0879
B	0.2	1.66	0.0768	1.24	0.0913	1.15	0.0869	−68.6	0.0929	−68.0	0.1094	−69.2	0.1037
B	0.3	1.76	0.0924	1.28	0.1055	1.19	0.1001	−66.3	0.1120	−70.0	0.1273	−69.1	0.1196
B	0.4	1.84	0.1079	1.33	0.1202	1.23	0.1135	−66.2	0.1322	−67.9	0.1455	−69.1	0.1358
B	0.5	2.02	0.1255	1.38	0.1350	1.27	0.1271	−66.1	0.1539	−67.9	0.1642	−69.0	0.1521
B	1.0	4.56	0.2527	1.69	0.2136	1.45	0.2776	−65.3	0.3303	−67.6	0.2618	−68.8	0.2362

$F^*_{\tau max}$ and $F^*_{\sigma max}$ are defined by

$$F^*_{\tau max} = (F_{\tau max})_{max}\sqrt{a/c} \tag{11}$$

$$F^*_{\sigma max} = (F_{\sigma max})_{max}\sqrt{a/c} \tag{12}$$

If $F^*_{\tau max}$ and $F^*_{\sigma max}$ are known, the maximum values, $(K_{\tau max})_{max}$ and $(K_{\sigma max})_{max}$, of $|K_{\tau max}|$ and $K_{\sigma max}$ can be calculated respectively by

$$(K_{\tau max})_{max} = F^*_{\tau max}\, p_0 \sqrt{\pi c} \tag{13}$$

$$(K_{\sigma max})_{max} = F^*_{\sigma max}\, p_0 \sqrt{\pi c} \tag{14}$$

In these calculations, when $F_I < 0$, $F^*_{\sigma max} = 0$ was assumed and crack growth in shear mode was also assumed.

From the numerical results in Table 1, we can make the following predictions:

1. An elliptical subsurface crack propagates mainly in shear mode, that is, in Mode II. For example, if it is assumed that $p_0 = 1$ GPa, $c = 2$ mm, and $a/c = 1.0$, $F^*_{\tau max}$ corresponding to $\Delta K_{\tau th} = 1.5$ MPa $\cdot \sqrt{m}$ [21] becomes 0.06 and the values in Table 1 satisfy Eq 9.
2. The direction of crack growth in shear mode at the left (trailing) tip (A in Fig. 11) is different from that of the right (leading) tip (B in Fig. 11). The trailing crack tip extends along the extended plane of the crack face irrespective of the magnitude of the surface traction. On the other hand, the leading tip extends also along the original crack plane if the surface traction is very small, but if the surface traction becomes larger, say $f > 0.1$, it extends towards the surface and its angle θ_0 increases with increasing crack size a/c and with increasing surface traction, that is, f.
3. Shear mode crack growth is most likely to occur under a relatively small surface traction, say $f \leqq 0.5$ and when the crack lies near the depth of $d/c = 0.5$. However, with increasing surface traction, say $f = 1.0$, shallower cracks satisfy the condition of shear mode crack growth.
4. When the crack size becomes relatively large and a high surface traction or a heavy load is applied, the fatigue crack growth in tensile mode (Mode I) occurs.
5. If a crack extends by tensile mode under low surface traction, both loading and trailing crack tips will propagate into the interior of the solid. However, if high surface traction is applied, the trailing tip extends towards the surface and the leading tip extends in opposite direction.
6. Generally, the magnitudes of $F^*_{\tau max}$ and $F^*_{\sigma max}$ decrease with the increase in the coefficient of friction, f_c, between crack faces. However, the value of $F^*_{\sigma max}$ at the trailing tip (A in Fig. 11) is scarcely influenced by f_c if the surface traction or f becomes larger.
7. The shear-mode crack growth angle at the leading tip of the crack increases with increasing f_c, especially under conditions of shallower crack depth d/c and larger crack size a/c.

Concluding Remarks

Two tribology problems were studied analytically from the viewpoint of fracture mechanics. One is an inclined semicircular surface crack under Hertzian contact loading, and the other is an elliptical crack embedded parallel to the surface under Hertzian contact loading. The stress-intensity factors and crack opening/closing behavior were numerically analyzed. In the latter problem, the experiments using two-dimensional model were conducted and the crack opening/closing behaviors were compared with the analytical results.

1. In the former problem, it was concluded by the analytical results that the oil seepage into a surface crack is the crucial factor which causes the pitting phenomena in rolling/sliding contact fatigue. From this point of view, the reason why pitting phenomena are more frequently observed on the follower surface than on the driver surface can be clearly explained.

The crack opening displacement is controlled mainly by surface traction, contact pressure, and oil hydraulic pressure. Both the direction and the magnitude of the surface traction govern the oil seepage into the crack. The oil hydraulic pressure is induced by two kinds of mechanism depending on the movement of the contact pressure. One is due to the contact pressure transmitted directly by the oil to the crack faces when the crack mouth is kept open. The other is due to the oil blocking phenomenon caused by the closure of the mouth of the crack and by squeezing the oil contained in the crack.

2. In the latter problem, it was predicted by analyses and was confirmed by experiments that a subsurface crack opens even under compressive surface loading (Hertzian contact loading with surface traction). The crack opening occurs not only near the trailing tip of the crack behind the contact load, but also near the leading tip in front of the contact load. The possible mechanisms of crack growth and of the direction of crack extension were also predicted on the basis of analytical results.

References

[*1*] Suh, N. P., *Wear,* Vol. 44, 1977, pp. 1–16.

[*2*] Fleming, J. R. and Suh, N. P., *Wear,* Vol. 44, 1977, pp. 39–56.

[*3*] Hills, D. A. and Ashelby, D. W., *Engineering Fracture Mechanics,* Vol. 13, 1980, pp. 69–78.

[*4*] Rosenfield, A. R., *Wear,* Vol. 72, 1981, pp. 245–254.

[*5*] Keer, L. M., Bryant, M. D., and Haritos, G. K., *Transactions,* American Society of Mechanical Engineers, *Journal of Lubrication Technology,* Vol. 104, 1982, pp. 347–351.

[*6*] Hearle, A. D. and Johnson, K. L., CUED/C-Mech/TR26, Cambridge University Research Report, Cambridge, U.K., 1983.

[*7*] Sin, H.-C. and Suh, N. P., *Transactions,* American Society of Mechanical Engineers, *Journal of Applied Mechanics,* Vol. 51, 1984, pp. 317–323.

[*8*] Murakami, Y., Kaneta, M., and Yatsuzuka, H., *Transactions,* American Society of Lubrication Engineers, Vol. 28, 1985, pp. 60–68.

[*9*] Kaneta, M., Yatsuzuka, H., and Murakami, Y., *Transactions,* American Society of Lubrication Engineers, Vol. 28, 1985, pp. 407–414.

[*10*] Kaneta, M., Murakami, Y., and Okazaki, Y., *Transactions,* American Society of Mechanical Engineers, *Journal of Tribology,* Vol. 108, 1986, pp. 134–139.

[*11*] Cheng, H. S., Keer, L. M., and Mura, T., SAE Technical Paper Series 841086, SP-584, Gear Design and Performance, Society of Automotive Engineers, 1984, pp. 27–35.

[*12*] Hahn, G. T., Bhragava, V., Yoshimura, H., and Rubin, C. A. in *Proceedings,* 6th International Conference on Fracture, Pergamon, Oxford, U.K., 1984, p. 295.

[*13*] O'Regan, S. D., Hahn, G. T., and Rubin, C. A., *Wear,* Vol. 101, 1985, p. 333.

[*14*] O'Regan, S. D., Hahn, G. T., and Rubin, C. A., *Transactions,* American Society of Mechanical Engineers, *Journal of Tribology,* Vol. 108, 1986, p. 540.

[*15*] Murakami, Y. and Nemat-Nasser, S., *Engineering Fracture Mechanics,* Vol. 17, 1983, pp. 193–210.

[*16*] Murakami, Y., *Engineering Fracture Mechanics,* Vol. 22, 1985, pp. 101–114.

[*17*] Nisitani, H., *Bulletin of the Journal of the Society of Mechanical Engineers,* Vol. 11, 1968, pp. 14–23.

[*18*] Mindlin, R. D., *Physics 7,* 1936, pp. 195–202.

[19] Erdogan, F. and Sih, G. C., *Transactions,* American Society of Mechanical Engineers, Ser. D. Vol. 85, 1963, pp. 519–527.
[20] Murakami, Y., *Transactions,* Japan Society of Mechanical Engineers, Vol. 46, 1980, pp. 729–738.
[21] Otsuka, A., Mori, K., and Miyata, T., *Engineering Fracture Mechanics,* Vol. 7, 1975, pp. 429–439.
[22] Kitagawa, H. and Takahashi, S., *Transactions,* Japan Society of Mechanical Engineers, Vol. 45, 1979, pp. 1289–1303.
[23] Kitukawa, M. and Jono, M., *Transactions,* Japan Society of Mechanical Engineers, Vol. 47, 1981, pp. 468–482.
[24] Murakami, Y. and Endo, M. in *The Behavior of Short Fatigue Cracks,* EGF Pub. 1, K. J. Miller and E. R. de los Rios, Eds., 1986, Mechanical Engineering Publishers, pp. 275–293.
[25] Taylor, D., *A Compendium of Fatigue Thresholds and Growth Rates,* Engineering Materials Advisory Service, Ltd., The Midlands, U.K., 1985.
[26] Way, S., *Transactions,* American Society of Mechanical Engineers, *Journal of Applied Mechanics,* Vol. 2, 1935, pp. A46–A58.
[27] Soda, N. and Yamamoto, T. in *Proceedings,* Japan Society of Lubrication Engineers/American Society of Lubrication Engineers, International Lubrication Conference, Tokyo, 1975, pp. 458–465.
[28] Ichimaru, K., Nakajima, A., and Hirano, F., *Transactions,* American Society of Mechanical Engineers, *Journal of Mechanical Design,* Vol. 103, 1981, pp. 482–491.
[29] Otsuka, A., Togo, K., and Matsuyama, H. in *Proceedings,* 3rd Fracture Mechanics Symposium, Society of Materials Science, Japan, 1985, pp. 46–50.

Indexes

Author Index

R

S

T

W

X

Y

Z

Subject Index

D

E

F